Ultra-Wideband, Short-Pulse Electromagnetics 3

Ultra-Wideband, Short-Pulse Electromagnetics 3

Edited by

Carl E. Baum
U.S.A.F. Phillips Laboratory
Albuquerque, New Mexico

Lawrence Carin
Duke University
Durham, North Carolina

and

Alexander P. Stone
The University of New Mexico
Albuquerque, New Mexico

Plenum Press • New York and London

Library of Congress Cataloging-in-Publication Data

On file

Proceedings of the Third International Conference on Ultra-Wideband, Short-Pulse Electromagnetics, held May 27 – 31, 1996, in Albuquerque, New Mexico

ISBN 0-306-45593-5

A Division of Plenum Publishing Corporation
233 Spring Street, New York, N. Y. 10013

http://www.plenum.com

10 9 8 7 6 5 4 3 2 1

Printed in the United States of America

PREFACE

The first two international conferences on Ultra-Wideband (UWB), Short-Pulse (SP) Electromagnetics were held at Polytechnic University, Brooklyn, New York in 1992 and 1994. Their purpose was to focus on advanced technologies for generating, radiating, and detecting UWB,SP signals, on mathematical methods, their propagation and scattering, and on current as well as potential future applications. The success of these two conferences led to the desirability of scheduling a third conference. Impetus was provided by the electromagnetics community and discussions led by Carl Baum and Larry Carin resulted in the suggestion that the UWB conferences be moved around, say to government laboratories such as Phillips Laboratory. Consequently the decision was made by the Permanent HPEM Committee to expand AMEREM '96 to include the Third Ultra-Wide Band, Short-Pulse (UWB,SP 3) with the Third Unexploded Ordnance Detection and Range Remediation Conference (UXO) and the HPEM/NEM Conference in Albuquerque, New Mexico during the period May 27-31, 1996. Planning is now underway for EUROEM '98 in June, 1998 in Tel Aviv, Israel. Joseph Shiloh is the conference chairman. A fourth UWB,SP meeting is planned as a part of this conference and Ehud Heyman will coordinate this part of the meeting.

The papers which appear in this volume, the third in the UWB,SP series, update subject areas from the earlier UWB,SP conferences. These topics include pulse generation and detection, antennas, pulse propagation, scattering theory, signal processing, broadband electronic systems, and buried targets.

The choice of the logo on the hard cover of this volume was motivated by the Impulse Radiating Antenna (IRA) on display at AMEREM '96. This display generated much excitement and interest, and in fact one of the papers discusses this IRA.

The editors wish to thank all of those involved in AMEREM '96, including those involved in the two related parallel conferences (UXO and HPEM/NEM), for their assistance and participation. We also acknowledge with gratitude the sponsorship of the Summa Foundation and the Permanent HPEM Committee. AMEREM '96 was hosted by the Advanced Weapons and Survivability Directorate of the USAF Phillips Laboratory. Cooperating institutions and agencies include the IEEE, URSI, AFOSR, ARL, Centre d'Etudes de Gramat (France), DNA, LANL, NRL, NSWC, SNL, Swedish Defense Materiel Administration, Swiss Federal Institute of Technology, TNO Physics and Electronics Laboratory (Netherlands), USAF Wright Laboratory, Yuma Proving Ground, and Duke University, the University of Illinois at Chicago, and the University of New Mexico.

Carl E. Baum
Lawrence Carin
Alexander P. Stone

CONTENTS

PULSE GENERATION AND DETECTION

Semiconductor Switching

The Time Evolution of Photonic Crystal Bandgaps 1
K. Agi, M. Mojahedi and K. J. Malloy

Optically Excited Photoconducting Antennas for Generating Ultra-Wideband Pulses 9
David W. Liu and Paul H. Carr

Ground Penetrating Radar Enabled by High Gain GaAs Photoconductive Semiconductor Switches 17
G. M. Loubriel, M. T. Buttram, J. F. Aurand and F. J. Zutavern

General

Ultrawideband Pulser Technology 25
David M. Parkes

High Power, Sub-Nanosecond Rising Waveforms Created by the Stacked Blumlein Pulsers 31
F. Davanloo, D. L. Borovina, J. L. Korioth, R. K. Krause, C. B. Collins, F. J. Agee, J. P. Hull, J. S. H. Schoenberg and L. E. Kingsley

The Problems of Picosecond Analog Devices Modeling and Creation 39
V. N. Ilyushenko, O. V. Stukach and B. I. Avdochenko

ANTENNAS

Impulse Radiating Antennas

Impulse Radiating Antennas, Part III 43
Everett G. Farr and Carl E. Baum

Transient Fields of Rectangular Aperture Antennas 57
Sergey P. Skulkin

Reflector Impulse Radiating Antennas

Temporal and Spectral Radiation on Boresight of a Reflector Type of Impulse Radiating Antenna (IRA) 65
D. V. Giri and Carl E. Baum

Coplanar Conical Plates in a Uniform Dielectric Lens with Matching Conical Plates for Feeding a Paraboloidal Reflector 73
Carl E. Baum, Joseph J. Sadler and Alexander P. Stone

Transient Fields of Parabolic Reflector Antennas 81
Sergey P. Skulkin and Victor I. Turchin

Use of the Synthesized Short Radio Pulse for Near-Field Antenna Measurements 89
Andrey V. Kalinin

Lens Impulse Radiating Antennas and TEM Horns

Design of the Low-Frequency Compensation of an Extreme-Bandwidth TEM Horn and Lens IRA 97
M. H. Vogel

A Radiating Structure Incorporating an Extended Ground Plane and a Brewster Angle Window 107
Jimmy Wells, Carl Baum, Norman Keator and William Prather

A TEM-Horn Antenna with Dielectric Lens for Fast Impulse Response 113
John F. Aurand

Optimized TEM Horn Impulse Receiving Antenna 121
Michael A. Morgan and R. Clark Robertson

Arrays

Transient Arrays 129
Carl E. Baum

Properties of Ultrawideband Arrays 139
Jodi Lisa Schwartz and Bernard D. Steinberg

General

Some Basic Properties of Antennas Associated with Ultrawideband Radiation 147
S. N. Samaddar and E. L. Mokole

Theorems on Time-Domain Far Fields 165
Arthur D. Yaghjian and Thorkild B. Hansen

Asymptotic Approximations for Optimal Conformal Antennas 177
T. S. Angell, R. E. Kleinman and B. Vainberg

Generation of Wideband Antenna Performance by [Z] and [Y] Matrix Interpolation in the Method of Moments 185
Kathleen L. Virga and Yahya Rahmat-Samii

Electromagnetic Analysis of Exponentially Tapered Coplanar Stripline Antennas Used in Coherent Microwave Transient Spectroscopy Technique 197
Valérie Bertrand, Michèle Lalande and Bernard Jecko

PULSE PROPAGATION AND GUIDANCE

Transient Dielectric Coefficient and Conductance in Dielectric Media in Nonstationary Fields 205
A. Gutman

The Short Pulses Propagation in the Dielectric Media 211
A. Gutman

Electromagnetic Pulse Propagation across a Planar Interface Separating Two Lossy, Dispersive Dielectrics 217
John A. Marozas and Kurt E. Oughstun

Time Domain Measurement of Material Permittivity and Permeability 231
Clifton Courtney, Tracey Bowen, Jane Lehr and Kami Burr

Measurements of Short-Pulse Propagation through Concrete Walls 239
John F. Aurand

Propagation of UWB Electromagnetic Pulses through Lossy Plasmas........ 247
Steven L. Dvorak, Donald G. Dudley and Richard W. Ziolkowski

SCATTERING THEORY, COMPUTATION, AND MEASUREMENTS

Early Time Signature Analysis of Dielectric Targets Using UWB Radar 255
Shane Cloude, Alec Milne, Chris Thornhill and Graeme Crisp

Conservation of Power in the Galerkin Approximation of the Electric Field Integral Equation 263
Stuart M. Booker

Scattering of Short Radar Pulses from Multiple Wires and from a Chaff Cloud........ 271
Herbert Überall and Yanping Guo

F.D.T.D. Method Applied to the Generation and Propagation of Short Pulse 279
F. Tristant, F. Torrès, P. Leveque, Pr. B. Jecko, D. Serafin, C. Cruciani and P. Noël

Short Pulse Scattering Measurements on Conducting Cylindrical Cavities........ 287
Marc Piette and David Perrot

RCS Determination from Localized Short-Pulse Scattering Measurements: Theory and Experiment 295
Morris P. Kesler, James G. Maloney, Eric J. Kuster, Paul G. Friederich and Brian L. Shirley

SIGNAL PROCESSING

Time-Frequency Analysis

Feature Extraction from Electromagnetic Backscattered Data Using Joint Time-Frequency Processing........ 305
L. C. Trintinalia and H. Ling

Classification of Buried Targets Using Time-Frequency Signatures Extracted by a Ground Penetrating Radar 313
H. C. Strifors, A. Gustafsson, S. Abrahamson and G. C. Gaunaurd

Short-Pulse Radar via Electromagnetic Wavelets 321
Gerald Kaiser

Spectral Techniques

The E-Pulse Technique for Dispersive Scatterers 327
S. Primak, J. LoVetri, Z. Damjanschitz and S. Kashyap

Spectral Correlation of Wideband Target Resonances........ 335
Vincent Sabio

Probabilistic Considerations

Robust Target Identification Using a Generalized Likelihood Ratio Test........ 343
Jon E. Mooney, Zhi Ding and Lloyd Riggs

New Methods of Designing Optimum Broad-Band Radar Signals........ 351
Jean-Philippe Ovarlez and Jacques Dulost

Ultra-Wideband Radar Detection in White Noise 361
M. Steiner, K. Gerlach and F. C. Lin

General

Error Correction in Transient Electromagnetic Field Measurements Using Deconvolution Techniques 373
Jian-Zhong Bao, Jonathan C. Lee, Michael E. Belt, David D. Cox, Satnam P. Mathur and Shin-Tsu Lu

BROADBAND ELECTRONIC SYSTEMS AND COMPONENTS

Systems and Components

Ultrawide Band Sources and Antennas: Present Technology, Future Challenges 381
W. D. Prather, C. E. Baum, F. J. Agee, J. P. O'Loughlin, D. W. Scholfield, J. W. Burger, J. Hull, J. S. H. Schoenberg and R. Copeland

A Device for Radiating High Power RF Fields from a Coaxial Source 391
Jimmy Wells, Clifton Courtney, Tracey Bowen, David Eckhardt, Norman Keator, Carl Noggle, Donald Voss, Gary Watt and Harvey Wigelsworth

High Voltage UWB Horn Antennas 397
P. D. Smith and C. J. Brooker

Antennas and Electric Field Sensors for Ultra-Wideband Transient Time-Domain Measurements: Applications and Methods 405
C. Jerald Buchenauer, J. Scott Tyo and Jon S. H. Schoenberg

Ultra-Wideband Radars

Dense Media Penetrating Radar 423
Kwang Min and Marcelious Willis, Jr.

First Achievement of Pump and Probe Experiments Involving an Optoelectronic Gigahertz Ultrashort Pulse Generator for Measurements of Transient Properties in Materials .. 431
Jean-François Eloy, Nicolas Breuil, Vincent Gerbe and Jean Hugues Trombert

Target Detection and Imaging Using a Stepped-Frequency Ultra-Wideband Radar 439
E. J. Rothwell, K. M. Chen, D. P. Nyquist, A. Norman, G. Wallinga and Y. Dai

Polarimetric Ultra-Wideband Radars

Polarimetry in Ultrawideband Interferometric Sensing and Imaging 447
Wolfgang-Martin Boerner and James Salvatore Verdi

Polarization Processing for UWB Radar 461
Shane R. Cloude

Implementation of the Optimal Polarization Contrast Enhancement Concept in Ultrawideband (Multispectral) POL-SAR Image Analysis 469
Harold Mott and Wolfgang-M. Boerner

Polarization Structure of Ultra-Wide-Band Radar Signals 477
V. A. Sarytchev and G. B. Katchalova

BURIED TARGETS

Analytic Methods for Pulsed Signal Interaction with Layered, Lossy Soil Environments and Buried Objects 485
Leopold B. Felsen

Short-Pulse Scattering from and the Resonances of Three-Dimensional Buried Targets 499
Lawrence Carin and Stanislav Vitebskiy

Comparative Analysis of UWB Underground Data Collected Using Step-Frequency, Short Pulse and Noise Waveforms 511
E. K. Walton and S. Gunawan

INDEX 517

Ultra-Wideband, Short-Pulse Electromagnetics 3

THE TIME EVOLUTION OF PHOTONIC CRYSTAL BANDGAPS

K. Agi, M. Mojahedi and K.J. Malloy

Center for High Technology Materials
University of New Mexico
Albuquerque, NM 87131

ABSTRACT

The concept of a scaled group delay time is applied to a finite one-dimensional periodic array of dielectrics as a means of obtaining a group velocity. The scaling factor is shown to be the physical distance and this derived group velocity is compared to the group velocity of an infinitely periodic structure. Joint time-frequency analysis is performed on the response of a one-dimensional structure and the time-to-formation of the pass bands is shown to be determined by the peak group velocity in a given band. These concepts are then extended to three-dimensional photonic crystals and shown to give good agreement.

INTRODUCTION

Photonic crystals (PCs) are three- or lower-dimensional periodic dielectric structures that exhibit pass- and stop-bands. The one-dimensional PC has a wide range of applications in the optical domain as reflectors, filters and anti-reflection coatings[1]. However, for lower frequency microwave/RF applications, conventional technology has limited the use of the one-dimensional PCs. On the other hand, the two- and three-dimensional PCs, such as frequency selective surfaces (two-dimensional) or photonic bandgap crystals (three-dimensional), have found some applications in the microwave domain such as substrates for narrowband antennas[2], filters[3], and frequency selective reflectors for high power microwave systems[4]. For ultra-wideband (UWB) systems, usage of PCs requires a better understanding of the time evolution of the pass- and stop-bands in the crystal. Fortunately, the ability to generate short electromagnetic pulses has made it possible to investigate the interaction of UWB signals with highly dispersive structures[5]. This paper addresses the issue of the band formation in PCs. Initially, one-dimensional structures are used to gain insight to the problem, and subsequently the ideas are extended into the experimental properties of a three-dimensional structure.

ONE-DIMENSIONAL PHOTONIC CRYSTALS

The analysis of one-dimensional PCs begins with the study of an infinitely periodic array of dielectric slabs. In order to study the evolution of the pass- and stop-bands, the group velocity of the system needs to be calculated. The group velocity is the inverse of the first derivative in the Taylor series expansion of the Bloch propagation constant (K) about a given frequency[6]. For this simple case, the required dispersion relation (ω vs. K) can be obtained analytically by applying periodic boundary conditions to the electric field[1].

To determine the evolution times in a finite periodic structure, a group velocity needs to be defined which should approach the group velocity of an infinitely periodic crystal in the

Ultra-Wideband, Short-Pulse Electromagnetics 3
Edited by Baum *et al.*, Plenum Press, New York, 1997

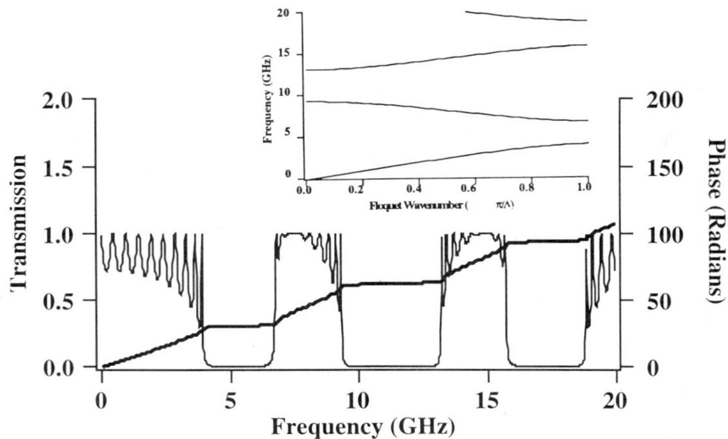

Figure 1. Magnitude (thin line) and unwrapped phase (thick line) of the transmission response of a 10 period multi-layer dielectric structure. d_1=d_2=0.635 cm, n_1=3.162, n_2=1. The inset is the corresponding dispersion curve obtained from the eigenvalue equation for the infinite structure.

limiting case. In order to discuss group velocity, the concept of group delay, which is simply the derivative of the phase of the transfer function with respect to frequency[7], is utilized. If the group delay is scaled by a length, the result is the desired group velocity.

In order to obtain the phase of the transfer function, a transmission line model is used. In this model, one period of the dielectric multi-layer is represented by two transmission lines with characteristic impedance Z_i, length d_i, and propagation constant k_i, for i=1,2, such that the overall ABCD matrix can be obtained[6]. The one period matrix is raised to the power of N, where N is the number of periods in the structure, and hence the transmission coefficient can be determined from the resultant matrix[8]. Figure 1 shows the transmission magnitude and unwrapped phase through the structure with the corresponding dispersion curve for the infinitely periodic structure shown as an inset.

To determine the scaling factor, consider an infinitely periodic structure. The relation between any field point and a field point NΛ away is given by Bloch's transformation theorem:

$$E(x+N\Lambda,K)=E(x,K)e^{iKN\Lambda}$$

where N is the number of periods and Λ is the physical length ($\Lambda=d_1+d_2$) of one period. The ratio of the two fields leads to a transfer function whose magnitude is 1 and whose phase, Φ, is KNΛ. The derivative of the phase with respect to frequency, which is the group delay, is given by

$$\frac{\partial\Phi}{\partial\omega}=\frac{N\Lambda}{v_g}$$

From the above it is clear that the scaling factor is the physical distance of the structure as opposed to the optical path length ($\Lambda=d_1+d_2$ vs. $n_1d_1+n_2d_2$, where n_i is the index of refraction). Figure 2 shows the comparison of the group velocity of the infinite structure (markers), calculated from the derivative of the dispersion curve, with a 10 period multi-layer (solid line), calculated from the scaled group delay. Away from the transition regions between the stop bands and the pass bands (i.e. band edges), the infinitely periodic result is approximately the average value of the finite structure. Near the band edges there is an insufficient number of periods to approximate the group velocity to any reasonable accuracy. However, the work here will be relying on the peak group velocity which occurs well away from the band edges.

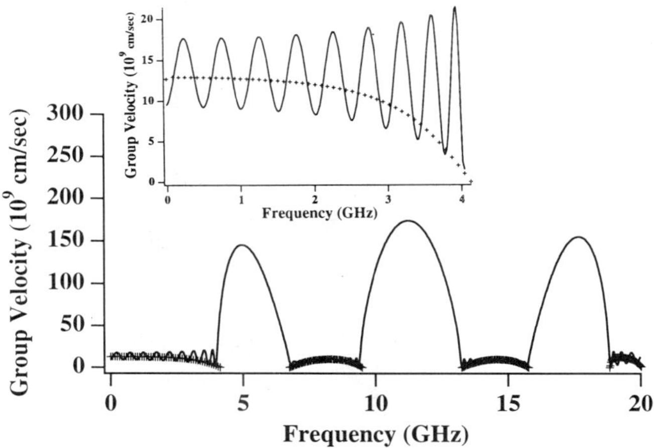

Figure 2. Group velocity calculated for the infinitely periodic array of dielectrics (markers) and a 10 period finite periodic structure (solid line). The inset is an expanded view of the first band.

The joint time-frequency analysis (JTFA) response can be obtained from the transmission response in Figure 1. Figure 3 shows the spectrogram using an adaptive, short-time fourier transform and Gabor algorithms[9], where in all cases, the vertical axis is time, the horizontal axis is frequency and the relative intensities are shown as the spectrogram. Independent of algorithm, the time-to-formation of the pass-bands, which is the start of the pulse to where the first wave appears, is governed by the peak group velocity. In other words, the first wave to appear is the undiffracted wave that is traveling at the peak group velocity for a given pass band. For all bands, there is good agreement between the delay time obtained from the JTFA spectrogram, the scaled group delay obtained from the phase of the transfer function and the derivative of the dispersion curve for the infinitely periodic structure. A summary of the results is given in Table 1.

Table 1. Summary of the group velocities obtained by scaling the JTFA delay time, scaling the group delay and the derivative of the dispersion curve for the infinitely periodic structure.

Band Number	JTFA (cm/s)	Group Delay (cm/s)	Infinite (cm/s)
1	1.41×10^{10}	1.29×10^{10}	1.308×10^{10}
2	1.12×10^{10}	0.97×10^{10}	0.99×10^{10}
3	1.01×10^{10}	0.94×10^{10}	0.97×10^{10}

On the other hand, the completion of the band, which is defined as the time from the start of the pass band to the end of the pass band, is difficult to deduce from the JTFA due to the algorithm dependence of the spectrograms. Hence it is difficult to differentiate between the real features and the extraneous ones. In other words, the decomposition of the time signal, to obtain the JTFA spectrogram, is dependent on the basis of the decomposition. This basis dependence creates cross-terms in the spectrogram which may be mistaken for real features. Hence, to avoid this dependence, the focus will be the formation time.

THREE-DIMENSIONAL PHOTONIC CRYSTALS

The concepts developed in the one-dimensional case are extended here. For the three-dimensional PC, a four-period face-centered-cubic structure is used. A detailed description of

Amplitude (Arb. Unl s)
-0.10
0.00
0.10
0.20
0
2
4
6
8
10
Time (ns)
Normalized Transmission
1.0
0.8
0.6
0.4
0.2
0.0
0
5
10
15
20
Frequency (GHz)

(a)

Amplitud e (Arb. Unl s)
-0.10
0.00
0.10
0.20
0
2
4
6
8
10
Time (ns)
Normalized Transmission
1.0
0.8
0.6
0.4
0.2
0.0
0
5
10
15
20
Frequency (GHz)

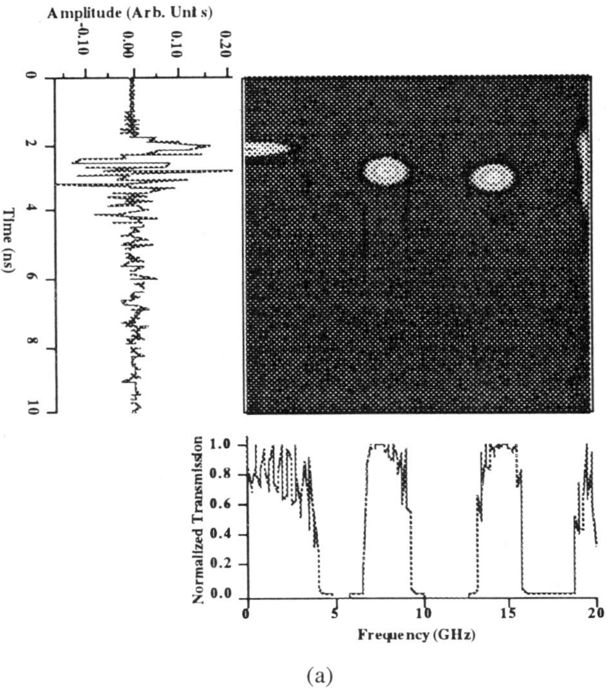

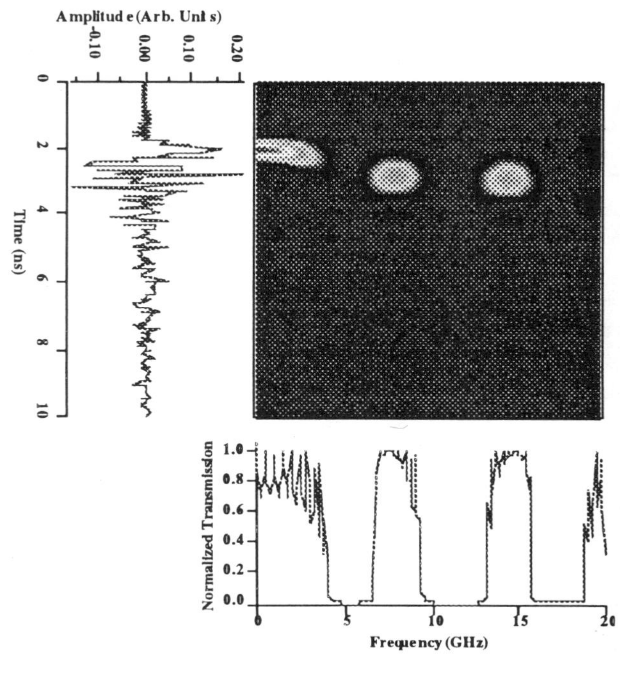

(b)

Figure 3.

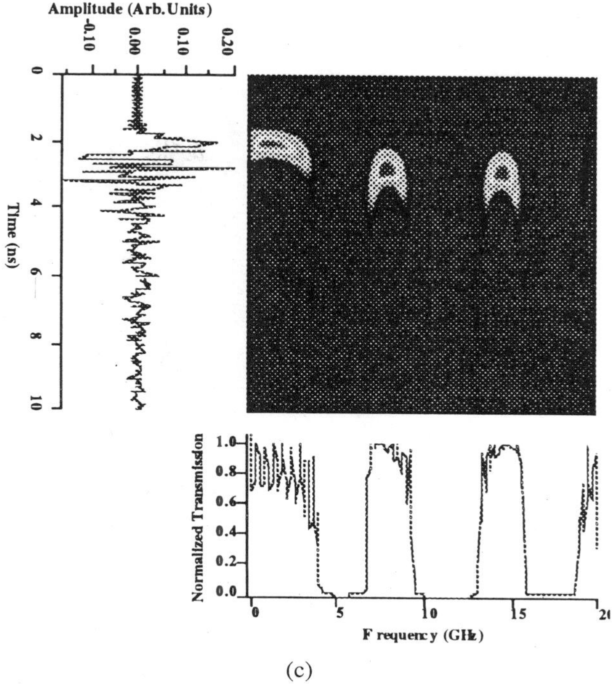

(c)

Figure 3. Joint time-frequency analysis using (a). adaptive algorithm, (b). short-time fourier transform and (c). Gabor transform. The time-to-formation of each pass band is determined by the undiffracted wave that is traveling at the peak group velocity for the particular band.

the structure can be found in the paper by Brown, et al[10]. The transmission response (S_{21}) of the PC is experimentally obtained using a vector network analyzer (HP 8510) from 15 to 25 GHz. Here the phase information is preserved, hence the group delay can be calculated. Figure 4 shows the magnitude and unwrapped phase for the frequency response of the crystal at normal incidence (L-point), obtained in the experiment. The points shown in Figure 4 are a linear curve fit to the data. This facilitates the determination of the slope and hence the group delay for the structure. In order to obtain the group delay, the slope of the phase curve is divided by 2π to scale the frequency into radian frequency correctly. Since there are only two pass bands that exist in the frequency range of the network analyzer, the calculations of the group delays will be limited to these two bands.

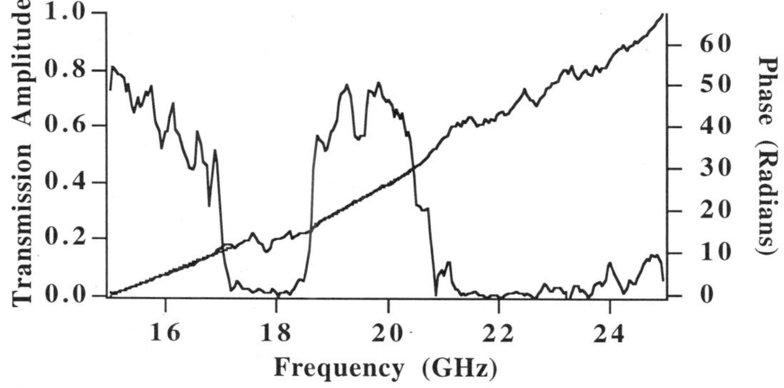

Figure 4. Magnitude and unwrapped phase of the experimental transmission response of a four-period, three-dimensional, face-centered-cubic photonic crystal. The points are a linear fit to the phase data which facilitate in the determination of the slope to obtain the group delay.

Table 2. Comparison of group delay and JTFA delay for various angles of incidence for the three-dimensional PC.

Crystal Direction	Band Number	Group Delay (ns)	JTFA Delay (ns)
L-point (0°)	1	0.844	0.9
	2	1.250	1.3
K-point (35.26°)	1	0.994	1.0
	2	2.020	1.9
W-point (39.2°)	1	0.943	1.0
	2	1.910	1.8

As in the one-dimensional case, the transmission response is inverse fourier transformed and the JTFA spectrogram is obtained. The adaptive algorithm is used to determine the spectrogram and is shown in Figure 5. Once again, the time-to-formation is determined from the delay in the spectrogram and calculated from the derivative of the phase. Here, the group delays can be compared directly using the two methods since there is no scaling factor in calculating delay times. For the three-dimensional PC, the transmission response for various high-symmetry directions (incident angles) are measured and the corresponding delays are calculated. The results are shown in Table 2. Good agreement is obtained with the two methods in determining the time-to-formation of the pass bands.

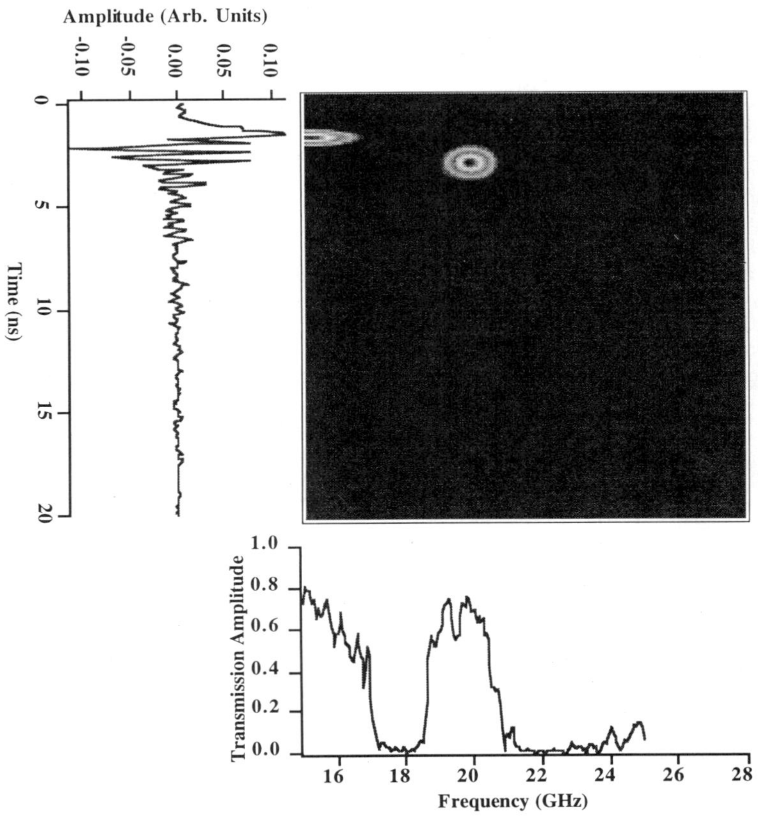

Figure 5. Joint time-frequency response, using the adaptive algorithm, of a four-period, three-dimensional face-centered-cubic photonic crystal at normal incidence. The delay shown is in good agreement with the group delay calculated by taking the derivative of the phase.

In summary, a group delay (velocity) is derived for a finite periodic structure. For the one-dimensional PCs, the group velocity is compared to the group velocity obtained by taking the derivative of the dispersion curve for an infinitely periodic structure. It is determined that the first wave to appear in all cases is the one travelling at the peak group velocity for each band. For the three-dimensional structure, the complex transmission is experimentally obtained and the phase is used, as in the one-dimensional case, to determine the group delay in the bands. The group delay is then compared to the results obtained using JTFA, where, the JTFA algorithm provides a method of pictorially obtaining a delay. Good agreement is obtained using the group delay and the JTFA algorithm for both the one-dimensional and three-dimensional structures.

ACKNOWLEDGEMENTS

This work was supported by the Air Force Office of Scientific Research in part through an AASERT grant.

REFERENCES

1. A. Yariv, P. Yeh, *Optical Waves in Crystals*, John Wiley and Sons, New York (1984).
2. E.R. Brown, C.D. Parker and E. Yablonovitch, Radiation properties of a planar antenna on a photonic-crystal substrate, JOSA B, 10:2 (1993).
3. T.K. Wu, *Frequcncy Selective Surfaces and Grid Arrays*, John Wiley and Sons, New York (1995).
4. K. Agi, L.D. Moreland, E. Schamiloglu, M. Mojahedie, K.J. Malloy and E.R. Brown, Photonic crystals: a new quasi-optical component for high-power microwaves, to appear in IEEE Trans. Plasma Sci.. June, 1996.
5. D. Kralj, L. Mei, T.T. Hsu and L. Carin, Short-pulse propagation in a hollow waveguide: analysis, optoelectronic measurement and signal processing.
6. D. Pozar, *Microwave Engineering*, Addison Wesley, Massachusetts (1990).
7. G.L. Matthaei, L. Young and E.M.T. Jones, *Microwave Filters, Impedance-Matching Networks, and Coupling Structures*, McGraw-Hill, New York (1964).
8. M. Born and E. Wolf, *Principles of Optics*, Pergamon Press, Oxford (1989).
9. L. Cohen, *Time Frequency Analysis*, Prentice Hall, New Jersey (1995).
10. E.R. Brown, K. Agi, C. Dill III. C.D. Parker and K.J. Malloy, A new face-centered-cubic photonic crystal for microwave and millimeter-wave applications, Microwave and Opt. Tech. Letters, 7:17 (1994).

OPTICALLY-EXCITED PHOTOCONDUCTING ANTENNAS FOR GENERATING ULTRA-WIDEBAND PULSES

David W. Liu, and Paul H. Carr

Rome Laboratory
Electromagnetics and Reliability Directorate (RL/ERAC)
Hanscom AFB, MA 01731-3010

INTRODUCTION

The generation of free-space ultrawideband (UWB) short-pulse (SP) electromagnetic radiation by laser pulses has been investigated vigorously in the past few years [1-8]. A variety of applications have been developed out of this innovative technique in the terahertz regime, such as, spectroscopic characterization of materials in the far-infrared region [9], transient nonlinear properties of dielectric materials at the submillimeter wavelengths [10], novel methods for characterizing semiconductor surfaces and interfaces [4], and very recently, terahertz time-domain imaging due to highly frequency-dependent absorption and dispersion for most chemicals [11]. When the photoconducting antennas are used in impulse radar communication and IFF system, the gigahertz (GHz) range is more applicable than the terahertz (THz), as electronic instrumentation is more mature.

Photoconducting antenna elements excited by 80 picosecond laser pulses are ultra-wideband sources of short-pulse electromagnetic radiation. In this work, 80 picosecond pulses from a frequency-doubled, modelocked, Q-switched YLF laser generate photoelectrons in dc-biased high resistivity semiconductor wafers. We have investigated InP:Fe, GaAs, and Low-Temperature grown GaAs for this application. The microwave radiation due to the dc-driven photocurrent is detected at the near field with an induction loop and at the far field by an impulse antenna. These signals are observed in real time with Tektronix 11802 sampling oscilloscope. We have studied nonlinearities in the microwave radiation for optical fluence as high as $300\mu J/cm^2$. This nonlinearity was analyzed and found to be consistent with the surface mobility of $1250 cm^2/V$-sec. We have also observed voltage nonlinearities at dc-bias fields as low as 12KV/cm.

A multi-element pulses antenna phased array can be implemented with multiple laser beams and the size of the radiating antenna can be varied by changing the diameter of the optical beam. Fiber optic feeds for individual elements make novel two-dimensional (2-D) and three-dimensional (3-D) phased array antennas possible. Spectrum shaping of the radiated waveform is being investigated for high-resolution-tracking-radar and noncooporative IFF systems.

EXPERIMENT

In this paper we demonstrate that 1-20 GHz microwave pulses are generated in free space by illuminating a semi-insulating semiconductor wafer with picosecond optical pulses at normal incidence. The experimental setup is illustrated schematically in Fig. 1. The mode locked, Q-switched, frequency-doubled YLF laser system provides an output of 100μJ pulse energy, 80ps pulse duration, and a center wavelength at 527nm. The laser pulses were selectively chosen by a Pockel cell with a repetition rate of 378Hz. After the frequency doubler (KDP crystal), the laser beam was split into two with approximately a 10:1 ratio. The weaker beam was fed into an ultrafast photodetector for triggering the TEK11802 sampling scope. The main beam was positioned to pass through an optical path delay line of about 15m (equivalent to a 50ns time delay) before reaching the sample in order to avoid the sampling scopes bandwidth limiting electrical delay line, allowing temporal resolution of 40GHz.

The electric pulses with an amplitude of 200V and width of 5μs were transformed by an automobile spark coil to provide up to 20KV bias voltage across the sample. The pulsed bias voltage was applied synchronously with the laser pulse to avoid the heat-load on the sample. The typical size of the GHz photoconducting antenna or the spacing between two electrodes was in the range of 1~5cm. The transient photocurrent generated in the voltage biased photoconductor served as the radiation source. The laser-illuminated region was deliberately positioned so as not to overlap the metallic electrode-semiconductor interface to prevent the possible Schottky barrier effect [4]. Semi-insulating GaAs, Fe-doped InP and MBE grown LTG-GaAs were investigated as the photoconducting materials. For the purpose of more directly characterizing the radiated electric field due to the induced photocurrent, measurements were made for the near field condition. A self-made loop-like wideband transducer- B-dot probe was placed about 5mm behind the sample.

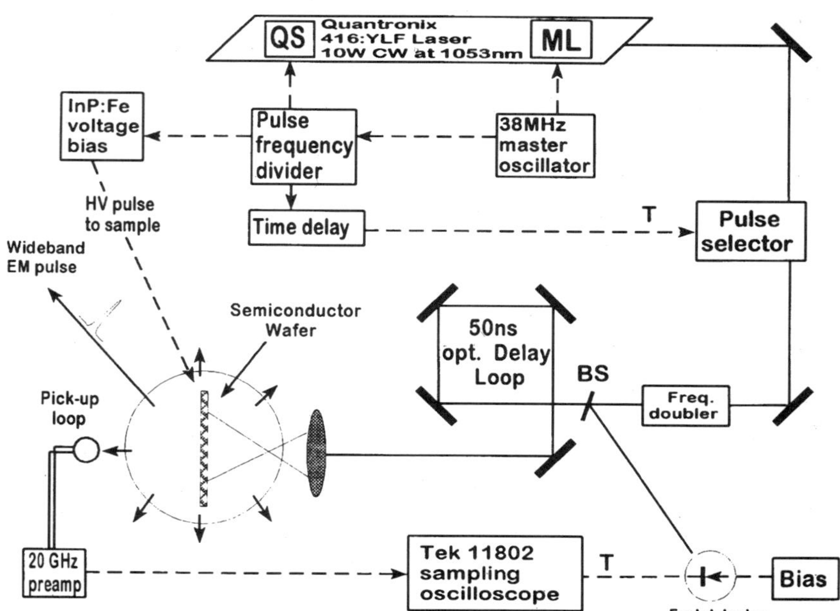

Figure 1. Schematic diagram for the experimental setup. EM, electromagnetic; T's, triggering; BS, beam splitter.

Since the output microwave pulse profile emulates the input optical signal, pulse-shaping is obtained by simply modulating the optical beams. Multi-element phased array antennas can be implemented with multiple laser beams. Fiber optic feeds for individual elements make novel 2-D and 3-D antenna arrays possible. The laser beam arrives at the Brewster angle incidence to avoid reflection loss. With the wavelength at 1.06μm, the beam will be partially absorbed and partially transmitted through the antenna element. Taking advantage of the nonmetallic feature of the antenna elements, a 3-D radiation source with a serial configuration was tentatively demonstrated.

POWER SCALING OF PHOTOCONDUCTING ANTENNA

A typical electromagnetic pulse received by the sampling oscilloscope from GaAs or InP sample is shown in Fig. 2 (a), the waveform is essentially a replica of the optical pulse since the photocarrier lifetime is much longer than the laser temporal width of 80ps (τ_c>1ns). On the other hand, when the photocarrier life time was much shorter than the laser temporal width, such as MBE low temperature grown LTG-GaAs (T_c<200fs) [12], a derivative waveform signal was displayed (Fig. 2(b)). The theoretical scope of the photoconducting antenna is primarily based upon a simple "current-surge model" [5,7,8]. The acceleration of the optically excited photocarriers under the applied electric field generates the electromagnetic radiation. The radiated electric field amplitude is proportional to the applied bias field for near field measurements. Field depletion under intensive optical fluence will occur and lead to a saturation phenomenon for the radiated electric field amplitude vs. the optical fluence. The experimental results generally follow the theoretical descriptions mentioned above [5-7].

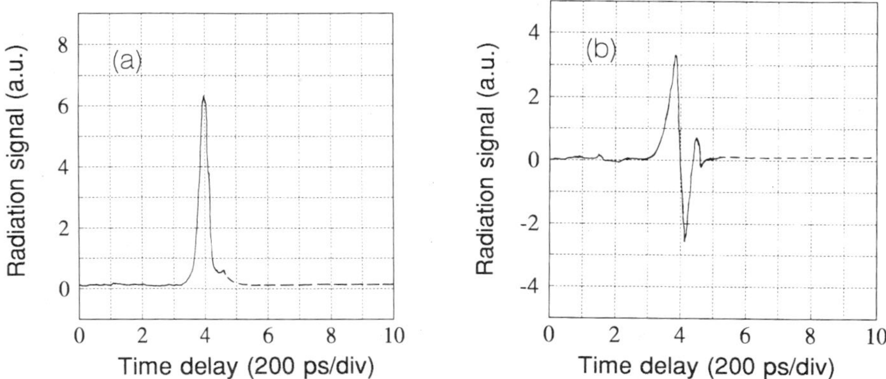

Figure 2. Picosecond microwave signal generated by 80 picosecond laser pulse on (a) semi-insulating GaAs, (b) MBE grown LTG-GaAs. Some artifacts after the main signal tail are not shown (in the dashed line region).

For the near field plane wave approximation, the inward radiated field **E** can be expressed in terms of the bias field, $\mathbf{E}_b$, and surface conductivity, σ, due to the required boundary conditions, as

$$\vec{E} = -\vec{E}_b \frac{\sigma_s(t)\eta_o}{\sigma_s(t)\eta_o + (1+\sqrt{\epsilon})} \qquad (1)$$

where η_o is the characteristic impedance of free space (377Ω) [5,7,8]. From equation (1) the radiated electric field amplitude is

$$E = E_b \frac{\sigma_{s,\max}\eta_o}{\sigma_{s,\max}\eta_o + (1+\sqrt{\epsilon})} \qquad (2)$$

The peak surface conductivity $\sigma_{s.max}$ can be expressed as

$$\sigma_{s.\max} = \frac{e(1-R)\mu_{tr}F_{op}}{\hbar\omega} \qquad (3)$$

where μ_{tr} is the transient carrier mobility at the point in time when the surface conductivity is maximum, R is the optical reflectivity, and F_{op} is the optical excitation fluence.

When the laser excitation fluence exceeds 20μJ/cm^2, as shown in Fig. 3, the radiation field is sublinear with respect to the optical excitation fluence. Strong saturation was observed when the optical fluence was above 100μJ/cm^2. Eqs. (1), (2), and (3), show that the radiated field amplitude, E, cannot exceed the bias field E_b, in magnitude and the radiated electric field is linearly proportional to F_{op} under the low excitation intensity (small $\sigma_{s,max}$ in comparison with $(1+\sqrt{\epsilon})/\eta_o$). Also, E saturates to E_b as $\sigma_{s,max}$ is much larger than the value of $(1+\sqrt{\epsilon})/\eta_o$. By adopting R=0.35 in eq.(3), we find that μ_{tr} =1250cm^2/(Vs) yields the best fit to the experimental data (Fig. 3). The photocarrier mobility obtained is approximately a factor of four higher than the corresponding value of the photoconducting antenna operating in the terahertz regime, which implying a better optical coupling efficiency for the generation of gigahertz electromagnetic pulses in this scheme.

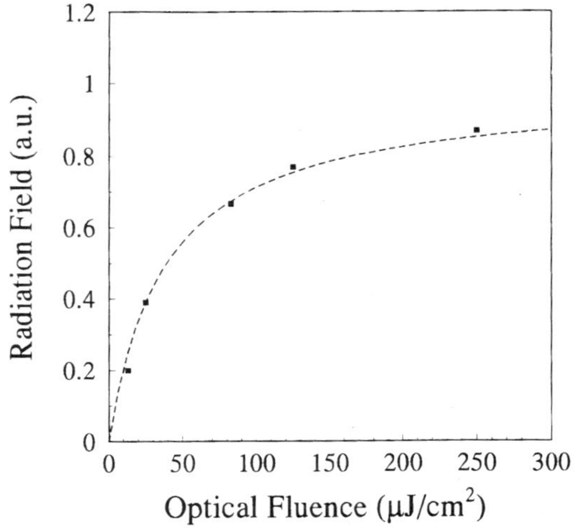

Figure 3. Microwave radiation intensity vs. optical excitation fluence at the bias field of 4KV/cm. Saturation effect due to the depletion of the electric field in GaAs was observed. The dashed curve is a best fit calculated with Eq. (2).

RADIATION FIELD vs. BIAS FIELD

We also observed a nonlinear relationship between the radiated field amplitude vs. the bias field in a gigahertz photoconducting antenna, which is dependent on the photocarrier population. It appears that "low" photocarrier population ($<3\times10^{16}/cm^3$) is the key for the saturation phenomenon, or even the possible "negative differential transient photoconductivity", to be observed. As shown in Fig. 4, under the low excitation ($<0.3\mu J/cm^2$ or less than $3\times10^{16}/cm^3$ photocarrier population), a departure from linearity begins at 4KV/cm for GaAs, and then increases monotonically a saturation value of approximately 6KV/cm (less deviation from linearity was found for InP). The characteristic curve is independent of the excitation intensity for optical fluence below $0.3\mu J/cm^2$. For some measurements, however, a slight drop (as much as 5%) in the peak amplitude was observed as the bias field was raised above 10KV/cm. Nevertheless, this so-called "negative differential photoconductivity" can not be confirmed by a lack of experimental precision. As the excitation density is increased, as shown in Fig. 5, the saturation phenomenon is less noticeable and gradually becomes sublinear before eventually approaching a linear relationship as the optical excitation fluence goes above $25\mu J/cm^2$.

According to equation (1), the departure from a linear behavior under low optical fluence implies that the maximum surface transient photoconductivity decreases as the bias field increases. Since the reflectivity was found to be insensitive to the bias field magnitude, the possibility of transient carrier velocity saturation as the bias field is above 6KV/cm for GaAs is suggested. For the energies of the photocarriers in excess of the intervalley separation of 0.31eV, intervalley scattering must be taken into consideration. With 2.35eV photon energy from our laser, more than 80% of the injected electrons can scatter into L valley within 200fs;

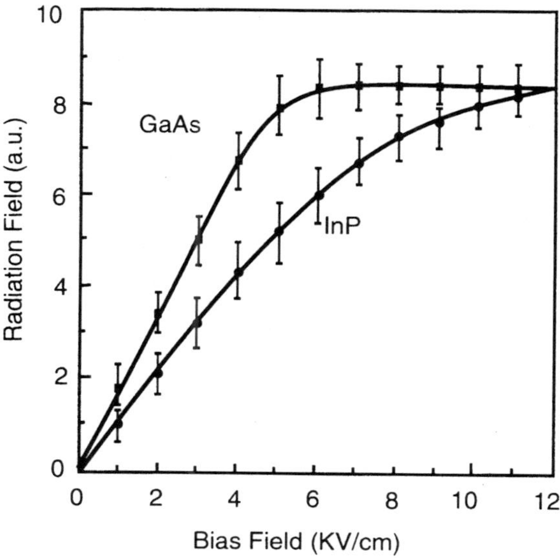

Figure 4. Microwave signal amplitude as a function of the bias field for GaAs, and InP with the optical fluence less than $0.3\mu J/cm^2$.

and relax back to the Γ valley approximately within 2~3 picosecond later [13,14]. Under a high electric field, the photocarriers get energized. Consequently, more electrons are likely to stay at the L valley, due to the Boltzmann distribution, with an electron temperature, T_e, much higher than the lattice temperature, T_L. The electrons at L valley have a higher effective mass and therefore a lower mobility, which is believed to be responsible for the saturation phenomenon observed (Fig. 4). Monte Carlo calculations also indicates a maximum carrier drift velocity at 5KV/cm electric field [14], which is in fair agreement with our experimental data.

The saturation becomes less visible and approaches a linear relationship when the optical fluence is greater than 20μJ/cm², as shown in Fig. 5. This is somewhat surprising if one considers that the temperature of hot carriers could be higher as the optical fluence is increased. Then, according to our previous qualitative analysis, the saturation phenomenon should be more prominent. This apparent discrepancy might be a result of the field depletion effect. According to equation (2), when $\sigma_{s,max}$ becomes comparable to $(1+\sqrt{\epsilon})/\eta_o$, the radiated electric field amplitude is no longer linearly proportional to $\sigma_{s,max}$, as in the low optical fluence case. The optical fluence of 20μJ/cm² will give $\sigma_{s,max}$ a significant value compared with $(1+\sqrt{\epsilon})/\eta_o$, manifested by Fig. 3, a deviation from a linear relationship for the radiated electric field amplitude vs. the optical fluence as the optical fluence is greater than 4μJ/cm². As indicated by equation (2), the value of E becomes much less sensitive to $\sigma_{s,max}$ as long as $\sigma_{s,max}$ is large with respect to $(1+\sqrt{\epsilon})/\eta_o$, which explains that the radiated field amplitude still goes higher monotonically when the bias field is above 5KV/cm at intense optical excitation. Another relevant aspect that might be considered is the electron-electron scattering mechanism, which can not be ignored with the carrier population higher than 10^{17}/cm³ [13,14]. As a result, the electron-electron scattering mechanism could "overwhelm" the contribution from the bias field and become the dominant factor responsible for the temperature evolution of the high density photocarriers. In retrospect, the result observed under intensive optical excitation gives us assurance that the earlier described saturation phenomenon could not be caused by the electrode degradation or field collapse at the high bias field. No further change

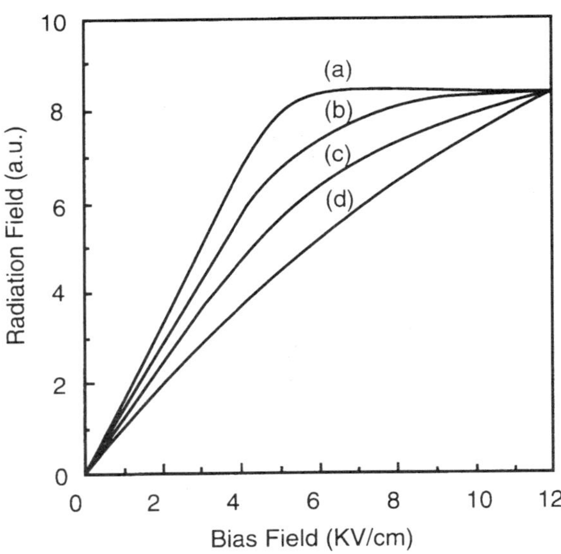

Figure 5. Microwave signal amplitude as a function of the bias field for GaAs at the optical excitation fluence of (a) 0.3, (b) 3.0, (c) 10.0, (d) 30.0, μJ/cm², respectively. All the curves are scaled to have the same radiation field magnitude at the bias field of 12KV/cm.

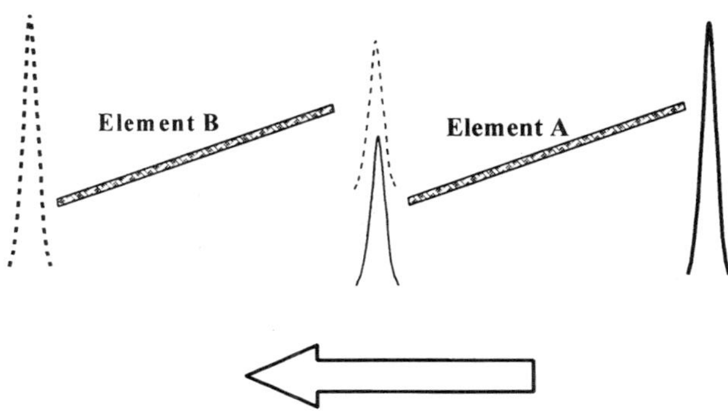

Figure 6. Establishment of a three-dimensional radiation source attributed by photoconducting antennas, as described in the text. Optical pulse and microwave pulse are depicted in solid line and dashed line respectively.

in the characteristic curve was observed when the photocarrier population is reduced to less than $3\times10^{16}/cm^3$. The band structure and field depletion effect are relevant for qualitatively describing the experimental data obtained.

THREE-DIMENSIONAL RADIATION SOURCE PERSPECTIVE

3-D radiation sources are attractive for having a compact device volume and with minimum aperture exposure area. It is known that 3-D sources are capable of radiating energetic short pulses in a selected (main) direction and low amplitude, long duration pulses in other directions. It is possible to optimize - for constant source volume and input energy - the radiation performance of general scalar, ultrawideband volume sources by a proper selection of the space distribution of the source [15].

The nonmetallic feature of photoconducting antenna makes a 3-D radiation source possible. As shown in Fig. 6, the 80 picosecond laser pulse at the wavelength of 1.06μm is employed as the optical source, which allows the laser beam to be partially transmitted through and partially absorbed in the sample. The optical pulse transmitted through goes together with the microwave pulse generated by the optical pulse absorbed in element A, which will activate element B. The optical incident angle can be positioned at the Brewster angle to avoid reflection loss and multiple-reflection between the elements. The elements also can be excited by different optical sources fed in from the side in a synchronous manner. In order to depressing the mutual disturbance or coupling between two adjacent elements, the bias field of the element B is set at the plateau region, as depicted in Fig. 4 for GaAs. As a result, the generation of the microwave pulse from the antenna B will not be affected by the presence of the microwave pulse from element A.

The establishment described above is beneficial in terms of delivering high angular-resolution microwave signals and compact device configuration. Parameters of the optical pulse profile, timing of each optical beam, and the bias field of individual antenna element provide the capability of spectrum-shaping the microwave signal output in real time.

CONCLUSION

The coherent nature of the signal from single GHz photoconducting antenna lays a groundwork for the fabrication of optical phased -array antenna, controlled by a fiber-optical network. Because the radiation sources are more localized, the schematic of such a phased-array antenna can significantly reduce the mutual disturbance, or so-called cross talk, between the individual antennas. The nonmetallic feature and fiber optic feeds for individual antenna elements make novel 2-D and 3-D antenna array possible. Such an establishment is expected to deliver high angular-resolution microwave signals out of compact device configuration. Due to the close assembly between the profile of the microwave signal and the optical signal, this technique provides the spectrum shaping capability for target recognition and tracking. The superposition nature of output microwave radiation controlled by the optical signals provides the degrees of freedom needed to steer and effectively modulate the radiation output.

REFERENCES

1. Ch. Fattinger and D. Grischkowsky, Terahertz beams, *Appl. Phys. Lett.* 54:490 (1989).
2. B.B. Hu, J.T. Darrow, X.-C. Zhang, and D.H. Auston, Optically steerable photoconducting antennas, *Appl. Phys. Lett.* 56:886 (1990).
3. X.-C. Zhang and D.H. Auston, Generation of steerable submillimeter waves from semiconductor surfaces by spatial light modulators, *Appl. Phys. Lett.* 59:768 (1991).
4. X.-C. Zhang, and D. H. Auston, Optoelectric measurement of semiconductor surfaces and interfaces with femtosecond optics, *J. Appl. Phys.* 71:326 (1992).
5. Justin T. Darrow, Xi-Cheng Zhang, David H. Auston, and Jeffrey D. Morse, Saturation properties of large-aperture photoconducting antennas, *J. of Quantum Electronics*, 28:1607 (1992).
6. D. You, R.R. Jones, P.H. Bucksbaum, and D.R. Dykaar, Generation of high-power sub-single-cycle 500-fs electromagnetic pulses, *Opt. Lett.* 18:290 (1993).
7. P.K. Benicewicz, J.P. Roberts, and A.J. Taylor, Scaling of terahertz radiation from large-aperture biased photoconductors, *J. Opt. Soc. Am. B*, 11:2533 (1994).
8. D.W. Liu, J.B. Thaxter, and D.F. Bliss, Gigahertz planar photoconducting antenna activated by picosecond optical pulses, *Opt. Lett.* 20:1544 (1995).
9. R.A. Cheville and D. Grischkowsky, Far-infrared terahertz time-domain spectroscopy of flames, *Opt. Lett.* 20:1646 (1995).
10. B.I. Greene, P.N. Saeta, Douglas R. Dykaar, S. Schmitt-Rink, and Shun Lien Chuang, Far-infrared light generation at semiconductor surfaces and its spectroscopic applications, *IEEE J. of Quantum Electron.* 28:2302 (1992).
11. B.B. Hu, and M.C. Nuss, Imaging with terahertz waves, *Opt. Lett.* 20:1716 (1996).
12. Gerald L. Witt, Optoelectronic properties of low-temperature III-Vs, in: *Proceedings of LEOS'94, 7th Annual Meeting*, 2:19 (1994).
13. Jagdeep Shah, Benoit Deveaud, T.C. Damen, W.T. Tsang, A.C. Gossard and P. Lugli, Determination of intervalley scattering rates in GaAs by subpicosecond luminescence spectroscopy, *Phys. Rev. Lett.* 59:2222 (1987).
14. G.M. Wysin, D.L. Smith, and Antonio Redondo, Picosecond response of photoexcited GaAs in a uniform electric field by Monte Carlo dynamics, *Phys. Rev. B*, 38:12514 (1988).
15. Edwin A. Marengo, Anthony J. Devaney, and Ehud Heyman, Analysis and characterization of ultrawideband scalar volume sources and the field they radiate, to be published in *IEEE Trans. Ant. & Propag.* (1996).

GROUND PENETRATING RADAR ENABLED BY HIGH GAIN GaAs PHOTOCONDUCTIVE SEMICONDUCTOR SWITCHES

G. M. Loubriel, M. T. Buttram, J. F. Aurand, and F. J. Zutavern

High Power Electromagnetics Department
Sandia National Laboratories, P. O. Box 5800, MS 1153,
Albuquerque, NM 87185-1153, (505) 845-7096

ABSTRACT

The ability of high gain GaAs Photoconductive Semiconductor switches (PCSS) to deliver fast risetime, low jitter pulses when triggered with small laser diode arrays makes them suitable for their use in ultrawide bandwidth (UWB), impulse transmitters. This paper will summarize the state-of-the-art in high gain GaAs switches and discuss how GaAs switches are being implemented in a transmitter for detection of underground structures. The advantage of this type of semiconductor switch is demonstrated operation at high voltages (100 kV) and repetition rates (1 kHz) with the potential for much higher repetition rates. The latter would increase the demonstrated average powers of 100 W to 1 kW and higher. We will also present an analysis of the effectiveness of different pulser geometries that result in transmitted pulses with varying frequency content. To this end, we have developed a simple model that includes transmit and receive antenna response, attenuation and dispersion of the electromagnetic impulses by the soil, and target cross sections.

INTRODUCTION (STATE-OF-THE-ART OF HIGH GAIN GaAs PCSS)

This research has focused on optically triggered, high gain GaAs switches for impulse sources for ultrawide bandwidth (UWB) transmitters. The practical significance of this high gain switching mode is that the switches can be activated with very low energy optical triggers, allowing for compact sources.[1] For example, we have shown[2] that a 90 nJ optical pulse has triggered switches that have delivered 48 MW in a 30-50 Ω system, and previously we have switched 6 MW for ~100 ns in a 0.25 Ω system.[3] The GaAs switches used in this experiment are lateral switches: they have two contacts on one side of a wafer separated by an insulating region of intrinsic material. At electric fields below 4 kV/cm, the

GaAs switches are activated by the creation of, at most, one electron hole pair per photon absorbed. This linear mode demands high laser power, and after the light is extinguished, the carrier density decays in 1- 10 ns. At higher electric fields these switches behave very differently. The high field induces carrier multiplication so that the amount of light required is reduced by as much as five orders of magnitude.[1,2] This high gain mode is characterized by fast current rise times (~430 ps). In the "on" state of the high gain switch there is a characteristic, constant field across the switch called the lock-on field. The switch current is circuit-limited provided the circuit maintains the lock-on field.[3] As the field increases, the switch risetime decreases and the trigger energy is reduced.[3] The PCSS are being tested for use in applications such as: UWB transmitters, firing sets for munitions, electro-optic modulators, current interrupters, active optical sensors, and pulsed power applications such as MV accelerators that operate at high repetition rates. Each of these applications imposes a different set of requirements on switch properties. Table I shows the best results obtained with the switches for these applications. The results are not all simultaneous. The work of many others has been presented at various conferences.[4]

Table 1. Results of tests with high gain GaAs switches. The first column is the best results obtained in various, independent tests. The second column are the results from a single system which can be used to produce ultra wideband pulses.

Table I Parameter	Best Individual Results*	Simultaneous Results (see next section)
Switch Voltage (kV)	155	100
Switch Current (kA)	7.0	1.26
Peak Power (MW)	120	48
Rise time (ps)	430	430
R-M-S jitter (ps)	150	150
Optical Trigger Energy (nJ)	90	180
Repetition Rate (Hz)	1,000	1,000
Electric Field (kV/cm)	100	67
Device Lifetime (# pulses)	6×10^6	5×10^4, (at 77 kV)

* Not all the results are simultaneous.

INITIAL IMPULSE SYSTEM

The feasibility of using GaAs switches to create voltage pulses suitable for driving UWB antennas has been previously demonstrated.[2] In that study we charged a nominally 1.0 ns long, 47 Ω, parallel plate transmission line to voltages of about 100 kV. This line was discharged with either one or two switches into a 30 Ω load. The circuit is shown in Figure 1. We measured the voltage on the transmission line and the current through the load. The voltage on the line rose to a peak value (100 kV in most cases, although voltages of 110 kV were also used) with a risetime of 210 ns. At peak voltage the laser diode arrays activated the switch and the line voltage dropped. If only switch one was triggered, the resulting load voltage was a monopulse. If both switches were triggered simultaneously the load current was a monocycle (bipolar pulse). This system operated in bursts of up to 5 pulses at a repetition rate of 1 kHz. The switches were fabricated from undoped GaAs with Ni-Ge-Au-Ni-Au metallization. Their insulating region separating the two contacts was 1.5 cm, the total contact width was 7.6 cm. Because of the high electric fields the switches were immersed in a dielectric liquid (Fluorinert®). To avoid corona and breakdown, the charged transmission line was in SF_6 gas. The laser diode arrays (with most of their electronics) are about 2" by 2" in size and were used in a configuration that allowed triggering the switches with as little as 90 nJ of energy.

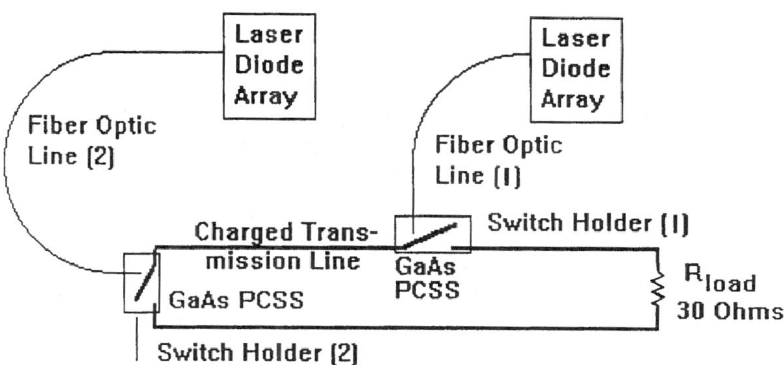

Figure 1. A short (1 ns), 47 Ω transmission line (the charge line) was charged to high voltage at a burst repetition rate of 1 kHz. Two switches were used on either side of the line to discharge the line into a 30 Ω load.

In one set of tests both laser diode arrays were used to activate switch one (in a configuration not shown in the figure) and obtain a monopulse. The highest current measured with this system is shown in Figure 2. The width of the current pulse and its peak value depend on the time delay between when the two laser diodes are triggered. When both diodes are triggered to produce simultaneous current pulses, the current is largest and the current pulse width is smallest. The highest current was 1.26 kA with a rise time of 430 ps and a pulse width of 1.4 ns. The peak power is 48 MW. With one laser diode activating one switch the current is about 1.0 to 1.1 kA with a rise time of about 770 ps and a pulse width of 1.7 -1.9 ns.

The second set of tests utilized both laser diodes, each triggering one switch, to produce a monocycle. Figure 3 shows the current waveform. In theory, with ideal switching, the monocycle should be composed of two monopulses of opposite polarity each with half the pulse width. Thus, we expect a monocycle composed of a negative and a positive pulse, each with a width of 0.9 ns. What we observe is a width of 1.0 ns for the negative pulse and 1.3 ns for the positive pulse. The reason for this is a timing error of about 200 ps. The minimum width should occur when both switches are triggered simultaneously. It is very important to trigger both switches at the same time to obtain full voltage and to obtain the proper waveform. In these tests, the switch jitter did not allow us to always reproduce the monocycle. As will be seen in the next sections, this monocycle is more effective than the monopulse for ground penetration. To reduce jitter it will be necessary to increase the laser energy. Fortunately, the laser diode array can be run, with electronics of the same size, at up to 1 μJ of energy. At this energy we expect the jitter to be sufficiently small to allow for effective switching.

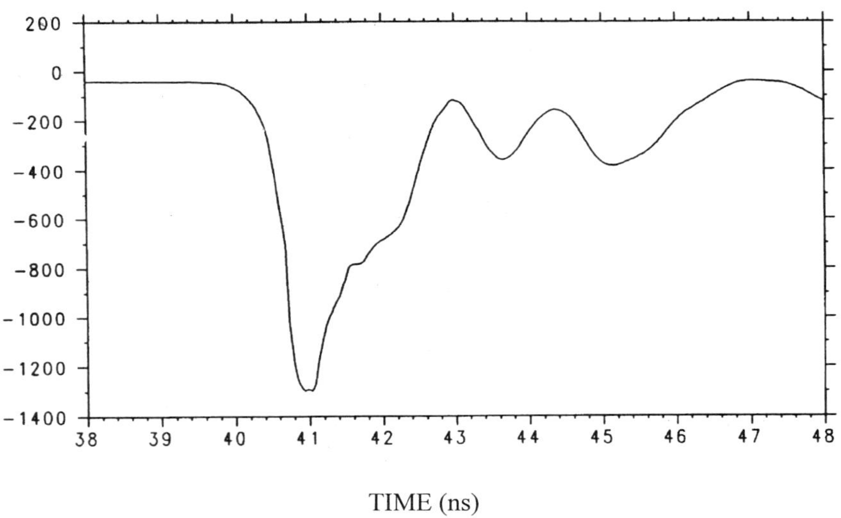

Figure 2. The current through the 30 Ω load when both laser diodes illuminate one switch: 1.26 kA peak (48 MW), 430 ps rise, 1.4 ns wide.

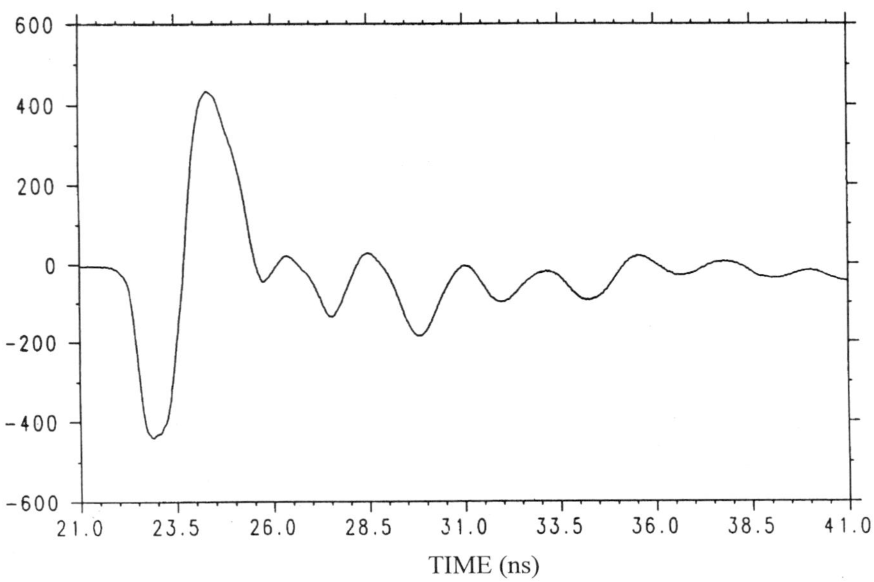

Figure 3. The current (in Amps) through the 30 Ω load when each of the two laser diodes is used to trigger one switch.

MODEL OF RADAR SYSTEM

To analyze the relative merits of different waveforms with varying frequency content we have developed a simple model that includes transmit and receive antenna response, attenuation and dispersion of the electromagnetic impulses by the soil, transmission at the air/soil boundary (upon entering and exiting), and target cross sections. The target that we will choose is a metallic, square plate (10 m by 10 m) buried at a depth D (10 m in this study). To examine the radar equation for a buried object, we use the simple radar equation for the frequency dependent amplitude of the transfer function [radar-to-target-to-receiver]. The beam is assumed to be normal to the ground which is flat and featureless. Figure 4 shows two different voltage waveforms (input to antenna): a monopulse and a monocycle (a bipolar pulse). The formulas used to calculate these waveforms are either a Gaussian (sigma = 0.8 ns) centered at 5 ns or the derivative of a Gaussian. Both were normalized to peak amplitude of 1 (as shown). These waveforms were meant to mimic the results obtained with the previous impulse system. Figure 5 shows the frequency spectrum of both of these waveforms. The monopulse, due to its non-zero integral, has peak frequency content at 0.0 Hz (dc). Most of its energy content is from 0.0 Hz to about 200 MHz. The monocycle, on the other hand, has no frequency content at 0.0 Hz and peaks at 200 MHz. Most of its energy content is from 100 MHz to 350 MHz.

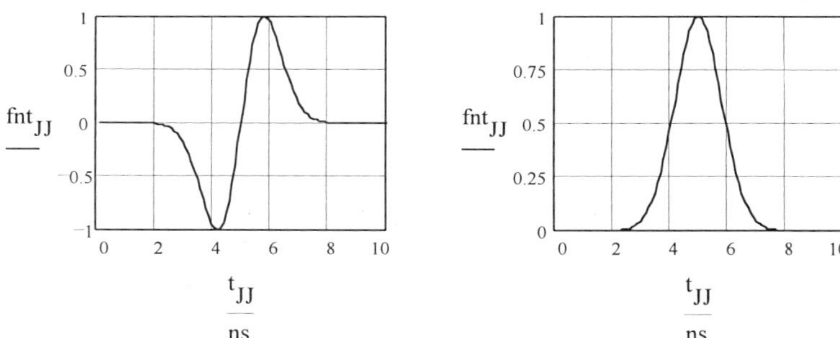

Figure 4. Voltage waveforms used in this study. On the left is a monocyle, on the right is a monopulse.

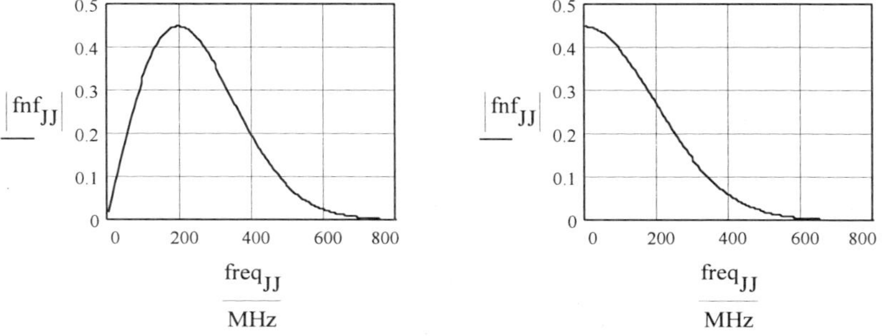

Figure 5. Frequency spectrum of the voltage waveforms (left: monocycle, right: monopulse).

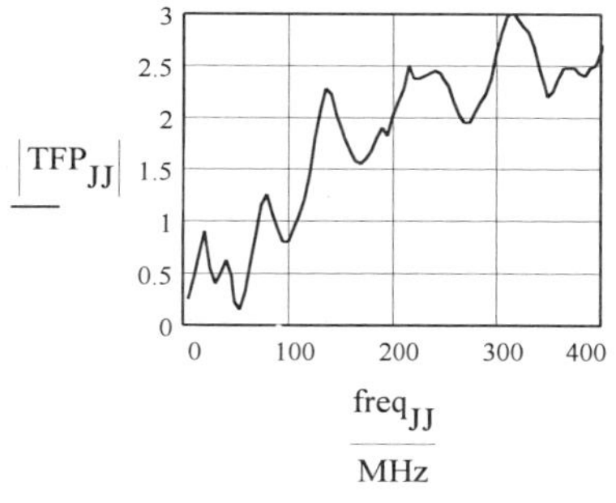

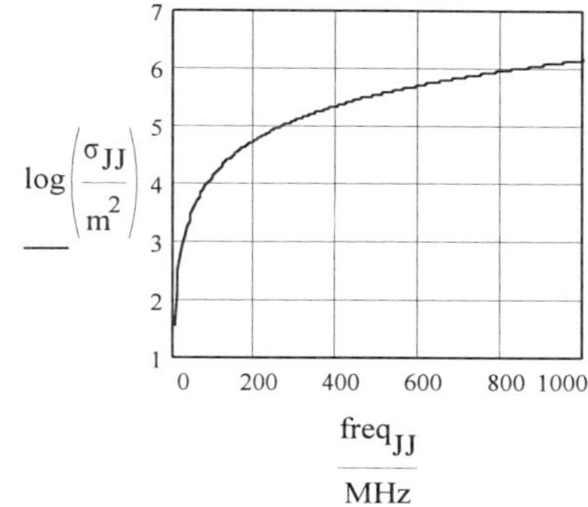

Figure 6. Antenna transfer function (left) and the cross section of a flat metal plate with dimensions of 10 m by 10 m.

An existing antenna which may be used for this work is a large TEM horn with flared aperture plates and a high-voltage inline coaxial 'zipper' balun (the 'adapter' in reference 5) as the input section. It has a transfer function whose magnitude is shown in figure 6. This was developed by measuring two time domain waveforms. One was a bipolar (monocycle) voltage pulse created by a custom pulse generator with a +/- 100 V amplitude, and primary spectral content from 50 to 400 MHz. The other was the main radiated E-field, at a range of 8.0 m. The transfer function was then formed by the complex ratio of the discrete Fourier transform of the radiated field divided by the transform of the input excitation voltage. The cross section (in Air) of the target is also shown in figure 6. Because the antenna transfer function and the target cross section are larger at higher frequencies, the monopulse waveforms will not radiate as efficiently as the monocycle. It is tempting to use waveforms that have very high frequency components. On the other hand, the attenuation by the soil is larger at higher frequencies. Figure 7 shows attenuation for two different types of soil: dry sand and San Antonio Clay Loam with a water content of 5%. The soil was modelled using the parameters in reference 6.

Using this information we can predict the receive waveform. Figures 8 and 9 show different receive electric field waveforms at the transmit site for a monocycle and monopulse voltage pulse. Figure 8 shows the return from penetration through 10 m of air and 10 m of soil. Note that the waveforms are similar to the derivative of the initial voltage pulse. Figure 9 shows the same type of results but with San Antonio clay loam with 5%

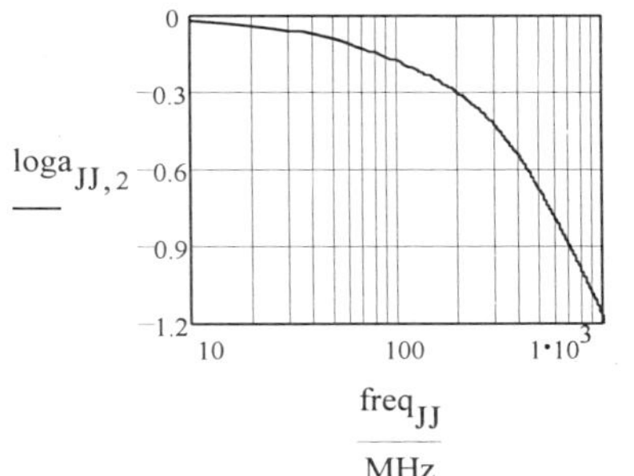

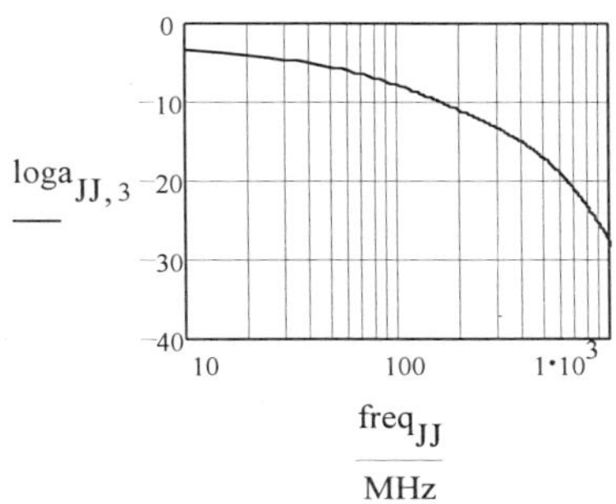

Figure 7. The log of the attenuation versus frequency for a penetration depth of 10 m for dry sand (left) and San Antonio clay loam, 5% water (right).

water content. The receive functions show two major effects. First, a large attenuation of about 10^5 in intensity. Second, that the temporal extent of the pulse is now about 15 ns. The pulse is expanded because the high frequencies are preferentially attenuated and the various frequency components are propagating through the soil at different speeds.

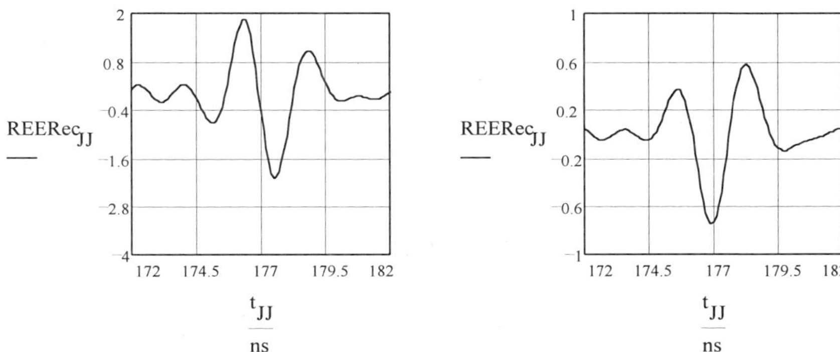

Figure 8. The electric field at the transmit site after transmission through the soil (dry sand) for the monocycle (left) and monopulse (right) waveforms.

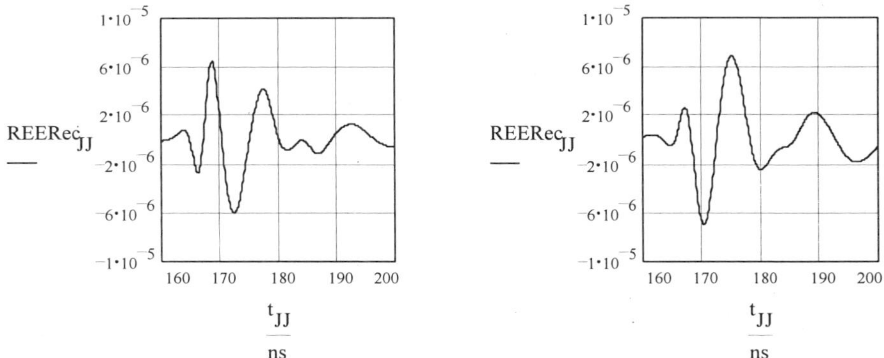

Figure 9. The electric field at the transmit site after transmission through the soil (San Antonio clay loam, 5% water) for the monocycle (left) and monopulse (right) waveforms.

CONCLUSION

We have developed a model that shows both the monocycle and monopulse waveforms should result in measurable penetration through different soils for the detection of large, deeply buried metal plates. There is little difference in the receive pulse amplitudes for these waveforms. That is surprising because the spectrum of the monopulse waveform has most of its energy at frequencies that are not transmitted by the antenna. This effect was cancelled by the attenuation of the soil. From a practical standpoint the production of a monopulse voltage waveform is much simpler than the monocycle and, thus, based solely on this result it may lead to the use of that type of waveform. Unfortunately, the monopulse may result in late time ringing of the antenna, since the energy that is not radiated stays trapped in the transmitter and may be radiated at later times. For this reason we will test both waveforms in future experiments to determine the transmitter efficiency and ringing.

The ultimate goal of the radar model is to assess different transmitter technologies for their effectiveness in detecting large, deeply buried structures. In particular, we need to know the average transmit power required for a given transmitter and target (of given size and depth). At this point in time, we know that we can produce 100 W average power, high peak power (48 MW) impulses in using laser diode triggered PCSS operated in the high gain mode in a configuration that allows for detection of large underground structures. The system needs to be optimized to allow for the highest possible average power, thus repetition rate is very important. The initial system was operated at a burst repetition rate of 1 kHz but the switches can be operated at much higher repetition rates and thus average powers of 1 kW could be used, if needed.

ACKNOWLEDGMENT

The authors want to thank W, D. Helgeson, D. J. Brown, M. W. O'Malley, B. Brock, A. G. Baca, H. J. Hjalmarson, T. A. Plut, and C. H. Sifford for their help. This work was supported by the United States Department of Energy under contract DE-AC04-94AL85000.

REFERENCES

1. G. M. Loubriel, M. W. O'Malley, and F. J. Zutavern, "Toward Pulsed Power Uses for Photoconductive Semiconductor Switches: Closing Switches," Proc. 6th IEEE Pulsed Power Conference, P. J. Turchi and B. H. Bernstein, eds., Arlington, VA, 1987, p. 145.
2. G. M. Loubriel, F. J. Zutavern, M. W. O'Malley, and W. D. Helgeson, "High Gain GaAs Photoconductive Semiconductor Switches for Impulse Sources," Proc. of SPIE Optically Activated Switching Conference IV, SPIE Proc. Series Vol. 2343, pp. 180-186, W. R. Donaldson, ed., Boston, MA, October 31-November 4, 1994.
3. F. J. Zutavern and G. M. Loubriel, "High Voltage Lateral Switches from Si or GaAs," High-Power Optically Activated Solid-State Switches, A. Rosen and F. J. Zutavern, Eds., Artech House, Boston, 1993, p. 245.
4. See Proceedings from: 6-8th IEEE Pulsed Power Conf., 1987, 1989, 1991; 18-20th IEEE Power Modulator Symposium, 1988, 1990, 1992; SPIE Optically Activated Switching I-III, (vol. 1378, 1632, 1873), 1990, 1992, 1993; and IEEE Trans. Elec. Devices, (vol. 37, 38), 1990, 1992.
5. R. S. Clark, L. F. Rinehart, M. T. Buttram, and J. F. Aurand, "An overview of Sandia National Laboratories' Plasma Switched, GigaWatt, Ultra-Wideband Impulse-Transmitter Program," Ultra-Wideband, Short Pulse Electromagnetics, L. Bertoni, L. Carin, and L. Felsen, eds., Plenum Press, New York, 1993, pp. 93- 98.
6. B. C. Brock and W. E. Patitz, "Optimum Frequency for Subsurface-Imaging Synthetic Aperture Radar," Sandia National Laboratories' report SAND93-0815 and B. C. Brock and K. W. Sorensen, "Electromagnetic Scattering from Buried Objects," report SAND94-2361.

ULTRAWIDEBAND PULSER TECHNOLOGY

Prof. David M. Parkes

DRA Malvern
St Andrews Road
Malvern
Worcestershire WR14 3PS
United Kingdom

INTRODUCTION

The advances in UWB pulser performance over the last decade have given rise to a large number of novel concepts covering a wide range of applications. Pulser characteristics are extremely varied, ranging from outputs of a few millivolts to megavolts and repetition frequencies of single shot to megahertz. Size, weight and prime power requirements vary significantly and it is now a major task for the engineer to be able to select the optimum pulser for his application. Research is still being conducted into new components and designs which now need to be tailored to a broader range of applications.

APPLICATIONS FOR UWB PULSERS

One of the major factors associated with the application of UWB pulsers is the range at which the device is expected to operate. This factor can easily be split into three regions, short range (< 1m), medium range (10-1000m) and long range (>1000m).

Short Range Applications

The short range applications cover the laboratory type of activity including the assessment of antenna performance, fast protection devices and system vulnerability as well as circuit analysis. Many of these applications can be undertaken in closed loop configuration and will require pulsers which have output voltages in the range of a few hundreds of millivolts to relatively high output voltages (>10kV), and have a high degree of repeatability. The repeatability is required so that consistent results can be obtained on large numbers of components under test, over a sustained period of time. It is important that the output impedance of these devices remains constant both during the pulse and over long periods, ensuring minimum reflections which could lead to corrupted data.

Ultra-Wideband, Short-Pulse Electromagnetics 3
Edited by Baum *et al.*, Plenum Press, New York, 1997

Other short range applications where the pulser is used to stimulate a radiative structure involve bounded wave simulators for EMC testing, simulators for medical research and chambers used for RCS measurement, including stealth material assessment.

Low voltage, fast risetime pulsers are already being used in medical applications for measuring heart beat. With the improvement in UWB signal processing it may be possible, in the near future, to provide some imaging of the body which could drastically reduce the hazards produced by equipments currently in use.

Mine detection is a very important short range application for UWB pulsers.[1] The ever increasing number of anti-personnel mines, both plastic and metal, is a major problem world wide. Detectors using UWB pulsers have been shown to work at short range in hand held systems. This technology needs to be extended to give at least 1m range to minimize the problems associated with booby trapped devices, and to possibly decrease the time required to clear given areas.

UWB pulsers are also being applied to the problems of void detection under major highways and to detecting the failure of reinforcing bars in concrete bridges and buildings.

Medium Range Applications

These applications cover areas where systems need to be platform mounted or bistatic ground based installations.

Airborne UWB SAR systems[2] have been used to detect a variety of targets on or below the ground surface. The advantages are that the radar has the ability to detect small objects partially obscured by vegetation and with care 3D images can be obtained through heavy foliage. The system is capable of operating in all weather conditions. Because of the low frequency content of the transmitted radiation it is possible that underground bunkers could be detected and this, associated with the normal attributes, could allow false targets to be identified.

In a bistatic ground configuration medium range intruder alarms can be produced which can give very accurate position location of targets entering a controlled area. This type of application could also give information about the type of system that has penetrated the area being protected.

There are a range of applications that could be applied to surface vehicle borne systems. Anti collision UWB systems could be utilised to inform car and heavy vehicle drivers of safe driving distances in inclement weather conditions (including fog, rain and snow). If utilised on ships, accurate distances could be provided to assist in safe docking procedures.

One of the major concerns for large aircraft pilots is maneuverability on the ground at major airports. Considerable damage does occur to aircraft when there is impact between the wing tips of aircraft that are taxiing and stationary aircraft. A UWB device fitted to the wing end section could give valuable early warning of impending contact. To ensure that there are no problems from interference, the system would only operate whilst the aircraft is on the ground.

Long Range Applications

These applications will almost certainly require pulsers with high voltage outputs or a number of lower voltage pulsers time locked to form an array so that a sufficient field strength can be achieved at long range.

Long range ground based UWB radars would be used in conjunction with conventional surveillance radars to assist on accurate target identification, where the high range resolution and system resonances would be utilised.

Space based UWB SAR systems could be designed to give precision data which when combined with the data obtained from other existing or new sensors could give enhanced information over a wide range of topics where accurate data is required.

UWB PULSER TECHNOLOGY

Research has been conducted over the last decade into improving Ultrawideband or Impulse Pulser Technology. Several new areas have been investigated and some older known technologies have been dramatically improved by incorporating modern techniques.

A large majority of the research has been aimed at specific applications and pulsers have tended to have been manufactured to meet customer generated specifications and have only been produced in small quantities.

Switches such as spark gaps and Hydrogen Thyratrons have improved because of techniques such as optical, electromagnetic and semiconductor triggering. High pressure Hydrogen thyratrons and ferrite line pulse sharpeners[3] are now used to produce fast risetimes which were not previously obtainable from conventional gas discharge devices.

Research has been undertaken into alternative methods of triggering high pressure switches to reduce the jitter which is inherent with gas discharge and a device (Polotron)[4] has been investigated which utilises the effect of RF initiation of a plasma discharge as the trigger mechanism. Since the initiation of the discharge occurs in times of <1ns there is the possibility of accurate triggering of many such devices operating in parallel or series.

Research into water cooled dielectric pulse sharpening ferrite lines has shown that high voltage operation at high repetition rates can now be achieved.

Semi conductor switch technology has been extensively researched and significant progress has been made with both discrete devices and optically triggered switches. The research has been undertaken virtually world wide in this area and has been widely reported.

Although many research institutes investigate these devices the number of companies who can produce off the shelf pulsers are very few.

PERFORMANCE ASSESSMENT

Once a specification for a UWB system has been defined the scientist or engineer has the problem of choosing the most appropriate pulser that will meet the requirements specified. It is very important at this time to know whether there is a pulser available from a manufacturer (there may be a need for large quantity orders of identical units) or whether there is a need to invest funding into research immediately to ensure that the technology can meet the requirement in the timescale envisaged.

So that these decisions can be made it is prudent that a review of the major suppliers of UWB pulse generators (both manufacturers and researchers) is available. From published data it is possible to ascertain many of the major parameters that are required, and it is then possible to produce a table which can form the basis of procurement or research objectives.

The major parameters which could be needed include the following: Output Voltage, Output Impedance, Pulse Risetime, Repetition Rate, Weight, Volume, Prime Power, degradation mechanism etc.

The system designer then needs to do an assessment of his requirement and provide a specification for both the ideal and acceptable UWB pulse generator for the application. Once this is completed an inspection of the performance review data will indicate either an immediate supplier or the area where research is required.

Table of Typical Pulser Data

Application & Parameters	Supplier A	Supplier B
Laboratory use	Yes	Yes
Proven Ground use	Yes	Yes
Proven Airborne use	Yes	Yes
Single Output	Yes	Yes
Multiple Output	Yes	Yes
Output Voltage	Pulser a)>2kV Pulser b) 10kV Pulser c) 18kV Pulser d) 12.5kV Pulser e) 23kV	Pulser a) 5kV Pulser b) 10kV Pulser c) 50kV Pulser d) 20kV Pulser e) 50kV
Risetime	<100ps (a, b, c & d) 130ps (e)	All<200ps
PRF<1kHz	Yes (d & e) 100Hz	Yes
PRF>1kHz	Yes to all (except d & e)	Yes to all
Phased Array application	Yes	Yes
Life >> 10^5 shots	Yes	Yes
Weight	Man portable	Man portable
Volume (cm)	a) 20x25x10 b) 30x35x20 c) 50x75x30 d) 50x75x30 e) 100x100x100	a) 100x75x30 b) 100x75x30 c) 100x75x30 plus 30x30x30 d) 1m3 + 1m3 e) TBA
Prime Power a) Battery b) Single Phase c) Three Phase d) 400Hz	Yes (a,b,c,d,e,)	Yes (a,b,c,d,e,)
Degradation a) Catastrophic b) Gradual	Yes	Yes
Spares a) Off Shelf b) Long lead c) Special	Yes	Yes

CONCLUSIONS

Large amounts of funding have been spent over the last decade in pulse generator research and in many areas there have been significant advances. Research, however, has tended to focus the development to uniquely defined objectives.

Research funding is now being reduced by many Governments and the amounts that are still available must be used to advantage in this area. To this end it is now very important that there is a direct link between the system designers and engineers and the research scientists involved with fast switch technologies that ensures an efficient "pull through" of the new and emerging technologies to the manufacturers of the generators.

Without this commitment, what could be a revolution in UWB radar technology will stagnate, and many of the advantages that could be exploited will be wasted.

REFERENCES

1. D. J. Daniels, Development in impulse radar technology for surface penetrating radar, *IEE Colloquium on 'Antenna and Propagation Problems of Ultrawideband Radar'* (1993).
2. R. S. Vickers, Dual polarized airplane: calibration and results, *SPIE 10th International Aerospace Symposium, 8-12 April 1996, Orlando* (1996).
3. J. E. Dolan, H. R. Bolton and A. J. Shapland, Development of a 30kV ferrite line pulse-sharpening system, *IEE Pulse Power Colloquium* (1996).
4. N. A. Ridge, P. F. Hirst, A. Maitland and D. M. Parkes, High Voltage Microwave-Triggered Switches, *to be published in IEEE Transactions on Plasma Science*.

HIGH POWER, SUB-NANOSECOND RISING WAVEFORMS CREATED BY THE STACKED BLUMLEIN PULSERS

Farzin Davanloo,[1] Dan L. Borovina,[1] Johnelle L. Korioth,[1] Raymond K. Krause,[1] Carl B. Collins,[1] Forrest J. Agee,[2] Jon P. Hull,[2] Jon S.H. Schoenberg,[2] and Lawrence E. Kingsley[3]

[1]Center for Quantum Electronics, University of Texas at Dallas
Richardson, TX 75083-0688

[2]U.S. Air Force Phillips Laboratory, WSR
Kirtland AFB, NM 87117-6008

[3]U.S. Army CECOM, S&TCD
Fort Monmouth, NJ 07703-5203

INTRODUCTION

The opportunity to develop a light, portable source of High Power Microwaves (HPM) is very clear. Advances in stacked Blumlein technology for voltage multiplication, together with the availability of fast photoconductive switches for commutation, would seem to make it unnecessary to continue to rely upon specialized vacuum or plasma tubes for HPM devices in some applications. At the 100 MW level of power, broad-band sources operating at kilo-Hertz repetition rates can be conceived which simply match compact pulse power devices to the radiation impedance of free space.

To fulfill the demand for pulse power sources producing several hundred kV pulses at moderately high repetition rates, the University of Texas at Dallas first introduced and implemented a new approach to combine the functions of pulse shaping and voltage multiplication using stacked Blumleins. This yielded the development of pulsers which consisted of several triaxial Blumleins stacked in series at one end.[1,2] The lines were charged in parallel and synchronously commuted with a single switch at the other end. This allowed switching to take place at a low charging voltage relative to the pulser output voltage. These pulsers have been extensively characterized by our group and their versatility has been demonstrated.[1-8] More recently, work in collaboration with the Phillips Laboratory has shown that, with slight modifications they can produce waveforms with fast risetimes and a wide range of pulse durations and peak values.

The stacked Blumlein pulsers have produced high power waveforms with risetimes and repetition rates in the range of 5-50 ns and 1-200 Hz, respectively, using a conventional thyratron or spark gap. To generate waveforms with sub-nanosecond risetimes at kilo-Hertz repetition rates, fast switching devices such as photoconductive switches offer significant advantages. This paper describes the feasibility of an intense

Ultra-Wideband, Short-Pulse Electromagnetics 3
Edited by Baum *et al.*, Plenum Press, New York, 1997

pulse power source for Ultra-Wideband HPM systems based upon stacked Blumlein technology by adapting the design for use with photoconductive switches. Significant progress in the development of the stacked Blumlein pulsers commuted with lateral GaAs switches in the avalanche mode is reported. These pulsers currently produce high power nanosecond wide pulses with risetimes on the order of 200-300 ps.

PHOTOCONDUCTIVE SWITCHES AND CHARGING SCHEMES

In recent years photoconductive semiconductor switches have gained much attention and become competitors to the conventional high power switches, such as spark gaps and thyratrons, for certain applications. These solid state devices operate jitter-free with switching speed as fast as the optical trigger pulse risetime or faster. Photoconductive switches are divided into the two categories of linear and avalanche type devices. The application of linear switches has been limited by the relatively high optical power required to perform the closure.[9-11] In the avalanche type switches the electron-hole pair produced by one photon is multiplied through an avalanche process, thus, reducing optical energy levels for the initiation of switch closure. The reliability and performance of such avalanche type semiconductor switches depend on charging mechanisms and the environment of operation.

To realize the capabilities of photoconductive switches in practical pulse power applications an intermediate module which might be described as a charging pulse compression (CPC) module was developed.[12] It conditions the output of the pulse charge power supply so that the lines in the final module and the photoconductive switch are subjected to high voltage for less than 100 ns. This supports the use of avalanche photoconductive switches by increasing their lifetime.

Design and construction of the CPC is similar to the single Blumlein pulse generators developed at UTD. Briefly, it consisted of two critical subassemblies: (1) a single Blumlein pulse forming line, and (2) a commutation system capable of operation at high repetition rates. The Blumlein was constructed from three copper plates 2.5 m long, potted with epoxy on outer surfaces to reduce corona, and separated by 1.65 mm thick laminated Kapton insulators. The Blumlein had a smooth taper which decreased the width of the line from 20 cm at the thyratron to 10 cm at the load. This combined the functions of Blumlein and tapered pulse transformer acting together to increase the Blumlein voltage gain at the output.

During operation, the CPC was resonantly charged with a pulse power supply capable of operating in the range of 3-75 kV at repetition rates of 1-1000 Hz. The middle conductor was charged to a positive high voltage and commutation was effected by a thyratron mounted in a grounded cathode configuration.[12] The output from this intermediate pulser was used to charge a Blumlein pulser designed for commutation by a photoconductive switch.

PHOTOCONDUCTIVE COMMUTATION OF STACKED BLUMLEIN PULSERS

Recently, we developed a low profile switching assembly for the operation of photoconductive switches with the stacked Blumlein pulsers. The base of the assembly contained two copper electrodes for holding the switch wafer and the means of coupling it to the pulse forming lines. The pressure cover was made from plexiglas and contained a chamber, into which the electrodes fit. To minimize the constriction of current flow to the switch from the pulse forming lines, the connections between the electrodes and the Blumlein were made using thin copper foils. The foils protruded from the base several centimeters to allow ample surface area for contact with the Blumlein and were passed through the cast material before it cured. The foils were joined to the electrodes by soldering them securely before immersion in the cast material.

Since the inductance of the switch assembly circuit was proportional to the area under the electrodes, this loop was decreased by minimizing the profile of the electrodes and by placing a laminated Kapton board 2.2 mm thick and 15 cm wide between the electrode leads. The board had a sufficient length to accommodate the stacked Blumlein pulser and was cast in place. The schematic drawing of this universal switch assembly is shown in Fig. 1 for reference.

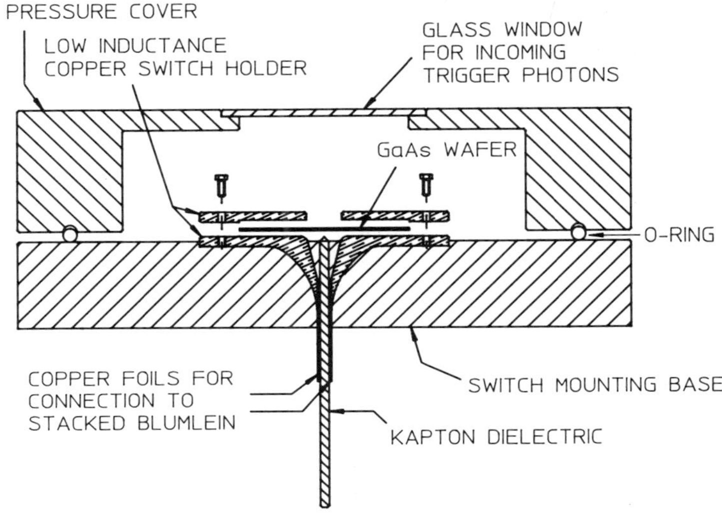

Figure 1. Schematic drawing of the cross-section of the universal switch assembly.

Switching of the Single Blumlein Pulser

It is interesting to note that the Blumlein may be the most proper pulse-forming line for use with photoconductive switches. It provides faster output pulse risetimes and reduce the percentage of stored energy deposited in the switch.[12] In this work, a single Blumlein with line length of 38 cm and line impedance of 100 Ω was connected to the universal switch assembly as shown schematically in Fig. 2. During operation, the CPC module was resonantly charged using our conventional pulse power supply. A schematic drawing of the power supply, the CPC module, laser and the timing circuits is shown in Fig. 3.

Synchronization between various components in the system was maintained through a series of timing pulses generated by a master oscillator. During operation, the master oscillator produced three pulses. The first, or the instant pulse, triggered the thyratron in the power supply and started the charging cycle for the CPC module. After a delay determined by the charging time of the CPC device, a second pulse triggered the thyratron in the CPC, producing short charging pulses for the main pulser. The stacked Blumlein pulser was fully charged in about 80 ns, after which a third trigger pulse Q-switched the Nd:YAG laser and provided trigger photons in a 40-ns (FWHM) pulse with a risetime of 20 ns for the GaAs photoconductive switch that commuted the pulser. Trigger laser photons were delivered to the switch by means of a diverging lens which was used to illuminate the active area uniformly. This allowed a few millijoules/pulse over the 2.5 cm x 1.3 cm surface area of the switch.

The GaAs switches were 50 mm diameter, 0.5 mm thick wafers, LEC grown with a resistivity of about 100 MΩ-cm. They were purchased from Litton Airtron. These switches did not have metal contacts, and connections to the electrodes in the switch assembly were done by pressure contact only. The photoconductive region for the switch, when installed, was about 2.5 cm wide by 1.3 cm long.

Figure 4 shows the charging voltage pulses for the single Blumlein pulser commuted with GaAs and Si switches. Closure of the GaAs switch is seen in Fig. 4(a) by a sharp fall time around the peak of waveform. The fall time in the charging pulse was about 5 ns which corresponded approximately to the two way transit time of the Blumlein. It should be noted that the fall time in the charging voltage represents the time it takes to deplete the charge stored in the Blumlein through the switch in each charging cycle. If switch risetime is faster than the one way transit time of the Blumlein, then the fall time in charging voltage must correspond to the two way transit time of the Blumlein. This seems to be the case as shown in Fig. 4(a).

If the switch risetime is slower than the one way transit time of the Blumlein, the fall time in the charging voltage should correspond to about twice the risetime of the switch (assuming the rise and fall times are equal). This was confirmed by replacing the GaAs switch with a highly resistive Si switch incapable of fast avalanche. Figure 4(b) shows a charging voltage for this case where switch closure is seen. The fall time was about 40 ns which corresponded approximately to twice the risetime of our laser pulse.

Results shown in Fig. 4 indicate that the GaAs switch risetime was faster than the one way transit time of the Blumlein of 2.5 ns. Since our laser pulse was relatively slow, the switch could not have been commuted in linear mode. Thus, in spite of a relatively long and energetic laser pulse, it was confirmed that the GaAs switch was operated in the avalanche mode.

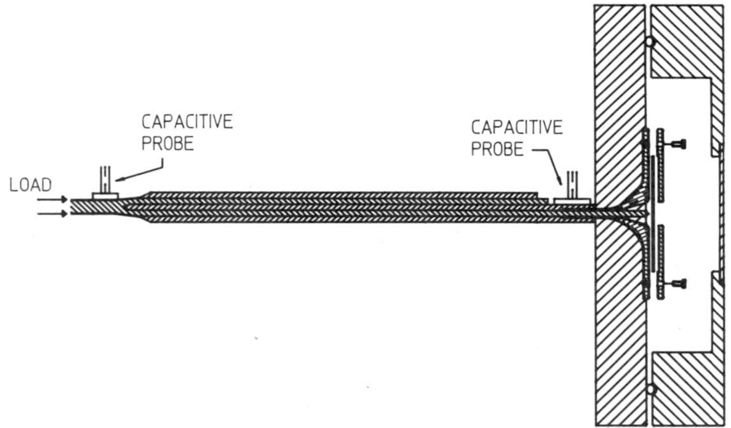

Figure 2. Schematic drawing of the single Blumlein connecting to the switch assembly. Two voltage probes were installed at the switch and the load sides to measure launching and output voltage waveforms.

Switching of the 2-line Stacked Blumlein Pulser

In this work, we designed and constructed a 2-line stacked Blumlein pulser for commutation by avalanche GaAs switches. For this device, the corresponding high voltage and ground copper plates for each of the Blumleins were connected and soldered together at the switch assembly. Blumleins with line impedances of 100 Ω were extended to the proximity of the stacking location, 12.7 cm from the switch assembly, where the center, high-voltage conductor was terminated. Then the lines angled toward each other along with the dielectric for another 5 cm. At this point, they were stacked directly on top of one another. The lines were joined in series for about 3.8 cm and the top and bottom plates were connected to a resistive load through a matched transmission line with a location for installing the capacitive probe. The resistive load was built from a stack of four 50-Ω, non-inductive carbon disc resistors. A top view is shown in the photograph of Fig. 5. The bottom Blumlein conductors were placed about 6 cm from the switch assembly, allowing the installation of a capacitive probe which was used to measure launching waveforms at the switch.

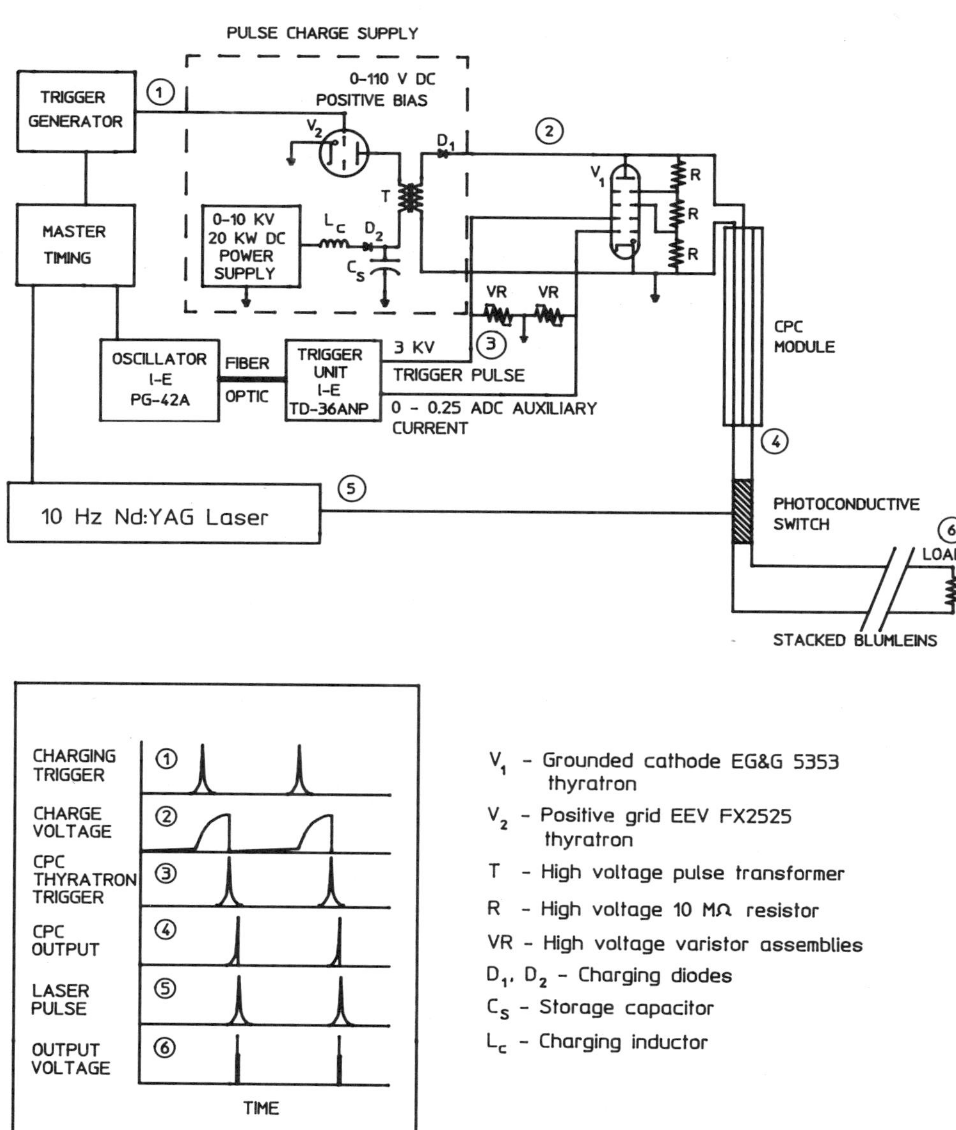

Figure 3. Schematic representation of the pulse power supply, CPC intermediate module and the Nd:YAG laser used to commute the Blumlein pulsers.

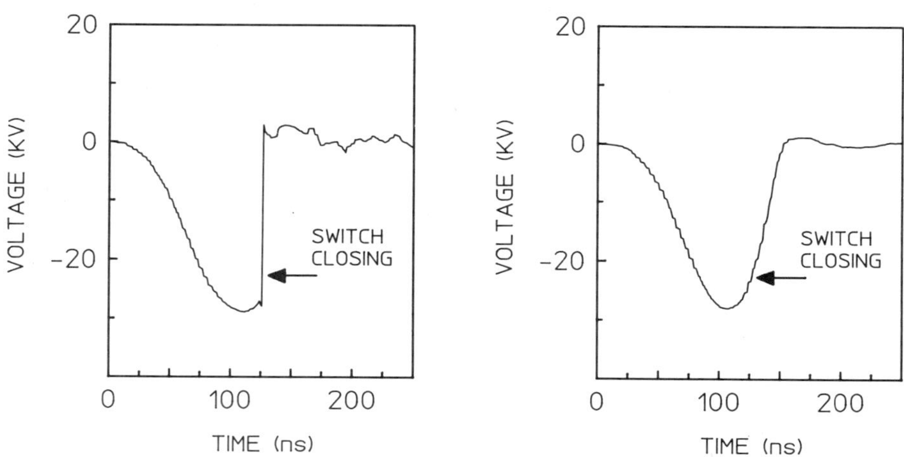

Figure 4. Charging voltage profiles for the single Blumlein pulser commuted by (a) GaAs switch in avalanche mode and (b) Si switch in linear mode. Switch closure is seen by respectively sharp and slow fall times of the pulses.

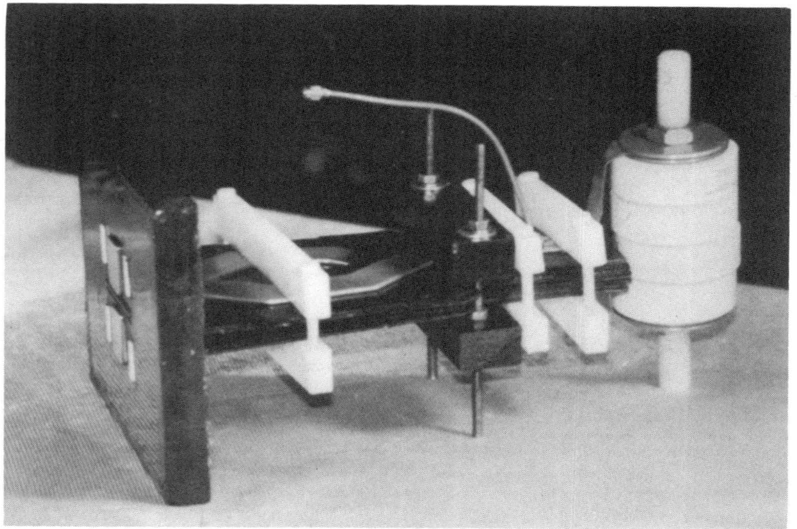

Figure 5. Photograph showing a top view of the 2-line stacked Blumlein pulser characterized in this work.

To prepare for operation, the 2-line pulser and the charging pulse compression (CPC) module were placed in separate RF shielded containers. The CPC module was resonantly charged using our conventional pulse power supply. The output from the CPC device was used to charge the stacked Blumlein pulser through a coaxial high-voltage cable in about 80 nsec after which a trigger pulse from the master oscillator Q-switched a Nd:Yag laser, providing trigger photons for the GaAs switch. We commuted the 2-line pulser with a GaAs switch in the avalanche mode with the same conditions of operation as described earlier for the single Blumlein pulser.

To probe sub-nanosecond risetimes expected of waveforms generated by our pulsers, two capacitive voltage probes were used. They were installed at the switch and the stack sides. Pulse heights were measured by a Tektronix 7912 transient digitizer to study the stacking function of the device. However, precise measurements of pulse-risetime were performed using a Tektronix SCD 5000 transient digitizer capable of recording waveforms with risetimes better than 80 ps.

The 2-line pulser, when operated at switch voltages of 29 kV or greater, generated pulses with risetimes in the range of 225 - 535 ps, and with an average risetime of about 385 ps. Pulse FWHM was in the range of 900 - 1230 ps with an average duration of about 1025 psec. Figure 6 presents a waveform with an average characteristic for reference. This waveform had a risetime of about 330 ps and FWHM of about 1030 ps. The pulse width corresponds to two way transit time of the Blumleins in the device, as expected. The voltage waveform of this figure indicates a voltage gain of about 1.8 which is consistent with our earlier results for the 2-line pulsers.[12]

The waveform presented in Fig. 6 exhibited a small step-like shape at about the middle of rising edge of the pulse. A design study of the 2-line device indicated following two possible reasons:

1. Waveforms reaching the stack location from each Blumlein have slightly different time profiles. This causes a step-shape in the resulting pulse as the two waveforms are added. A small difference in the characteristics of each Blumlein can produce such timing difference.
2. Waveforms generated in each Blumleins are slightly degraded by the difference in the length of the transmission line components.

It should be emphasized that the 2-line pulser characterized in this work is an early prototype, and design improvements are expected to enhance its performance in the near future.

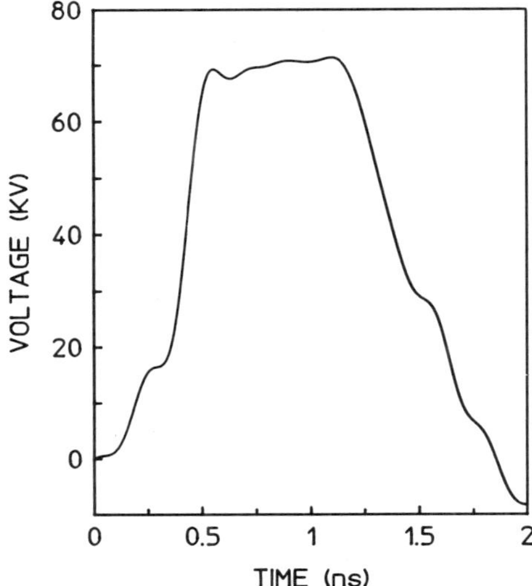

Figure 6. Voltage waveform obtained at the stack of the 2-line pulser by commuting the device with a GaAs switch in the avalanche mode. This particular waveform corresponded to a charging voltage of about 40 kV.

CONCLUSIONS

In this work, a single GaAs switch was used to commute a stacked Blumlein prototype pulser in the avalanche mode. The device was successfully operated at the peak power levels in the range of 50-70 MW. Advances in stacked Blumlein technology for voltage multiplication, together with the results obtained in this study, would seem to indicate the feasibility of an intense stacked Blumlein pulser commuted by photoconductive switches in the avalanche mode. In recent months, our efforts have been focused on activating the GaAs switch in our devices with the laser diode arrays. Proper avalanche operation at the power levels similar to those reported in this work have been obtained with laser pulse energies as low as 0.5 μJ. Results will be reported at a future date.

ACKNOWLEDGMENTS

This work was supported by the U.S. Air Force Phillips Laboratory, WSR, and U.S. Army Research Laboratory, PSD under contract DAAL01-95-K-3502.

REFERENCES

1. F. Davanloo, J.J. Coogan, T.S. Bowen, R.K. Krause, and C.B. Collins, "Flash X-ray Source Excited by Stacked Blumlein Generators," Rev. Sci. Instrum. 59, 2260 (1988).
2. J. J. Coogan, F. Davanloo, and C.B. Collins, "Production of High Energy Photons from Flash X-ray Sources Powered by Stacked Blumlein Generators," Rev. Sci. Instrum. 61, 1448 (1990).
3. F. Davanloo, R.K. Krause, J.D. Bhawalkar, and C.B Collins, "A Novel Repetitive Stacked Blumlein Pulse Power Source," in Proceedings of the 8th International Pulsed Power Conference, 1991, pp. 971-974.
4. F. Davanloo, J. D. Bhawalkar, C.B. Collins, F. J. Agee, and L. E. Kingsley, "High Power, Repetitive Stacked Blumlein Pulse Generators Commuted by a Single Switching Element," in Conference Record of the 1992 Twentieth Power Modulator Symposium, 1992, pp. 364-367.
5. J. D. Bhawalkar, F. Davanloo, C.B. Collins, F. J. Agee, and L. E. Kingsley, "High Power, Repetitive Blumlein Pulse Generators to Drive Lasers," in Proceedings of the International Conference on Lasers '92, edited by C.P. Wang (STS Press, McLean, VA, 1993) pp. 360-364.
6. J. D. Bhawalkar, F. Davanloo, C.B. Collins, F. J. Agee, and L. E. Kingsley, "High Power Repetitive Stacked Blumlein Pulse Generators Producing Waveforms with Pulse Durations Exceeding 500 nsec," in Proceedings of the 9th International Pulsed Power Conference, 1993, pp. 857-860.
7. J. D. Bhawalkar, D.L. Borovina, F. Davanloo, C.B. Collins, F. J. Agee, and L. E. Kingsley, "High Power Repetitive Stacked Blumlein Pulse Generators," in Proceedings of the International Conference on Lasers '93, edited by V.J. Corcoran and T.A. Goldman (STS Press, Mclean, VA, 1994) pp. 712-717.
8. F. Davanloo, D.L. Borovina, J. D. Bhawalkar, C.B. Collins, F. J. Agee, and L. E. Kingsley, "High Power Repetitive Waveforms Generated by Compact Stacked Blumlein Pulsers," in Conference Record of the 1994 Twenty-First Power Modulator Symposium, 1994, pp. 201-205.
9. M.D. Pocha and R.L. Druce, "35-KV GaAs Subnanosecond Photoconductive Switches," IEEE Trans. Electron Devices, 37, 2486 (1990).
10. F.J. Zutavern, G.M. Loubriel, W.D. Helgeson, M.W. O'Malley, R.R. Gallegos, A.G. Baca, T.A. Plut, and H.P. Hjalmarson, "Fiber-optic Control of Current Filaments in High Gain Photoconductive Semiconductor Switches," in Conference Record of the 1994 Twenty-First Power Modulator Symposium, 1994, pp. 116-119.
11. W. C. Nunnally, "Photoconductive Pulse Power Switches," in Proceedings of the 4th International Pulsed Power Conference, 1983, pp. 620-623.
12. D.L. Borovina, R.K. Krause, F. Davanloo, C.B. Collins, F. J. Agee, and L. E. Kingsley, "Switching the Stacked Blumlein Pulsers: Status and Issues," in Proceedings of the 10th International Pulsed Power Conference, 1995 (in press).

THE PROBLEMS OF PICOSECOND ANALOG DEVICES MODELING AND CREATION

V.N. Ilyushenko, O.V. Stukach, and B.I. Avdochenko

Tomsk State Academy of Control Systems and Radioelectronics (TACSR),
40 Lenin Avenue, Tomsk, 634050, Russia

In the last years the works on development picosecond pulse range are intensity developed. The great successes in high-voltage picosecond pulse generation, formation, amplification, control of parameters and registration are achieved. However the theoretical and applied bases of creation of the amplifiers, devices of amplitude picosecond pulse controlling and other devices in working frequency band from a zero or 1 ...10 kHz up to 1...10 GHz are not enough developed. The complexity of their creation is stipulated extremely by high significance of a frequency range factor. In the most of practical problems it resulted to necessary of the element characteristics realization, near to limiting (potential). As a result essentially difficulties of the analysis and synthesis of parameters grow because the classical models in the basis R,L,C- elements and operator function has high order. The active and passive elements not always satisfy to the requirements on duration of processable signals because of change of reactive conductivity character in a frequency band or incompatibility of electrical or design-technical characteristics. Besides those element sizes become commensurable with of working range wave length, that causes dependence of pulse signal parameters (delay, amplitude, form) from the element geometry sizes and their connections. Additional transfer parasitic channels, transformation and formation of pulses are creating. A multi-channeling becomes the characteristic structural attribute of the systems. It results to necessary of research the signal transformation and formation processes at a structurally functional level and to necessary of the highspeed problem complex decisions, based on development of the system theory.

In this connection the questions of search and development of new concepts and approaches to research are urgent, based on structural construction laws of complex systems and devices. They should be also the basis for understanding of the system characteristics formation processes and general principles of system behaviour. Key and most difficult position in these problems is the problem of modeling. Offering to decide it on the system approach basis, enabling to exclude circuitry aspects at a certain designing stage. Besides it gives possibility to investigate the complex objects (systems) on the basis of ready and base functional units connecting on determine rules with beforehand investigate or known characteristics. The system approach includes the following: the definition of the

Ultra-Wideband, Short-Pulse Electromagnetics 3
Edited by Baum *et al.*, Plenum Press, New York, 1997

characteristics, in the most complete describing objects; operator designing, construction of generalized structural system model.

In the classical circuit theory in mathematical models the integration operators are the most frequently used. Using of these operators as a unique means of models construction of picosecond systems for revealing of laws of their construction is not effectively because of high order of transfer functions. The computing complexities, stipulating by necessity of high accuracy of the characteristics approximation are also essential.

We offered the approach to research and modeling of systems, which permits partially to decide a series of considered problems. It is based on the description of systems by following:

$$K(w)=K_n(w)+\Delta K(w); \qquad h(t)=h_n(t)+\Delta h(t);$$
$$\varphi(w)=\varphi_n(w)+\Delta\varphi(w); \qquad g(t)=g_n(t)+\Delta g(t), \tag{1}$$

where $K_n(w)$, $\varphi_n(w)$ are the amplitude-frequency and phase-frequency characteristic of the initial system, $h_n(t)$, $g_n(t)$ are the transient and pulsing characteristic of this system; $K(w)$, $\varphi(w)$, $h(t)$, $g(t)$ are the appropriate characteristics of researching (simulating) system, $\Delta K(w)$, $\Delta\varphi(w)$, $\Delta h(t)$, $\Delta g(t)$ are the characteristics deviation from initial and researching systems. Normalizing frequency w and time t and using a formula for the transient characteristic:

$$h(t)=0{,}5+1/\pi\int_0^\infty K(w)\sin[wt+\varphi(w)]/w\,dw, \tag{2}$$

at $K(w)<0{,}1...0{,}2$, $\varphi(w)<\pi$ we receive:

$$h(t)=h_n(t)+\Delta h_k(t)+\Delta h_\varphi(t), \tag{3}$$

where $\Delta h_k(t)$, $\Delta h_\varphi(t)$ are the transient characteristic deviations, stipulated by amplitude-frequency and phase-frequency characteristics deviations accordingly. Approximating of the frequency characteristics deviations by Fourie series, transforming them in the temporal domain the following mathematical models has been founded.

For non-minimum phase systems:

$$h(t)=h_n(t)+\Delta h_\varphi(t)=h_n(t)+0{,}5\sum_{i=1}^{m} b_i[h_n(t+T_i)-h_n(t-T_i)];$$
$$h(t)=h_n(t)+\Delta h_k(t)=h_n(t)+0{,}5\sum_{i=0}^{m} a_i[h_n(t+T_i)+h_n(t-T_i)]. \tag{4}$$

For minimum phase systems:

$$h(t)=h_n(t)+\Delta h_{k\varphi}(t)=h_n(t)+0{,}5a_0h_n(t)+\sum_{i=1}^{m} a_ih_n(t-T_i),$$
$$h(t)=h_n(t)+\Delta h_{\varphi k}(t)=h_n(t)-\sum_{i=1}^{m} b_ih_n(t-T_i)]. \tag{5}$$

Here $\Delta h_{k\varphi}(t)$, $\Delta h_{\varphi k}(t)$ are deviation in the transient characteristic of a minimum phase system, caused either K(w) and Hilbert transformation connected with it φ(w), or on the contrary in $\Delta h_{\varphi k}(t)$; a,b,T are the decomposition factors and periods in a Fourie series of the frequecy characteristics deviations.

Therefore, as following from (1)-(5), the basis of operators affecting on the initial system response includes the decomposition of initial system on m responses, algebraical weighing, delay and summation. These operators can be realized by picosecond functional circuits relative simply, in physical multichannel (including matrix) model of a system. Such mathematical models, based on structural representations of the frequency and tempo-ral characteristics deviations evidently reflects the physical processes of the signal distortion (formation) in the temporal domain and nature of these distortions in the frequency domain. The developed models reflects also such feature picosecond systems as multi-channeling signal transfer, stipulated or system structure or existence of parasitic channels.

Besides of basic functional circuits realizing the elementary operations, the developed models structure includes an initial system having beforehand realized or investigated characteristics. Such system, in general case of any complexity essentially expands the opportunity of research high-speed processes and laws of structural construction of picosecond pulsing technique systems and devices. These models as a basis for creation of a generalized multichannel system structure, variants of its are distribute amplifiers, amplifiers with direct connection and feedback, multichannel correctors and their single-channel analogues on the non-uniform lines, controllable attenuators, pulse shapers, phase shifters, delay lines and other devices are developed by us.

The generalized structures application to the linear systems (in particular, to amplifiers) has made possible to generate a new sight on feedbacks. Besides of classical feedback and direct connection is offered to use for increase of speed operation the following: the frequency-dependent feedbacks and direct connections, simultaneously in a working band, the feedbacks in a part of frequency band with inversion in direct connections at the high frequency, that has allowed to increase speed of operation, gain and stability.

On the basis of theoretical and experimental results obtained, the following hybrid-integrated devices are developed:

1. The amplifiers of picosecond electrical signals having working frequency bands 80 kHz ...7 GHz, 80 kHz...6 GHz, 0...5,6 GHz, with gain accordingly 14, 40 and 20 dB and with a rise time of the transient characteristics of 50 and 70 ps.

2. The gigabyte sequence pulse generators with reiterative frequency up to 3 Ghz.

Thus, as a result of researches the picosecond devices are developed. Such devices are applicable in subsurface radar, oscillography, in broadcast systems, experimental physics, measuring engineering, and in other fields of science and engineering.

IMPULSE RADIATING ANTENNAS, PART III

Everett G. Farr[1] and Carl E. Baum[2]

[1]Farr Research, Inc.
Albuquerque, NM 87123
[2]Phillips Laboratory,
Kirtland AFB, NM 87117

ABSTRACT

In this paper we continue our general discussion of Impulse Radiating Antennas (IRAs), which has been carried on during the first two Ultra-Wideband, Short-Pulse Electromagnetics conferences. IRAs are a class of antennas that consist of a TEM feed section and either a lens or reflector to focus the aperture field. We summarize here much of the more recent information, including new antenna designs, new calculation methods, and an optimization of the impedance of the lens IRA.

First, we explore a wide variety of new IRA designs that include two reflecting or refracting surfaces. By using two surfaces, one can achieve considerable additional flexibility in design. This flexibility allows very compact designs, and also allows additional choices of feed impedance.

Next, we consider the optimal feed impedance for long TEM horns and lens IRAs. We consider both TEM horns whose plates are flat, and whose plates are confined to a circular arc. We also consider both infinite apertures and circular apertures of finite radius. The optimal impedance is determined as the impedance that provides the highest radiated field for a given input power.

Finally, we calculate the field radiated from a four-wire aperture, both on- and off-boresight. This is an approximation to the aperture field of a four-armed reflector IRA.

I. INTRODUCTION

The theory of Impulse Radiating Antennas has been building for some time [1,2]. Simple models are now available for the radiated field for both lens and reflector designs. Sketches of lens and reflector IRAs are shown in Figure 1.1.

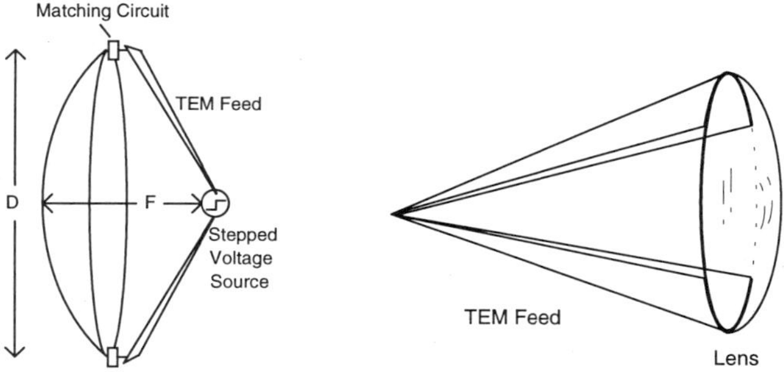

Figure 1.1. A reflector IRA (left) and a lens IRA (right).

In this paper we extend the theory in several key areas. First, we consider a broad range of new designs which make use of two reflecting or refracting surfaces. Second, we calculate the optimal impedance for infinitely long TEM horns and lens IRAs. Finally, we calculate the radiated field from a four-wire aperture on- and off-boresight.

II. IRAS WITH TWO REFLECTING OR REFRACTING SURFACES

Several new antenna designs can be fabricated with a combination of two reflecting and/or refracting surfaces[3]. A ReLIRA is an IRA consisting of both a reflector and a lens, and an example is shown in Figure 2.1. It consists of a TEM feed and a planar reflector, all embedded in a dielectric medium, with a prolate spheroidal lens[4] to focus the rays. Note that a prolate spheroid is just an ellipse that has been rotated around its major axis.

The lens for this antenna forces all the high-frequency rays to arrive at some aperture plane concurrently, as shown in Figure 2.2. To satisfy the condition, the lens shape is described by

$$\sqrt{\varepsilon_1}\,\ell \;=\; \sqrt{\varepsilon_1}\,r \;+\; \sqrt{\varepsilon_2}(-z) \tag{2.1}$$

where all the symbols are shown in Figure 2.2. This results in a ellipse of revolution that is described by

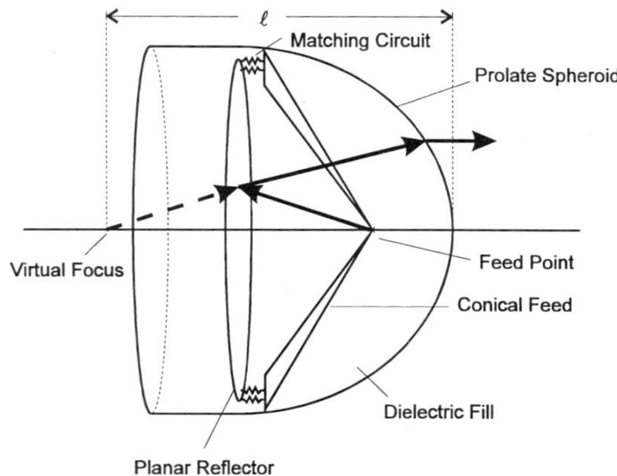

Figure 2.1. ReLIRA with Planar Reflector and Prolate Spheroidal Lens, two-arm version.

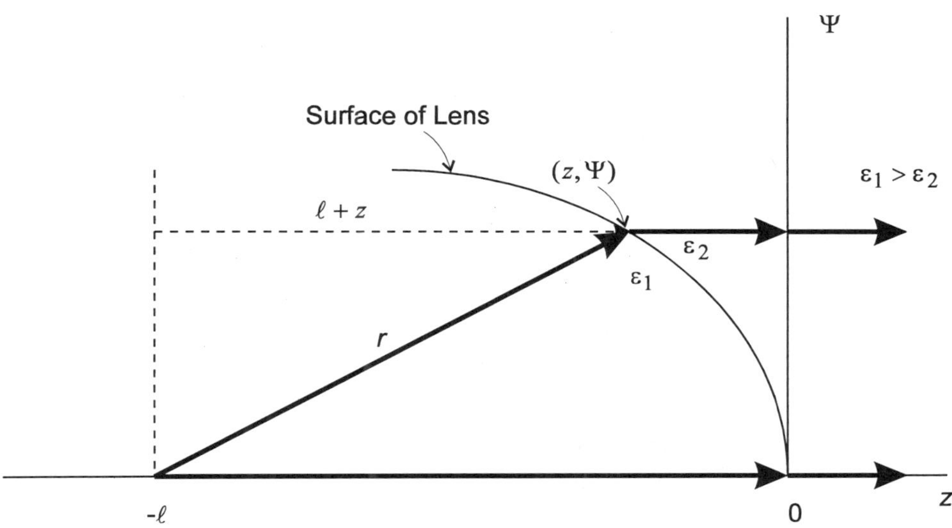

Figure 2.2. Geometry of the prolate spheroid dielectric interface (an ellipse when projected into the Ψ- z plane).

$$\frac{(z+a)^2}{a^2} + \frac{\Psi^2}{b^2} = 1 \quad , \quad a = \frac{\ell}{1+q} \quad , \quad b = \ell\sqrt{\frac{1-q}{1+q}} \quad , \quad q = \sqrt{\varepsilon_2 / \varepsilon_1} \qquad (2.2)$$

where a and b are the major and minor axes of the ellipse. Details of the derivation are provided in [3].

There are a number of variations of this antenna, the first of which is the Solid Dielectric Lens IRA. This combines a TEM horn embedded in a dielectric, with a prolate spheroidal lens interface. A sketch of this configuration is shown in Figure 2.3. Note that the feed point of the TEM horn is located at a focus of the prolate spheroid. Other variations might include using either a hyperboloidal reflector (with a prolate spheroidal lens) or a paraboloidal reflector (with a planar lens). These two configurations are shown in Figure 2.4. While the version with a paraboloidal reflector may seem to be a trivial case, it may have some use in applications requiring high mechanical stability (shock-hardened) or in applications requiring a lower feed impedance.

Yet another variation is the Split IRA (SPIRA), an example of which is shown in Figure 2.5. This is a technique for placing two half IRAs in close proximity, in order to implement separate transmit and receive antennas in a compact design. In this design, each of the two half IRAs share a thick ground plane, through which feed cables are run. Any of the designs previously discussed in this paper, or any of the classical reflector or lens IRA designs could be implemented in such a fashion. An example of such a design, using an ReLIRA with a planar reflector and prolate spheroidal lens is shown in Figure 2.5.

An interesting feature of the Split IRA is that one can achieve a very low feed impedance. By using two arms for each half, the feed impedance is typically 100 Ω in air. If the dielectric material has a relative dielectric constant of four, then the input impedance is 50 Ω. This is a convenient impedance for matching to a source. In addition, this is a single-ended impedance, so it matches well to a coaxial cable input.

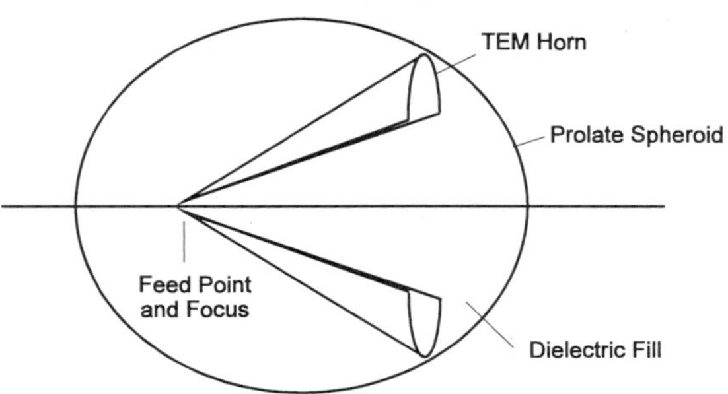

Figure 2.3. A solid dielectric lens IRA.

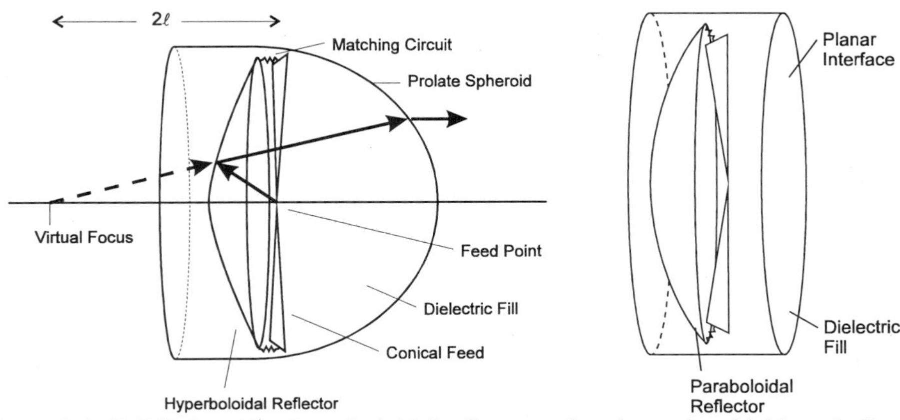

Figure 2.4. ReLIRAs with a hyperboloidal reflector and prolate spheroidal lens (left) and with a paraboloidal reflector and planar lens (right). (Two-arm versions are shown.)

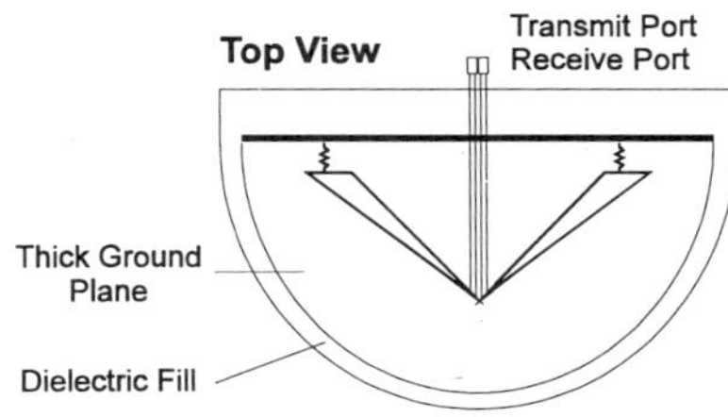

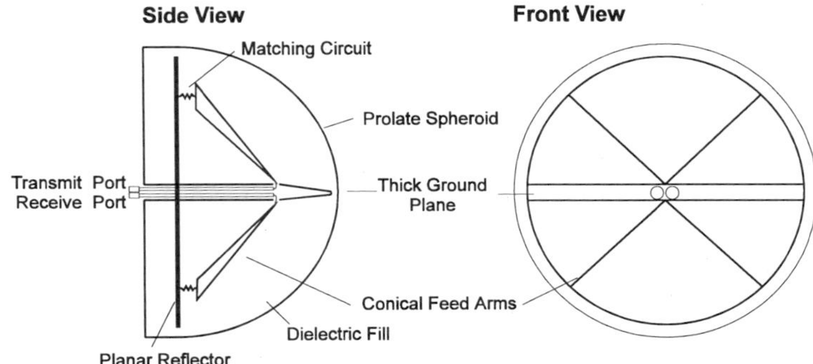

Figure 2.5. An example of a SPIRA, using a flat plate reflector and a prolate spheroidal lens. (Four-arm version is shown.)

III. IMPEDANCE OPTIMIZATION OF LONG TEM HORN AND LENS IRA

In previous papers[2,5], we demonstrated a method for optimizing the impedance of two-wire and four-wire apertures. These apertures corresponded to the aperture fields for two-arm and four-arm reflector IRAs. Let us now optimize the impedance of long TEM horns, or lens IRAs. This material originally appeared in [6].

We wish to optimize the impedance of the flat plates of a TEM horn, under the constraint that they lie within a circle of radius a_o, as shown in Figure 3.1. Thus, we search for the aspect ratio, b/a, which optimizes the radiated field for constant input power. We treat two cases here, one with the flat plates located in free space, and the other with a conducting plate that blocks fields outside the circle. We optimize the impedance in an early-time sense, using the static fields in the aperture.

To optimize the impedance, we must find the quasi-static electric field in the aperture between two plates. First, we calculate the characteristic impedance of the feed. The impedance is determined from [7,8]

$$\begin{aligned} f_g &= \frac{K(m_1)}{K(m)}, \quad m_1 = 1-m, \quad \sin(\phi_o) = \sqrt{\frac{1}{m}\left(1-\frac{E(m)}{K(m)}\right)} \\ \frac{a}{b} &= \frac{2}{\pi}\left[K(m)E(\phi_o|m) - E(m)\,F(\phi_o|m)\right] \end{aligned} \tag{3.1}$$

where $K(m)$ and $E(m)$ are the complete elliptic integrals of the first and second kind, and $F(\phi_o|m)$ and $E(\phi_o|m)$ are the incomplete elliptic integrals of the first and second kind. Furthermore, $f_g = Z_c/Z_o$, where $Z_o = 376.727\ \Omega$. To find f_g for a given value of b/a, one must solve numerically the above set of equations. At low impedances, an asymptotic form is used for the impedance [8],

$$f_g = \frac{b/a}{1 + \frac{b/a}{\pi}\left[1+\ln\left(\frac{2\pi}{b/a}\right)\right]} \tag{3.2}$$

This form was used for $b/a < 0.3$. Since the aperture is constrained to be of radius a_o, the relationship between a and b is expressed as

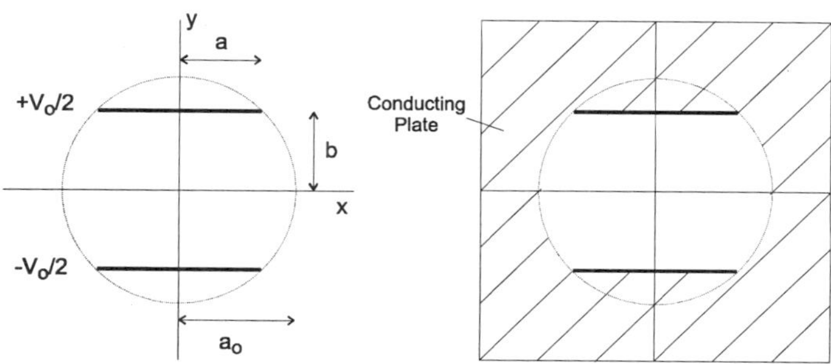

Figure 3.1. Two cases for optimizing the radiated field of a flat TEM horn, infinite aperture (left) and blocked aperture (right).

$$\frac{a^2}{a_o^2} + \frac{b^2}{a_o^2} = 1 \ , \quad b = \frac{a_o}{\sqrt{1+(a/b)^2}} \ , \quad a = \frac{a_o}{\sqrt{1+(b/a)^2}} \tag{3.3}$$

To obtain the radiated fields, we find the static fields in the aperture. Thus, we need to find a potential function in the form

$$\begin{aligned} \zeta &= x + j y \\ w(\zeta) &= u(\zeta) + j v(\zeta) \ , \qquad x, y, u, \text{ and } v \text{ are all real} \end{aligned} \tag{3.4}$$

When cast into this form, the aperture field and feed impedance are [7,8]

$$E_y(x,y) = -\frac{V_o}{\Delta u}\frac{\partial u(x,y)}{\partial y} \ , \qquad f_g = \frac{\Delta u}{\Delta v} \tag{3.5}$$

where Δu is the change in u from one conductor to the other, and Δv is the change in v as one goes around one conductor. We now need a suitable complex mapping to fit the problem.

The potential function that describes the aperture is [7, eqn. 2.15]

$$\frac{\zeta}{b} = \frac{2j}{\pi}\left[K(m)\,E(w|m_1) + w \times (E(m) - K(m))\right] \tag{3.6}$$

This is the simplest form. Another form is used in [8,9], but it is less convenient because the conductors are located on surfaces of constant v instead of the more customary constant u. As a cautionary note, we point out that most numerical packages expect the first argument of the incomplete elliptic function, $E(w|m)$, to be in the form of an angle. Thus, the incomplete elliptic integral has to be cast into the form of $E(\text{am}(w|m)|m)$, where $\text{am}(w|m) = \arcsin(\text{sn}(w|m))$ is the Jacobian amplitude function. A plot of the resulting complex mapping appears in Figure 3.2.

The figure of merit we use for the radiated field is the transient power gain normalized to constant input power, as defined in [2,5,6,10]. Thus, we have

$$G_p = \frac{h_a}{\sqrt{f_g}} \tag{3.7}$$

where h_a is the normalized integral over the aperture field [11],

$$h_a = -\frac{f_g}{V_o}\iint_{S_a} E_y(x',y')\,dx'\,dy' \tag{3.8}$$

and S_a is the total surface over which radiation occurs. This integral will be calculated for the two configurations as a function of Z_c.

Let us pause for a moment, to consider whether the figure of merit, G_p, makes sense. To see why our expression of gain in (3.7) is reasonable, we recall that the fast part of the field radiated from an aperture on boresight is [2,5,6]

$$\begin{aligned} E_{rad}(t) &= -\frac{h_a}{2\pi r c f_g}\frac{dV(t)}{dt} \\ E_{rad}(t) &= -\frac{G_p}{2\pi r c}\frac{d\left(V(t)/\sqrt{f_g}\right)}{dt} \end{aligned} \tag{3.9}$$

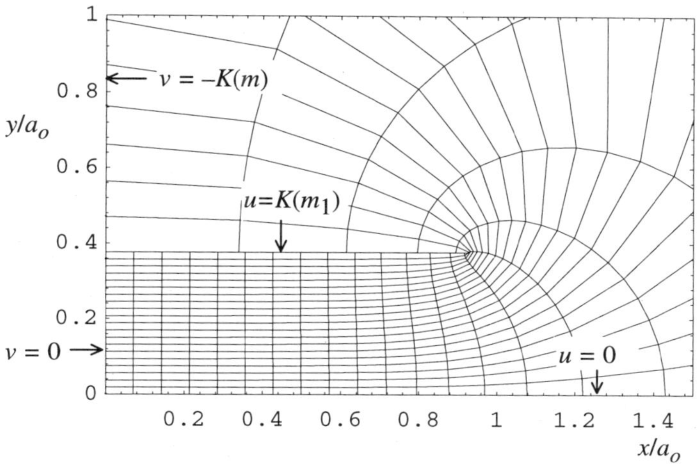

Figure 3.2. Complex potential map of the parallel plate configuration, for $Z_c = 100\ \Omega$.

Furthermore, the received voltage for an incident field on boresight is

$$\begin{aligned} V_{rec}(t) &= -h_a\ E_{inc}(t) \\ \frac{V_{rec}(t)}{\sqrt{f_g}} &= -G_p\ E_{inc}(t) \end{aligned} \tag{3.10}$$

Thus, both the radiated field and the received voltage are proportional to power gain. Note also that all the voltages are cast into the form of square root of power, by dividing by $\sqrt{f_g}$. On this basis, our definition of the figure of merit seems reasonable.

A. Flat Plates with Infinite Aperture

We now calculate h_a and G_p for two cases. First, we consider the case of an infinite aperture, as shown on the left in Figure 3.1. In this case, it is particularly simple to calculate h_a, since it can alternatively be expressed in terms of the dipole moment of the charge in the aperture [11]. Thus,

$$h_a = \frac{1}{2} \frac{\iint q(x,y)\ y\ dA}{\iint q(x,y)\ dA} \tag{3.11}$$

where $q(x, y)$ is the charge density on the conductors, and the integrals are carried out over all the conductors in the aperture. Since all the charge is located at $y = \pm\ b$, the value of h_a is calculated trivially as

$$h_a = b \tag{3.12}$$

We have plotted h_a/a_o as a function of Z_c in Figure 3.3, on the left. As expected, at high impedances it approaches unity asymptotically. This is the expected result because for a pair of thin wires, we know from [11] that $h_a = a_o$. Finally, we have plotted the power gain, $G_p = h_a / \sqrt{f_g}$, in Figure 3.3, on the right. The peak occurs at $Z_c = 242.3\ \Omega$, where the power gain is $1.09 \times a_o$. At this point, $b/a = 1.82$.

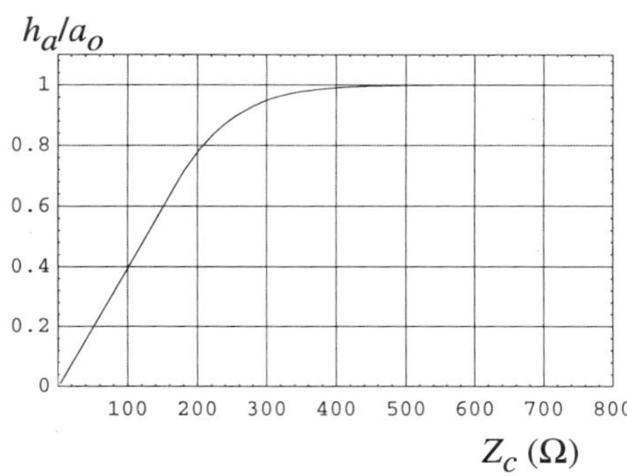

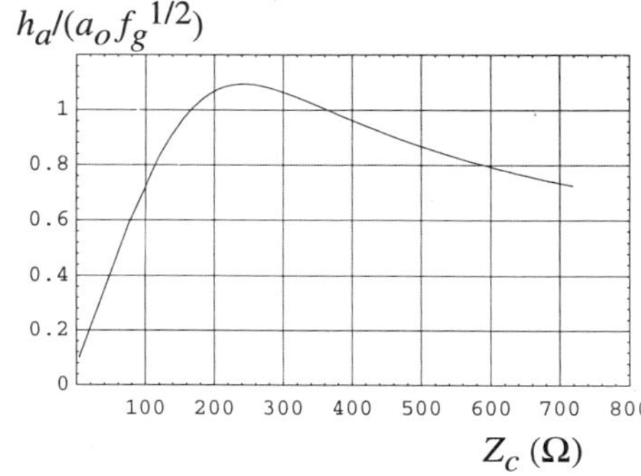

Figure 3.3. Normalized aperture height for the infinite aperture with flat plates (left) and the corresponding figure of merit (right).

B. Flat Plates with Blocked Aperture

Next, we calculate h_a for the case of a blocked aperture, as shown on the right in Figure 3.1. This is the normal case of interest to the antenna designer. Only a portion of the aperture field is included in the aperture integral, because only that portion is focused.

To calculate h_a , we express it as a contour integral [2,5,11], i.e.,

$$h_a = -\frac{4}{\Delta v} \oint_{C_a} v(\zeta)\, dy \tag{3.13}$$

where Δv is the change in v around a conductor. Furthermore, C_a is a contour consisting of four segments which surround one quadrant of the exposed portion of the aperture, as shown in Figure 3.4. To calculate this integral, we note that the integrals over C_1 and C_3 are identically 0, since there is no change in y. Furthermore, the integral over C_4 is also zero, because $v = 0$ there. Therefore, we need only calculate the integral over C_2.

Calculating the integral over C_2 presents us with a challenge, because there is no easy way to calculate $v(\zeta)$. Normally, one would find v from the complex contour mapping, i.e., equation (3.6). But in the present case there is no simple way to invert the contour mapping to find v in terms of ζ. Thus, one would have to solve (3.6) numerically for $v(\zeta)$ at each location on the arc. Since we must also calculate the integral along this path, this becomes computationally expensive.

A simpler method is to just assume that the contour C_2 is along a contour of constant v, instead of being along a portion of a circular arc. This is rigorously true in the limit of high impedances, $b/a \to \infty$, because the conductors become thin wires at that point. Since we know the optimal values will occur at relatively high values of b/a, the approximation is reasonable. Thus, we search numerically for the value of v that intersects the circle at $(x, y) = (a_o, 0)$, and use that for our contour of constant v_o. From (3.6), we solve numerically the following for v_o,

$$\frac{a_o}{b} = \frac{2j}{\pi}\left[K(m)\, E(j v_o | m_1) + j v_o \times (E(m) - K(m))\right] \tag{3.14}$$

After finding v_o, and using the fact that $\Delta v = 2K(m)$, we find the aperture height to be

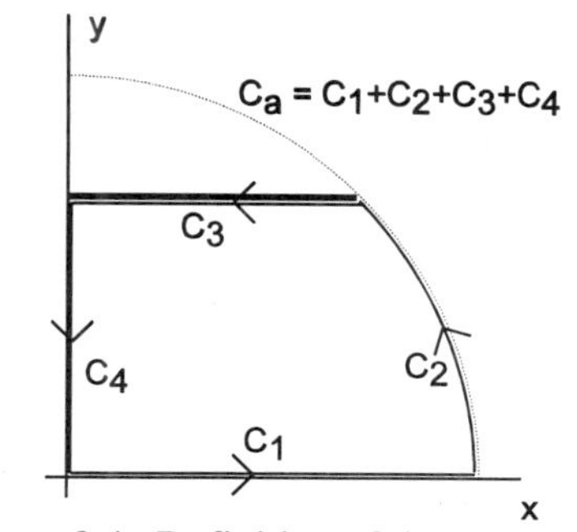

Figure 3.4. Definition of the contour C_a.

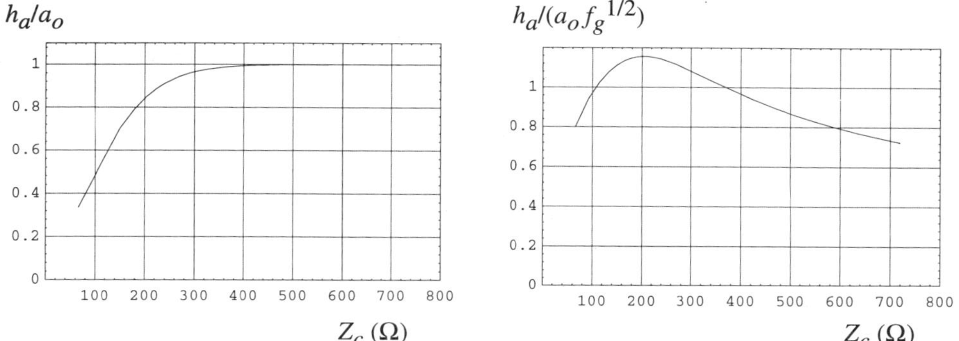

Figure 3.5. Normalized aperture height for the blocked aperture with flat plates (left), and the normalized figure of merit (right).

$$h_a = -\frac{2bv_o}{K(m)} \tag{3.15}$$

Note that v_o is a negative number, so h_a is positive, as it must be. We estimate the error in this procedure to be a few percent.

The results for h_a are plotted in Figure 3.5 (left), and the power gain is plotted on the right of the same figure. The peak gain occurs at $Z_c = 203.7\ \Omega$, where the power gain is $1.16 \times a_o$. At this point, $b/a = 1.28$.

It is interesting to compare the results for the blocked aperture to those with the infinite aperture. One might expect that an infinitely large aperture would have better performance, but that is not the case. The blocked aperture actually has a slightly better power gain, with a value of $1.16 \times a_o$, compared to $1.09 \times a_o$ for the infinite aperture. The likely reason for this is that we have blocked out fields that have a negative contribution to the total radiated field, just above the top plate and just below the bottom plate. Therefore, if one wanted to improve the performance of a simple long TEM horn, one would block out the portion of the aperture fields just above and below the top and bottom plates.

A similar case with curved plates confined to a circular arc was treated in [6]. The optimal impedance for that case was $f_g = 0.5$, or $Z_c = 376.727 / 2\ \Omega$, which occurs with plates with angular widths of 90°. This is true for both the blocked aperture and for the infinite aperture. It was demonstrated in [12] that these two cases are equivalent. At this impedance, we have $h_a = 0.85 \times a_o$, and $G_p = 1.20 \times a_o$, where a_o is the aperture radius.

IV. RADIATION FROM FOUR-WIRE APERTURES

Finally, we consider the radiated field pattern of a four-wire aperture. This is an approximation to the four-arm reflector Impulse Radiating Antenna. In previous papers [2,13] we considered the radiated field for a two-wire aperture, so the technique used here is essentially the same. The only change is that the aperture field is described by a different aperture potential [14]. Note that only the fast part of the radiated field is considered here.

When the conical geometry is projected onto a plane, we have an aperture field that is created from four conductors, as shown in Figure 4.1. The potential function is calculated by adding the potential for two two-wire problems, including translation. The potential function for a single pair of wires, where the charge centers are located at (x=0, y/a = 1), is

$$w_2(\zeta) \;=\; 2\,j\,\mathrm{arccot}(\zeta\,/\,a) \;=\; \ln\!\left(\frac{\zeta - j\,a}{\zeta + j\,a}\right) \tag{4.1}$$

where a is the aperture radius. Here, $\zeta = x + j\,y$ is the location in the Cartesian coordinate space. This potential function is plotted in [13, Figure 2]. The complex potential for the four-wire case is just a sum of two two-wire potentials that have been shifted and rescaled,

$$w_4(\zeta) \;=\; w_2\big((\zeta\,/\,a + \sqrt{2})\,/\,\sqrt{2}\big) \;+\; w_2\big((\zeta\,/\,a - \sqrt{2})\,/\,\sqrt{2}\big) \tag{4.2}$$

This function is complex, i.e., has both real and imaginary parts. Let us therefore set

$$u(\zeta) \;=\; \mathrm{Re}(w_4(\zeta)) \quad , \qquad v(\zeta) \;=\; \mathrm{Im}(w_4(\zeta)) \tag{4.3}$$

We can plot contours of constant u and v, and these are shown in Figure 4.2, for the upper right quadrant. The conductors correspond to a contour of constant u.

To calculate the radiated field, we need the aperture fields and the normalized aperture potentials. The aperture field is

$$E_y(x,y) \;=\; \frac{-(2V_o)}{\Delta u}\frac{\partial\, u(x,y)}{\partial\, y} \tag{4.4}$$

where $2V_0$ is the voltage between the top and bottom conductors, and Δu is the change in u between the two conductors.

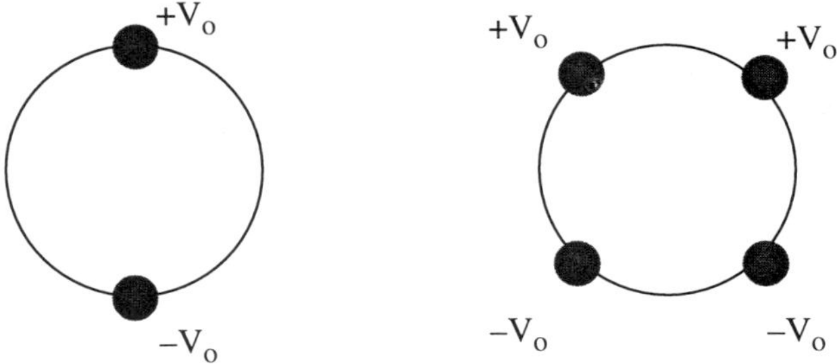

Figure 4.1. The apertures for a two-wire and four-wire configuration.

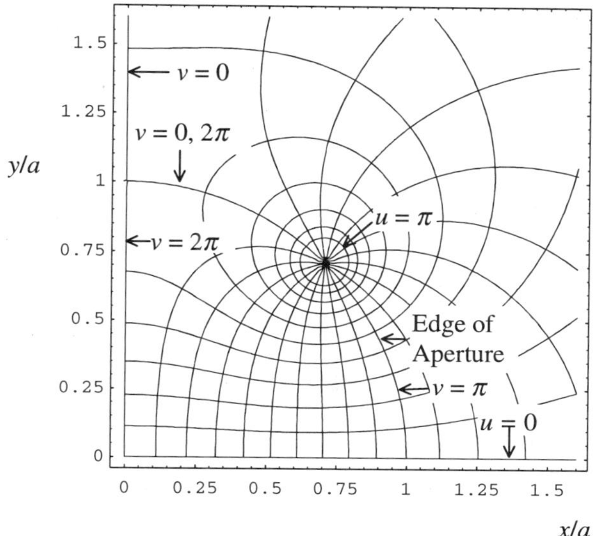

Figure 4.2. Contour map for $w_4(\zeta)$. Increments for u and v are $\pi/10$.

The normalized potentials are integrals over linear paths in the aperture field. We need to calculate these because the radiated field is proportional to them. The normalized potentials for the H-plane calculation is

$$\Phi^{(h)}(x) \quad = \quad -\frac{1}{(2V_o)}\int_{C_1(x)} E_y \, dy \tag{4.5}$$

where the contour $C_1(x)$ is a vertical line cut through the aperture plane, as shown in Figure 4.3. To simplify this H-plane integral, one substitutes (4.4) into (4.5), generating

$$\Phi^{(h)}(x) \quad = \quad \frac{1}{\Delta u}\int_{C_1(x)} \frac{\partial u}{\partial y}\, dy \quad = \quad \frac{2}{\Delta u}\, u\left(x, \sqrt{a^2 - x^2}\right) \tag{4.6}$$

We can now calculate $u(x,y)$ as the real part of the potential function given in (4.2). Note that the value of $u(x,y)$ is a maximum when it cuts through the conductors. At this point, the value of $u(x,y)$ is $u_o = \pi f_g$, where f_g is the relative impedance for a single pair of arms located on opposite sides of the circle (typically f_g = 400 Ω / 377 Ω). Note also that for values of x that cut through the conductors, the normalized potential is unity. This normalized potential function is plotted in Figure 4.4 (left), for a few different values of f_g.

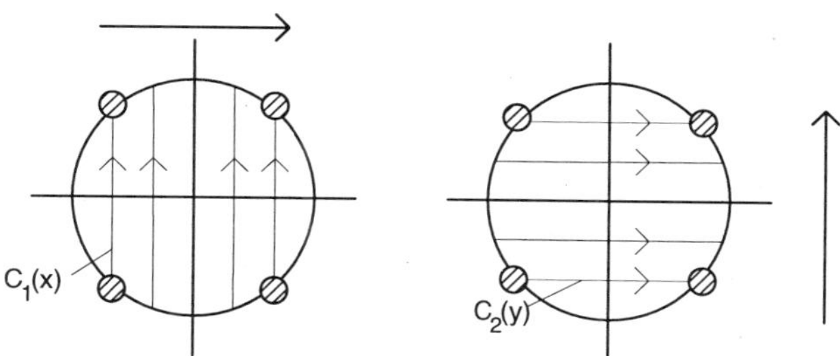

Figure 4.3. Locations of $C_1(x)$ and $C_2(y)$.

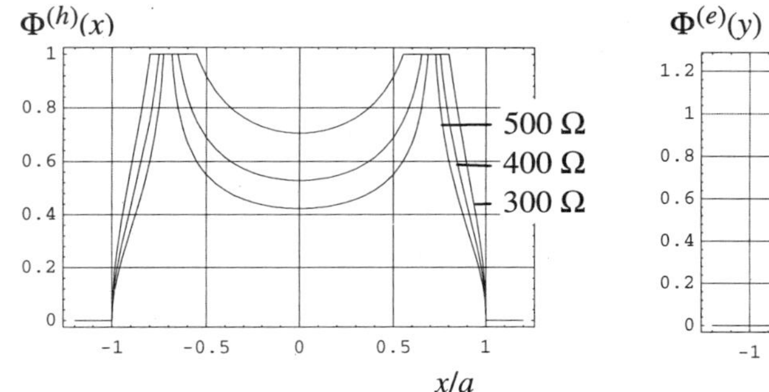

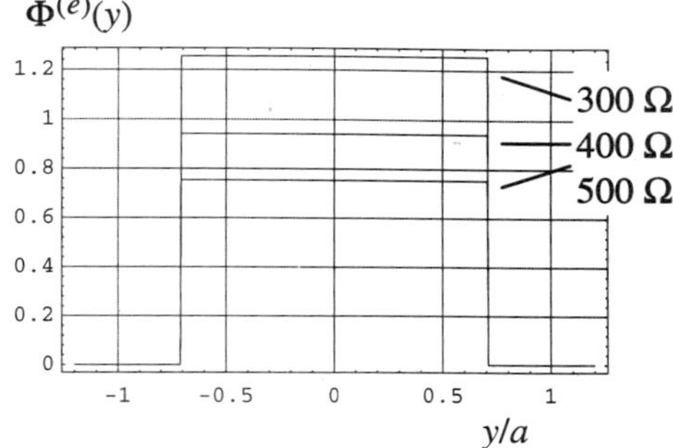

Figure 4.4. The normalized potential functions, $\Phi^{(h)}(x)$ and $\Phi^{(e)}(y)$, for a few different impedances.

The normalized potential for the E-plane is expressed as

$$\Phi^{(e)}(y) = -\frac{1}{(2V_o)}\int_{C_2(y)} E_y\,dx = \frac{1}{\Delta u}\int_{C_2(y)}\frac{\partial u}{\partial y}\,dx \qquad (4.7)$$

where $C_2(y)$ is a horizontal linear cut through the aperture plane, as shown in Figure 4.3. To evaluate this, we require the Cauchy-Riemann relation for analytic functions,

$$\frac{\partial u}{\partial y} = -\frac{\partial v}{\partial x} \qquad (4.8)$$

which allows us to recast the integral as

$$\Phi^{(e)}(y) = \frac{-2}{\Delta u}\left[v\left(\sqrt{a^2-y^2},\,y\right) - v(0,y)\right] \qquad (4.9)$$

This is a particularly simple form, because the edges of the circular aperture are also lines of constant v. Thus, the normalized potential is evaluated analytically as

$$\Phi^{(e)}(y) = \begin{cases} 1/f_g & |y|/a < 1/\sqrt{2} \\ 0 & \text{else} \end{cases} \qquad (4.10)$$

Note that we show a very abrupt transition between the two values, but it is actually more smooth. This transition occurs as $C_2(y)$ passes through the two wires, and if the wire is thin, an abrupt transition is an excellent approximation. We have plotted the normalized potentials for a few impedances in Figure 4.4 (right).

With the normalized potentials calculated, we can now calculate the radiated field as a function of angle off boresight in the H and E-planes. The H-plane and E-plane are the planes that are perpendicular and parallel to the dominant radiated field on boresight, respectively. In the H-plane and E-plane, the field radiated by a step voltage of magnitude $2V_o$ across the aperture is [13]

$$\begin{aligned} \vec{E}_{step}^{(h)}(r,\theta,t) &= \vec{1}_y\left(\frac{-(2V_o)}{r}\right)\frac{\cot(\theta)}{2\pi}\,\Phi^{(h)}\left(\frac{ct}{a\sin(\theta)}\right) \\ \vec{E}_{step}^{(e)}(r,\theta,t) &= \pm\vec{1}_\theta\left(\frac{-(2V_o)}{r}\right)\frac{1}{2\pi\sin(\theta)}\,\Phi^{(e)}\left(\frac{ct}{a\sin(\theta)}\right) \end{aligned} \qquad (4.11)$$

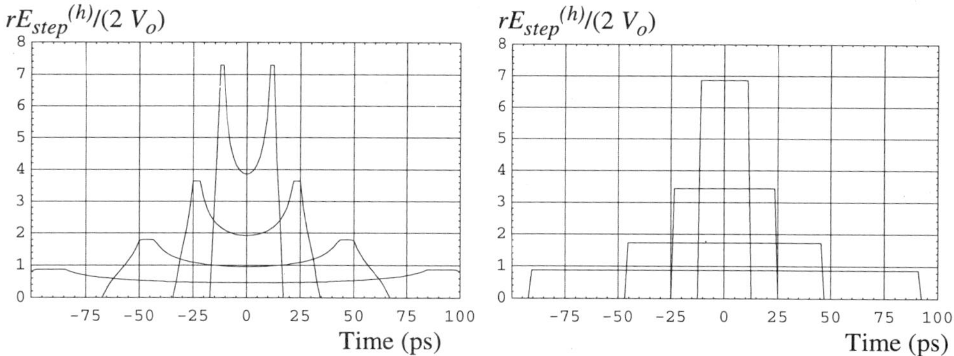

Figure 4.5. Step response of the 4-armed reflector IRA in the H-plane (left) and the E-plane (right), at 2.5, 5, 10, and 20 degrees off-boresight.

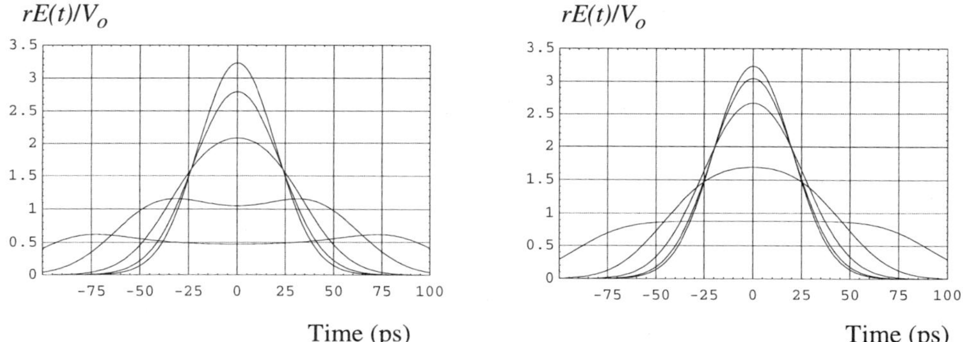

Figure 4.6. Fast part of the H-plane (left) and E-plane (right) radiated field for the reflector IRA at $\theta = 0^o$, 2.5^o, 5°, 10° and 20° away from boresight.

In Figure 4.5 we have plotted these for the case of diameter = $2\ a$ = 22.9 cm, for a four-arm feed with 200 Ω input impedance. Time plots of these two step responses are shown for a few different values off-boresight in the H and E-planes. Note that we cannot plot the step response at 0°, because it is a delta function there, with infinite magnitude and zero width.

To obtain a radiated field, we convolve the step responses with the derivative of the driving voltage. To drive the antenna, we assume an integrated Gaussian with a peak magnitude of V_o and a risetime of $t_d = 50$ ps. We have plotted the time response at $\theta = 0^o$, 2.5^o, 5°, 10° and 20° away from boresight in the E and H planes, in Figure 4.6.

VI. CONCLUSION

We have considered here a number of extensions to IRA theory, including an assortment of new IRA designs using two reflecting or refracting surfaces. We have also considered the optimal impedance for a long TEM horn or lens IRA, and we have calculated the off-boresight field radiated from a four-wire aperture.

ACKNOWLEDGMENTS

Dr. Charles A. Frost, of Pulse Power Physics, first suggested the Solid Dielectric Lens IRA. Dr. Gary D. Sower, of EG&G MSI, suggested an early version of the Split IRA.

REFERENCES

1. C. E. Baum and E. G. Farr, "Impulse Radiating Antennas," pp. 139-147 in H. L. Bertoni et al (eds.), *Ultra Wideband/Short-Pulse Electromagnetics*, Plenum Press, New York, 1993.

2. E. G. Farr, C. E. Baum, and C. J. Buchenauer, "Impulse Radiating Antennas, Part II, pp. 159-178 in L. Carin et al (eds.), *Ultra Wideband/Short-Pulse Electromagnetics 2*, Plenum Press, New York, 1995.

3. E. G. Farr and C. E Baum, Impulse Radiating Antennas With Two Refracting or Reflecting Surfaces, Sensor and Simulation Note 379, May 1995.

4. C. E. Baum, J. J. Sadler, and A. P. Stone, Uniform Isotropic Dielectric Equal-Time Lenses for Matching Combinations of Plane and Spherical Waves, Sensor and Simulation Note 352, December 1992.

5. E. G. Farr, Optimizing the Feed Impedance of Impulse Radiating Antennas, Part I: Reflector IRAs, Sensor and Simulation Note 354, January 1993.

6. E. G. Farr, Optimization of the Feed Impedance of Impulse Radiating Antennas, Part II: TEM Horns and Lens IRAs, Sensor and Simulation Note 384, November 1995.

7. C. E. Baum, D. V. Giri, and R. D. Gonzalez, Electromagnetic Field Distribution of the TEM Mode in a Symmetrical Two-Parallel-Plate Transmission Line, Sensor and Simulation Note 219, April 1976.

8. C. E. Baum, Impedances and Field Distributions for Parallel Plate Transmission Line Simulators, Sensor and Simulation Note 21, June 1966.

9. P. Moon and D. E. Spencer, Field Theory Handbook, second edition, Springer-Verlag, Berlin, 1971.

10. E. G. Farr and C. E. Baum, Extending the Definitions of Antenna Gain and Radiation Pattern Into the Time Domain, Sensor and Simulation Note 350, November 1992.

11. C. E. Baum, Aperture Efficiencies for IRAs, Sensor and Simulation Note 328, June 1991.

12. E. G. Farr and C. E. Baum, Radiation from Self-Reciprocal Apertures, Chapter 6 in C. E. Baum and H. N. Kritikos, *Electromagnetic Symmetry*, Taylor and Francis, 1995.

13. E. G. Farr and C. E. Baum, The Radiation Pattern of Reflector Impulse Radiating Antennas: Early-Time Response, Sensor and Simulation Note 358, June, 1993.

14. E. G. Farr and C. A. Frost, Development of a Reflector IRA and a Solid Dielectric Lens IRA, Part I: Design, Predictions and Construction, Sensor and Simulation Note 396, April 1996.

TRANSIENT FIELDS OF RECTANGULAR APERTURE ANTENNAS

Sergey P. Skulkin

Radiophysical Research Institute
Nizhny Novgorod, Russia

INTRODUCTION

Properties of transient field of impulse radiating antennas (IRA) were described predominantly for reflector antennas with circular aperture[1-9]. However, the shape of the aperture changes the spatial-temporal field distribution, especially, in the near-field region. The goals of the paper are to describe the transient fields of rectangular aperture and to compare these results with those obtained for circular aperture[9].

The properties of ultra wide band (UWB) antennas can be completely characterized by the radiated field's spatial-frequency dependence $\vec{E}(\omega, \vec{r})$, where $\vec{r}$ is the radius- vector of the observation point and ω is the circular frequency. In contrast to the monochromatic antennas where ω is a constant, for the UWB antennas the functional dependence $\vec{E}(\omega, \vec{r})$ as a function of ω is significant because this dependence determines the form of the radiating pulse $\vec{E}(t, \vec{r})$ (where t is time variable) at the point $\vec{r}$. The latter is the Fourier transform of the former:

$$\vec{E}(t, \vec{r}) = \int \vec{E}(\omega, \vec{r}) e^{-i\omega t} d\omega \tag{1}$$

If $\vec{E}(\omega, \vec{r})$ is calculated for a unit complex exciting amplitude at every ω, then $\vec{E}(t, \vec{r})$ can be considered as the totality of the pulse radiating characteristics (PRC) of an "antenna-free space" system for each point $\vec{r}$ of the space around the antenna[9]. In this case the radiating antenna transient field for the polarization characterized by the unit vector $\vec{e}$ can be defined as a convolution:

$$S_e(t, \vec{r}) = \vec{e} \cdot E(t, \vec{r}) \bigotimes S_0(t), \tag{2}$$

where $S_0(t)$ is the antenna input impulse. Respectively, $E(t, \vec{r})$ is the antenna transfer function in the time domain. We note that the antenna transfer function in the time domain is calculated more simply than in the frequency domain, where, in the best case, the antenna near-field can be represented analytically by a series of special functions[10,11]. For large antennas, the transfer function calculation can be organized

Ultra-Wideband, Short-Pulse Electromagnetics 3
Edited by Baum *et al.*, Plenum Press, New York, 1997

into two independent parts. The first is the transfer function of the primary radiator. The calculated results for such antennas are given, for example, in [12]. The second issue is the calculation of the transfer function of the aperture itself, formed by the reflector or array of primary radiators. In the case of a reflector antenna the primary radiator pattern can be considered as the aperture illumination function $g(\vec{r}_a, \omega)$ (here $\vec{r}_a$ is the radius-vector of a point on aperture plane). In the case of an antenna array the primary radiator pattern is introduced into the integral as the integrand factor, provided we ignore the coupling between array elements etc. For large antennas the spatial field distribution is defined, first of all, by the aperture shape and size. Usually the main lobe of the primary radiator pattern is quite wide and approximated within the scope of aperture antenna theory by model functions[10,11]. In this approximation the electrical field projection $\vec{E}(\omega, \vec{r})$ on the unit vector $\vec{e}$ can be represented as:

$$E_e(\omega, \vec{r}) = \frac{i\omega}{2\pi c} \iint_{S_a} g(\vec{r}_a, \omega)\gamma(\vec{r}, \vec{r}_a)\frac{e^{i\frac{\omega}{c}|\vec{r}-\vec{r}_a|}}{|\vec{r} - \vec{r}_a|} dS_a, \tag{3}$$

where S_a is the aperture region, $g(\vec{r}_a, \omega)$ is the aperture illumination function, $\gamma(\vec{r}, \vec{r}_a)$ is the factor defined by the polarization relations. In particular, for a long focal length parabolic antenna we approximately consider the aperture surface S_a as a plane, and for the basic polarization we assume:

$$g(x', y', \omega) \sim f(\frac{x'}{F}, \frac{y'}{F}, \omega)$$

where x', y' are the Cartesian coordinates on the aperture plane, F is the focal distance, $f(\xi, \eta, \omega)$ is the pattern of the illuminating radiator, ξ, η are the direction cosines; and

$$\gamma(\vec{r}, \vec{r}_a) \simeq (\vec{n}\vec{r}_0)^2 \tag{4}$$

where $\vec{r}_0 = (\vec{r} - \vec{r}_a)/|\vec{r} - \vec{r}_a|$, $\vec{n}$ is the normal to the aperture[12]. Here, for a wideband illuminating radiator, we will assume that the aperture illumination function and the shape of the illuminator's radiation pattern are independent of the frequency:

$$g(\vec{r}_a, \omega) \simeq g(\vec{r}_a) w_{ill}(\omega). \tag{5}$$

Based on (5) the radiating antenna transient field $S_e(t, \vec{r})$ for polarization $\vec{e}$ can be represented as:

$$S_e(t, \vec{r}) = S_0(t) \otimes S_{ill}(t) \otimes (\frac{d}{dt}\tilde{E}_e(t, \vec{r})), \tag{6}$$

where $S_{ill}(t)$ is the PRC of the illuminating radiator, it is defined as the Fourier transformation of $w_{ill}(\omega)$, $d\tilde{E}_e/dt$ is the Fourier transformation of $i\omega\vec{E}_e(\omega, \vec{r})$. Taking into account (3) and (6) we can write:

$$\tilde{E}_e(t, \vec{r}) = \frac{1}{2\pi c} \iint_{S_a} \frac{g(\vec{r}')\gamma(\vec{r}, \vec{r}')\delta(t - \frac{1}{c}|\vec{r} - \vec{r}_a|)}{|\vec{r} - \vec{r}_a|} dS_a, \tag{7}$$

where δ is the Dirac delta function. It is essential for representation (7) of E_e that (3) and (5) are correct for all frequencies $-\infty < \omega < \infty$, whereas (3) is suitable for field calculations of antennas with the aperture diameter D_a much larger than the wavelength, and (5) does not hold for all frequencies. To overcome this impediment we will assume that the energy of pulse signal S_0 is concentrated essentially in the frequency band $\omega_{min} < \omega < \omega_{max}$ and $D_a \gg 2\pi c/\omega_{min}$.

Although formally, (7) is not strictly correct; in this case we will obtain the right result for $\tilde{E}_e(t, \vec{r})$.

BASIC FORMULAS

To calculate the PRC let us take an formula for an integral, containing a δ-function of complex argument:

$$\iint\limits_S f(x,y)\delta[\varphi(x,y)]dxdy = \int\limits_\Gamma \frac{f(x(\gamma),y(\gamma))d\gamma}{|grad\varphi|_{,x=x(\gamma),y=y(\gamma)}}, \tag{8}$$

where Γ is the curve determined from the equation $\varphi(x,y) = 0$; $x = x(\gamma)$, $y = y(\gamma)$ is a parametric representation of Γ, $d\gamma$ is the element of length of Γ. It is supposed here, that the solution of the equation $\varphi(x,y) = 0$ for $x, y \in S$ exists and determines a unique curve Γ. If for all $x, y \in S$ $\varphi > 0$ or $\varphi < 0$, then the integral (8) equals zero. For the integral (8) the equation determining the curve Γ, is

$$|\vec{r} - \vec{r}_a| = ct. \tag{9}$$

In three-dimensional space (9) describes the sphere with its center at the point $\vec{r}$ and radius ct (c is the speed of light). The sphere crosses the aperture plane for $ct > z$, where z is the distance from the point $\vec{r}$ to the aperture plane, it is obvious for $ct < z$, that $\tilde{E}_e = 0$. Curve Γ (a locus of intersection of the sphere and the aperture plane) is the circle with radius $b = \sqrt{(ct)^2 - z^2}$ and the center at the point $\vec{\rho}$, where $\vec{\rho}$ is the projection of vector $\vec{r}$ on the aperture plane (see Fig.1). For this circle $|\vec{r} - \vec{r}_a|| grad(\frac{1}{c} | \vec{r} - \vec{r}_a|) |_{\vec{r}_a \in \Gamma} = b/c$, hence,

$$\tilde{E}_e(t,\vec{r}) = \begin{cases} 0, & (a); \\ \frac{1}{2b}\int\limits_{\Gamma_a} \gamma(\vec{r},\vec{r}_a{}')g(\vec{r}_a \in \Gamma_a)d\gamma & (b); \end{cases} \tag{10}$$

Γ_a is the part of Γ, belonging to S_a. In the case (a) we have $\Gamma_a \notin S_a$.

PRC OF RECTANGULAR PLANE APERTURE

For the constant amplitude distribution over the aperture $g(\vec{r}_a) = 1$, there follows the elementary formula for $\tilde{E}_e$:

$$\tilde{E}_e(t,x,y,z) = \gamma(\vec{r},\vec{r}_a{}')(2\pi - \sum_1^8 \phi_{m,n}), \tag{11}$$

Here for simplicity we will consider $\gamma(\vec{r},\vec{r}_a{}') = 1$.

$$\phi_{mn} = \begin{cases} 0, & ct < \sqrt{z^2 + l_m^2}; \\ \arccos\frac{|l_m|}{\sqrt{(ct)^2 - z^2}}, & \sqrt{z^2 + l_m^2} < ct < \sqrt{z^2 + l_m^2 + l_n^2}; \\ \arccos\frac{|l_m|}{\sqrt{l_m^2 + l_n^2}}, & \sqrt{z^2 + l_m^2 + l_n^2} < ct; \end{cases} \tag{12}$$

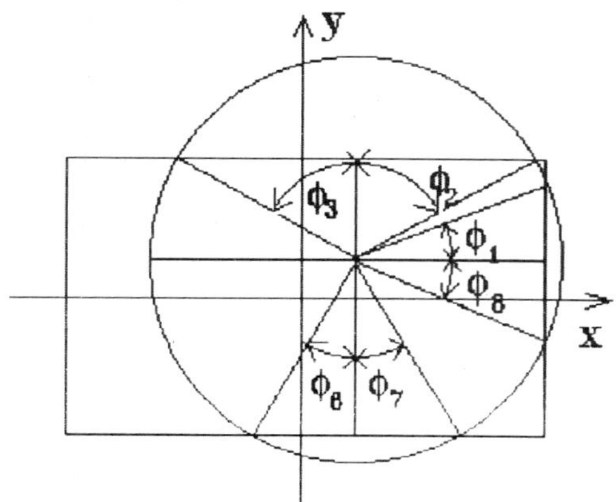

Fig. 1. The curve of the intersection of the sphere with radius (ct') and the aperture plane.

$m = 1 \ldots 4$ is the number of the aperture boundary; n is the number of the boundary perpendicular to that with the number m, l_m, l_n is the distance from the projection of the observation point on the aperture plane up to the boundaries of apertures with numbers m, n, respectively.

Such a transformation in the physical sense can be explained the following way. If each element of the aperture radiates a δ-pulse at the moment $t = 0$, then, at the moment $t' > 0$ the field at the point $\vec{r}$ is defined only by elements lying on the circle or its part, where the sphere of ct' radius and center at $\vec{r}$ point crosses the aperture plane. The field amplitude is defined by the weighted integral over the given circle.

The form of the field $E_a(t, x = 0, y < a_y, z)$ for $a_x = a_y$ is given in Figure 2.

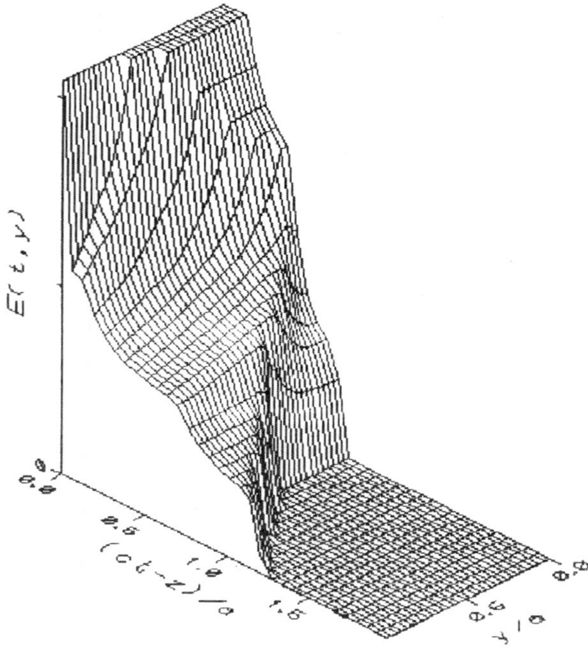

Fig. 2. The transient characteristics $\tilde{E}_e(t, x = 4, y < a, z = a)$. Here for simplicity we will concider $\gamma(\vec{r}, \vec{r_a}') = 1$.

In contrast to PRC of circular plane aperture at $\rho = 0$, the duration of the trailing edge is always not zero and contains several zones defined by different character of intersection between the circles and the aperture boundary. With increase of the distance from the axis, the duration of the plane zone at the top of PRC is decreased, and the total duration of the pulse is increased. The form of the trailing edge for the rectangular aperture will depend also on the relation of a_x/a_y. We give expressions for the field outside the limits of the projector area for two cases: when only one projection is outside the aperture limits (13) and when both projections are outside the aperture limits (see Figure 5, expression (17)).

$$\tilde{E}_e(t, x, y, z) = \phi_{right} + \phi_{left}, \tag{13}$$

$$\phi_{right,left} = |\phi_2 - \phi_1|, \tag{14}$$

$$\phi_1 = \begin{cases} 0, & ct < \sqrt{z^2 + l_i^2}; \\ \arcsin \dfrac{|l_i|}{\sqrt{(ct)^2 - z^2}}, & \sqrt{z^2 + l_i^2} < ct < \sqrt{z^2 + l_i^2 + l_{j,n}^2}; \\ 0, & ct > \sqrt{z^2 + l_i^2 + l_{j,n}^2}. \end{cases} \tag{15}$$

$$\phi_2 = \begin{cases} 0, & ct < \sqrt{z^2 + l_k^2}; \\ \arcsin \dfrac{l_i}{\sqrt{(ct)^2 - z^2}}, & \sqrt{z^2 + l_k^2} < ct < \sqrt{z^2 + l_k^2 + l_j^2}; \\ \arccos \dfrac{l_j}{\sqrt{(ct)^2 - z^2}}, & \sqrt{z^2 + l_k^2 + l_j^2} < ct < \sqrt{z^2 + l_i^2 + l_j^2}; \\ 0, & \sqrt{z^2 + l_i^2 + l_j^2} < ct; \end{cases} \tag{16}$$

where $\phi_{right,left}$ are angles in the right-hand and left-hand of half-plane, the boundaries of half-planes are defined by the normal to two boundaries passing through the projection of the observation point, l_k, l_i is the distance from the projection of the observation point on the aperture plane up to the nearest aperture boundary and to the parallel one; j, n is the number of the aperture boundary perpendicular to that with number i.

$$\tilde{E}_e(t, x, y, z) = \phi_{max} - \phi_{min}, \tag{17}$$

where $\phi_{max} > \phi_{min} > 0$, two others $\phi_m = 0$

$$\phi_m = \begin{cases} 0, & ct < \sqrt{z^2 + l_m^2 + l_n^2}; \\ \arccos \dfrac{l_m}{\sqrt{(ct)^2 - z^2}}, & \sqrt{z^2 + l_m^2 + l_n^2} < ct < \sqrt{z^2 + l_m^2 + l_k^2}, \\ & \text{for } m = 1, 2; \\ \arcsin \dfrac{l_m}{\sqrt{(ct)^2 - z^2}}, & \sqrt{z^2 + l_m^2 + l_n^2} < ct < \sqrt{z^2 + l_m^2 + l_k^2}, \\ & \text{for } m = 3, 4; \\ 0, & \sqrt{z^2 + l_m^2 + l_k^2} < ct; \end{cases} \tag{18}$$

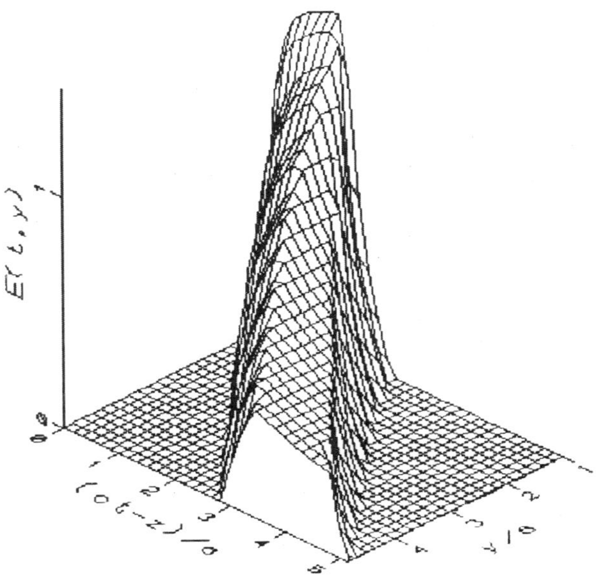

Fig. 3. The transient characteristics $\tilde{E}_e(t, x = 4, y > a, z = a)$.

where l_m is the distance from the projection of the observation point up to one of the aperture boundaries with number m; $m = 1 \ldots 4$ is the number of the aperture boundary; n, k are the numbers of boundaries perpendicular to that with number m, here $l_k > l_n > l_m$.

At infinity the field at each moment of time will be defined as the integral over the line of two plane intersection, the first is the aperture plane and the plane (the sphere of the infinite radius) inclined by the angle θ to the aperture plane (polar angle) and will be presented in the form: $\tilde{E}_e(t, \theta, r, \varphi) \to \frac{1}{r} f(t, \theta, \varphi)$, where φ- azimuth angel, θ- polar angle (here $x = r \sin\theta \cos\varphi$, $y = r \sin\theta \sin\varphi$, $z = r\cos\theta$, $ct' = ct - r$, it being known that $a_x \sin\theta\cos\varphi < a_y \sin\theta \sin\varphi$, $\sin\theta\cos\varphi > 0, \sin\theta\sin\varphi > 0$).

$$f(t, \theta, \varphi) = \begin{cases} \dfrac{2a_y c}{\sin\theta\cos\varphi}, & \sin\theta(a_x\cos\varphi + a_y \sin\varphi) < |ct'|; \\ \dfrac{c(a_x\cos\varphi - a_y\sin\varphi - |ct'|)}{\sin\theta\sin\varphi\cos\varphi}, & \sin\theta(a_x\cos\varphi - a_y\sin\varphi) < |ct'| < \\ & < \sin\theta(a_x\cos\varphi + a_y\sin\varphi); \\ 0, & \sin\theta(a_x\cos\varphi + a_y\sin\varphi) < |ct'|. \end{cases} \tag{19}$$

CONCLUSIONS

A rectangular aperture response in the near-field region is defined by the derivative of $\tilde{E}_e(t, x, y, z)$ and consists of several pulses, whereas for circular aperture antennas this number of pulses is limited by two[9]. For the both aperture shapes the initial pulse is shortest and has the maximal magnitude.

The method proposed is also useful for the monochromatic field calculation, because for field computations we use the one-dimensional Fourier transformation of signals, given by the strict formulas which are expressed in elementary functions. This allows to simplify the estimation of the temporal sampling and provides any precision required.

REFERENCES

1. C. E. Baum, "Impulse radiating antennas",in book Ultra-Wideband, Short-Pulse Electromagnetics, ed. by Bertroni et al., Plenum Press, 1993.
2. S. P. Skulkin, V. I. Turchin, et al., "The time-pulse method of measuring antenna characteristics in near zone," Radiophysics and quantum electronics, vol.32, no.1, pp. 61-70, July 1989.
3. C. E. Baum, "Focused aperture antennas", Sensor and Simulation note 306, May 19 1987.
4. E. G. Farr and C. E. Baum, "Prepulse associated with the TEM feed of an impulse radiating antennas", Sensor and Simulation note 337, March 1992.
5. D.V. Giri and C. E. Baum, "Reflector IRA Design and Boresight Temporal Waveforms", Sensor and Simulation note 365, February 2 1994.
6. E. G. Farr and C. E. Baum, "A canonical scatterer for transient scattering range calibration", Sensor and Simulation note 342, June 1992.
7. C. E. Baum and E. G. Farr, "Hyperboloidal scatterer for spherical TEM waves", Sensor and Simulation note 342, June 1992.
8. C. E. Baum, "Circular Aperture Antennas in Time Domain", Sensor and Simulation note 351, November 2 1992.
9. S. P. Skulkin, V. I. Turchin, "Radiation of nonsinusoidal waves by aperture antennas," Proc. EUROEM '94 Symposium, Bordeaux, France, part2, pp.1498-1504, May 1994.
10. R. C. Hansen, Microwave Scanning Antennas. vol.1, New York and London: Academic Press, 1964.
11. M.Born, E.Wolf, Principles of Optics, Pergamon Press, 1964.
12. V. Borovikov, B. Kinber, Geometrical Theory of Diffraction, Moscow, Svaz', 1978 (in Russian).

TEMPORAL AND SPECTRAL RADIATION ON BORESIGHT OF A REFLECTOR TYPE OF IMPULSE RADIATING ANTENNA (IRA)

D.V. Giri[1] and Carl E. Baum[2]

[1]Pro-Tech, 47 Lafayette Circle, #364, Lafayette, CA 94549-4321
[2]Phillips Laboratory/WSQ, 3550 Aberdeen Avenue S.E., Kirtland AFB, NM 87117

INTRODUCTION

The IRA considered here is a parboloidal reflector fed by conical TEM lines. Such an IRA has been analyzed in the past for its performance characteristics such as the prepulse or feed step in the impulse, assuming a step function excitation. In this paper, we extend the analysis to include the diffracted fields from the launcher plates and the circular rim of the reflector. The diffraction from the launcher plates can be viewed to consist of two parts. One has the plate edge diffraction followed by diffraction associated with the total currents on the plates. The leading terms in all of these diffracted signals arriving at an observer in the far field have been determined. In addition, the predominant portion of the radiated field consisting of the feed step and the impulse has been analytically Fourier transformed to get the boresight radiated spectrum. Experimental results from a prototype IRA system are also included.

BORESIGHT RADIATION

The reflector IRA under consideration is schematically shown in Figure 1. It consists of a parboloidal reflector fed by a pair of coplanar conical TEM lines that are terminated in their characteristic impedances. The basic principles of this type of IRA are well analyzed and understood[1,2]. When this form of IRA was originally proposed[1], the

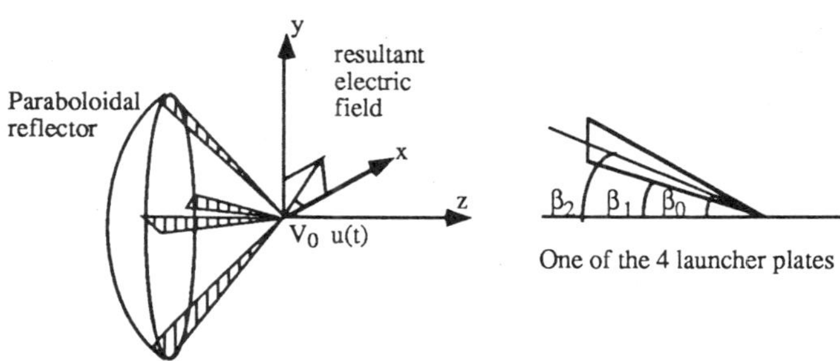

Figure 1. An example of a reflector fed by a pair of coplanar conical transmission lines.

Ultra-Wideband, Short-Pulse Electromagnetics 3
Edited by Baum *et al.*, Plenum Press, New York, 1997

boresight radiation was predicted to consist of a feedstep followed by an impulse-like behavior, for a step function source. We can build on the past analyses, and conceptually list the various temporal elements of the boresight radiation as follows, assuming that the step-function pulse generator is switched on at $t = 0$, and the observer is at a distance x(=z) to the right of the focal point of the reflector. These temporal elements are:

A. Prepulse — 1) feed step

B. Main pulse of interest — 2) impulse

C. Postpulse — 3) feed plate diffraction consisting of two parts
  a) plate edge of finite width>>compared to wavelength
  b) plate of finite width<<wavelength
4) edge diffraction from the circular rim of the reflector

D. Constraints on entire pulse — 5) low-frequency dipole moment radiation and no radiation at zero frequency (dc).

We now discuss each of the above elements.

Feed Step or Prepulse

This is a direct radiation from the source or the "switch" towards the observer. It starts at a time $t_r = (r/c)$and lasts for a duration approximately = (2F/c). The feed step is negative for the assumed sign of voltages on the launcher plates. For a pair of narrow plates (plate width << plate separation, with applied voltages of $\left[\pm(V_0/2)u(t)\right]$, the feed step in the far field may be written as[2],

$$E_{y1}(r,t) = -\frac{V_0}{4\pi f_g}\frac{1}{2F}\left[u(t-t_r)-u(t-t_r-T)\right] \text{ with } T = 2F/c \tag{1}$$

where $V_0 \equiv$ potential difference between the two feed plates, $D \equiv$ diameter of the reflector = 2b, $f_g \equiv$ feed impedance/free space impedance, $F \equiv$ focal length and $c \equiv$ speed of light. (Later this and other expressions are multiplied by $\sqrt{2}$ to account for the fact that four plates are used, using symmetry conditions.)

Impulse

The impulse-like radiation occurs at a time $t = \left[t_r + (2F/c)\right]$ for an observer location on axis. The impulse amplitude is given for narrow plates by[1,2],

$$E_{y2}\ (r,t) = \frac{V_0}{r}\frac{D}{4\pi c f_g}\delta\left(t - t_r - \frac{2F}{c}\right) \tag{2}$$

Feed-plate Diffraction

The diffraction from the feed plates may be classified into two parts. As the spherical TEM wave launched by the feed plates scatters off the paraboloidal surface, it encounters the plate edge and then the plates themselves. Consequently, the feed-plate diffraction occurs from the plate edge at the early time. At early times, when the finiteness of the plate edge is unobservable, this problem can be modeled by the classical diffraction from a half-plane[3]. Keller and Blank[3] have treated a general problem of diffraction from a wedge, which can be simplifed and specialized[4] to a half-plane, by making the wedge angle equal to zero. The resultant far field on axis due to the diffraction from the plate edge at early times is given by[4],

$$E_{y3a}(r,t) = -\frac{V_0}{r}\frac{1}{\sqrt{\sin(\beta_1)}}\left(\frac{2}{3\pi}\right)\sqrt{\frac{a}{ct-(r+2F)}}\,u\left(t-\frac{r}{c}-\frac{2F}{c}\right) \tag{3}$$

recalling that the source is turned on at t=0, and the origin of coordinates is the focal point of the reflector, which is also the true apex of the launcher plates. The above equation is valid for a time <(a/c), on average after this diffraction process is initiated at a time $t=\left[t_r+(2F/c)\right]$, where a is the average value of the launcher plate width (half the maximum since it is a flate-plate cone).

For times larger than (a/c) and smaller than (b/c), only the total current on the plates is important, and the scattered wave is diffracted from the plates. We can estimate this by modeling the plates by equivalent round conductors and quantifying the diffraction[3]. The resultant leading term in this portion of the diffracted field is given by[4] (assumed initially for one TEM conical line),

$$E_{y4b}(r,t) \cong -\frac{V_0}{r}\frac{1}{4\sin(\beta_0)}\frac{1}{\left[\frac{1}{\Gamma_0}\left\{\frac{2(ct-r)}{a_e\sin(\beta_0)}+1\right\}\right]}+\ldots\ldots \tag{4}$$

where a_e ≡equivalent radius of the plates $\cong a/4$, and Γ_0 = exponential of Euler's constant = 1.7810... It should be remarked that we have estimated the diffraction in two different time regimes. It is not clear how the two time regime solutions connect up. The diffraction from the launcher plates is of opposite sign compared to the impulse-like radiation and intially falls off like $\left(1/\sqrt{t}\right)$and then asymptotically behaves like $\left(1/\ell n\sqrt{t}\right)$.

Diffracted Field from the Reflector Edge

Immediately following the impulse, one also has the diffracted signal from the circular rim of the reflector. This can be estimated exactly in closed form, since the incident field on the reflector rim is the TEM field of the conical line, which is well known. In addition, this incident field is parallel[6] to the reflector rim. We have done this in the frequency domain and were able to analytically Fourier invert the spectral domain result[4]. With reference to the geometry in figure 2, the diffracted spectrum turns out to be

$$E_{y4} = -\frac{e^{-jk(r+2F)}}{4\pi r}\tan\left(\frac{\pi-\beta_0}{4}\right)\oint_C \tilde{E}_y d\ell \tag{5}$$

The integral represents the integration of the incident field along the circular rim of the reflector (contour C). The integrand is the spherical TEM field of a conical line and is well known in closed form, on the contour C.

After performing the integral and taking the inverse Fourier transform, one gets[4],

$$E_{y4}(r,t) = \frac{V_0}{2}\left(\frac{\pi}{2}\right)\frac{1-\sin(\beta_o/2)}{\cos(\beta_o/2)}\frac{1}{m^{1/4}K(m)}\left[1-\frac{2}{\pi}\text{are}\sin\left(\frac{1-\sqrt{m}}{1+\sqrt{m}}\right)\right]u\left(t-\frac{r}{c}-\frac{2F}{c}\right) \tag{6}$$

The above expression is the leading term in the y-directed, rim-diffracted field on axis at a ristance r from the foacl point for the case of one pair of coplanar plates. It is noted that the

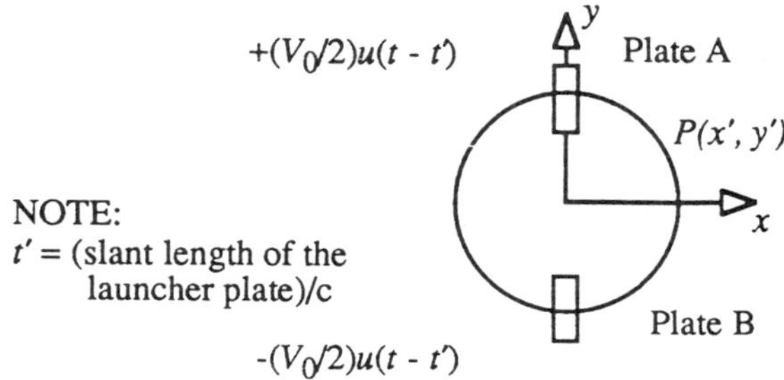

Figure 2. Circular rim of the reflector in the aperture plane

above expression is asymptotically exact since we have used the exact expression for the TEM incident field (ϕ directed) which is parallel to the cicular rim of the reflector. This result is a step function in an early-time sense and it is of the opposite sign compared to the diffracted fields from the launcher plates.

Constraints on the Entire Pulse

The radiation at low frequncies may be characterized by a set of electric and magnetic dipole moments. Let us denote these late-time dipole moments by $\vec{p}(t)$ and $\vec{m}(t)$ respectively. With reference to figure 2, it is observed that the electric dipole moment is oriented along the y-axis and the magnetic dipole moment is oriented along the negative x-axis, so that the resultant radiation from the two dipole moments is in the z direction. The low-frequency radiation from this pair of balanced ($|\vec{m}| = c|\vec{p}|$)dipole moments is a cardioid pattern that peaks along the boresight and has a null in the negative z direction, behind the reflector. The complete time integral of the radiated waveform must be zero and the low-frequency radiation is proportional to $\vec{p}(\infty)$ or $\vec{m}(\infty)$which are the late-time dipole moments. The constraints on the radiated far field are as follows

$$\int_{-\infty}^{\infty}\int_{-\infty}^{t}\vec{E}_f(r,t')dt'dt = -\frac{\mu_0}{4\pi r}\left[\vec{p}(\infty) - \frac{\vec{1}_x}{c}\times\vec{m}(\infty)\right] \text{ and } \int_{-\infty}^{\infty}\vec{E}_f(r,t) = 0 \tag{7}$$

Gathering various temporal elements of the boresight radiation discussed above, one arrives at figure 3 for the radiated field from a reflector IRA fed by one set of coplanar plates.

Recall that the radiated field outlined above is for a step function $u(t)$ excitation. In practial situations, one never has an ideal step function and so $u(t)$ may be replaced by a general $f(t)$. A typical example is $(e^{-\beta t} - e^{-\alpha t})u(t)$ with $\alpha >> \beta \equiv$ fast rising, slowly decaying double-exponential function, for which the step function would be replaced by $f(t)$ and the impulse $\delta(t)$ would be replaced by $(\partial f / \partial t)$. The leading terms of the time-domain radiated field on axis described above are useful in the design and evaluation of IRAs of this type.

BORESIGHT RADIATED SPECTRUM

It is observed from the results of the previous section that the radiated field is dominated by the feed-step and the impulse. It has also been established in the literature that these two terms, by themselves result in a net zero area for the time integral. The post pulse terms are small and have a net zero area. Considering the dominant terms of the boresight

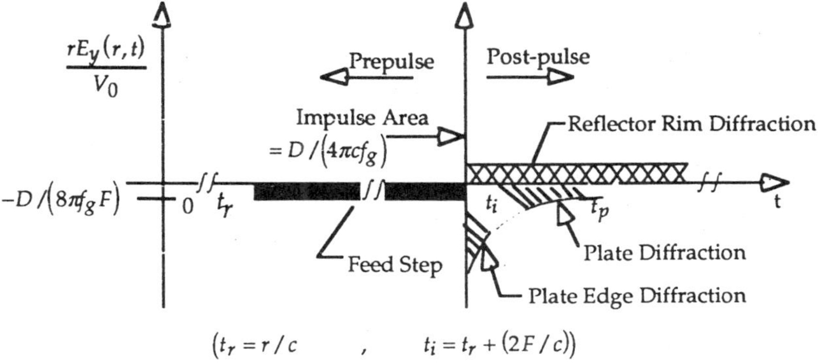

Figure 3. On-axis radiation from a reflector IRA fed by a single coplanar conical TEM line

radiated field for a general $V(t)$ excitation, we have

$$\vec{E}(r,t) = \frac{D}{4\pi c f_g}\frac{1}{r}\left\{\frac{\partial V}{\partial t}(t-T) - \frac{1}{T}\left[V(t) - V(t-T)\right]\right\} \tag{8}$$

Consider a two-sided Laplace transform of the above expression with the complex frequency s denoted by $s = \Omega + j\omega$. Upon setting $\Omega = 0$, one then gets

$$\vec{E}(r,\omega) = \frac{D}{4\pi c f_g}\frac{1}{r}\left\{e^{-j\omega\tau}\left[j\omega\tilde{V}(\omega) - V\left(0^+\right)\right] - \frac{1}{T}\left[\tilde{V}(\omega) - e^{j\omega T}\tilde{V}(\omega)\right]\right\} \tag{9}$$

Setting the initial condition $V(0^+) = 0$, without any loss of generality, denoting $\omega T = x$, and then considering the spectral magnitude, we get

$$\left|\vec{E}(r,\omega)\right| = \tilde{E}_M(r,\omega) = \frac{D}{4\pi c f_g}\frac{1}{r}\left\{1 - \frac{2\sin(x)}{x} + \frac{\sin^2(x/2)}{(x/2)^2}\right\}^{1/2} \omega\tilde{V}_M(\omega) \tag{10}$$

where $\tilde{V}_M(\omega)$ is the spectral magnitude of the voltage input to the IRA. For step and a double exponential function inputs, we have the radiated spectra given by[6]

$$\tilde{E}_M^{(s)}(r,\omega) = \left[\frac{D}{4\pi c f_g}\frac{V_0}{r}\right]\tilde{f}_s(\omega) \quad , \quad \tilde{E}_M^{(d.e.)}(r,\omega) = \left[\frac{D}{4\pi c f_g}\frac{V_0}{r}\right]\tilde{f}_{d.e.}(\omega)$$

$$\tilde{f}_s(\omega) = \left\{1 - \frac{2\sin(x)}{x} + \frac{\sin^2(x/2)}{\left(x/2^2\right)}\right\}^{1/2} \quad , \quad \tilde{f}_{d.e.}(\omega) = \tilde{f}_s(\omega) = \frac{\omega(\alpha-\beta)}{\sqrt{\beta^2+\omega^2}\sqrt{\alpha^2+\omega^2}} \tag{11}$$

are respectively the normalized radiated spectra for step and double exponential inputs.

PROTOTYPE IRA SYSTEM DESCRIPTION

The prototype reflector IRA that has been fabricated and tested is shown in figure 4. It consists of a solid surface, spun aluminum, paraboloidal reflector of diameter

Figure 4. Prototype reflector IRA

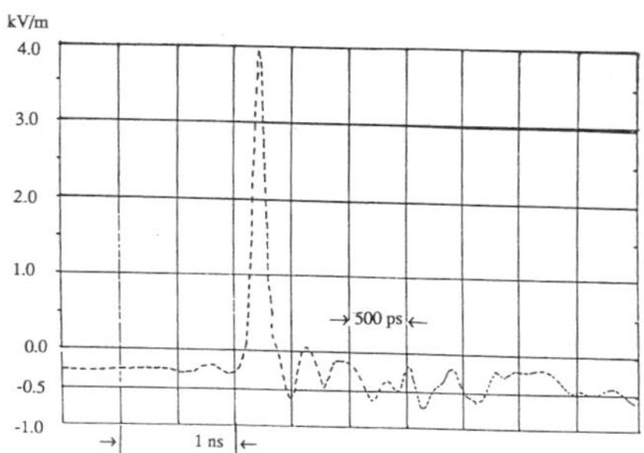

Figure 5. Measured boresight temporal e-field at r=304m

D = 3.66 m, forcal length F= 1.21m and F/D = 0.33. There are two pairs of conical lines, each with a TEM characteristic impedance of 400 Ω with a net impedance of 200 Ω. The terminating impedance (an RC combination) was experimentally optimized by a TDR measurement. A uniform dielectric, electromagnetic lens[7,8] is present in the launch region to ensure an approximate spherical TEM wave onto the conical transmission lines. Outside the lens, it is an approximate spherical TEM wave centered at the prime focus of the reflector. The source can be represented by a switch inside the lens. The closure of the switch at $t = 0$, generates a fast pulse, which may be represented by a double exponential as follows.

$$V(t) = V_0\left(e^{-\beta t} - e^{-\alpha t}\right)u(t) \text{ with } V_0 = 10^5 V, \alpha = 2.2\times10^{10}/s \text{ and } \beta = 5\times10^7/s \quad (12)$$

For this source, we estimate the near field to have a peak amplitude of about 34 kV/m which is in agreement with a measurement of the incident electric field at the vertex of the reflector. This electric field incident on the reflector is obtained by a direct measurement of the total magnetic field which is twice the incident magnetic field. The prepulse duration = $(2F/c)$ is about 8ns and the impulse amplitude was estimated to be about 4.4 kV/m at a distance of 304m in the far field. These estimated are obtained by multiplying the foregoing

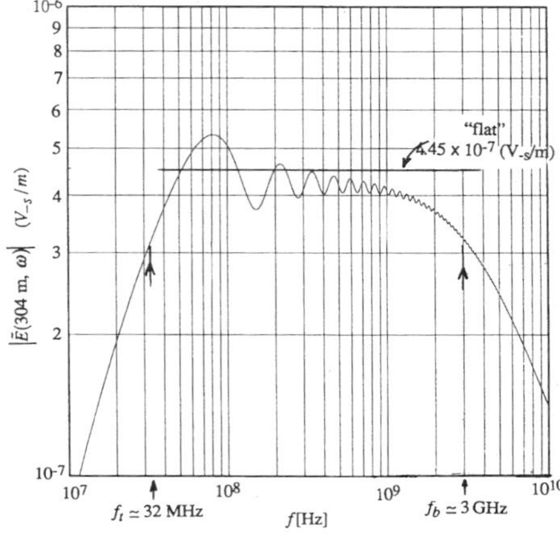

Figure 6. Calculated boresight radiated e-field spectrum at r=304m

Table 1. Summary of the reflector IRA measurements[9]

Peak Electric Field on Boresight at r = 305 meters	4.2 kV/m with FWHM of 130 ps
Boresight Elec. Field (10-90%) risetime, r = 305 m	99 ps
Boresight Electric Field Spectrum, r = 305 meters	<12 dB varia. over 50 Mhz - 4 Ghz
Main Beam Scan	
FWHM θ -beam width	-1.77°, +1.45°
FWHM ϕ - beam width	-0.98°, +2.31°
Azimuthal or H-plane Pattern	
FWHM ϕ - beam width	3.18°
-3 dB peak power beam width	1.8°
Incident Electric Field at the Center of the Dish	~34 kV/m
(10-90%) Risetime at the Center of the Dish	126 ps*
$V_{far} = r\ E_{far}$	~1281 kV
*(instrumentation limited sensor ~ 100 ps, scope ~ 58 ps)	

formulas by a factor of $\sqrt{2}$ for vector addition of the fields associated with the two pais of launcher plates. Note that each pair of plates is orthogonal to the fields produced by the other pair (non interacting). The far field, for such a pulsed antenna, starts at a distance when the clear time t_c is $<< t_r$. The clear time t_c is the differential time of arrival from the edge of the radiating aperture and the center of the aperture, at the observer. Typically $t_c \leq t_r/5$ could define the far field where t_r is the (10-90)% risetime of the incident pulse. The measured far field at a distance of 304m is shown in figure 5. It is observed that $V_{far} = r\ E_{far} \cong 1281$ kV, the prepulse is observable and the "impulse" duration is about 130ps. An extensive set of measurements have been made[9] and the results are available. The radiated spectrum at r = 304m is shown in figure 6, which is obtained[9] from a calculation using (11). A summary of the prototype IRA performance is shown in Table 1. The calculated [from (11)] bandratio br (3 GHz/32 MHz) of this antenna is 93 and the Fourier transformation of the measured temporal waveform had a slightly smaller br highlighting the problems at low frequencies associated with the measurement.

SUMMARY

Since it was first proposed in 1989[1] paraboloidal reflectors fed by TEM transmission lines have received a lot of attention, owing to their main attractive property of extremeley wide bandwidth, without the adverse effects of dispersion. The bandwidth associated with time-domain antennas is to be distinguished from the approximately 10 to 1 bandwidth of the so-called frequency independent antennas such as the log-periodic antenna, which is highly dispersive. Reflector IRAs overcome this problem and even have equivalent electric and magnetic dipole moments characterizing the low-frequency performance. Even the dipolar radiation at low frequencies is along the optical axis of the reflector. Such pulsed antennas have been called ultra-wideband antennas, when percent bandwidths exceed 25%. The prototype IRA antenna described in this paper has a percent bandwidth of about 190 out of a possible maximum of 200. A better way to characterize such antennas is by their bandratio br = (f_h / f_ℓ) where f_h and f_ℓ are the upper and lower 3dB rolloff frequencies. We are approaching a bandratio of 100 [or bandratio decades (brd) of 2]. Future systems may even have a br value of 1000 (or brd of 3). Consequently, the reflector type of an impulse radiating antenna is an example of a **d**ispersionless, **e**lectromagnetic, **l**arge-bandratio, **t**ransient **a**ntennas and **s**ystems (DELTAS) which may be denoted by δs, with δ being the symbol for Dirac-delta function.

Based on past analytical efforts, an optimal reflector IRA has been designed, fabricated and tested. It produces an impulse-like electric field waveform with a peak amplitude of ~ 4.2 kV/m and a FWHM ~ 130 ps at a distance of 305 meters from the antenna. This results in a bandratio br of about 93 (or brd of 1.97) and $V_{far} = r\ E_{far}$ of~1281 kV.The experimental results reported here testify to the usefulness and desirability of such antennas, which can be built to different sizes and frequency ranges tailored to specific application, such as hostile target identification, buried object detection and identification, high power jamming, transmission/reception of various signals separated in frequency and/or time, etc.

REFERENCES

1. C.E. Baum, Radiation of Impulse-Like Transient Fields, *Sensor and Simulation Note* 321, 25, Nov. (1989).
2. C.E. Baum and E.G. Farr, Impulse Radiating Antennas in *Ultra-Wideband, Short-Pulse Electromagnetics*, H.L. Bertoni et al. ed., pp 130-147, Plenum Press, NY 1993, and E.G. Farr, C.E. Baum and C.J. Buchenauer, Impulse Radiating Antennas, Part II in *Ultra-Wideband, Short-Pulse Electromagnetics 2*, edited by L. Carin and L.B. Felsen., pp 159-170, Plenum Press, NY (1995).
3. J.B. Keller and A. Blank, Diffraction and Reflection of Pulses by Wedges and Corners, *Communications on Pure and Applied Mathematics*, Volum 4, 1951, pp 75-94.
4. D.V. Giri and C.E. Baum, Reflector IRA Design and Boresight Temporal Waveforms, *Sensor and Simulation Note 365*, 2 February (1994).
5. P.R. Barnes, The Axial Current Induced on an Infinitely Long Perfectly Conducting, Circular Cylinder in Free Space by a Transient Electromagnetic Plane Wave, *Interaction Note 64*, March (1971).
6. E.G. Farr and C.E. Baum, Radiation from Self-Reciprocal Apertures, Chapter 6, pp 281-308 in *Electromagnetic Symmetry*, C.E. Baum and H.N. Kritikos, eds., Taylor and Francis, (1995)
7. D.V. Giri, Radiated Spectra of Impulse Radiating Antennas (IRAs), *Sensor and Simulation Note 386*, 23 November (1995).
8. C.E. Baum, J.J. Sadler, and A.P. Stone, A Uniform Dielectric for Launching A Spherical Wave Into A Paraboloidal Reflector, *Sensor and Simulation Note 360*, 22 July (1993).
9. C. Courtney, et al., Measurement and Characterization of the Impulse Radiating Antenna, *Prototype IRA Memo 5*, 6 September (1995).

COPLANAR CONICAL PLATES IN A UNIFORM DIELECTRIC LENS WITH MATCHING CONICAL PLATES FOR FEEDING A PARABOLOIDAL REFLECTOR

Carl E. Baum and Joseph J. Sadler

Phillips Laboratory
PL/WSR, KAFB
Albuquerque, NM 87117

Alexander P. Stone

Department of Mathematics and Statistics
University of New Mexico
Albuquerque, NM 87131

INTRODUCTION

One form of an impulse radiating antenna (IRA) consists of a paraboloidal reflector fed by conical transmission lines that propagate a spherical TEM wave, which originates from the focal point of the reflector. A diagram which depicts such an IRA is shown schematically in Figure 1. The design considerations of a uniform dielectric lens useful in launching a spherical TEM wave onto such a paraboloidal reflector have been investigated in earlier work[1, 2, 3], and the geometrical parameters for the lens design are indicated in Figure 2. The lens design in essence consists of specifying certain parameters. These include the cylindrical radius, h, of the outermost ray which intercepts the reflector,the relative permittivity ϵ_r of the lens medium, and the angle, $\theta_r^{(out)}$, from the apex of the conical transmission line (focal point) to the edge of the reflector. This angle is determined by the reflector geometry through the ratio F/D, where F is the focal distance from the center of the reflector and D is the reflector diameter. We also need to specify the angle $\theta_r^{(in)}$ made by the line inside the lens with respect to the directions of launch.

There are several reasons for employing dielectric lenses in such an application, and these may be briefly summarized as follows. First, the ideal point source at the focal point is physically realized by a closing switch. The high voltage between the switch electrodes dictates the use of a dielectric medium surrounding the switch enclosure. Secondly, a dielectric medium can also be used to ensure a spherical TEM wave launch inside the lens. All rays emanating from the switch center should exit the lens boundary and end up on a spherical wavefront centered on the focal point at the same instant. The "equal transit-time" condition, which provides the basis for the lens design, generates an optimal spherical wave (not truly TEM at very early times due to different transmission coefficients at different points on the lens surface) launch onto the reflector. Note that $\theta_r^{(in)}$ in[1] is taken as 90° and defines the center (symmetry plane) of the conical plate, thereby maximizing the conical-transmission line impedance inside the lens, and the plate spacing for high voltage reasons. The frequency dependence of the dielectric

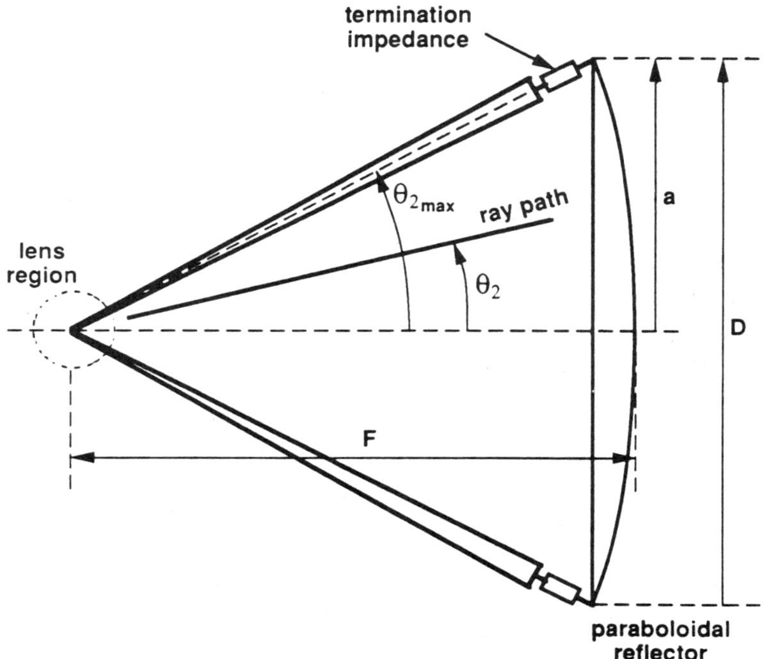

Figure 1: Schematic diagram of Prototype IRA-2

constant of the lens medium is also an important consideration in the choice of the material. The dielectric constant chosen is approximately 2.26, which is the dielectric constant for materials such as transformer oil or polyethylene and is approximately frequency independent over the frequency range of interest.

In this article we are concerned with impedance calculations for a lens geometry, which we take as a coplaner conical plate geometry as investigated in [2]. The lens impedance as a function of the half-angle, α', (Figure 3) of the interior conical plates, as well as the impedance outside of the lens, is calculated on the basis of this coplaner conical geometry. The reflector geometry, specified by the F/D ratio, determines the acceptable range for the angle α'. The notation in Figure 3 is the same as that which appears in [2]. For consistency, changes have been made in the notation which appears in [1]. For example, θ_1 becomes $\beta_1^{(in)}$ or $\beta_2^{(in)}, \theta_2$ becomes $\beta_1^{(out)}$ or $\beta_2^{(out)}$, while θ_{1max} is replaced by $\theta_r^{(in)}$ and θ_{2max} is replaced by $\theta_r^{(out)}$. The lens characteristic impedance is denoted by Z_{in} while the characteristic impedance of the region outside of the lens is denoted by Z_{out}. These impedances are calculated and tabulated for various F/D ratios in the later sections of this article.

THEORETICAL CONSIDERATIONS IN LENS DESIGN

Let us recall the equal-time condition for a diverging spherical wave in a medium with permittivity ϵ_1 going into another diverging spherical wave in a second medium with permittivity ϵ_2. The equation which describes this equation is derived in [4] and it also appears in [3] in the form

$$\sqrt{\epsilon_1}\ \{[(z_b - \ell_2 + \ell_1)^2 + \Psi_b^2]^{1/2} - \ell_1\} = \sqrt{\epsilon_2}\ \{[z_b^2 + \Psi_b^2]^{1/2} - \ell_2\} \tag{1}$$

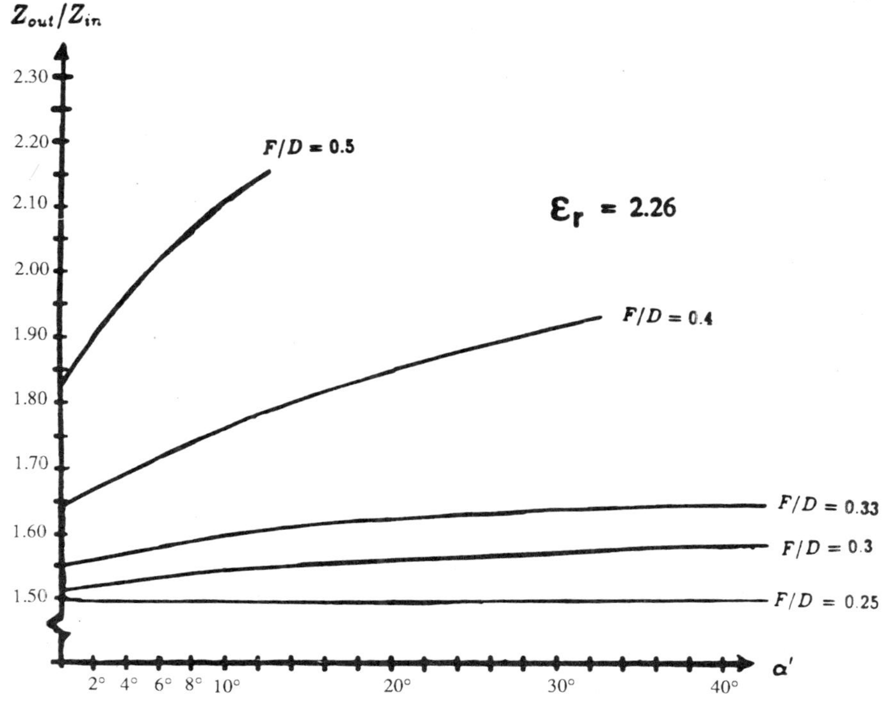

Figure 2: Lens Design Parameters

where ℓ_1, ℓ_2, z_b, and Ψ_b are as they appear in Figure 2. The coordinates (z_b, Ψ_b) are coordinates of a point on the lens boundary. This equation may then be used to obtain an analytical expression for the lens boundary curve, which expresses the angle $\theta^{(out)}$ as a function of $\theta^{(in)}$. We may compute $\cos(\theta^{(out)})$ by calculating

$$\frac{AB\sin^2(\theta^{(in)}) + |B\cos(\theta^{(in)}) - A\epsilon_r^{\frac{1}{2}}|\sqrt{\left[B^2 - 2AB\epsilon_r^{\frac{1}{2}}\cos(\theta^{(in)}) + A^2\epsilon_r\right] - A^2\sin^2(\theta^{(in)})}}{B^2 - 2AB\epsilon_r^{\frac{1}{2}}\cos(\theta^{(in)}) + A^2\epsilon_r}$$

where $A = (\ell_2/\ell_1) - 1$ and $B = (\ell_2/\ell_1) - \epsilon_r^{\frac{1}{2}}$. The parameter ℓ_2/ℓ_1 is determined once $\theta_r^{(in)}, \theta_r^{(out)}$, are ϵ_r are chosen. If $\theta_r^{(in)} = 90°$, the formula for ℓ_2/ℓ_1 is, as in [1],

$$\frac{\ell_2}{\ell_1} = \epsilon_r^{\frac{1}{2}} \frac{[\cos(\theta_r^{(out)}) + \sin(\theta_r^{(out)})] - 1}{\cos(\theta_r^{(out)}) + \epsilon_r^{\frac{1}{2}}\sin(\theta_r^{(out)}) - 1} \tag{2}$$

Thus when $\theta_r^{(in)} = 90°$, the parameters $\ell_2/\ell_1, A$, and B may be expressed in terms of ϵ_r and F/D. The results are:

$$A = \frac{(\epsilon_r^{\frac{1}{2}} - 1)}{2}\left[\frac{16(F/D)^2 - 1}{4\epsilon_r^{\frac{1}{2}}(F/D) - 1}\right] \tag{3}$$

$$B = \frac{(\epsilon_r^{\frac{1}{2}} - 1)}{2}\left[\frac{16(F/D)^2 - 8\epsilon_r^{\frac{1}{2}}(F/D) + 1}{4\epsilon_r^{\frac{1}{2}}(F/D) - 1}\right] \tag{4}$$

$$\frac{\ell_2}{\ell_1} = \frac{\frac{1}{2}\left\{(\epsilon_r^{\frac{1}{2}} - 1)[16(F/D)^2 + 8(F/D) - 1] + 2[4(F/D) - 1]\right\}}{4\sqrt{\epsilon_r}(F/D) - 1} \tag{5}$$

We may also express $\theta_r^{(out)}$ in terms of F/D by the formulas

$$\sin(\theta_r^{(out)}) = \frac{8(F/D)}{16(F/D)^2 + 1} \quad \text{and} \quad \cos(\theta_r^{(out)}) = \frac{16(F/D)^2 - 1}{16(F/D)^2 + 1} \tag{6}$$

The lens boundary curve is now specified by first choosing an appropriate F/D ratio which determines the angle $\theta_r^{(out)}$ of Figure 1. The lens parameters h, ϵ_r, and $\theta_r^{(in)}$ are then chosen (consistent with certain constraints which are discussed in [1]) so that the lens parameters ℓ_1/h and ℓ_2/h are obtained and a boundary curve is generated with

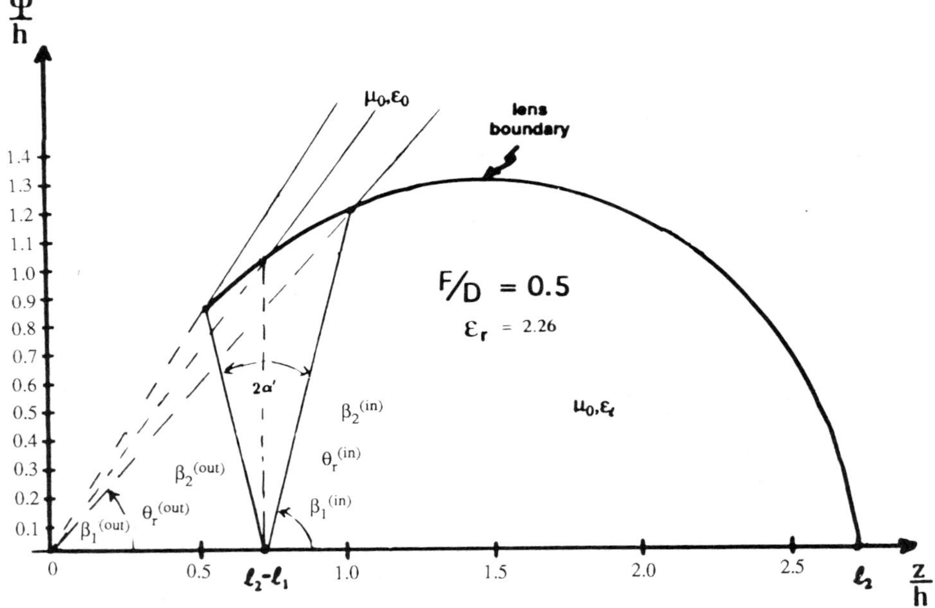

Figure 3: Coplanar Conical Plate Geometry

$0 \leq \theta^{(in)}$ and $0 \leq \theta^{(out)}$. Note that $\theta_r^{(out)}$ is not the largest $\theta^{(out)}$ on the lens boundary, but that $\theta^{(out)}$ which corresponds to the edge of the reflector and similarly for $\theta_r^{(in)}$. In the cases which we investigate the angle $\theta_r^{(in)}$ will be 90°. For the chosen value of F/D, a family of lens designs may then be generated by various choices of $\theta_r^{(in)}$, subject to the constraints mentioned in [1]. In Figure 4, lens boundary curves for $\theta_r^{(in)} = 90°$ and F/D values of 0.25, 0.3, 0.33, 0.4, and 0.5 are displayed. Note that larger values of F/D lead to larger lenses, while the value of $\theta_r^{(out)}$ decreases with increasing F/D. The limiting or trivial case occurs when $F/D = 0.25$. In this case, unlike the cases where $F/D > 0.25$, reflections from the lens surface (spherical) do occur, but the TEM wave is not distorted. There are also some limitations on how small $\theta^{(out)}$ can be for a given $\theta_r^{(in)}$. These are based on the slope of the lens boundary matching the ray direction.

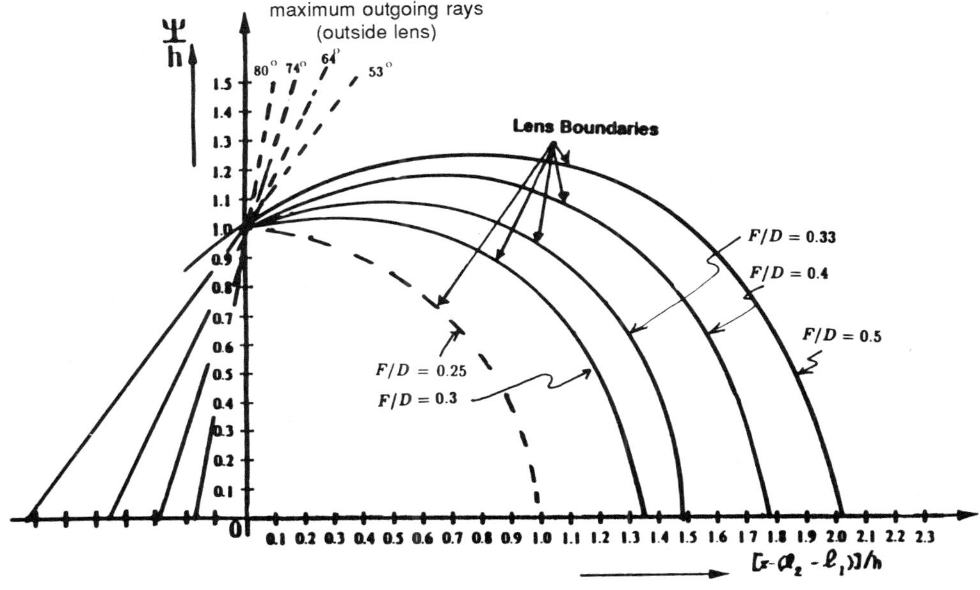

Figure 4: Lens Shapes with $\theta_r^{(in)} = 90°, \epsilon_r = 2.26$ (Cylindrical Coordinates)

IMPEDANCE CALCULATIONS

We begin by fixing the F/D ratio which determines the angle $\theta_r^{(out)}$ of Figure 1. Since the angle $\theta_r^{(in)}$ is 90° for all chosen values of F/D, and $\epsilon_r = 2.26$, the range of the angle α' is determined. Thus for all values of F/D considered, the value of Z_{in} depends only on α'. The lens characteristic impedance Z_{in} can then be calculated by the procedure described in this section. Note that $2\alpha' = \beta_2^{(in)} - \beta_1^{(in)}$ and so the value of α' determines the angles $\beta_i^{(in)}, i = 1, 2$. The angles $\beta_i^{(out)}$ are then found from equation (2) and thus the lens shape will be determined.

The problem of calculating the characteristic impedance of the lens region and the region exterior to the lens can now be considered. The geometry, that of coplanar conical plates, is described in Figure 3. A formula for the characteristic impedance of such a geometry appears in [2] and is given by

$$Z_{in} = \frac{Z_0}{\sqrt{\epsilon_r}} \frac{K(m_{in})}{K'(m_{in})} \tag{7}$$

where ϵ_r is the relative permittivity for the lens material and the parameter m_{in} is given by

$$m_{in} = \frac{\tan^2(\beta_1^{(in)}/2)}{\tan^2(\beta_2^{(in)}/2)} \tag{8}$$

The impedance Z_0 is 376.73 ohms. The quantities $K(m_{in})$ and $K'(m_{in})$ are complete elliptic integrals of the first kind. Formulas for these integrals appear in many places (for example, in [5]) and are given by

$$K(m_{in}) = \int_0^{\pi/2} (1 - m_{in}\sin^2(\theta))^{-\frac{1}{2}} d\theta \tag{9}$$

$$K'(m_{in}) = \quad K(1 - m_{in}) \equiv K(m_1^{(in)}) \tag{10}$$

where $0 \leq m_{in} < 1$ and $m_1^{(in)} = 1 - m_{in}$. We may also introduce a geometric impedance factor $f_g^{(in)}$ as

$$f_g^{(in)} = \frac{K(m_{in})}{K'(m_{in})} \tag{11}$$

and thus rewrite (9) in the form

$$Z_{in} = f_g^{(in)} Z_0/\sqrt{\epsilon_r} \tag{12}$$

Similar expressions hold for the region exterior to the lens. Thus we take for the characteristic impedance of this region

$$Z_{out} = Z_0 \frac{K(m_{out})}{K'(m_{out})} \tag{13}$$

where

$$m_{out} = \frac{\tan^2(\beta_1^{(out)}/2)}{\tan^2(\beta_2^{(out)}/2)} \quad \text{and} \quad K'(m_{out}) = K(1 - m_{out}). \tag{14}$$

Likewise a geometric factor, $f_g^{(out)}$, is given by

$$f_g^{(out)} = \frac{K(m_{out})}{K'(m_{out})}. \tag{15}$$

The elliptic integrals which depend on the parameters m_{in} and m_{out} may be evaluated by the tables in [5] or by use of the Mathematica program.

When m_{in}, or m_{out}, is near 0 or 1, there are expressions for $K(m)/K'(m)$ which may be derived. One may then verify that

$$\frac{Z_{out}}{Z_{in}} = \sqrt{\epsilon_r}\frac{K(m_{out})}{K'(m_{out})} \cdot \frac{K'(m_{in})}{K(m_{in})} \sim \sqrt{\epsilon_r}\, \frac{\ell n\left[\frac{16}{1-(1+a(1-m_{in}))^2}\right]}{\ell n\left[\frac{16}{1-m_{in}}\right]} \longrightarrow \epsilon_r^{\frac{1}{2}} \tag{16}$$

as $m_{in} \longrightarrow 1$.

Armed with the above formulas we may now calculate the impedances Z_{in} and Z_{out}, and also the ratio Z_{out}/Z_{in} for various practical values of F/D. In our impedance calculations we have chosen F/D ratios of 2.5, 3.0, 3.3, 4.0 and 5.0. Graphical results are presented in Figure 5, which gives plots of Z_{out}/Z_{in} versus α' , for each of the chosen values of F/D. The special case where $F/D = 0.25$ was mentioned previously. Note that for our choice of $\epsilon_r = 2.26$ and for the limiting value of $F/D = 0.25$, we have $Z_{out}/Z_{in} = \sqrt{\epsilon_r} = 1.503$.

CONCLUSIONS

In Figure 5 we observe that as α' increases, the ratio Z_{out}/Z_{in} increases. Small values of α' correspond to larger values of Z_{in}, and so as $\alpha' \longrightarrow 0$, Z_{in} increases and smaller values of Z_{out}/Z_{in} imply a better match at the lower frequencies (radian wavelengths of the order of h or larger) for which a transmission-line approximation is appropriate.

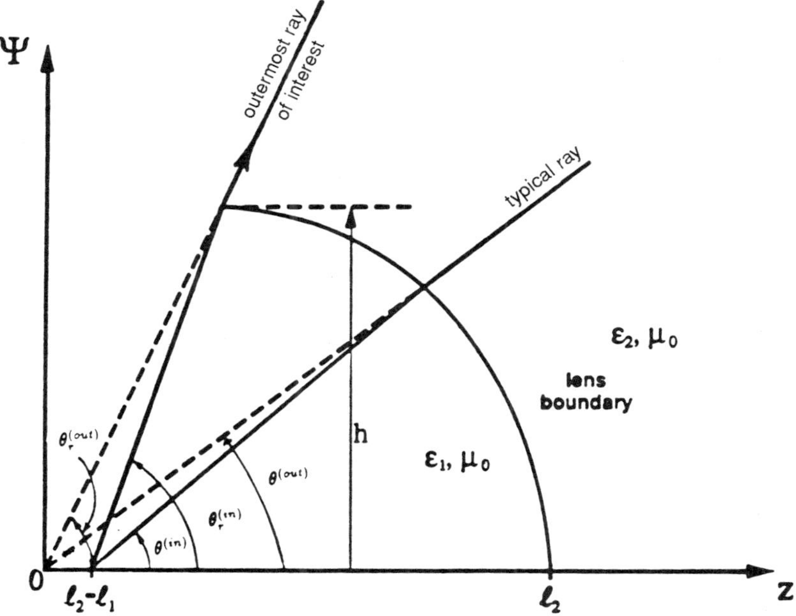

Figure 5: Impedance Ratio, Z_{out}/Z_{in}, versus half-angle, α'

As $\alpha' \longrightarrow 0$, $Z_{out}/Z_{in} \longrightarrow \epsilon_r^{\frac{1}{2}} = 1.503$, which is the ratio of the wave impedances of the media. For a ray propagating along the z axis (in the positive z direction) this represents a transmission coefficient of

$$T_\epsilon = \frac{2\epsilon_r^{\frac{1}{2}}}{\epsilon_r^{\frac{1}{2}} + 1} = 1.201 \tag{17}$$

while the smaller values of α' make the transmission-line transmission coefficient

$$T_Z = \frac{2\frac{Z_{out}}{Z_{in}}}{\frac{Z_{out}}{Z_{in}} + 1} \tag{18}$$

approach the same value. Note that for $F/D = 0.25$ the two transmission coefficients are the same for all α'.

The special case of $F/D = 0.25$ (a spherical lens) has the property that waves transmitted and reflected via the lens boundary are spherical TEM waves. However the reflected wave then is focused on the apex of the interior conical transmission line where it reflects back toward and through the lens boundary (with a fast rise time) unless there is a matched load at this apex. Larger F/D values make this wave reflected from the lens boundary dispersed in the lens and exiting the lens giving a smoother transition from the initially-transmitted fast-rising wave to the late-time behavior. It would then appear that F/D should be larger than 0.25, but not excessively so, since Z_{out}/Z_{in} increases significantly above $\sqrt{\epsilon_r}$.

We also note that for F/D values larger than 0.337, the values of α' cut off at an angle determined by the choice of F/D. When $F/D = 0.4$, this value of α' is approximately 33°, while for $F/D = 0.5$, the maximum value of α' is approximately 12°.

References

1. C. E. Baum, J. J. Sadler, and A. P. Stone, A uniform dielectric lens for launching a spherical wave into a paraboloidal reflector, *Sensor and Simulation Note* 360, (July 1993).

2. E. G. Farr and C. E. Baum, Prepulse associated with the TEM feed of an impulse radiating antenna, *Sensor and Simulation Note* 337, (March 1992).

3. D. V. Giri, Design considerations of a uniform dielectric lens for launching a spherical TEM wave on the the protype IRA, *Protype IRA Memos*, Memo 3, (May 1994).

4. C. E. Baum, J. J. Sadler, and A. P. Stone, Uniform isotropic dielectric equal-time lenses for matching combinations of plane and spherical waves, *Sensor and Simulation Note* 352, (December 1992).

5. M. Abramowitz and I. A. Stegun, Handbook of mathematical functions, in: *National Bureau of Standards*, AMS-55, Chapters 16 and 17, (June 1964).

TRANSIENT FIELDS OF PARABOLIC REFLECTOR ANTENNAS

Sergey P. Skulkin[1], Victor I. Turchin[2]

[1]Radiophysical Research Institute
Nizhny Novgorod, Russia
[2]Institute of Applied Physics Russian Academy of Science
Nizhny Novgorod, Russia

INTRODUCTION

Requirements to increase the information content of radar and communication systems result in a band width increase and are a reason to investigate and describe the transient fields from ultra wideband (UWB) large antennas[1]. Furthermore, a need in such an analysis occurs in near field time domain antenna measurements[2]. In these cases not only the transient far field is of interest. The spatial-temporal near-field distribution is useful too to the optimal antenna arrangement on the complex objects. Mention must be made that many issues of the theory of an impulse radiating antenna (IRA) are discussed by Carl E. Baum, Everett G. Farr and D. V. Giri[3-8]. The transient field from the circular focused aperture with uniform field distributions was considered in [3] . Much attention is given to IRA which consists of a large angle conical TEM feed that attaches to a reflector antenna. It was shown that radiated field includes three distinct parts. The first of these is the direct radiation from the feed structure (prepulse), which is of rather low magnitude, but lasts for a fairly long time. This is followed by an impulse, which lasts for a brief time and is high in amplitude. Finally, there is a tail expected after the impulse. Closed-form expressions for prepulse were presented in [4]. Analysis was extended to include the diffracted fields from the launcher plates and the circular rim of the reflector[5]. It was shown that there is an exact time-domain match of the spherical and planar TEM waves (inhomogeneous) at the reflector[4-7]. The exact expressions for the acoustic fields on the z axis for a circular aperture were obtained in [8]. While the above-mentioned works describe basic properties of transient fields of an IRA, many questions are yet to be investigated.

In this paper, the exact expressions for the transient fields radiated from a parabolic reflector antenna are obtained for each point of the half-space in front of the antenna.

Ultra-Wideband, Short-Pulse Electromagnetics 3
Edited by Baum *et al.*, Plenum Press, New York, 1997

BASIC FORMULAS

If the system is fed by the input signal $U(t)$, then the electric field $\vec{E}(\vec{r},t)2$ at any point defined by the radius-vector $\vec{r}$ can be written as

$$\vec{E}(\vec{r},t) = \vec{E}_a(\vec{r},t) \bigotimes U(t), \tag{1}$$

where $\bigotimes$ denotes the convolution over the time and $\vec{E}_a(\vec{r},t)$ represents the transient field: the response on the input signal $\delta(t)$ at the point $\vec{r}$. The exact calculation and analysis of $\vec{E}_a(\vec{r},t)$ from any reflector antenna need solving the nonstationary diffraction problem, that is a heavy task, but we suppose that the most part of energy of the input signal spectrum is concentrated in the interval $[\omega_{min}, \omega_{max}]$ and

$$\lambda_{max} = 2\pi c/\omega_{min} \ll F, a, |\vec{r}| \tag{2}$$

where F is the focal distance, a is the radius of the reflector. In this case, regular approximations can be used. In particular, the current on the reflector surface $\vec{j}_S$ can be represented as

$$\vec{j}_S \simeq 2\vec{n} \times \vec{H} \tag{3}$$

where $\vec{H}$ is the magnetic field on the reflector surface, $\vec{n}$ is the normal to the surface. If the input signal $U(t)$ is defined as $U = \frac{di}{dt}$, where i is the current of the dipole, the current on the reflector surface can be found as

$$\vec{j}_S \simeq \frac{1}{2\pi c r_1} \vec{r}_1' \times \vec{n} \times \vec{e}\delta(t - r_1/c), \tag{4}$$

where $\vec{e}$ is the unit vector oriented along the dipole, $r_1 = |\vec{r}_S - \vec{r}_F|$, vectors $\vec{r}_S$ and $\vec{r}_F$ denote the coordinates of the reflector surface and focus respectively, $\vec{r}_1' = (\vec{r}_S - \vec{r}_F)/r_1$. Thereafter the transient field $\vec{E}_a$ can be written in the following form

$$\vec{E}_a(\vec{r},t) \simeq \frac{Z_0}{4\pi c}\frac{d}{dt} \iint_S \vec{q}\, \frac{\delta(t - r_2/c - r_1/c)}{r_1 r_2} dS \tag{5}$$

where

$$\vec{q} = \vec{r}_2' \times \vec{r}_1' \times \vec{n} \times \vec{e} \times \vec{r}_2', \tag{6}$$

$r_2 = |\vec{r} - \vec{r}_S|$, $\vec{r}_2' = (\vec{r} - \vec{r}_S)/r_2$, S is the reflector surface and Z_0 is the impedance of the free space. Also, here we suppose that vector $\vec{e}$ is oriented along y axis.

For each polarization, the temporal dependencies of $\vec{E}_a$ can be obtained from (5) immediately, using the integration technique for a δ-function of complex argument[9]. For any functions f and φ the integral over the surface S can be represented as the integral over the curve L:

$$\iint_S f(x,y)\delta[\varphi(x,y)]dxdy = \int_{L_S} f_L \frac{dl}{G_L}, \tag{7}$$

where L_S is a part of L, belonging to S ($L_S \in S$). L is determined by the equation $\varphi(x,y) = 0$, f_L represents the function f on L, dl is an element of length in L, and

$$G = \sqrt{(\frac{\partial\varphi}{\partial x})^2 + (\frac{\partial\varphi}{\partial y})^2}\ |_{x,y \in L}, \tag{8}$$

We suppose here that there is only one solution of the equation $\varphi = 0$ and there exists the first derivative of φ.

Consider the parabolic reflector antenna with the center in the origin of Cartesian coordinate system (x, y, z). The reflector focus is located at the point $(0, 0, F)$. The surface of the paraboloid is described by: $z = \rho^2/4F$, where $\rho^2 = x^2 + y^2$, and

$$r_1 = F + \frac{\rho^2}{4F}, \tag{9}$$

$$r_2 = \sqrt{(x - x_0)^2 + (y - y_0)^2 + (z_0 - \frac{\rho^2}{4F})^2}, \tag{10}$$

where (x_0, y_0, z_0) is the coordinates of the observation point. The curve L is determined from the equation

$$ct = r_1 + r_2. \tag{11}$$

We find that L is the circle

$$(x - x_{0,t})^2 + (y - y_{0,t})^2 = a_t^2, \tag{12}$$

with the center at the point $(x_{0,t}, y_{0,t})$ and the radius a_t, where

$$x_{0,t} = \frac{x_0}{1 + \beta_t}, \quad y_{0,t} = \frac{y_0}{1 + \beta_t}, \tag{13}$$

$$\beta_t = \frac{ct' - z_0}{2F},$$

$$a_t = \frac{1}{1 + \beta_t}\sqrt{a_{t,0}^2 1 + \beta_t - \rho_0^2 \beta_t},$$

$$a_{t,0} = \sqrt{(ct')^2 - z_0^2},$$

$$t' = t - F/c, \; \rho_0^2 = x_0^2 + y_0^2.$$

For the field of the paraboloid we can write

$$\vec{E}_a = \frac{Z_0}{2F}\frac{d}{dt'}\vec{E}_\delta, \quad \vec{E}_\delta = \frac{F}{2\pi}\int\limits_{L_a} \vec{q}_L \frac{J_L dl}{G_L r_{1,L} r_{2,L}}, \tag{14}$$

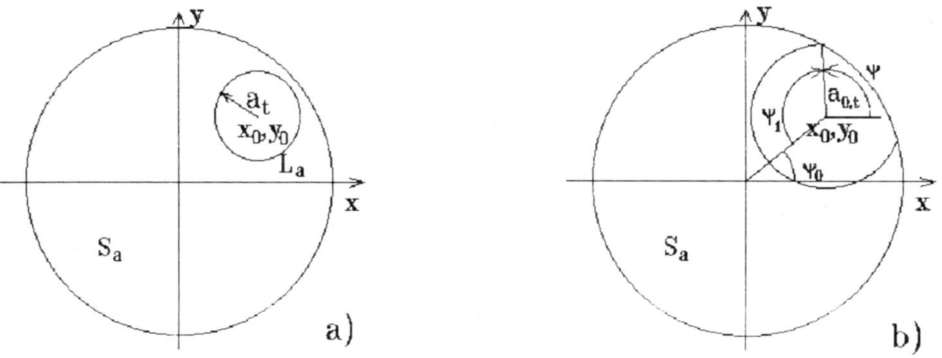

Fig. 1. The curves of the intersection of the sphere with radius (ct') and the aperture plane.

where J is the function that describes the surface element of the paraboloid:

$$dS = Jdxdy, \quad J_L = (1 + \frac{\rho_L^2}{4F^2})^{-1/2},$$

here subscript L ndicates that functions take their arguments on the contour L. Going over to the planar polar coordinates $x - x_{0,t} = a_t \cos\psi$, $y - y_{0,t} = a_t \sin\psi$ and using the polar coordinates $x_0 = \rho_0 \cos\psi_0, y_0 = \rho_0 \sin\psi_0$ (see figure 1), we get

$$\rho_L^2 = \rho_{0,t}^2 + a_t^2 + 2\rho_{0,t} a_t \cos(\psi - \psi_0),$$

$$\rho_{0,t} = \frac{\rho_0}{1 + \beta_t}.$$

Using (13) we have:

$$G_L = \frac{a_t}{r_{2,L}\sqrt{1 + \beta_t}}. \tag{15}$$

Conclusively we can write

$$\vec{E}_\delta = \frac{1}{2\pi\sqrt{1 + \beta t}} \int\limits_{\pi+\psi 1+\psi_0}^{\pi-\psi 1+\psi_0} \vec{q}_L \frac{d\psi}{(1 + \frac{\rho_L^2}{4F^2})^{3/2}}, \tag{16}$$

Consider the planar aperture ($F \to \infty$). Here:

$$r_1 \to F, \; r_2 \to \sqrt{(x - x_0)^2 + (y - y_0)^2 + z_0^2},$$

$$\vec{e} = \vec{y}_0, \; \vec{n} \to \vec{z}_0, \; \vec{r}_1 \to -\vec{z}_0, \; \vec{q} \to \vec{y}_0 - \vec{r}_2(\vec{r}_2, \vec{y}_0),$$

where $\vec{y}_0, \vec{z}_0$ are the unit vectors which directed along y and z axes respectively. In this case, the equation of the circle (10) becomes:

$$(x - x_0)^2 + (y - y_0)^2 = a_{t,0}^2. \tag{17}$$

Taking into account that $r_2 = ct'$ and $G = \frac{a_{t,0}}{ct'}$ on contour L, we get

$$\vec{E}_\delta = \frac{1}{2\pi} \int\limits_{\pi+\psi 1+\psi_0}^{\pi-\psi 1+\psi_0} \vec{q}_L d\psi \tag{18}$$

where $\vec{q}_L$ is the vector $\vec{q}$ on the circle (19) and L_a is the part of L, belonging to S_a. In the polar coordinate system ρ, ψ

$$x - x_0 = a_{t,0} \cos\psi, \; y - y_0 = a_{t,0} \sin\psi,$$

and the projections $q_{L,x}, q_{L,y}, q_{L,z}$ of $\vec{q}_L$ can be presented as follows

$$q_{L,x} = -\frac{1}{2}(1 - \frac{z_0^2}{(ct')^2}) \sin 2\psi, \tag{19}$$

$$q_{L,y} = \frac{1}{2}[(1 + \frac{z_0^2}{(ct')^2}) + (1 - \frac{z_0^2}{(ct')^2}) \cos 2\psi], \tag{20}$$

$$q_{L,z} = -\frac{1}{2} \frac{z_0\sqrt{(ct')^2 - z_0^2}}{(ct')^2} \sin\psi. \tag{21}$$

Prior to integrate, we consider conditions under which either the contour L or its part belongs to aperture S_a, that is $\vec{E}_a$ is not equal to 0. For $\rho_0 < a$

$$z_0 < ct' < l_2. \tag{22}$$

For $\rho_0 > a$

$$l_1 < ct' < l_2, \tag{23}$$

where $l_{1,2} = \sqrt{z^2 + (a \mp \rho_0)^2}$. For $\rho_0 < a$, interval (22) partitions into two intervals, the first one is

$$z_0 < ct' < l_1, \tag{24}$$

and the second one is defined by (23). For the time moments satisfying (24), the whole contour L belongs to aperture S_a, and $\psi_1 = \pi$. In the case of (23), only part of L belongs to S_a and angle ψ ranges as follows

$$\pi - \psi_1 + \psi_0 \le \psi \le \pi + \psi_1 + \psi_0, \tag{25}$$

$$\psi_1 = \arccos\left(\frac{\rho^2 - a^2 + a_{t,0}^2}{2\rho_0 a_{t,0}}\right). \tag{26}$$

For $\rho_0 > a$ integrating limits over ψ are defined by (25), however they are different for $\rho_0 < a$ and $\rho_0 > a$. When $\rho_0 < a$ and $a - \rho_0 < a_{t,0} < a + \rho_0$ (that is equivalent to ranging ct' in limits defined by (23)) angle ψ_1 decreases from π to 0 monotonically (see figure 1a). For $\rho_0 > a$ and $a - \rho_0 < a_{t,0} < a + \rho_0$, angle ψ_1 is increased from 0 to $\arcsin(a/\rho_0)$ at first and then is decreased to 0 (see figure 1b). From the above reasoning y component of $\vec{E}_A$ (the basic polarization) is determined by:

$$\vec{E}_{\delta,y} = \begin{cases} 0, & 0 < ct' < z_0,\ for\ \rho_0 < a\ or\ ct' < l_1\ for\ \rho_0 > a; \\ & and\ ct' > l_2; \\ \frac{1}{4\pi}(1 + \frac{z_0^2}{(ct')^2}) & l_1 < ct' < l_1\ for\ \rho_0 < a; \\ \frac{1}{4\pi}\left[(1 + \frac{z_0^2}{(ct')^2})\psi_1 - (1 - \frac{z_0^2}{(ct')^2})\cos 2\psi_0 \sin 2\psi_1\right], & l_1 < ct' < l_2; \end{cases} \tag{27}$$

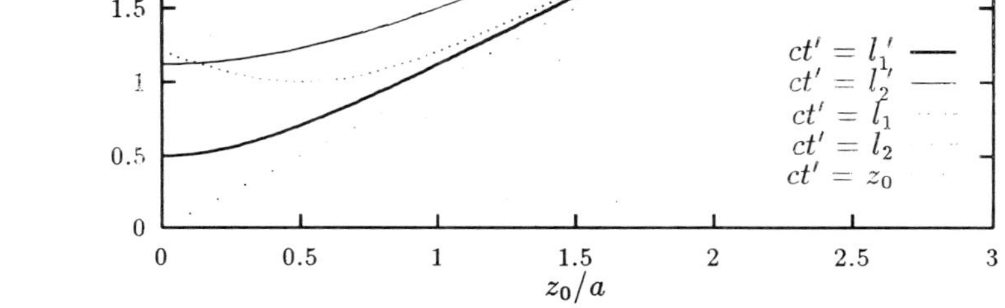

Fig. 2. The pulse limits for parabolic and planar antennas.

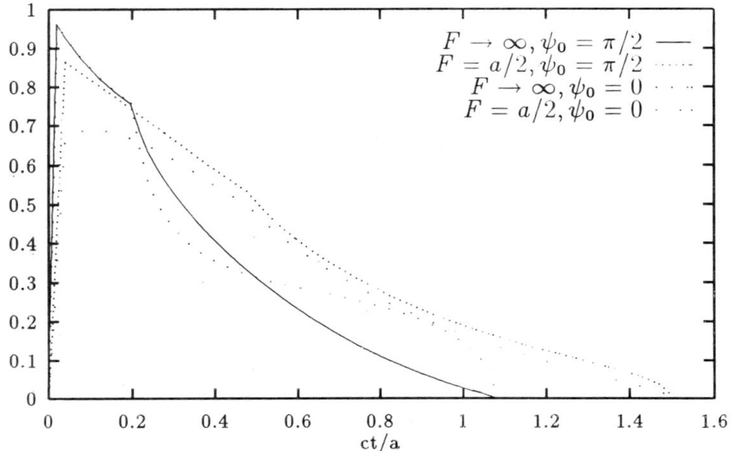

Fig. 3. The transient characteristics $E_{\delta,y}$ for parabolic and planar antennas.

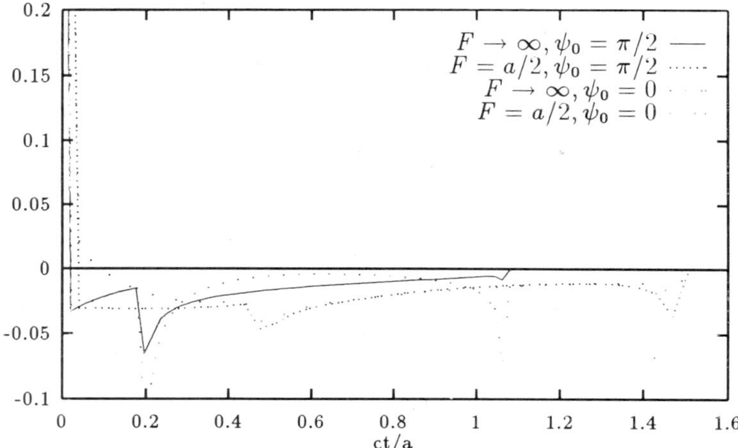

Fig. 4. The transient fields $E_{a,y}$ for parabolic and planar antennas.

RESULTS

Figure 2 shows the pulse limits at different z_0. Here l_1, l_2 are the pulse limits for the planar aperture, $l_1{}', l_2{}'$ are the pulse limits for the parabolic reflector (F=a/2).

From this figure we observe that the pulse radiated from the parabolic reflector is more durable than the pulse radiated from the planar aperture.

Figure 3 shows the time domain dependencies of $E_{\delta,y}$ for parabolic and planar antennas. Here $\rho_0 = z_0 = a/2, F_1 = a/2, F_2 = \infty$, If $\psi_0 = 0$, the projection of the observation point is on the axis x. If $\psi_0 = \pi/2$, then the projection of the observation point is located on the axis y.

Figure 4 shows the time domain dependencies of $E_{a,y}$, the derivative of $E_{\delta,y}$. From this figure we see that the amplitudes of additional "parasitic" pulses for parabolic reflector are less than those for planar aperture.

CONCLUSIONS

The transient fields for parabolic reflector antenna have been obtained for each point of the half-space in front of the aperture. The formulas obtained turn out to be simpler than those for aperture antennas in the monochromatic case[10,11].

We illustrated that the structure of the spatial-temporal field distribution is quit complex, especially in the near-field region. We show that in this region there is a difference between transient fields of parabolic antennas and plane aperture.

The various temporal dependencies are presented. It is shown that in the case, when the observation point $\vec{r}$ is placed inside the projection of the aperture, the transient field is represented by two pulses. This result was confirmed in experiment[2].

REFERENCES

1. C. E. Baum, "Impulse radiating antennas",in book Ultra-Wideband, Short-Pulse Electromagnetics, ed. by Bertroni et al., Plenum Press, 1993.
2. S. P. Skulkin, V. I. Turchin, et al., "The time-pulse method of measuring antenna characteristics in near zone," Radiophysics and quantum electronics, vol.32, no.1, pp. 61-70, July 1989.
3. C. E. Baum, "Focused aperture antennas", Sensor and Simulation note 306, May 19 1987.
4. E. G. Farr and C. E. Baum, "Prepulse associated with the TEM feed of an impulse radiating antennas", Sensor and Simulation note 337, March 1992.
5. D.V. Giri and C. E. Baum, "Reflector IRA Design and Boresight Temporal Waveforms", Sensor and Simulation note 365, February 2 1994.
6. E. G. Farr and C. E. Baum, "A canonical scatterer for transient scattering range calibraion", Sensor and Simulation note 342, June 1992.
7. C. E. Baum and E. G. Farr, "Hyperboloidal scatterer for spherical TEM waves", Sensor and Simulation note 342, June 1992.
8. C. E. Baum, "Circular Aperture Antennas in Time Domain", Sensor and Simulation note 351, November 2 1992.
9. S. P. Skulkin, V. I. Turchin, "Radiation of nonsinusoidal waves by aperture antennas," Proc. EUROEM '94 Symposium, Bordeaux, France, part2, pp.1498-1504, May 1994.
10. R. C. Hansen, Microwave Scanning Antennas. vol.1, New York and London: Academic Press, 1964.
11. M.Born, E.Wolf, Principles of Optics, Pergamon Press, 1964.

USE OF THE SYNTHESIZED SHORT RADIO PULSE FOR NEAR-FIELD ANTENNA MEASUREMENTS

Andrey V. Kalinin

Radiophysical Research Institute (NIRFI)
Bolshaya Pecherskaya 25, Nizhny Novgorod
603600, Russia

INTRODUCTION

This paper considers a mirror parabolic antenna near-field, which is synthesized by measurements at a discrete number of frequencies with the following Fourier transformation of data into time domain, where separation of the signal components with different propagation occurs. Identification of these components taking into account the geometry of the measurement facility and the inverse Fourier transformation of a part of them allows us to define more accurately the antenna near field, as well as to define the interference field. The efficiency of such measurements is defined first of all by the band and the discrete of the frequency variation choosen.

Feasibilities of the above methodics were investigated for planar near field measurement facility, which has been created in NIRFI as a model of the facility for ground tests of the transformed space antennas. Synthesized signal of the antenna near field was experimentally investigated in the time range up to 500 ns (in space — up to 150 m) with the resolution about one ns (30 cm), that permits us to identify the paths and to determine levels of the signal being multireflected between the antenna and scanner elements.

These method can be used for increasing the accuracy of antenna measurements either for determining the reflection level in facility.

THE METHOD OF MEASUREMENTS

The near-field measurements in a frequency band with following Fourier transformation of data into time domain allow to separate the components of measured signal with various propagation times. It gives an opportunity to separate the test antenna field from the facility scattered field, if the time duration of antenna signal is shorter than the delay of multireflections.

Figure 1 shows the geometry of facility for near field measurements

Ultra-Wideband, Short-Pulse Electromagnetics 3
Edited by Baum *et al.*, Plenum Press, New York, 1997

of mirror parabolic antenna. According to Huygens principle the impulse transient characteristic of such antenna may be written in following form

$$h(t) = \int_0^{D/2}\int_0^{2\pi} \delta\left(t - \frac{r_1 + r_2}{c}\right) \frac{1}{r_1 r_2}\, a_1(\rho)\sqrt{1 + \frac{\rho^2}{4F^2}}\, \rho\, d\rho\, d\varphi, \tag{1}$$

where

$$r_1 = F + \rho^2/4F, \qquad r_2^2 = (H - \rho/4F)^2 + R^2 + \rho^2 - 2R\rho\cos(\phi - \varphi),$$

$a_1(\rho)$ — paraboloid feeding distribution; H — distance from mirror apex to measurement plane; F — paraboloid focal length; R, ϕ — measuring probe polar coordinates on measurement plane; ρ, φ — mirror surface integrate coordinates; c — velocity of light; $D, F \gg \lambda$.

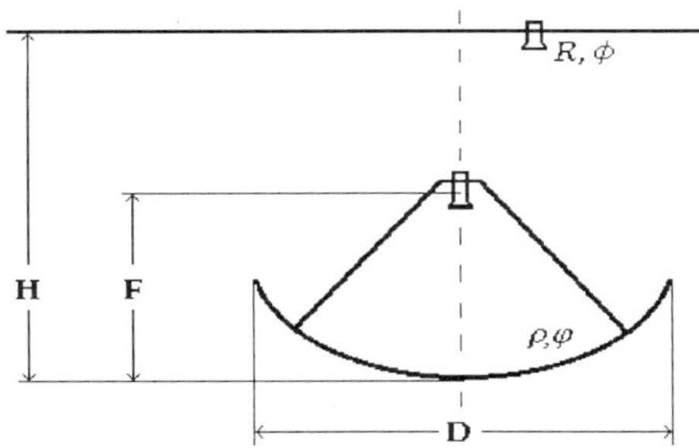

Figure 1. Geometry of facility for near field measurements

The time domain response $S(t)$ is the convolution of the synthesized pulse $I(t)$ and impulse transient characteristic of antenna $h(t)$

$$S(t) = \int_{-\infty}^{\infty} I(\tau)\, h(t - \tau)\, d\tau, \qquad I(t) = \int_{\omega_l}^{\omega_h} a_2(\omega)\, e^{i\omega t}\, d\omega, \tag{2}$$

where $\omega_l - \omega_h$ — frequency retune band; $a_2(\omega)$ — feed frequency characteristic.

An analysis of integral (1) shows that synthesized in time domain near field response of the mirror parabolic antenna is concentrated in following time interval

$$\begin{gathered} F + H < ct < F + A^2/4F + \sqrt{(H - A^2/4F)^2 + (A + R)^2}, \qquad R < A, \\ F + A^2/4F + \sqrt{(H - A^2/4F)^2 + (R - A)^2} < ct \\ < F + A^2/4F + \sqrt{(H - A^2/4F)^2 + (a + R)^2}, \quad R > A \end{gathered} \tag{3}$$

Figure 2 shows calculated results of antenna impulse transient characteristic $h(t)$, modules of the synthesized pulse $I(t)$ and the time domain response of antenna $S(t)$ for point of measurement with coordinate $R = 0$.

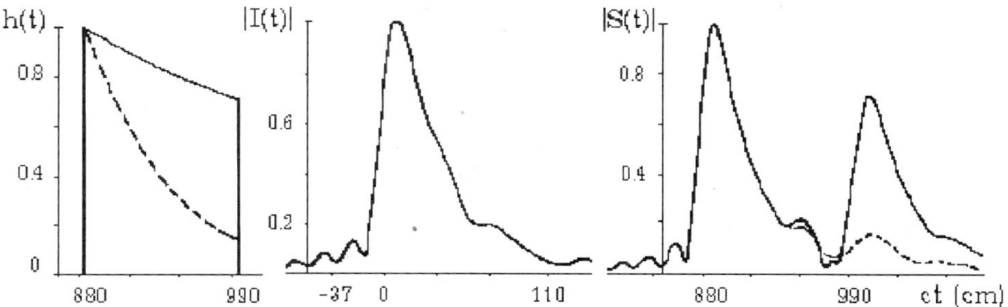

Figure 2. Impulse transient characteristic $h(t)$, synthesized pulse $I(t)$ and antenna signal $S(t)$ for uniform (solid line) and fall down (dotted line) feeding $a_1(\rho)$.

In practice the measurements are executed at a finite set of discrete frequencies $f_n = f_0 + \Delta f\, n, \quad n = 0, 1, \ldots, N-1$, after that the response $S(t)$ at the time moments $t_m = \Delta t\, m, \quad m = 0, 1, \ldots, N-1$ is defined using the discrete Fourier transformation.

In this case the near field antenna response is not equal to zero in the time range

$$c\,t_0 < c\,t < c\,t_0 + \Delta R, \tag{4}$$

where

$$c\,t_0 = \Delta r_{min} - (c\,\Delta f)\, entire(\Delta r_{min} \Delta f / c),$$

Δr_{min} — minimum difference of path between reference and measured signals; ΔR — time duration of antenna response.

For geometry of Figure 1

$$\begin{gathered} \Delta r_{min} = \Delta r_f + \sqrt{(H-R)^2 + R^2}, \\ \Delta R = \sqrt{(H - A^2/4F)^2 + (R+A)^2} + F + A^2/4F - \sqrt{(H-F)^2 + R^2}, \end{gathered} \tag{5}$$

Δr_f — difference of electrical lengths in reference and measured pathfeeds.

THE MEASUREMENT FACILITY

The facility for near field antenna measurements (see Figure 3) was designed in NIRFI (Nizhny Novgorod) under supervision of prof. N.Tseitlin[1,2]. This facility was intended to model the facility for ground tests of transformed space antennas[3]. In USA this measuring scheme was developed by prof.Y.Rahmat-Samii and called "bi-polar"[4].

In NIRFI facility the test antenna is directed to zenith and is rotated slowly around azimuth axis. The probe is placed at the end of horizontal arm of the scanner and is rotated continuously over circular arc above the antenna axis with velocity up to 15 circles per minute.

The probe rotation plane is about 15 meters over the ground. Length of the scanner arm is 7.5 meters. The probe is used in radiation mode while the test antenna receives this signal.

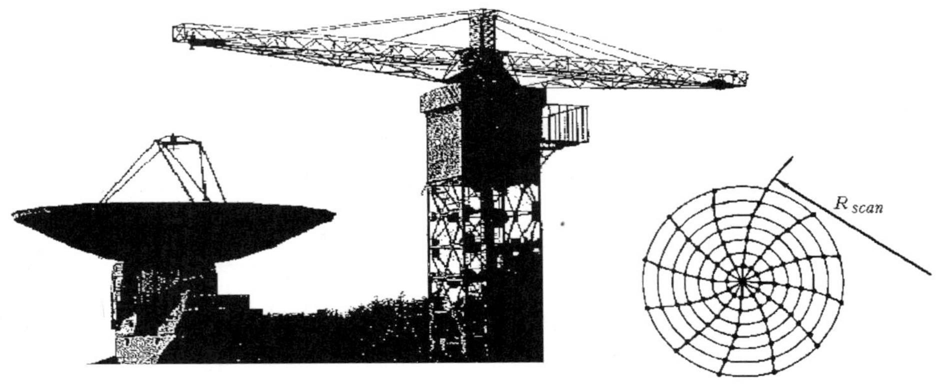

Figure 3. HIRFI near field facility and measuring grid

Oscillator of the type G4–81 (frequencies from 4. GHz through 5.6 GHz) manufactured by Russian industry is used as signal source at the 6–cm range. The receiver-ampliphasemeter FK2–24 has a frequency band 3.2–5.6 GHz. The similar equipment is used for measuring at the 18–cm range. Computer implements of antenna rotation control, probe position and near field complex data collection and far-field pattern restoration.

The near field measurement grid formed by this facility is shown at Figure 3. For restoration of far-field pattern directly from this grid the modified Jacobi-Bessel algorithm is used[5].

An advantage of above mechanical scheme is the stability of antenna state during measurements through its elevation immobility. Besides, an absence of VHF-cable twists allows to obtain high accuracy of phase measurements.

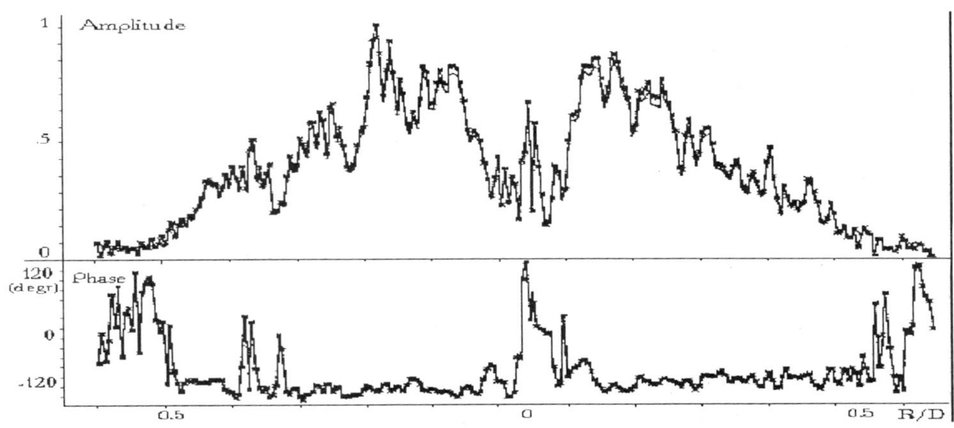

Figure 4. Near field cross section of the 7–meter antenna at $\lambda = 6\,\mathrm{cm}$.

The random error of phase measurement is less than one degree at speed about 2–4 circles per minute and about 3 degrees at 12–15 circles per minute. Figure 4 shows two implementations of the same near field cross sections measured data. There is good coincidence of two implementations. The small scale fringe appreciable on this illustration is caused by presence of some interference components at the measured field. The above method was used for investigation of these components.

EXPERIMENTAL RESULTS

First of all the responses synthesized in time domain were investigated at several points of the measuring plane.

Figure 5 shows the frequency dependencies of signal amplitude and phase for coordinate $R = 1$ m. The amplitude downfall at the frequencies about 4500 MHz is associated with feed characteristic. The amplitude oscillations are caused by interference of the measured field components.

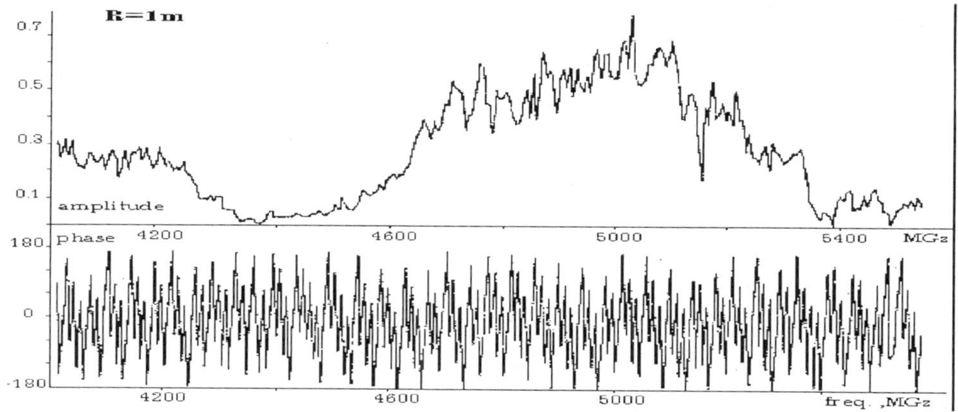

Figure 5. Frequncy dependency of near field signal.

The time domain signal synthesized with the frequency band 254 MHz and the discrete 2 MHz is shown on Figure 6. This distribution obtained at the time range 0–500 ns demonstrates the concentration of the measured components at the window about 100 ns (30 meters). This result allows us to increase frequency variation discrete for obtaining higher resolution with smaller frequency number and thus to transfer the components investigation to a small delay range.

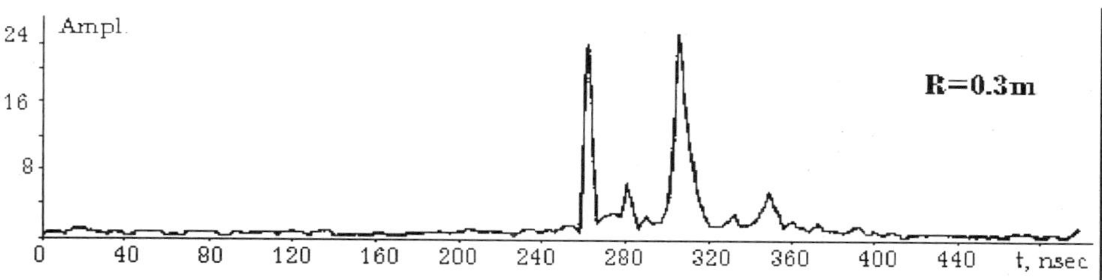

Figure 6. Time domain signal with frequency discrete 2 MHz.

Figure 7 shows the amplitude of the time domain signal synthesized in different points of measurement plane with the frequency band about 1.5 GHz and discrete 6 MHz. The resolution obtained by this measurements is about 1 ns (30 sm), that is practically the limit for this facility.

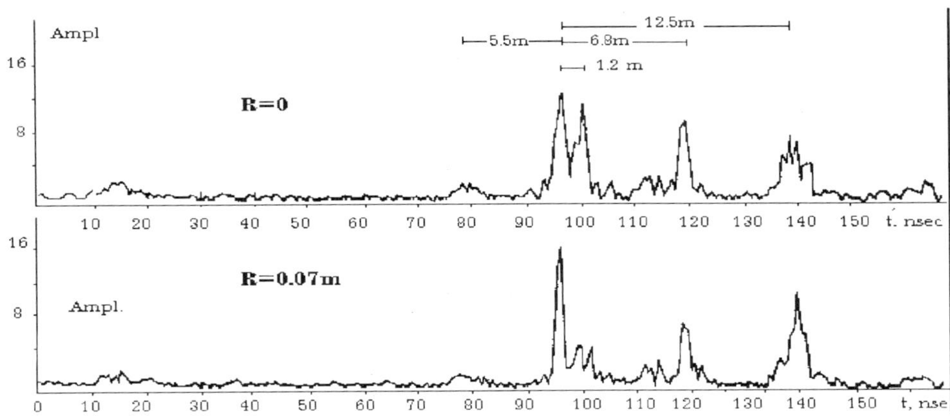

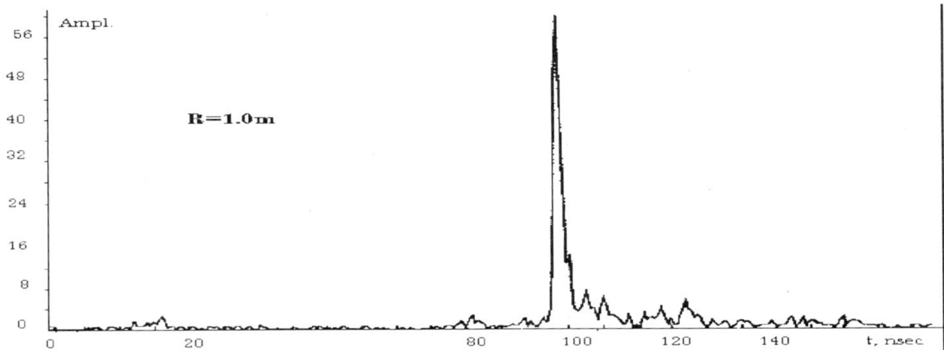

Figure 7. Time domain signals with frequency discrete 6 MHz.

A time domain separation of the measured field components is obvious at Figure 7. In according to the (5) the test antenna components are concentrated at the interval 75–107 ns. Among them the components with delay from 94 to 101 ns are propagated along path the "probe-mirror-feed". The splitting of this response is in according to modeling results showed at Figure 2. The components with delays near 78 ns are caused by reception through backlobes of the antenna feed.

Among interference background the components with delays near 119, 139 and 13 ns are distinguished. The delay near 119 ns corresponds to doubled path between probe and antenna focus. Therefore, these components are caused by reflection between the feed support elements and scanner arm. Similarly, the components with delays near 139 may be identified as corresponding to the propagation "probe-mirror-scanner-mirror-feed", and the components with delay near 13 ns — as having propagated twice.

CONCLUSION

The time domain investigation of the measured antenna near field allows us to define the signal components origin, levels and space distributions. This measurements show that the reflections between test antenna and scanner are essential for probe position near the aperture center. Using the inverse Fourier transformation of proper part of the time signal, the antenna near field was restored more precisely as well as the plane distribution of scattered field at the 6 cm wavelength. These distributions were used for estimation of pattern distortion caused by scattering. At this facility the scattering distorts insignificantly the main beam and first sidelobes of antenna far field pattern. With using of this method the estimation was made of the expedience of radioabsorbing cover for this scanner.

REFERENCES

1. N.M. Tseitlin, V.I. Turchin, A.V. Kalinin et.al., Radio holographic methods for measurements of radio telescope antenna characteristics, in: *Proc. of the Int. Workshop Holography Testing of Large Radio Telescopes,* Nauka, Leningrad (1991).
2. N.M. Tseitlin, Yu.I. Belov, A.V. Kalinin, et.al., The planar testing specialities of the antennas directed to zenith,in: *Proc. of Fourteenth ESA Workshop on Antenna Measurements,* ESTEC, Noordwijk (1991).
3. A.V. Kalinin, Yu.I. Belov and V.I. Altunin. *Radiotechnical testing of the Radioastron Antenna,* IKI Rep. 1244, Acad.of Science, Moscow (1987).
4. Y.Rahmat-Samii, Modern concepts in analysis, synthesis and measurements of antennas, in: *Modern Radio Science 1993,* H. Matsumoto, ed., Oxford University Press, New York (1993).
5. Yu.I. Belov, A.V. Kalinin and E.E. Kalinina, Use of the Jacobi-Bessel series for reconstructing of antenna far-field from near-field data, measured in quasi-radial grid, *Izv.VUZov Radiophisics,* 30, 10 (1987).

DESIGN OF THE LOW-FREQUENCY COMPENSATION OF AN EXTREME-BANDWIDTH TEM HORN AND LENS IRA

M.H. Vogel

Phillips Laboratory / WSQ ,
3550 Aberdeen SE, Kirtland AFB, NM 87117-5776, U.S.A.
on assignment from
TNO Physics and Electronics Laboratory
P.O. Box 96864, 2509 JG The Hague, NETHERLANDS.

1. INTRODUCTION

Many applications require radiation of a very short (i.e. extreme-bandwidth) pulse of electromagnetic energy out to large distances. Short pulses for which the ratio between the highest and the lowest frequencies in the spectrum (at the -3 dB points) are of the order of 100:1 without dispersion (i.e. with pulse fidelity) are often desirable. To radiate such an extreme-bandwidth pulse, a TEM horn can be used.

A diagram of a TEM horn is shown in figure 1.1. It consists of a TEM transmission line of almost constant impedance. A lens may be included at the horn aperture to obtain an improvement in the boresight directivity for the high frequencies. In the latter case, it is called a lens impulse radiating antenna (lens IRA or LIRA).

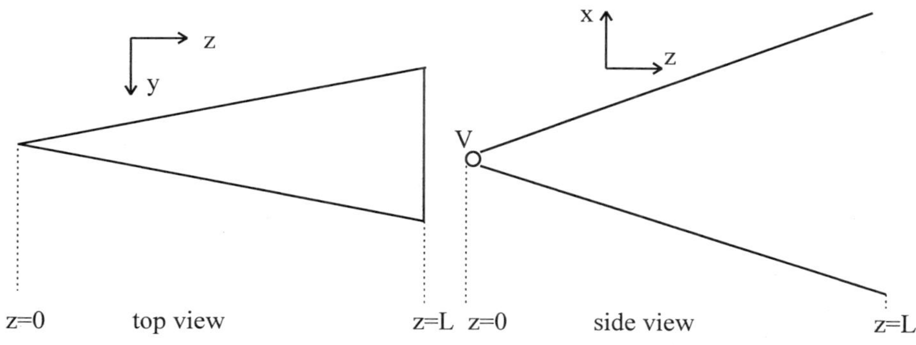

Figure 1.1. TEM horn

In this paper, we concentrate on the low-frequency behaviour of the TEM horn. As a TEM horn can be modeled as a transmission line, it presents an open circuit for the low-frequency part of the pulse[1]. As a result, a large part of the energy will be reflected towards the source and may damage it. A remedy is to connect a resistive termination to the horn, so that it will no longer act as an open circuit for the low frequencies. Preferably, the resistance of this termination is matched to the impedance of the horn, so that it behaves as a matched load to a transmission line. The physical shape of this resistive termination is important, as it significantly affects the low-frequency performance. This has been pointed out by Baum[2,3], and will be summarized in the next section.

2. DESIGN CONSIDERATIONS FOR THE TERMINATION

In the low-frequency limit, we are dealing with a quasistatic problem. Then, because of the voltage difference between the plates, there will be positive charge on the upper antenna plate and negative charge on the lower plate. As a consequence, the antenna (in combination with the terminating loop) has an electric dipole moment and there is a toroidal electric-field distribution around it. Furthermore, the current that flows through the antenna and the terminating loop produces a magnetic dipole moment, and there is a magnetic field around the antenna, the properties of which are determined by the size and shape of the terminating loop and the magnitude of the current. It is desirable to have the magnitudes and directions of both dipole moments matched in such a way that they combine to orient the low-frequency radiation in the forward direction and cancel the low-frequency radiation in the backward direction.

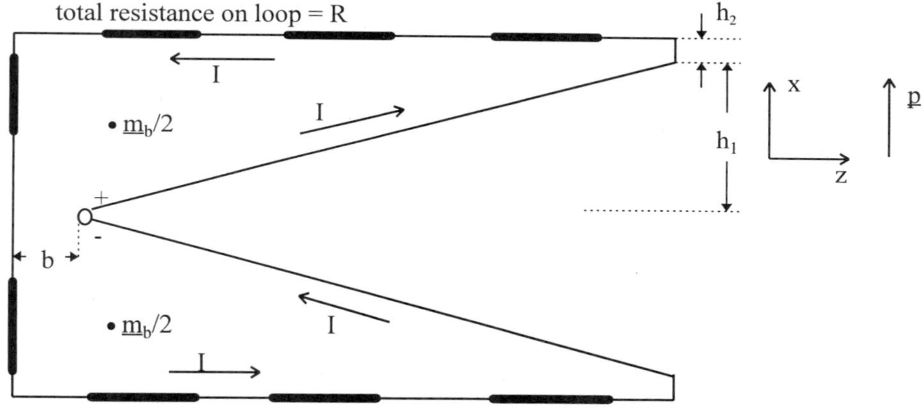

Figure 2.1. Possible design of the resistive termination

Consider figure 2.1. In this design the electric and magnetic dipole moment will, if their magnitudes are matched, combine to direct the Poynting vector for low-frequencies forward. A cardioid antenna pattern will result for these low frequencies, with a null in the backward direction. The matching condition, as explained by Baum[2], is

$$m_y = p_x c\ ,$$
$$m_x = m_z = 0,\ \ p_y = p_z = 0\ , \tag{2.1}$$

in which m_x, m_y and m_z are the components of the magnetic dipole moment $\underline{m}$ in Am^2, p_x, p_y and p_z are the components of the electric dipole moment $\underline{p}$ in Cm, and c is the speed of light in m/s. From figure 2.1, we conclude that, to optimize our design with a rectangular loop shape, we can vary

- h_2, the height above the aperture rim,
- b, the distance behind the apex of the antenna,
- the cross section of the loop in the x and y directions,
- the distribution of the resistors along the loop.

As loop cross sections, we have performed calculations on designs with one wire and with two parallel wires, in both cases with variable radii, as well as on strip-like structures. We will present the results in Section 4. For a constant loop area, the variation of the loop cross section doesn't affect the magnetic dipole moment. The way the electric dipole moment is affected depends on how the total charge on the loop is affected. The effect can be both positive and negative. The distribution of resistors along the loop doesn't affect the magnetic dipole moment. It does affect the charge distribution, and therefore electric dipole moment. In general, when we move part of the resistors closer to the antenna aperture, p_x decreases significantly. When we move part of the resistors farther away from the aperture, p_x increases significantly.

3. ANALYTICAL APPROACH

In our analytical approach, we assume that the antenna and the compensating loop are both very long, so that end effects can be neglected. As a start, we take b=0, i.e. the loop doesn't extend behind the source. The design procedure then consists of the following steps:

1. Specify the distribution of resistors along the loop and specify h_2.
2. Knowing the distribution of resistors, solve for the voltage distribution along the loop. The voltage on the upper antenna plate is V_0, the voltage on the lower antenna plate is $-V_0$.
3. Solve for the charge distribution on the entire structure (antenna plates and loop).
4. Knowing the charge distribution, calculate the electric dipole moment $\underline{p}$.
5. Knowing the current and the loop shape, calculate $\underline{m}$ and compare $\underline{m}$ to $\underline{p}c$ (cf. (2.1)). If a modification is needed, go back to step 1 and repeat the procedure.

Obviously, the main challenge is in step 3. An efficient method to obtain the charge distribution on a cylindrical structure (a two-dimensional problem) has been presented by Clements, Paul, et al.[4,5]. Consider the cross section of our antenna+loop structure as given in figure 3.1. The cross section has been taken at an arbitrary location between source and aperture (i.e. arbitrary z-coordinate). In this figure, W is the width of the plates at the aperture plane, and L is the distance between apex and aperture plane (cf. fig. 1.1). Note that, as we have assumed the structure to be very long, we can, for each z, solve for the charge distribution as if the geometry were two-dimensional. We just need to obtain a relationship between the charge density distribution and the voltage distribution for the two-dimensional case depicted in figure 3.1. The charge density distribution will be a function of z of course, as the distances between the conductors, the voltages on the loop and the widths of the plates are all functions of z.

Because of symmetry, the problem reduces to the following equations:

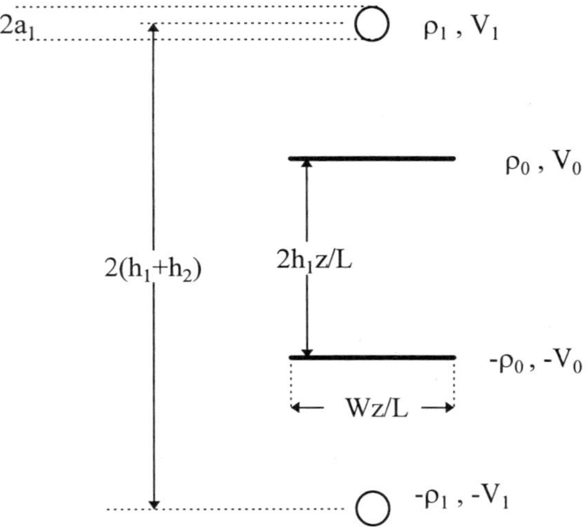

Figure 3.1. Cross section of antenna+loop structure at arbitrary z

$$\rho_1 = c_{11}V_1 + c_{12}V_0 \ ,$$
$$\rho_0 = c_{21}V_1 + c_{22}V_0 \ , \tag{3.1}$$

in which $\rho_{1,0}$ denotes the charge per unit length on the upper half of the loop or on the upper antenna plate, respectively, and $V_{1,0}$ denotes the voltages on these structures. All are functions of z. To obtain the coefficients c_{11}, c_{12}, c_{21}, c_{22}, we invoke the inverse equations

$$V_1 = d_{11}\rho_1 + d_{12}\rho_0 \ ,$$
$$V_0 = d_{21}\rho_1 + d_{22}\rho_0 \ . \tag{3.2}$$

Once matrix $\underline{\underline{D}}=(d_{n,m})$ has been found, matrix $\underline{\underline{C}}=(c_{n,m})$ can be obtained by taking the inverse of $\underline{\underline{D}}$. To obtain the coefficients of $\underline{\underline{D}}$, we use the relation between voltage and charge density in two dimensions as given by Smythe[6]

$$V(\underline{r}_p) = \frac{-1}{2\pi\varepsilon_0}\int \rho_s(\ell)\ln\left|\underline{r}_p - \underline{r}(\ell)\right|^2 d\ell \ , \tag{3.3}$$

in which $\underline{r}_p$ denotes the point of observation, $\underline{r}(\ell)$ denotes the point on the surface where the surface charge density is given by $\rho_s(\ell)$, and ℓ is a local coordinate on the surface (at constant z) over which the integration is carried out.

Using (3.3) while making simplifying assumptions about the charge density distribution $\rho_s(\ell)$ (for constant z) over the surface of each conductor, we establish linear relations for each z between the charge densities ρ_1 and ρ_0 and the voltages V_1 and V_0, i.e. we have obtained the coefficients of matrix D. The coefficients of C are obtained by taking the inverse of D. The resulting relation between charge densities and voltages is

$$\rho_1(z) = \frac{2\pi\varepsilon_0}{\ln(f)\ln(g) - \ln^2(k(z))}\left(\ln(g)V_1(z) - \ln(k(z))V_0\right), \tag{3.4a}$$

$$\rho_0(z) = \frac{2\pi\varepsilon_0}{\ln(f)\ln(g) - \ln^2(k(z))}\left(\ln(f)V_0 - \ln(k(z))V_1(z)\right), \tag{3.4b}$$

in which

$$f = \frac{2h}{a_1}, \quad g = \frac{8h_1}{W}, \quad k(z) = \frac{hL + h_1 z}{hL - h_1 z}. \tag{3.4c}$$

In (3.4c), a_1 is the radius of the wire, W is the width of the antenna plate at the aperture, h is the height of the horizontal section of the wire above the apex of the antenna, L is the length of the antenna, h_1 is half the vertical dimension of the aperture (at z=L). For the entire antenna+loop p_x is subsequently calculated by means of (3.5):

$$p_x = \int_{z=0}^{L} \left(2\rho_0(z)\frac{zh_1}{L} + 2\rho_1(z)h\right)dz. \tag{3.5}$$

We have calculated an expression for p_x for the case of a uniform distribution of resistors along the loop, i.e. a linear voltage distribution. For V_0=1.5 MV, Z=240 Ω, L=137 cm, W=50 cm, h_1=50 cm, h_2=10 cm, b=0, a_1=2 cm, we obtain

$$p_x = 28\ \mu\text{Cm}, \quad m_y = 7.7\ \text{kAm}^2.$$

Hence, p_x multiplied by the speed of light c is equal to 8.4 kAm^2, which is 10% larger than m_y. As our goal is to match $p_x c$ and m_y, this result indicates that just a small decrease in p_x or a small increase in m_y will give us the desired result.

We can now proceed to adjust b, h_2 or a_1 and repeat the process, or we can derive an expression for p_x resulting from another resistor distribution, and match $p_x c$ and m_y. However, we prefer to proceed with numerical calculations, in which the analytical result obtained thus far will serve as a useful starting point and as a necessary check on the first numerical results.

4. NUMERICAL APPROACH

As pointed out by Harrington[7], the Method of Moments is very well suited to calculate the charge distribution on an arbitrarily shaped three-dimensional perfectly conducting structure, once the voltages are known everywhere on this structure. In three dimensions, the relation between the electrostatic potential V at the point of observation denoted by $\underline{r}_p$ and the surface charge density distribution $\rho_s(\underline{r})$ on a metal object is given by[7]

$$V(\underline{r}_p) = \iint_{surface} \frac{\rho_s(\underline{r})}{4\pi\varepsilon_0\left|\underline{r}_p - \underline{r}\right|} dS. \tag{4.1}$$

On the antenna and the loop, the voltage distribution is known while the charge density distribution is to be calculated. To this aim, the structure is divided into many planar quadrilateral panels. As has been proven by King[8], a wire of radius a_1 can be modeled as a strip of width $4a_1$. When we number the panels 1 to N, N being the total number of panels, we have for the voltage on each panel the following equation:

$$V_n = \sum_{m=1}^{N} L_{nm}\rho_{s,m}, \tag{4.2}$$

where, for combinations of panels that are not very close[7],

$$L_{nm} = \frac{A_m}{4\pi\varepsilon_0 R_{nm}} , \tag{4.3}$$

in which A_m is the area of panel m and R_{nm} is the distance between the centers of panels n and m. The self term for m=n is given by[7]

$$L_{nn} = \frac{0.282\sqrt{A_n}}{\varepsilon_0} . \tag{4.4}$$

For panels close together, special measures have to be taken, such as a further subdivision of the panels. Finally, a matrix equation results:

$$\underline{V} = \underline{\underline{L}}\,\underline{\rho}_s , \tag{4.5}$$

where $\underline{V}=(V_n)$, $\underline{\underline{L}}=(L_{nm})$, $\underline{\rho}_s=(\rho_{s,m})$, in which $\underline{\rho}_s$ is unknown. The matrix equation is solved by LU-decomposition.

For the 1.37 meter antenna+loop, the characteristic linear dimension of the panels in the antenna and in the loop is 5 cm, with a maximum of 9 cm. The maximum linear dimension of the panels in the ground plane (to be introduced later) is 12 cm. Doubling the number of panels typically changes the calculated value for p_x by 3%.

Note that the antenna, as depicted in figure 2.1, is symmetrical in the vertical direction. Therefore, a ground plane (V=0) can be introduced and the lower part of the antenna+loop omitted. The subsequent calculations have been performed for an antenna with a finite ground plane (fig. 4.1), the reason being that the antenna will actually be built with a finite-sized ground plane. The size and shape of the ground plane provide extra variables in the antenna design. In all cases, we choose its minimum width to be 0.5 m at z=-0.25 m, and its maximum width to be 1.5 m. The extension of the ground plane in the positive z-direction is kept as a variable. This mainly influences p_z, which we assume to satisfy the condition $p_z=0$ when its calculated absolute value is less than 0.2 μCm. All calculations with the finite-sized ground plane have been performed subject to the condition that the total charge on the entire structure be zero.

We start with the case where the compensating loop consists of one wire (figure 4.1). The final design for this case has a uniform distribution of resistors along the entire wire (including the vertical sections). Further, h_2=2.5 cm (this means the loop hardly takes more space in the vertical direction than the antenna itself), b=15 cm (this means the loop extends 15 cm behind the source, which is usually no problem), and the wire radius a_l=1 cm. The ground plane extends to z=1.65 m. For this design, p_x=19 μCm and m_y=5.7 kAm^2, while the other components are zero. Hence, the matching condition (2.1) has been satisfied.

The sensitivity of the result to variation in some parameters is illustrated by the following examples:

When h_2=5 cm instead of 2.5 cm, p_x increases by 3% while m_y increases by 8%.
When b =17.5 cm instead of 15 cm, p_x increases by 1% while m_y increases by 3%.
When a_l=2 cm instead of 1 cm, p_x increases by 2% while m_y remains constant.
When the resistors are distributed uniformly over the horizontal section of the loop only, p_x decreases by 9 % while m_y remains constant.
When the ground plane extends to z=1.75 m instead of z=1.65 m, p_z=3 μCm while p_x decreases by 1%.

We proceed with the case where the compensating loop consists of two wires (figure 4.2). In the final design for this case the wires are 25 cm apart, while again h_2=2.5 cm, a_1=1 cm, and the resistors are uniformly distributed. The loop now extends 20 cm behind the source (i.e. b=20 cm), and the ground plane extends to z=1.60 m. For this design, p_x=20 μCm and m_y=6.0 kAm^2, while p_y, p_z, m_x, and m_z are zero. Hence, the matching condition (2.1) has again been satisfied.

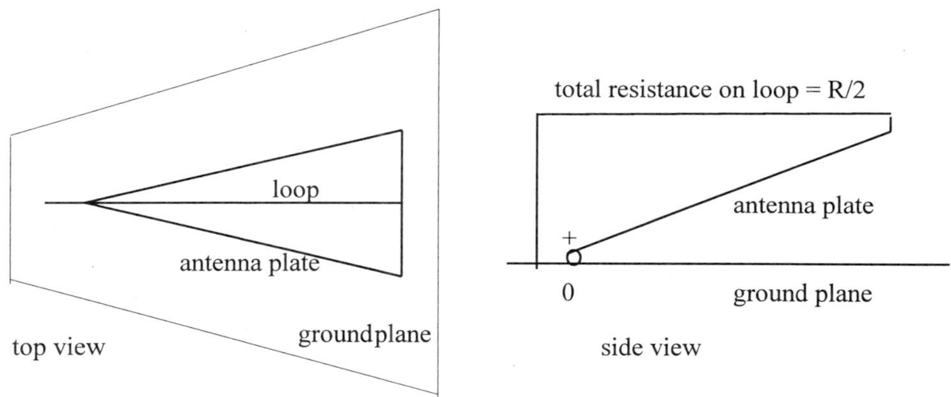

Figure 4.1. TEM horn with one-wire compensating loop

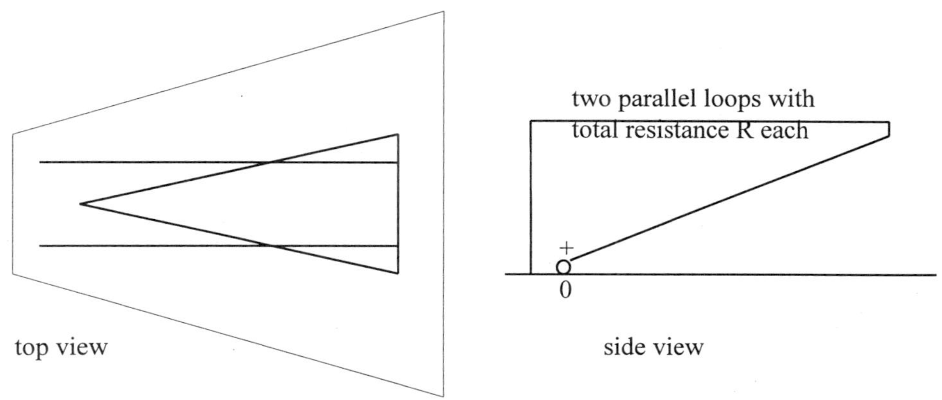

Figure 4.2. TEM horn with two-wire compensating loop

At this point, we have two designs for the low-frequency compensation of the TEM horn that are both easy to build and that both satsfy all requirements for the low frequencies. For the intermediate and higher frequencies however, note that the fields associated with the charges in the loop and on the plate side facing the loop will radiate in unwanted directions. It is desirable to have a loop design that suppresses the radiation of the TEM fields above the TEM horn. This observation leads us to the design depicted in figure 4.3. We call it the TEM horn with the one-triangle compensating loop.

The idea behind this design is as follows. Fields propagating forward in the space between the antenna plate and the loop will in this design be propagating in a structure that looks like a *receiving TEM horn* with a ground plane. With a matched load at the end (i.e. at the attachment point), their energy will be dissipated in the resistor. A 120 Ω resistor is a matched load for this case, which is exactly the total resistance required in the loop. Hence, in this case there are no further resistors along the loop; all resistance is concentrated in the attachment point. The voltage along the loop is zero. Unfortunately, with the one-triangle design we are far from satisfying the matching condition $p_x c = m_y$. The reason is that by bringing all resistance as far forward as possible, we are allowing a significant negative charge on the loop, resulting in a low electric dipole moment. We cannot avoid this by moving the resistors, as all resistance has to be concentrated in the attachment point in order to act as a matched load. Hence, we cannot use the one-triangle design.

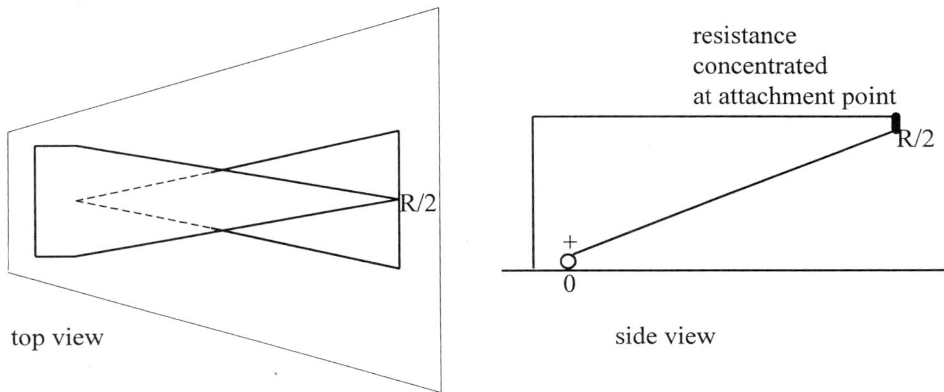

Fig. 4.3. TEM horn with one-triangle compensating loop

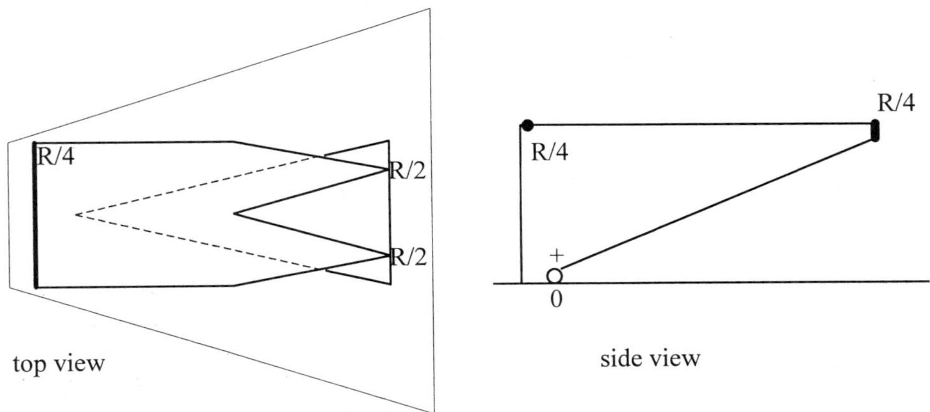

Figure 4.4. TEM horn with two-triangle compensating loop

An alternative design, in which the idea of the matched load can be used as well, is depicted in figure 4.4. Now there are two attachment points on the antenna plate. We call this the TEM horn with the two-triangle loop. Note that radiation traveling in the space between an antenna plate and the loop will still encounter, in the frontal section, a structure that looks like a receiving TEM horn with a ground plane. In each attachment point, the matched load is provided by a 120 Ω resistor. An important difference with the previous design is that we now have some resistance left to place along the loop at will! Two parallel 120 Ω resistors form together a 60 Ω resistor, and we have another 60 Ω left to place along the loop. We can use this freedom to satisfy the matching condition. It turns out that this is achieved with b=10 cm, h_2=2.5 cm, and with the 60 Ω placed at the point where where the loop makes an angle (as indicated in the figure). Further, the ground plane extends to z=1.75 m. In this design, p_x= 18 μCm and m_y= 5.4 kAm^2.

5. CONCLUSION

We have presented several designs for a low-frequency compensated TEM horn which does not reflect low-frequency radiation towards the source and which, for the lowest frequencies, does not radiate a toroidal pattern but a cardioid pattern in the forward direction. The compensating loop can be achieved with one or two wires or with adequately shaped plates, with the appropriate distribution of resistors along it. The TEM horn with the two-triangle compensating loop is expected to have the best directivity for the intermediate and higher frequencies. Therefore, we have chosen this design to be fabricated and tested. Its properties for the entire frequency spectrum will the subject of further research, both experimental and computational.

REFERENCES

1. E.G. Farr and C.E. Baum, A simple model of small-angle TEM horns, *Sensor and Simulation Note* 340 (1992).
2. C.E. Baum, Low-frequency compensated TEM horn, *Sensor and Simulation Note* 377 (1995).
3. C.E. Baum, Some characteristics of electric and magnetic dipole antennas for radiating transient pulses, *Sensor and Simulation Note* 125, (1971).
4. J.C. Clements, C.R. Paul and A.T. Adams, Computation of the capacitance matrix for systems of dielectric-coated cylindrical conductors, *IEEE Trans. Electromagnetic Compatibility*, 17:4 (1975).
5. C.R. Paul and A.E. Feather, Computation of the transmission-line inductance and capacitance matrices from the generalized capacitance matrices, *IEEE Trans. Electromagnetic Compatibility*, 18:4 (1976).
6. W.R. Smythe, *Static and Dynamic Electricity*, Third Edition, Chapter 4, Hemisphere Publishing Corporation (Taylor and Francis Group), New York (1989).
7. R.F. Harrington, *Field Computation by Moment Methods*, First Printing, Chapter 2, Macmillan Company, New York (1968).
8. R.W.P. King, *The Theory of Linear Antennas*, Chapter I-7, Harvard University Press, Cambridge, Massachusetts (1956).

A RADIATING STRUCTURE INCORPORATING AN EXTENDED GROUND PLANE AND A BREWSTER ANGLE WINDOW

Jimmy Wells,[1] Carl Baum,[2] Norman Keator,[3] William Prather[2]

[1]Fiore Industries Inc.
1009 Bradbury Drive SE
PO Box 9243
Albuquerque, NM 87119-9243

[2]USAF Phillips Laboratory
3550 Aberdeen Ave.
Albuquerque, NM 87117

[3]Voss Scientific
416 Washington SE
Albuquerque, NM 87106

INTRODUCTION

The H-series of ultra wideband (UWB) pulsers is designed in a coaxial geometry. Therefore, in order to effectively radiate the energy, it is necessary to have a mode convertor and a non-dispersive radiating structure. This paper presents such a combination which was designed and built specifically for the H-3 high power ultra wideband source [1]. It incorporates the Point Geometry Convertor (PGC) and a TEM horn with a Brewster angle window providing the oil to air interface.

OVERALL ANTENNA DESIGN

Figure 1 shows the top and side views of the entire mode conversion and radiating structure. Note that this is what we call an unbalanced antenna design. Since the output of the H-series machines is in a coaxial mode, it is not possible to generate a true differential mode over all frequencies, at least not with the mode convertors we have. If one tries to use the antenna in a differential mode, some of the energy is lost to common mode radiation or may cause ringing on the structure if it is not terminated. Therefore, this antenna system is built to be unbalanced, that is, a half TEM horn over a symmetry

Ultra-Wideband, Short-Pulse Electromagnetics 3
Edited by Baum *et al.*, Plenum Press, New York, 1997

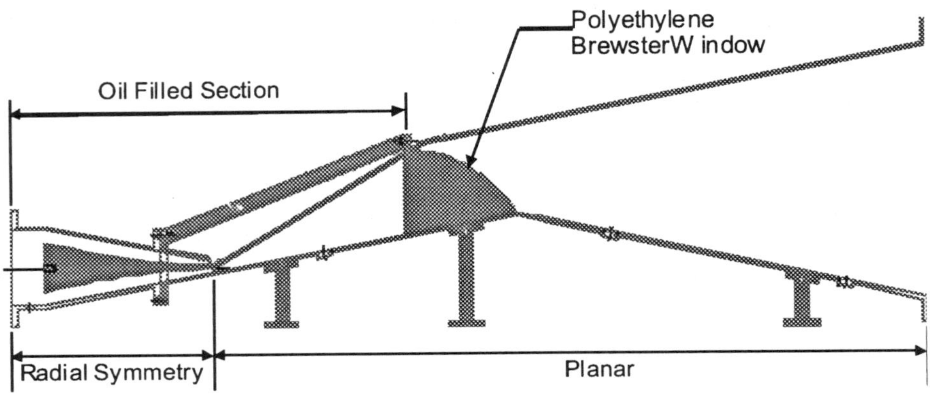

Figure 1. TEM Horn with Balun and Brewster Angle Lens

plane. When mated to the output of the H-3 high power microwave source, this antenna system is designed to radiate transient electromagnetic energy with peak electric fields in excess of 100 kV/m at 5 m and risetimes on the order of 100 picoseconds.

POINT GEOMETRY CONVERTOR

The feed from the coaxial output of the H-3 source is shown on the left hand side. In the center is the PGC balun (mode convertor) which matches the coaxial feed to an unbalanced conical transmission line structure [2]. This then becomes the feed to the TEM horn antenna. The entire high voltage output section of the H-3 is insulated with transformer oil as is the PGC and the narrow part of the TEM horn structure.

BREWSTER ANGLE LENS

The TEM horn has a dielectric window midway up its length to transition the wave from oil to air. The window is made of polyethylene which has a dielectric constant $\varepsilon_r \approx 2.3$ which is very close to that of transformer oil. It uses a Brewster angle window which is curved only in the **E** plane. In the **H** plane, the window is not curved. That is, the radius of curvature is set to infinity. The impedance of the device transitions from 20 Ω at the output of the H-3 to 60 Ω at the highest field stress section of the PGC. The impedance then increases linearly to 120 Ω at the Brewster angle window. This is described in more detail in References 3 and 5.

FOCUSING LENS

The final design of this radiating structure will include a polyethylene focusing lens at the output of the horn [3, 4]. This part is currently being designed and has not been produced at the time of this writing. The purpose of the focusing lens is two-fold. First, it will concentrate the energy into a narrower beam in order to get more power on target. Second, it will straighten out the wavefront from a spherical to a plane wave, at least on boresight. This will reduce the spatial dispersion which stands in the way of getting extremely fast risetimes of less than 100 ps [4].

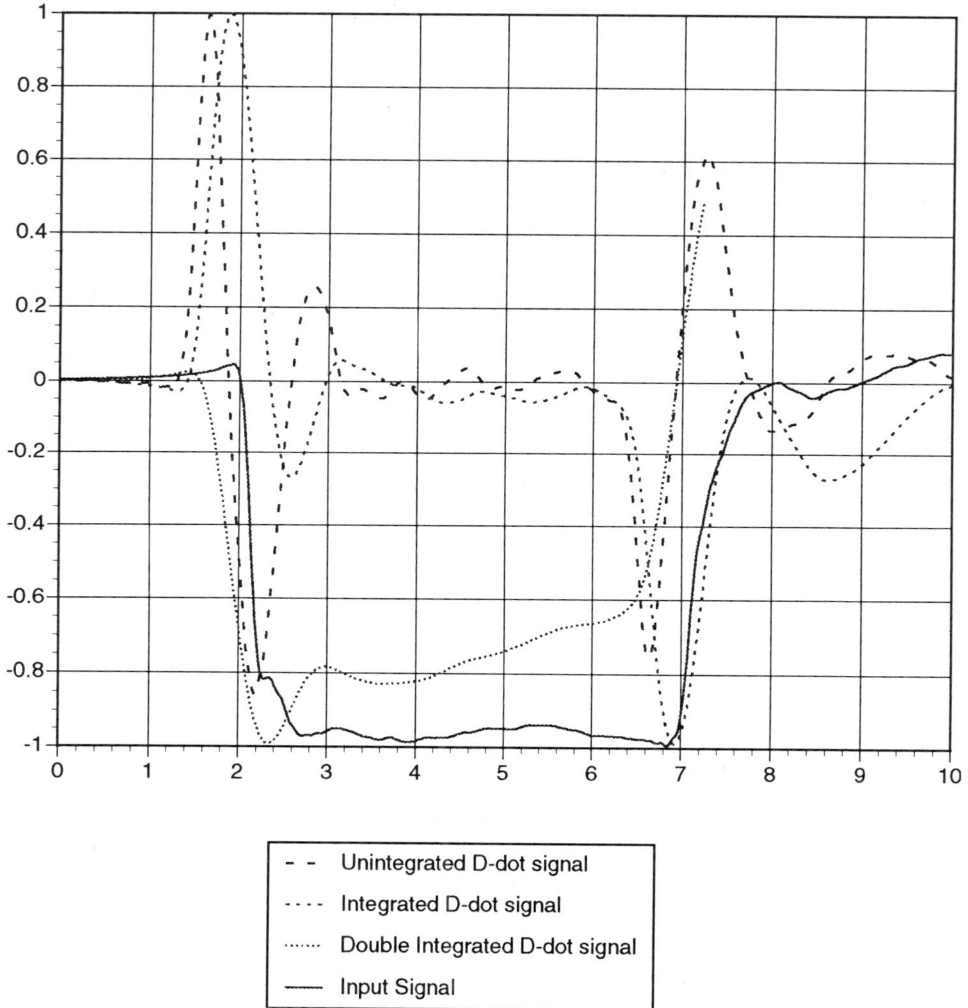

Figure 2 Long pulse input signal with radiated D-dot signals

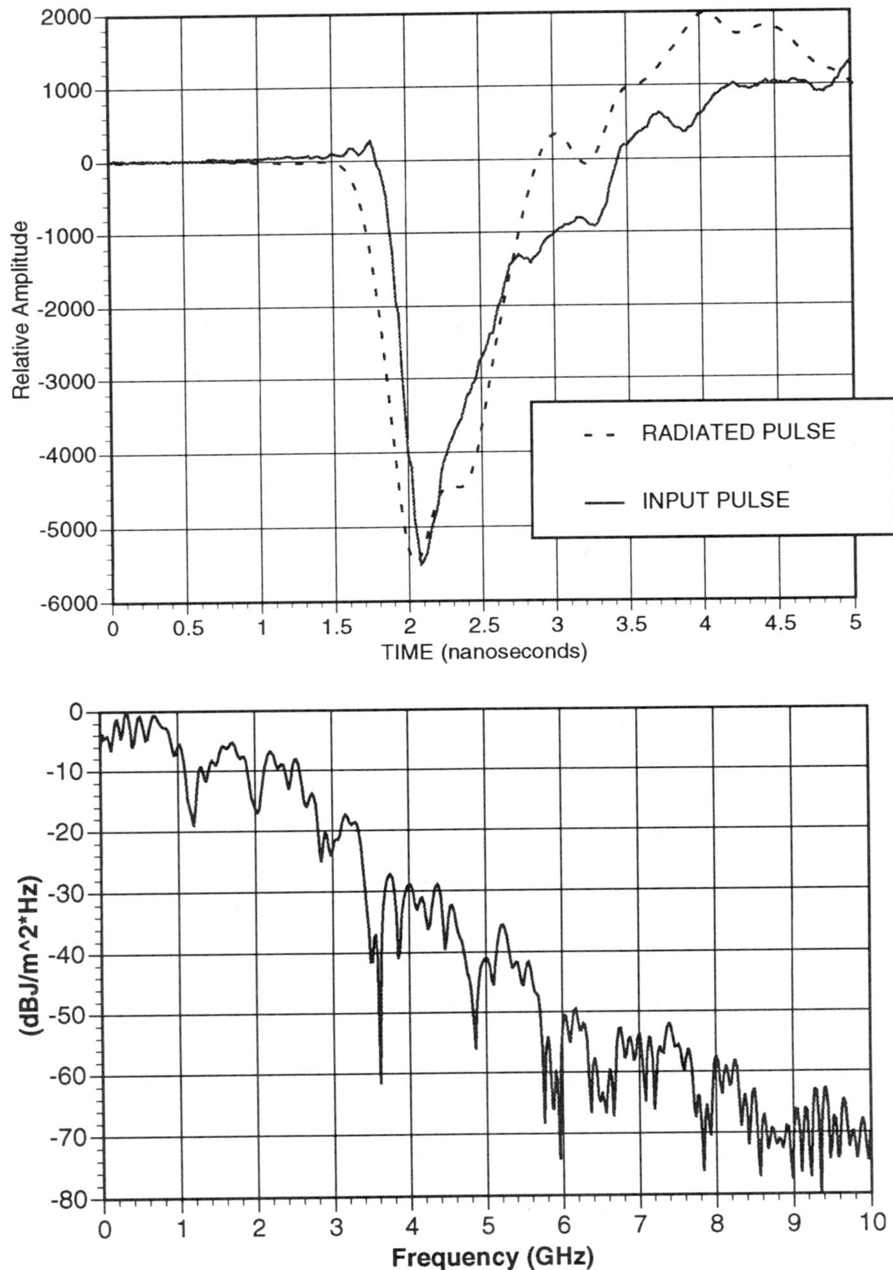

Figure 3. Measured Fast pulse Input, Radiated, and Spectral Waveforms

LOW FREQUENCY ENHANCEMENT

This radiator is also designed to incorporate a **p x m** loop to enhance the low frequency radiation [3]. The initial design has been completed and is found in the paper by Martin Vogel [5]. This design will also be turned into hardware during the early part of FY97 and subsequently tested at the Phillips Laboratory antenna range.

MEASURED RESULTS

The initial measured results which we have from this assembly are shown in Figure 2. The configuration which was measured was as shown in Figure 1 except that it included the **p x m** loop. That is, it includes the PGC balun, the TEM horn with the Brewster angle lens and the **p x m** loop. The focusing lens was not present. This will be included in subsequent measurements, but was not available for this test.

These initial short pulse measurements were made with a Kentech 6 kV pulse generator in order to be able to assess the fast transient behavior of the system. Moreover, the Kentech pulser while not having very high voltage output, does have a risetime of 90 picoseconds which is faster than the H-3, so it will give a better evaluation of the transient response of the system. The risetime of the Kentech was degraded to about 110 picoseconds using a 75 foot length of RG141 semi-rigid signal cable. Following this, measurements will be made with the H-3 pulser in order to evaluate the full voltage and risetime performance. The H-3 Source is a 20 Gigawatt, 300 picosecond pulse width, 150 picosecond risetime UWB HPM source which uses two 1500 psi hydrogen spark gaps for pulsecompression.

Cold test measurements of the radiated field were taken at a distance of seven meters from the antenna. Two lower voltage pulsers were used for these tests. One had 300 picosecond pulse width, 90 picosecond risetime and amplitude of six kilovolts the other had six nanoosecond pulse width, 200 picosecond risetime and amplitude of 1.5 kilovolts. Both pulsers have SMA output connections at 50Ω impedance and were fitted to the extended ground plane structure via a linear impedance transformer which transfers from the 50Ω SMA to the 20Ω at six inch diameter input to the extended ground plane. The pulser and PGC are fitted with fast, self-integrating electric field sensors with sufficient band width for measuring the waveshape as the pulse propagates along the length of the structure. D-dot sensors are also placed along the ground plane in the throat of the horn. Measurements of the near radiated field were made with free-field sensors placed 7 meters in front of horn, in an anechoic chamber. This will be sufficient to assess the performance of the system. However, when the focusing lens is added, we will have to use the outdoor antenna range in order for evaluation. The performance of such structures will just not show up accurately in a small anechoic chamber. Figure 2 shows the results of the long pulse application to the structure. Note that in radiating the extended ground plane structure also differentiates the pulse and thus we must perform two integration's of the D-dot signal to recover the input waveform. Figure 3 shows the results of the short pulse application to the structure. Note that both the risetime and the pulse shape are reproduced at the free field sensor with good accuracy considering sensor calibration error and the double integration.

REFERENCES

1. W.D. Prather, et al., "Ultrawide Band Sources: Present Technology, Future Challenges," *Ultra Wideband/Short-Pulse Electromagnetics 3*, this volume.

2. J. Wells, et al., "A Device for Radiating High Power RF Fields From a Coaxial Source," *Ultra Wideband/Short-Pulse Electromagnetics 3*, this volume.

3. C.E. Baum, "Low Frequency Compensated TEM Horn," Sensor and Simulation Note 377, January 1995.

4. E.G. Farr, C.E. Baum and C.J Buchenauer, "Impulse Radiating Antennas, Part 2," *Ultra Wideband/Short-Pulse Electromagnetics 2,* Plenum Press, New York, 1994.

5. C.E. Baum, "Brewster Angle Interface Between Flat-Plate Conical Transmission Lines," Sensor and Simulation Note 389, November 1995.

6. M.H. Vogel, "Design of the Low Frequency Compensation of an Extreme Bandwidth TEM Horn and Lens IRA," Sensor and Simulation Note 391, February 1996.

A TEM-HORN ANTENNA WITH DIELECTRIC LENS FOR FAST IMPULSE RESPONSE

Dr. John F. Aurand

Sandia National Laboratories [1]
High-Power Electromagnetics Department 9323
P.O. Box 5800, MS-1153
Albuquerque, NM 87185-1153

INTRODUCTION

We recently designed and built a TEM-horn antenna with a dielectric aperture lens in order to achieve faster transient pulse response. TEM (*t*ransverse *e*lectro*m*agnetic) horns are commonly used for wideband time-domain work because they offer minimum dispersion as a traveling-wave endfire structure (which can be made fairly nonresonant). However, even carefully designed TEM horns have an inherent pulse-smearing geometrical effect due to spherical wavefront propagation within the structure. A dielectric planar-convex aperture lens is used to accomplish this plane-wave to spherical-wave conversion, by collimating the wavefront between the plates in order to improve the impulse response.

The antenna consists of a conventional TEM-horn configuration (two flat, long triangularly-shaped conducting plates with a constant separation angle), and an additional solid Teflon™ lens placed at the aperture end of the plates. A 91-cm-long antenna was designed and built. Two different schemes were employed for the plate configuration: the first version utilizes single-sided etched copper traces on low-loss printed-wiring boards, and the other version utilizes solid copper plates. In both configurations, expanded polystyrene is employed as a solid structural supporting material between the plates, and the dielectric planar-convex lens is located at the aperture end of the plates.

The *printed-board configuration* is designed with stepped resistive loading at the aperture end of the traces in order to minimize ringing antenna currents, and a custom transition from the parallel-plate antenna structure to coaxial feedpoint. The *solid-plate configuration* was then developed because the impulse response of the printed-board topology wasn't good enough. The resulting step-equivalent risetime (10-90%) of the solid-plate version is 20 ps, the fastest TEM-horn we have designed and built to date. This paper describes our antenna design for both plate configurations, and measurements of the

[1] This Work was Supported by the U.S. DOE under Contract DE-AC04-94AL85000.

resulting performance for two nominally identical antennas. This type of antenna offers very good short-pulse operation, and is highly recommended for wideband time-domain antenna work.

PRINTED-BOARD ANTENNA CONFIGURATION

The printed-board configuration was the first design effort for our next-generation of time-domain antennas, with two primary features: 1) incorporate a planar-convex dielectric lens to collimate the internal spherical-wave behavior of a TEM horn, resulting in much faster transient response (for either transmit or receive operation), and 2) incorporate resistive loading on the conducting plates of the TEM horn in order to reduce the undesired ringing of antenna currents between the feedpoint and the aperture.

Figure 1 shows one of two identically-built TEM horn antennas which utilizes the printed-board construction. Most of the body consists of solid expanded polystyrene, with a dielectic constant very close to free-space (for minimal perturbation of the electromagnetic fields in the antenna structure). The lens is seen positioned between the aperture end of the two printed-wiring boards, with single-sided copper traces on the outside surfaces of the two boards. The aperture plate (and lens) width is 30.5 cm, and the plate separation (lens height) is 12.7 cm. The length of the flat-plate conductors is 80.0 cm from the focal point to the center of the aperture end of each conductor. The resulting plate-separation angle is 9°, and the plate-width angle is 22°. Because the primary goal was to minimize the duration of the time-domain impulse response, the plate-separation angle is shallow compared to typical TEM-horn designs of 15-20°. The surge impedance along the antenna was allowed to increase from the feedpoint level of 50 Ω up to about 95 Ω at the aperture, in order to increase the main-beam transmit gain and/or receive sensitivity with a given aperture width.

Figure 1. Printed-board configuration of TEM-horn antenna with dielectric lens.

Figure 2 shows the top and bottom printed-wiring boards (of the two antennas - one is turned upside-down). These boards are Rogers RT/Duroid 5880 material, with the lowest dielectric constant and loss commercially available. At the aperture end (or right side) of the boards, there are six rectangular sections of resistive loading, consisting of nichrome thin-film material. Their surface resistivity increases in a stepped exponential fashion, to approximate a continuously-increasing loss to the antenna plate currents.

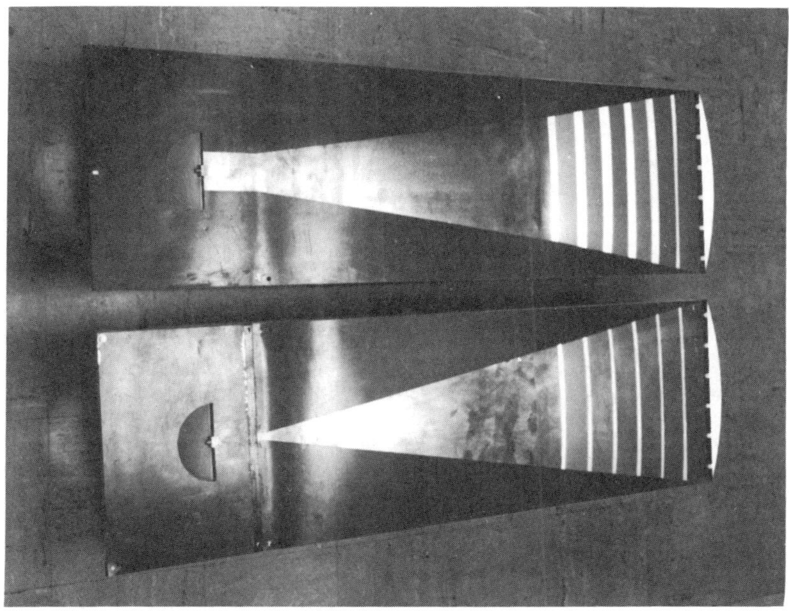

Figure 2. Top surface (lower antenna shown) and bottom surface of printed-board configuration of TEM-horn antenna.

At the feedpoint region, the bottom antenna has a small butterfly-shaped copper trace which is an integral portion of the custom transmission-line transition from the endfire-coax input (SMA female) to the balanced parallel-plate cross-section of the antenna proper. This transition was carefully designed and is a unique succession of endfire coax, balanced stripline, unbalanced microstrip, and finally balanced parallel-plate topology for most of the antenna flared region. It was expressly designed to provide a smooth transition from the coaxial port to balanced antenna currents on the antenna plates, with minimal variation in surge impedance through the feed transition region.

The Teflon™ (polytetrafluoroethylene) lens was designed with straight-forward geometrical equal-path ray tracing, with a planar inner face and a convex aperture or outer face. Then a 3D spherical curve-fit was performed so that the resulting planar-spherical lens could be easily made on a CNC machine.

After two nominally identical printed-board antennas were assembled, the time-domain radiation performance was measured in a boresight transmit-receive configuration, with step generator and 20-GHz sampling oscilloscope. Compared to our older TEM horns, this new antenna offered several times better gain or sensitivity, but the step-equivalent risetime was much too slow (on the order of 100 ps). The increased main-beam gain is directly due to the increased plate separation or aperture height, which increases the aperture area of the antenna. The poor time-domain response is surmised to be due to pulse smearing of the currents in the antenna conductors due to the direct presence of the printed-wiring board dielectric substrate. This has the effect of slowing the propagation velocity of the antenna current on the substrate side (inner surface) of the copper traces. On the other hand, the current on the outer/air side of the copper traces propagates at the speed of light, and the resulting combination of these two current components suffers pulse spreading due to different overall propagation times along the antenna conducting plates.

SOLID-PLATE ANTENNA CONFIGURATION

After we found out that the transient response of the printed-board configuration was too slow, a simpler version was designed and retrofitted into the same polystyrene antenna body. This second configuration consists of solid metal plates in place of the single-sided printed-wiring boards, in the attempt to remove the board substrate and its' pulse-spreading effect. Figure 3 shows both TEM horns, after retrofitting with the metal-plate configuration for the antenna conductors.

Figure 3. Solid-plate configuration of TEM-horn antennas.

Figure 4 shows the top and bottom conductors of the solid-plate configuration, with a very clean geometry. The triangular conductors are full length, with no resistive loading; the lenses are the same ones used for the printed-board configuration. Figure 5 then shows a close-up view of the feedpoint region. This consists of a simple, but carefully constructed transition from transverse-fed coaxial line to unbalanced parallel-plate which smoothly transforms to the balanced parallel-plate antenna structure. In past antenna design efforts, a variety of coax-to-parallel-plate transitions have been constructed and evaluated. This transverse-fed topology has worked the best for short-duration time-domain operation at low voltage levels. Attempts were made to build both antennas with identical feedpoint characteristics, and step time-domain reflectometry was used to empirically adjust the

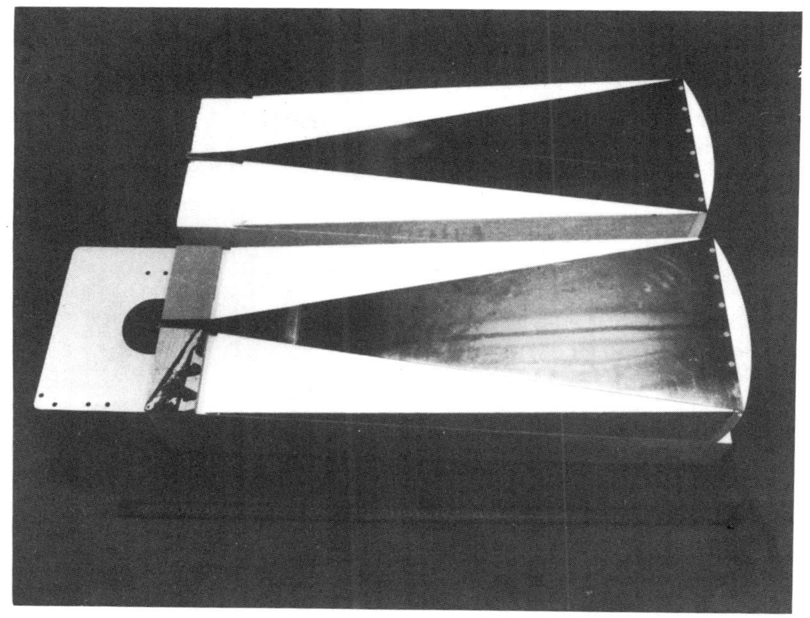

Figure 4. Top and bottom conductors of solid-plate antenna configuration.

Figure 5. Close-up view of feedpoint region of solid-plate antenna configuration.

feedpoint conductor spacing in order to have the same surge impedance variation in each antenna.

After this pair of TEM-horn antennas was assembled, they were extensively used for time-domain electromagnetic measurements. After about seven months of use, we re-measured the boresight/main-beam time-domain performance of the antenna pair in a transmit-receive configuration. Figure 6 shows the indoor antenna setup, with both TEM horns boresight aligned for main-beam response. The transmitting apparatus is at the rear of the photo, aimed at the receiving equipment, which is nearest to the camera. The aperture separation was 2.00 m, and the transmitting antenna was excited with a very fast step waveform. The receiving system consisted of a 20-GHz sampling oscilloscope. The measurement system step response had a 10-90% transition duration (risetime) of 29 ps.

Figure 6. Experimental setup to measure boresight time-domain response of solid-plate antennas.

Figure 7 shows the two measured waveforms for this characterization. The step waveform is that of the measurement system, without the antennas. The impulse waveform is the received waveform with the double-antenna configuration; this clearly shows the derivative transmitting response of a TEM-horn antenna (the receive antenna has a replica response - the time-integral of the transmitting response). The time alignment of the two waveforms is arbitrary, as there is a 2-m propagation delay between the antennas.

Figure 8 shows the same system step waveform and a time-integration of the antenna response waveform. Using simple quadrature analysis of the transition duration (10-90% risetime), based on the assumption of Gaussian system components, the degradation or slowing of the transition duration can be unfolded from the response risetime. The observed or measured risetime, $t_r(mmt)$, is the squared sum of the each antenna's 'step-equivalent' risetime, $t_r(ant)$, and the risetime of the measurement system, $t_r(sys)$:

$$t_r^2(mmt) = 2\, t_r^2(ant) + t_r^2(sys).$$

This assumes that the two antennas are very similar in radiation performance, which has been verified. Solving for the antenna step-equivalent transition duration,

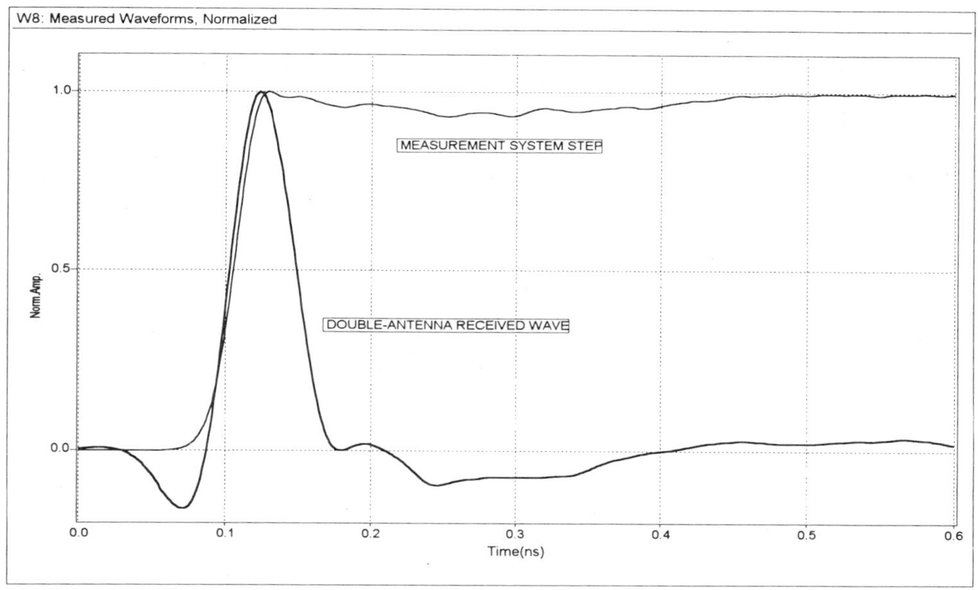

Figure 7. Boresight time-domain response of solid-plate antennas.

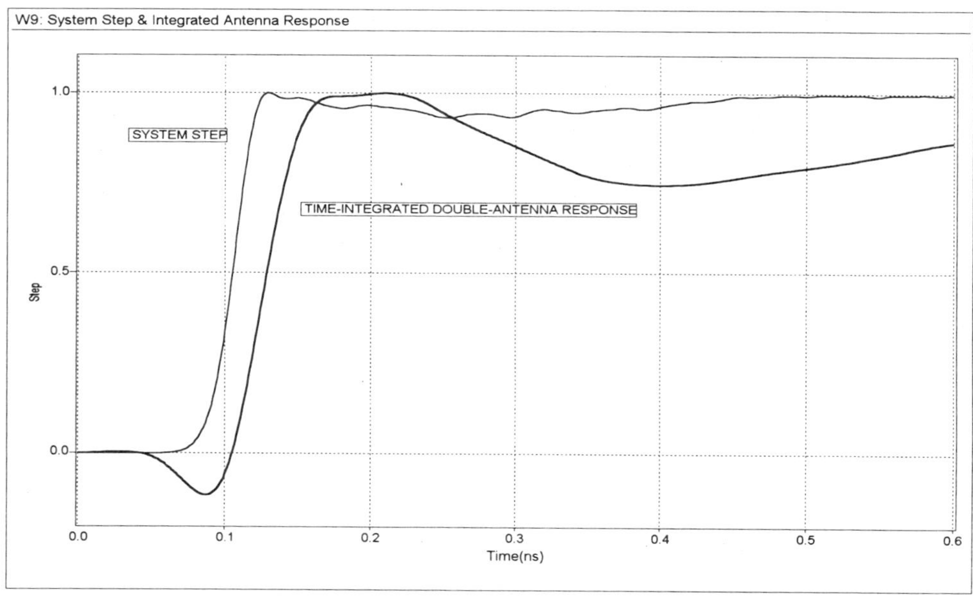

Figure 8. Time-Integrated 'Step-Equivalent' boresight time-domain response of solid-plate antennas.

$$t_r(ant) = \sqrt{\tfrac{1}{2}\left[t_r^2(mmt) - t_r^2(sys)\right]}$$
$$= \sqrt{\tfrac{1}{2}\left[(40.7ps)^2 - (28.9ps)^2\right]}$$
$$= 20.3ps.$$

This is the fastest transient response of any time-domain antenna which we have designed and built in our department.

SUMMARY

We designed and constructed a pair of TEM-horn antennas specifically for very fast time-domain boresight response. Two physical topologies were made: a printed-board configuration, and a solid-plate configuration. The printed-board configuration has much slower transient response, which we think is due to pulse-smearing of the antenna currents in the dielectric substrate of the printed-wiring boards. The solid-plate version has a 20-ps transition-duration response in the main-beam endfire (boresight) direction, which is the fastest (or highest bandwidth) we have seen to date. And, since the antenna has a round-trip antenna-current propagation time of 6 ns, it offers clean radiated electromagnetic-field measurement capability with a clear time of several ns. The printed-board version has resistive loading at the aperture end of the conductors, which should offer better low-frequency performance (time has not permitted studying this issue however). The dielectric lens certainly does improve the transient performance of the TEM horn, and was simple to design. In the future, we plan to study possible enhancements to either configuration version.

OPTIMIZED TEM HORN IMPULSE RECEIVING ANTENNA

Michael A. Morgan and R. Clark Robertson

Electrical and Computer Engineering Department
Naval Postgraduate School
Monterey, CA 93943-5121

INTRODUCTION

Impulse receiving antennas are usually designed for ultra-wideband applications involving high-fidelity sensing of incident impulsive electromagnetic fields. These antennas may be utilized in advanced radar and information warfare systems, such as in the receive portion of impulse radar systems, or as stand-alone sensors for ultra-wideband direction finding. Impulse receiving antennas can also be used for measurement of field strength and waveform type being produced by high-power microwave sources or by natural and man-made EMP generators.

Due to their dispersive characteristics conventional wideband antennas, such as log-periodic dipole arrays and spiral type antennas, perform poorly as high-fidelity sensors for wideband impulsive waveforms.[1] Loaded dipole antennas were one of the earliest impulse sensors to be developed for omnidirectional situations.[2] The traveling wave TEM horn forms a natural structure for low-dispersion launching and receiving of ultra-wideband impulses in applications where sensitivity and direction-of-arrival are important. Extensive engineering design of loaded TEM horns has been conducted since the late 1970's at the National Bureau of Standards, later renamed National Institute of Standards and Technology (NIST).[3,4]

Results from the NIST research were used at the Naval Postgraduate School (NPS) in 1993 to design a high-performance loaded TEM horn antenna which we will denote as TEM-1. Specifics of the TEM-1 design, its measured performance, and detailed descriptions of a new measurement procedure and laboratory facility were presented at the Ultra-Wideband Short-Pulse EM-2 Symposium.[5]

New requirements drove a redesign of the antenna to achieve the following:

(1) Design to fit within a 18cm $\times$ 23cm $\times$ 76cm cubical volume;
(2) Display a usable passband of 10 MHz to 8 GHz for $Z_{in} = 50\Omega$;
(3) Provide high-fidelity reception of impulsive waveforms having duration <100pS;
(4) Optimize directivity and sensitivity given the size limitation in (1).

Second-generation receiving antennas, named TEM-2A and TEM-2B, were designed, constructed in pairs, and tested. The A-model employs conventional discrete chip resistor loading while the B-model uses tapered resistive coatings, as illustrated in Fig. 1. Both versions have a tapered impedance flare section and offer improved low-frequency performance, bandwidth, and sensitivity beyond that of the linear-tapered TEM-1 antenna. A 2:1 differential balun, which is not shown in the figure, is connected to the terminal feed points at the apex of the horn antennas. This balun transitions the nominal 100Ω balanced configuration of the horn feed to a 50Ω coaxial feed.

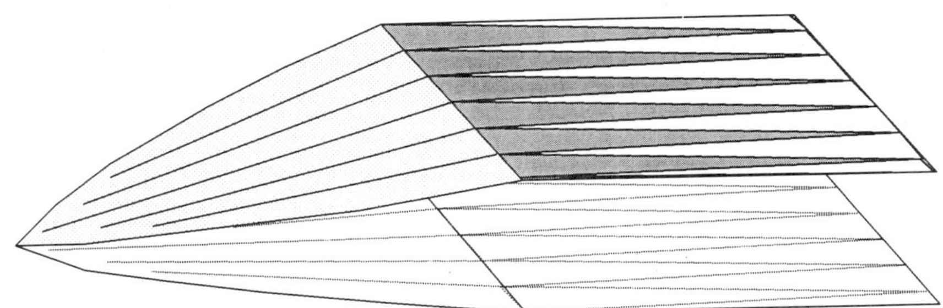

Figure 1. Second-Generation TEM-2B Horn Antenna

In preparation for discussion of design considerations for high-fidelity impulsive signal reception, we will now review the definition of antenna "sensitivity."

ANTENNA SENSITIVITY

The sensitivity function of a receiving antenna can be considered as the frequency domain transfer function relating terminal load voltage to incident electric field. Sensitivity is determined by the frequency dependence of the antenna's effective length and the impedance match it has with the receiver input impedance.

To introduce effective length, first consider the transmitting case, whereby the far-zone $\overline{E}$-field is being produced at a distance r in the direction of the unit vector $\hat{r}$ by a referenced input current having Fourier transform $I_{in}(\omega)$. The transmitted field can be used to define a complex effective length $\overline{L}_T(\hat{r}, \omega)$, given by[6]

$$\overline{E}(\overline{r}, \omega) = \frac{jk_0\eta_0}{4\pi r} e^{-jk_0 r}\, I_{in}(\omega)\, \overline{L}_T(\hat{r}, \omega) \tag{1}$$

where $k_0 = \omega/c$ is the wavenumber and $\eta_0 = 120\pi\Omega$.

Consider now the same antenna operating as a receiver, as shown in Fig. 2. In this role, the antenna acts as a transducer, converting an incident plane-wave field having $\overline{E}^i(\omega)$ into an open-circuit voltage, $V_{oc}(\omega)$ across the terminals. This V_{oc} drives the series antenna and load impedances in a Thevenin circuit voltage divider. The receiving complex effective length, $\overline{L}_R(\hat{r}, \omega)$, is defined as the linear transfer function which relates V_{oc} to $\overline{E}^i$

$$V_{oc}(\omega) = \overline{L}_R(\hat{r}, \omega) \cdot \overline{E}^i(\omega) \tag{2}$$

The Lorentz reciprocity relationship and definitions above can be used to show that[6]

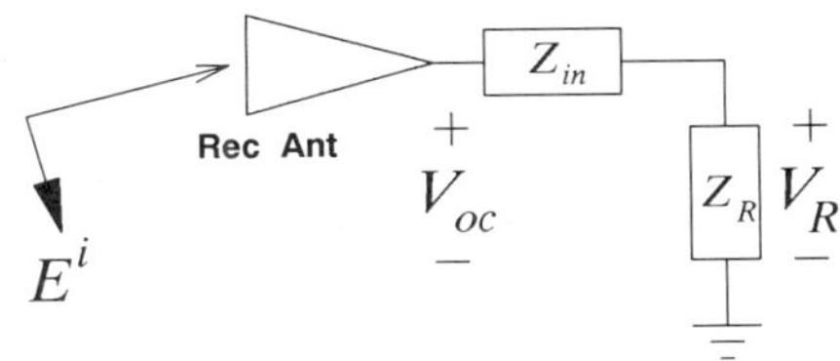

Figure 2. Receiving Antenna

$$\overline{L}_R(\hat{r},\omega) = \overline{L}_T(\hat{r},\omega) = \overline{L}(\hat{r},\omega) \tag{3}$$

The equality of transmitting and receiving effective lengths is used as the basis for deriving the more common reciprocity relationships between the antenna power transfer quantities of effective area, A_e, and power gain, G. Denoting R_{in}=Re$\{Z_{in}(\omega)\}$ as the antenna input resistance, this relationship becomes, with λ_0=wavelength,

$$A_e(\hat{r},\omega) = \frac{\eta_0}{4R_{in}} \, |\overline{L}(\hat{r},\omega)|^2 = \frac{\lambda_0^2}{4\pi} \, G(\hat{r},\omega) \tag{4}$$

The antenna effective area provides the power transfer between incident Poynting vector power flux density P^i and received power W_{rec} into a matched load, Z_L=Z_{in}^*, as given by W_{rec}=$A_e \; P^i$

Consider again the situation shown in Fig. 2, where the receiving antenna is connected to a receiver having load impedance Z_R. Antenna sensitivity, $\overline{H}_R$, is defined as the transfer function relating incident $\overline{E}^i$ and voltage across the receiver input impedance, Z_R,

$$V_R(\omega) = \overline{H}_R(\hat{r},\omega) \cdot \overline{E}^i \tag{5}$$

where, in terms of effective length

$$\overline{H}_R(\hat{r},\omega) = \frac{Z_R(\omega)}{Z_R(\omega) + Z_{in}(\omega)} \, \overline{L}(\hat{r},\omega) \tag{6}$$

DESIGN CONSIDERATIONS

Some important facts result from the relationships in the previous section:

(1) High-fidelity impulse reception, according to (5), requires a non-dispersive sensitivity, where the magnitude of the complex $\overline{H}_R(\hat{r},\omega)$ is nominally constant within the passband of the impulse while its phase is a linear function of frequency (having the form $-\omega t_0$, with t_0 representing a time delay).

(2) Conventional wideband CW antennas such as log-periodic arrays, spirals, and ridged horns are usually designed to provide constant gain, $G(\hat{r},\omega)$, and with a dispersive, non-linear phase characteristic.

(3) An ideal impulse receiving antenna will exhibit gain, via (4), which increases with the square of the frequency. In the time-domain, an ideal impulse receiving antenna will act as a differentiating impulse transmitting antenna.

The traveling-wave configuration of the resistively-loaded TEM horn forms a natural vehicle for design of ultra-wideband impulse receiving antennas. Longitudinal insulating slots, as shown in Fig. 1, are used to suppress inadvertent creation (due to imperfect construction) of TE modes having transverse directed currents.

A parallel-plate loading section serves to dampen large variations of the sensitivity at lower frequencies caused by standing wave resonances. An impedance profile which increases with distance from the flared section is employed to gradually absorb remnant current which remains beyond the radiation process taking place in the flare section. A tapered resistive coating is used with TEM-2B, in lieu of the more conventional discrete resistor loading. This has advantages of ruggedness, ease of construction, and more uniform absorption.

Another unique aspect of both TEM-2 antennas is the flare section which forms a smooth impedance transition to match the 100Ω input feed to a 158Ω (under lossless conditions) parallel plate loading section. This design feature was used to maximize the plate spacing within the volume allocated, and thus optimize sensitivity, while meeting the width constraint.

Although empirically-based formulas developed at NIST[3] were employed in the design of first-generation TEM horns previously reported[1], it became important to have an accurate numerical model for discerning the effects of new features and for optimizing performance while honoring the size constraints imposed. A custom, large-memory version of a commercial method-of-moments wire-grid analysis program was employed.[7] The plan view of the 3-D wire-grid array for TEM-2 is shown in Fig. 3. A useful feature of the modeling software is stepped-frequency operation with adaptive subdomain definition and ability to use discrete or distributed loads.

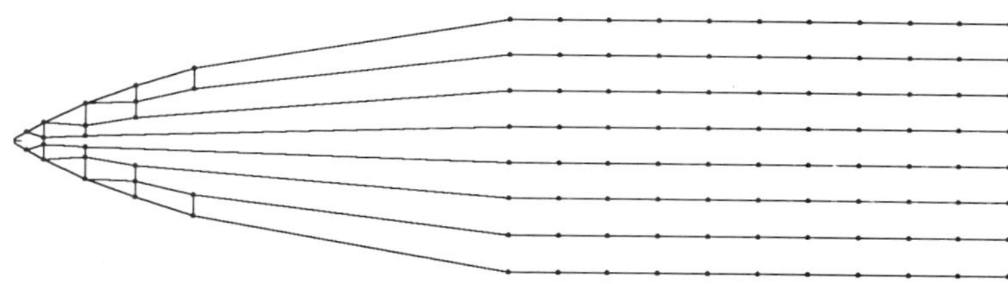

Figure 3. TEM-2 Wire-Grid Geometry Plan View

Stepped-frequency computations spanning 10 MHz to 8 GHz were used to estimate the antenna sensitivity on boresight when feeding a hypothetical Z_R=100Ω balun load impedance. Computed sensitivity performance, including low and high-frequency roll-offs, was monitored as geometrical dimensions and resistive load tapering were varied about those found using the baseline empirical design procedure employed with TEM-1. This iterative procedure was used to approach an optimized design meeting the specifications.

MEASUREMENT PROCEDURES

Prototype horns (TEM-2A with chip resistors and TEM-2B with resistive coatings) were each fabricated in matched pairs. Initial $Z_{in}(\omega)$ estimates were made using measured time-domain reflectometer (TDR) waveforms corresponding to short-circuit, matched load (50Ω) and antenna load (with balun connected). To quantify the influence of the parallel-plate load section, measurements were also made using partially constructed TEM-2B antennas which lacked the loaded sections. Figure 4 shows overlays of TDR-based $Z_{in}=R_{in}+jX_{in}$ measurements for the unloaded and loaded TEM-2B antenna. As expected, the load section primarily influences Z_{in} at low frequencies, where it acts to absorb remnant currents which have not been radiation damped in the flare section. This tends to mitigate reflections which cause resonant undulations of the input impedance and radiation/reception properties. Note the diminishing effect of the load section at higher frequencies, where efficient wave launching (and corresponding radiation damping of currents) takes place in the flare section.

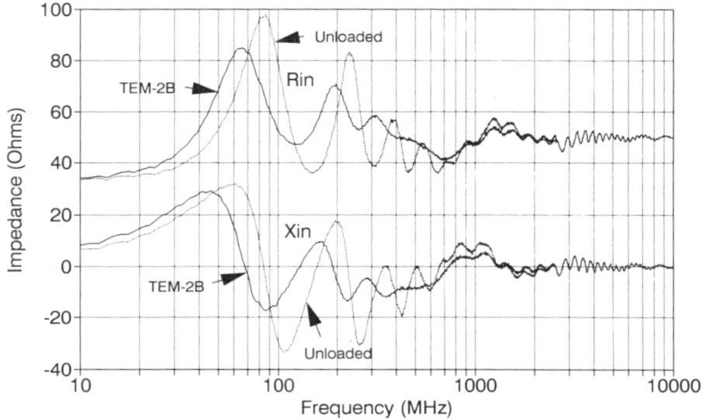

Figure 4. TDR Measured Input Impedance for Loaded and Unloaded TEM-2B

Receiving antenna performance evaluations were conducted in an anechoic chamber using a low-power dc-50 GHz step-waveform source and a 50 GHz bandwidth sampling oscilloscope receiver, as depicted in Fig. 5. Measurements were made using one of the identical antennas in each pair as the transmitter and the other antenna in that same pair as the receiver. Antenna apertures were separated by r= 3.89 m.

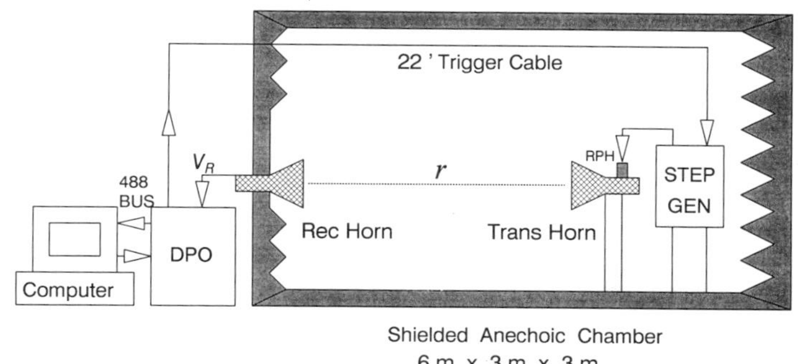

Figure 5. Impulse Antenna Laboratory

The generator's remote pulse head (RPH) is triggered by the 0.25V step output of the digital processing oscilloscope (DPO). The RPH drives the transmitting antenna with a –9 volt, negative step waveform, $v_T(t)$, shown in Fig. 6.

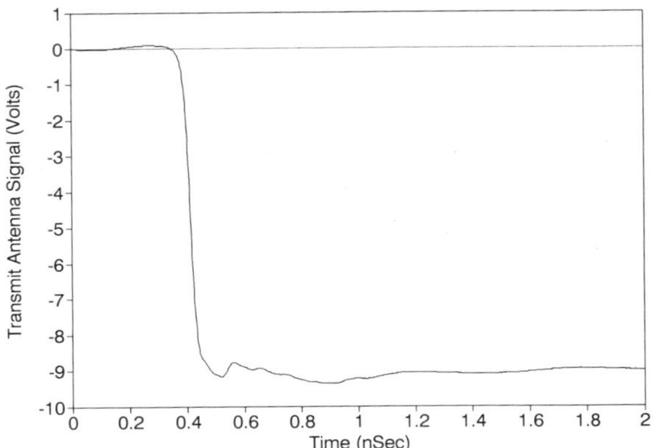

Figure 6. Measured Transmit Antenna Input Signal

DPO sampled waveforms are ensemble averaged 256 times and sent by IEEE-488 bus to the controller (i486-80 based computer). Received signals, $v_R(t)$, using TEM-2B transmitting and receiving antennas are displayed in Figs. 7 and 8. Received waveforms for E-plane (vertical plane in Fig. 1) incident field directions of 10°, 20° and 30° are compared in Fig. 7 to those of 0° (boresight) and 180° (backfire incidence). Figure 8 overlays measured waveforms for boresight and H-plane incidence of 10°, 20°, 30° and 90° (broadside). Note the expected progressive degradation of received pulse amplitude and high-frequency performance (via increased pulsewidth) with increasing arrival angles away from boresight. Degradation versus off-boresight arrival angle appears to be similar for both planes of incidence for the TEM-2 design which has a close to square aperture face.

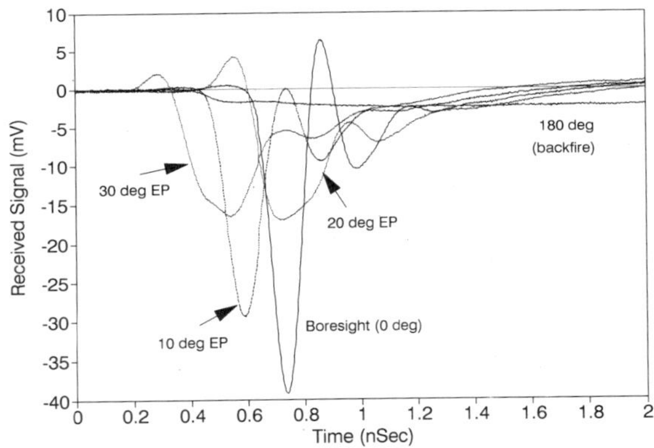

Figure 7. Measured TEM-2B Receive Signals for E-Plane Incident Angles

Acquired waveforms were post-processed using custom MatLab software to estimate the receive antenna sensitivity. The estimation employs a new formulation for sensitivity which was derived in our previous work.[1,5] For the special case being considered, which assumes identical transmitting and receiving antennas, the result becomes especially simple,

$$H_R(\omega) = \sqrt{\frac{2\pi c r Z_0}{\jmath\omega\eta_0}} \sqrt{\frac{V_R(\omega)}{V_T(\omega)}} e^{\jmath\frac{\omega r}{2c}} \quad . \tag{7}$$

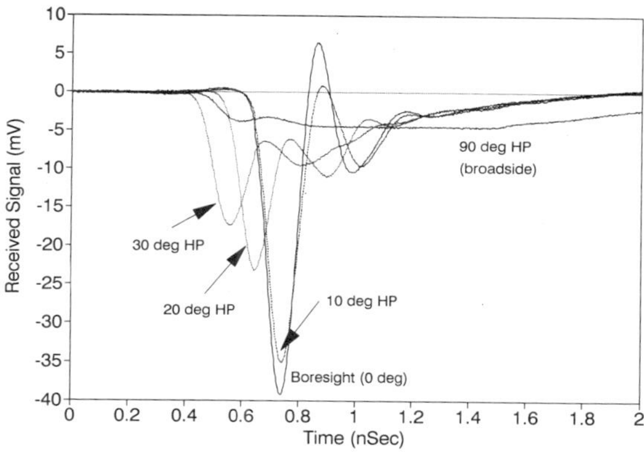

Figure 8. Measured TEM-2B Receive Signal for H-Plane Incident Angles

where $V_T(\omega)$ and $V_R(\omega)$ are the computed spectra of transmit and receive voltages when measured feeding a standard Z_0=50Ω load. The exponential term provides a phase shift to compensate for the time shift implied in the square-root of the voltage ratio.

Processed estimates for antenna sensitivity are overlayed in Fig. 9 for the partially constructed unloaded version of TEM-2B and the two loaded TEM-2 versions. The chip resistor (A) and resistive coating (B) models provide similar performance over the lower portion of the passband shown. The A-version displays notches near 500 MHz and 600 MHz, but also appears to offer slightly better high-frequency performance beyond 8 GHz. Improved high-end performance of TEM-2A may be due to lot differences in the custom baluns used in the construction of the two antenna pairs that took place months apart. As with input impedance, the major effect of loading is to improve low frequency performance. This can be seen clearly in the figure. The unloaded TEM-2 also displays a deep performance notch at about 380 MHz. Beyond 2 GHz the loaded and unloaded TEM-2B antennas are virtually identical. This indicates that loading is not an important factor in high-frequency performance.

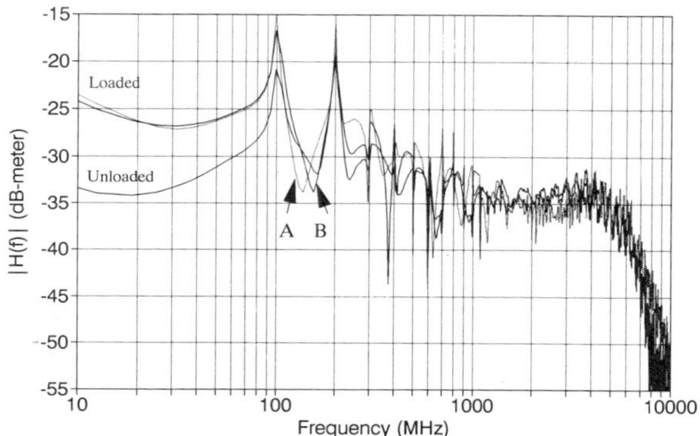

Figure 9. Measured TEM-2 Boresight Sensitivity: Comparing Unloaded Antenna to Loaded Versions A (Chip Resistor) and B (Resistive Coating)

CONCLUSIONS

Second-generation impulse receiving antennas which meet challenging specifications have been designed, constructed and tested. These antennas employ tapered flare transitions to allow maximum aperture height and thus optimize sensitivity under constrained physical dimensions. A chip resistor model was first constructed and evaluated, followed by a new design employing tapered resistive coatings.

Wire-grid numerical modeling was employed to optimize the physical design, including flare transition and loading profiles for chip resistors and resistive coatings in the two models.

Physical intuition and measurements indicate that loading has a substantial effect in improving low frequency performance of these traveling wave structures by absorbing currents which would reflect at the antenna aperture.

Further performance gains in the low frequency regime, where these antennas are electrically small, may be possible through use of active impedance transformation. Research is required to design such active devices to provide this characteristic at low frequencies while maintaining a low-loss direct path for high frequencies.

REFERENCES

1. R. C. Robertson and M. A. Morgan, "Ultra-wideband impulse antenna study and prototype design," *Naval Postgraduate School Technical Report*, NPSEC-93-010, March 1993.
2. M. Kanda, "A relatively short cylindrical broadband antenna with tapered resistive loading for picosecond pulse measurements," *IEEE Trans. on Antennas Propagation*, vol. AP-26, pp. 439–447, May 1978.
3. A. R. Ondrejka, J. M. Ladbury, and H. W. Medley, "TEM horn antenna design guide," National Institute of Standards and Technology, unpublished report.
4. M. Kanda, "The effects of resistive loading of 'TEM' horns," *IEEE Trans. on Electromagnetic Compatibility*, vol. EMC-24, pp. 245–255, May 1982.
5. R. C. Robertson and M. A. Morgan, "Ultra-wideband impulse receiving antenna design and evaluation," *Ultra-Wideband Short-Pulse Electromagnetics – 2*, L. Carin and L.B. Felsen, eds., New York: Plenum, 1994.
6. J. D. Kraus, *Antennas, Second ed.* New York: McGraw-Hill, 1988.
7. A. R. Djordjevic, M. B. Bazdar, G. M. Vitosevic, T. K. Sarkar, and R. F. Harrington, *Analysis of Wire Antennas and Scatterers,* Software and Manual, Boston: Artech House, 1990.

TRANSIENT ARRAYS

Carl E. Baum

Phillips Laboratory
3550 Aberdeen Ave SE
Kirtland AFB, NM 87117-5776

1. INTRODUCTION

Analogous to phased arrays with narrowband excitation one can have timed arrays for radiating transient pulses, the role of phase being replaced by the time shifts between the application of temporal waveforms (or one common waveform) to the various array elements. Our concern here is for such arrays to operate in transmission and/or reception over very large band ratios (ratio of upper frequency to lower frequency of interest), similar to other forms of impulse radiating antennas (IRAs) using reflectors or lenses[29, 30]. One might think of such transient arrays as array IRAs.

Such arrays are an extension of those studied, and in some cases realized, for simulation of the nuclear electromagnetic pulse (EMP)[13, 27]. Such an array has been referred to as a distributed source[3, 4] or a distributed switch[14]. In this case, the approach consists of synthesizing the TEM mode (planar or spherical, in general inhomogeneous) over some aperture surface serving as the electrical source for a cylindrical or conical transmission line. By use of such a technique one can suppress the generation of higher order (E and H) modes up to frequencies limited, not by wavelengths of the order of (or larger than) the transmission-line-conductor spacing and width, but by wavelengths of the order of (or larger than) the element spacing in the array forming the distributed source. The reader should note that such arrays are comprised of interconnected elements which allow for current continuity through the array, this being essential for adequate low-frequency performance. In contradistinction to the case of many narrow-band arrays in which the mutual interaction of the array elements is made (or assumed) small, the present arrays are designed so that the mutual interaction is strong and is an integral part of the array operation, at least for frequencies with wavelengths of the order of and larger than the element spacing.

The motivation for developing such arrays to drive EMP simulators has been the desire to go to higher and higher voltages while retaining a sufficiently small risetime in the pulse. As one goes to higher voltages (MV) on a single switch the risetime increases to the point where one considers using multiple switches at lower voltages. These switches then need to be distributed over an appropriate aperture to synthesize the desired TEM wave. Note that the risetime is influenced not only by the switch size, but also by how these switches are integrated into the array-element design, and how small is the jitter in the timing of these switches as compared to the ideal (desired) switching time.

An important approach to synthesis of a transient array is an array of flat-plate conical transmission lines each launching a spherical TEM wave from a small source, these waves combining on an aperture plane feeding a parallel-plate cylindrical (or conical) transmission line[1] illustrated in fig. 1. Some improvement in the aperture synthesis is obtained as the individual wave launchers are lengthened to make the spherical waves better approximate a plane wave on the aperture plane. One can have individual pulsers at the apices of the individual-element conical transmission lines, or one can feed various numbers of such source points from one or more common pulsers via transmission lines[1] as indicated by the example in fig. 2. Various combinations of series and parallel connections with appropriate matched delays (transit times in the transmission lines) are possible. Later papers[2–4, 6–10, 12–16, 23] have considered more details of such transient arrays in the context of EMP simulators.

Such arrays have been realized in EMP simulators. The large ATLAS I for testing large aircraft[13, 27] has a two-element array in a series configuration (fig. 1) with the two adjacent conical plates connecting at the aperture plane being larger than the two outermost so as to form a central-ground-plane wedge and prevent coupling between the two launchers until they meet at the aperture plane[1]. The SIEGE simulation concept[13]

Ultra-Wideband, Short-Pulse Electromagnetics 3
Edited by Baum *et al.*, Plenum Press, New York, 1997

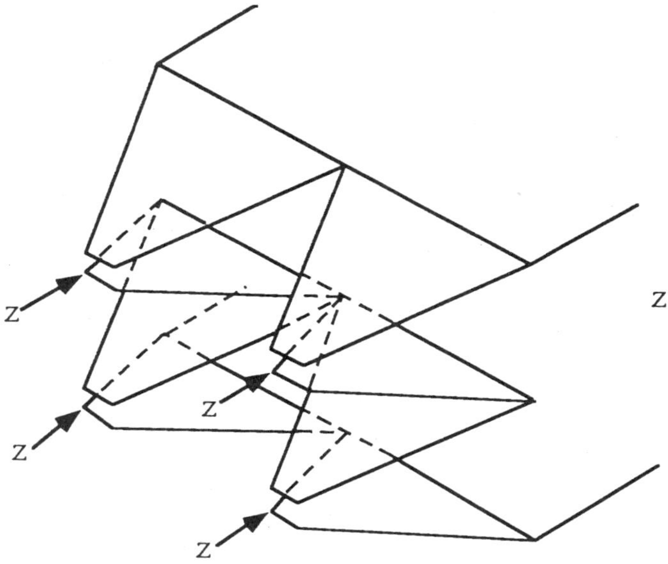

Figure 1. Multiple Conical Transitions.

replaces one of the plates of a two-parallel-plate waveguide by the earth surface for testing buried systems. This has been realized using a four-parallel-element array in which the long conical plates are bent in a contour to account for mutual interaction before the aperture plane and thereby maintain a constant characteristic impedance along each of the four wave launchers[2]. Another type of simulator for buried systems is DISCUS[4, 13] in which the array is attached to the ground surface for driving the fields into the earth, thereby introducing additional matching problems at the ground surface. Such wave launchers have been designed and constructed involving Brewster-angle and transmission-line techniques[15, 23]. In one experiment a twelve-element array (100 kV pulser per element, dimensions 1 m in the direction of the electric field and 2 m in the direction of the magnetic field), connected in series with fiber-optic signals to trigger each pulser module, produced about 70 kV/m with 7 ns risetime in the soil[16].

2. EARLY CONSIDERATIONS FOR RADIATING TRANSIENT ARRAYS

There was early recognition[5] that such arrays were also suitable for radiating transient pulses, i.e., without an additional waveguiding structure (cylindrical or conical transmission line). In this case, the object was to see what kinds of pulses could be sent to distances far from the radiating source array using the kinds of pulse power technology in EMP simulators.

Approximating large arrays as infinite for initial considerations (theoretical simplification) one can pay

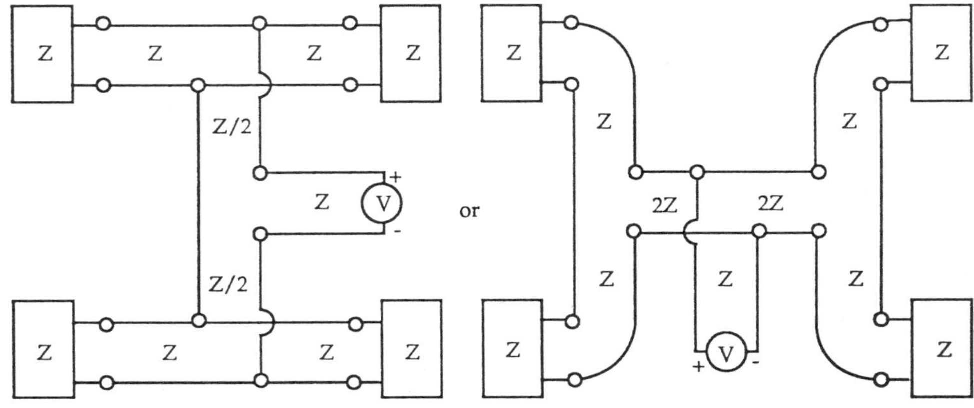

Figure 2. Single Signal Feed Transmission Line Networks for Multiple Conical Feed Points.

attention to the details of the unit cells (individual elements) in an array of identical unit cells which is periodic in two dimensions. Thus one can have waves propagating in each cell, including interaction with other cells, without including effects associated with array truncation (edge effects). A section of such an array with rectangular unit cells[5] (height w_1, width w_2) comprised of conical transmission lines is illustrated in fig. 3. This is but one of various types of unit cells that one might consider for array elements. The two-dimensional translation group T_2 admits five kinds of parallelogram systems for the boundaries of the unit cells[25, 31]. One can adjoin compatible rotations and reflections in the unit cells to give the two-dimensional space groups E_2 with a richer symmetry structure. Some of the more interesting types of unit cells are illustrated in fig. 4 based on squares, regular hexagons, and equilateral triangles[5]. In this case, the unit cells are configured such that by changing electrical connections to the sources one can achieve multiple polarizations.

The early-time (or, equivalently, high-frequency) performance of an infinite array of conical wave launchers can be found by first considering the same performance regime of a single conical element[11, 17]. As indicated in fig. 5 consider a rectangular array of elements with source points (conical apices) on the $z = 0$ plane. Letting one element have conical apex at $\vec{r} = \vec{0}$ with voltage excitation $V_o\, u(t)$, the early-time field is described by the TEM field

$$\vec{E}(\vec{r},t) = -\frac{V_o}{r}\,\vec{F}(\theta,\phi)\,u\!\left(t-\frac{r}{c}\right) \quad , \quad \vec{F}(\theta,\phi) = \nabla_{\theta,\phi}\, f(\theta,\phi) \tag{2.1}$$

where $f(\theta,\phi)$ is the potential function and $\nabla_{\theta,\phi}$ is the gradient on the unit sphere in the usual spherical coordinate system. The form that $f(\theta,\phi)$ takes depends on the detailed shape of the conical-transmission-line elements. Detailed calculations have been performed[11, 17] for planar bicones and flat-plate cones such as in fig. 3.

As one would expect as one makes the length ℓ of the conical transmission line larger than the opening w_1 (plate separation at the aperture plane where the individual plates connect to adjacent ones) the field at some distance r continues to increase. On the center line (the z axis) we have

$$\vec{F}(\theta = 0\,,\ \phi) \simeq \frac{\ell}{w_1}\,\vec{E}_{rel} \tag{2.2}$$

as the angle between the plates decreases, $\vec{E}_{rel}$ being the electric field at the aperture plane on the z axis for 1 Volt between the plates (now approximately parallel). For wide plate ($w_2 >> w_1$) $\left|\vec{E}_{rel}\right| \simeq 1/w_1$. Note, however, that as ℓ / w_1 increases the time for which (2.1) is valid decreases due to the earlier arrival of the diffraction from the end of the conical plate (at the aperture plane). Furthermore, the behavior as in (2.1) being restricted to angles (θ,ϕ) lying between the plates (for far fields due to the presence of adjacent wave launchers), then large ℓ / w_1(or ℓ / w_2) restricts θ to angles near 0, an important consideration in the context of array scanning.

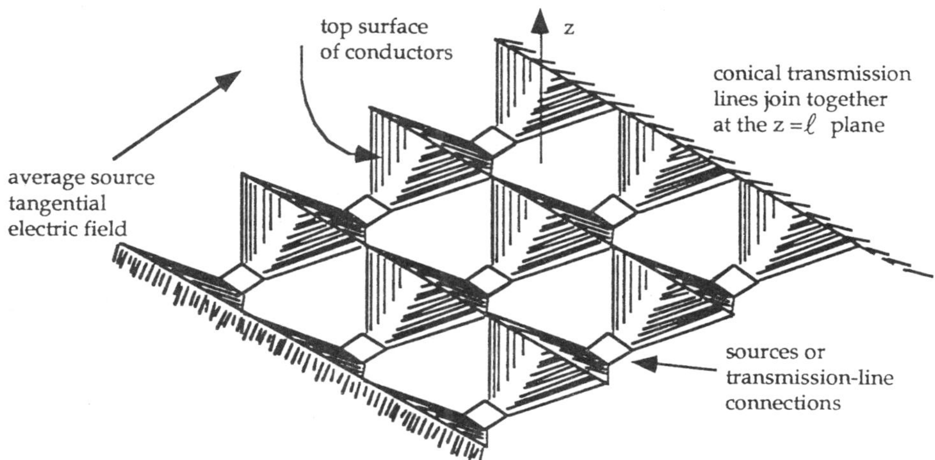

Figure 3. Non-Planar Conical-Transmission-Line Array.

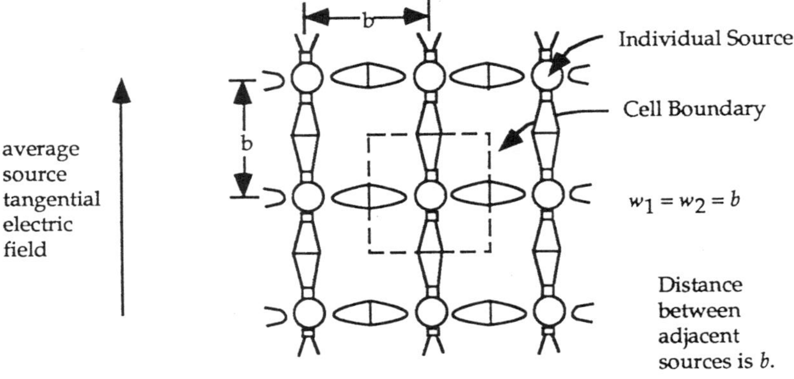

A. Square Cell Geometry

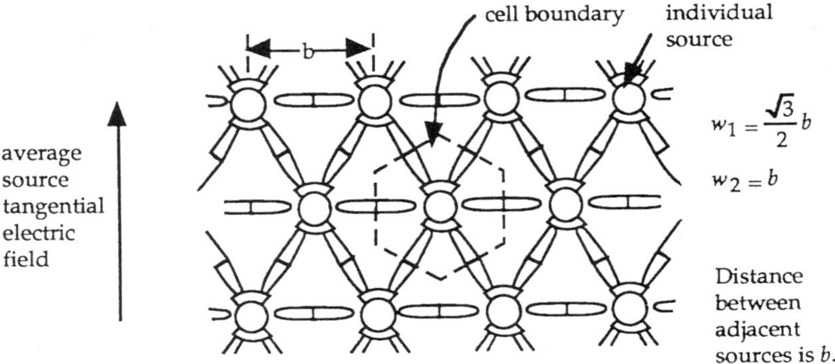

B. Regular hexagonal cell geometry

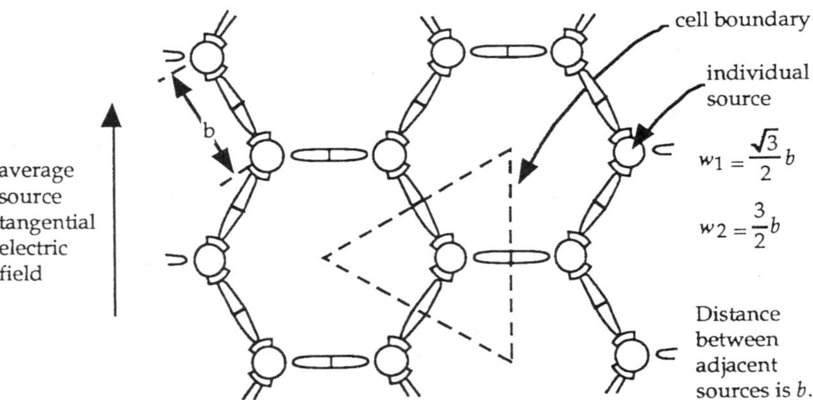

C. Equilateral triangular cell geometry

Figure 4. Arrays For Changing Polarization: Top Views.

Now, assuming an infinite array, we need to sum over the early-time signals of the individual elements. Let $\vec{1}_1$ define the direction propagation of a plane wave and define retarded time by

$$t_r \equiv t - \frac{\vec{1}_1 \cdot \vec{r}}{c} \quad , \quad c = \text{speed of light} \tag{2.3}$$

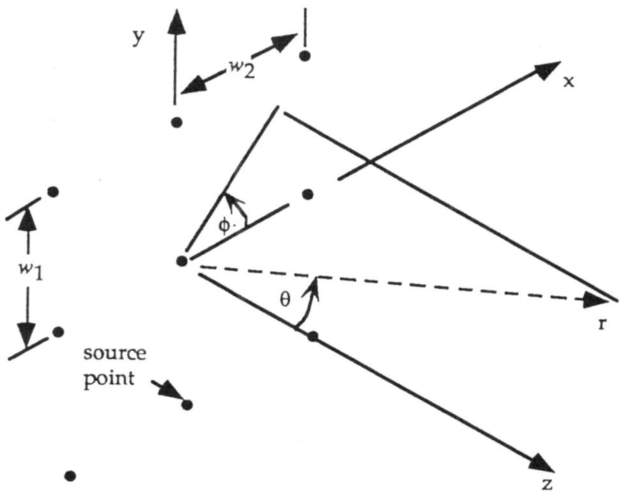

Figure 5. Rectangular Array of Spherical TEM Elements.

The individual source points in the $z = 0$ plane are turned on (with $V_0\ u(t_r)$) at zero retarded time at each source point. Then consider the field propagating in the $\vec{1}_1$ direction for large r with

$$\cos(\theta_1) = \vec{1}_1 \cdot \vec{1}_z > 0 \tag{2.4}$$

Restrict $\vec{1}_1$ (make θ_1 small enough) that (2.1) is valid for the individual elements for a window of retarded time

$$0 \le t_r < t_{c_f} \equiv \text{clear time in the far field (after which finite launcher dimensions can be observed)} \tag{2.5}$$

where t_{c_f} is a function of (θ_1, ϕ_1), being maximum (for symmetrical wave launchers) with $\theta_1 = 0$. By considering some t_r slightly greater than zero and letting $r \to \infty$ in the $\vec{1}_1$ direction more and more sources are seen by the observer. Summing these[11] over the expanding ellipse on the aperture plane (as $r \to \infty$) gives a far electric field for early times as

$$\begin{aligned} \vec{E}_{f_0} &= -\frac{2\pi V_o}{A_e \cos(\theta_1)} \vec{F}(\theta_1, \phi_1)\ ct_r\ u(t_r) \ \text{ for } t_r < t_{c_f} \\ A_e &= w_1 w_2 \equiv \text{area of unit cell of array element} \end{aligned} \tag{2.6}$$

Note that in the limit the step has become a ramp function. This result applies to infinite arrays for which the far field does not decrease with r for constant retarded time. (This will be modified later for finite-size arrays.) The above result applies not only to rectangular arrays but other shapes in fig. 4 as well, with w_1 and w_2 appropriately interpreted.

The array performance can also be calculated for late times or low frequencies for which the wavelengths are large compared to element dimensions[11]. In this case one considers the average tangential electric field along the array as

$$\vec{E}_t = -\vec{1}_s \frac{V_o}{w_1} \tag{2.7}$$

where $\vec{1}_s$ can be considered as $\vec{1}_y$ in fig. 5 for convenience. Then with appropriate element symmetry[11] the late-time far field is (for step excitation)

$$\vec{E}_{f\infty} = \begin{cases} -\dfrac{V_o}{w_1 \cos(\theta_1)} \vec{1}_{\theta_1} \text{ in } E \text{ plane } (\phi_1 = \pi/2) \\ -\dfrac{V_o}{w_1} \vec{1}_y \text{ in } H \text{ plane } (\phi_1 = 0) \end{cases} \tag{2.8}$$

Equating this to the early-retarded-time result in (2.6), one can extrapolate the ramp to the late-time value to give an effective time constant for the rise of the far field as

$$ct_1 = \begin{cases} \dfrac{w_2}{2\pi c} \left| \vec{F}\left(\theta_1, \dfrac{\pi}{2}\right) \right|^{-1} \text{ in } E \text{ plane} \\ \dfrac{w_2}{2\pi c} \left| \vec{F}(\theta_1, 0) \right|^{-1} \cos(\theta_1) \text{ in } H \text{ plane} \end{cases} \tag{2.9}$$

For $\ell >> w_1$ and w_2, we can have t_1 arbitrarily small, except that θ_1 is restricted to smaller and smaller values.

Considering impedances, the value for early-time (high-frequency) considerations is tabulated for planar bicones[11], and is given by $Z_0/2$ for square unit cells (a self-complementary case) with $Z_0 \simeq 377\,\Omega$ for free space. There are approximate values as well as more detailed calculations for non-planar flat-plate conical wave launchers[17]. For late times (low frequencies) the impedance appropriate to an individual unit cell is

$$Z_\ell = \begin{cases} \dfrac{Z_o}{2} \dfrac{w_1}{w_2} \cos(\theta_1) \text{ in } E \text{ plane} \\ \dfrac{Z_o}{2} \dfrac{w_1}{w_2} \cos^{-1}(\theta_1) \text{ in } H \text{ plane} \end{cases} \tag{2.10}$$

To minimize reflections, one can try to match early- and late-time results, but as these formulae indicate this is a function of $\vec{1}_1$, albeit not a severe one if one restricts the variation of $\vec{1}_1$ to not-too-large scan angles.

As ℓ/w_1 and ℓ/w_2 are increased (>> 1) there is a significant interaction of the fields on an element with adjacent elements before the wave reaches the aperture plane. This can be partly accounted for by considering quasi TEM waves as on a multiconductor transmission line together with the symmetry conditions for a periodic structure. Allowing for this mutual coupling one can shape the wave-launching conductors to be no longer conical so as to optimize the wave transport to the aperture plane and the impedance presented to the sources. For the case of $\theta_1 = 0$ some analytic solutions have been obtained[18–21].

3. MODERN CONTEXT

In applying these array concepts to even faster transient electromagnetic pulses (picoseconds for the fastest components) the first thing to observe is the usual electromagnetic scaling of time and frequency according to physical dimensions. The array elements basically need to be smaller, which for a given array size means many more elements. Of course, the switching for launching sufficiently fast-risetime pulses (with sufficiently small jitter) on the wave launchers needs to be incorporated as well.

Now we consider finite arrays. On the aperture plane (now $z = 0$) there is some electric field with tangential components $\vec{E}_t(\vec{r}_s, t)$ giving a far field[22, 29]

$$\vec{E}_f(\vec{r}, t) = \frac{1}{2\pi c r} \left[\vec{1}_z \cdot \vec{1}_r \overleftrightarrow{1} - \vec{1}_z \vec{1}_r \right] \cdot \frac{\partial}{\partial t} \int_{S_a} \vec{E}_t \left(\vec{r}_s', t_r + \frac{\vec{1}_r \cdot \vec{r}_s'}{c} \right) dS'$$

$$\overleftrightarrow{1} \equiv \vec{1}_x \vec{1}_x + \vec{1}_y \vec{1}_y + \vec{1}_z \vec{1}_z \equiv \text{identity dyadic} \tag{3.1}$$

For the simple case that the observer is in the $\vec{1}_z$ direction (i.e., $\vec{1}_r = \vec{1}_z$) we have

$$\vec{E}_f(r\,\vec{1}_z,t) = \frac{1}{2\pi c r}\frac{\partial}{\partial t}\int_{S_a}\vec{E}_t(\vec{r}'_s,t_r)\,dS' \tag{3.2}$$

Furthermore, if we have a step-function tangential field on S_a (corresponding to a plane wave propagating in the $\vec{1}_1$ direction) we have

$$\begin{aligned}\vec{E}_t(\vec{r}_s,t) &\equiv \vec{E}_{t0}(\vec{r}_s)\,u\left(t-\frac{\vec{1}_1\cdot\vec{r}'_s}{c}\right)\\ \vec{E}_f(\vec{r},t) &= \frac{1}{2\pi c r}\left[\vec{1}_z\cdot\vec{1}_r\overleftrightarrow{1}-\vec{1}_z\,\vec{1}_r\right]\cdot\frac{\partial}{\partial t}\int_{S_a}\vec{E}_{t0}(\vec{r}'_s)\,u\left(t_r+\frac{\vec{1}_r\cdot\vec{r}'_s}{c}-\frac{\vec{1}_1\cdot\vec{r}'_s}{c}\right)dS'\end{aligned} \tag{3.3}$$

which on the beam center ($\vec{1}_r = \vec{1}_1$) reduces to

$$\begin{aligned}\vec{E}_f(r\,\vec{1}_1,t) &= \frac{1}{2\pi c r}\left[\vec{1}_z\cdot\vec{1}_r\vec{1}-\vec{1}_i\,\vec{1}_i\right]\cdot\delta_a(t_r)\int_{S_a}\vec{E}_{t0}(\vec{r}'_s)\,dS'\\ \vec{E}_f(r\,\vec{1}_z,t) &= \frac{1}{2\pi c r}\,\delta_a(t_r)\int_{S_a}\vec{E}_{t0}(\vec{r}'_s)\,dS' \quad \text{for } \vec{1}_1=\vec{1}_z\end{aligned} \tag{3.4}$$

where δ_a is the approximate delta function[22]. Note the change in form from the infinite-array result. There is a 1/r dependence and the introduction of a time derivative in going from the near to the far field. The aperture S_a is assumed to be of finite linear dimensions with area A_a.

In applying the aperture integral to a transient array one can also see the effect of the field distribution on the unit cells for individual array elements. Instead of a simple plane-wave step excitation as in (3.3) one can view the actual aperture field as a deviation from this. Consider the simple case that the observer is in the $\vec{1}_z$ direction and all elements are turned on at the same time. If each element is a symmetrical conical transmission line of length ℓ, such as in the staggered cell arrangement in fig. 6, then, with a step-function TEM wave launched from the source, the wave does not arrive at the aperture plane all at the same time.

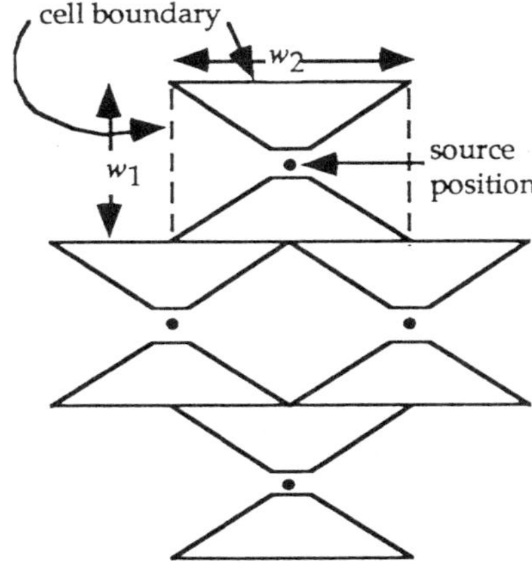

Figure 6. Staggered Array of Flat-Plate Conical Wave Launchers in a Symmetrical Configuration.

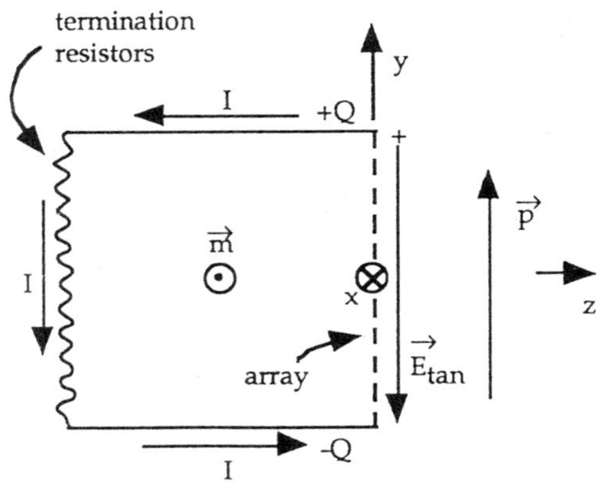

Figure 7. Additions to Array for Balanced Low-Frequency Electric- and Magnetic-Dipole Moments.

There is a dispersion distance[1] or dispersion time which gives the difference in time of arrival of the field on each element unit cell of the aperture plane (for rectangular cells) as

$$d_e^{(1)} \equiv ct_e^{(1)} = \left[\ell^2 + \frac{w_1^2 + w_2^2}{4}\right]^{\frac{1}{2}} - \ell \to \frac{w_1^2 + w_2^2}{8\ell} \quad \text{as} \quad \frac{\left[w_1^2 + w_2^2\right]^{\frac{1}{2}}}{\ell} \to 0 \tag{3.5}$$

Using this in (3.2), then $t_e^{(1)}$ approximates the radiated pulse width due to the time derivative of the spatial integral of the field on the aperture plane. Note for large ℓ that $d_e^{(1)} \to 0$ for constant w_1, w_2. However, as discussed previously, large ℓ also introduces significant coupling of the wave to adjacent cells before reaching the aperture plane, thereby modifying the fields that reach the aperture plane.

If, however, the array is to scan the beam over some angular domain of $\vec{1}_1$, then $t_e^{(1)}$ is not the only dispersion of interest for the unit cells[24]. As indicated in (3.3) the ideal field distribution on the aperture plane does not turn on all at once, but sweeps across the aperture, and hence across each unit cell in the array. The time difference in the ideal turn-on-time across the unit cell then gives another dispersion distance and time as

$$d_e^{(2)} = c\, t_e^{(2)} = w \sin(\theta_1) \tag{3.6}$$

where w represents the width of the unit cell in the direction of wave propagation across the array. For the rectangular array, this can be w_1 and w_2 for particular directions, with $\sqrt{w_1^2 + w_2^2}$ as the largest value achieved by w. If the individual element is designed to minimize $t_e^{(1)}$, there is still $t_e^{(2)}$ with which to reckon. This dispersion time also represents the pulse width in the far field so that $\delta_a(t_r)$ in (3.4) is replaced by a pulse of approximate width $t_e^{(2)}$, which we see increases with increasing θ_1 (by steering the beam off boresight).

One may combine these two dispersion times to obtain some effective total dispersion and effective far-field pulse width. If one limits θ_1 to some range $0 \le \theta_1 \le \theta_{1_{max}}$ centered on boresight, then $t_e^{(2)}$ is limited to some $t_{e_{max}}^{(2)}$. Looking again at $t_e^{(1)}$, there is not much point in making this too much smaller than $t_{e_{max}}^{(2)}$. So a certain consistency in design is called for in which these two time dispersions are roughly comparable.

As with the reflector[22] and lens[26] IRAs, an array IRA can also be designed to optimize its low-frequency performance [24]. As indicated in fig. 7, additional conductors can be added behind the array. The low-frequency (or late time) voltage across the array (summing the voltages of the series elements) induces a charge +Q on the top conductors and -Q on the bottom conductors thereby giving an electric-dipole moment $\vec{p}$ for low frequencies. With an array of terminating resistors in the back, the low-frequency voltage produces a current I around a closed loop (including the source array), thereby giving a magnetic-dipole moment $\vec{m}$. With appropriate symmetry, these two moments are perpendicular and we have by proper choice of the termination resistors (controlling V/I)

$$\vec{p} = p\,\vec{1}_y \;,\; \vec{m} = -m\,\vec{1}_x \;,\; p = \frac{m}{c} \;,\; \vec{p} \times \vec{m} = p\, m\,\vec{1}_z \tag{3.7}$$

This gives low-frequency directionality (a cardoid pattern) with maximum in the $\vec{1}_z$ direction (boresight) and a null in the $-\vec{1}_z$ direction. The example in fig. 7 is for a single polarization (the $\vec{1}_y$ direction with unit cells such as in fig. 3 or 6). One can also extend this low-frequency performance to the case of dual polarization[24] by the addition of conducting wires or strips (parallel to $\vec{1}_z$) on the sides as well as top and bottom. The termination resistors then form a two-dimensional grid in the back. The unit cells then take a form as in fig. 4A with two sources at each source point for the two orthogonal polarizations (or one source switched between the two orthogonal conical transmission lines).

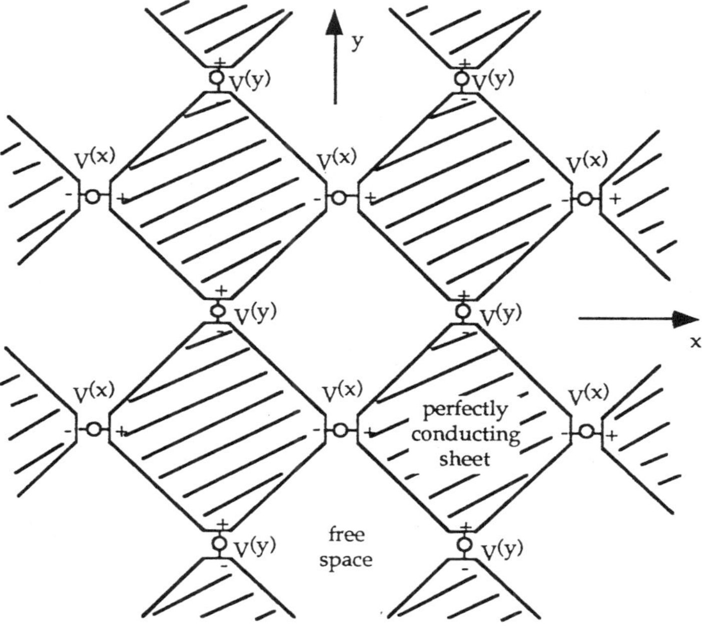

Figure 8. Planar Array with C_{2ac} Symmetry and Dual Polarization.

One can also apply the concept of self-complementary antennas to the design of transient arrays[25, 28]. In this case, not only the unit cells, but also the array as a whole (assumed infinite) is self complementary. This applies strictly only to planar arrays, but can apply approximately to non-planar arrays of planar bicones if ℓ is not very long; the characteristic impedance of the conical transmission line does vary much from the ideal value of $Z_0/2$ appropriate to square unit cells. The self-complementarity principle[31] is a symmetry in which the conducting sheets are replaced by free space, and conversely, and dyadic admittance sheets are replaced by other special sheets (the complement), but the structure remains the same except for a point symmetry operation (rotation, reflection), and now a translation as well. While this applies most simply to a square version of fig. 3 with square unit cells, other versions are possible with various types of impedance loading and other types of unit cells (e.g., as in fig. 4). The sources also enter into the consideration of self complementarity. The simplest case has identical sources, all triggered simultaneously so that $\vec{1}_1 = \vec{1}_z$ (boresight). These can be for a single polarization (as in fig. 3), or for two polarizations as in fig. 8 with two separate sets of sources with voltages $V^{(x)}$ and $V^{(y)}$ for the two polarizations[25, 28]. By including resistors of value $Z_0/2$ at these various connection points for sources one can retain self complementarity with sources that differ from each other (as in a scanning array) in the sense that an individual source (with all other sources zero) drives an impedance of $Z_0/2$ for all frequencies (or Z_0 if there is a $Z_0/2$ resistor with this source)[25]. However, when there is more than one source operating, one source does send currents through the other sources, so that one needs to allow for this.

4. CONCLUDING REMARKS

Transient arrays can be designed to be impulse radiating antennas (IRAs) in a sense similar to reflector and lens IRAs for both high- and low-frequency performance. The array IRA is much more complicated than the other kinds due to the numerous array elements and associated timing requirements. However, one gains the ability to electronically scan. So it all depends on what function one wishes it to perform.

REFERENCES

1. C. E. Baum, The Conical Transmission Line as a Wave Launcher and Terminator for a Cylindrical Transmission Line, Sensor and Simulation Note 31, January 1967.
2. G. W. Carlisle, Matching the Impedance of Multiple Transitions to a Parallel-Plate Transmission Line, Sensor and Simulation Note 54, April 1968.

3. C. E. Baum, The Distributed Source for Launching Spherical Waves, Sensor and Simulation Note 84, May 1969.
4. C. E. Baum, A Distributed-Source Conducting-Medium Simulator for Structures Near and Below the Ground Surface, Sensor and Simulation Note 87, July 1969.
5. C. E. Baum, Some Characteristics of Planar Distributed Sources for Radiating Transient Pulses, Sensor and Simulation Note 100, March 1970.
6. P. R. Barnes, Pulse Radiation by an Infinitely Long, Perfectly Conducting, Cylindrical Antenna in Free Space Excited by a Finite Cylindrical Distributed Source Specified by the Tangential Electric Field Associated with a Biconical Antenna, Sensor and Simulation Note 110, July 1970.
7. C. E. Baum, General Principles for the Design of ATLAS I and II, Part IV: Additional Considerations for the Design of Pulser Arrays, Sensor and Simulation Note 146, March 1972.
8. C. E. Baum, General Principles for the Design of ATLAS I and II, Part V: Some Approximate Figures of Merit for Comparing the Waveforms Launched by Imperfect Pulser Arrays onto TEM Transmission Lines, Sensor and Simulation Note 148, May 1972.
9. Z. L. Pine and F. M. Tesche, Pulse Radiation by an Infinite Cylindrical Antenna with a Source Gap with a Uniform Field, Sensor and Simulation Note 159, October 1972.
10. F. M. Tesche and Z. L. Pine, Approximation to a Biconical Source Feed on Linear EMP Simulators by Using N Discrete Voltage Gaps, Sensor and Simulation Note 175, May 1973.
11. C. E. Baum, Early Time Performance at Large Distances of Periodic Planar Arrays of Planar Bicones with Sources Triggered in a Plane-Wave Sequence, Sensor and Simulation 184, August 1973.
12. T. K. Liu, Admittances and Fields of a Planar Array with Sources Excited in a Plane Wave Sequence, Sensor and Simulation Note 186, October 1973.
13. C. E. Baum, EMP Simulators for Various Types of Nuclear EMP Environments: An Interim Categorization, Sensor and Simulation Note 240, January 1978, and IEEE Trans. Antennas and Propagation, 1978, pp. 35-53, and IEEE Trans. EMC, 1978, pp. 35-53.
14. C. E. Baum and D. V. Giri, The Distributed Switch for Launching Spherical Waves, Sensor and Simulation Note 289, August 1985.
15. Y.-G. Chen, S. Lloyd, R. Crumley, C. E. Baum, and D. V. Giri, Design Procedures for Arrays which Approximate a Distributed Source at the Air-Earth Interface, Sensor and Simulation Note 292, May 1986.
16. T. M. Flanagan, C. E. Mallon, R. Denson, R. Leadon, and C. E. Baum, A Wide-Bandwidth Electric-Field Sensor for Lossy Media, Sensor and Simulation Note 297, January 1987.
17. D. V. Giri and C. E. Baum, Early Time Performance at Large Distances of Periodic Arrays of Flat-Plate Conical Wave Launchers, Sensor and Simulation Note 299, April 1987.
18. C. E. Baum, Coupled Transmission-Line Model of Period Array of Wave Launchers, Sensor and Simulation Note 313, December 1988.
19. D. V. Giri, Impedance Matrix Characterization of an Incremental Length of a Periodic Array of Wave Launchers, Sensor and Simulation Note 316, April 1989.
20. C. E. Baum, Canonical Examples for High-Frequency Propagation on Unit Cell of Wave-Launcher Array, Sensor and Simulation Note 317, April 1989.
21. D. V. Giri, A Family of Canonical Examples for High Frequency Propagation on Unit Cell of Wave-Launcher Array, Sensor and Simulation Note 318, June 1989.
22. C. E. Baum, Radiation of Impulse-Like Transient Fields, Sensor and Simulation Note 321, November 1989.
23. Y.-G. Chen, S. Lloyd, and R. Crumley, Low Voltage Experiments Concerning a Section of a Pulser Array Near the Air-Earth Interface, Sensor and Simulation Note 322, February 1990.
24. C. E. Baum, Timed Arrays for Radiating Impulse-Like Transient Fields, Sensor and Simulation Note 361, July 1993, and Proc. SPIE Automatic Object Recognition V, Orlando Florida, April 1955, Vol. 2485, pp. 168-174.
25. C. E. Baum, Self-Complementary Array Antennas, Sensor and Simulation Note 374, October 1994.
26. C. E. Baum, Low-Frequency-Compensated TEM Horn, Sensor and Simulation Note 377, January 1995.
27. C. E. Baum, From the Electromagnetic Pulse to High-Power Electromagnetics, System Design and Assessment Note 32, June 1992, and Proc. IEEE, 1992, pp. 789-817.
28. N. Inagaki, Y. Isogai, and Y. Mushiake, Self-Complementary Antenna with Periodic Feeding Points, Trans. IECE, Vol. 62-B, No. 4, 1979, pp. 388-395, in Japanese. Translation NAIC-ID(RS)T-0506-95, 11 December 1995, National Air Intelligence Center.
29. C. E. Baum and E. G. Farr, Impulse Radiating Antennas, pp. 139-147, in H. Bertoni et al (eds.) *Ultra-Wideband, Short-Pulse Electromagnetics*, Plenum Press, 1993.
30. E. G. Farr, C. E. Baum, and C. J. Buchenauer, Impulse Radiating Antennas, Part II, pp. 159-170 in L. Carin and L. B. Felsen (eds.), *Ultra-Wideband, Short-Pulse Electromagnetics 2*, Plenum Press, 1955.
31. C. E. Baum and H. N. Kritikos, Symmetry in Electromagnetics, ch. 1, pp. 1-90, in C. E. Baum and H. N. Kritikos (eds.), *Electromagnetic Symmetry*, Taylor & Francis, 1995.

PROPERTIES OF ULTRAWIDEBAND ARRAYS*

Jodi Lisa Schwartz and Bernard D. Steinberg

University of Pennsylvania
202 S. 33rd Street
Philadelphia, PA 19104

INTRODUCTION

Highly thinning the elements required in a filled $\lambda/2$-element spaced array can greatly reduce the quantity of electronics in the array and the data-handling requirements of an imaging system for a fixed number of array channels, greatly enlarge the aperture, and correspondingly improve its resolving power. The cost is a dramatic decrease in dynamic range. The resulting sparse array generally leaves the shapes of the main lobe and near-in sidelobes intact, but the loss in absolute gain implies that main lobe energy has been redistributed into the side radiation region. In conventional narrowband arrays, periodic thinning produces grating lobes of similar shape and strength to the main lobe. Aperiodic thinning destroys the coherent sidelobe buildup in the grating lobes but not the grating lobe energy, which becomes distributed throughout the visible region in a manner determined by the particular thinning procedure. It is common knowledge that sidelobe statistics are similar for a wide variety of thinning procedures, both deterministic and random, with a few notable exceptions[1]. Thus many proposed designs for high resolution two-dimensional arrays are based upon a random distribution of elements[2].

Current technology in ultrasound and electromagnetics is now able to construct radiating array elements that can transmit a high energy pulse of only two or three cycles. The corresponding arrays are called ultrawideband (UWB) arrays. In a highly thinned UWB array, the distribution of side energy is very different from conventional narrowband (NB) arrays. Due to the UWB nature of the pulse, the radiated waveform in space varies in time. As a result, the waveform has an extra dimension of time, with respect to a NB waveform, across which undesired side energy can be distributed. This extra degree of freedom enables UWB arrays to be highly thinned and achieve a much lower side radiation level than NB arrays. An ultrasparse ultrawideband array is capable of reducing the number of elements conventionally required by $\lambda/2$-element spacing to approximately the square root of that number ($\sqrt{N}$ array). This paper presents a method for characterizing ultrasparse ultrawideband one- and two-dimensional arrays. Contrary to conventional wisdom, this analysis shows that in a highly thinned UWB array, a periodic array is preferred over a random array. Experimental analysis performed using an UWB ultrasonic array system is provided to demonstrate the theory.

---

* Supported by NSF under grant BCS92-09680. AMEREM, Albuquerque, New Mexico May 1996.

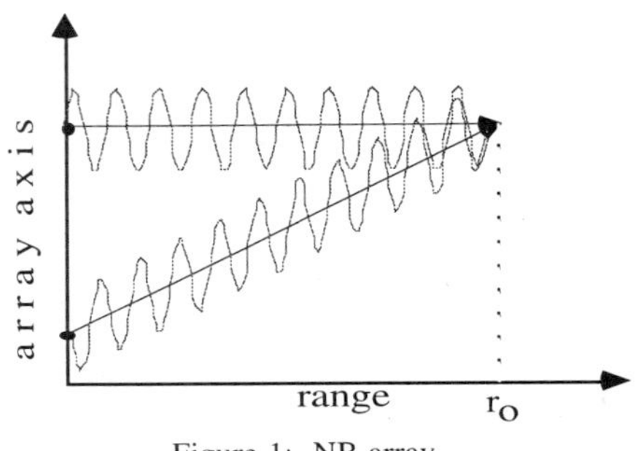

Figure 1: NB array.

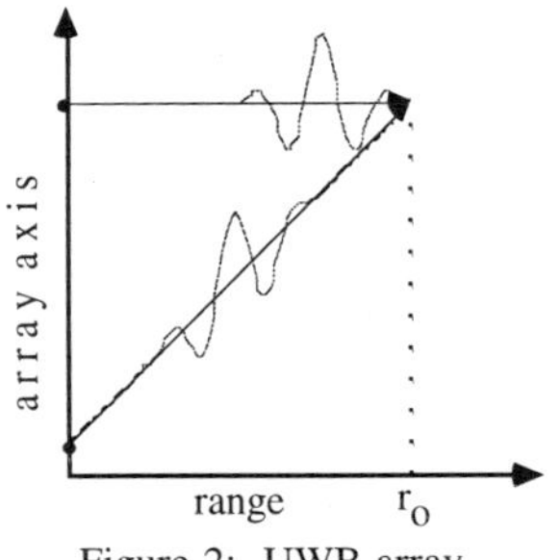

Figure 2: UWB array.

CONVENTIONAL NARROWBAND ARRAYS

Conventionally, radiating elements in an array transmit a narrowband pulse, which is very long in duration. When an array of elements transmits narrowband pulses, the radiated waveform in the field will exhibit interference due to all the pulses from the array (Figure 1). The interference is the result of the pulses existing simultaneously in space. Because of the condition of simultaneity and the interference that results, the dimensionality of this radiated waveform can be reduced by the use of Fourier transforms between time and frequency and between the aperture plane and the image plane. Thus, conventional NB theory, which typically involves Fourier transforms, is based on the assumption that all pulses from the radiating elements are continuously interfering in space. The resulting waveform or radiation pattern in the far-field becomes a function of only the angular dimensions and can be represented by a three-dimensional isometric plot.

An example of periodically and randomly thinned NB radiation patterns are shown in Figure 3. In the thinned, periodic NB array, where the spacing between elements of arrays becomes too large, an ambiguity results in distinguishing the direction from which the energy reflected by a scatterer is received. This ambiguity occurs due to the creation of multiple mainlobes or grating lobes which are aliasing effects caused by the periodically undersampled aperture. In NB, this problem is often remedied by randomly thinning the array. However, the random array redistributes the grating lobe energy into the side radiation region which results in a high sidelobe floor, where the average sidelobe level in power (ASL) = 1/N and N is the number of elements in the array[1].

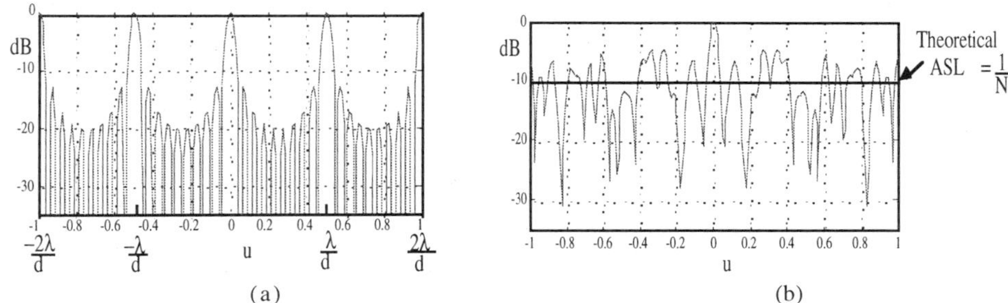

Figure 3: NB radiation patterns of (a) a periodically thinned 10-element, 2λ-spaced array and (b) a randomly thinned 10-element array with a 2λ average element spacing. All arrays are linear and have the same length of 18λ. The reduced angle is $u=\sin\theta$, where θ is the angle off the normal to the array axis.

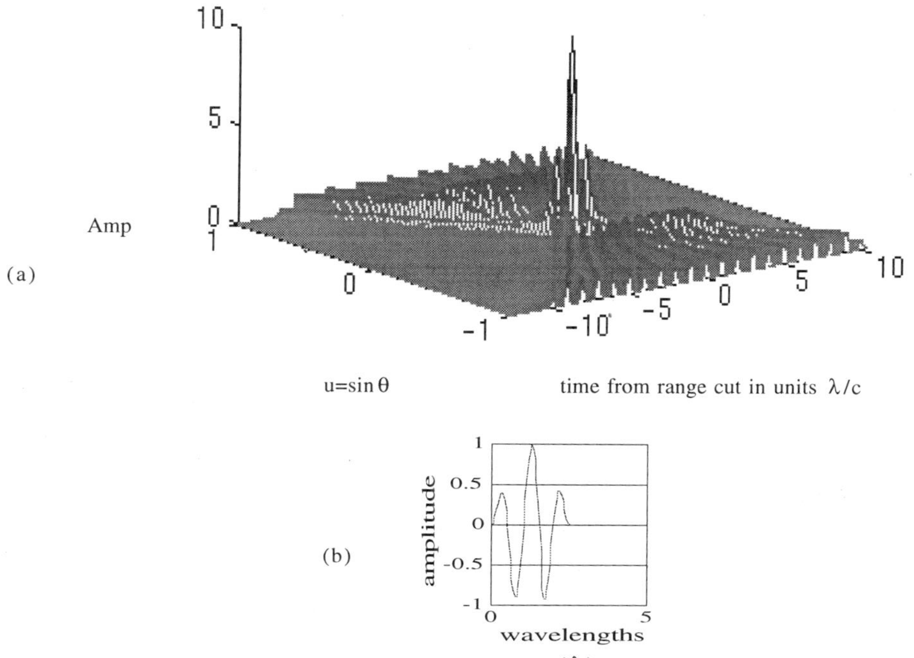

Figure 4: (a) 3-D waveform of an UWB ultrasparse periodic array consisting of 10 elements with an array length L= 18 λ and an interelement spacing of d=2 λ, in which the array elements radiate the Q~2 pulse shown in (b).

ULTRASPARSE, ULTRAWIDEBAND ARRAYS

In highly-thinned ultrawideband arrays, because the pulse duration is so short, a situation can arise where the interelement spacing, d, is greater than the length of the pulse cT, where c is the speed of propagation and T is the duration of the pulse. This type of array can be termed ultrasparse with the condition for being ultrasparse is an array where the average interelement spacing, $\bar{d}$, is greater than the length of the pulse. In an ultrasparse array, there are regions in space, as in Figure 2, where the pulses from the array do not always exist simultaneously at a field point; therefore, they do not interfere. Because there is no interference, NB analysis is no longer applicable and Fourier transforms are no longer useful tools for reducing the dimensionality of the radiated waveform. In the far-field, this waveform varies in the angular dimensions and in time. For a two-dimensional array, the waveform varies in four dimensions and for a one-dimensional array in three dimensions. An example of the radiated waveform of a linear, one-dimensional UWB ultrasparse array at a cut in range is shown in Figure 4.

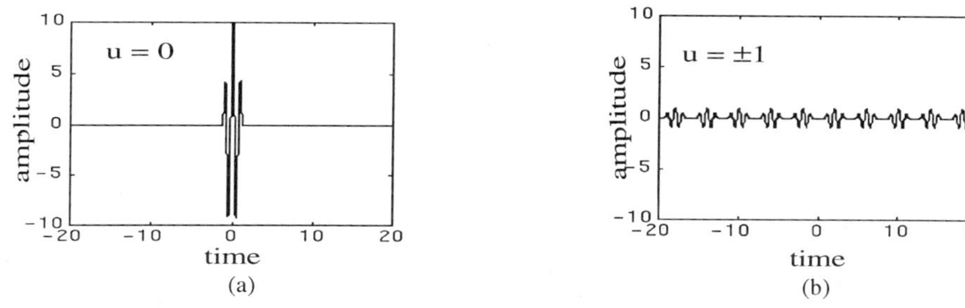

Fig. 5: Cuts of the waveform in Figure 4a. (a) IR at u=0(broadside) and (b) NIR at u=±1(horizon).

The radiation from UWB arrays forms a time-varying waveform where the pulses can spread out in time and not interfere with each other. Thus, UWB arrays are effective in distributing side energy through variations in the time dimension so there is no build up in the energy that causes high sidelobes. Consequently, a thinned, periodic UWB array does not form grating lobes; the grating lobe energy is spread out in the time domain and only forms a low level grating plateau where NB grating lobes would form.

To demonstrate that the side radiation level of an UWB ultrasparse pattern is much lower than in the NB array, the dimensionality of the UWB pattern needs to be reduced to that of the NB array in order to form a comparison. One method for simplifying this three- or four-dimensional pattern is to collapse or project the time dimension of the waveform onto its angular dimensions and treat the maximum envelope of this projection as a beampattern. This method is called maximum projection. Examples of maximum projections from periodic and random ultrasparse UWB arrays are shown in Figure 6. Analysis of maximum projections can be simplified by defining two regions in azimuth, an interfering region (IR), where the pulses are primarily interfering, and a non-interfering region (NIR), where most of the pulses are separated in time and are not interfering with the other pulses from the array. Figure 5 shows the radiated waveform in Figure 4a at cuts in angle in the IR and the NIR, respectively. Since the IR involves interference, its structure is similar to a NB array. In the NIR, the side radiation spreads out in time and, from the maximum projection, has a minimum level which is the height of one pulse, corresponding to a side radiation power level of $1/N^2$. This may be compared to the ASL of the NB random array which is $1/N$; thus, for a large number of elements, the ultrasparse UWB array is able to achieve a much lower side power level. Once the minimum side radiation power level in the NIR is reached, the aperture length can be increased, thereby increasing angular resolution, without any increase in side power level[3].

For comparison with the NB patterns in Figure 3, examples of maximum projections from the periodic and random UWB thinned arrays are shown in Figure 6. The most significant difference from NB is the periodically thinned array. As stated before, the periodically thinned array has no grating lobes, only grating plateaus. Thus, its pattern is not only unambiguous, but also maintains a lower side radiation power level (SL) than the randomly thinned UWB array in Figure 6b. Notice that in the IR near the mainlobe of the random UWB array where most of the transmitted pulses are fully interfering, the pattern does behave similarly to NB random array theory. Outside the mainlobe region in the IR, the pulses start to spread out in time and only partially interfere; thus, this region in the IR has a side radiation level lower than that predicted by NB theory.

The periodic array gives a lower inner side radiation level in the maximum projection than the random array. This is because destructive interference between pulses from adjacent elements occurs in a more continuous fashion in periodic arrays than in random, and when a maximum projection is performed on its radiation pattern, lower side radiation levels are observed. Thus, in a maximum projection, the periodic array has a better side radiation profile than the random array. In addition, the entire periodic

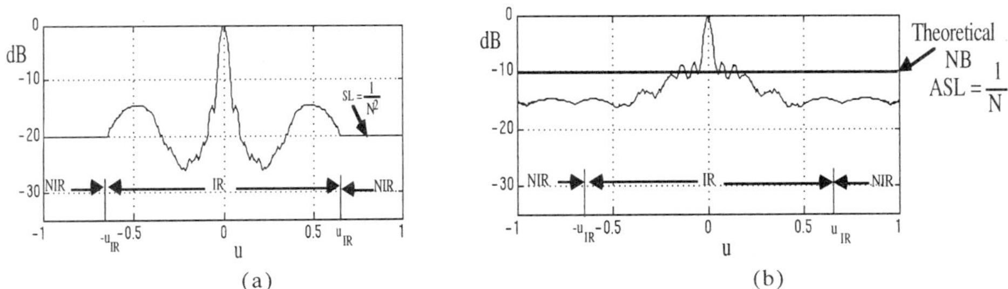

Figure 6: UWB maximum projection patterns of (a) a periodically thinned 10-element, 2λ-spaced array and (b) a randomly thinned 10-element UWB array with an average element spacing of 2λ. All arrays are linear and have the same array length of 18λ.

maximum projection is lower in side radiation level than the random projection because energy is not conserved in the maximum projection as it is in the radiation pattern of a NB array.

TWO-DIMENSIONAL ULTRASPARSE ARRAYS

In the case of a two-dimensional array, as in Figure 7a, the maximum projection varies not only in azimuth (as for the one-dimensional array) but also in elevation. The projection slice method (PSM) allows analysis of the two-dimensional array to be performed as a set of one-dimensional arrays[1]. PSM projects the elements of the two-dimensional array onto a one-dimensional axis at the observation angle ϕ and then finds the maximum projection from that linear array. The resulting pattern represents the maximum projection from the two-dimensional array at that ϕ cut. When the elements of the two-dimensional array are projected, they often shadow each other and result in an effective projected aperture of fewer elements (Figure 7b). The one-dimensional aperture at $\phi = 0$ has only four elements, whereas at $\phi = \pi/8$ it has all 16 elements. The maximum projections of shadowed projections have higher side energy levels than projections in which there is no element shadowing. Based on the analysis of the one-dimensional array and shadowing, the lowest side energy level in a two-dimensional aperture results from a two-dimensional grid in which the effective projected aperture at each ϕ cut has no element shadowing and the elements on that projection are periodic.

EXPERIMENTAL RESULTS

A synthetic array operating in a water tank was used to demonstrate the above theory. The experiment was performed in a one-way transmit-receive mode. A hemispherical source, which behaves much like a point source, was used as the transmitter,

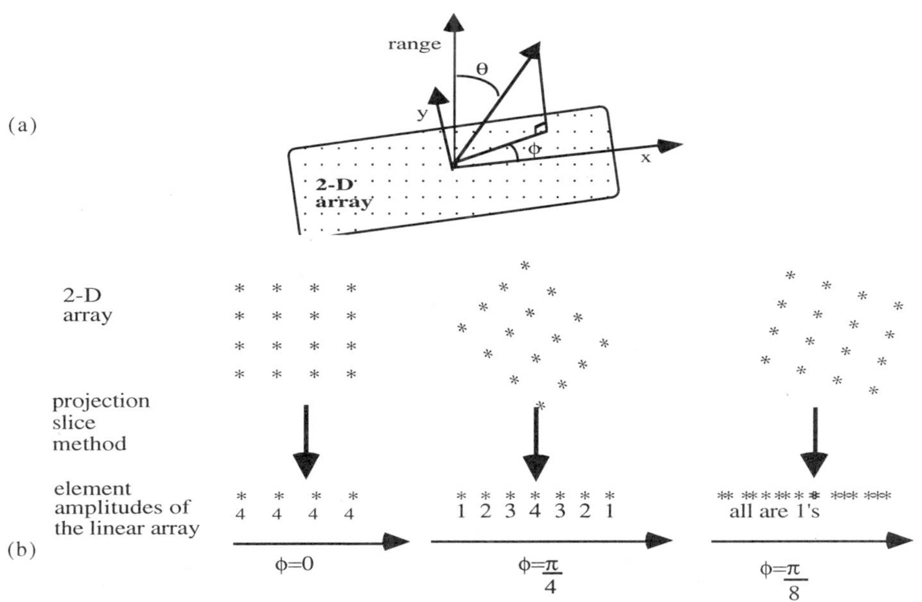

Figure 8: (a) 2-D geometry. (b) The projection slice method applied to a 16-element array.

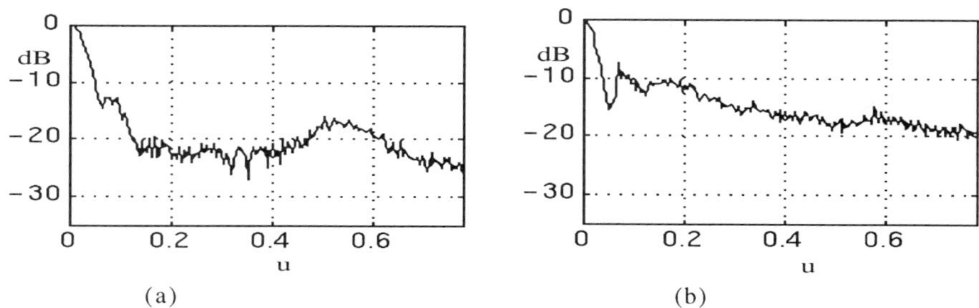

Figure 8: The maximum projection of the image as received by a 10-element array of length L= 18 λ and average interelement spacing of $\bar{d}=2\lambda$. (a) Periodic array. (b) Random array. Theoretical predictions of the maximum projection from the same periodic array as in (a) and from a random array with similar statistics as in (b) are shown in Figure 6.

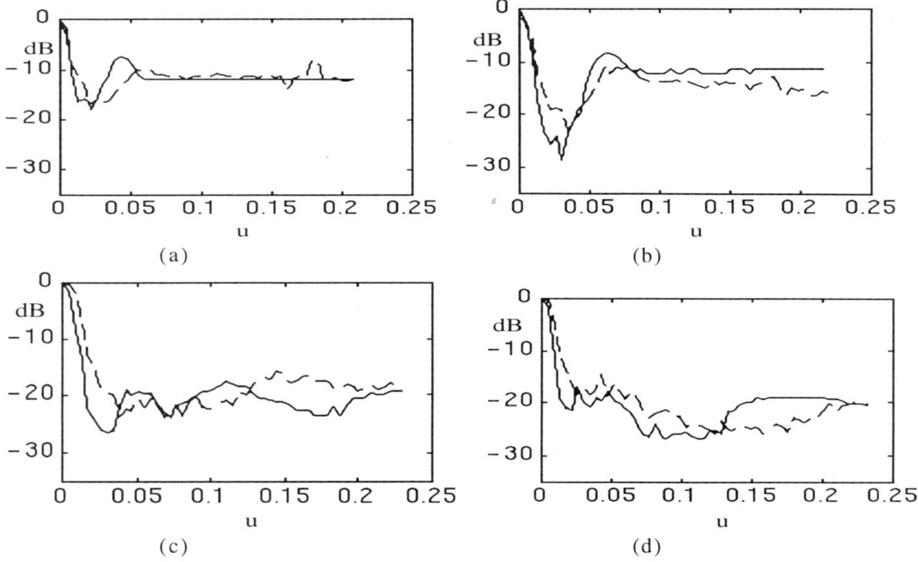

Figure 9: The maximum projections at (a) $\phi = 0$, (b) $\phi = \frac{\pi}{4}$, (c) $\phi = \frac{\pi}{8}$, and (d) $\phi = 0.25$ of the images as received by a 4 x 4 element square grid array of length L= 25 mm x 25 mm with an average interelement spacing in x and y of $\bar{d}= 15.5\lambda$. The theoretical patterns are solid lines and the measured are dashed lines.

operating at a center frequency of 3.75 MHz, and a 0.6 mm hydrophone as the receiver. The hemispherical source transmitted a $Q \approx 2$ pulse. The hydrophone was moved in the xy-plane, perpendicular to the direction of propagation, by a computer-controlled positioning system. The instantaneous voltage of each received waveform was measured and stored in an HP digital oscilloscope. The data were then appropriately steered using time delays and noise filtered in post-processing, and the image of the source was formed by backpropagation. A maximum projection was performed on the image to form the beampattern.

A one-dimensional periodic array with the same parameters as in Figure 6a and a random array with similar statistics as Figure 6b were tested and their maximum projections are shown in Figure 8. Notice that the maximum projections behave very closely to the expected theoretical patterns. In the periodic array, the side radiation level in the NIR is slightly lower than the theoretical level of $SL = 20\log(1/N) = 20\log(1/10) = -20dB$ due to the tapering effect of the element pattern of the 0.6 mm hydrophone. The inner side radiation level in the random array is approximately at the NB ASL, which is ASL=1/N. In accord with the theoretical analysis, the measured side radiation level of the random array is higher than that of the periodic array. Thus, the periodic array outperforms the random array in side radiation level.

The two-dimensional array was verified by a 4x4 element, 15.5λ-spaced in x and y, square grid UWB array. The corresponding measured and calculated maximum projections at several ϕ cuts are in close agreement as is shown in Figure 9.

CONCLUSIONS

The ultrasparse UWB array is capable of achieving high resolution and a low side radiation level with very few elements. Analysis of an UWB array is different from a NB array because NB array analysis is based on the assumption that the pulses from the array elements are always interfering in space. Although the UWB arrays are more difficult to analyze, UWB arrays are advantageous for highly-thinned arrays because they exhibit an extra dimension over NB across which undesired side radiation energy could be distributed. The dimensionality of the radiated waveform produced by an UWB ultrasparse array is reduced by taking a maximum projection of the waveform. The angular dimension of the maximum projection of ultrasparse UWB arrays can be separated into two regions, an IR and an NIR, that are useful when describing the performance of each array. The performance of these regions has been verified by experimentation. A periodic arrangement of elements is shown both theoretically and experimentally to give a lower side radiation level than the random array. In a two-dimensional array, the best configuration of elements is an arrangement where all the projected linear arrays have no element shadowing and are periodic. Two potential two-dimensional structures for achieving a lower side radiation level than the square grid array are the spiral and multiple ring arrays. These arrays are currently being analyzed.

REFERENCES

1. B.D. Steinberg, Principles of Aperture and Array System Design: Including Random and Adaptive Arrays, John Wiley and Sons, Inc., New York, (1976).
2. D.H. Turnbull, P.K. Lum, A.T. Kerr, and F. S. Foster, "Simulation of B-Scan Images from Two-Dimensional Transducer Arrays: Part 1: Methods and Quantitative Contrast Measurements, " Ultrasonic Imaging, 14, pp. 323-343 (1992).
3. F. Anderson, Fullerton, L., Christensen, W., and B. Kortegaard, "Wideband Beam Patterns for Sparse Arrays," Ultra-Wideband Radar: Proceedings of the First Los Alamos Symposium, (1991).

SOME BASIC PROPERTIES OF ANTENNAS ASSOCIATED WITH ULTRAWIDEBAND RADIATION

S. N. Samaddar and E. L. Mokole

Naval Research Laboratory
Radar Division
4555 Overlook Avenue, S. W.
Washington, DC 20375-5336

INTRODUCTION

The purpose of this paper is to present in a tutorial fashion some of the fundamental properties and characteristics of antenna elements that are responsible for very broadband or ultrawideband (UWB) behavior. Before selecting, designing, or analyzing UWB radiators, one must understand the how and the why of the broadband nature of antenna elements that are thought to be UWB. In an ideal situation, when excited by an UWB input signal, an UWB antenna should be capable of radiating a distortionless and diminutive replica of that signal. Although UWB signals come in two broad classes, temporally short and long pulses like the single-cycle sine and linear FM, respectively, short-pulse signals are sufficient for characterizing UWB antennas.

For the purposes of this discussion, the following definitions are used. Let f_L and f_U be the lower and upper frequencies of the passband of the power spectral density of a signal or of the square magnitude of an antenna's transfer function, then the fractional bandwidth B_F is given by

$$B_F = \frac{f_U - f_L}{(f_U + f_L)/2} 100\%. \tag{1}$$

A signal or antenna is categorized in terms of its fractional bandwidth as[1]

a. narrowband if $0\% \leq B_F \leq 1\%$,
b. wideband if $1\% < B_F \leq 25\%$, and
c. ultrawideband if $25\% < B_F < 200\%$.

Ultra-Wideband, Short-Pulse Electromagnetics 3
Edited by Baum *et al.*, Plenum Press, New York, 1997

Since $0 < f_L/f_U \leq 1$, B_F is a strictly monotonically decreasing function of f_L/f_U. The maximum value of 200% for B_F is not achievable because it corresponds to $f_L = 0$, and the minimum value of 0% corresponds to a monochromatic signal ($f_L = f_U > 0$) or to an antenna that only supports radiation at a single frequency.

The analytical representations of UWB exciting voltages is a matter of some importance, and one can choose from several options: Morlet wavelet, Gaussian modulated sine, double exponential, linear FM, single-cycle sine, linear period modulation, etc. A canonical waveform for analyzing conventional antennas is a long-duration, truncated sine function at a single carrier frequency, where the number of cycles is large (continuous wave). Hence, a natural choice for a canonical UWB waveform is a single cycle of the same sine function. Specifically, if one uses the half-power criterion to determine the passband of a single cycle of $\sin(\omega_0 t)$, then $f_L \doteq 0.411 f_0$ and $f_U \doteq 1.309 f_0$, where $f_0 = (2\pi)^{-1}\omega_0$ and f_L and f_U are calculated from $[f_0^2 \sin^2(\pi f/f_0)]/[\pi^2(f_0^2 - f^2)^2] = 0.5$. Consequently, the fractional bandwidth of the single-cycle sine is approximately 104%, which is UWB by the preceding definition. The single-cycle sine is used as the excitation in all calculations presented here.

The definitions of an UWB signal and an UWB antenna are based on the square magnitude of the excitation's voltage spectrum and the antenna's transfer function, respectively. That begs the question, why not define an UWB signal or antenna so that the definition includes the effect of the pertinent phase spectrum? For example, one might say that an antenna is UWB if an UWB input signal is minimally distorted, that is, if the antenna's transfer function has nearly flat magnitude and nearly linear phase across its passband.[2] One way of mitigating signal distortion is to load the antenna's discontinuities, except at the desired point of radiation. Alternately, an antenna can be made broadband by appropriately shaping it so that the discontinuities in the antenna's current and voltage cause negligible radiation and reflection. Although shaped antennas are considered broadband for continuous-wave (CW) operation, they may not be suitable for UWB applications, because they have frequency-dependent phase centers. On the other hand, the classical thin half-wave dipole, usually thought to be a narrowband antenna, may be suitable for UWB radar applications, since the phase of its transfer function is approximately linear and the square magnitude of its transfer function may be improved considerably (made more flat) through loading.[3,4]

Initially, instead of investigating a complex antenna structure, the behavior of the radiated and received pulses associated with the classical, thin, half-wave dipole under a zero-order approximation to the induced current was investigated[5,6] to gain an understanding of the fundamental characteristics of UWB radiation. Because the thin dipole has some desirable properties, the authors felt that analyzing it would provide a sense of how UWB antennas behave in general. For example, an analytical treatment of this dipole is mathematically tractable when its induced current distribution is represented by a zero-order approximation. Specifically, closed-form expressions for the dipole's transfer function and its time- and frequency-domain radiated/received fields are available. Although the magnitude of the dipole's transfer function fluctuates significantly, its phase is approximately linear. Moreover, when the thin dipole is excited by a single-cycle sine, the distortion of the radiated far field is moderate; that is, the duration of the single cycle is increased by 50%, and its number of zero crossings increases from 3 to 5.

The information gained from the dipole studies is summarized in the second section and is applied to explain the mechanisms of radiation and reception associated with other simple antennas that are attributed to be very wideband or UWB. In the remaining sections, this background is used to explain the expected broadband behavior of these antennas from their performance with CW excitations. Specifically, the loaded dipole, flared and tapered radiators, spirals, fat antennas, the log-periodic antenna, and wide-angle bicones are briefly discussed. Although spiral, flared- or tapered-notch, and log-periodic antennas are considered broadband when excited by a CW signal, they have phase centers that vary with frequency and significantly distort short pulses. One possible method for mitigating the migration of the phase center with frequency is mentioned. Another section addresses the behavior of wide-angle biconical antennas to illuminate why finite antennas like the TEM horn, triangular plates, the bow-tie, etc. have UWB characteristics. Finally, the possibility of radiating and receiving circularly polarized sinusoidal pulses involving a pair of orthogonal dipoles is discussed.

DIPOLES

Although a thin dipole, a low-gain radiating element, usually is not considered to be an ultrawideband antenna, a familiarity with its radiation mechanism and characteristics might lead one to understand better how wideband and UWB radiators should perform. In a previous study,[5] by using a simplified model of the thin dipole which carries only a zero-order current, it was found that radiation takes place from the discontinuities of the antenna.

The dipole under consideration is a center-fed, cylindrical antenna with radius a and length $2h$ that is oriented along the z axis (Figure 1). The parameter θ measures the angle from the positive z axis to the observation vector, and r is the distance from the origin to the observation point. For a single-cycle input voltage,

$$V_g(t) = V_1 \sin(\omega_0 t)[U(t) - U(t - T_1)], \tag{2}$$

the radiated electric far field can be expressed as

$$\begin{aligned} E_\theta(r,\theta,t) = \frac{1}{2\Omega r \sin\theta}\Bigg[& V_g(t^*) - V_g\left(t^* - \frac{h}{c}(1-\cos\theta)\right) \\ & - V_g\left(t^* - \frac{h}{c}(1+\cos\theta)\right) + V_g\left(t^* - \frac{2h}{c}\right)\Bigg], \end{aligned} \tag{3}$$

where

$$t^* = t - \frac{r}{c}, \quad T_1 = \frac{2\pi}{\omega_0}, \quad \Omega = 2\ln\frac{2h}{a}. \tag{4}$$

The quantities $U(t)$ and c respectively are the unit step function and the speed of light in free space, and $U(t)$ is defined by

$$U(t) = \begin{cases} 1, & \text{for } t \le 0, \\ 0, & \text{for } t > 0. \end{cases} \tag{5}$$

The thin and half-wave conditions correspond to $2h/a \gg 1$ and $h/c = \pi/(2\omega_0)$, respectively. Although the source impedance is usually a function of frequency, it is assumed equal to the characteristic impedance $Z_0 = \Omega\zeta_0/(2\pi)$ (ζ_0 is the free-space impedance) in the derivation of Eq. (3); that is, the dipole is perfectly matched to the feed network.

At the observation point, the field is not observed until the retarded time t^* exceeds 0 or, equivalently, until the field has propagated the distance r ($t > r/c$). In

Eq. (3), the first term of the far field is radiated directly from the feed-point ($z = 0$) of the dipole and resembles very much the sinusoidal input voltage; however, it is delayed by the propagation time r/c. The second and third terms represent radiation from the upper ($z = h$) and lower ($z = -h$) ends, respectively, of the dipole. Both endpoints radiate replicas of the feed point's radiation h/c seconds later, but they arrive at the observation point at different times because the corresponding distances to the observation point are unequal unless $\theta = \pi/2$. The fourth term can be identified as a second radiation from the feed point that results from currents which are reflected from both endpoints. In other words, since the dipole is fed symmetrically at the mid-point, one half of the last term of Eq. (3) is contributed by the current which travels from $z = 0$ to $z = h$ and back to $z = 0$, and the other half is due to the current traveling from $z = 0$ to $z = -h$ to $z = 0$. This representation of the field indicates that radiation emerges only from the discontinuities of the antenna. In an actual situation, the radiated field will have considerably more terms with attenuated amplitudes than the four signals in Eq. (3). The reason that Eq. (3) has only four terms is a consequence of using the zero-order approximation to the current, which allows the transmitter current to be reflected only once from each endpoint.

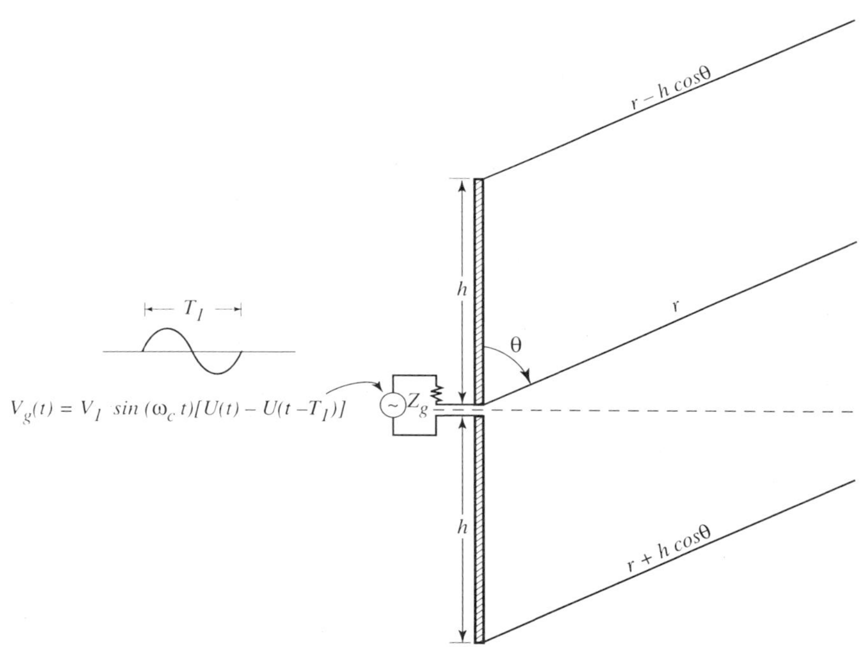

Figure 1. Thin half-wave dipole excited by single-cycle sine.

Instead of a single-cycle sinusoidal voltage, one could have taken any other UWB pulse and still have obtained similar observations on the mechanism of radiation. The choice of the finite-cycle sine as the source for the study of radiation or propagation problems is not new and has been used by many researchers, including Sommerfeld and Brillouin,[7] because it provides a reasonably accurate characterization of physical phenomena. On the other hand, a finite-cycle sinusoidal pulse probably is not possible to generate in actual practice without distorting the leading edge of the first cycle and the trailing edge of last cycle.[2] Evidently, as a consequence of its finite duration, an arbitrary input voltage $V_g(t)$ must at least satisfy the following end conditions,

$$V_g(0) = 0 = V_g(T), \qquad V_g'(0) = 0 = V_g'(T), \qquad V_g''(0) = 0 = V_g''(T), \tag{6}$$

where T is the duration of the pulse that emits at $t = 0$ and the prime denotes differentiation with respect to time t. Clearly, an ideal signal like the finite-cycle sine does not satisfy the second and third conditions of Eq. (6), which real signals like the one generated in Ref. [2] apparently obey.

To illustrate why real signals should satisfy the conditions of Eq. (6), consider a thin, very short (Hertzian) dipole. Under the short assumption $(t^* - T \gg h/c)$, the current is spatially invariant along dipole's length. Suppose that

$$V_g(t) = F(t)[U(t) - U(t-T)] \tag{7}$$

represents a real signal for some appropriately well behaved function F. According to Ref.[2], Eq. (3) is still valid, and the radiated field can be obtained from Eq.(3) by expanding the four terms in second-order Taylor series in powers of t^*. Upon ignoring terms involving powers of h/c that are greater than two, the far field for the short dipole is approximately

$$E_\theta(r, \theta, t) \doteq \frac{h^2}{c^2} \frac{\sin\theta}{2r\Omega} \frac{d^2}{dt^{*2}} V_g(t^*). \tag{8}$$

Implicit in expanding the four expressions involving V_g in second-order Taylor series is the assumption that V_g, hence F, is a twice differentiable function on the real line. The fact that E_θ must be continuous with respect to t further implies that the second derivatives of V_g and F are continuous. By using basic calculus and the continuity of V_g and F at $t = r/c$ and $t = (r/c) + T$, one can argue rigorously that F and V_g satisfy Eq. (6). When V_g is specified by Eq. (2), the second derivative in Eq. (8) includes a Dirac delta $\delta(t^*)$ and its derivative $\delta'(t^*)$, which cannot exist with a realizable signal. Therefore, although the single-cycle sine is a useful analytical tool for studying UWB radiation, it only approximates a real signal, which is much smoother at its start and finish. Thus, when real UWB signals are applied to the well known representation of the radiated far field for a short dipole, these signals must satisfy Eq. (6).

Now suppose the thin dipole acts as a receiving antenna. Under the assumptions that lead to Eq. (3), an incident field induces voltage pulses at the input terminal of the receiving dipole in four distinct ways.[6] The energy that is incident to the feed point divides into two components. Half of the energy is injected as a voltage pulse directly into the input terminal, which is located at the dipole's midpoint. The other half of the energy is equally split between induced currents on each half of the dipole that travel towards the associated endpoints, reflect from them, and travel back towards the terminal, where each current induces a second voltage pulse at the terminal. The remaining two components enter at each endpoint, and travel along the dipole to the terminal point, where they induce two more voltage pulses. Analogous to radiation case, the total received induced voltage is stretched in time by the superposition of the four voltages.

The following is a list of observations that are gleaned from studying dipoles when they are excited by a short-pulse voltage and that are relevant to UWB antennas.

(1) Radiation takes place only from the discontinuities in the current and voltage along the surface of the antenna.

(2) Spatial separation of the discontinuities generates multiple radiations at different intervals of time. Therefore the radiated pulse is stretched in time, thereby decreasing the bandwidth of the radiated field.

(3) An exact replica of the exciting pulse is impossible to radiate due to the finite size of any antenna.

(4) The open-circuit voltage received by a short dipole is approximately a replica of the incident field.

(5) The open-circuit voltage received by a half-wave dipole is proportional to a time integral of the input voltage of the transmitting dipole.

(6) In general, the time variation of the received voltage depends on how the receiving dipole is matched.

(7) If by some artifice, the reflections and radiations from a dipole's endpoints could be prevented, the exciting input pulse ideally would radiate without distortion; that is, a relatively narrowband dipole would perform as an UWB radiator.

(8) As one knows from a familiarity with the frequency-domain analysis of a given narrowband antenna, the primary reason for said antenna's narrowband designation is the rapid variation of its input impedance with frequency. One, therefore, may infer that such variations are caused mainly by reflections of the exciting signal from the discontinuities of the current and voltage distributions of the radiator, assuming that the antenna is properly matched to the feed network.

Since distortionless radiation of an UWB exciting signal by a real radiator is theoretically impossible due to the radiator's finite size, the foregoing observations suggest that one should be content with minimally distorted radiated and received signals. To accomplish this goal, the antenna should be selected and designed appropriately. The next few sections attempt to explain whether specific UWB antennas are expected to minimize the distortion of the radiated field when the antenna is excited by UWB inputs.

LOADED DIPOLES

If nature would permit an infinitely long dipole to exist, only the first term of Eq. (3), which is an exact replica of the exciting voltage, would appear in the radiated field. One then might argue that since an infinite dipole is not practical, consider a long dipole of length $2h$ which is excited by a very short pulse of duration T. If the radiated pulse is observed at a distance r in the time interval $0 \leq t^* \leq \min\{h(1 \pm \cos\theta)/c\}$ (Figure 1), the radiated field will appear to be distortionless insofar as these observation times are concerned. Employing this kind of argument to draw inferences about UWB radiation may have validity for some special applications; however, this line of reasoning is not followed here. Since loading can play an important role in reducing reflections from discontinuities, improving the behavior of the dipole so that its performance is more broadband is discussed in terms of loading.

The use of loaded dipoles for the prevention of reflections is not a new idea. Apparently, the notion was first suggested and tested by Hallen,[8] which might have prompted Wu and King[3] to provide the appropriate theory. As noted below, the reflectionless behavior of the antenna makes it a traveling-wave type radiator, which is in general broadband.[9] The introduction of lossy materials inevitably lowers the efficiency, however the overall efficiency in a transmitting system is often sacrificed for improvements in its broadband and directional properties. It is also recognized that efficiency may not be the most important factor in a receiving antenna for which broadbandedness, directivity, and the simplicity of the system are the most desirable features.

Most of existing work on loaded antennas is confined to resistive loading, which reduces efficiency. The possibility of reactive loading (in particular, capacitive loading), although mentioned by Hallen, lacks appropriate attention except the work reported in Ref. 10. This aspect of antenna loading should be explored, since apparently no loss of energy is incurred. Consequently, the efficiency of a reactively loaded antenna is expected to increase when compared to a resistively loaded one.

Some recent works[4,11] show both experimentally and theoretically that a resistively loaded antenna, excited by a short pulse, is capable of radiating the same pulse with no appreciable distortion. For a resistively loaded dipole, the theory[3,4] assumes that the antenna is made of resistive material so that its internal impedance Z_i per unit length is proportional to $1/(h - |z|)$, where z is the distance measured along an antenna having total length $2h$. Wu and King[3] have shown that the current distribution on such a loaded dipole behaves like a traveling wave of the form $(h - |z|)e^{-ik|z|}$, where $k = \omega/c = 2\pi/\lambda$. By using this analytical representation of the current, they approximately computed the radiated field as well as the power. With this form, the current distribution vanishes at the dipole's endpoints, but the voltage, although small, does not vanish. However, Hallen claimed that his experimental results for a capacitively loaded dipole showed that both current and voltage vanish near the end of the dipole.

For an arbitrary antenna, it is then natural to think that any discontinuous locations which emit undesirable radiation may be resistively loaded to make the antenna in question more broadband. For example, experiments show[12] that the radiation pattern of a horn antenna consists of a forward lobe, which may be called the desirable lobe or gain lobe, and some undesirable backward lobes. The latter might be generated by the horn's discontinuous edge. However, when the end surface of the horn was "extended by lossy tapered metalized mylar strips," the backlobes disappeared without distorting the size and shape of the forward lobe. This shows that the antenna gain was not reduced significantly. Thus it may be inferred that the antenna was matched to free space by the mylar strips. This method should also be tried on other antennas that are excited by short-pulse inputs, and the associated experiments should determine gain, beamwidth, efficiency, etc.

FLARED AND TAPERED RADIATORS

In the second section, it is analytically argued for the thin dipole that radiation emanates from geometric discontinuities (feed point and endpoints) and from reflections back to the feed point from the end discontinuities.[5] That raises the question, how can one characterize where radiation occurs on an antenna? Clearly, a break in an antenna's surface causes radiation, but radiation also takes place at locations where breaks do not occur. For example, all cross sections of a given surface may be continuous curves, and yet radiation will occur at points where the surface's tangent plane changes abruptly. Quantifying the location of the radiation in terms of geometric discontinuities is probably inadequate. To obtain a complete quantification, perhaps one also should characterize radiation with electrical quantities like the current and voltage distributions on an antenna's surface. In particular, the authors conjecture that discontinuities in the first partial derivatives with respect to the spatial variables of an antenna's current and voltage distributions or places on the antenna's surface where the tangent plane does not exist are the locations of all radiations. A plausibility argument in support of this conjecture follows momentarily.

As a general rule, it seems evident that the less sharp an antenna's surface the smaller the reflections and the broader its bandwidth. Therefore, increased bandwidth can be achieved by shaping the antenna to a smoother surface so that the current and voltage distributions are smoother functions with respect to spatial variables, thereby reducing reflections to a large extent. In reality, no antenna can be made so that the tangent plane varies continuously because imperfections, which cause radiation, will be introduced by the manufacturing process. Consequently, the best one can do in practice is to mitigate the effects of imperfections. However the discontinuities in the current and voltage distributions of a shaped antenna are manifested, they cause its phase center to vary with frequency over bandwidths that are appropriate for UWB radiation. This in turn results in antennas with dispersive characteristics. How much such antennas will distort the radiated signals associated with short pulse excitations depends on how slowly the tangent to the antenna surface changes within the longest wavelength it is required to radiate. One may view the process of antenna shaping as an empirical methodology for relating an antenna's surface geometry and the behavior of the induced current and voltage distributions.

Examples of shaping are antennas that widen (flare) or narrow (taper) from the feed, of which the flared notch is a special case. A V-antenna is the simplest example of a flared radiatior. Such shaped antennas are made from waveguides, wires, tapered slots, etc. For the flared notch, the feed point has the narrowest gap and supports radiation at the highest frequency; whereas the open end has the largest separation and radiates the lowest frequency. Discussions of other radiators of this category can be found in Ref. 13. Sengupta and Ferris[14] show the performance of such an antenna under the name, "Rudimentary Horn," where the theory was developed for a symmetrical rudimentary horn having exponentially curved radiating elements. Although the experimental work of Sengupta and Ferris is satisfactory, their theory is inadequate according to them.

Generally, the geometric shape of flared and tapered radiators makes mathematical analysis difficult. To provide a heuristic argument for the aforementioned conjecture on the locations of an antenna's radiation, consider a one-dimensional antenna geometry (a wire antenna), where the curve Γ denotes the physical location of the radiator. The origin of the xyz coordinate frame is placed at the antenna's feed point, and $\mathbf{r}$ and $\mathbf{r}'$ denote the vectors from the origin to the observation point (x, y, z) and to the point (x', y', z') on Γ, respectively. Also note that (r, θ, ϕ) are spherical coordinates. For simplicity, Γ is chosen to lie in the yz plane and is specified by the arc length s for $0 \leq s \leq s^*$, with $s = 0$ corresponding to the feed point. The frequency-domain far field $\hat{\mathbf{E}}$ is proportional to the vector potential $\hat{\mathbf{A}}$ ($\hat{\mathbf{E}} = -i\omega\hat{\mathbf{A}}$), and $\hat{\mathbf{A}}$ can be expressed as[15]

$$\hat{\mathbf{A}} = \frac{\mu_0}{4\pi r} e^{-i\omega r/c} \int_{\Gamma} \hat{\mathbf{T}}(\mathbf{r}')\hat{I}(s,\omega)e^{i\omega(\mathbf{r}'\cdot\hat{\mathbf{r}})/c}\, ds, \tag{9}$$

where $\hat{\mathbf{r}} = \mathbf{r}/r$, $r = |\mathbf{r}|$, μ_0 is the permeability of free space, and $\hat{I}$ is the current distribution along Γ. The quantity $\hat{\mathbf{T}}$ is the direction of current flow (unit tangent vector) at (x', y', z') and is given by

$$\hat{\mathbf{T}} = \frac{dy'}{ds}\hat{\mathbf{y}} + \frac{dz'}{ds}\hat{\mathbf{z}}. \tag{10}$$

A sufficient condition for the existence of the integral in Eq. (9) is that Γ has a parametric representation and that both $\hat{\mathbf{T}}$ and $\hat{I}$ are continuous in a region D containing Γ. Suppose the parametric equations of Γ are

$$x' = 0, \quad y' = \eta(\tau), \quad z' = \nu(\tau), \qquad a \leq \tau \leq b, \tag{11}$$

such that η and ν have continuous derivatives with respect to τ on (a, b). The line integral becomes

$$\hat{\mathbf{A}} = \frac{\mu_0}{4\pi r}e^{-i\omega r/c}\left\{\hat{\mathbf{y}}\int_a^b G_1(\tau,\omega,\theta,\phi)\,d\tau + \hat{\mathbf{z}}\int_a^b G_2(\tau,\omega,\theta,\phi)\,d\tau\right\}, \tag{12}$$

where

$$G_1(\tau,\omega,\theta,\phi) = \hat{I}(s(\tau),\omega)\frac{d\eta}{d\tau}(\tau)e^{i\omega(\mathbf{r}'\cdot\hat{\mathbf{r}})/c}, \tag{13a}$$

$$G_2(\tau,\omega,\theta,\phi) = \hat{I}(s(\tau),\omega)\frac{d\nu}{d\tau}(\tau)e^{i\omega(\mathbf{r}'\cdot\hat{\mathbf{r}})/c}, \tag{13b}$$

$$\mathbf{r}'\cdot\hat{\mathbf{r}} = \eta(\tau)\sin\theta\sin\phi + \nu(\tau)\cos\theta. \tag{13c}$$

Since each integrand of Eq. (12) is continuous, an antiderivative exists for each indefinite integral and is denoted Φ_α for $\alpha = 1, 2$. Moreover, the partial derivative of Φ_α with respect to τ is G_α[16], and

$$\hat{\mathbf{A}} = \frac{\mu_0}{4\pi r}e^{-i\omega r/c}\left\{\hat{\mathbf{y}}[\Phi_1(b,\omega,\theta,\phi) - \Phi_1(a,\omega,\theta,\phi)] + \hat{\mathbf{z}}[\Phi_2(b,\omega,\theta,\phi) - \Phi_2(a,\omega,\theta,\phi)]\right\}. \tag{14}$$

Suppose further that the behavior of $\hat{I}$ and the voltage distribution $\hat{V}$ can be represented by a transmission line model for $0 \leq s \leq s^*$. Then $\hat{I}$ and $\hat{V}$ satisfy

$$\frac{\partial\hat{V}}{\partial s}(s,\omega) = -Z(s,\omega)\hat{I}(s,\omega) \qquad \text{and} \qquad \frac{\partial\hat{I}}{\partial s}(s,\omega) = -Y(s,\omega)\hat{V}(s,\omega), \tag{15}$$

where Z and Y are the known continuous impedance and admittance, respectively, of the transmission line. Equation (15) and the continuity of $\hat{I}$ imply that $\hat{I}$ and $\hat{V}$ are continuously differentiable on (a, b). Clearly, the far field under these conditions depends only on the feed point ($\tau = a$) and the endpoint ($\tau = b$) of the radiator; that is, radiation occurs only at these two points, which coincidentally happen to be the only places where the current and voltage distributions have discontinuous derivatives.

Next, at some d in the open interval (a, b), suppose either $\hat{I}$ fails to be continuously differentiable, which implies $\hat{I}$ is discontinuous by Eq. (15), or suppose $\hat{\mathbf{T}}$ fails to exist. In either case, each integral in Eq. (12) must be split into a sum of integrals over $[a, d]$ and $[d, b]$, and

$$\begin{aligned}\hat{\mathbf{A}} = \frac{\mu_0}{4\pi r}e^{-i\omega r/c}\Big\{&\hat{\mathbf{y}}[\Phi_1(b,\omega,\theta,\phi) - \Phi_1(d+,\omega,\theta,\phi) + \Phi_1(d-,\omega,\theta,\phi) - \Phi_1(a,\omega,\theta,\phi)] \\ &+ \hat{\mathbf{z}}[\Phi_2(b,\omega,\theta,\phi) - \Phi_2(d+,\omega,\theta,\phi) + \Phi_2(d-,\omega,\theta,\phi) - \Phi_2(a,\omega,\theta,\phi)]\Big\}.\end{aligned} \tag{16}$$

As with Eq. (14), Eq. (16) indicates that radiation comes from the discontinuities of $\hat{\mathbf{T}}$ and the first spatial derivatives of $\hat{I}$ and $\hat{V}$.

One could extend the preceding argument to a finite number of discontinuities in $\hat{\mathbf{T}}$, $\hat{I}$, and $\hat{V}$ and to three-dimensional radiators by rotating Γ about the y axis, assuming the excitation is symmetric about the y axis. Although the conjecture has not been argued for arbitrarily shaped antennas, the authors feel that it is valid and should be the subject of future research.

FAT ANTENNAS

Although a thin wire (thin cylindrical conductor) is a narrowband antenna, a fat cylindrical dipole behaves like a wideband radiator. In general, fat antennas are broadband. In addition to the fat cylindrical dipole, members of the fat-dipole family include wide-angle conical (or biconical), spherical, and spheroidal antennas. Theoretical analysis of a finite, fat cylindrical dipole and a wide-angle cone are very difficult; whereas exact analytical field expansions for spherical and spheroidal antennas can be obtained with less effort.[17,18] Unfortunately, even these exact solutions do not necessarily provide the underlying physical understanding of the antenna's characteristics for an UWB range of frequencies.

Those who are familiar with the problem of scattering from spheres and spheroids know that the representation of the scattered field converges very slowly as the frequency increases. Analogous mathematical difficulties are encountered with spherical and spheroidal antennas. In scattering problems, resonance scattering may help in identifying whether the scattering object resembles a sphere or spheroid. Similarly, a radiator may be identified from its resonance radiations. Thus resonance radiations are important phenomena for investigation. Most scattering problems are worked out in the frequency domain, because transient scattering and radiation problems, which require inverse Fourier transforms, are more difficult to handle. However, some attempts[19,20] have been made for some limiting cases, such as for very low and high frequencies. Since a larger number of resonance radiations are associated with fat antennas than their thin counterparts, one may infer that these large number of resonances are responsible for making a fat antenna broadband.

POSSIBLE REDUCTION OF PHASE VARIATION IN SOME ANTENNAS

Spirals and log-periodic antennas are known to be independent of frequency. In practice, however, each has a very large bandwidth which may have a frequency ratio on the order of 100:1 or more. Unfortunately, such frequency-independent antennas have phase centers (like the tapered and flared antennas) that depend on frequency, and therefore their responses to a short pulse are highly dispersive. One may then inquire whether ways exist to modify such structures so that their short-pulse characteristics become less dispersive.

The exact answer to this question is unknown. However, some investigations along this direction have appeared in the Russian literature,[21] where Yatskevich and Fedosenko considered log-periodic antennas which consisted of a number of parallel linear dipoles placed in a certain manner. It would be a worthwhile effort to develop their technique so that it could be extended to other types of dispersive radiators.

WIDE-ANGLE BICONICAL ANTENNAS

When one searches the recent literature for the names of very-wideband or ultrawideband (UWB) antenna elements that are nondispersive, biconical antennas are most frequently cited. Axially symmetric bicones, excited by a symmetrically fed source at the apex, radiate axially symmetric electromagnetic fields. This property of a bicone may not make it a good candidate for radar applications. However, analyses of TEM horns, V-conical antennas, triangular plates, and bow-tie antennas, which are often used for launching impulse radiations, require some knowledge of a biconical antenna's characteristics. Furthermore, some conical radiators that are geometrically asymmetrical, such as when the axes of two cones are inclined to each other, radiate axially non-uniform electromagnetic waves. These unsymmetrically radiating cones may find their applications in a modern UWB radar antenna system. However, analyses of such asymmetrical conical antennas will not be easy to treat.

It should be noted that TEM-fed antennas play a dominant role in most of the UWB systems that have been studied and built. Such systems roughly consist of a source, a parabolic reflector, and feeds that connect the source and reflector. In particular, Baum and his associates[22] have developed and analyzed impulse radiating antennas (IRAs), where the source-fed elements may be circular cones, TEM horns of different shapes, etc. They have a newly designed system, located at Phillips Laboratory's High Energy Research Technology Facility (HERTF), which consists of four feeds (two pairs) that connect a source-fed lens and a parabolic reflector.

The literature dealing with the TEM-fed antennas considers them to be infinitely long when analyzing the TEM wave launching properties. Although a TEM wave is the dominant mode, a real finite antenna of this type can also support TM waves both within and without the antenna region. This possibility is hardly mentioned in the literature and leads one to ask why the TM waves inside the antenna region, in particular, are neglected in such investigations. The answer to this question is found in the literature on wide-angle biconical antennas. Specifically, this point is made clear in the works of Smith[23] and Tai.[24–26]

Since wide-angle biconical antennas play an important role in understanding TEM-fed antennas, some of the important characteristics of wide-angle bicones are now discussed. Both Smith and Tai obtained analytical expressions of the bicone's input impedance and showed it also depends on TM modes. Tai[26] was also able to cast the expression for the input impedance in a variational form. This variational form indicates that the influence of TM waves becomes negligible for the wider cone angles together with increasing frequencies, which then suggests that finite wide-angle bicones behave like infinite cones for high frequencies. In other words, in the high-frequency limit, the input impedance of a finite cone approaches the characteristic impedance of the biconical antenna. Samaddar and Mokole[27,28] have extended the work of Smith[23] and Tai[26] to include bicones having unequal cone angles (Figure 2), where expressions for the input impedance, its variational form, and the field are obtained. Related research on this subject is found in Refs. 29–32. Moreover, the work of Harrison and Williams[32] is generalized in Ref. 28.

The following are highlights of the research on the generalized biconical antenna in Refs. 27 and 28.

(1) The radiation pattern of an electrically very small ($ka \ll 1$ where $k = 2\pi/\lambda$), wide-angle cone of length a is proportional to $\sin\theta$. This dependence of the pattern on $\sin\theta$ is similar to the pattern for a short dipole (Table 1). The angle θ is measured from the axis of symmetry of the cone to the direction of the observation. In this case, maximum radiation takes place at $\theta = \pi/2$.

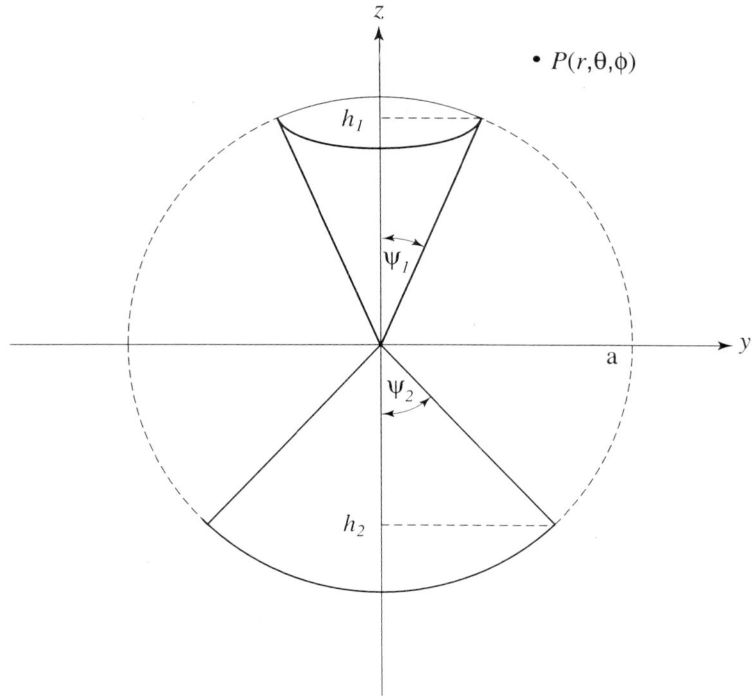

Figure 2. Spherically capped biconical antenna with unequal cone angles.

Table 1. Behavior of the radiated field by wide-angle bicones, where r is the distance of the observation point from the origin, Z_0 is the free space impedance, Z_c is the characteristic impedance of the bicone, C_{in} is the equivalent input capacitance of an electrically small bicone, and $t^* = t - (r/c)$ is retarded time.

Antenna Type	Normalized Effective Height $\frac{\omega}{c} h_c(\theta,\omega)$	Antenna Input Impedance Z_{in}	Generator Impedance Z_g	Radiated Field $E_\theta(\theta,t)$
Wide-Angle Small Cone $[ka << 1]$	$\sim \frac{\omega a}{c}\sin\theta$	$\sim \frac{-iZ_c}{(\omega/c)a} = \frac{1}{i\omega C_{in}}$	$Z_c << \frac{1}{\omega C_{in}}$	$\sim \frac{Z_0 C_{in}}{r}\sin\theta \frac{d^2}{dt^2}V_g(t^*)$
Wide-Angle Large Cone $[ka >> 1]$	$\sim \frac{-i}{\sin\theta}$	$\sim Z_c$	Z_c	$\sim \frac{Z_0}{Z_c r}\frac{V_g(t^*)}{\sin\theta}$

(2) On the other hand, the radiation pattern of an electrically very long ($ka \gg 1$), wide-angle cone is proportional to $1/\sin\theta$. For such cones, the maximum radiation occurs near the radial tangent to the cone's surface.

The transient behaviors of the transmitting and receiving wide-angle cones under different matching conditions with their associated feed networks are summarized in Tables 1 and 2, where the transmitting and receiving antennas are modeled by the equivalent circuits of Figure 3. The time-domain representation of the radiated field clearly depends on the passband of the input voltage $V_g(t)$ relative to the bicone's passband and on the match between the antenna and generator impedances. In particular, for frequencies satisfying $ka \ll 1$, the far field is proportional to the second derivative of V_g; whereas for high frequencies ($ka \gg 1$), the far field behaves like a time-delayed replica of V_g.

In contrast, the received load voltage depends on the match between the antenna and load impedances, as well as the relation between the passbands of V_g and the bicone. Moreover, the load impedance can be resistive, capacitive, or a combination, and Table 2 displays how the received voltage V_L varies accordingly. For $ka \ll 1$, V_L is proportional to $\sin\theta$ times the incident field E^{inc} or the temporal derivative of E^{inc},depending on the matching conditions. Similarly, for $ka \gg 1$, V_L is proportional to $1/\sin\theta$ times E^{inc} or the time integral of E^{inc}.

RADIATING CIRCULARLY POLARIZED WAVES FROM ORTHOGONAL DIPOLES

This last topic addresses the issue of whether an UWB antenna can transmit a circularly polarized radiated field. To ascertain if this is feasible, a pair of thin, coplanar, half-wave, electromagnetically uncoupled, orthogonal dipoles in free space that are excited by a finite-cycle sinusoidal voltage has been investigated.[33] The coordinate frame is chosen so that the x-axis and z-axis point along the axes of the dipoles (Figure 4). Each dipole has length $2h$, and the frequency f_0 ($= \omega_0/(2\pi)$) of the input voltage is related to the length of the dipoles by $h/c = 1/(4f_0)$. The dipole along the z-axis is excited first with voltage S_1, and the excitation S_2 to the dipole along the x-axis is delayed by a quarter cycle. Specifically,

$$S_1(t) = V_1 \sin(\omega_0 t)\, p_T(t)\,, \tag{17a}$$

$$S_2(t) = V_1 \sin\left[\omega_0\left(t - \frac{T_0}{4}\right)\right] p_T\left(t - \frac{T_0}{4}\right), \tag{17b}$$

$$p_T(t) = U(t) - U(t-T), \tag{17c}$$

where U is the unit step function of Eq. (5), $T = N/f_0$ is the duration of the finite-cycle sine with period $T_0 = 1/f_0$, and N is the number of cycles of the sine.

One obtains a closed-form expression for the far field by using a zero-order approximation of the current distribution along each dipole. Furthermore, mutual coupling between the dipoles is ignored, and each dipole is assumed to be perfectly matched to its feed network. Applying Eq. (3) to each excitation in Eq. (17) and summing the resulting vector fields yield the total field. When the observation direction is broadside ($\theta = \phi = \pi/2$), the radiated field reduces to

$$\vec{\mathbf{E}}(\vec{\mathbf{r}}; t^*) = K f_1(t^*, \pi/2)\widehat{\theta} + K f_3(t^*, \pi/2, \pi/2)\widehat{\phi} \tag{18a}$$

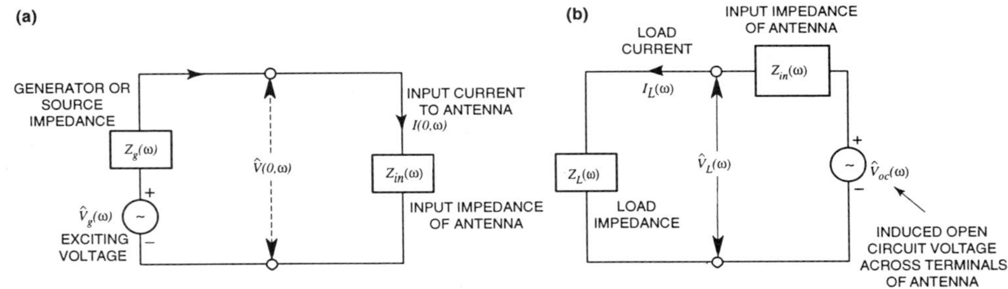

Figure 3. (a) Equivalent circuit of transmitting antenna. (b) Equivalent circuit of receiving antenna.

Table 2. Behavior of received voltage by wide-angle bicones, where Z_c is the characteristic impedance of the bicone, C_{in} is the equivalent input capacitance of the bicone, R_L is the real part of Z_L, and C_L is the capacitance associated with Z_L.

Antenna Type	Normalized Effective Height $\frac{\omega}{c}h_e(\omega,\theta)$	Antenna Input Impedance Z_{in}	Load Impedance Z_L	Received Voltage $V_L(t)$
Wide-Angle Small Cone $[ka \ll 1]$	$\sim \frac{\omega a}{c}\sin\theta$	$\sim \frac{1}{i\omega C_{in}}$	$Z_L = R_L \gg \frac{1}{\omega C_a}$	$\sim a\sin\theta\, E^{inc}(t)$
			$Z_L = R_L + \frac{1}{i\omega C_L}$ with $R_L \ll \frac{1}{\omega C_L}$	Voltage across Z_L is $\sim a\sin\theta\left(\frac{C_{in}}{C_{in}+C_L}\right)E^{inc}(t)$ Voltage across R_L is $\sim a\sin\theta\left(\frac{R_L C_L C_{in}}{C_L+C_{in}}\right)\frac{d}{dt}E^{inc}(t)$
			$Z_L = Z_c \ll \frac{1}{\omega C_{in}}$	$\sim aZ_c C_{in}\sin\theta\frac{d}{dt}E^{inc}(t)$
Wide-Angle Large Cone $[ka \gg 1]$	$\sim \frac{-i}{\sin\theta}$	$\sim Z_c$	$Z_L = R_L + \frac{1}{i\omega C_L}$ with $Z_c + R_L \ll \frac{1}{\omega C_L}$	Voltage across Z_L is $\sim \frac{c}{\sin\theta}\int_{-\infty}^{t}E^{inc}(t')dt'$ Voltage across R_L is $\sim cR_L C_L\frac{E^{inc}(t)}{\sin\theta}$
			$Z_L = Z_c$	$\sim \frac{c}{\sin\theta}\int_{-\infty}^{t}E^{inc}(t')dt'$

$$f_1(t^*, \pi/2) = V_1\left\{\sin(\omega_0 t^*)p_T(t^*) + \sin\left[\omega_0\left(t^* - \frac{T_0}{2}\right)\right]p_T\left(t^* - \frac{T_0}{2}\right)\right.$$
$$\left. - 2\sin\left[\omega_0\left(t^* - \frac{T_0}{4}\right)\right]p_T\left(t^* - \frac{T_0}{4}\right)\right\} \qquad (18b)$$

$$f_3(t^*, \pi/2, \pi/2) = V_1\left\{\sin\left[\omega_0\left(t^* - \frac{T_0}{4}\right)\right]p_T\left(t^* - \frac{T_0}{4}\right) + \sin\left[\omega_0\left(t^* - \frac{3T_0}{4}\right)\right]\right.$$
$$\left. \times p_T\left(t^* - \frac{3T_0}{4}\right) - 2\sin\left[\omega_0\left(t^* - \frac{T_0}{2}\right)\right]p_T\left(t^* - \frac{T_0}{2}\right)\right\} \qquad (18c)$$

On inspecting Eq. (18), one observes that the time-domain radiated field is extended in time with respect to the input signal's duration N/f_0. In particular, the duration of the field is $(N + .75)/f_0$. As noted in the dipole section, this temporal extension arises from radiations that occur at the ends of each dipole, as well as from delayed radiations at the feed points that are caused by current being reflected from the ends of the dipoles. The field is initially vertically polarized for $.25/f_0$, is elliptically polarized with two different orientations for a total time of $.5/f_0$, is circularly polarized for time $(N - .75)/f_0$, is elliptically polarized again for $.5/f_0$, and finally is horizontally polarized for a time of $.25/f_0$. For a single-cycle sine ($N = 1$), the radiated field's duration is 1.75 times the duration of the input, and circular polarization occurs for only one-quarter of the period of the input signal. As one allows the number of input cycles to increase, circular polarization is predominant. Usually when someone speaks of circular polarization, they are considering a CW signal and do not bother with the beginning and ending of that signal.

These results indicate that the familiar observations of circularly polarized fields associated with CW waves are made in the interval between the transient leading and trailing edges, since the so-called CW field in reality has a start and a finish.

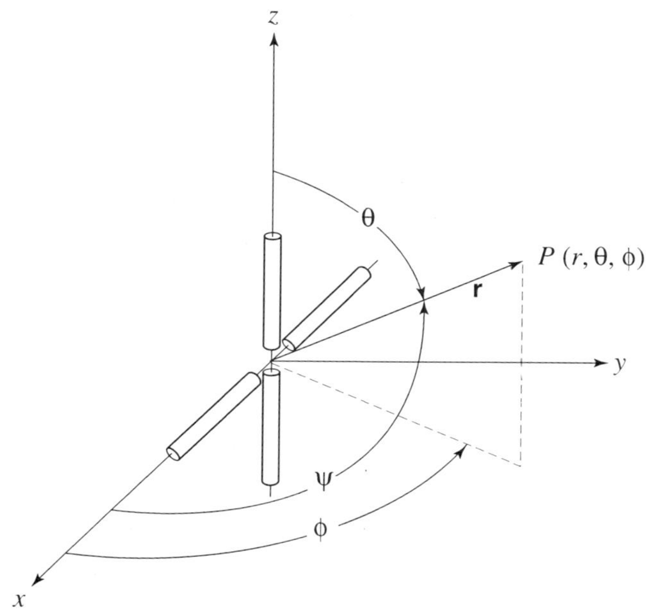

Figure 4. Coplanar, orthogonal, thin, half-wave dipoles.

Observers appear to be unaware of the existence of the different polarizations at the start and finish of the radiated field. Further, these results establish the feasibility of radiating circularly polarized fields in the broadside direction.

CONCLUSIONS

Studies of the dipole indicate that the temporal distortion and stretching of an UWB excitation are caused by radiation from the discontinuities at the dipole's endpoints and from a delayed radiation at its feed point induced by reflections from the endpoints. These conclusions also are supported by the behavior of other simple radiators like the wide-angle bicone, which can be viewed as a fat dipole. Moreover, as a result of further heuristic analysis, the authors arrive at the following conjecture: an antenna radiates from locations where discontinuities occur in the first partial derivatives with respect to the spatial variables of the current and voltage distributions along the antenna surface or where the tangent plane to the antenna's surface does not exist. The preceding observations suggest: if by suitable shaping or by some artificial means like resistive and capacitive loading the radiations and reflections from an antenna's discontinuities can be eliminated or at least minimized, the input excitation would be radiated with minimal distortion. Consequently, such an antenna would be UWB.

The behavior of the radiated field pattern of an electrically small, wide-angle bicone is $\sin\theta$ (like a short dipole); whereas for an electrically large, wide-angle bicone, the field pattern varies as $1/\sin\theta$, where θ is the angle from the bicone's axis of symmetry to the observation direction. In addition, the existence of circularly polarized radiated fields for UWB signals in a preferential direction appears to be feasible. Specifically, when N cycles of a sine pulse are used to excite a pair of orthogonal dipoles with the vertical dipole excited first, a circularly polarized field exists at broadside for $N - 0.25$ cycles. The leading edge of the field is vertically polarized for the first quarter cycle and then takes two distinct elliptical polarizations, each for a quarter cycle. On the other hand, the trailing edge of the radiated field consists of two elliptically polarized portions of a half-cycle total duration, followed by a horizontally polarized component that lasts one quarter cycle.

REFERENCES

1. J. G. Siambis and R. E. Symons, "Ultrawideband clustered-cavity^{TM} klystron," in: *Ultra-Wideband, Short-Pulse Electromagnetics*, H. L. Bertoni, L. Carin, and L. B. Felsen, eds., Plenum, New York (1993), p. 121.

2. G. F. Ross, "A time domain criterion for the design of wideband radiating elements," *IEEE Trans. Antennas Propagat.* **AP-16** (3):355 (1968).

3. T. T. Wu and R. W. P. King, "The cylindrical antenna with non-reflecting resistive loading," *IEEE Trans. Antennas Propagat.* **13**:369 (1965).

4. J. G. Maloney and G. S. Smith, "A study of transient radiation from the Wu-King resistive monopole – FDTD analysis and experimental measurements," *IEEE Trans. Antennas Propagat.* **41**:668 (1993).

5. S. N. Samaddar, "Transient radiation of single-cycle sinusoidal pulse from a thin dipole," *J. Franklin Inst.* **329** (2):259 (1992).

6. S. N. Samaddar, "Behavior of a received pulse radiated by a half-wave dipole excited by a single-cycle sinusoidal voltage," *J. Franklin Inst.* **330** (1):17 (1993).

7. L. Brillouin, *Wave Propagation and Group Velocity*, Academic Press, New York (1960).

8. E. Hallen, *Electromagnetic Theory*, Wiley, New York (1962), pp. 501-502.

9. L. C. Shen and T. T. Wu, "Cylindrical antenna with tapered resistive loading," *Radio Sci.* **2**:191 (1967).

10. J. B. L. Rao, J. E. Ferris, and W. E. Zimmerman, "Broadband characteristics of cylindrical antennas with exponentially tapered capacitive loading," *IEEE Trans. Antennas Propagat.* **AP-17**:145 (1969).

11. J. G. Maloney and G. S. Smith, "Optimization of a conical antenna for pulse radiation: an efficient design using resistive loading," *IEEE Trans. Antennas Propagat.* **41**:940 (1993).

12. M. Parent, private communication (1995).

13. J. D. Kraus, *Antennas* (2nd ed.), McGraw-Hill, New York (1988).

14. D. L. Sengupta and J. E. Ferris, "Investigation of the rudimentary horn," Final Report AD857364, The University of Michigan, Electrical Engineering Dept., Radiation Laboratory, Ann Arbor MI (1969).

15. R. F. Harrington, *Field Computation by Moment Methods*, Krieger, Malabar FL (1968), p. 64.

16. P. Franklin, *Treatise on Advanced Calculus*, Wiley, New York (1940), p. 202.

17. L. J. Chu and J. A. Stratton, "Steady-state solutions of electromagnetic field problems. Forced oscillations of a conducting sphere," *J. Appl. Phys.* **12** (12):236 (1941).

18. L. J. Chu and J. A. Stratton, "Steady-state solutions of electromagnetic field problems. Forced oscillations of a prolate spheroid," *J. Appl. Phys.* **12** (12):241 (1941).

19. G. Franceschetti, "A canonical problem in transient radiation – the spherical antenna," *IEEE Trans. Antennas Propagat.* **AP-26**:551 (1978).

20. O. M. Bucci and G. Franceschetti, "Input admittance and transient response of spheroidal antennas in dispersive media," *IEEE Trans. Antennas Propagat.* **AP-22**:526 (1974).

21. V. A. Yatskevich and L. L. Fedosenko, "Antennas for radiation of very-wide-band signal," *Izvestiya Vuz. Radioelektronika* **29** (2):69 (1986).

22. C. E. Baum and E. G. Farr, "Impulse radiating antennas," in: *Ultra-Wideband, Short-Pulse Electromagnetics*, H. L. Bertoni, L. Carin, and L. B. Felsen, eds., Plenum, New York (1993), pp. 139,147.

23. P. D. P. Smith, "The conical dipole of wide-angle," *J. Appl. Phys.* **19**:11 (1948).

24. C. T. Tai, "On the theory of biconical antennas," *J. Appl. Phys.* **19**:1155 (1948).

25. C. T. Tai, "A study of the E.M.F. method," *J. Appl. Phys.* **20**:717 (1949).

26. C. T. Tai, "Application of variational principle to biconical antennas," *J. Appl. Phys.* **20**:1076 (1949).

27. S. N. Samaddar, "Preliminary study of some antenna elements which have potential usefulness as ultrawideband radiators," NRL Memorandum Report, NRL/MR/5341-95-7794, Nov. 24, 1995, Naval Research Laboratory, Washington DC (1995).

28. S. N. Samaddar and E. L. Mokole, "Biconical antennas with unequal cone angles," submitted to *IEEE Trans. Antennas Propagat.* (1996).

29. C. H. Papas and R. W. P. King, "Input impedance of wide-angle conical antennas fed by a coaxial line," *Proc. IRE* **37**:1269 (1949).

30. C. H. Papas and R. W. P. King, "Radiation from wide-angle conical antennas fed by a coaxial line," *Proc. IRE* **39**:49 (1951).

31. S. S. Sandler and R. W. P. King, "Compact conical antennas for wide-band coverage," *IEEE Trans. Antennas Propagat.* **AP-42**:436 (1994).

32. C. W. Harrison and C. S. Williams, "Transients in wide-angle conical antennas," *IEEE Trans. Antennas Propagat.* **AP-13**:230 (1965).

33. A. K. Choudhury, E. L. Mokole, and S. N. Samaddar, "Radiated field from thin half-wave uncoupled orthogonal dipoles excited by finite-cycle sinusoids," presented at the National Radio Science Meeting, Boulder CO (9–13 January 1996).

THEOREMS ON TIME-DOMAIN FAR FIELDS

Arthur D. Yaghjian and Thorkild B. Hansen

RL/ERCS, Hanscom AFB, MA 01731

INTRODUCTION

The far-field characteristics of classical electromagnetic fields satisfying Maxwell's equations have been investigated quite thoroughly for sources radiating at a single frequency, that is, for frequency-domain or time-harmonic fields [1]. For example, frequency-domain far fields radiated by integrable sources in a volume of finite extent decay as $1/r$ or faster as the distance r to the far field approaches infinity. In addition, these far fields are entire analytic functions of their angular variables θ and ϕ. Using a plane-wave decomposition, frequency-domain near fields can be expressed as an integral of the far-field pattern and its analytic continuation to complex angles of observation [2],[3].

The characteristics of time-domain far fields, that is, the far fields of electromagnetic sources that turn on and off in a finite time interval, have not been investigated nearly as thoroughly as those of frequency-domain far fields. Thus, the purpose of this paper is to derive some of the characteristics of time-domain far fields and to concentrate on those characteristics that are unique to the time domain.

Specifically, conditions are given on the source current that ensure the time-domain far fields will decay as $1/r$ or faster as $r \to \infty$. When these conditions are not satisfied, slower than $1/r$ far-field decay can occur and, in particular, the necessary conditions are derived for the existence of "electromagnetic missiles," far-field pulses with energy decay slower than $1/r^2$.

The time-domain near fields are expressed as integrals of time-domain far-field functions (plane-wave spectra) over real and complex angles of observation. After all sources are turned off, it is shown that the time-domain fields throughout all space can be written as an integral of the far-field pattern over real angles of observation only.

It is shown that time-domain far fields with zero sidelobes cannot be excited by sources in a finite region of space. Under certain conditions, satisfied, for example, by bandlimited fields, the time-domain far-field pattern, like the frequency-domain far-field pattern, is part of an analytic function of complex variables θ and ϕ. However, the analytic continuation of the time-domain far-field pattern to complex angles of observation does not, in general, equal the time-domain evanescent far-field function (time-domain plane-wave evanescent spectrum).

It is proven that there are also restrictions on the time dependence of electromagnetic far fields. Namely, the integral over all time of the far-field pattern must be zero for sources that turn on and off in a finite time interval.

Many of the results contained in this paper for electromagnetic fields, and the corresponding results for acoustic fields, may also be found in [3] and [4].

Ultra-Wideband, Short-Pulse Electromagnetics 3
Edited by Baum *et al.*, Plenum Press, New York, 1997

FAR FIELDS IN TERMS OF CURRENT

The far-field equations derive rigorously from the vector-potential solution to Maxwell's equations in the time domain outside the source region [5]

$$\mathbf{H}(\mathbf{r},t) = \frac{1}{\mu}\nabla \times \mathbf{A}(\mathbf{r},t) \tag{1}$$

$$\mathbf{E}(\mathbf{r},t) = \frac{1}{\epsilon}\nabla \times \int_{t_0}^{t} \mathbf{H}(\mathbf{r},t')dt', \quad \mathbf{r} \notin V \tag{2}$$

where

$$\mathbf{A}(\mathbf{r},t) = \frac{\mu}{4\pi}\int_V \frac{\mathbf{J}(\mathbf{r}',t-R/c)}{R}dV' \tag{3}$$

and $\mathbf{R} = \mathbf{r} - \mathbf{r}'$ ($R = |\mathbf{r} - \mathbf{r}'|$). $\mathbf{E}(\mathbf{r},t), \mathbf{H}(\mathbf{r},t), \mathbf{A}(\mathbf{r},t)$ and $\mathbf{J}(\mathbf{r},t)$ denote the electric field, magnetic field, vector potential, and current density at the position $\mathbf{r}$ and time t. The constants μ and ϵ are the permeability and permittivity of free space, respectively. The current in (3) lies in a volume V of finite extent, and $\mathbf{J}(\mathbf{r}',t-R/c)$ is assumed integrable over this finite source volume V (at least for observation points $\mathbf{r}$ outside the source region) [6]. Inserting $\mathbf{A}$ from (3) into (1), and bringing the curl operator under the integral sign produces the following expression for the magnetic field

$$\mathbf{H}(\mathbf{r},t) = -\frac{1}{4\pi}\int_V \left[\frac{\mathbf{R}\times\frac{\partial}{\partial t}\mathbf{J}(\mathbf{r}',t-R/c)}{cR^2} + \frac{\mathbf{R}}{R^3}\times\mathbf{J}(\mathbf{r}',t-R/c)\right]dV' \tag{4}$$

where we have used the identity [8, app.1, eq.186] $\nabla \times \frac{\mathbf{J}(\mathbf{r}',t-R/c)}{R} = -\frac{\mathbf{R}\times\frac{\partial}{\partial t}\mathbf{J}(\mathbf{r}',t-R/c)}{cR^2} - \frac{\mathbf{R}}{R^3}\times \mathbf{J}(\mathbf{r}',t-R/c)$. The interchange of differentiation and integration required to obtain (4) is valid outside the source region if the time derivative of the current exists and is finite, or more generally, $|\frac{\partial}{\partial t}\mathbf{J}(\mathbf{r}',t)| \leq \alpha(\mathbf{r}')$, where $\alpha(\mathbf{r}')$ is a function, independent of time, that is integrable over the finite source region V [7, sec.246, p.355]. (This inequality implies that $\mathbf{J}(\mathbf{r}',t)$ is also bounded by an integrable function; namely $|\mathbf{J}(\mathbf{r}',t)| \leq |t-t_0|\alpha(\mathbf{r}')$, where t_0 is the time at which the sources are turned on.)

Multiplying (4) by r, then taking the absolute value and limit of (4) as $r \to \infty$ gives

$$\lim_{r\to\infty}|r\mathbf{H}(\mathbf{r},t)| = \frac{1}{4\pi}\lim_{r\to\infty}\left|\int_V r\left[\frac{\mathbf{R}\times\frac{\partial}{\partial t}\mathbf{J}(\mathbf{r}',t-R/c)}{cR^2} + \frac{\mathbf{R}}{R^3}\times\mathbf{J}\right]dV'\right| \leq \frac{1}{4\pi c}\int_V \alpha(\mathbf{r}')dV'. \tag{5}$$

Bringing the limit inside the last integral of (5) is allowed by standard theorems on limits of integrals [7, p.324]. To obtain the analogous results for the electric field, insert $\mathbf{H}$ from (4) into (2) to get

$$\mathbf{E}(\mathbf{r},t) = -\tfrac{1}{4\pi}\sqrt{\tfrac{\mu}{\epsilon}}\int_V \left[\tfrac{1}{cR}\tfrac{\partial}{\partial t}\mathbf{J}(\mathbf{r}',t-R/c) - \tfrac{1}{cR^3}\left(\mathbf{R}\cdot\tfrac{\partial}{\partial t}\mathbf{J}(\mathbf{r}',t-R/c)\right)\mathbf{R}\right.$$
$$\left. -\tfrac{3}{R^4}\mathbf{R}\cdot\mathbf{J}\mathbf{R} + \tfrac{\mathbf{J}}{R^2} - \tfrac{3c}{R^5}\mathbf{R}\cdot\left(\int_{t_0}^{t-R/c}\mathbf{J}(\mathbf{r}',t')dt'\right)\mathbf{R} + \tfrac{c}{R^3}\int_{t_0}^{t-R/c}\mathbf{J}(\mathbf{r}',t')dt'\right]dV'. \tag{6}$$

And as $r \to \infty$ one obtains the far-field result similar to (5), namely

$$\lim_{r\to\infty}|r\mathbf{E}(\mathbf{r},t)| \leq \frac{1}{4\pi c}\sqrt{\frac{\mu}{\epsilon}}\int_V \alpha(\mathbf{r}')dV'. \tag{7}$$

The inequalities (5) and (7) prove that the far magnetic and electric fields decay as $1/r$ (or faster) as $r \to \infty$, provided the first time derivative of the current is bounded by an integrable function in the finite source region V.

If, in addition, the first time derivative of the current is a continuous function of time, the limit of (4) and (6) as $r \to \infty$ can be brought inside the integrals [7, p.324] to get the following explicit expressions for the far magnetic and electric fields

$$\mathbf{H}(\mathbf{r},t) \sim -\frac{1}{4\pi cr}\hat{\mathbf{r}}\times\int_V \frac{\partial}{\partial t}\mathbf{J}(\mathbf{r}',t-r/c+\hat{\mathbf{r}}\cdot\mathbf{r}'/c)dV' + O\left(\frac{1}{r^2}\right) \tag{8}$$

$$\mathbf{E}(\mathbf{r},t) \sim \sqrt{\frac{\mu}{\epsilon}} \frac{1}{4\pi cr} \hat{\mathbf{r}} \times \left(\hat{\mathbf{r}} \times \int_V \frac{\partial}{\partial t} \mathbf{J}(\mathbf{r}', t - r/c + \hat{\mathbf{r}} \cdot \mathbf{r}'/c) dV' \right) + O\left(\frac{1}{r^2}\right) \tag{9}$$

which satisfy the far-field relationship

$$\mathbf{H}(\mathbf{r},t) \sim \sqrt{\frac{\epsilon}{\mu}} \hat{\mathbf{r}} \times \mathbf{E}(\mathbf{r},t), \qquad r \to \infty. \tag{10}$$

The far-field expressions in the time domain, (8) and (9), check with the far-field expressions in the frequency domain [3] in that they are Fourier transforms of each other. Here we have given conditions on the current that assure the validity of the time-domain far-field expressions (5), (7), (8) and (9). In the section on "electromagnetic missiles," we consider time-domain far fields radiated by current sources that do not satisfy these conditions.

NEAR FIELDS IN TERMS OF FAR FIELDS

In this section we derive expressions for the time-domain near fields of a radiating source distribution in terms of time-domain far-field functions. Begin the derivation by defining the time-domain and frequency-domain far-field patterns as

$$\mathcal{F}(\theta, \phi, t) = \lim_{r \to \infty} r\mathbf{E}(\mathbf{r}, t + r/c) \tag{11}$$

$$\mathcal{F}_\omega(\theta, \phi) = \lim_{r \to \infty} r\, e^{-i\omega r/c} \mathbf{E}_\omega(\mathbf{r}) \tag{12}$$

where the frequency-domain electric field $\mathbf{E}_\omega(\mathbf{r})$ is the Fourier transform of the time-domain electric field $\mathbf{E}(\mathbf{r}, t)$, and thus the frequency-domain far-field pattern $\mathcal{F}_\omega(\theta, \phi)$ is simply the Fourier transform of the time-domain far-field pattern $\mathcal{F}(\theta, \phi, t)$:

$$\mathcal{F}_\omega(\theta, \phi) = \frac{1}{2\pi} \int_{-\infty}^{+\infty} \mathcal{F}(\theta, \phi, t) e^{i\omega t} dt \tag{13}$$

$$\mathcal{F}(\theta, \phi, t) = \int_{-\infty}^{+\infty} \mathcal{F}_\omega(\theta, \phi) e^{-i\omega t} d\omega. \tag{14}$$

From (9) and (13) the far-field patterns satisfy the relationships $\hat{\mathbf{r}} \cdot \mathcal{F} = 0$ and $\hat{\mathbf{r}} \cdot \mathcal{F}_\omega = 0$. The time-domain and frequency-domain electromagnetic fields are solutions to Maxwell's time-dependent and time-harmonic equations, respectively. It is assumed that the time-domain electromagnetic fields are sufficiently well behaved that the time- and frequency-domain fields, including the far fields, can be expressed as Fourier transforms of each other. (The theory of generalized functions allows the rigorous application of Fourier transform theory to all but the most pathological time dependence, certainly to all physically realizable fields [8, app.6]).

The frequency-domain electric field can be written as a superposition of propagating and evanescent plane waves [2],[3]

$$\begin{aligned} \mathbf{E}_\omega(\mathbf{r}) = {} & \frac{1}{2\pi c} \int_0^{2\pi} \int_0^{\pi/2} i\omega \mathcal{F}_\omega(\sigma_\theta, \sigma_\phi) e^{i\omega c^{-1}(x \cos\sigma_\phi \sin\sigma_\theta + y \sin\sigma_\phi \sin\sigma_\theta + z \cos\sigma_\theta)} \sin\sigma_\theta d\sigma_\theta d\sigma_\phi \\ & - \frac{i}{2\pi c} \int_0^{2\pi} \int_0^{+\infty} i\omega \mathcal{F}_\omega(\pi/2 - i\tfrac{\omega}{|\omega|}\alpha, \sigma_\phi) e^{i\omega c^{-1}(x \cos\sigma_\phi \cosh\alpha + y \sin\sigma_\phi \cosh\alpha)} \frac{\omega}{|\omega|} e^{-(z/c)|\omega| \sinh\alpha} \cosh\alpha d\alpha d\sigma_\phi \end{aligned} \tag{15}$$

where the $\exp(-i\omega t)$ harmonic time dependence has been suppressed. The plane-wave formula (15) holds for all frequencies $(-\infty < \omega < +\infty)$ and for all observation points $\mathbf{r} = x\hat{\mathbf{x}} + y\hat{\mathbf{y}} + z\hat{\mathbf{z}}$ in the source-free half space $z > z_0$. It is assumed that the sources lie within a volume of finite extent in the half space $z < z_0$. The evanescent spectrum, given by $\mathcal{F}_\omega(\pi/2 - i\frac{\omega}{|\omega|}\alpha, \sigma_\phi)$ in (15) for $0 < \alpha < \infty$, is the frequency-domain far-field pattern $\mathcal{F}_\omega(\sigma_\theta, \sigma_\phi)$ analytically continued to complex values of σ_θ. Taking the inverse Fourier transform of (15), applying the convolution rule along with the identity

$$-i \int_{-\infty}^{+\infty} \frac{\omega}{|\omega|} e^{-(z/c)|\omega| \sinh\alpha} e^{-i\omega t} d\omega = -\frac{2t}{(zc^{-1}\sinh\alpha)^2 + t^2} \tag{16}$$

and defining the time-domain evanescent far-field function (time-domain plane-wave evanescent spectrum) as

$$\mathcal{F}_e(\pi/2 - i\alpha, \phi, t) = \int_{-\infty}^{+\infty} \mathcal{F}_\omega(\pi/2 - i\frac{\omega}{|\omega|}\alpha, \phi)e^{-i\omega t}d\omega. \tag{17}$$

one finds that the time domain electric field is given by

$$\mathbf{E}(\mathbf{r},t) = -\frac{1}{2\pi c}\int_0^{2\pi}\int_0^{\pi/2}\frac{\partial}{\partial t}\mathcal{F}(\sigma_\theta,\sigma_\phi,t-\mathbf{r}\cdot\hat{\boldsymbol{\sigma}}_{\theta\phi}/c)\sin\sigma_\theta d\sigma_\theta d\sigma_\phi \tag{18}$$
$$+\frac{1}{2\pi^2 c}\int_0^{2\pi}\int_0^{+\infty}\int_{-\infty}^{+\infty}\frac{\partial}{\partial t}\mathcal{F}_e(\pi/2-i\alpha,\sigma_\phi,t-t'-\mathbf{r}\cdot\hat{\boldsymbol{\sigma}}_\phi c^{-1}\cosh\alpha)\frac{t'}{(zc^{-1}\sinh\alpha)^2+t'^2}dt'\cosh\alpha d\alpha d\sigma_\phi$$

where the unit vectors are given by $\hat{\boldsymbol{\sigma}}_{\theta\phi} = \hat{\mathbf{x}}\cos\sigma_\phi\sin\sigma_\theta + \hat{\mathbf{y}}\sin\sigma_\phi\sin\sigma_\theta + \hat{\mathbf{z}}\cos\sigma_\theta$ and $\hat{\boldsymbol{\sigma}}_\phi = \hat{\mathbf{x}}\cos\sigma_\phi + \hat{\mathbf{y}}\sin\sigma_\phi$.

For sources in a finite region of the half space $z < z_0$, the formula (18) gives the time-domain field everywhere in the half space $z > z_0$ in terms of the time-domain far-field pattern $\mathcal{F}(\theta,\phi,t)$ and the time-domain evanescent far-field function $\mathcal{F}_e(\pi/2 - i\alpha,\phi,t)$. If only the far-field pattern $\mathcal{F}$ is known, (18) cannot in general be used to determine the near field. However, after all sources have been turned off so that $\mathbf{E}(\mathbf{r},t)$ satisfies the homogeneous wave equation ($\nabla\times\nabla\times\mathbf{E} - \partial^2\mathbf{E}/(\partial ct)^2 = 0$ and $\nabla\cdot\mathbf{E} = 0$) throughout *all* space, one may derive a formula for the near field, first derived by Moses *et al.* [9], involving only the far-field pattern evaluated at real angles of observation.

We shall derive this formula directly from Green's second identity using the advanced Green's function. The frequency-domain electric field $\mathbf{E}_\omega(\mathbf{r})$ and the advanced Green's function $G_\omega^a(\mathbf{r},\mathbf{r}') = e^{-ik|\mathbf{r}-\mathbf{r}'|}/(4\pi|\mathbf{r}-\mathbf{r}'|)$ satisfy the following inhomogeneous wave equations throughout all space

$$\nabla^2\mathbf{E}_\omega(\mathbf{r}) + k^2\mathbf{E}_\omega(\mathbf{r}) = -\mathbf{Q}_\omega(\mathbf{r}) \tag{19}$$

$$\nabla^2 G_\omega^a(\mathbf{r},\mathbf{r}') + k^2 G_\omega^a(\mathbf{r},\mathbf{r}') = -\delta(\mathbf{r}-\mathbf{r}') \tag{20}$$

where $\delta(\mathbf{r})$ is the three-dimensional delta function, and $k = \omega/c$. The frequency-domain source function $\mathbf{Q}_\omega(\mathbf{r})$ can be written in terms of the frequency-domain current and charge densities as

$$\mathbf{Q}_\omega(\mathbf{r}) = i\omega\mu\mathbf{J}_\omega(\mathbf{r}) - \nabla\rho_\omega(\mathbf{r})/\epsilon. \tag{21}$$

Although the advanced Green's function G_ω^a obeys the same wave equation as the retarded Green's function, it does not satisfy the outgoing radiation condition.

Inserting $\mathbf{E}_\omega$ and G_ω^a into Green's second identity and making use of (19) and (20) produces the expression

$$\mathbf{E}_\omega(\mathbf{r}) = \frac{1}{4\pi}\int_V\frac{\mathbf{Q}_\omega(\mathbf{r}')e^{-ik|\mathbf{r}-\mathbf{r}'|}}{|\mathbf{r}-\mathbf{r}'|}dV' + \int_S\left[G_\omega^a(\mathbf{r},\mathbf{r}')\frac{\partial}{\partial n'}\mathbf{E}_\omega(\mathbf{r}') - \mathbf{E}_\omega(\mathbf{r}')\frac{\partial}{\partial n'}G_\omega^a(\mathbf{r},\mathbf{r}')\right]dS' \tag{22}$$

where V is a volume of finite extent that contains the sources and S is a surface that encloses V and the observation point $\mathbf{r}$. Let S be the sphere of radius r' and let $r' \to \infty$. Then

$$\frac{\partial G_\omega^a}{\partial n'} \sim -ikG_\omega^a \sim -\frac{ike^{-ikr'}e^{ik\mathbf{r}\cdot\hat{\mathbf{r}}'}}{4\pi r'}\,, \quad \frac{\partial\mathbf{E}_\omega}{\partial n'} \sim ik\mathbf{E}_\omega \sim \frac{ik\mathcal{F}_\omega(\mathbf{r}')}{r'} \tag{23}$$

and we find from (22) that

$$\mathbf{E}_\omega(\mathbf{r}) = \frac{1}{4\pi}\int_V\frac{\mathbf{Q}_\omega(\mathbf{r}')e^{-ik|\mathbf{r}-\mathbf{r}'|}}{|\mathbf{r}-\mathbf{r}'|}dV' + \frac{1}{2\pi}\int_0^{2\pi}\int_0^{\pi} ik\mathcal{F}_\omega(\sigma_\theta,\sigma_\phi)e^{ik\mathbf{r}\cdot\hat{\boldsymbol{\sigma}}_{\theta\phi}}\sin\sigma_\theta d\sigma_\theta d\sigma_\phi \tag{24}$$

where the surface integral has been rewritten in terms of spherical coordinates.

Taking the inverse Fourier transform of (24) gives

$$\mathbf{E}(\mathbf{r},t) = \frac{1}{4\pi}\int_V\frac{\mathbf{Q}(\mathbf{r}',t+|\mathbf{r}-\mathbf{r}'|/c)}{|\mathbf{r}-\mathbf{r}'|}dV' - \frac{1}{2\pi c}\int_0^{2\pi}\int_0^{\pi}\frac{\partial}{\partial t}\mathcal{F}(\sigma_\theta,\sigma_\phi,t-\mathbf{r}\cdot\hat{\boldsymbol{\sigma}}_{\theta\phi}/c)\sin\sigma_\theta d\sigma_\theta d\sigma_\phi \tag{25}$$

throughout all space and time. The time-domain source function $\mathbf{Q}(\mathbf{r},t)$ can be found in terms of the time-domain current and charge densities by taking the inverse Fourier transform of (21) to get

$$\mathbf{Q}(\mathbf{r},t) = -\mu\frac{\partial \mathbf{J}(\mathbf{r},t)}{\partial t} - \nabla\rho(\mathbf{r},t)/\epsilon. \tag{26}$$

After all sources are turned off, that is, $\mathbf{J}(\mathbf{r},t) = 0$ and $\rho(\mathbf{r},t) = 0$ so that $\mathbf{Q}(\mathbf{r},t) = 0$ for all $\mathbf{r}$ and $t > t_1$, the advanced source function $\mathbf{Q}(\mathbf{r}', t + |\mathbf{r} - \mathbf{r}'|/c)$, and thus the volume integral, is zero. Consequently, for $t > t_1$ (25) reduces to simply

$$\mathbf{E}(\mathbf{r},t) = -\frac{1}{2\pi c}\int_0^{2\pi}\int_0^{\pi}\frac{\partial}{\partial t}\boldsymbol{\mathcal{F}}(\sigma_\theta,\sigma_\phi,t - \mathbf{r}\cdot\hat{\boldsymbol{\sigma}}_{\theta\phi}/c)\sin\sigma_\theta d\sigma_\theta d\sigma_\phi, \quad t > t_1 \tag{27}$$

where t_1 is the time after which all sources are zero. Similarly, the magnetic field corresponding to (27) is given by

$$\mathbf{H}(\mathbf{r},t) = -\frac{1}{2\pi c}\sqrt{\frac{\epsilon}{\mu}}\int_0^{2\pi}\int_0^{\pi}\hat{\boldsymbol{\sigma}}_{\theta\phi}\times\frac{\partial}{\partial t}\boldsymbol{\mathcal{F}}(\sigma_\theta,\sigma_\phi,t - \mathbf{r}\cdot\hat{\boldsymbol{\sigma}}_{\theta\phi}/c)\sin\sigma_\theta d\sigma_\theta d\sigma_\phi, \quad t > t_1. \tag{28}$$

(The formulas (27) and (28) can also be derived directly in the time domain using the advanced free-space time-domain Green's function. This alternative derivation avoids the use of the Fourier transform and does not require the existence of the Fourier transform of the time-domain functions.)

Note that (27) and (28) are valid for all observation points $\mathbf{r}$ as long as t is greater than t_1. Also, the integrations in (27) and (28) cover the entire far-field sphere, whereas the integration over the real angles in (18), that is, the first term on the right side of (18), covers only the far-field hemisphere in the half space $z > z_0$ in which (18) is valid. Equating (18) and (27) shows that after all sources are turned off, and for $z > z_0$, the integral over the evanescent time-domain far-field function (second term on the right hand side of (18)) equals the integral of the time-domain far-field pattern over the opposite hemisphere ($z < z_0$).

In summary, the electromagnetic near fields, for observation points outside the source region and for every instant of time, can be expressed as integrals of far-field functions (plane-wave spectra) over both real and complex angles of observation. Moreover, the near fields, at all points of observation and at times after all sources have been turned off, can be expressed in terms of an integral of the far-field pattern over only real angles of observation. This latter, rather surprising result was first derived by Moses *et al.* [9] for both acoustic and electromagnetic fields.

ANALYTICITY OF THE FAR-FIELD PATTERN

It may be tempting to insert an arbitrary desired far field into (27) or (28) to compute the near field that will generate this far field. However, care must be taken to choose a time-domain far-field pattern that is compatible with sources in a finite region of space. Specifically, the time-domain far-field pattern of sources confined to a finite region of space cannot have zero sidelobes, and must, under certain conditions, be part of an analytic function of complex angles θ and ϕ.

To prove these results, express the time-domain far-field pattern in terms of the frequency-domain far-field pattern, that is, as the inverse Fourier transform in (14)

$$\boldsymbol{\mathcal{F}}(\theta,\phi,t) = \int_{-\infty}^{+\infty}\boldsymbol{\mathcal{F}}_\omega(\theta,\phi)e^{-i\omega t}d\omega. \tag{29}$$

From Theorem 29 of Müller [1], the frequency-domain far-field pattern $\boldsymbol{\mathcal{F}}_\omega(\theta,\phi)$ of sources in a finite region of space is an entire analytic function of complex (θ,ϕ).

The impossibility of obtaining zero-sidelobe far fields in the time domain from sources in a volume of finite extent follows directly from (29). Specifically, taking the inverse Fourier transform of (29) in an angular region where $\mathcal{F}(\theta, \phi, t)$ is chosen to be zero shows that $\mathcal{F}_\omega(\theta, \phi)$ equals zero in the same angular region. Since analyticity of $\mathcal{F}_\omega(\theta, \phi)$ for sources in a volume of finite extent implies $\mathcal{F}_\omega(\theta, \phi)$ cannot equal zero throughout an angular region (unless it is zero everywhere), it follows that $\mathcal{F}(\theta, \phi, t)$ cannot be chosen zero throughout an angular region, unless it is zero everywhere. In particular, "electromagnetic bullets" (time-domain fields with zero far fields outside an angular cone) [9],[11] cannot be excited by sources in a finite region of space.

If, unknowingly, a zero sidelobe far-field pattern is inserted into (27) or (28), one obtains a near-field distribution that cannot be generated by sources in a finite region of space but requires a source of infinite extent. However, it is emphasized that (18) and (27) or (28), like their frequency-domain counterparts, can still be useful for determining near fields and source distributions that produce very low sidelobes.

The analyticity of $\mathcal{F}_\omega(\theta, \phi)$ in (29) also implies the analyticity of $\mathcal{F}(\theta, \phi, t)$ with respect to θ and ϕ in a complex domain D under the following conditions [12]:
Throughout the domain D

1. the integral in (29) converges,

2. the first derivatives of $\mathcal{F}_\omega(\theta, \phi)$ with respect to θ and ϕ are continuous functions of ω,

3. and the Fourier integrals over ω of the first derivatives converge uniformly in θ and ϕ.

These restrictions on $\mathcal{F}_\omega(\theta, \phi)$ are satisfied for all complex (θ, ϕ) by fields that have continuous, bandlimited frequency spectra. We note that $\mathcal{F}(\theta, \phi, t)$ for complex θ and ϕ is not necessarily a real-valued function. Also note from (17) and (29) that even when $\mathcal{F}(\pi/2 - i\alpha, \phi, t)$ exists (that is, (29) converges for $\theta = \pi/2 - i\alpha, 0 < \alpha < \infty$), it does not, in general, equal the evanescent spectrum $\mathcal{F}_e(\pi/2 - i\alpha, \phi, t)$ defined in (17). (In [13] we failed to make this distinction between the time-domain far-field functions, $\mathcal{F}_e(\pi/2 - i\alpha, \phi, t)$ and $\mathcal{F}(\pi/2 - i\alpha, \phi, t)$.)

FAR FIELDS INTEGRATED OVER TIME

There are also restrictions on the time dependence of the far fields. In particular, we shall show that the time-domain far-field pattern integrated over all time is zero, if the sources are located in a finite region of space and are turned on and off in a finite time period.

The behavior of the electromagnetic far fields integrated over all time can be found directly from the expressions (4) and (6) for the fields in terms of the current density $\mathbf{J}(\mathbf{r}, t)$ located in a volume V of finite extent. Since the current is in a volume of finite extent, (4) and (6) can be integrated over all time, and the integrations over time and space can be interchanged to prove

$$\int_{-\infty}^{+\infty} \mathbf{H}(\mathbf{r}, t) dt \overset{r \to \infty}{\sim} O\left(\frac{1}{r^2}\right) \tag{30}$$

$$\int_{-\infty}^{+\infty} \mathbf{E}(\mathbf{r}, t) dt \overset{r \to \infty}{\sim} O\left(\frac{1}{r^2}\right) \tag{31}$$

that is, the time-integrated electric and magnetic fields approach zero as $1/r^2$ or faster as $r \to \infty$, if the current turns on and off in a finite time interval. In terms of the time-domain far-field pattern

$$\int_{-\infty}^{+\infty} \mathcal{F}(\theta, \phi, t) dt = 0 \tag{32}$$

if the current exists only during a finite time interval.

Only if the current remains nonzero as $t \to \infty$ can the far-field pattern integrated over all time be nonzero. For example, if the time-domain current is zero for $t < t_0$ and approaches a nonzero static value as $t \to \infty$

$$\lim_{t \to \infty} \mathbf{J}(\mathbf{r}, t) = \mathbf{J}_0(\mathbf{r}) \tag{33}$$

the integrals of the far-field pattern over all time can be written from (9) and (11) as

$$\int_{-\infty}^{+\infty} \mathcal{F}(\theta, \phi, t)dt = \sqrt{\frac{\mu}{\epsilon}}\frac{1}{4\pi c}\hat{\mathbf{r}} \times \hat{\mathbf{r}} \times \int_V \mathbf{J}_0(\mathbf{r}')dV'. \tag{34}$$

With $\mathbf{J}_0$ inserted from the vector identity, $\nabla\cdot(\mathbf{r}\mathbf{J}_0) = \mathbf{r}(\nabla\cdot\mathbf{J}_0)+\mathbf{J}_0$, the equation (34) becomes

$$\int_{-\infty}^{+\infty} \mathcal{F}(\theta, \phi, t)dt = -\sqrt{\frac{\mu}{\epsilon}}\frac{1}{4\pi c}\hat{\mathbf{r}} \times \hat{\mathbf{r}} \times \int_V \mathbf{r}'\frac{\partial}{\partial t}\rho_0(\mathbf{r}', \infty)dV' \tag{35}$$

where use has been made of the divergence theorem and the continuity equation

$$-\frac{\partial}{\partial t}\rho_0(\mathbf{r}, \infty) = -\lim_{t\to\infty}\frac{\partial}{\partial t}\rho_0(\mathbf{r}, t) = \nabla \cdot \mathbf{J}_0(\mathbf{r}). \tag{36}$$

The divergence of $\mathbf{J}_0(\mathbf{r})$ is not a function of time in (36) so that the charge distribution $\rho_0(\mathbf{r}, t)$ increases linearly with time as $t \to \infty$, that is

$$\rho_0(\mathbf{r}, t) \sim -\nabla \cdot \mathbf{J}_0(\mathbf{r})t, \quad t \to \infty. \tag{37}$$

Thus, the electromagnetic far-field pattern integrated over all time can be nonzero when the current approaches a static value as $t \to \infty$, but only if the static charge distribution, or more precisely, the electric dipole moment or higher order multipole moments of the charge distribution, grows linearly with time as $t \to \infty$. (Of course, the continuity equation valid for all times implies that the *total* charge for sources in a volume of finite extent remains zero because it is assumed zero for $t < t_0$.)

In general, the time-domain near fields at a point in space do not integrate over all time to zero. However, an integration over all time of the expression (18) for the near electric field (and the analogous expression for the magnetic field) shows that it is the evanescent part of the fields only that contribute to the integration over all time t, provided the sources (and thus the far fields) are nonzero only during a finite time interval. The integration over all time t of the propagating part of the fields, that is, the first integral on the right side of (18), can be interchanged with the *finite* spatial integration to give zero for far fields that are zero outside a finite time interval. The integration over all time t of the evanescent part of the near fields, that is, the second integral on the right side of (18), cannot be rigorously interchanged with the *infinite* spatial and t' integrations, and is not, in general, equal to zero.

ELECTROMAGNETIC MISSILES

In 1985 Wu published a paper [14] in which he showed that the energy in an electromagnetic pulse could decay more slowly than $1/r^2$ as r approaches infinity. He proved these results mathematically from the classical Maxwell equations for sources in a finite region of space radiating a finite amount of energy. He called these unusual far-field pulses "electromagnetic missiles." In this section we shall give the necessary conditions in both the time and frequency domains for the time-domain far fields to decay more slowly than $1/r$, and for the far-field energy to decay more slowly than $1/r^2$, that is, for the existence of an electromagnetic missile. (The necessary conditions for an electromagnetic missile are more restrictive than for the time-domain far fields to decay slower than $1/r$, because the time duration of the far-field pulse may approach zero as r becomes infinite.)

We shall derive these necessary conditions for two kinds of sources: first, sources composed of a finite number of moving point charges, and secondly, for a continuum of charge-current.

EM Missiles for Accelerating Point Charges

For a finite number of separate point charges, each point charge q that moves with velocity $\mathbf{u}$ and acceleration $\dot{\mathbf{u}}$ has far fields given by [5, p.475],[15]

$$\mathbf{E}(\mathbf{r}, t) \sim \frac{\mathcal{F}(\theta, \phi, t - r/c)}{r} = \frac{q\gamma_r^3}{4\pi\epsilon rc^2}\hat{\mathbf{r}} \times [(\hat{\mathbf{r}} - \mathbf{u}'/c) \times \dot{\mathbf{u}}'] = \frac{q\gamma_r^2}{4\pi\epsilon rc^2}|\dot{\mathbf{u}}'| \left[\gamma_r(\hat{\mathbf{r}} - \mathbf{u}'/c)(\hat{\mathbf{r}} \cdot \hat{\dot{\mathbf{u}}}') - \hat{\dot{\mathbf{u}}}'\right] \tag{38}$$

$$\mathbf{H}(\mathbf{r},t) \sim \sqrt{\frac{\epsilon}{\mu}}\hat{\mathbf{r}} \times \mathbf{E}(\mathbf{r},t) \sim \sqrt{\frac{\epsilon}{\mu}}\hat{\mathbf{r}} \times \frac{\mathcal{F}(\theta,\phi,t-r/c)}{r}, \quad r \to \infty \tag{39}$$

where $\gamma_r = 1/(1 - \hat{\mathbf{r}} \cdot \mathbf{u}'/c)$ and the primes on the velocity and acceleration indicate that they are evaluated at the retarded time $t' = t - |\mathbf{r} - \mathbf{r}_q(t')|/c$. The vector $\mathbf{r}_q(t')$ is the position of the charge at the retarded time t'. The origin of the coordinate system, and thus that of the asymptotically large vector $\mathbf{r}$, is chosen in the vicinity of the charge during the retarded times of interest.

Equations (38) and (39) reveal that the far electric and magnetic fields of the point charge decay as $1/r$ (or faster) unless the acceleration of the charge is infinite at some point (or points) in time. Thus, we conclude that *the far fields of a finite number of moving point charges decay more slowly than $1/r$ only if the acceleration of a least one of the point charges becomes infinite at some point in time.* (Of course, it is assumed that each point charge q is finite.) Also, the square brackets in (38) cannot be zero (for nonzero $\dot{\mathbf{u}}$) in all directions of observation $\hat{\mathbf{r}}$. Thus, for a single point charge, the above necessary condition for slower than $1/r$ far-field decay is a sufficient condition as well.

For rectilinear motion, these points of infinite acceleration must be isolated points in time, and must integrate over time to a value less than c, because the speed of each point charge must be less than the speed of light.

The magnitude of the time-domain acceleration can be written in terms of the Fourier transform of the velocity spectrum, to show

$$|\dot{\mathbf{u}}(t)| = \left|\int_{-\infty}^{+\infty} -i\omega \mathbf{u}_\omega e^{-i\omega t} d\omega\right| \leq \int_{-\infty}^{+\infty} |\omega \mathbf{u}_\omega| d\omega \tag{40}$$

which is finite if $|\mathbf{u}_\omega|$ decays as $|\omega|^{-(2+\alpha)}, \alpha > 0$, as $|\omega| \to \infty$. Using the even and odd functional dependence on ω of the real and imaginary parts of the velocity spectrum $\mathbf{u}_\omega$, and assuming that $\dot{\mathbf{u}}(t)$ is zero before some initial time, it can also be shown that the condition $|\mathbf{u}_\omega| \sim 1/\omega^2$ as $|\omega| \to \infty$ produces a finite acceleration. Thus, we conclude that *the far fields of a moving point charge can decay more slowly than $1/r$ only if the frequency spectrum of the velocity of the point charge decays slower than $1/\omega^2$ as $|\omega| \to \infty$.* In addition, if $|\mathbf{u}_\omega| \sim |\omega|^{-(2-\alpha)}$ as $|\omega| \to \infty$, examples of $\mathbf{u}_\omega$ can be found that produce infinite acceleration at some point in time.

The energy radiated per unit area in a far-field pulse can be expressed from (38) and (39) as

$$\int_{pulse} (\mathbf{E} \times \mathbf{H}) \cdot \hat{\mathbf{r}} dt \sim \frac{1}{r^2}\sqrt{\frac{\mu}{\epsilon}} \int_{pulse} |\mathcal{F}(\theta,\phi,t)|^2 dt \sim O\left(\frac{1}{r^2}\right) \int_{pulse} |\dot{\mathbf{u}}|^2 dt = 2\pi O\left(\frac{1}{r^2}\right) \int_{-\infty}^{+\infty} \omega^2 |\mathbf{u}_\omega|^2 d\omega\,. \tag{41}$$

(Only $|\dot{\mathbf{u}}|^2$ is retained in the second integral of (41) because the expression in the square brackets of the last part of (38) is never infinite but is nonzero for most far-field directions. Because $dt = dt'/\gamma_r$ and $1/\gamma_r$ is never infinite or zero, it also has been omitted from the time integration of $|\dot{\mathbf{u}}|^2$ in (41).) Thus, *the necessary condition for the finite number of point charges to radiate an electromagnetic missile (energy that decays slower than $1/r^2$) is that at least one point charge have infinite acceleration at some time, and in addition*

$$\int_{pulse} |\dot{\mathbf{u}}|^2 dt = 2\pi \int_{-\infty}^{+\infty} \omega^2 |\mathbf{u}_\omega|^2 d\omega = \infty \tag{42}$$

which implies that the magnitude of the velocity spectrum $|\mathbf{u}_\omega|$ *decays as $1/|\omega|^{3/2}$ or slower as $|\omega| \to \infty$.* Furthermore, examples of velocities (such as $t^\alpha \exp(-t)$, $t \geq 0$, and zero for $t \leq 0$) are easily constructed that produce electromagnetic missiles ($0 < \alpha \leq 1/2$), or no electromagnetic missile yet slower than $1/r$ far-field decay ($1/2 < \alpha < 1$).

We can also find conditions on the frequency-domain far-field pattern required to have slower than $1/r$ decay in the time-domain far field, and slower than $1/r^2$ energy decay in the far-field pulse. The time-domain far-field pattern is expressed in (29) as the inverse Fourier transform of the frequency-domain far-field pattern. If the time-domain far field is to decay slower than $1/r$ in some direction, the far-field pattern must be infinite in that direction. Thus, from (29) we have

$$|\mathcal{F}(\theta,\phi,t)| = \left|\int_{-\infty}^{+\infty} \mathcal{F}_\omega(\theta,\phi)e^{-i\omega t}d\omega\right| \leq \int_{-\infty}^{+\infty} |\mathcal{F}_\omega(\theta,\phi)|\, d\omega = \infty \tag{43}$$

or that the magnitude of the frequency-domain far-field pattern must decay slower than $1/|\omega|$ as $|\omega| \to \infty$ for slower than $1/r$ far-field decay in the time domain. (By the same reasoning explained after (40), when the magnitude of the frequency-domain far-field pattern decays as $1/|\omega|$, the time-domain far-field decays as $1/r$.)

Similarly, in order for the energy in the time-domain far-field pulse to decay slower than $1/r^2$, we must have

$$\int_{pulse} |\mathcal{F}(\theta,\phi,t)|^2\, dt = \infty \tag{44}$$

or from (29) and Parseval's theorem

$$\int_{-\infty}^{+\infty} |\mathcal{F}_\omega(\theta,\phi)|^2\, d\omega = \infty. \tag{45}$$

Equation (45) implies that the magnitude of the frequency-domain far-field pattern must decay as $1/|\omega|^{1/2}$ or slower as $|\omega| \to \infty$ for the energy in the far-field pulse to decay slower than $1/r^2$, that is, for an electromagnetic missile to exist.

For rectilinear motion, any one of the conditions (44), (45), or (42) also implies from (38) and (39) that the point charge radiates an infinite amount of energy because the energy radiated per unit area decays slower than $1/r^2$ over most of the far-field sphere.

EM Missiles for a Continuum of Charge-Current

Now assume the source region consists of a continuum of current rather than a finite number of moving point charges. We proved above that the electromagnetic far fields decay as $1/r$ (or faster) provided the first time derivative of the source current exists and is bounded by an integrable time-independent function in the finite source region V. Therefore, *only if the first time derivative of the current in a region of space (or secant slope if the time derivative does not exist [16]) is infinite at some point in time [18] can the far fields decay more slowly than* $1/r$. Expressing the first time derivative of the current as a Fourier transform of the current spectrum, shows that

$$\left|\frac{\partial}{\partial t}\mathbf{J}(\mathbf{r},t)\right| = \left|\int_{-\infty}^{+\infty} -i\omega \mathbf{J}_\omega(\mathbf{r})e^{-i\omega t}d\omega\right| \leq \int_{-\infty}^{+\infty} |\omega \mathbf{J}_\omega(\mathbf{r})|\, d\omega \tag{46}$$

which implies that the necessary condition in the frequency domain for slower than $1/r$ time-domain far-field decay is that *the magnitude of the frequency spectrum of the current* $|\mathbf{J}_\omega(\mathbf{r})|$ *for some* $\mathbf{r}$ *decays more slowly than* $1/\omega^2$ *as* $|\omega| \to \infty$.

As in the case of the accelerating point charge, an additional condition on the current is required to excite an electromagnetic missile. Namely, the energy carried by the far-field pulse must decay slower than $1/r^2$. From equations (8) and (9), one finds that the energy radiated per unit area in a far-field pulse behaves as

$$\int_{pulse}(\mathbf{E}\times\mathbf{H})\cdot\hat{\mathbf{r}}dt \sim \frac{1}{r^2}\sqrt{\frac{\mu}{\epsilon}}\int_{pulse}|\mathcal{F}(\theta,\phi,t)|^2\, dt \sim O\left(\frac{1}{r^2}\right)\int_V\int_V\left[\int_{pulse}\frac{\partial}{\partial t}\mathbf{J}(\mathbf{r}',t')\cdot\frac{\partial}{\partial t}\mathbf{J}(\mathbf{r}'',t'')dt\right]dV'dV'' \tag{47}$$

where $t' = t - r/c + \hat{\mathbf{r}}\cdot\mathbf{r}'/c$ and $t'' = t - r/c + \hat{\mathbf{r}}\cdot\mathbf{r}''/c$. (For surface current the volume integrals in (47) are replaced by surface integrals.)

Equation (47) implies that the far-field energy in the pulse cannot decay more slowly than $1/r^2$ unless the time integral is infinite over some region of $(\mathbf{r}',\mathbf{r}'')$. Specifically

$$\int_{pulse}\frac{\partial}{\partial t}\mathbf{J}(\mathbf{r}',t')\cdot\frac{\partial}{\partial t}\mathbf{J}(\mathbf{r}'',t'')dt = \infty \tag{48}$$

for some $\mathbf{r}'$ and $\mathbf{r}''$. If we let $\mathbf{r}$ be the value of $\mathbf{r}'$ or $\mathbf{r}''$ in (48) that has the largest time derivative of the current, (48) implies that there must be some $\mathbf{r}$ for which

$$\int_{pulse} \left| \frac{\partial}{\partial t} \mathbf{J}(\mathbf{r}, t) \right|^2 dt = 2\pi \int_{-\infty}^{+\infty} \omega^2 |\mathbf{J}_\omega(\mathbf{r})|^2 d\omega = \infty \,. \tag{49}$$

Thus, in addition to the first time derivative of the current being infinite at some point in time, (49) is a necessary condition on the singularity of the current to produce an electromagnetic missile. Specifically, *the first time derivative of the current in a region of space must be infinite at some point in time, and the time integral of the magnitude squared of the first time derivative of the current must be infinite to produce an electromagnetic missile. In terms of the frequency spectrum of the current,* $|\mathbf{J}_\omega(\mathbf{r})|$ *for some* $\mathbf{r}$ *must decay as* $1/|\omega|^{3/2}$ *or slower as* $|\omega| \to \infty$. For linearly polarized current, the singularities (infinities) of the first time derivative of the current must be isolated and integrate over time to a finite value if the current itself is to remain a finite function of time, even though the square of the first time derivative of current must integrate to an infinite value to produce an electromagnetic missile. The lower bound of $1/|\omega|^{3/2}$ as $|\omega| \to \infty$ for the frequency spectrum of the current producing an electromagnetic missile was also found by Myers [19] for current on a circular disk.

For a continuum of current, the far-field patterns must also satisfy (43) and (44)-(45) for slower than $1/r$ and $1/r^2$ decay, respectively, in the time-domain far field and energy in the far-field pulse. Thus, the magnitude of the frequency-domain far-field pattern $|\mathcal{F}_\omega(\theta, \phi)|$ in a direction (θ, ϕ) must decay slower than $1/|\omega|$ as $|\omega| \to \infty$ for slower than $1/r$ decay of the time-domain far field in the direction (θ, ϕ). And $|\mathcal{F}_\omega(\theta, \phi)|$ must decay slower than $1/|\omega|^{1/2}$ as $|\omega| \to \infty$ for an electromagnetic missile to exist in the direction (θ, ϕ).

Recall that the frequency-domain far-field pattern is defined as the limit as $r \to \infty$ in (12). Thus, when referring to the limit of the far-field pattern as $|\omega| \to \infty$, it is assumed that the limit as $r \to \infty$ has already been taken.

The exact expressions for the time-domain fields radiated by a circular current disk, derived in Blejer *et al.* [17], demonstrate that a spatial distribution of current can be found that will produce an electromagnetic missile for any given time or frequency dependence satisfying (49). It is emphasized that the condition (49) or (44)-(45) does not imply that the current source radiates an infinite amount of energy, because a continuum of current, unlike an accelerating point charge, can have infinite directivity. Therefore, the energy in the far-field pulse that decays slower than $1/r^2$ can be confined to a finite transverse area as $r \to \infty$. Conversely, if the total energy radiated by the sources is finite, electromagnetic missiles can occur only in isolated directions, that is, over a zero solid angle (namely points and curves) on the far-field sphere.

Also it is emphasized that (8)-(9) and (47) may no longer be valid expressions for evaluating the far fields and far-field energy when $\partial \mathbf{J}/\partial t$ becomes infinite at some point in time. To find the exact behavior of the far fields when $\partial \mathbf{J}/\partial t$ becomes infinite, one can evaluate the expressions (1) and (2), or (4) and (6), before letting $r \to \infty$, as is done in Blejer *et al.* [17].

Concluding Remarks on Electromagnetic Missiles

In principle, a pulse can be sent to infinity with finite or even infinite energy through a finite area. For example, equations (26) and (27) of Blejer *et al.* [17] show that a current source with delta-function time dependence on a circular disk radiates two pulses with infinite energy to an indefinitely large distance in the direction of the z axis, which is normal to the center of the disk.

In general, however, one can argue that a current source of finite energy in a finite region of space cannot radiate an electromagnetic missile with nondecaying energy to an infinitely large distance from the source [14]. The argument goes as follows. Because a source pulse of finite bandwidth cannot generate an electromagnetic missile, any electromagnetic missile must maintain itself on the energy in the increasingly higher frequencies. If the total energy radiated by the sources is finite, the energy in the frequency spectrum at frequencies higher than any finite value must approach zero.

Nonetheless, finite energy current sources are mathematically possible that will produce a pulse that travels without energy decay arbitrarily far from the source region. To see this, let the surface current $\mathbf{K}(t)$ on a current disk of fixed radius a have a time dependence given by the function

$$\mathbf{K}(t) = \hat{\mathbf{x}} \begin{cases} \frac{1}{\sqrt{t_1}}, & 0 \leq t \leq t_1 \\ 0, & t < 0 \text{ and } t > t_1 \end{cases} \tag{50}$$

with $t_1 > 0$. This current radiates a finite energy since $\int_0^{t_1} |\mathbf{K}(t)|^2 dt = 1$. Now equations (26) and (27) of Blejer *et al.* [17] show that this current will radiate two separate nondecaying finite-energy pulses to the far field along the z axis provided $a^2/(2cz) \geq t_1$. In other words, the pulses will travel with finite nondecaying energy to a distance z given by

$$z = \frac{a^2}{2ct_1}. \tag{51}$$

Since, in principle, t_1, the width of the current pulse, can be made arbitrarily small, the excited pulses can be made to travel arbitrarily far along the z axis without any loss in energy. Of course, for a fixed nonzero value of t_1, the energy in the pulse will begin to decay beyond the distance z given in (51).

In summary, classical sources of finite charge and current in a finite region of space can produce classical Maxwellian far-field pulses that decay slower than $1/r$ and carry energy that decays slower than $1/r^2$. However, these anomalous pulses described by Maxwell's equations require either infinite acceleration of point charges or infinite first time derivatives in the macroscopic continuum of current such that the frequency spectrum of the current decays as $1/|\omega|^{3/2}$ or slower as $|\omega| \to \infty$. The infinite acceleration of a point charge is presumably not possible, and all known sources have frequency spectra that decay much faster than $1/|\omega|^{3/2}$ as $|\omega| \to \infty$. Thus, electromagnetic missiles described by Maxwell's equations cannot be generated in the laboratory today nor in the foreseeable future. Of course, the greater the serviceable bandwidth of the generator and the larger the dimensions of the radiator, the further out one can push the distance to the boundary between the near and far fields.

References

[1] C. Müller, *Foundations of the Mathematical Theory of Electromagnetic Waves* (Springer-Verlag, New York, 1969).

[2] D.M. Kerns, *Plane-Wave Scattering-Matrix Theory of Antennas and Antenna-Antenna Interactions* (NBS Monograph 162, U.S. Gov. Printing Off., Washington, D.C., 1982).

[3] T.B.Hansen and A.D. Yaghjian, *Formulation of time-domain planar near-field measurements without probe correction* (Rome Laboratory Technical Report RL-TR-93-210, Hanscom AFB, MA 01731, 1993).

[4] T.B.Hansen and A.D. Yaghjian, IEEE Trans. Antennas Propagat., *42*, 1280-1300 (1994).

[5] J.A. Stratton, *Electromagnetic Theory* (McGraw-Hill, New York, 1941).

[6] The interchange of the time integration with the curl operator on the right side of (2) can be proven valid by interchanging the space and time integrations in the integral form of Maxwell's second equation to get $\epsilon \int_S \mathbf{E} \cdot \mathbf{dS} = \oint_C \left(\int_{t_0}^t \mathbf{H} dt'\right) \cdot \mathbf{dl}$, where no point of the surface S or the curve C is in the source region V. Then Stokes' theorem is applied to obtain (2). This interchange of space and time integrations is permitted by standard theorems of integration [7, sec.237] under the very weak condition that $|\mathbf{H}(\mathbf{r}, t)|$ is integrable over any finite space-time domain (C, t), where no point of the

curve C is in the source region V. Alternatively, one could simply take the above integral equation as the fundamental form of Maxwell's second equation.

[7] E.W. Hobson, *The Theory of Functions of a Real Variable* (Dover, New York, 1957), Vol. II.

[8] J. van Bladel, *Electromagnetic Fields* (McGraw-Hill, New York, 1964).

[9] H.E. Moses, R.J. Nagem and G.V.H. Sandri, J. Math. Phys., **33**, 86-101 (1992).

[10] J.W. Dettman, *Applied Complex Variables* (Dover, New York, 1984).

[11] H.E. Moses and R.T. Prosser, SIAM J. Appl. Math., **50**, 1325-1340 (1990).

[12] E.T. Whittaker and G.N. Watson, *A Course of Modern Analysis* (Cambridge University Press, London, 1952), sec.5.32.

[13] A.D. Yaghjian and T.B. Hansen, J. Appl. Phys., **79**, 2822-2830 (1996).

[14] T.T. Wu, J. Appl. Phys., **57**, 2370-2373 (1985).

[15] A.D. Yaghjian, *Relativistic Dynamics of a Charged Sphere: Updating the Lorentz-Abraham Model* (Springer-Verlag, New York, 1992).

[16] If the first time derivative does not exist, bringing the curl operator under the integral sign to obtain (8) and (9) is no longer a valid interchange. In that case, the derivatives operating on the vector potential integral can be expressed in terms of their defining limits to show that the far fields decay as $1/r$ unless the secant slope, $[\mathbf{J}(\mathbf{r},t+\Delta t) - \mathbf{J}(\mathbf{r},t)]/\Delta t$, becomes infinite for some t and Δt [17],[7, sec.246, p.355]. Of course, if the limit of the secant slope exists as $\Delta t \to 0$, the limit equals the time derivative. These concepts are illustrated in Blejer *et al.* [17] for the specific case of the fields radiated by the current on a disk.

[17] D.J. Blejer, R.C. Wittmann and A.D. Yaghjian, in *Ultra-Wideband, Short-Pulse Electromagnetics* (Plenum Press, New York, 1993), pp. 285-292.

[18] Note that the first time derivative of the current $\frac{\partial}{\partial t}\mathbf{J}(\mathbf{r},t)$ must become infinite as a function of time t (not just position $\mathbf{r}$) to generate an electromagnetic missile. For example, the current parallel to a perfectly conducting sharp edge is infinite right at the edge. Yet this singularity is a function of the spatial coordinates and will not generate an electromagnetic missile because the singularity is integrable with respect to the spatial coordinates.

[19] J.M. Myers, *Developmental Study of Electromagnetic Missiles* (Gordon McKay Laboratory Annual Progress Report No. 3, Harvard University, Cambridge, MA 02138, January 1989), sec.5.

ASYMPTOTIC APPROXIMATIONS FOR OPTIMAL CONFORMAL ANTENNAS

T. S. Angell and R.E. Kleinman[1]
Department of Mathematical Sciences
University of Delaware
Newark, Delaware 19716, U.S.A.

and

B. Vainberg[2]
Department of Mathematical Sciences
University of North Carolina
Charlotte, North Carolina 28223, U.S.A.

Abstract:

In earlier work, we have given constructive methods to compute the surface currents on a conformal antenna which is required to radiate a maximal amount of energy into a predetermined sector of the far field. More recently, the authors have used asymptotic methods to compute approximate optimal surface currents in the time harmonic two-dimensional electromagnetic case (the Helmholtz equation with impedance boundary condition), for the case of high frequency.

In the present work, we extend these asymptotic results to the full three-dimensional time-harmonic electromagnetic case. We obtain a representation of the suboptimal current which explicitly shows the dependence on the total curvature, $\kappa \mathbf{x}$ of the radiating structure at each point $\mathbf{x}$.

1 Introduction and Notation

The problem of choosing feedings which optimize antenna performance has been studied for many years, usually with reference to a particular type of antenna. Likewise, different situations call for different measures of performance with respect to which the optimization is to be accomplished. The object of our current discussion is to examine a class of problems involving closed three-dimensional bodies. Rather than considering elementary radiators mounted on conducting bodies, we treat the entire body as an antenna and address the question of determining the surface current distribution which optimizes a particular antenna performance measure. It is usually the case that such distributions are subject to constraints which reflect physical limitations on the currents that can be produced. A more complete discussion of this type of problem has been presented in a recent paper by Angell et al. [1].

[1]Effort sponsored by the Air Force Office of Scientific Research, Air Force Materiel Command, USAF, under grant number F9620-96-1-0039. The U.S. Governnment is authorized to reproduce and distribute reprints for Governmental purposes notwithstanding any copyright notation thereon.

[2]Effort partially supported by funds from the University of North Carolina, Charlotte.

Here, we consider one of the particular problems presented in [1], namely that of choosing surface currents on a conformal antenna which is required to radiate a maximal amount of energy into a predetermined sector of the far field. As explained in [1], this problem is closely related to the familiar directivity problem that has been widely discussed. We have presented an analysis of and numerical results for this problem before (see e.g. [2],[8]). In the first section we will give a very brief summary of those earlier results. The main body of the paper is devoted to asymptotic methods for computation of optimal currents in the high frequency regime for a perfectly conducting body in three dimensions. Earlier work on the two–dimensional case was discussed by the authors in [4].

The analysis is complicated by the fact that we are dealing simultaneously with two large parameters, both the wave number k and the position $\mathbf{x}$. In order to avoid these difficulties, we use the time dependent problem which is related to the original one by the limiting amplitude principle. Instead of taking the limit as $t \to \infty$ to get the limiting amplitude at high frequency we show that one can use the averaging with respect to time over a finite interval of time. This approach reduces the high frequency analysis to the analysis of the time dependent problem in which we have only one large parameter, namely the high frequency. By applying an appropriate form of the principle of limiting amplitude, we succeed in obtaining the high frequency characterization of the Green's function for the problem and then an explicit form for an approximation to the optimal current.

We begin by fixing some notation. Let Ω be the exterior of a strongly convex bounded obstacle $B \subset \mathbb{R}^n$ with infinitely smooth boundary $\partial\Omega$. We denote points by their position vectors $\mathbf{x}$ and $\mathbf{y}$, and write $r = |\mathbf{x}|$ for the radial variable in hyperspherical coordinates and $\hat{\mathbf{r}} = \mathbf{x}/r$ as the unit radial vector.

We consider the time–harmonic Maxwell equations and the problem of finding a solution, in the region exterior to Ω, of the boundary value problem:

$$\begin{cases} ik\mathbf{E} = \nabla \times \mathbf{H}, & \\ ik\mathbf{H} = -\nabla \times \mathbf{E}, & \\ \operatorname{div}\mathbf{E} = \operatorname{div}\mathbf{H} = 0, & x \in \Omega, \\ \hat{\mathbf{n}} \times \mathbf{H} = \mathbf{h}, & x \in \partial\Omega, \end{cases} \tag{1.1}$$

together with the Silver–Müller radiation condition

$$\begin{cases} \hat{\mathbf{r}} \times \nabla \times \mathbf{E} + ik\mathbf{E} = o(1/r) & \text{and,} \\ \hat{\mathbf{r}} \times \nabla \times \mathbf{H} + ik\mathbf{H} = o(1/r) & \text{as } r \to \infty, \end{cases} \tag{1.2}$$

where $\hat{\mathbf{n}}$ is the unit normal vector which is directed into the exterior region Ω, and the prescribed vector function $\mathbf{h}$ does not depend on the wave number k.

Calderón [6] has shown that this problem has a unique solution for every $\mathbf{h} \in \mathbf{L}^2_t(\partial\Omega)$ where $\mathbf{L}^2_t(\partial\Omega)$ is the set of all vector functions, defined in $\partial\Omega$, whose normal component vanishes and whose magnitude is square integrable. Denote this unique solution corresponding to the boundary data h by $\{\mathbf{E}_h, \mathbf{H}_h\}$.

This unique solvability of the boundary value problem (1.1)–(1.2) allows us to define an "$\mathbf{H} \to \mathbf{E}$"–map

$$T : \mathbf{h} \to \hat{\mathbf{n}} \times \mathbf{E}_{\mathbf{h}}, \tag{1.3}$$

which relates the electric surface current $\hat{\mathbf{n}} \times \mathbf{H}_{\mathbf{h}}$ to the tangential component of the electric field $\hat{\mathbf{n}} \times \mathbf{E}_{\mathbf{h}}$ on the boundary $\partial\Omega$. We remark in passing that this mapping is a generalization of the more familiar Dirichlet–to–Neumann map used, for example, by [9]. Indeed, if we consider the transverse electric case in which the field $\mathbf{E}$ is assumed to be polarized normal to the z–axis and $\mathbf{H}$ is taken parallel to the z–axis, then

$$\mathbf{H} = \hat{\mathbf{z}}\, u(p), \; p \in \Omega,$$

and the Maxwell equations then imply that the scalar function u satisfies the two–dimensional Helmholtz equation. Then the electric current on $\partial\Omega$ is given by

$$\hat{\mathbf{n}} \times \mathbf{H} = -\hat{\mathbf{t}}\, u$$

where $\hat{\mathbf{t}}$ is the unit vector tangent to the cross–section of the three dimensional cylinder in the plane of the cross–section. Then

$$\hat{\mathbf{n}} \times \mathbf{E} = \hat{\mathbf{n}} \times (\nabla \times \mathbf{H}) = \hat{\mathbf{n}} \times (\nabla \times u\hat{\mathbf{z}}) = -\frac{\partial u}{\partial \mathbf{n}}\hat{\mathbf{z}}.$$

The behavior of the antenna is usually described in terms of its **far field** or **radiation** pattern. Indeed, the fields $\mathbf{E}$ and $\mathbf{H}$ are known [7] to have the following asymptotic behavior at infinity:

$$\begin{cases} \mathbf{E}_{\mathbf{h}}(\mathbf{x}) = \left(\frac{e^{ikr}}{r^{(n-1)/2}}\right) \mathbf{F}_{\mathbf{h}}(\hat{\mathbf{r}})(1 + o(1)), & \text{as } r \to \infty \\ \mathbf{H}_{\mathbf{h}}(\mathbf{x}) = \left(\frac{e^{ikr}}{r^{(n-1)/2}}\right) \hat{\mathbf{r}} \times \mathbf{F}_{\mathbf{h}}(\hat{\mathbf{r}})(1 + o(1)), & \text{as } r \to \infty. \end{cases} \tag{1.4}$$

The vector function $\mathbf{F}_{\mathbf{h}}$, which has no radial component, is called the **radiation pattern**.

Let $a = a(\hat{\mathbf{r}})$ be a piecewise continuous non-negative function on the unit sphere, and define the functional J by

$$J(\mathbf{h}) = \int_{S^2} |\mathbf{F}_{\mathbf{h}}(\hat{\mathbf{r}})|^2 a(\hat{\mathbf{r}}) dS_{\hat{\mathbf{r}}}, \quad \mathbf{h} \in \mathbf{L}^2_t(\partial\Omega), \tag{1.5}$$

where dS is the element of surface area. The problem we consider is that of finding current distributions, $\mathbf{h}$ defined on $\partial\Omega$, which maximize the radiated power in the far field weighted by the function α which could, for example, be the characteristic function of an angular sector. We are interested in determining the maximum value of the functional J on the set U of functions $\mathbf{h}$ in $\mathbf{L}^2_t(\partial\Omega)$ which is closed and convex in that Hilbert space. Moreover, for specific choices of the set U, we wish to characterize the functions, $\mathbf{h} \in U$, where the functional J attains its maximum. The *existence* of such functions $\mathbf{h}$ follows from the results of [2], [1], [8].

2 Maximizing Power Radiated in a Sector

For the problem of maximizing the function J given above over the unit ball in $\mathbf{L}^2_t(\partial\Omega)$, the approach in our previous work consists of reducing the problem to a generalized eigenvalue problem i.e., an eigenvalue problem of the form

$$R\,\mathbf{h} = \lambda Q\,\mathbf{h}.$$

This problem is then solved approximately by projection onto a finite dimensional subspace of suitably large dimension.

Indeed, the relationship between the surface current $\mathbf{h} \in \mathbf{L}^2_t(\partial\Omega)$ and the far–field pattern $\mathbf{F} \in \mathbf{L}^2_t(S^2)$ is given by a compact operator $\mathcal{K} : \mathbf{L}^2_t(\partial\Omega) \to \mathbf{L}^2_t(S^2)$ so that the constrained optimization problem can be written as

$$\max_{\mathbf{h}\in U} \int_{S^2} \alpha(\hat{\mathbf{r}})|\mathcal{K}\mathbf{h}(\hat{\mathbf{r}})|^2 \, dS \tag{2.1}$$

where we define the constraint set U to be the unit ball in the Hilbert space $\mathbf{L}^2_t(\partial\Omega)$:

$$U = \{\mathbf{h} \in \mathbf{L}^2_t(\partial\Omega) \mid \|\mathbf{h}\|_{\mathbf{L}^2_t(\partial\Omega)} \leq 1\}. \tag{2.2}$$

Since $\alpha(\hat{\mathbf{r}})$ is real, we may rewrite (2.1) in terms of the usual inner product on the Hilbert space $\mathbf{L}^2_t(\partial\Omega)$ as

$$\max_{\mathbf{h}\in U} \, (\mathcal{K}^\star \alpha \mathcal{K}\mathbf{h}, \mathbf{h})_{\mathbf{L}^2_t(\partial\Omega)}, \tag{2.3}$$

$\mathcal{K}^\star$ being the adjoint of the compact operator $\mathcal{K}$. This characterization of the cost functional not only leads to a relatively simple proof of the existence of optimizers, but also allows the development of a simple computational method. Indeed, the optimal

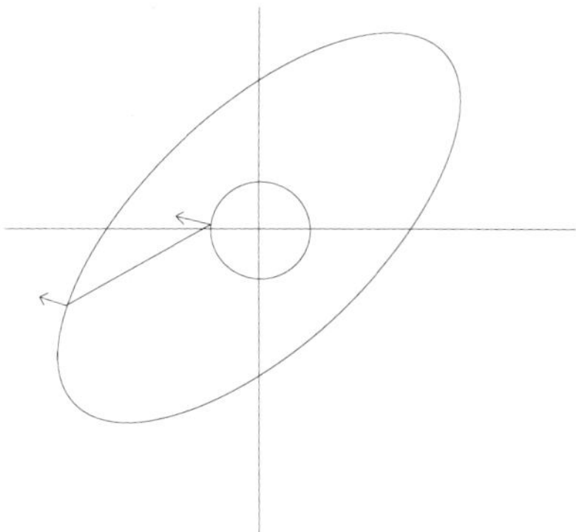

Figure 1: The Map P

value of the cost functional is just the maximum eigenvalue of the compact operator $\mathcal{K}^\star\alpha\mathcal{K}$ while the corresponding normalized eigenvector represents the coefficients of the optimal current distribution. The proofs for the full electromagnetic case in [8] follow almost exactly the earlier proofs in [2] [3] given for the three–dimensional acoustic and two–dimensional electromagnetic cases. Moreover, the computational procedure has been carried out in [8] for both spherical and ellipsoidal surfaces.

3 High Frequency Results

In order to establish the asymptotic results, we modify the constraint set. Thus, instead of maximizing the functional (2.1) over the unit ball in $\mathbf{L}^2_t(\partial\Omega)$ as before, we introduce a new norm on this space of tangential vectors as follows

$$|||\mathbf{h}||| := \left(\|\hat{\mathbf{n}}\times\mathbf{H}_{\mathbf{h}}\|^2_{\mathbf{L}^2_t(\partial\Omega)} + \|\hat{\mathbf{n}}\times\mathbf{E}_{\mathbf{h}}\|^2_{\mathbf{L}^2_t(\partial\Omega)}\right)^{\frac{1}{2}} \tag{3.1}$$

Notice that this is just a type of graph norm of the operator T. We then define the constraint set U_1 to be the set of elements which are bounded by 1 in this new norm:

$$U_1 := \{\mathbf{h}\in\mathbf{L}^2_t(\partial\Omega) \mid |||\mathbf{h}||| \le 1\}, \tag{3.2}$$

which is a closed and convex subset of $\mathbf{L}^2_t(\partial\Omega)$.

Certainly, since $\mathbf{h} = \hat{\mathbf{n}}\times\mathbf{H}_{\mathbf{h}}$, we have

$$\|\mathbf{h}\| \le |||\mathbf{h}|||,$$

and so $U_1 \subset U$. We therefore expect that

$$\max_{\mathbf{h}\in U} J \ge \max_{\mathbf{h}\in U_1} J$$

In order to formulate the main result we need to introduce some notation. Let the mapping $P : \Gamma \to S^{n-1}$ transfer each point $\mathbf{x}\in\partial\Omega$ into the point $\hat{\theta}\in S^{n-1}$ for which $\hat{\mathbf{n}} = \hat{\theta}$, where $\hat{\mathbf{n}}$ is the unit exterior normal to $\partial\Omega$ at the point $\mathbf{x}$. In general, $\hat{\theta} = P(\mathbf{x}) \ne \hat{\mathbf{r}}$ but the assumption that the obstacle is strongly convex ensures that the mapping is one–to–one (see Fig.1).

Note that we will use the symbol $\hat{\theta}$ to denote both a point on the unit sphere and the position vector of that point, but $\hat{\theta}$ is *not* the angular unit basis vector in spherical coordinates of $\mathbf{x}$.

For an arbitrary $\epsilon > 0$ we construct a function $\mathbf{g}_\epsilon = \mathbf{g}_\epsilon(\hat{\theta})$ such that

$$\int\limits_{S^{n-1}} |\mathbf{g}_\epsilon(\hat{\theta})|^2 dS = 1, \quad \int\limits_{S^{n-1}} |\mathbf{g}_\epsilon(\hat{\theta})|^2 a(\hat{\theta}) dS \geq \sup a(\hat{\theta}) - \epsilon. \tag{3.3}$$

It is obvious that we can take, for example,

$$\mathbf{g}_\epsilon = \varphi / \Big(\int\limits_{S^{n-1}} |\varphi|^2 dS \Big)^{1/2}$$

where φ is *any* function on S^{n-1} with support in a region where $a(\hat{\theta}) > \sup a(\hat{\theta}) - \epsilon$.

Let $\kappa(\mathbf{x})$ be the total curvature (product of the principle curvatures) of $\partial\Omega$ at the point $\mathbf{x} \in \partial\Omega$. The main result can be expressed as:

Theorem 3.1 *1. If* $\mathbf{h}$ *is such that* $|||\mathbf{h}||| = 1$, *then*

$$0 \leq J(\mathbf{h}) \leq \frac{1}{2} \sup a(\hat{\mathbf{r}}). \tag{3.4}$$

2. Let $\epsilon > 0$ *be an arbitrary positive number,* $\mathbf{g}_\epsilon$ *be a fixed function (independent of* k*) which satisfies the relations (3.3) and*

$$\mathbf{h}_\epsilon := \mathbf{h}_\epsilon(\mathbf{x}) = \frac{\sqrt{2}}{2} \left(\mathbf{g}_\epsilon(P\mathbf{x}) \sqrt{\kappa(\mathbf{x})} \right), \quad \mathbf{x} \in \partial\Omega. \tag{3.5}$$

Then

$$|||\mathbf{h}||| = 1 + 0(k^{-1}), \ k \to \infty \tag{3.6}$$

and there exists $k_0 = k_0(\epsilon)$ *such that*

$$J(\mathbf{h}_\epsilon) > \frac{1}{2} \sup a(\hat{\theta}) - 2\epsilon \tag{3.7}$$

if $k \geq k_0$.

From this theorem it follows that if

$$\tilde{\mathbf{h}}_\epsilon = \frac{\mathbf{h}_\epsilon}{|||\mathbf{h}_\epsilon|||}$$

then $J(\tilde{\mathbf{h}}_\epsilon)$ differs from its maximum value on the set $U_1 = \{\mathbf{h} \in \mathbf{L}^2_t(\partial\Omega) \mid |||\mathbf{h}||| = 1\}$ by not more than 3ϵ if k is sufficiently large.

The proof of Theorem 1 is based on the high frequency asymptotic expansions of solutions to problem (1.1) and on the asymptotic behavior of the corresponding far field amplitudes. We will discuss these expansions elsewhere [5].

4 Indications of Proofs

It is comparitively easy to prove the first assertion of Theorem 3.1 using the following form of Green's theorem.

Lemma 4.1 *Let* $\Omega_1 \subset \Omega$ *be a bounded domain exterior to the body* B. *Then it is true that*

$$\int_{\partial\Omega_1} -\hat{\mathbf{n}} \times \mathbf{H} \cdot \overline{\mathbf{E}} + \hat{\mathbf{n}} \times \mathbf{E} \cdot \overline{\mathbf{H}} \, dS = 0. \tag{4.1}$$

Proof:
Note first that $\overline{\mathbf{E}} \cdot \hat{\mathbf{n}} \times \mathbf{H} = \hat{\mathbf{n}} \cdot (\mathbf{H} \times \overline{\mathbf{E}})$ and $\overline{\mathbf{H}} \cdot \hat{\mathbf{n}} \times \mathbf{E} = \hat{\mathbf{n}} \cdot (\mathbf{E} \times \overline{\mathbf{H}})$. Then, for any bounded $\Omega_1 \subset \Omega$, the vector Green's theorem yields:

$$\int_{\partial\Omega_1} \hat{\mathbf{n}} \cdot \left[\mathbf{E} \times \overline{\mathbf{H}} - \mathbf{H} \times \overline{\mathbf{E}}\right] dS = \int_{\Omega_1} \nabla \cdot \left(\mathbf{E} \times \overline{\mathbf{H}} - \mathbf{H} \times \overline{\mathbf{E}}\right) dV \quad =$$

$$\int_{\Omega_1} \left[\overline{\mathbf{H}} \cdot (\nabla \times \mathbf{E}) - \mathbf{E} \cdot (\nabla \times \overline{\mathbf{H}}) - \overline{\mathbf{E}} \cdot (\nabla \times \mathbf{H}) + \mathbf{H} \cdot (\nabla \times \mathbf{E})\right] dV \quad =$$

$$(ik) \int_{\Omega_1} (\overline{\mathbf{H}} \cdot \mathbf{H} + \overline{\mathbf{E}} \cdot \mathbf{E} - \overline{\mathbf{E}} \cdot \mathbf{E} - \mathbf{H} \cdot \overline{\mathbf{H}})\, dV = 0. \tag{4.2}$$

Having established this simple form of Green's theorem, we can use the definition of the far field (1.4) and the radiation conditions to establish the first estimate, (3.4) as follows:
First, note that by starting with the condition

$$\hat{\mathbf{r}} \times (\nabla \times \mathbf{E}) + ik\mathbf{E} = o(\frac{1}{r}), \quad \text{as } r \to \infty,$$

and using the Maxwell equations to substitute for $\nabla \times \mathbf{E}$, we see that

$$\hat{\mathbf{r}} \times \mathbf{H} = \mathbf{E} + o(\frac{1}{kr}), \quad \text{as } r \to \infty. \tag{4.3}$$

Moreover, it follows from the definition of the far field pattern (1.4), that

$$|r\mathbf{E}| = |\mathbf{F}(\hat{\mathbf{r}})(1 + o(1))|, \quad \text{as } r \to \infty. \tag{4.4}$$

Now let Ω_1 be the region exterior to B and interior to a ball, B_R, of radius R which contains B in its interior. Then using the preceeding lemma

$$\begin{aligned} \int_{\partial\Omega} \left[(\hat{\mathbf{n}} \times \mathbf{H}) \cdot \overline{\mathbf{E}} - (\hat{\mathbf{n}} \times \mathbf{E}) \cdot \overline{\mathbf{H}}\right] dS \quad &= \\ \int_{B_R} \left[-(\hat{\mathbf{n}} \times \mathbf{H}) \cdot \overline{\mathbf{E}} + (\hat{\mathbf{n}} \times \mathbf{E}) \cdot \overline{\mathbf{H}}\right] dS \quad &= \\ \int_{B_R} \left[-(\hat{\mathbf{n}} \times \mathbf{H}) \cdot \overline{\mathbf{E}} - (\hat{\mathbf{n}} \times \overline{\mathbf{H}}) \cdot \mathbf{E}\right] dS \quad &= \\ 2\int_{B_R} (-\hat{\mathbf{r}} \times \mathbf{H}) \cdot \overline{\mathbf{E}}\, dS = 2\int_{B_R} (\mathbf{E} + o(\frac{1}{kR})) \cdot \overline{\mathbf{E}}\, dS, \quad \text{as } R \to \infty. \end{aligned} \tag{4.5}$$

Taking the limit of this last expression as $R \to \infty$, it follows from (4.4) that

$$\begin{aligned} \frac{1}{2} \int_{\partial\Omega} \left[(\hat{\mathbf{n}} \times \mathbf{H}) \cdot \overline{\mathbf{E}} - (\hat{\mathbf{n}} \times \mathbf{E}) \cdot \overline{\mathbf{H}}\right] dS \quad &= \\ \lim_{R\to\infty} \int_{\partial B_R} |\mathbf{E}|^2\, dS \quad &= \\ \int_{|\mathbf{r}|=1} |\mathbf{F}(\hat{\mathbf{r}})|^2\, dS. \end{aligned} \tag{4.6}$$

From this last equation, we arrive at the inequalities

$$\int_{|\mathbf{r}|=1} |\mathbf{F}(\hat{\mathbf{r}})|^2\, dS \leq \int_{\partial\Omega} |\mathbf{E_h}||\mathbf{H_h}|\, dS \leq \int_{\partial\Omega} \frac{1}{2} \left(|\mathbf{E_h}|^2 + |\mathbf{H_h}|^2\right) dS, \tag{4.7}$$

and therefore

$$J(\mathbf{h}) \leq \frac{1}{2} \sup\ a(\hat{\theta}), \quad \text{if } |||h||| = 1. \tag{4.8}$$

This, then, establishes the first inequality of the theorem.

The other conclusions of the theorem follow from the corresponding results in [4] and the asymptotic expansion of the solution of the exterior radiation problem. The details needed for the establishment of the expansions will be presented elsewhere. [3]

5 References

[1] T.S. ANGELL, A. KIRSCH, AND R.E. KLEINMAN, Antenna control and optimization, *Proc. IEEE*, 79:1559–1568, (1991).

[2] T. S. ANGELL AND R. E. KLEINMAN, Generalized exterior boundary value problems and optimization for the Helmholtz equation, *J. Optimization Theory Appl.*, 37:469-497, (1982).

[3] T.S. ANGELL AND R.E. KLEINMAN, A new optimization mentod for antenna design, *Ann. des Telecommunications*, 40: 341–349, (1985).

[4] T.S. ANGELL, R.E. KLEINMAN, AND B. VAINBERG, Asymptotic methods for an optimal antenna problem, submitted to *SIAM J. Appl. Math.*

[5] T.S. ANGELL, R.E. KLEINMAN, AND B. VAINBERG, to appear.

[6] A. CALDERÓN, Multipole expansion of radiation fields, *J. Rat. Mech. Anal.*, 3:523–537, (1954).

[7] D.L. COLTON AND R. KRESS, *Integral Equation Methods in Scattering Theory*, Wiley Interscience, New York, (1983).

[8] S. FAST, *An Optimization Method for Solving a Radiation Direction Problem*, Ph.D. Thesis, University of Delaware, Newark, Delaware, (1988).

[9] G. ULHMANN, Inverse boundary value problems and applications, *Asterisque*, 207:153–211, (1992).

---

[3] DISCLAIMER: The views and conclusions contained herein are those of the authors and should not be interpreted as necessarily representing the official policies or endorsements, either expressed or implied, of the Air Force Office of Scientific Research or the U.S. Government.

GENERATION OF WIDEBAND ANTENNA PERFORMANCE BY [Z] AND [Y] MATRIX INTERPOLATION IN THE METHOD OF MOMENTS

Kathleen L. Virga and Yahya Rahmat-Samii

Department of Electrical Engineering
University of California, Los Angeles
Los Angeles, CA 90095-1594

INTRODUCTION

Designing antennas for modern radar and communications applications often requires the evaluation of the antenna's ultra-wide band (UWB) operation capabilities. Identifying the appropriate electromagnetic modeling tools for UWB antennas can be challenging, since such antennas come in a wide-variety of configurations that range from thin-wire types to complex structures such as spirals, bow-ties, etc. The triangular surface patch method of moments (MoM) formulation[1,2] is one popular modeling approach. The surface mesh allows flexibility in modeling detailed antenna features. Since the elements of the MoM impedance matrix, or [Z], must be recomputed for each new frequency, the computation of antenna performance over an wide frequency range can take a long time.

This chapter discusses the utilization of an efficient method to compute the performance of wideband antennas using *frequency* interpolation of the [Z] matrix[3-5]. This method is used with the method of moments in order to significantly reduce the computation time required to evaluate the elements of [Z] at each frequency. Fig. 1 outlines the [Z] matrix interpolation procedure as well as a comparable [Y] matrix interpolation procedure. Both methods incorporate knowledge of the *frequency* characteristics of their corresponding matrix elements in a way that reduces the time it takes to compute the elements at each frequency. The [Z] matrix interpolation method has particular attributes that allows it to be easily applied to a different of antennas. In this method, the elements of only a few [Z] matrices, evaluated at relatively large frequency intervals, are directly computed. These matrix elements are used to interpolate the elements of the [Z] matrices at the intermediate frequencies. The method significantly reduces the time it takes to compute the antenna performance over a wide frequency band and is straightforward to apply.

Methods to reduce the computation time of the MoM emphasize the use of numerical and geometrical approximations to quickly fill [Z] or accentuate efficient matrix inversion and solution algorithms[6,7]. Some researchers have implemented methods that use *spatial* interpolation to fill the impedance matrix[8-10]. This process requires special attention since

Ultra-Wideband, Short-Pulse Electromagnetics 3
Edited by Baum *et al.*, Plenum Press, New York, 1997

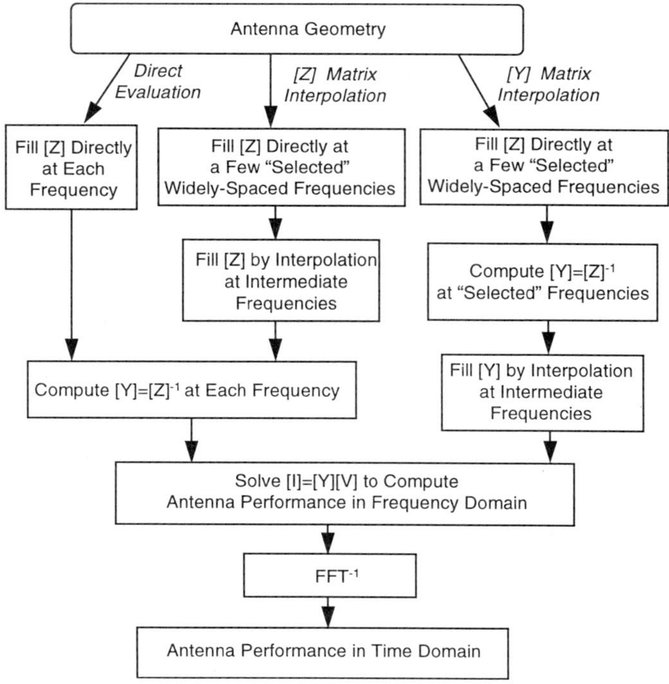

Fig. 1 Comparison of [Z] and [Y] matrix interpolation methodologies

the interpolation sampling depends on the antenna geometry and on the structure of the surface mesh. A technique has been implemented to compute rational function approximations for the transfer functions of antenna performance parameters[11]. The transfer functions are determined for one parameter and thus do not provide any additional information on the overall antenna performance. The Cauchy technique has been used to calculate rational function approximations for the surface currents on a conducting cylinder with a slit[12]. The coefficients used in the rational function and Cauchy techniques are determined by frequency samples of the response or from the response and higher-order derivatives at one or two frequencies. Computation of the derivatives requires access to the MoM source code and can be difficult when complex basis or testing functions are used.

The method of [Z] matrix interpolation with frequency is based upon an algorithm that

a) accurately constructs the antenna surface currents, impedance, patterns, etc.,
b) is independent of the angle of incidence or the excitation,
c) applies to a wide variety of complex antenna configurations,
d) utilizes simple interpolation functions that require only a few coefficients, and
e) is easily implemented to an existing method of moments computer code.

The method is applied to the performance analysis of a circular helix antenna on an infinite ground plane, Fig. 2a, and a planar inverted F antenna (PIFA), Fig. 2b. The circular helix is a complex thin-wire antenna. The helix with dimensions shown in Fig. 2a is in the axial mode near 3 GHz. The MoM model consists of 88 wire subsections. The antenna is fed by a delta-gap source located at the base of helix. The wire subsections are numbered consecutively from 1 (at the base of the helix) to 88 (at the top of the helix.) The PIFA is a compact low-profile antenna that consists of an air-suspended rectangular patch element, small ground plane, and a shorting plate. The overall surface mesh and wire section model consists of 527 unknowns. The antenna is fed by a delta-gap source that is placed between the base of the feed wire and the ground plane.

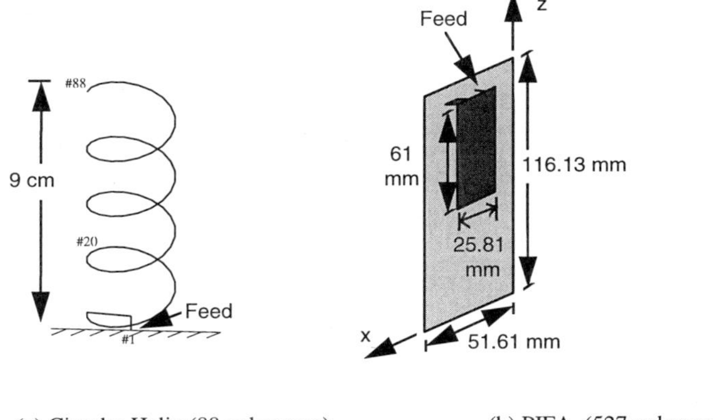

(a) Circular Helix (88 unknowns) (b) PIFA (527 unknowns)

Fig. 2 Antennas for the application of [Z] Matrix interpolation

One particular application of the [Z] matrix interpolation method is the analysis of antenna characteristics of short pulse radiation. The time domain response can be determined by performing an inverse Fourier transform on the frequency domain data. The simulation run times of the direct [Z] matrix computation and the [Z] matrix interpolation computation are compared. The discussion includes examining the characteristics of some of the elements of the impedance matrix with respect to the corresponding elements of the admittance matrix, [Y]. An in-depth investigation of the effects of different interpolation parameters, such as the interpolation scheme and the size of interpolation frequency step, on the computed antenna performance is presented. Guidelines on implementing the [Z] matrix interpolation method for different antenna structures are given.

[Z] MATRIX INTERPOLATION METHODOLOGY

Triangular Surface Patch Method of Moments Implementation

In the triangular surface patch method of moments formulation for antenna radiation problems, the antenna surfaces are partitioned into N sufficiently small subsections. From this, the solution the system of equations [Z][I]=[V] determines the N surface currents on the antenna, where [Z] is the N x N impedance matrix, [I] is the N x 1 current coefficient matrix to be determined, and [V] is the N x 1 voltage or excitation matrix. This formulation uses the electric field integral equation (EFIE) for perfect electric conductors. Using this condition and expressing the total radiated field in terms of potential functions allows one to write

$$\left[j\omega \frac{\mu}{4\pi} \int_S \vec{J}_S(\vec{r}\,') \frac{e^{-jkR}}{R} dS' - \frac{1}{j4\pi\omega\varepsilon} \nabla \int_S \nabla' \cdot \vec{J}_S(\vec{r}\,') \frac{e^{-jkR}}{R} dS' \right]_{tan} = \vec{E}^{i}_{tan}(\vec{r}) \qquad (1)$$

where *tan* refers to the vector component that is tangential to the conductor surface S, $R = |\vec{r} - \vec{r}\,'|$ is the distance between the observation point $\vec{r}$ and source point $\vec{r}\,'$ on S, λ is the wavelength, $k=2\pi/\lambda$, and μ and ε are the permittivity and permeability, respectively of the medium. An $e^{j\omega t}$ time convention is used, where $\omega=2\pi f$ and f denotes frequency.

These equations and the form of the basis and testing functions determine the frequency characteristics of the elements of [Z]. The basis functions in the triangular surface patch formulation depend only upon the geometrical parameters of the subsection[2]. Significant computational effort is required to fill the N^2 elements in [Z]. This effort increases when techniques, such as higher order basis functions[13, 14] are used.

Equation (1) reveals that the term e^{-jkR} dominates the frequency behavior of the [Z] elements. For each [Z] element, R is equal to $r_{mn} = |\vec{r}_m - \vec{r}_n|$ where $\vec{r}_m$ is the observation location and $\vec{r}_n$ is the source location. When the observation and source are close to each other, r_{mn} is small, and $e^{-jkr_{mn}}$ varies slowly with frequency. When they are far from each other, r_{mn} is large, and $e^{-jkr_{mn}}$ fluctuates rapidly with frequency.

[Z] Matrix versus [Y] Matrix Interpolation

The interpolation of the elements of the [Y] matrix over frequency would significantly reduce the time required to compute the antenna performance over many frequencies. It would eliminate having to invert the [Z] matrix at each intermediate frequency. Unfortunately, while the elements of the [Z] matrix are practically unaffected by the resonant characteristics of the antenna, the elements of the [Y] matrix are strongly influenced by the resonant behavior. Each element of [Z] depends upon the relative spacing of the two subsections, whereas the elements of [Y] strongly depend upon the *overall* behavior of the entire antenna structure. Fig. 3 compares the behavior of some of the [Z] and [Y] matrix elements for the circular helix. The self-term [1,1] as well a several other matrix terms are shown. The [Z] matrix elements in Fig. 3a vary slowly with frequency while the [Y] matrix elements in Fig. 3b fluctuate rapidly with frequency. The elements of [Z] can be evaluated over a frequency range by low-order interpolation functions, such as a quadratic. The evaluation of the elements of [Y] over a frequency range via interpolation requires complex interpolation functions that must be tailored for each different antenna.

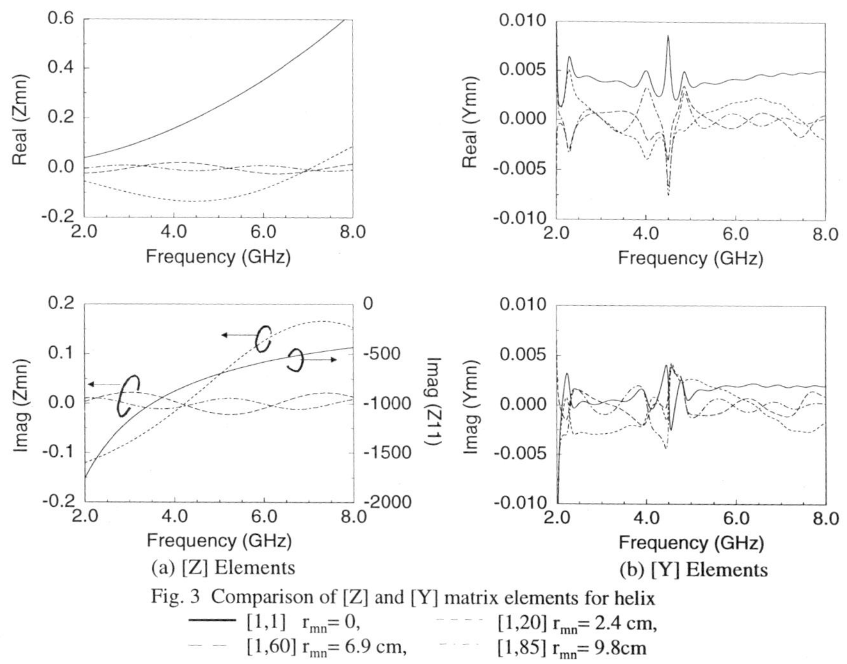

(a) [Z] Elements (b) [Y] Elements

Fig. 3 Comparison of [Z] and [Y] matrix elements for helix
—— [1,1] r_{mn}= 0, - - - [1,20] r_{mn}= 2.4 cm,
— — [1,60] r_{mn}= 6.9 cm, - · - [1,85] r_{mn}= 9.8cm

[Z] Matrix Interpolation

The [Z] matrix interpolation process begins by partitioning the entire frequency band of interest into steps and defining several "selected" frequencies. The interval between adjacent "selected" frequencies defines the interpolation frequency step size. Three neighboring "selected" frequencies define a frequency range. The elements of the [Z] matrices for the first three selected frequencies are directly computed by evaluating the potential integrals in (1). The elements of [Z] for the intermediate frequencies are approximated by a quadratic function

$$Z_{mn}(f) = A_{mn}f^2 + B_{mn}f + C_{mn} \tag{2}$$

where A_{mn}, B_{mn}, and C_{mn} are the *m,nth* elements of the complex coefficient matrices [*A*], [*B*], and [*C*], respectively. Equation (2) can be cast into a system of three equations and three unknowns. These equations along with the elements of the directly computed [Z] matrices calculated at three selected frequencies are used to determine the coefficient matrices. If the frequency band of interest is especially wide, it may be necessary to divide the band into several interpolation frequency sub-bands.

Improved [Z] Matrix Interpolation

The [Z] matrix interpolation method can be improved upon by incorporating the frequency behavior of the elements of [Z] into the process[5]. This is done to increase the accuracy in evaluating the terms of [Z] at the intermediate frequencies and to allow larger interpolation frequency ranges. Quadratic interpolation works best for antennas with surfaces that are smaller than 0.5λ. For larger antennas, the rapid frequency variation of the factor $e^{-jkr_{mn}}$ dominates the frequency variation of the [Z] matrix elements that have r_{mn} greater than 0.5λ. The quantity

$$Z'_{mn} = \frac{Z_{mn}}{e^{-jkr_{mn}}} \tag{3}$$

changes slowly with frequency and can be interpolated with low-order interpolation functions. When r_{mn} is large with respect to λ, the improved computation of Z_{mn} is evaluated in a two steps. First Z'_{mn} is computed by quadratic interpolation using the corresponding elements of the directly computed $\lfloor Z' \rfloor$matrices. Z_{mn} is determined by multiplying the resultant value by the known quantity, $e^{-jk_{mn}}$. Fig. 4 compares the behavior of the elements of [Z] and $\lfloor Z' \rfloor$ for the circular helix. Fig. 4a shows how the elements of [Z] oscillate with frequency and how the number of oscillations over the frequency range increase as the distance between the *mth* and *nth* element, r_{mn}, increases. The elements of $\lfloor Z' \rfloor$, shown in Fig. 4b are much smoother over the entire frequency band, and therefore are easier to interpolate. Note that the concept of factoring out the $e^{-jkr_{mn}}$ term does not apply to [Y] matrix interpolation. The interpolation can also be improved by incorporating the known form of the singular and closely spaced terms of [Z] into the process[5].

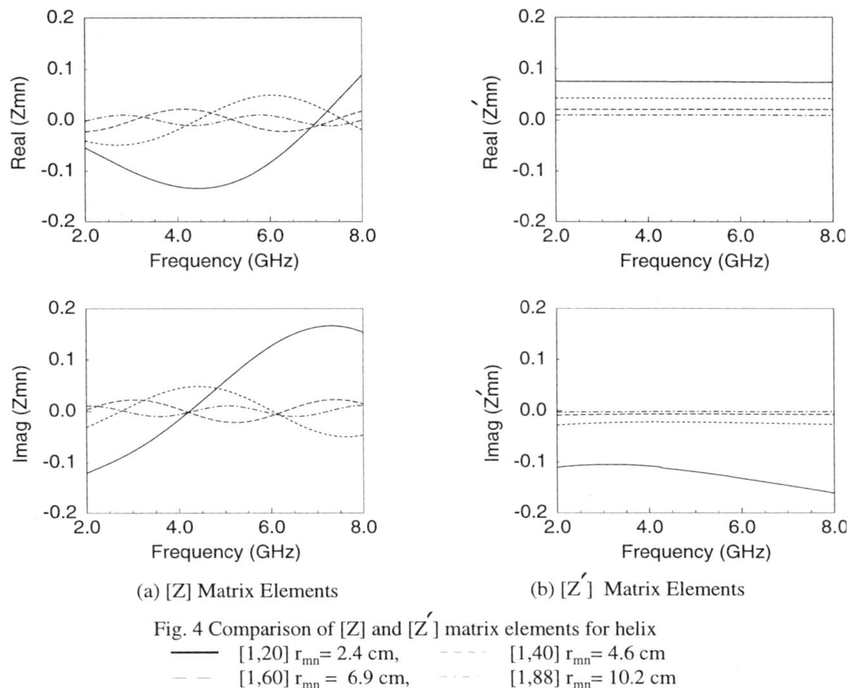

(a) [Z] Matrix Elements (b) [Z′] Matrix Elements

Fig. 4 Comparison of [Z] and [Z′] matrix elements for helix
[1,20] r_{mn}= 2.4 cm, [1,40] r_{mn}= 4.6 cm
[1,60] r_{mn} = 6.9 cm, [1,88] r_{mn}= 10.2 cm

REPRESENTATIVE ANTENNA EXAMPLES

The quadratic and improved [Z] matrix interpolation schemes were used to compute the input impedance of a 4 turn circular helix antenna on an infinite ground plane. The input impedance computed by direct [Z] evaluation and by both interpolation methods is compared in Fig. 5. The input impedance in each case is computed and plotted at every 20 MHz. The selected frequencies used for the interpolation are denoted by the five stars, '$*$'. The impedance is calculated by interpolation over the entire 4:1 frequency band using only 5 directly computed [Z] matrices, i.e., three sub-bands are used. Both interpolation approaches reconstruct the input impedance well. Even the rapid impedance changes between 4 and 5 GHz only slightly differ from the directly computed results. This is significant since the closest selected frequencies are at 3.5 GHz and 5 GHz and lie outside this range.

The quadratic [Z] matrix interpolation scheme has been applied to the computation of the input impedance of the PIFA[15]. Fig. 6 examines the impact of the specific location of the selected frequencies on the results. The input impedance in this figure is computed by direct [Z] evaluation and by two different [Z] matrix interpolation simulations. Both interpolation cases use a 250 MHz interpolation frequency step. In one case the three selected frequencies are 0.6, 0.85 and 1.1 GHz while in the other the selected frequencies are 0.53, 0.78 and 1.03 GHz. Both interpolation cases predict the resonant behavior at 0.78 GHz even though there is no selected frequency specifically located at this point.

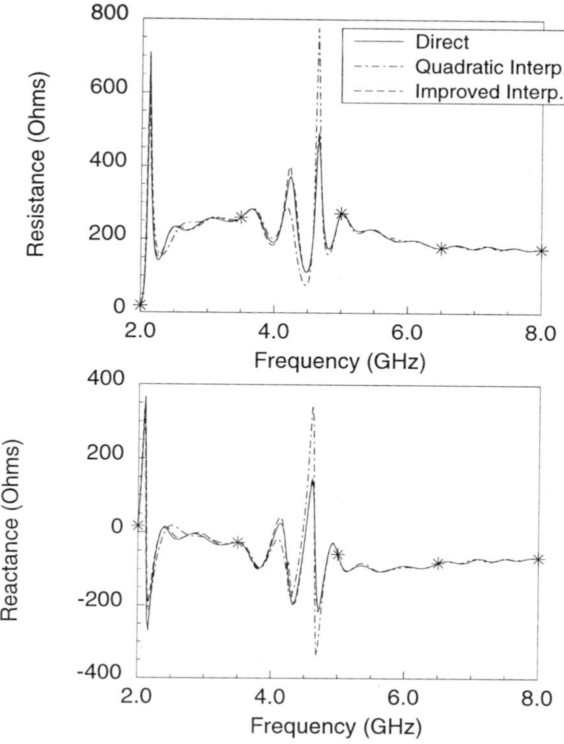

Fig. 5 Circular helix input impedance comparison of quadratic and improved interpolation
* denote "selected" frequencies where [Z] is directly computed

Fig. 7 compares the far-field pattern of the PIFA at 0.90 GHz computed by the direct MoM evaluation and by [Z] matrix interpolation with selected frequencies at 0.53, 0.78 and 1.03 GHz. The computation of the far-field involves the integration of all of the currents on the surfaces of the conductors, whereas the input impedance only depends upon the current at the antenna feed input. These results show that the interpolation method can also be used to accurately predict the far-field characteristics of the antenna.

Fig. 8 compares the computation time of direct [Z] evaluation and [Z] matrix interpolation. All the computations were performed on an IBM RS/6000 530H workstation. The timing comparison for the circular helix involve the computation of the helix input impedance for all 301 frequencies (including matrix fill and inversion.) The timing comparison of the PIFA involves the time to fill [Z] for each frequency.

IMPLEMENTATION GUIDELINES

In order to compute the broadband frequency performance of the antenna as rapidly as possible, the interpolation frequency step must be carefully chosen. A small interpolation frequency step means that [Z] matrices are directly computed and filled for many frequencies and that new quadratic coefficients must be recomputed many times. A very large interpolation frequency step results in poor reconstruction of the elements in [Z]. In order to choose a suitable interpolation frequency step size, one must review the behavior of a few [Z] matrix elements. A plot of a few of the elements of [Z] that represent the full range (closest, farthest and mid-range) of r_{mn} values can be used to quickly indicate how they vary with frequency. Such an approach can be used to ensure the accuracy of the simulated results without knowing the antenna resonant behavior. The [Z] matrix elements shown in Figs. 3 and 4 are examples of such plots. These simulation guidelines were used in determining the interpolation approaches for the helix and the PIFA. The computation time required to compute a few [Z] matrix values is significantly smaller than computing and inverting the overall [Z] matrix for many frequencies.

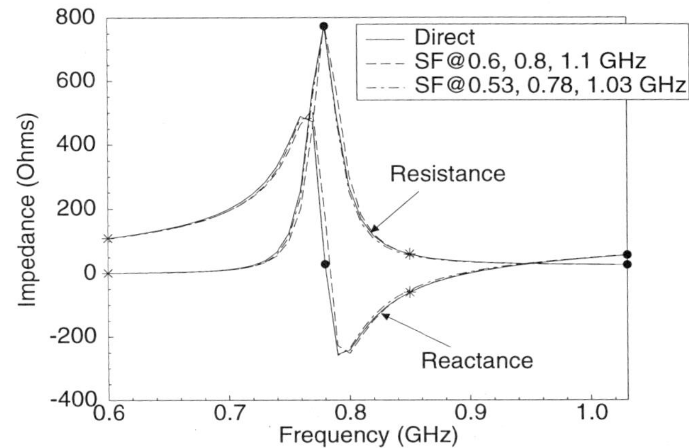

Fig. 6 PIFA input impedance computed from different sets of "selected" frequencies

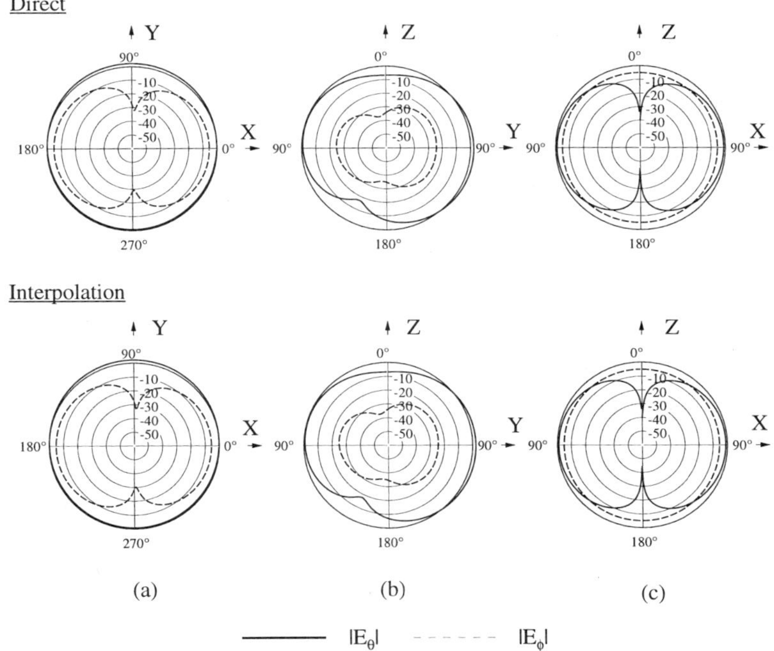

Fig. 7 Comparison of far-field patterns of PIFA at 0.90 GHz computed by direct moment method and by [Z] matrix interpolation (a) |E| dB vs. ϕ, $\theta=90^\circ$, (b) |E| dB vs. θ, $\phi=90^\circ$, (c) |E| dB vs. θ, $\phi=0^\circ$. Selected frequencies are at 0.53, 0.78, 1.03 GHz

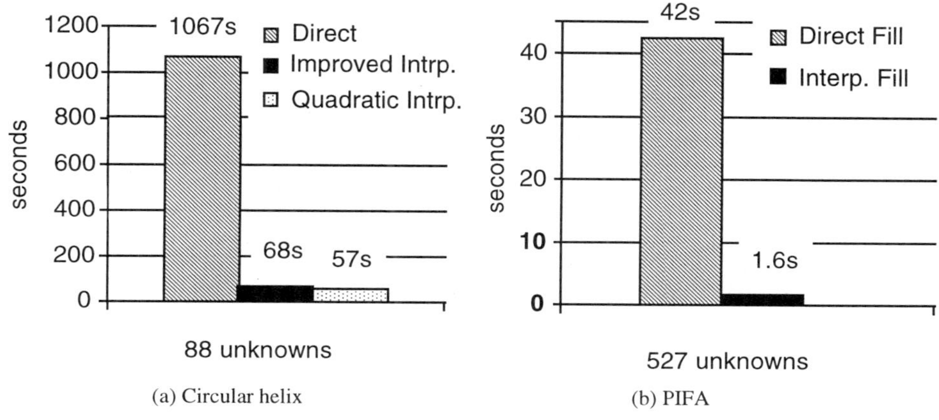

(a) Circular helix (b) PIFA

Fig. 8 Timing comparison of direct [Z] evaluation and [Z] matrix interpolation

RATIONAL FUNCTION APPROXIMATION OF [Y] MATRIX ELEMENTS

Transfer functions expressed as rational functions provide an alternate way to represent the behavior of a specified performance parameter over some bandwidth. The use of rational functions to represent the frequency transfer functions of the elements of the MoM admittance matrix, [Y], is suggested in [11]. This method provides an overall representation of antenna performance, since the surface currents on the antenna at a particular frequency can be directly computed from the corresponding [Y] matrix.

The rational functional form can be written as a ratio of two polynomials given as,

$$F(X) = \frac{\sum_{i=0}^{p} N_i X^i}{\sum_{i=0}^{d} D_i X^i} \tag{4}$$

The feasibility of [Y] matrix interpolation was investigated by applying it to the evaluation of a [Y] matrix element for the helix. Fig. 9 compares $Y_{1,60}$ as a function of frequency computed three different ways: (a) by directly inverting the entire [Z] matrix and extracting $Y_{1,60}$ at each frequency, (b) by approximating the value of $Y_{1,60}$ by the ratio of two 4th order polynomials, and (c) by approximating the value of $Y_{1,60}$ by the ratio of two 10th order polynomials. The coefficients for the ratio of 4th order polynomials were computed by evaluating the elements of [Y] (via the computation and inversion of [Z]) at 9 frequencies. The coefficients for the ratio of 10th order polynomials were computed by evaluating the elements of [Y] at 21 frequencies.

Fig. 9 shows that the ratio of two high-order (greater than 10th degree) polynomials are required to reconstruct the behavior of the $Y_{1,60}$ matrix element. The ratio of polynomials representation for each element of [Y] requires the computation and inversion of [Z] at 21 frequencies as well at the storage of more than 21 [88 x 88] coefficient matrices. The rational function approximation of the elements of [Y] must represent the resonant characteristics of the antenna. Different polynomial orders that depend upon the resonant nature of the structure and the bandwidth of the evaluation frequencies are required for different types of antennas. This is an undesirable feature, since it makes the task of routinely applying [Y] matrix interpolation to new problems rather cumbersome.

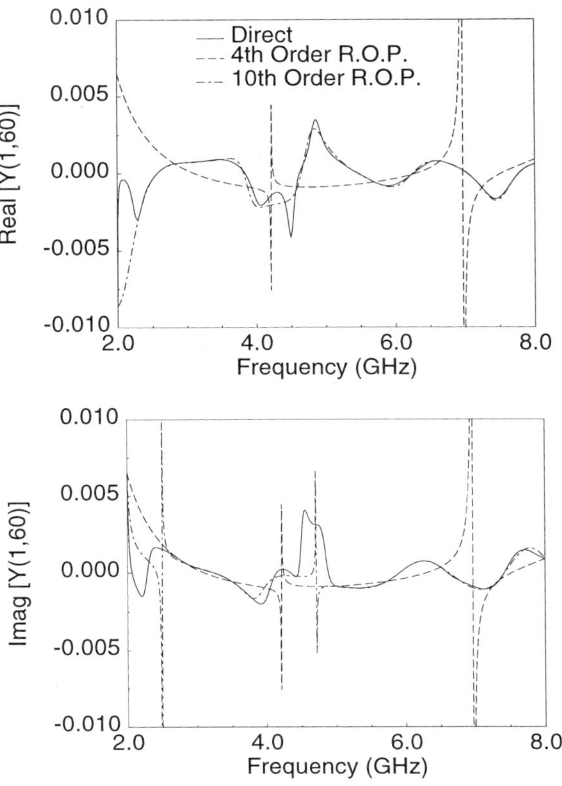

Fig. 9 Comparison of ratio of polynomials approximation of Y1,60 for circular helix computed by direct [Y] evaluation and the ratio of polynomials (R.O.P)

SHORT-PULSE TIME-DOMAIN ANTENNA PERFORMANCE ANALYSIS

The [Z] matrix interpolation method has been applied to the analysis of the source current of the circular helix antenna shown in Fig. 2a for short-pulse excitation. The input voltage source is the time derivative of a Gaussian pulse. The temporal behavior of the source voltage impressed at the connection between the base of the helix wire and the infinite ground plane is[16]

$$V_s(t) = 2a^2(t - t_{max})^2 \, exp[-a^2(t - t_{max})^2] \quad \text{volts} \tag{5}$$

where $a = 1.5 \times 10^9$ and $t_{max} = 1.43 \times 10^{-3}$ sec. The temporal step size is $\delta t = 6.26 \times 10^{-11}$ sec. Fig. 10 shows the input admittance of the helix as a function of the scaled frequency, F/c, where c = 3×10^8 m/s. This figure compares the input admittance, Y_{in}, at 256 frequency points computed by the directly filling the elements of [Z] and by interpolating the elements of [Z]. Fourteen selected frequencies, denoted by '*', are used to interpolate the elements of [Z] across the 256:1 bandwidth. When there is a high density of selected frequencies at the lower frequency range, the values of Y_{in} computed by [Z] matrix interpolation agree well with the valued computed by direct [Z] computation. Recall that the imaginary part of the elements of [Z] for the self and closely spaced segments have logarithmic behavior with frequency. More points are needed at the lower frequencies to adequately reconstruct this behavior. Fig. 11 compares the time-domain response of the source current computed by direct evaluation of [Z] and by [Z] matrix interpolation.

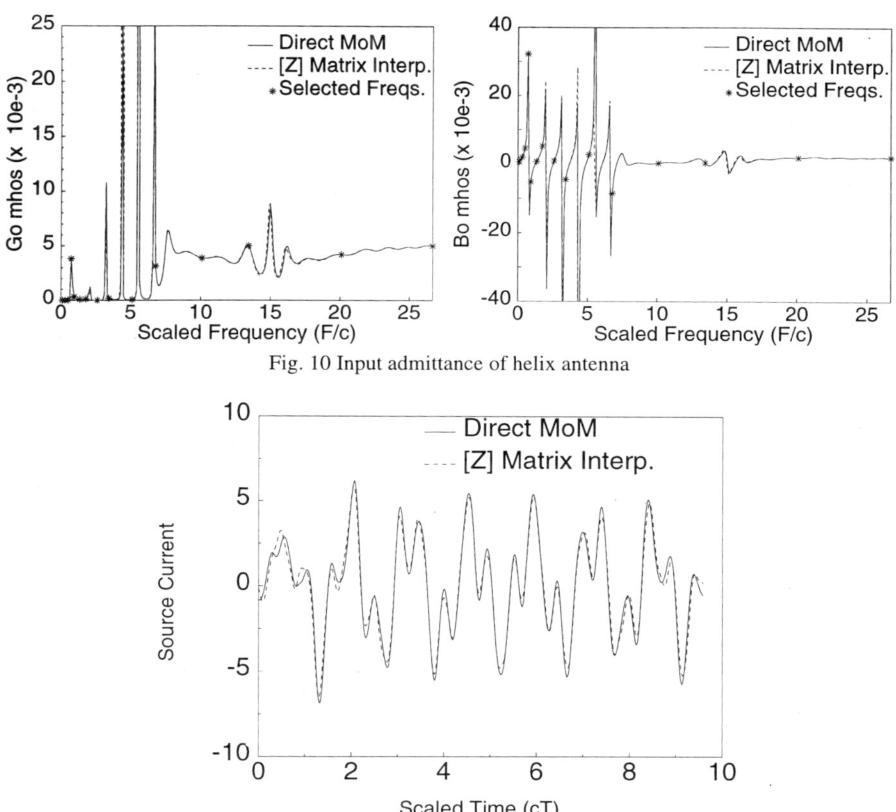

Fig. 10 Input admittance of helix antenna

Fig. 11 Time -domain response of source current for helix antenna for time derivative of a Gaussian voltage source excitation

CONCLUSIONS

A detailed look into why the [Z] matrix interpolation method works well has been conducted. This includes a comparison of the elements of the impedance matrix [Z] with the elements of the admittance matrix [Y]. The elements of [Z] vary slowly with frequency, while the elements of [Y] typically vary quite rapidly. Guidelines that can be used to determine the parameters for the successful implementation of [Z] matrix interpolation for the MoM analysis of other types of antennas have been given.

The [Z] matrix interpolation methodology has been used with the method of moments to significantly reduce the computation time required for the wideband performance evaluation of antennas. The substantial time savings of simulations using this method allows detailed performance evaluation and sensitivity studies to be performed. The method has the advantage in that it computes the entire surface currents, which can be used to compute the antenna performance parameters such as near-field and far-field patterns, and input impedance. The value of the method was illustrated by its application to some different types of antennas used for various applications. The examples clearly demonstrate the accuracy and time savings that are achievable with the [Z] matrix interpolation method.

Acknowledgment. This work was supported by Rockwell International and the UC MICRO.

REFERENCES

1. S. M. Rao, D. R. Wilton, and A. W. Glisson, "Electromagnetic scattering by surfaces of arbitrary shape," *IEEE Trans. Antennas Propagat.,* vol. AP-30, pp. 409-418, May 1982.
2. S.-U. Wu and D. R. Wilton, "Electromagnetic scattering and radiation by arbitrary configurations of conducting bodies and wires," Technical Document 1325, University of Houston. Applied Electromagnetics Laboratory, Dept. of Electrical Engineering, Aug. 1988.
3. E. H. Newman and D. Forrai, "Scattering from a microstrip patch," *IEEE Trans. Antennas Propagat.*, vol. AP-35, pp. 245-251, Mar. 1987.
4. K. L Virga and Y. Rahmat-Samii, "Wide-band evaluation of communications antennas using [Z] matrix interpolation with the method of moments," *1995 IEEE AP-S Intl. Symposium Digest*, pp. 1262-1264, Newport Beach, CA, June 1995.
5. E. H. Newman, "Generation of wide-band data from the method of moments by interpolating the impedance matrix," *IEEE Trans. Antennas Propagat.*, AP-36, pp. 1820-1824, Dec. 1988.
6. T. K Sarkar, K. R. Siarkiewicz, and R. F. Stratton, "Survey of numerical methods for solution of large systems of linear equations for electromagnetic field problems," *IEEE Trans. Antennas Propagat.,* vol. AP-29, pp. 847-856, Nov. 1981.
7. F. X. Canning, "Direct solution of the EFIE with half the computation," *IEEE Trans. Antennas Propagat.,* vol. AP-39, pp. 118-119, Jan. 1991.
8. G. F. Herrmann, "Note on interpolational basis functions in the method of moments," *IEEE Trans. Antennas Propagat.*, vol. AP-38, pp. 134-137, Jan. 1990.
9. T. W. Nuteson, K. Naishadham, and R. Mittra, "Spatial interpolation of the moment matrix in electromagnetic scattering and radiation problems," *1993 IEEE AP-S Intl. Symposium Digest*, pp. 860-863, Ann Arbor, MI, June 1993.
10. G. Vecchi, P. Pirinoli, L. Matekovits, and M. Orefice, "Reduction of the filling time of method of moments matrices," *11th Annual Review of Progress in Applied Computational Electromagnetics*, pp. 600-605, Monterery, CA, Mar. 1995.
11. G. J. Burke, E. K. Miller, S. Chakrabarti, and K. Demarest, "Using model-based parameter estimation to increase the efficiency of computing electromagnetic transfer functions," *IEEE Trans. Magn.*, vol. 25, pp. 2807-2809, July 1989.
12. K. Kottapalli, T. K. Sarkar, Y. Hua, E. K. Miller, and G. J. Burke, "Accurate computation of wide-band response of electromagnetic systems utilizing narrow-band information," *IEEE Trans. Microwave Theory Tech.*, vol. 39, pp. 682-687, Apr. 1991.
13. A. F. Peterson, "Higher-order surface patch basis functions for EFIE formulations," *1994 IEEE AP-S Intl. Symposium Digest*, pp. 2162-2165, Seattle, WA, June 1994.
14. D. R. Wilton, "Review of current status and trends in the use of integral equations in computational electromagnetics," *Electromagnetics*, vol. 12, pp. 287-341,1992.
15. K. Hirasawa and M. Haneishi, *Analysis, Design and Measurement of Small Low Profile Antennas*, Artech House, 1992.
16. A. J. Poggio, E. K. Miller, and G. J. Burke, "An integro-differential equation technique for the time-domain analysis of thin-wire structures," *Journal of Computational Physics*, vol. 12, pp. 210-233, 1973.

ELECTROMAGNETIC ANALYSIS OF EXPONENTIALLY TAPERED COPLANAR STRIPLINE ANTENNAS USED IN COHERENT MICROWAVE TRANSIENT SPECTROSCOPY TECHNIQUE

Valérie Bertrand, Michèle Lalande, and Bernard Jecko

Institut de Recherche en Communications Optiques et Microondes
CNRS - URA n° 356
IRCOM Limoges, 123 avenue Albert Thomas - 87065 Limoges (France)

INTRODUCTION

With the advent of reliable sources of ultrashort optical pulses, photoconductive switches and photoconductive sampling have become usual methods that allow to generate and to detect ultra-fast electric pulses. Recent experiments have shown that commuting devices of optoelectronic pulses associated with appropriate antennas can be used to generate, control and detect picosecond bursts of electromagnetic radiation. It leads to various applications[1] such as the transient radiation properties of antennas[2], the scattering of radiation by three-dimensional objects[3] or the characterization of material by the coherent microwave transient spectroscopy technique[4].

In this paper, an electromagnetic analysis by the Finite Difference Time Domain (FDTD) method of exponentially tapered coplanar stripline (ETCS) antennas used for the material characterization by the spectroscopy technique is reported. The effect of substrate thickness on radiation properties of ETCS antennas is studied. The computation of the field in the far zone allows to know radiation patterns and the received signal in the time domain.

EXPERIMENT

The experimental configuration is shown schematically in Figure 1. The transmitter is biased with a DC voltage, while the receiver is connected to low-frequency signal-averaging electronics. The optical beam is focussed on the feedline of the transmitter, where an ultra-fast electric pulse is generated. All or part of the spectrum contained in this pulse is radiated, depending on the characteristics of the transmitting antenna. The field incident on the receiver induces a transient voltage across its feedline which is photoconductively sampled as a function of time by the variably delayed probe beam. For material characterization, a sample is located between the two antennas. Generally, exponentially tapered coplanar stripline antennas are employed in this experiment because of their broad bandwidth, directivity and transmission-line nature[5].

Ultra-Wideband, Short-Pulse Electromagnetics 3
Edited by Baum *et al.*, Plenum Press, New York, 1997

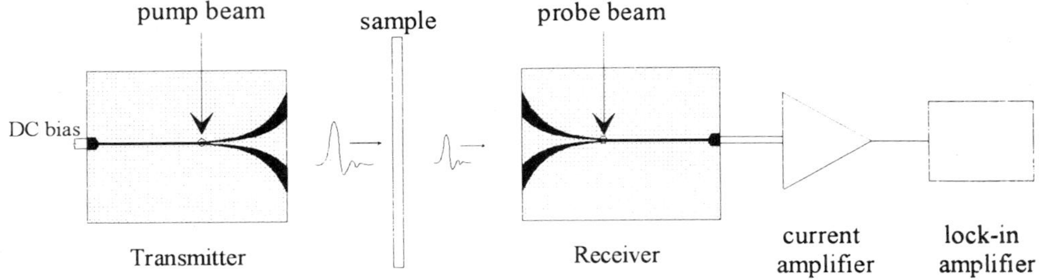

Figure 1. Schematic of the coherent microwave transient spectroscopy experiment.

EFFECT OF SUBSTRATE THICKNESS

This paper deals with an electromagnetic analysis of these ETCS antennas. In particular, the purpose of this electromagnetic study is to understand the radiation mechanism of this type of antenna and to optimize it. For this analysis and for the considered frequency bands, the FDTD method is a suitable method[6] because it takes all the propagation and radiation modes into account. The FDTD method based on a discrete form of Maxwell's equations leads directly to the time evolution of the radiated field.

The time evolution of the electric field is observed in free space at different points in the propagation direction (Figure 2) for two antenna geometric designs. The strip width and spacing of the feedline are respectively 5 µm and 10 µm. The antenna length is 3.5 mm and the end opening width measured from the outside edges of metallization is 2 mm. A 1V bias is supplied to the transmitter. The relative permittivity of the substrate is equal to 9.95. The substrate thickness is the only characteristic that varies between the two antennas. For the first antenna the substrate thickness is 3.34mm, for the second the substrate thickness is 340µm. The excitation signal is a Gaussian pulse with Full Width at Half Maximum (FWHM) of 3.5 picoseconds. The results (Figure 3) show that for the thinner substrate the peak height of the radiated field is larger indicating increased radiation efficiency. Less apparent, the width of the central transient is more narrow for the thinner substrate antenna. These results are in agreement with [7].

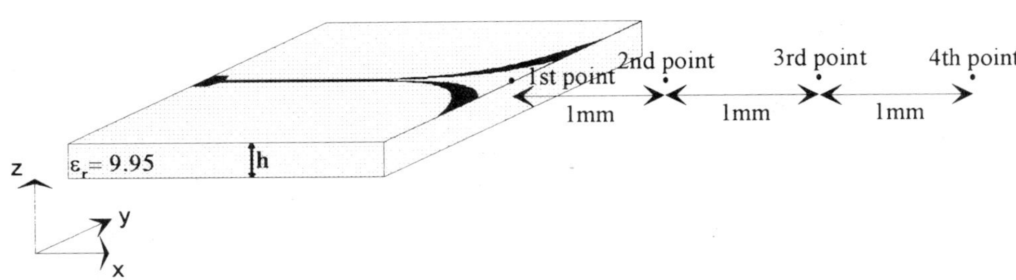

Figure 2. Observation points.

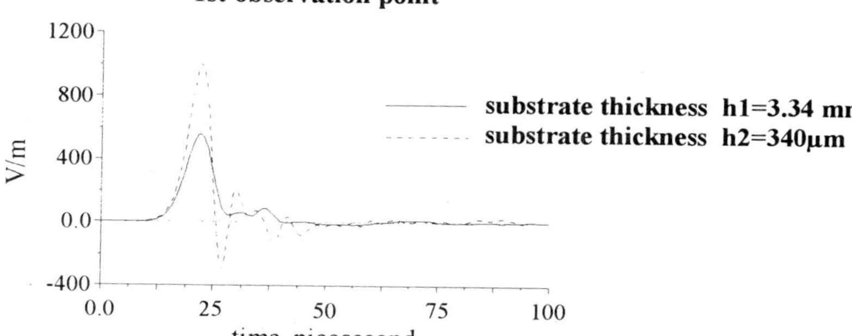

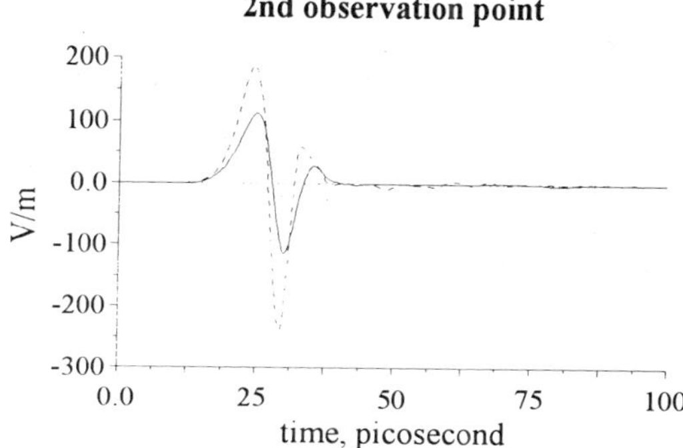

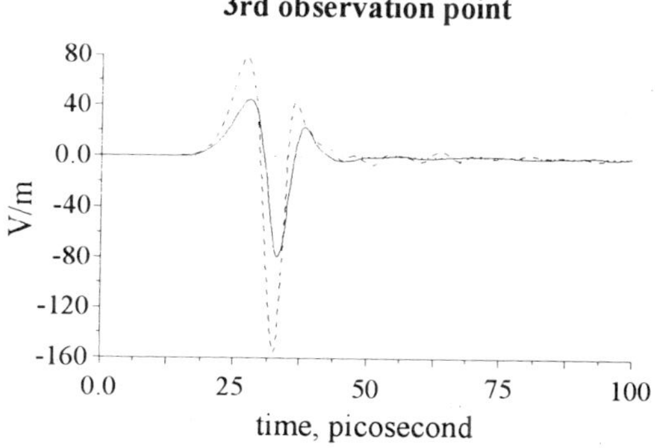

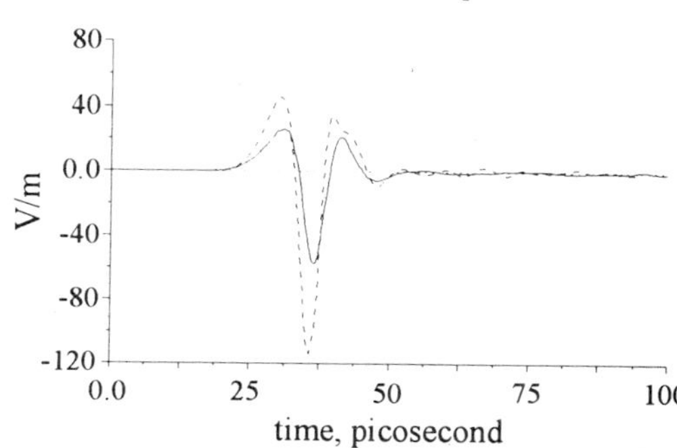

Figure 3. Time evolution of y electric field component.

FAR FIELD

The spectroscopy measurements require the material sample and the receiving antenna to be located in the far zone. So, the numerical analysis of the received signals leads to the physical characteristics of the material. Then, an electromagnetic study in the far field is necessary. The first far-field characteristic presented is the radiation pattern.

Radiation patterns

For the computation of antenna patterns, it is necessary to know the field in the far zone. The electromagnetic form of Huygen's principle is used to transform the near field obtained by the FDTD method on a surface surrounding the ETCS antenna into the far field. Equivalent magnetic and electric surface currents are defined from the near field. A surface integration of these equivalent currents, weighted by the free space Green's function is then calculated. The two angular components of far electric fields E_θ and E_ϕ are deduced from this integration. Different observation angles require a recalculation of the integral.

The E- and H-plane configuration is shown in Figure 4. The E- and H-plane radiation patterns of the ETCS antenna at 50 GHz and 75 GHz are shown in Figures 5a and 5b respectively. The E-plane patterns are symmetric because of the perfectly symmetric structure modeling and symmetric feed conditions. At 0°, the co-polarization is maximum whereas the cross-polarization is minimum. The cross-polarization maximum level is about -20 dB at ±90 degrees for the frequency of 50 GHz and is equal to -20.7 dB at ±35° for the frequency of 75 GHz.

For the H-plane, only the co-polarization is represented. In fact, the cross-polarization is quite negligible. No symmetry is noticed for the H-plane patterns. The main lobe is directed into the substrate[8] and is peaked at an angle of 95° for the frequency of 50 GHz. This angle is wider at 75 GHz and is about 108°. Generally in the spectroscopy experiment, microwave lenses are used to intercept and collimate the lobe of transient radiation emitted from the ETCS antennas[9].

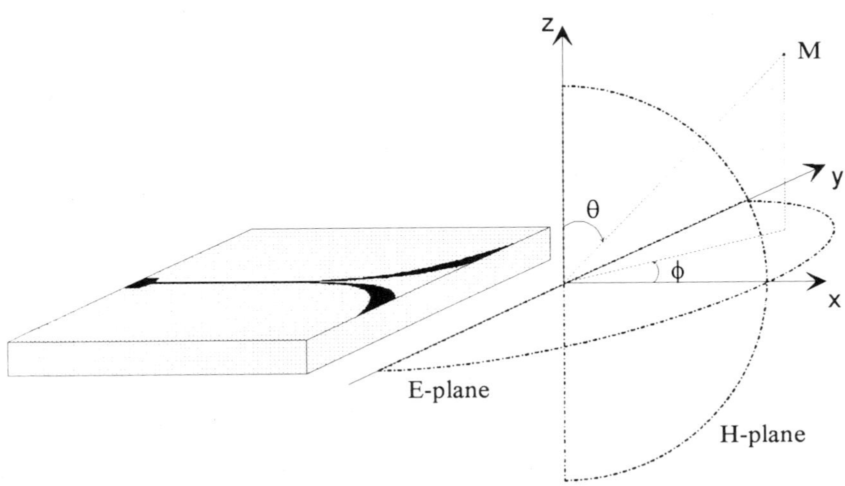

Figure 4. E- and H-plane configuration.

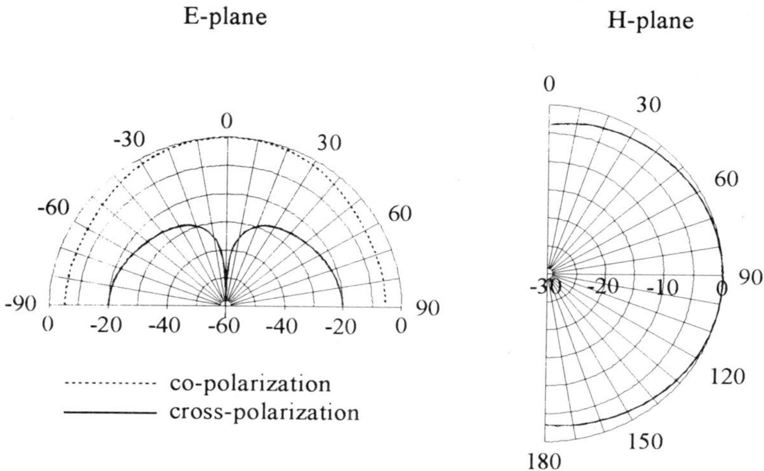

Figure 5a. Radiation patterns of the ETCS antenna at 50 GHz.

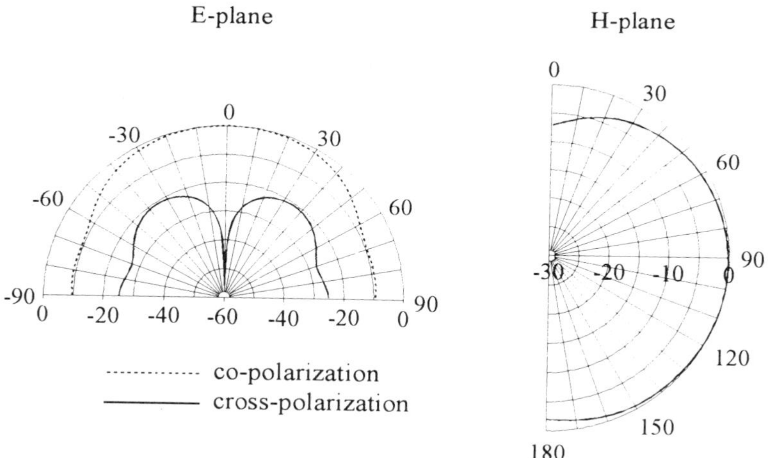

Figure 5b. Radiation patterns of the ETCS antenna at 75 GHz.

Received signal

In the previous paragraph, we explain that the far field is obtained by using Huygen's principle. The same principle is applied but in time domain form : equivalent magnetic and electric surface currents are introduced into the time domain integral to obtain the time evolution of the electric field in the far zone.

Figure 6a represents the time evolution of the radiated field in the propagation direction x 5 centimeters away, for an excitation gaussian with FWHM of 3.5 picoseconds. Figure 6b shows the spectrum obtained by a numerical Fourier transform of the radiated field shown in Figure 6a. The spectrum contains a range of frequency components from 25 GHz up to 150 GHz at -10 dB. The peak of the spectrum is at 75 GHz, which is the resonnant frequency of the end opening of the antenna.

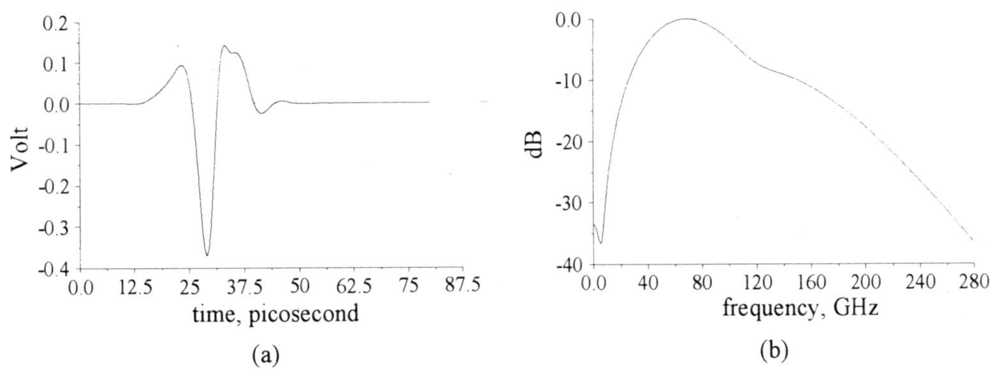

Figure 6. Radiated field (a) time evolution (b) normalized spectrum

In the second step, the received signal is determined by using Huygen's principle again. Time electric and magnetic surface current densities deduced from the radiated field are imposed on a surface surrounding the receiving antenna. These equivalent densities induce a transient voltage across the gap of the receiving antenna.

The received current is represented Figure 7a. The current magnitude obtained corresponds to an excitation signal with a maximum level of 1V and a distance between the two antennas of 5 centimeters. The frequency components of the received signal (Figure 7b) are contained between 25 GHz and 140 GHz at -10 dB.

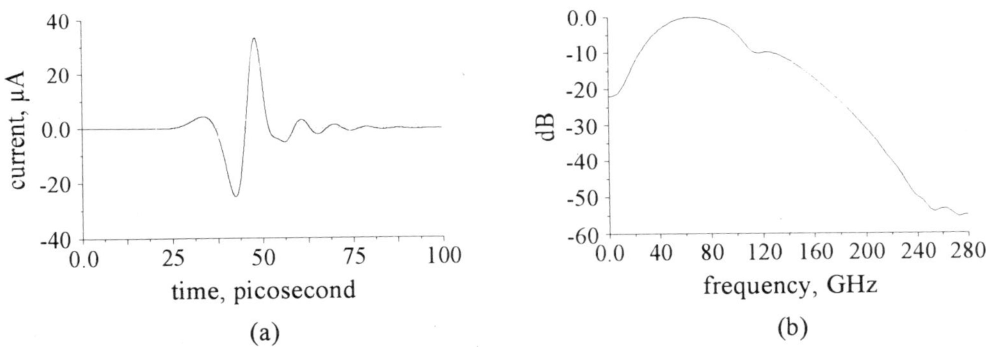

Figure 7. Received current (a) time evolution (b) spectrum

THEORY VERSUS EXPERIMENT

In Figure 8, the experimental received signal[10] and our theoretical received signal are represented. At first, the two signals are similar but the theoretical signal decreases more rapidly.

In fact, the two excitation signals are different, which induces differences between the two time waveforms. The theoretical signal is a gaussian pulse; the rise time of the experimental signal created by photoconductive effect is similar to the gaussian pulse but it decreases more slowly.

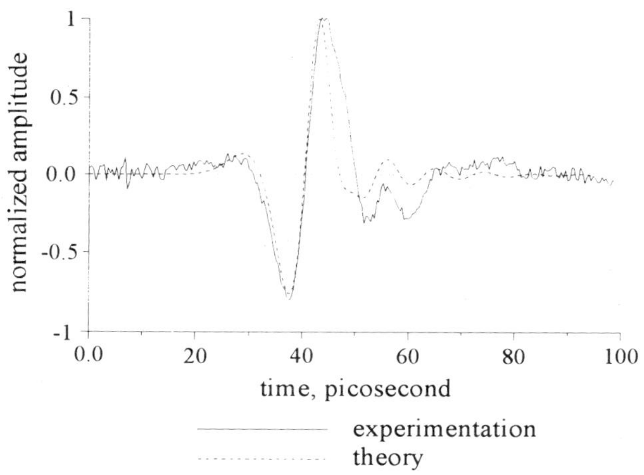

Figure 8. Comparison between experimental and theoretical received signals

CONCLUSION

To know the validity domain of material characterization by the spectroscopy technique, it's essential to have a precise knowledge of the transient and spectral evolution of fields where the material sample will be positioned. The main advantage of this electromagnetic analysis is that it leads to the time and harmonic evolution of radiated fields at every point in free space.

So, the coherent microwave transient spectroscopy technique can be optimized by using thin substrate antennas. The radiation patterns show that the ETCS antenna radiates into the substrate. This effect can be reduced by reducing the substrate thickness. But in practice, hemispherical microwave lenses are commonly used to collimate the transient radiation diverging from the transmitter and to focus it onto the receiver. The comparison between theory and experiment is quite satisfactory, which shows the validity of our numerical method.

ACKNOWLEGMENTS

This work has been performed in collaboration with CEA/CESTA ("Commissariat à l'Energie Atomique / Centre d'Etudes Scientifiques et Techniques d'Aquitaine").

The authors would like to thank I.D.R.I.S. ("Institut du Developpement et des Ressources en Informatique Scientifique") for their help with data processing (CRAY C98).

REFERENCES

1. G. Arjavalingam, Y. Pastol, J.-M. Halbout and W. M. Robertson, "Optoelectronically-pulsed antennas : characterization and applications", *IEEE Antennas and Propagation Magazine*, pp. 6-11, Feb. 1991.
2. M. M. Gitin, F. W. Wise, G. Arjavalingam, Y. Pastol and R. C. Compton, "Broad-band characterization millimeter-wave log-periodic antennas by photoconductive sampling", *IEEE Trans. on Antennas and Propagation*, Vol. 42, no. 3, March 1994.
3. W. M. Robertson, G. V. Kopcsay and G. Arjavalingam, "Picosecond time domain electromagnetic scattering from conducting cylinders", *IEEE Microwave and Guided Wave Letters*, Vol. 1, no.12, pp 249-251, Dec. 1991.

4. G. Arjavalingam, Y. Pastol, J.-M. Halbout and G. V. Kopcsay, "Broad-band microwave measurements with transient radiation from optoelectronically pulsed antennas", *IEEE Trans. Microwave Theory Tech.*, vol. 38, no. 5, pp.615-621, May 1990.
5. W. M. Robertson, *Optoelectronic Techniques for Microwave and Millimeter-Wave Engineering*, Artech House, Boston-London, 1995.
6. A. Taflove, "Advances in finite-difference time-domain methods for engineering electromagnetics", in : *Ultra-Wideband, Short-Pulse, Electromagnetics 2*, L. Carin and L. B. Felsen, Plenum Press, New York and London, 1995.
7. M. Heimlich, W. M. Robertson, G. Arjavalingam and J.-M. Halbout, "Effect of substrate thickness on radiation properties of coplanar strip antennas", *Electronics Letters*, vol. 28, no. 3, Jan. 1992.
8. Y. Pastol, G. Arjavalingam, J.-M. Halbout and G. V. Kopcsay, "Characterisation of an optoelectronically pulsed broadband microwave antenna", *Electronics Letters*, vol. 24, no 21, pp 1318-1319, Oct. 1988.
9. W. M. Robertson, G. Arjavalingam and G. V. Kopcsay, "Microwave diffraction and interference in reflection using transient radiation from optoelectronically pulsed antennas", Appl. Phys. Lett. 57 (19), Nov. 1990.
10. J. F. Eloy, V. Gerbe, J. H. Trombert, "Mise au point d'une expérience optoélectronique de caractérisation des propriétés transitoires de matériaux", to be presented at the *4ème Journées de Caractérisation Microonde et Matériaux*, Chambéry (France), April 1996.

TRANSIENT DIELECTRIC COEFFICIENT AND CONDUCTANCE IN DIELECTRIC MEDIA IN NONSTATIONARY FIELDS

A.Gutman

Department of Physics
Voronezh State Forestry Engineering Academy
Timirjasev St.,8 Russia 394613

INTRODUCTION

The establishment process of medium parameters becomes an essential one for the pulse propagation if the pulse duration and character times of medium processes are commensurable. The polarization, dielectric coefficient and conductance are these medium parameters in the case of the electromagnetic pulse propagation. The paper contains their investigation in the medium with elastic tied charges. The establishment process also depends on the electromagnetic pulse form. Therefore the following three pulse formes of the exciting field are considered here: 1) the function of a single step (Heviside function), 2) a linearly growing function of time, 3) the pulse function of special form. The dielectric response from Heviside function excitation can be used for constructing the response from other ones, but it is found that their direct calculation is more convenient.

The real pulse changes its form during the propagation in the dielectric media. Therefore it is necessary to obtain the general correlation between the nonstationary electromagnetic field and the polarization vector. In the paper this correlation is obtained for the Laplace transformations of the field and the polarization. The result is used for constructing wave equation in another paper [1].

For solving these problems according to the classic notions a molecule is simulated by one or several linear harmonic oscillators. Each of the oscillators corresponds to the normal electron oscillations in the molecule. The molecule demensions are considerably less than electromagnetic wave length in wide frequency range. Therefore one may suppose that the field acting on the elementary dipole is a homogeneous one. In this case the correlation between the field, the polarization vector and the average macroscopic field is simple. Then an integral equation for the polarization vector may be obtained from the harmonic oscillator motion equation. Its solution is obtained by the Laplace transformation use.

CALCULATION

In the flat field

$$\vec{E} = E(t)\vec{e}_x \tag{1}$$

Ultra-Wideband, Short-Pulse Electromagnetics 3
Edited by Baum *et al.*, Plenum Press, New York, 1997

the motion equation of an elementary oscillator has the form:

$$\ddot{x}+2\alpha\,\dot{x}+\omega_0^2 x=\frac{e}{m}\left[E(t)+\frac{4\pi}{3}P(t)\right] \tag{2}$$

where m, e are electrons mass and charge, ω_0 is a resonance frequence of normal oscillations, α is an absorption coefficient of the oscillations, P is a polarization of dielectric

$$P=Nex, \tag{3}$$

$E+\dfrac{4\pi}{3}P$ is the effective field, N is a molecular concentration.

The solutions of equation (2) at the initial conditions $\dot{x}(0)=x(0)=0$ are

$$x=\frac{e}{m|\omega|}\int_0^t\left[E(z)+\frac{4}{3}\pi\,P(z)\right]e^{\alpha(z-t)}\cdot\begin{cases}\sin\omega(t-z) & \omega_0^2>\alpha^2,\\ sh|\omega|(t-z) & \omega_0^2<\alpha^2,\end{cases}dz,\quad \omega=\sqrt{\omega_0^2-\alpha^2} \tag{4}$$

$$x=\frac{e}{m}\int_0^t\left[E(z)+\frac{4}{3}\pi\,P(z)\right]e^{\alpha(z-t)}\cdot(t-z)\,dz,\quad \omega=0,\ \omega_0=\alpha \tag{5}$$

Three integral equantions for $P(t)$ may be obtained, using (3), (4), (5). We shall consider below the first one:

$$aP(t)=f(t)+\int_0^t P(z)e^{-\alpha(t-z)}\sin\omega\,(t-z)\,dz \tag{6}$$

where

$$a=3\frac{\omega}{\omega_p^2},\quad \omega_p^2=\frac{4\pi\,Ne^2}{m}, \tag{7}$$

$$f(t)=\frac{3}{4\pi}\int_0^t E(z)e^{-\alpha(t-z)}\sin\omega\,(t-z)\,dz. \tag{8}$$

Let us apply the Laplace transformation to equation (6). We shall also use the convolution theorem:

$$\int_0^t P(z)\varphi(t-z)\,dz\to\overline{P}(\xi)\overline{\varphi}(\xi), \tag{9}$$

where a line over a function means its Laplace transformation.
In the integrals (6) and (8)

$$\varphi(z)=e^{-\alpha z}\sin\omega\,z\ \to\ \overline{\varphi}(\xi)=\frac{\omega}{(\xi+\alpha)^2+\omega^2}, \tag{10}$$

The transformation for polarization function is given now by

$$\overline{P}(\xi)=\frac{\omega_p^2\,\overline{E}(\xi)}{4\pi\left[\Omega^2+(\xi+\alpha)^2\right]},\quad \Omega^2=\omega^2-\omega_p^2/3. \tag{11}$$

The equation (11) is the general correlation between the Laplace transformation of the polarization and nonstationary electric field by the zeroes initial conditions. If the initial conditions are nonzeroes, it is necessary to add them by solutions of equation (2).

The inverse transformations and the corresponding solutions of equation (6) are obtained for the following modes of $E(t)$ and their Laplace transformations by the use of the decomposition theorem and the table formulas:
for the function of a single step (Heviside function)

$$E_1(t)=\left.\begin{matrix}0, & t<0\\ 1, & t\geq 0\end{matrix}\right\}\quad \overline{E}_1(\xi)=\frac{1}{\xi}, \tag{12}$$

for the linearly growing function of the time

$$E_2(t)=\left.\begin{matrix}0, & t\leq 0\\ At, & t\geq 0\end{matrix}\right\}\quad \overline{E}_2(\xi)=\frac{A}{\xi^2}, \tag{13}$$

for the pulse function

$$E_3(t)=\left.\begin{matrix}0, & t\leq 0\\ Bte^{-\beta t}, & t\geq 0\end{matrix}\right\}\quad \overline{E}_3(\xi)=\frac{B}{(\xi+\beta)^2}. \tag{14}$$

The corresponding polarization functions have the form :

$$P_1=\frac{\omega_p^2}{4\pi\left(\omega_0^2-\omega_p^2/3\right)}\left[1-e^{-\alpha t}\left(\frac{\alpha}{\Omega}\sin\Omega t+\cos\Omega t\right)\right], \tag{15}$$

$$P_2=\frac{A\omega_p^2}{4\pi\left(\omega_0^2-\omega_p^2/3\right)}\left[t+\frac{e^{-\alpha t}}{\Omega}\sin\left(\Omega t+2arctg\frac{\Omega}{\alpha}\right)-\frac{2\alpha}{\omega_0^2-\omega_p^2/3}\right], \tag{16}$$

$$P_3=\frac{\beta\omega_p^2}{4\pi\left(\Omega^2+\gamma^2\right)}\left[\frac{e^{-\alpha t}}{\Omega}\sin\left(\Omega t+2arctg\frac{\Omega}{\gamma}\right)+e^{-\beta t}\left(t-\frac{2\gamma}{\Omega^2+\gamma^2}\right)\right],\gamma=\alpha-\beta \tag{17}$$

The dielectric coefficient ε and conductance σ, created by the polarization losses, are determined from the correlations :

$$P=\frac{\varepsilon-1}{4\pi}E,\quad j=\dot{P}(t)=\sigma E, \tag{18}$$

where j is the polarization current density.

From (15), (16), (17), (18) we get :

$$\varepsilon_1 = 1 + \frac{\omega_p^2}{\omega_0^2 - \omega_p^2/3}\left[1 - e^{-\alpha t}\left(\cos\Omega t - \frac{\alpha}{\Omega}\sin\Omega t\right)\right], \tag{19}$$

$$\sigma_1 = \frac{1}{4\pi}\frac{\omega_p^2}{\Omega}e^{-\alpha t}\left|\sin\Omega t\right|, \tag{20}$$

$$\varepsilon_2 = 1 + \frac{\omega_p^2}{\omega_0^2 - \omega_p^2/3}\left[1 + \frac{e^{-\alpha t}\left(\alpha^2 - \Omega^2\right)}{\omega_0^2 - \omega_p^2/3}\frac{\sin\Omega t}{\Omega t} + \frac{2\alpha\left(e^{-\alpha t}\cos\Omega t - 1\right)}{\left(\omega_0^2 - \omega_p^2/3\right)t}\right], \tag{21}$$

$$\sigma_2 = \frac{1}{4\pi}\frac{\omega_p^2}{\omega_0^2 - \omega_p^2/3}\frac{1}{t}\left[1 - e^{-\alpha t}\left(\frac{\alpha}{\Omega}\sin\Omega t + \cos\Omega t\right)\right], \tag{22}$$

$$\varepsilon_3 = 1 + \frac{\omega_p^2}{\Omega^2 + \gamma^2}\left\{1 - \frac{1}{t}\left[\frac{2\gamma}{\Omega^2 + \gamma^2} - \frac{e^{-\gamma t}}{\Omega}\sin\left(\Omega t + 2\,arctg\frac{\Omega}{\gamma}\right)\right]\right\} \tag{23}$$

$$\sigma_3 = \frac{1}{4\pi}\frac{\omega_p^2}{\Omega^2 + \gamma^2}\frac{1}{t}\left\{1 - e^{-\gamma t}\left[\frac{\alpha}{\Omega}\sin\left(\Omega t + 2arctg\frac{\Omega}{\gamma}\right) - \cos\left(\Omega t + 2arctg\frac{\Omega}{\gamma}\right)\right] + \right.$$
$$\left. + \beta\left(\frac{2\gamma}{\Omega^2 + \gamma^2} - t\right)\right\} \tag{24}$$

The correlations obtained are in keeping with the oscillating mode of the P, ε, σ establishment. Analogous correlations may be obtained by the aperiodic mode of the P, ε, σ establishment from the integral equations in the form (6) for $\omega_0^2 \le \alpha^2$.

The established values of ε_1, σ_1 and ε_2, σ_2 obtained from (19) - (22) are equal to the known ones for the static field in the ideal dielectric [2] :

$$\varepsilon_{1,2} = 1 + \frac{\omega_p^2}{\omega_0^2 - \omega_p^2/3}, \quad \sigma_{1,2} = 0. \tag{25}$$

The establishment process for ε_1, σ_1 follows exponential law $\left(e^{-\alpha t}\right)$, for ε_2, σ_2 alongside with the exponential addend there arises one that decreases according to the law t^{-1}.

The established values of ε_3, σ_3, correspond to the pulse train :

$$\varepsilon_3 = 1 + \frac{\omega_p^2}{\Omega^2 + \gamma^2} = 1 + \frac{\omega_p^2}{\omega_0^2 - \omega_0^2/3 + \beta^2 - 2\alpha\beta}; \quad \sigma_3 = \frac{1}{4\pi}\left|\frac{\beta\omega_p^2}{\Omega^2 + \gamma^2}\right|. \tag{26}$$

Note that a finite established value of σ_3 corresponds only to the ratio j/E. The polarisation losses jE aspire to zero for big t.

CONCLUSION

There are three motives for research in this paper: to find the transient polarization, dielectric coefficient and conductance in dielectric medium with elastic tied charges, to clarify their dependence on the microparameters of the medium and on the form of the pulse, to prepare the investigation of the pulse propagation in the medium.

The transient electromagnetic parameters obtained are presented in the correlations (15) - (17), (19) - (24). The establishment process has exponential decay periodical or aperiodical character depending upon microscope parameters of medium ($\alpha, \omega_0, \omega_p$). The pulse form stipulates the possibility and the time of achivement of the stationary values.

The general correlation between the Laplace transformation of the polarization and the electrical field (equation (11)) opens the possibility to the Laplace transformation for the wave equation. This is the start to the pulse porpagation investigation. The estimation of the pulse absorption can be obtained from the formula (20) for the conductance σ_1, if it is possible to neglect the pulse dispersion.

REFERENCES

1. A.Gutman. *Short Pulses Propagation in the Dielectric Media*, "Ultra-Wideband, Short-Pulse Electromagnetics 3", US (1996).
2. M.Vinogradova, O.Rudenko, A.Sukhorukov. *The Wave Theory*, Moscow (1979).

THE SHORT PULSES PROPAGATION IN THE DIELECTRIC MEDIA

A.Gutman

Department of Physics
Voronezh State Forestry Engineering Academy
Timirjasev St.,8 Russia 394613

INTRODUCTION

The analysis of the short pulses propagation in the dielectric medium with the elastic tied elementary charges has intention to determine the absorption and dispersion of the pulses. How will the pulse absorption decrease in this medium if the pulse duration decreases and the conductance of the medium does not reach its stationary value during the pulse duration? How will the pulse form be changed when all pulse parts have different velocities because the dielectric coefficient is not established? These questions excite great theortical and practical interest.

To answer these questions we shall use the transient conductance obtained in[1] and the Laplace transformation method to the wave equation solution. The nonstationary processes investigation with the help of the Laplace transformation has a long time history. The difficulties of its use here are stipulated by the wave equation evolution: the pulse velocity is changed during the propagation in the medium and these changes are determined by the changes of the pulse form. The wave equation for "self-coordinative" field is constructed in the paper for the Laplace transformation of the field. The general correlation between the Laplace transformations of the field and polarization obtained in the previous paper [1] is also used here. The solution of the equation for the Laplace transformation field is obtained. But a precise general inverse Laplace transformation of the solution is impossible. Therefore two other ways are considered in the paper: 1) the use of the general approximate methods for the inverse Laplace transformation, 2) the analysis of the particular examples where it is possible to obtain the precise inverse Laplace transformation.

The approximate estimation of the pulse absorption can be obtained without solving the wave equation, because in this case it is possible to use the average conductance of the medium for the pulse.

THE ABSORPTION OF THE PULSE FIELD IN DIELECTRIC MEDIUM

The results obtained [1] are used to estimate the nonstationary absorption of short electromagnetic pulses in dielectric media. The estimation of absorption is quite simple, if it

Ultra-Wideband, Short-Pulse Electromagnetics 3
Edited by Baum *et al.*, Plenum Press, New York, 1997

is possible to neglect the pulse dispersion. The pulse penetration through thin dielectric layer is a suitable example.

The plasma frequency of the dielectric – ω_p, the oscillation frequency of the elementary oscillator – ω_0, the attenuation constant of this oscillator – α are necessary for the calculation.

They may be obtained for the following correlations [2] between these and macroscopic parameters of dielectric:

$$\varepsilon'(\omega) = 1 + \frac{\omega_p^2\left(\omega_0^2 - \omega_p^2/3 - \omega^2\right)}{\left(\omega_0^2 - \omega_p^2/3 - \omega^2\right)^2 + 4\omega^2\alpha^2}, \tag{1}$$

$$\varepsilon''(\omega) = \frac{2\omega_p^2\alpha\omega}{\left(\omega_0^2 - \omega_p^2/3 - \omega^2\right)^2 + 4\omega^2\alpha^2}, \tag{2}$$

$$\varepsilon(0) = 1 + \frac{\omega_p^2}{\omega_0^2 - \omega_p^2/3} \tag{3}$$

where $\varepsilon(\omega) = \varepsilon'(\omega) + i\,\varepsilon''(\omega)$ – is the dielectric coefficient in the stationary absorption band, ω is the central frequency of this band, $\varepsilon(0)$ is the dielectric coefficient in the static field.

Now the effective absorption by an arbitrary form pulse can be calculated by means of the convolution between the function of the pulse versus time and the transient conductance. The correlation for the conductance according to [1] is

$$\sigma = \frac{1}{4\pi}\frac{\omega_p^2}{\Omega}e^{-\alpha t}\left|\sin\Omega t\right|, \quad \Omega^2 = \omega_0^2 - \alpha^2 - \omega_p^2/3 \tag{4}$$

In the case of a rectangular pulse the transient conductance does not change the law of establishing during the pulse duration. Therefore the effective conductance for a rectangular pulse can be defined as an average one during the pulse duration:

$$\sigma_{eff} = \frac{1}{4\pi}\frac{\omega_p^2}{\Omega}\frac{1}{\tau}\int_0^{\tau} e^{-\alpha t}\left|\sin\Omega t\right|dt \tag{5}$$

where τ is the pulse duration.

The conductance coefficient in the stationary field is

$$\sigma_{st} = \frac{\omega\varepsilon''}{4\pi} \tag{6}$$

The ratio σ_{eff}/σ_{st} is a suitable value for the estimation of the pulse absorption. The ratio is calculated for $\varepsilon(\omega) = 6 + i\,0.6$, $\omega = 2\pi\cdot 10^{10}\,s^{-1}$, $\varepsilon(0) = 2$.

The calculation results are represented in table 1.

Table 1.

$\tau(s)$	10^{-10}	10^{-11}	10^{-12}	10^{-13}
σ_{eff}/σ_{st}	~ 1	~ 0,5	~ 0,05	~ 0,005

It is obvious from the table that the ratio can be considerably less than unity if the pulse duration is much shorter than the duration of the establishment of the conductance.

GENERALIZED WAVE EQUATION AND PULSE DISPERSION

The generalized wave equation may be obtained for the nonstationary field, using the well-known correlation

$$\vec{D} = \vec{E} + 4\pi\vec{P} \tag{7}$$

It is true for any nonstationary field, because $div\,\vec{P} = -\overline{\rho}$, where $\overline{\rho}$ is a real (microscopic) charge density, and $\vec{P}$ is a real electrical moment of volume unity. Therefore the equation for the flat field is

$$\frac{\partial^2 E}{\partial x^2} - \frac{1}{c^2}\frac{\partial^2}{\partial t^2}(E + 4\pi P) = 0 \tag{8}$$

The Laplace transformation of the equation for zero initial conditions is

$$\frac{\partial^2 \overline{E}(x,\xi)}{\partial x^2} - \frac{\xi^2}{c^2}\left(1 + \frac{\omega_p^2}{(\xi+\alpha)^2 + \Omega^2}\right)\overline{E}(x,\xi) = 0 \tag{9}$$

where a line above the function means the Laplace transformation and the general expression for $\overline{P}(\xi)$ is taken from [1].

The solution of (9) at the boundary conditions:

$$\overline{E}(x,\xi) \underset{x\to\infty}{\to} 0, \quad \overline{E}(x,\xi)\big|_{x=0} = \overline{E}(0,\xi) \tag{10}$$

has the form:

$$\overline{E}(x,\xi) = \overline{E}(0,\xi)e^{-\frac{\xi x}{V}}, \quad V \equiv c\left(1 + \frac{\omega_p^2}{(\xi+\alpha)^2 + \Omega^2}\right)^{-\frac{1}{2}} \tag{11}$$

Let us consider the possibilities of the method using the example of the single step field propagation. We shall examine the case of the dielectric with permittivity in static field close to unity. It is a simple case.

Then it can be assumed that

$$\left| \frac{\omega_p^2}{(\xi+\alpha)^2+\Omega^2} \right| << 1 \tag{12}$$

The boundary condition for the single step field is $\bar{E}(0,\xi) = \frac{1}{\xi}$ (13)

Now equation (11) takes the form:

$$\bar{E}(x,\xi) = \frac{1}{\xi} e^{-\frac{\xi x}{c}} \left(1 - \frac{\xi x}{c} \frac{\omega_p^2}{2\left[(\xi+\alpha)^2+\Omega^2\right]} + \cdots \right) \tag{14}$$

The inverse Laplace transformation for (14) obtained by the theorem of the decomposition is

$$E(x,t) = \begin{cases} 0, & t < \frac{x}{c}; \\ 1 - \frac{\omega_p^2}{2\Omega} \frac{x}{c} e^{-\alpha\left(t-\frac{x}{c}\right)} \sin\Omega\left(t - \frac{x}{c}\right) + \cdots, & t > \frac{x}{c}. \end{cases} \tag{15}$$

The expression (15) has a clear physical sense. It can be used for the determination of a pulse dispersion.

The single step dispersion presented (Fig.1) corresponds to (15) with $\omega_p^2 x / 2\Omega c = \frac{1}{2}$. The rectangular pulse dispersion presented (Fig.2) with relative pulse duration 2π and $\alpha/\Omega = 1$ corresponds to two single steps of different signs.

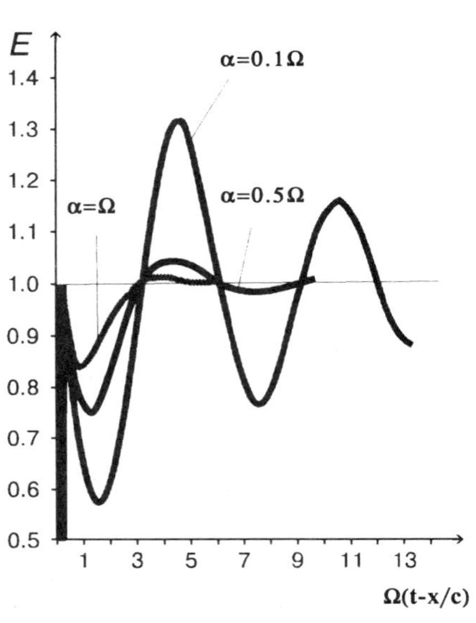

Figure 1.

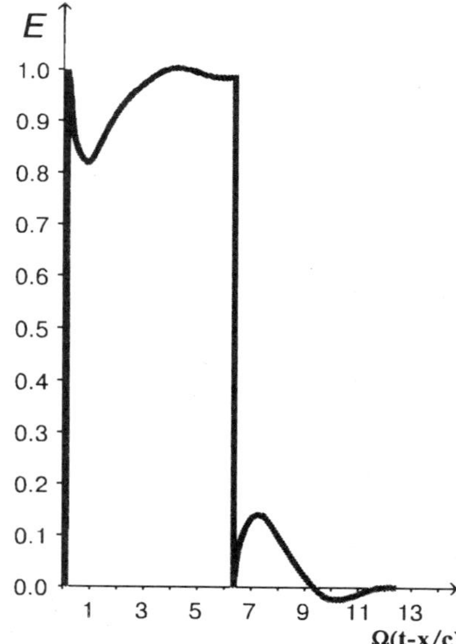

Figure 2.

One can see that the increase of x can stimulate the pulse destruction. But one cannot use the formulas (14), (15) for big x values.

To obtain the inverse Laplace transformation of (11) by the arbitrary form of $\overline{E}(0,\xi)$ it is necessary to use the approximate methods and the modern computer techniques. For our purpose the method of the Laplace transformation inversion with the generalized Lagger polynomials sets[3] is suitable. The application of the method for the function $\overline{E}(\xi)$ is possible if an unknown function $E(t)$ satisfies the condition:

$$\int_0^\infty e^{-t} \left|E(t)\right|^2 dt < \infty \tag{16}$$

Then it is possible to present the function $g(t) = t^{-\lambda} E(t)$ by the set:

$$g(t) = \sum_{k=0}^{\infty} a_k \frac{k!}{\Gamma(k+\lambda+1)} L_k^{(\lambda)}(t), \tag{17}$$

where $L_k^{(\lambda)}(t)$ are generalized Lagger polynomials. The coefficient a_k are determined from the equation:

$$a_k = \frac{(-1)^k}{k!} \frac{d^k}{dz^k} \left\{ \frac{1}{z^{\lambda+1}} \overline{E}\left(\frac{1}{z}\right) \right\}_{z=1} \tag{18}$$

CONCLUSION

Two problems are examined in this paper:

1. the search of the general methods for the investigation of the short pulse propagation in the dielectric medium with elastic tied charges,
2. the estimation of the short pulse dispersion and absorption in the medium for the simple situation, without using special computer programms.

The generalized wave equation is presented in (8), (9), where general connection between the Laplace transformation of the field and polarization is used. The general solution for the Laplace transformation of the field by the typical boundary conditions (10) is given by (11). Thus all difficulties of the problem come down to the Laplace transformation inversion. The general approximate method for the inversion is suggested in (17), (19). Its application is possible for the wide class of the fields, which satisfy the condition (16).

The estimation technique of the pulse absorption is given by the formulas (1) - (6), the calculation results are represented in table 1. The relative pulse absorption decrease can be considerable with the pulse duration decrease.

The pulse dispersion in the simple example is illustrated by (15) and Fig.1 and Fig.2.

It is clear that there can be essential change of the pulse form.

REFERENCES

1. A. Gutman. *Transient Dielectric Coefficient and Conductance in Dielectric Media in Nonstationary Fields*. "Ultra-Wideband, Short-Pulse Electromagnetics 3", US (1996).
2. M.Vinogradova, O.Rudenko, A.Sukhorukov. *The Wavetheory*. Moscow (1979).
3. W.A.Ditkin, A.P.Prudnikov. *Operational calculus*. Moscow (1975).

ELECTROMAGNETIC PULSE PROPAGATION ACROSS A PLANAR INTERFACE SEPARATING TWO LOSSY, DISPERSIVE DIELECTRICS

John A. Marozas and Kurt E. Oughstun

College of Engineering and Mathematics
University of Vermont
Burlington, VT 05405-0156

INTRODUCTION

Our understanding of the dynamics of pulse propagation in a homogeneous, isotropic, locally linear, temporally dispersive lossy medium is a fundamental problem in electromagnetic wave theory[1-6] . As a plane wave pulse propagates through a dispersive, lossy medium, each spectral component propagates with its own characteristic phase velocity $v_p(\omega) = \omega/\beta(\omega)$ and is attenuated at a rate that is characterized by its own attenuation coefficient $\alpha(\omega) = \Im\{\tilde{k}(\omega)\}$, where $\tilde{k}(\omega) = \beta(\omega) + i\alpha(\omega)$ is the complex wavenumber of a time-harmonic plane wave of angular frequency ω. The relative phases and amplitudes of the spectral components of the initial pulse then change by differing amounts as the propagation distance increases, thereby giving rise to the observed distortion of the propagated pulse. The seminal analysis of dispersive pulse propagation in a linear dielectric filling the half-space $z \geq 0$ was provided by Sommerfeld[1] and Brillouin[2,4] in 1914 using the asymptotic method of steepest descents. The recent analysis of Oughstun and Sherman[7-9] using modern asymptotic expansion techniques has resulted in significant quantitative improvements in the entire description of the propagated field as well as a correct description of the signal velocity in causally dispersive media. This analysis has also led to a clear physical interpretation of the dynamical pulse evolution in a given dispersive medium in terms of the frequency dispersion of the attenuation coefficient and the energy transport velocity for a time-harmonic field[9-11]. A fundamental extension of this theory is provided by the problem of the reflection and transmission of a pulsed electromagnetic beam field that is incident upon the planar interface separating two half-spaces containing different dispersive media. Previous treatments of this problem have either focused on time-harmonic beam fields in lossless media[12,13] or on pulsed plane wave fields when the incident medium is vacuum and the second medium is a dispersive dielectric[14,15]. The analysis presented in this paper describes the general situation in which both media are dispersive and absorptive and the incident field is a pulsed electromagnetic beam field. The results of this mathematically rigorous analysis have immediate application in integrated and fiber optic device technologies, the analysis and design of low-observable surfaces, and the analysis of bioelectromagnetic effects in stratified tissue[16,17].

GENERAL FORMULATION

The electromagnetic field vectors in a homogeneous, isotropic, locally linear, temporally dispersive medium with the constitutive relations

$$\mathbf{D}(\mathbf{r},t) = \int_{-\infty}^{t} \hat{\varepsilon}(t-\tau)\mathbf{E}(\mathbf{r},\tau)d\tau \quad , \tag{1a}$$

$$\mathbf{B}(\mathbf{r},t) = \mu\mathbf{H}(\mathbf{r},t) \quad , \tag{1b}$$

where μ is the constant magnetic permeability and where $\hat{\varepsilon}(t)$ denotes the real-valued dielectric permittivity response function, satisfy the vector Helmholtz equations

$$\left(\nabla^2 + \tilde{k}^2(\omega)\right)\tilde{\mathbf{E}}(\mathbf{r},\omega) = \mathbf{0} \quad , \tag{2a}$$

$$\left(\nabla^2 + \tilde{k}^2(\omega)\right)\tilde{\mathbf{B}}(\mathbf{r},\omega) = \mathbf{0} \quad , \tag{2b}$$

where $\left\{\mathbf{E}(\mathbf{r},t),\tilde{\mathbf{E}}(\mathbf{r},\omega)\right\}$ and $\left\{\mathbf{B}(\mathbf{r},t),\tilde{\mathbf{B}}(\mathbf{r},\omega)\right\}$ each form temporal Fourier transform pairs. Here

$$\tilde{k}(\omega) = \beta(\omega) + i\alpha(\omega) = \frac{\omega}{c}n(\omega) \tag{3}$$

is the complex wavenumber of the electromagnetic disturbance with angular frequency ω that is propagating in the medium with complex index of refraction

$$n(\omega) = \left[\frac{\mu\varepsilon(\omega)}{\mu_0\varepsilon_0}\right]^{1/2} \quad , \tag{4}$$

where $c = (\mu_0\varepsilon_0)^{-1/2}$ is the speed of light in vacuum.

With reference to some arbitrarily oriented rectangular coordinate system (u,v,w) with corresponding unit vectors $\left(\hat{\mathbf{1}}_u,\hat{\mathbf{1}}_v,\hat{\mathbf{1}}_w\right)$, let the electromagnetic field vectors be specified on the plane $w = w_0 = 0$ as

$$\mathbf{E}(\mathbf{r}_T,w_0,t) = \mathbf{E}_0(\mathbf{r}_T,t) \quad , \tag{5a}$$

$$\mathbf{B}(\mathbf{r}_T,w_0,t) = \mathbf{B}_0(\mathbf{r}_T,t) \quad , \tag{5b}$$

where $\mathbf{E}_0(\mathbf{r}_T,t) = \mathbf{E}_0(u,v,t)$ and $\mathbf{B}_0(\mathbf{r}_T,t) = \mathbf{B}_0(u,v,t)$ are known functions of time and the transverse position vector

$$\mathbf{r}_T = \hat{\mathbf{1}}_u u + \hat{\mathbf{1}}_v v \quad , \tag{6}$$

whose two-dimensional spatial Fourier transform in the transverse coordinates and temporal Fourier-Laplace transform exist and are given by

$$\tilde{\tilde{\mathbf{E}}}_0(\mathbf{k}_T,\omega) = \int_{-\infty}^{\infty} dt \int_{-\infty}^{\infty}\int_{-\infty}^{\infty} dudv \mathbf{E}_0(\mathbf{r}_T,t) e^{-i(\mathbf{k}_T\cdot\mathbf{r}_T-\omega t)} \quad , \tag{7a}$$

$$\tilde{\tilde{\mathbf{B}}}_0(\mathbf{k}_T,\omega) = \int_{-\infty}^{\infty} dt \int_{-\infty}^{\infty}\int_{-\infty}^{\infty} dudv \mathbf{B}_0(\mathbf{r}_T,t) e^{-i(\mathbf{k}_T\cdot\mathbf{r}_T-\omega t)} \quad , \tag{7b}$$

where

$$\mathbf{k}_T = \hat{\mathbf{1}}_u k_u + \hat{\mathbf{1}}_v k_v \tag{8}$$

is the transverse wavevector. The inverse transforms of Eqs. (7a,b) are also assumed to exist and are given by

$$\mathbf{E}_0(\mathbf{r}_T,t) = \frac{1}{(2\pi)^3}\int_C d\omega \int_{-\infty}^{\infty}\int_{-\infty}^{\infty} dk_u dk_v \tilde{\tilde{\mathbf{E}}}_0(\mathbf{k}_T,\omega) e^{i(\mathbf{k}_T\cdot\mathbf{r}_T-\omega t)} \quad , \tag{9a}$$

$$\mathbf{B}_0(\mathbf{r}_T,t) = \frac{1}{(2\pi)^3}\int_C d\omega \int_{-\infty}^{\infty}\int_{-\infty}^{\infty} dk_u dk_v \tilde{\tilde{\mathbf{B}}}_0(\mathbf{k}_T,\omega) e^{i(\mathbf{k}_T\cdot\mathbf{r}_T-\omega t)} \quad . \tag{9b}$$

If the initial time dependence of the field vectors at the plane $w = w_0$ is such that both field vectors vanish for all $t < t_0$ for some finite value of t_0, then the corresponding time-frequency transform pairs appearing in Eqs. (7) and (9) are both Laplace transforms and the contour C is the straight line path $\omega = \omega' + ia$ with $\omega' = \Re\{\omega\}$ extending from negative to positive infinity, and with a being greater than the abscissa of absolute convergence for the initial time evolution of the field; if not, then they are both Fourier transforms. The propagated field vectors in either the positive half-space $w > w_0$ or the negative half-space $w < w_0$ are then given by the angular spectrum representation[9]

$$\mathbf{E}(\mathbf{r},t) = \frac{1}{(2\pi)^3}\int_C d\omega \int_{-\infty}^{\infty}\int_{-\infty}^{\infty} dk_u dk_v \tilde{\tilde{\mathbf{E}}}_0(\mathbf{k}_T,\omega) e^{i[\tilde{\mathbf{k}}^{\pm}(\omega)\cdot\mathbf{r}-\omega t]} \quad , \tag{10a}$$

$$\mathbf{B}(\mathbf{r},t) = \frac{1}{(2\pi)^3}\int_C d\omega \int_{-\infty}^{\infty}\int_{-\infty}^{\infty} dk_u dk_v \tilde{\tilde{\mathbf{B}}}_0(\mathbf{k}_T,\omega) e^{i[\tilde{\mathbf{k}}^{\pm}(\omega)\cdot\mathbf{r}-\omega t]} \quad . \tag{10b}$$

Here

$$\tilde{\mathbf{k}}^{\pm}(\omega) = \hat{\mathbf{1}}_u k_u + \hat{\mathbf{1}}_v k_v \pm \hat{\mathbf{1}}_w \gamma(\omega) \tag{11}$$

is the complex wavevector, where $\tilde{\mathbf{k}}^{+}$ is used for propagation into the positive half-space $w > w_0$, while $\tilde{\mathbf{k}}^{-}$ is used for propagation into the negative half-space $w < w_0$, where $\gamma = \gamma(\omega)$ is defined as the principal branch of the expression

$$\gamma = \left[\tilde{k}^2(\omega) - k_T^2\right]^{1/2} \quad , \tag{12}$$

with $k_T^2 = k_u^2 + k_v^2$, and where

$$\tilde{k}(\omega) = \left(\tilde{\mathbf{k}}^{\pm} \cdot \tilde{\mathbf{k}}^{\pm}\right)^{1/2} = \frac{\omega}{c} n(\omega) \tag{13}$$

is the associated complex wavenumber [cf. Eq.(3)]. Finally, the spatio-temporal spectra of the electromagnetic field vectors at the plane $w = w_0$ are related by the transversality conditions

$$\tilde{\tilde{\mathbf{E}}}_0(\mathbf{k}_T, \omega) = -\frac{\|c\|}{\omega\mu\varepsilon(\omega)} \tilde{\mathbf{k}}^{\pm} \times \tilde{\tilde{\mathbf{B}}}_0(\mathbf{k}_T, \omega) \quad , \tag{14a}$$

$$\tilde{\tilde{\mathbf{B}}}_0(\mathbf{k}_T, \omega) = \frac{\|c\|}{\omega} \tilde{\mathbf{k}}^{\pm} \times \tilde{\tilde{\mathbf{E}}}_0(\mathbf{k}_T, \omega) \quad , \tag{14b}$$

$$\tilde{\mathbf{k}}^{\pm} \cdot \tilde{\tilde{\mathbf{E}}}_0(\mathbf{k}_T, \omega) = \tilde{\mathbf{k}}^{\pm} \cdot \tilde{\tilde{\mathbf{B}}}_0(\mathbf{k}_T, \omega) = 0 \quad . \tag{14c}$$

Notice that both Gaussian (cgs) and MKS units are employed here through use of a conversion factor that appears inside the double brackets of an affected equation, as in Eqs.(14a,b). If this factor is included in the equation it is then in cgs units provided that one also sets $\varepsilon_0 = \mu_0 = 1$, while if this factor is replaced by unity the equation is then in MKS units. Finally, if no such factor appears, then that equation is correct in both systems of units.

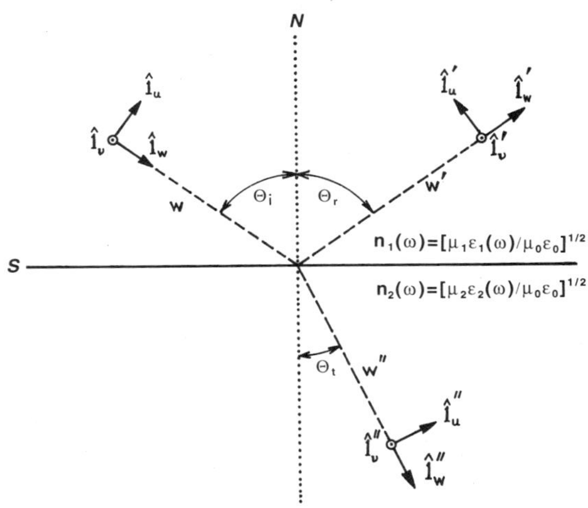

Figure 1. Incident $(\hat{\mathbf{1}}_u, \hat{\mathbf{1}}_v, \hat{\mathbf{1}}_w)$, reflected $(\hat{\mathbf{1}}'_u, \hat{\mathbf{1}}'_v, \hat{\mathbf{1}}'_w)$, and transmitted $(\hat{\mathbf{1}}''_u, \hat{\mathbf{1}}''_v, \hat{\mathbf{1}}''_w)$ coordinate systems at a planar dielectric interface *S* with normal *N*.

Consider now the reflection and transmission of a pulsed electromagnetic beam field that is incident upon a planar interface separating two different dielectric media. Let the medium in which the incident and reflected fields reside be described by the complex-valued dielectric permittivity $\varepsilon_1(\omega)$ and constant magnetic permeability μ_1, and let the medium in which the transmitted field resides be described by the complex-valued dielectric permittivity $\varepsilon_2(\omega)$ and constant magnetic permeability μ_2. Let the incident field be along the direction $\hat{\mathbf{1}}_w$ which is at the angle Θ_i with respect to the normal to the interface S, the reflected field be along the direction $\hat{\mathbf{1}}'_w$ which is at the angle Θ_r with respect to the normal, and let the transmitted field be along the direction $\hat{\mathbf{1}}''_w$ which is at the angle Θ_t with respect to the normal to S, as indicated in Fig.1. The right-handed rectangular coordinate systems $(\hat{\mathbf{1}}_u,\hat{\mathbf{1}}_v,\hat{\mathbf{1}}_w)$, $(\hat{\mathbf{1}}'_u,\hat{\mathbf{1}}'_v,\hat{\mathbf{1}}'_w)$, and $(\hat{\mathbf{1}}''_u,\hat{\mathbf{1}}''_v,\hat{\mathbf{1}}''_w)$ are then defined along each of these directions such that the unit vectors $\hat{\mathbf{1}}_v,\hat{\mathbf{1}}'_v,\hat{\mathbf{1}}''_v$ are each directed out of the plane of incidence that is defined by the unit vector $\hat{\mathbf{1}}_w$ of the incident field and the normal N to the interface, as indicated in Fig.1.

Let the incident field vectors be specified as in Eq.(5) on the plane $w = w_0$ that is a distance $w > 0$ from the interface along the $\hat{\mathbf{1}}_w$ direction. The electromagnetic field vectors incident upon the interface S are then obtained from Eq.(10) as

$$\mathbf{E}^{(i)}(\mathbf{r},t) = \frac{1}{(2\pi)^3}\int_C d\omega \int_{-\infty}^{\infty}\int_{-\infty}^{\infty} dk_u dk_v \tilde{\tilde{\mathbf{E}}}_0(k_u,k_v,\omega)e^{i[k_u u + k_v v + \gamma_1(\omega)w - \omega t]} \quad , \tag{15a}$$

$$\mathbf{B}^{(i)}(\mathbf{r},t) = \frac{1}{(2\pi)^3}\int_C d\omega \int_{-\infty}^{\infty}\int_{-\infty}^{\infty} dk_u dk_v \tilde{\tilde{\mathbf{B}}}_0(k_u,k_v,\omega)e^{i[k_u u + k_v v + \gamma_1(\omega)w - \omega t]} \quad , \tag{15b}$$

where $\gamma_1^2(\omega) = \tilde{k}_1^2(\omega) - k_T^2$ with $\tilde{k}_1(\omega) = \omega n_1(\omega)/c$. The propagated plane wave spectra of the incident field at the interface are then seen to be given by $\tilde{\tilde{\mathbf{E}}}_0(\mathbf{k}_T,\omega)\exp(i\gamma_1 w)$ and $\tilde{\tilde{\mathbf{B}}}_0(\mathbf{k}_T,\omega)\exp(i\gamma_1 w)$, so that the corresponding reflected plane wave spectra at the interface are given by $\vec{\mathbf{r}}(\mathbf{k}_T,\omega)\tilde{\tilde{\mathbf{E}}}_0(\mathbf{k}_T,\omega)\exp(i\gamma_1 w)$ and $\vec{\mathbf{r}}(\mathbf{k}_T,\omega)\tilde{\tilde{\mathbf{B}}}_0(\mathbf{k}_T,\omega)\exp(i\gamma_1 w)$, and the corresponding transmitted plane wave spectra at the interface are given by $\vec{\mathbf{t}}(\mathbf{k}_T,\omega)\tilde{\tilde{\mathbf{E}}}_0(\mathbf{k}_T,\omega)\exp(i\gamma_1 w)$ and $\vec{\mathbf{t}}(\mathbf{k}_T,\omega)\tilde{\tilde{\mathbf{B}}}_0(\mathbf{k}_T,\omega)\exp(i\gamma_1 w)$, where $\vec{\mathbf{r}}(\mathbf{k}_T,\omega) = \vec{\mathbf{r}}(k_u,k_v,\omega)$ is the amplitude reflection matrix and $\vec{\mathbf{t}}(\mathbf{k}_T,\omega) = \vec{\mathbf{t}}(k_u,k_v,\omega)$ is the amplitude transmission matrix for monochromatic plane wave reflection and transmission at the planar interface S. The reflected electromagnetic field at the $u'v'$-plane of the reflected coordinate system located a distance w' from the interface along the $\hat{\mathbf{1}}'_w$ direction is then given by

$$\mathbf{E}^{(r)}(\mathbf{r},t) = \frac{1}{(2\pi)^3}\int_C d\omega \int_{-\infty}^{\infty}\int_{-\infty}^{\infty} dk_u dk_v \vec{\mathbf{r}}(k_u,k_v,\omega)\tilde{\tilde{\mathbf{E}}}_0(k_u,k_v,\omega)e^{i[k_u u' + k_v v' + \gamma_1(\omega)(w+w') - \omega t]} \quad , \tag{16a}$$

$$\mathbf{B}^{(r)}(\mathbf{r},t)=\frac{1}{(2\pi)^3}\int_C d\omega\int_{-\infty}^{\infty}\int_{-\infty}^{\infty}dk_u dk_v \tilde{\mathbf{r}}(k_u,k_v,\omega)\tilde{\tilde{\mathbf{B}}}_0(k_u,k_v,\omega)e^{i[k_u u'+k_v v'+\gamma_1(\omega)(w+w')-\omega t]} \quad . \tag{16b}$$

The transmitted electromagnetic field at the $u''v''$-plane of the transmitted coordinate system located a distance w'' from the interface along the $\hat{\mathbf{1}}''_w$ direction is given by

$$\mathbf{E}^{(t)}(\mathbf{r},t)=\frac{1}{(2\pi)^3}\int_C d\omega\int_{-\infty}^{\infty}\int_{-\infty}^{\infty}dk_u dk_v \tilde{\mathbf{t}}(k_u,k_v,\omega)\tilde{\tilde{\mathbf{E}}}_0(k_u,k_v,\omega)e^{i[k_u u''+k_v v''+\gamma_1(\omega)w+\gamma_2(\omega)w''-\omega t]}, \tag{17a}$$

$$\mathbf{B}^{(t)}(\mathbf{r},t)=\frac{1}{(2\pi)^3}\int_C d\omega\int_{-\infty}^{\infty}\int_{-\infty}^{\infty}dk_u dk_v \tilde{\mathbf{t}}(k_u,k_v,\omega)\tilde{\tilde{\mathbf{B}}}_0(k_u,k_v,\omega)e^{i[k_u u''+k_v v''+\gamma_1(\omega)w+\gamma_2(\omega)w''-\omega t]}, \tag{17b}$$

where $\gamma_2^2(\omega)=\tilde{k}_2^2(\omega)-k_T^2$ with $\tilde{k}_2(\omega)=\omega n_2(\omega)/c$.

GENERALIZED FRESNEL EQUATIONS

Application of the electromagnetic field vector boundary conditions at the planar interface S separating medium 1, a lossy dielectric whose frequency dispersion is described by the complex-valued dielectric permittivity $\varepsilon_1(\omega)=\varepsilon_1'(\omega)+i\varepsilon_1''(\omega)$, from medium 2, another lossy dielectric whose frequency dispersion is described by the complex-valued dielectric permittivity $\varepsilon_2(\omega)=\varepsilon_2'(\omega)+i\varepsilon_2''(\omega)$, results in both the generalized laws of reflection and refraction as well as in the generalized Fresnel equations. Consider the inhomogeneous plane wave spectral components of the incident electromagnetic field vectors given in Eqs.(15a,b), each with the complex wave vector

$$\tilde{\mathbf{k}}_1^+(\omega)=\hat{\mathbf{1}}_u k_u+\hat{\mathbf{1}}_v k_v+\hat{\mathbf{1}}_w\gamma_1(\omega) \tag{18a}$$

which has the complex direction cosine representation[9]

$$\tilde{\mathbf{k}}_1^+(\omega)=\tilde{k}_1(\omega)\left(\hat{\mathbf{1}}_u p+\hat{\mathbf{1}}_v q+\hat{\mathbf{1}}_w m_1\right) \tag{18b}$$

with

$$\tilde{k}_1(\omega)=\beta_1(\omega)+i\alpha_1(\omega)=\frac{\omega}{c}n_1(\omega) \quad , \tag{19}$$

where $p=k_u/\tilde{k}_1(\omega), q=k_v/\tilde{k}_1(\omega)$, and $m_1=\gamma_1(\omega)/\tilde{k}_1(\omega)$. Since k_u,k_v must both be real-valued quantities, then $p=p'+ip'', q=q'+iq''$ must both be complex-valued with

$$p''=-\frac{\alpha_1(\omega)}{\beta_1(\omega)}p' \quad , \tag{20a}$$

$$q''=-\frac{\alpha_1(\omega)}{\beta_1(\omega)}q' \quad . \tag{20b}$$

Under this transformation to the complex direction cosine representation, the k_u, k_v spatial frequency integrals appearing in Eqs.(15a,b) become p,q direction cosine integrals taken over the respective straight line contours C_p, C_q which pass through the origin and are both at the angle $-\tan^{-1}(\alpha_1(\omega)/\beta_1(\omega))$ to the p',q' axis, where p',q' extends from negative to positive infinity. The complex direction cosine $m_1 = m_1' + i m_1''$ is then given by

$$m_1 = \left(1 - p^2 - q^2\right)^{½} = \left[1 - \left(1 - \frac{\alpha_1^2(\omega)}{\beta_1^2(\omega)}\right)\left(p'^2 + q'^2\right) + 2i\frac{\alpha_1(\omega)}{\beta_1(\omega)}\left(p'^2 + q'^2\right)\right]^{½} \quad , \qquad (21)$$

where the principal branch of the square root expression is to be taken [see Ref.9, Section 4.1.1]. With these results, the spatial phase term appearing in the exponential factor of the angular spectrum of plane waves representation (15a,b) is seen to be given by

$$\begin{aligned}\tilde{\mathbf{k}}_1^+ \cdot \mathbf{r} &= k_u u + k_v v + \gamma_1(\omega) w \\ &= \beta_1(\omega)\left(1 + \frac{\alpha_1^2(\omega)}{\beta_1^2(\omega)}\right)(p'u + q'v) + \left[\beta_1(\omega)m_1' - \alpha_1(\omega)m_1''\right]w + i\left[\alpha_1(\omega)m_1' + \beta_1(\omega)m_1''\right]w\end{aligned} \qquad (22)$$

Similar expressions hold for both the reflected and transmitted electromagnetic waves. This then represents the spatial part of an inhomogeneous plane wave of angular frequency ω whose surfaces of constant amplitude $w = constant$ are, in general, different from the surfaces of constant phase

$$\left(1 + \frac{\alpha_1^2(\omega)}{\beta_1^2(\omega)}\right)(p'u + q'v) + \left(m_1' - \frac{\alpha_1(\omega)}{\beta_1(\omega)}m_1''\right)w = constant \quad .$$

The attenuative part of the complex wave vector for each inhomogeneous plane wave spectral component is directed along the $\hat{\mathbf{1}}_w$ direction, while each inhomogeneous plane wave phase front propagates in the direction that is specified by the real-valued vector

$$\mathbf{s} = \left(1 + \frac{\alpha_1^2(\omega)}{\beta_1^2(\omega)}\right)\left(p'\hat{\mathbf{1}}_u + q'\hat{\mathbf{1}}_v\right) + \left(m_1' - \frac{\alpha_1(\omega)}{\beta_1(\omega)}m_1''\right)\hat{\mathbf{1}}_w \qquad (23a)$$

with magnitude

$$s = \left[\left(1 + \frac{\alpha_1^2(\omega)}{\beta_1^2(\omega)}\right)^2\left(p'^2 + q'^2\right) + \left(m_1' - \frac{\alpha_1(\omega)}{\beta_1(\omega)}m_1''\right)^2\right]^{½} \quad . \qquad (23b)$$

This direction is then completely specified by the set of real-valued direction cosines

$$\{\xi, \psi, \zeta\} = \left\{\frac{1}{s}\left(1 + \frac{\alpha_1^2(\omega)}{\beta_1^2(\omega)}\right)p', \frac{1}{s}\left(1 + \frac{\alpha_1^2(\omega)}{\beta_1^2(\omega)}\right)q', \frac{1}{s}\left(m_1' - \frac{\alpha_1(\omega)}{\beta_1(\omega)}m_1''\right)\right\} \quad , \qquad (24)$$

as illustrated in Fig.2. These expressions then yield both the angle and plane of incidence of each inhomogeneous plane wave spectral component that is incident upon the dielectric interface S.

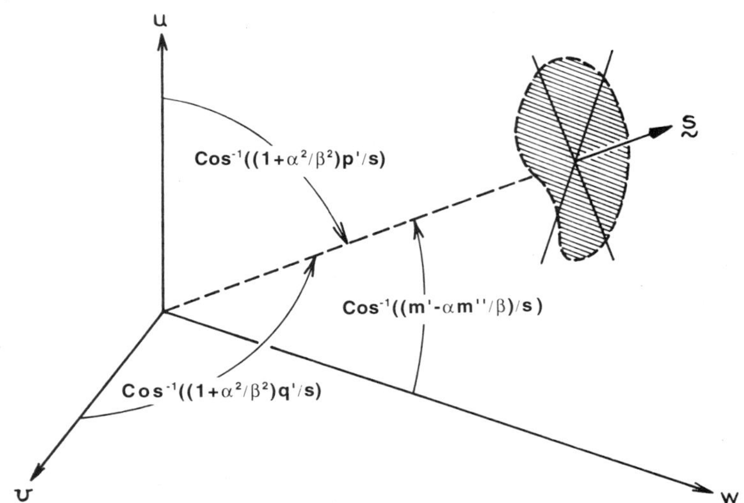

Figure 2. Inhomogeneous plane wave phase front propagating in the direction that is specified by the vector **s**. The attenuative part of the complex wave vector is directed along the positive w-axis.

Let the normal to the phase front of the incident inhomogeneous plane wave spectral component be at the angle θ_i with respect to the normal to the interface, as specified by its direction cosines relative to the positive w-axis which itself is incident at the angle Θ_i with respect to the normal to the interface (see Fig.1). The generalized law of reflection[18] then shows that the reflected coordinate axis in the direction specified by the unit vector $\hat{\mathbf{1}}'_w$ lies in the α-plane of incidence that is defined by the incident unit vector $\hat{\mathbf{1}}_w$ and the normal N to the interface S and is at the angle

$$\Theta_r = \Theta_i \tag{25a}$$

with respect to the normal to the interface, while the reflected inhomogeneous plane wave spectral components' phase front normal lies in the β-plane of incidence that is defined by the vector **s** and the normal N to the interface S and is at the angle

$$\theta_r = \theta_i \tag{25b}$$

relative to the positive w'-axis of the reflected coordinate system. The generalized law of refraction[18] shows that the refracted coordinate axis in the direction specified by the unit vector $\hat{\mathbf{1}}''_w$ lies in the α-plane of incidence and is at the angle Θ_t with respect to the normal to the interface (see Fig.1) that is given by

$$\alpha_t \sin(\Theta_t) = \alpha_i \sin(\Theta_i) \quad , \tag{26a}$$

while the refracted inhomogeneous plane wave spectral components' phase front normal lies in the β-plane of incidence and is at the angle θ_t with respect to the refracted coordinate axis and is given by

$$\beta_t \sin(\theta_t) = \beta_i \sin(\theta_i) \quad , \tag{26b}$$

where the normal component of β_t is given by the real part and the normal component of α_t is given by the imaginary part of the expression

$$\beta_{t_n}(\omega) + i\alpha_{t_n}(\omega) = \left[n_2^2(\omega)k_0^2 - \left(\beta_{i_\tau}(\omega) + i\alpha_{i\tau}(\omega)\right)^2 + \alpha_{i_v}^2(\omega)\right]^{1/2} \quad , \tag{26c}$$

with $k_0 = \omega/c$, where the principal branch of the square root expression is to be taken.

The tangential components of the reflected and transmitted inhomogeneous plane wave spectral component electric field vectors are related to those of the incident field by the generalized Fresnel equations[18]

$$\tilde{\tilde{\mathbf{E}}}_r^{(\tau)}(\omega) = \vec{\mathbf{r}}_\tau(\omega)\tilde{\tilde{\mathbf{E}}}_i^{(\tau)}(\omega) \quad , \tag{27a}$$

$$\tilde{\tilde{\mathbf{E}}}_t^{(\tau)}(\omega) = \vec{\mathbf{t}}_\tau(\omega)\tilde{\tilde{\mathbf{E}}}_i^{(\tau)}(\omega) \quad , \tag{27b}$$

where the generalized Fresnel reflection matrix is given by

$$\vec{\mathbf{r}}_\tau(\omega) = \left[\vec{\mathbf{Y}}_i(\omega) + \vec{\mathbf{Y}}_t(\omega)\right]^{-1}\left[\vec{\mathbf{Y}}_i(\omega) - \vec{\mathbf{Y}}_t(\omega)\right] \quad , \tag{28}$$

and where the generalized Fresnel transmission matrix is given by

$$\vec{\mathbf{t}}_\tau(\omega) = \left[\vec{\mathbf{Y}}_i(\omega) + \vec{\mathbf{Y}}_t(\omega)\right]^{-1}\left[2\vec{\mathbf{Y}}_i(\omega)\right] \quad . \tag{29}$$

Notice that the reflection and transmission matrices $\vec{\mathbf{r}}(k_u, k_v, \omega)$ and $\vec{\mathbf{t}}(k_u, k_v, \omega)$ appearing in Eqs.(16) and (17), respectively, are linearly related to the generalized Fresnel reflection and transmission matrices that are given in Eqs.(28) and (29), respectively. The incident and transmitted complex admittance matrices appearing here are given by

$$\vec{\mathbf{Y}}_i(\omega) = \frac{1}{\eta_0 k_0 \tilde{k}_{i_n}(\omega)} \begin{pmatrix} -i\alpha_{i_v}(\omega)\tilde{k}_{i_\tau}(\omega) & \tilde{k}_{i_\tau}^2(\omega) - n_1^2(\omega)k_0^2 \\ n_1^2(\omega)k_0^2 + \alpha_{i_v}^2(\omega) & i\alpha_{i_v}(\omega)\tilde{k}_{i_\tau}(\omega) \end{pmatrix} \quad , \tag{30}$$

$$\vec{\mathbf{Y}}_t(\omega) = \frac{1}{\eta_0 k_0 \tilde{k}_{t_n}(\omega)} \begin{pmatrix} -i\alpha_{t_v}(\omega)\tilde{k}_{t_\tau}(\omega) & \tilde{k}_{t_\tau}^2(\omega) - n_2^2(\omega)k_0^2 \\ n_2^2(\omega)k_0^2 + \alpha_{t_v}^2(\omega) & i\alpha_{t_v}(\omega)\tilde{k}_{t_\tau}(\omega) \end{pmatrix} \quad , \tag{31}$$

respectively, where $\eta_0 = \sqrt{\mu_0/\varepsilon_0}$ is the intrinsic impedance of free space. Here the subscript τ denotes the tangential component defined by the plane of incidence and the interface plane, while the subscript ν denotes the component the lies in the interface plane but is normal to the plane of incidence, and are such that the unit normal vector to the interface that is directed from medium 1 into medium 2 is given by $\hat{\mathbf{n}} = \hat{\mathbf{1}}_\tau \times \hat{\mathbf{1}}_\nu$.

The frequency dispersion of the angular dependence of the real part of the Fresnel reflection coefficient for a TM mode field incident upon the planar interface separating two double resonance Lorentz model dielectrics whose frequency dependent dielectric permittivitys are given by

$$\varepsilon_j(\omega) = \varepsilon_j + \sum_{l=0,2} \frac{b_{jl}^2}{\omega^2 - \omega_{jl}^2 + 2i\delta_{jl}\omega} \quad , \tag{32}$$

$j = 1,2$, is illustrated in Fig.3 for the special case when the $\alpha-$ and $\beta-$ planes of incidence coincide. For simplicity, the two Lorentz model dielectrics considered here differ only in their limiting high frequency values ε_1 and ε_2, where $\varepsilon_1 > \varepsilon_2$ and consequently $\varepsilon_1(\omega) > \varepsilon_2(\omega)$ for all real-

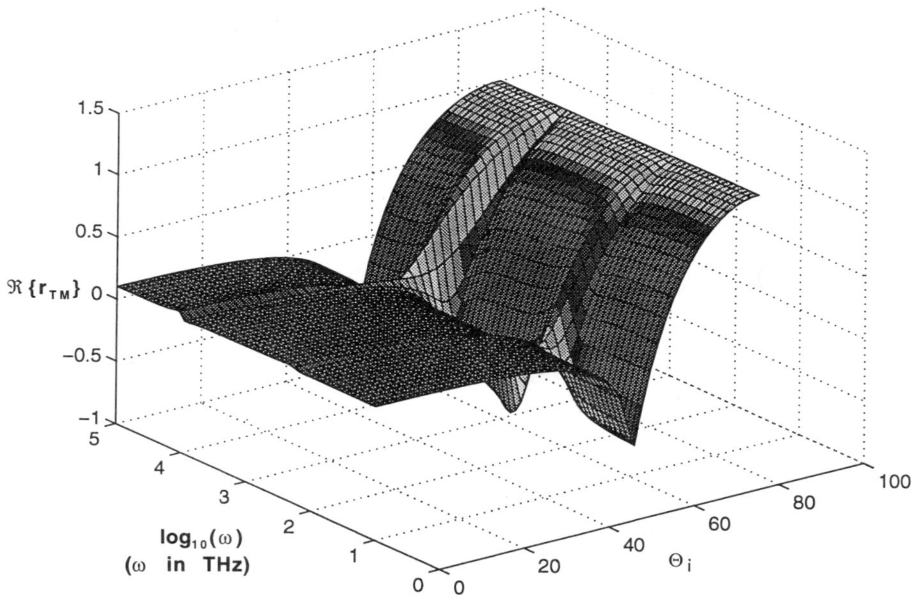

Figure 3. Real part of the Fresnel reflection coefficient as a function of both the angular frequency and the angle of incidence of a TM mode field incident upon the planar interface separating two double resonance Lorentz model dielectrics. The medium parameters for both medium 1 and 2 are $\omega_{10} = \omega_{20} = 174.12THz$, $\delta_{10} = \delta_{20} = 49.555THz$, $b_{10} = b_{20} = 121.55THz$ for the lower resonance line and $\omega_{12} = \omega_{22} = 9144.8THz$, $\delta_{12} = \delta_{22} = 1424.1THz$, $b_{12} = b_{22} = 6719.8THz$ for the upper resonance line, where medium 1 and medium 2 have the respective high-frequency limits $\varepsilon_1 = 2.9938$ and $\varepsilon_2 = 1.9938$.

valued ω. Notice that the frequency dependence of the Fresnel reflection coefficient is greatest near the critical angle and decreases as the angle of incidence approaches normal incidence while it becomes negligibly small as the angle of incidence approaches grazing incidence.

ASYMPTOTIC DESCRIPTION OF THE DYNAMICAL FIELD EVOLUTION

Let the time dependence of the initial field at the input plane $w = w_0$ be such that its Fourier-Laplace spectrum is ultrawideband, and let the initial carrier frequency ω_c of the pulse lie between the two resonance frequencies of the double resonance Lorentz model dielectric, so that $\omega_{10} < \omega_c < \omega_{12}$. If the distance w to the interface S is large in comparison to the absorption depth at the carrier frequency of the initial pulse, then each component of either the electric or magnetic field vector that is incident upon the interface has an asymptotic representation[19,20] that may be expressed either in the form

$$A(\mathbf{r},t) = A_S(\mathbf{r},t) + A_B(\mathbf{r},t) + A_m(\mathbf{r},t) + A_c(\mathbf{r},t) \quad , \tag{33}$$

or in a somewhat more complicated form that is given by a linear superposition of fields that are themselves expressed in the form given in Eq.(33). The reflected and transmitted fields each will also have the same form for their asymptotic representations as either $w' \to \infty$ or $w'' \to \infty$, respectively.

The asymptotic behavior of the component field $A_S(\mathbf{r},t)$ is due to the above-resonance frequency components $|\omega| \geq \sqrt{\omega_{j2}^2 + b_{j2}^2 - \delta_{j2}^2}$ in the initial pulse spectrum and is referred to as the first or Sommerfeld precursor field[7-9], where *j=1* for the reflected field and *j=2* for the transmitted field. The front of the Sommerfeld precursor oscillates at or just below an infinite instantaneous oscillation frequency and travels at the speed of light in vacuum. As the Sommerfeld precursor field evolves at a fixed propagation distance, its amplitude rapidly builds to a peak value soon after its arrival and thereafter decays as the attenuation factor monotonically increases and the instantaneous oscillation frequency chirps downward and approaches the value $\sqrt{\omega_{j2}^2 + b_{j2}^2 - \delta_{j2}^2}$, where *j=1* for the reflected field and *j=2* for the transmitted field.

The asymptotic behavior of the component field $A_B(\mathbf{r},t)$ is due to the below-resonance frequency components $|\omega| \leq \sqrt{\omega_{j0}^2 - \delta_{j0}^2}$ that are present in the initial pulse spectrum and is referred to as the Brillouin precursor field[7-9], where *j=1* for the reflected field and *j=2* for the transmitted field. As the Brillouin precursor evolves at a fixed propagation distance , its amplitude rapidly builds to a maximum value that is at or very near to its zero attenuation point that propagates at the velocity $c/n_j(0)$ through the dispersive medium; this point in the dynamical field evolution only decreases in amplitude as $z^{-1/2}$ and so will eventually dominate the entire propagated field structure for a sufficiently large propagation distance for either the reflected (*j=1*) or transmitted (*j=2*) fields. The Brillouin precursor field is quasi-static over its initial rise to its peak value, and thereafter becomes oscillatory with an instantaneous oscillation frequency that chirps upward and approaches the value $\sqrt{\omega_{j0}^2 - \delta_{j0}^2}$ while its amplitude decreases as its attenuation factor monotonically increases from zero.

The asymptotic behavior of the component field $A_m(\mathbf{r},t)$ is due to the frequency components in the intermediate frequency domain $\sqrt{\omega_{j0}^2 + b_{j0}^2 - \delta_{j0}^2} < |\omega| < \sqrt{\omega_{j2}^2 - \delta_{j2}^2}$ that lies between the two absorption bands of the double resnance Lorentz medium, where *j=1* for the reflected field and *j=2* for the transmitted field. A condition for the appearance of this additional precursor field in a double resonance Lorentz model dielectric may be found in Ref.19. Its dynamical evolution, if present in the propagated field structure, occurs just prior to the signal arrival when the input pulse carrier frequency lies in this intermediate frequency domain.

The final contribution $A_c(\mathbf{r},t)$ to the asymptotic description (33) of the propagated field is due to the poles (if any) of the initial pulse envelope spectrum whose real coordinate locations occur at the input pulse carrier frequency at $\omega = \omega_c$. This contribution to the asymptotic behavior of the propagated field describes the steady state behavior of the propagated signal[7-9] that oscillates at $\omega = \omega_c$.

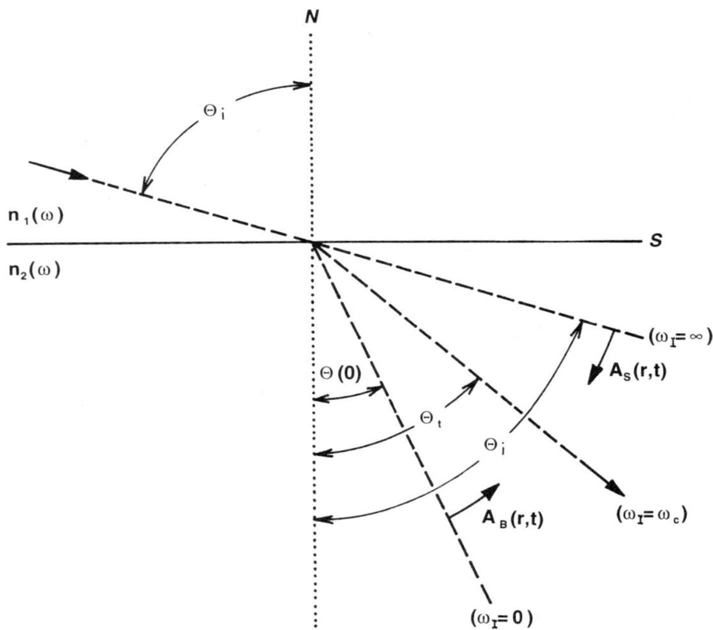

Figure 4. Simplified depiction of the dynamical evolution of the space-time structure of the refracted field due to an incident Heaviside-step-function modulated plane wave field with carrier frequency ω_c. The steady-state angle of refraction Θ_t is for the main signal at $\omega = \omega_c$. The Sommerfeld precursor front is at the angle of incidence Θ_i and, as the Sommerfeld precursor evolves in time, its angle of refraction sweeps to smaller angles as its oscillation frequency chirps downward, as indicated in the figure. The Brillouin precursor front is refracted at the quasistatic angle of refraction at $\omega = 0$ and, as the Brillouin precursor evolves in time, its angle of refraction sweeps to larger angles as its oscillation frequency chirps upward, as indicated in the figure.

Several uniquely interesting phenomena appear in the dynamical evolution of the transmitted field. Since the instantaneous angular frquency of oscillation of the Sommerfeld precursor field decreases monotonically from infinity as it evolves, the real part of the index of refraction presented to this transient field component will decrease and the angle of refraction will therefore dynamically change as the field evolves so that the Sommerfeld precursor will spatially fan out from the angle of incidence to larger angles as it crosses the interface. A similar effect will occur for the Brillouin precursor field, but in the opposite angular direction and beginning at the quasi-static angle of refraction. This spatio-temporal coupling is due to the combined effects of angular dispersion at the interface and temporal dispersion in the transmission medium as well as (but to a lesser extent) temporal dispersion in the incident medium. Notice that this effect will allow one to spatially separate the individual precursor fields from the main body of the pulse in a well-designed experiment, as depicted in Fig.4.

ACKNOWLEDGEMENTS

The research presented in the paper has been supported, in part, by the United States Air Force Office of Scientific Research under Grant No. F49620-94-1-0430 and by the Patricia Harris Fellowship Foundation.

REFERENCES

1. A. Sommerfeld, Über die fortpflanzung des lichtes in disperdierenden medien, Ann. Phys. 44:177 (1914).
2. L. Brillouin, Über die fortpflanzung des licht in disperdierenden medien, Ann. Phys. 44:203 (1914).
3. H. Baerwald, Über die fortpflanzung von signalen in disperdierenden medien, Ann. Phys. 7:731 (1930).
4. L. Brillouin. "Wave Propagation and Group Velocity," Academic, New York (1960).
5. J.A. Stratton. "Electromagnetic Theory," McGraw-Hill, New York (1941), pp.333-340.
6. J.D. Jackson. "Classical Electrodynamics," 2nd.ed., Wiley, New York (1975), ch.7.
7. K.E. Oughstun and G.C. Sherman, Propagation of electromagnetic pulses in a linear dispersive medium with absorption (the Lorentz medium), J. Opt. Soc. Am. B 5:817 (1988).
8. K.E. Oughstun and G.C. Sherman, Uniform asymptotic description of electromagnetic pulse propagation in a linear dispersive medium with absorption (the Lorentz model), J. Opt. Soc. Am. A 6:1394 (1989).
9. K.E. Oughstun and G.C. Sherman. "Electromagnetic Pulse Propagation in Causal Dielectrics," Springer-Verlag, Berlin (1994).
10. G.C. Sherman and K.E. Oughstun, Description of pulse dynamics in Lorentz media in terms of the energy velocity and attenuation of time-harmonic waves, Phys. Rev. Lett. 47: 1451 (1981).
11. G.C. Sherman and K.E. Oughstun, Energy-velocity description of pulse propagation in absorbing, dispersive dielectrics, J. Opt. Soc. Am. B 12:229 (1995).
12. B.R. Horowitz and T. Tamir, Unified theory of total reflection phenomena at a dielectric interface, Appl. Phys. 1:31 (1973).

13.C.C. Chen and T. Tamir, Beam phenomena at and near critical incidence upon a dielectric interface, J. Opt. Soc. Am. A 4:655 (1987).
14.E. Gitterman and M. Gitterman, Transient processes for incidence of a light signal on a vacuum-medium interface, Phys. Rev. A 13:763 (1976).
15.J.G. Blaschak and J. Franzen, Precursor propagation in dispersive media from short-rise-time pulses at oblique incidence, J. Opt. Soc. Am. A 12:1501 (1995).
16.R. Albanese, J. Penn, and R. Medina, Short rise-time microwave pulse propagation through dispersive biological media, J. Opt. Soc. Am. A 6:1441 (1989).
17.R. Albanese, Ultrashort electromagnetic signals: biophysical questions, safety issues, and medical opportunities, Aviat. Space Environ. Med. 65:116 (1994).
18.J.A. Marozas, "Asymptotic Description of Ultrawideband Electromagnetic Pulse Propagation in Lossy, Dispersive Dielectric Waveguides," Ph.D. dissertation, University of Vermont (in progress).
19.S. Shen and K.E. Oughstun, Dispersive pulse propagation in a double resonance Lorentz medium, J. Opt. Soc. Am. B 6:948 (1989).
20.K.E. Oughstun, Dynamical structure of the precursor fields in linear dispersive pulse propagation in lossy dielectrics, in: "Ultra-Wideband, Short-Pulse Electromagnetics 2," L. Carin and L.B. Felsen, eds., Plenum, New York (1995),pp.257-272.

TIME DOMAIN MEASUREMENT OF MATERIAL PERMITTIVITY AND PERMEABILITY

Clifton Courtney[1], Tracey Bowen[2], Jane Lehr[3], and Kami Burr[4]

[1]Voss Scientific, 412 Washington St. SE, Albuquerque, NM
[2]Phillips Laboratory / WSM, KAFB, NM
[3]Fiore Industries, 1009 Bradbury Dr. SE, Albuquerque, NM
[4]Science and Engineering Assoc., 6100 Uptown Blvd. NE, Albuquerque, NM

INTRODUCTION

Accurate knowledge of material complex relative permittivity ($\varepsilon_r = \varepsilon_r' - j\varepsilon_r''$) and permeability ($\mu_r = \mu_r' - j\mu_r''$) is required for just about any application utilizing the electromagnetic properties of materials. Applications that need precise information of the frequency dependence of ε_r and μ_r include design of radar absorbing material and RAM geometry, design of transmission line circuits on microwave substrates, and simulation and analysis of the propagation of electromagnetic waves in and through complex media. Frequency domain measurement of material properties are well known, and can be accomplished in a number of ways. These include lumped circuit and balanced bridge methods at low frequencies, and waveguide, TEM transmission line and resonant cavity methods for high frequencies[1,2], and optical techniques. A comprehensive overview of material electromagnetic properties measurement techniques was recently presented by Afsar[3]. These techniques typically are conducted at low voltage and low electric field strengths, and assume that the material properties are independent of the field strength.

In the late 60's and early 70's, before the advent of the automatic network analyzer, time domain material measurement methods were popular[4,5]. These techniques utilized the spectral content of a fast rise time pulse to determine the frequency dependence of the complex ε_r and μ_r of materials. However, the method described by Nicolson and Ross[4] relied on approximations that may not be valid for some cases of interest. Though the scheme reported here is an extension of previously described methods, the hardware design and data reduction techniques used in the present method eliminate some of the approximations and sources of potential error of the earlier work. In addition, the hardware costs associated with this technique are quite lower than that required of a frequency domain measurement.

THEORY OF REDUCTION OF MEASURED WAVEFORMS TO MATERIAL PARAMETERS

This section reviews the theory of TEM propagation in a two-media region, derives the relationship between the time harmonic transmission and reflection coefficients (or scattering parameters s_{21} and s_{11}) and complex ε_r and μ_r material parameters, and describes how the scattering parameters can be determined from a partial time history of a transient waveform.

Time Harmonic TEM Propagation in a Two Material Region

Consider the situation depicted in Figure 1a. There, a TEM wave is shown incident on a planar interface between Region 1 characterized by (ε_0, μ_0) and Region 2 characterized by (ε_2, μ_2). The normal to the interface between semi-infinite Regions 1 and 2 is opposite to the direction of the incident wave, and the thickness of Region 2 is d; the coordinate convention is indicated in the figure. A second planar interface between Regions 2 and 3, Region 3 characterized as well by (ε_0, μ_0), is indicated also. The wave interactions of the geometry of Figure 1a can be modeled by the coaxial transmission line shown in Figure 1b. Note the transit times and observation positions indicated. For observation locations in Regions 1 and 3, the waveforms of interest include $\mathbf{E}^{inc}$ (assumed known), $\mathbf{E}^{ref}$ = the total reflected waveform in Region 1, and $\mathbf{E}^{tran}$= the total transmitted waveform in Region 3. The usual microwave s-parameters can be written for time-harmonic excitation as

$$s_{11}(\omega)=\frac{\mathbf{E}^{ref}}{\mathbf{E}^{inc}},\ \text{and}\ s_{21}(\omega)=\frac{\mathbf{E}^{tran}}{\mathbf{E}^{inc}} \tag{1}$$

In terms of the wave impedances of each region, the above can be written as

$$s_{11}(\omega)=\frac{\left(Z_2^2-Z_0^2\right)\cdot\left(1-e^{-j2\beta_2 d}\right)}{\left(Z_2+Z_0\right)^2-\left(Z_2-Z_0\right)^2\cdot e^{-j2\beta_2 d}},\ \text{and} \tag{2a}$$

$$s_{21}(\omega)=\frac{4\left(Z_0 Z_2\right)\cdot e^{-j\beta_2 d}}{\left(Z_2+Z_0\right)^2-\left(Z_2-Z_0\right)^2\cdot e^{-j2\beta_2 d}}. \tag{2b}$$

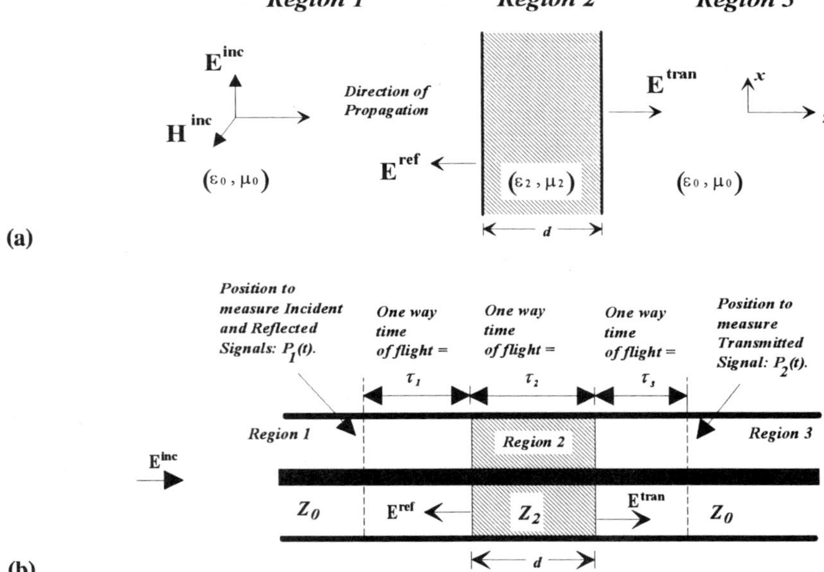

Figure 1. Geometry of interest: (a) a TEM wave is incident on the interface of Region 1 characterized by (ε_0, μ_0) and Region 2 characterized by (ε_2, μ_2); and (b) a coaxial transmission line can be used to model the geometry of (a).

Harrington[6] writes the complex material parameters as: $\varepsilon_2 = \varepsilon_0 \cdot (\varepsilon'_{r2} - j\varepsilon''_{r2})$; $\mu_2 = \mu_0 \cdot (\mu'_{r2} - j\mu''_{r2})$; $Z_0 = \sqrt{\mu_0 / \varepsilon_0}$; $Z_2 = \sqrt{\mu_2 / \varepsilon_2}$; and $\beta_2 = \frac{\omega\sqrt{\varepsilon_{r2} \cdot \mu_{r2}}}{c_0}$, where c_0 = speed of light in vacuum. Note that Eqns. 2a and 2b assume the evaluation is at the interface between Regions 1 and 2, and 2 and 3 respectively. Evaluation positions at other locations in Regions 1 and 3 require modification of the expressions by an appropriate amount of phase shift corresponding to propagation in Regions 1 and 3.

Frequency Dependent Material Properties from Knowledge of s_{11} and s_{12}

Given knowledge of the s-parameters s_{11} and s_{12} associated with a specific material, perhaps measured in a coaxial fixture similar to that shown in Figure 1b, the material properties ε_r and μ_r can be determined. A procedure for extracting the values is given below and follows the algorithm presented by Nicolson and Ross[4], and summarized by Lederer[7]. Equations 2a and 2b can be rewritten in more compact and convenient forms as

$$s_{11}(\omega) = \frac{(1 - z^2) \cdot \Gamma_{12}}{1 - \Gamma_{12}^2 \cdot z^2}, \text{ and} \tag{3a}$$

$$s_{21}(\omega) = \frac{(1 - \Gamma_{12}^2) \cdot z}{1 - \Gamma_{12}^2 \cdot z^2}; \tag{3b}$$

where $\Gamma_{ab} = \frac{Z_b - Z_a}{Z_b + Z_a}$ = the reflection coefficient for a wave passing from the semi-infinite Region a into the semi-infinite Region b, and $z = e^{-j\beta_2 d}$. The following relation then can be shown to hold:

$$z = \frac{s_{11} + s_{21} - \Gamma_{12}}{1 - (s_{11} + s_{21}) \cdot \Gamma_{12}}. \tag{4}$$

Now recall from above $z = e^{-j\beta_2 d} = e^{-j(\omega / c_0) \cdot \sqrt{\mu_{r2}\varepsilon_{r2}}}$; taking logarithms of both sides[8] yields

$$ln(z) = \mathbf{Ln}|z| + j\,arg(z) = \mathbf{Ln}\left| e^{-j(\omega / c_0) \cdot \sqrt{\mu_{r2}\varepsilon_{r2}}} \right| + j\,arg\left(e^{-j(\omega / c_0) \cdot \sqrt{\mu_{r2}\varepsilon_{r2}}} \right). \tag{5}$$

If we assume that Region 2 is not too lossy, then

$$|z| = \left| e^{-j(\omega / c_0) \cdot \sqrt{\mu_{r2}\varepsilon_{r2}}} \right| \approx 1, \text{ and} \tag{6}$$

$$\ln(z) = 0 - j(\omega / c_0) \cdot \sqrt{\mu_{r2}\varepsilon_{r2}}. \tag{7}$$

Then the product of the material parameters is

$$\mu_{r2}\varepsilon_{r2} = -\left\{ \left(\frac{c_0}{d \cdot \omega} \right) \cdot \ln(z) \right\}^2 = C_1. \tag{8}$$

Furthermore, $\Gamma_{12} = \frac{Z_2 - Z_1}{Z_2 + Z_1} = \frac{\sqrt{\mu_{r2} / \varepsilon_{r2}} - 1}{\sqrt{\mu_{r2} / \varepsilon_{r2}} + 1}$, and solving for the ratio of $\mu_{r2} / \varepsilon_{r2}$ yields

$$\frac{\mu_{r2}}{\varepsilon_{r2}} = \left(\frac{1+\Gamma_{12}}{1-\Gamma_{12}}\right)^2 = C_2 . \tag{9}$$

With the relations expressed in Eqns. 8 and 9, the material parameters can be separated as

$$\mu_{r2} = \sqrt{C_1 \cdot C_2} \text{ and } \varepsilon_{r2} = \sqrt{C_1 / C_2} . \tag{10}$$

The parameters C_1 and C_2 then are defined completely in terms of the s-parameters, and the material constants ε_{r2} and μ_{r2} are likewise determined.

Derivation of s_{11} and s_{12} from a Transient Time Domain Measurement

The previous section demonstrated that the material parameters ε_{r2} and μ_{r2} could be determined from the time-harmonic s-parameters. In this section, we will show how the required s-parameters can be expressed in terms of the partial time history of a transient waveform.

Consider the bounce diagram[9] shown in Figure 2. The diagram represents the geometry of Figure 1, and depicts the transient wave behavior one would observe if a wave $e^{inc}(t)=e_0(t-\tau_1)$ were incident on the Region 1 - Region 2 interface from the left. Note that the reference is at position P_1 (see Figure 1b), but the waveform incident on the interface is $e^{inc}(t)$. In the figure

$$\Gamma_{12} = \frac{Z_2 - Z_0}{Z_2 + Z_0} = \text{ reflection coefficient at Region 1 / 2 interface,} \tag{11a}$$

$$\Gamma_{21} = \frac{Z_0 - Z_2}{Z_2 + Z_0} = -\Gamma_{12} = \text{ reflection coefficient at Region 2 / 1 interface,} \tag{11b}$$

$$T_{12} = 1 + \Gamma_{12} = \text{ transmission coefficient at Region 1 / 2 interface, and} \tag{11c}$$

$$T_{21} = 1 + \Gamma_{21} = \text{ transmission coefficient at Region 2 / 1 interface.} \tag{11d}$$

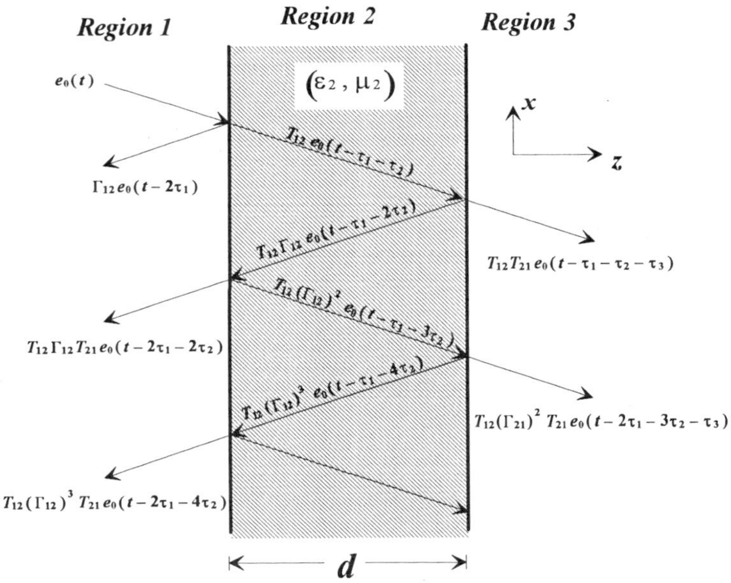

Figure 2. Bounce diagram for the geometry depicted in Figure 1; a transient wave undergoes an infinite number of transmissions and reflections at the Region 1-2 / 2-1 and Region 2-3 interfaces.

Also, τ_1 = propagation time from the point of observation in Region 1 to the Region 1/2 interface, τ_2 = propagation time across Region 2, and τ_3 = propagation time from the point of observation in Region 3 to the Region 2/3 interface (see Figure 1b).

In general, the reflected (transmitted) wave is given by the convolution of the incident wave with the time domain reflection (transmission) transfer function, or

$$e^{ref}(t) = e^{inc}(t) * s_{11}(t) = \int_0^\infty e^{inc}(\lambda) s_{11}(t-\lambda)\, d\lambda\,, \tag{12}$$

where $e^{ref}(t)$ is the reflected waveform at the interface in Region 1, $e^{inc}(t)=e_0(t\text{-}\tau_1)$ is the waveform incident on the interface, and $e^{inc}(t)=0$ for $t<0$. In the frequency domain, this relationship becomes simple multiplication

$$E^{ref}(\omega) = E^{inc}(\omega)\cdot s_{11}(\omega)\cdot e^{-j\omega\tau_1} = E_0(\omega)\cdot s_{11}(\omega)\cdot e^{-j2\omega\tau_1}\,, \tag{13}$$

where we have now let the reference location be position P_1 as shown in Figure 1b (characterized by the multiplying term $e^{-j2\omega\tau_1}$, where τ_1 is the one-way transit time from position P_1 to the Region 1-2 interface). Other ways to view the time domain interaction of the electromagnetic field with the dielectric have been presented[10,11,12]. Note that although the convolution operation is implied in Figure 2, it is not explicitly shown. Examination of Figure 2 reveals that the reflected and transmitted waves can be written as

$$e^{ref}(t) = e_0\left(t-2\tau_1\right)*\Gamma_{12} + \sum_{n=0}^{\infty} T_{21}(t)*\left\{(\Gamma_{21}*)^{2n+1}\left\{T_{12}*e_0\left(t-2\tau_1-2(n+1)\tau_2\right)\right\}\right\},\ \text{and} \tag{14a}$$

$$e^{tran}(t) = \left\{e_0\left(t-\tau_1-\tau_2-\tau_3\right)*T_{12}\right\}*T_{21} + \sum_{n=0}^{\infty} T_{21}*\left\{(\Gamma_{21}*)^{2(n+1)}\left\{T_{12}*e_0\left(t-\tau_1-(3+2n)\tau_2-\tau_3\right)\right\}\right\}. \tag{14b}$$

The notation $(\Gamma_{21}*)^m$ means that the expression to the right of this term is to undergo m convolutions with Γ_{21}, and the explicit time dependence is not shown. Also, for convenience and notational simplicity the time lag in the right-hand side has been grouped together in the term $e_0(\cdot)$. Fourier transforms of both sides of Eqns. 14a and 14b then give an equivalent frequency domain description. That is

$$E^{ref}(\omega) = E_0(\omega)\cdot\Gamma_{12}(\omega)\cdot e^{-j2\omega\tau_1} + \sum_{n=0}^{\infty} T_{21}(\omega)\cdot T_{12}(\omega)\cdot(\Gamma_{21}(\omega))^{2n+1} E_0(\omega) e^{-j2\omega\left(\tau_1+(n+1)\tau_2\right)},\ \text{and} \tag{15a}$$

$$\begin{aligned} E^{tran}(\omega) = {} & E_0(\omega)\cdot T_{12}(\omega)\cdot T_{21}(\omega)\cdot e^{-j\omega\left(\tau_1+\tau_2+\tau_3\right)} \\ & + \sum_{n=0}^{\infty} T_{12}(\omega)\cdot T_{21}(\omega)\cdot\left(\Gamma_{21}(\omega)\right)^{2(n+1)}\cdot E_0(\omega)\cdot e^{-j\omega\left(\tau_1+(3+2n)\tau_2+\tau_3\right)}. \end{aligned} \tag{15b}$$

The scattering coefficients s_{11} and s_{21} can be expressed in the frequency domain as

$$s_{11}(\omega) = \frac{E^{ref}(\omega)}{E_0(\omega)} = e^{-j2\omega\tau_1}\Gamma_{12}(\omega) + \sum_{n=0}^{\infty} T_{21}(\omega)\cdot T_{12}(\omega)\cdot\left(\Gamma_{21}(\omega)\right)^{2n+1}\cdot e^{-j2\omega\left(\tau_1+(n+1)\tau_2\right)},\ \text{and} \tag{16a}$$

$$s_{21}(\omega)=\frac{E^{tran}(\omega)}{E_0(\omega)}=T_{12}(\omega)\cdot T_{21}(\omega)\cdot e^{-j\omega(\tau_1+\tau_2+\tau_3)}+\sum_{n=0}^{\infty}T_{12}(\omega)\cdot T_{21}(\omega)\cdot\left(\Gamma_{21}(\omega)\right)^{2(n+1)}\cdot e^{-j\omega(\tau_1+(3+2n)\tau_2+\tau_3)}. \tag{16b}$$

The important point to note about Eqns. 16a and 16b is that they can be completely determined in terms of $\Gamma_{12}(\omega)$ and $T_{12}(\omega)\cdot T_{21}(\omega)$ ($\Gamma_{21}(\omega)$ can be defined in terms of $\Gamma_{12}(\omega)$ through Eqn. 11b).

Referring back to the bounce diagram, Figure 2, one notes that in the time domain the first component observed in Region 1 of the (total) reflected waveform is

$$p_1(t)=\Gamma_{12}*e_0(t-2\tau_1), \tag{17}$$

then the time domain reflection coefficient Γ_{12} (= the reflection coefficient at the Region 1/2 interface when the width of Region 2, $t\to\infty$) can be determined directly from a measurement of $p_1(t)$. Similarly, the first transmitted component observed in Region 3 is

$$p_2(t)=T_{12}*T_{21}*e_0(t-\tau_1-\tau_2-\tau_3), \tag{18}$$

and the time domain product of the transmission coefficients $T_{ab}\cdot T_{ba}$ (where T_{ab} = the transmission coefficient from Region a to Region b when the width of Region b, $d\to\infty$) can be determined directly from a measurement of $p_2(t)$.

The values of $\Gamma_{12}(\omega)$ and $T_{12}(\omega)\cdot T_{21}(\omega)$ will be found from Fourier transforms of the measured quantities $p_1(t)$ and $p_2(t)$ (Eqns. 17 and 18), and division by the incident waveform. The scattering coefficients s_{11} and s_{21} can be determined next from Eqns. 16a and 16b, and finally values of material ε_2 and μ_2 can be extracted in the manner described previously.

Measurement Procedure and Reduction to Material Properties

Here, a description of the physical process one would use to determine the material properties ε_2 and μ_2 using the algorithmic process described above. The procedure is as follows. A voltage pulse is introduced in a coaxial test fixture that contains the material sample, and measured at Position P_1. To allow uncorrupted measurement of the incident, reflected and transmitted pulses, the incident pulse width must not exceed a maximum value. The incident pulse, $e_0(t)$, is measured at location P_1. Then, the first component of the reflected pulse, $p_1(t)$ is also measured at P_1, and finally the first component of the transmitted pulse, $p_2(t)$, is measured at position P_2 in the coaxial test fixture. The measurements must be tied to the same time reference. Note that values of τ_1, τ_2, and τ_3 will need to be determined. This can be done in a number of ways; for example direct measurement on the scope, or TDR-type measurements.

Once the raw data has been collected, the transforms of $p_1(t)$ and $p_2(t)$ are formed and used to evaluate $\Gamma_{12}(\omega)$ and $T_{12}(\omega)\cdot T_{21}(\omega)$ via Eqns. 17 and 18. Next, form the quantities s_{11} and s_{21} via Eqns. 16a and 16b. Though originally written in terms of an infinite series, there is an easier way to evaluate the quantities s_{11} and s_{21}. Consider the well-known infinite series[13]

$$\sum_{n=0}^{\infty}x^n=\frac{1}{1-x},\quad |x|<1. \tag{20}$$

Then, Eqn. 20 can be used to express Eqns. 16a and 16b in forms more easily evaluated numerically. Equation 16a becomes

$$s_{11}(\omega)=e^{-j2\omega\tau_1}\Gamma_{12}(\omega)+\frac{T_{21}(\omega)\cdot T_{12}(\omega)\cdot\Gamma_{21}(\omega)\cdot e^{-j2\omega(\tau_1+\tau_2)}}{1-\Gamma_{21}(\omega)^2\cdot e^{-j2\omega\tau_2}} \tag{21a}$$

and 16b becomes

$$s_{21}(\omega) = T_{12}(\omega)\cdot T_{21}(\omega)\cdot e^{-j\omega(\tau_1+\tau_2+\tau_3)} + \frac{T_{12}(\omega)\cdot T_{21}(\omega)\cdot \Gamma_{21}(\omega)^2 \cdot e^{-j\omega(\tau_1+3\tau_2+\tau_3)}}{1-\Gamma_{21}(\omega)^2\cdot e^{-j\omega 2\tau_2}} \quad (21b)$$

The expressions for the scattering coefficients given in Eqns. 21a and 21b are then easily evaluated from knowledge of $\Gamma_{12}(\omega)$ and $T_{12}(\omega)\cdot T_{21}(\omega)$ determined from measurements, Fourier transforms of $p_1(t)$ and $p_2(t)$, and simple mathematical manipulation as described above.

MEASURED PROPERTIES OF NORYL

A bulk sample of Noryl EN265 was machined to a length of 6.000 ± 0.001 inches, and with cross sectional dimensions compatible with a coaxial line. A differentiating E-field probe was used to measure the time derivative of the incident, reflected and transmitted waveforms. These waveforms are shown in Figure 3. Since the reduction procedure considers ratios of quantities in the frequency domain, the derivative response is sufficient (no time integration of the waveform is required). Note that just the first component of the reflected and transmitted waveforms is measured (and shown in the figure). The reduction algorithm forms the complete response from just these partial components. These waveforms then are analyzed to yield the required values of propagation times. To recall, τ_1 = propagation time from the first sensor to the front face of the sample, τ_2 = one-way propagation time through the sample, and τ_3 = propagation time from the rear face of the sample to the second sensor. From the data shown in Figure 3 these values were found to be: τ_1 = 1.828 ns; τ_2 = 0.8825 ns; and τ_3 = 0.381 ns. The physical length of the sample, the time of flight values, and the waveforms of Figure 3 were submitted to automated data reduction software; the reduced values of relative permittivity and permeability for Noryl EN265 are given in Figures 4a-d for the frequency range 50 MHz - 8 GHz, a greater than 7½ octave span. Note that both the raw measured data and a linear interpolation are given in each graph. One sees that the real part of the relative dielectric constant is essentially constant over the band ($\varepsilon_r' \approx 2.7$), while the imaginary (loss) component varies from $-0.06 < \varepsilon_r'' < 0$ over the frequency band. Figures 4c and 4d show the Noryl is a non-magnetic material with $\mu_r' \approx 1.0$ and $\mu_r'' \approx 0.0$. GE data sheets[1] give the relative dielectric constant of Noryl EN265 as 2.69 and dissipation factor (loss tangent) as 0.0007, both at 60 Hz. These values are consistent with the results reported here.

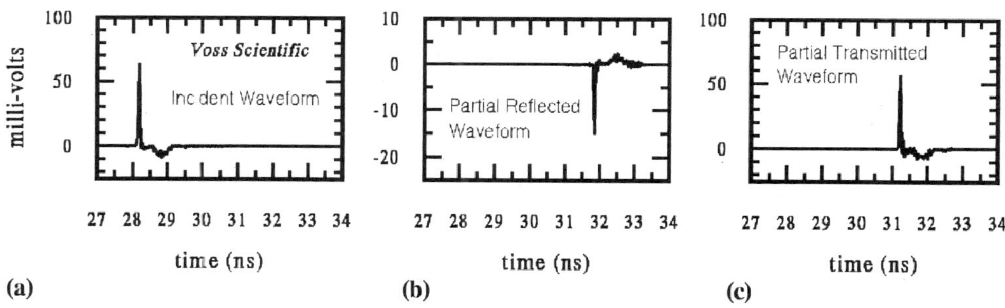

Figure 3. The waveforms used to determine NORYL sample material properties: (a) time derivative of the incident; (b) first component of the reflected; and (c) and first component of the transmitted waveforms.

[1] http://www.ge.com/datasheets/NORYLEN265.html

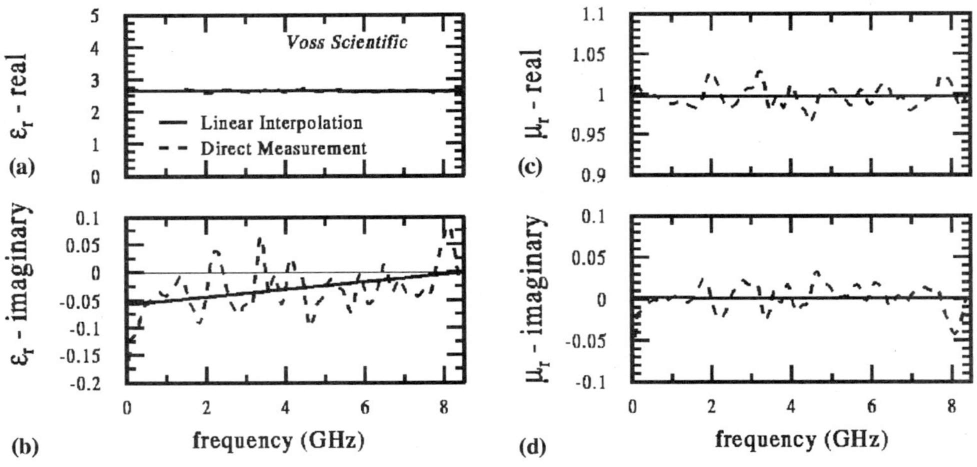

Figure 4. Reduced values of relative permittivity and permeability for Noryl (GE EN265): (a) real component of the relative permittivity; (b) imaginary component of the relative permittivity; (c) real component of the relative permeability; and (d) imaginary component of the relative permeability.

REFERENCES

1. A. von Hippel, *Dielectrics and Waves*, Second Edition, ISBN 0-89006-803-8, Artech House, Boston (1995).
2. A. von Hippel, editor, *Dielectric Materials and Applications*, Second Edition, ISBN 0-89006-805-4, Artech House, Boston (1995).
3. N. Afsar, J.R. Birch, and R.N. Clarke, edited by G.W. Chantry, The measurement of the properties of materials, *Proceedings of the IEEE*, vol. 74, no. 1, pp. 183 - 199 (1986).
4. M. Nicolson and G.F. Ross, Measurement of the intrinsic properties of materials by time-domain techniques, *IEEE Trans. on Instr. and Meas.*, vol. IM-19, no. 4, pp. 377 - 382 (1970).
5. M. Nicolson, P.G. Mitchell, R.M. Mara, and A.M. Auckenthaler, Time domain measurement of microwave absorbers, *Final Technical Report AFAL-TR-71-353*, Air Force Avionics Laboratory, Air Force Systems Command, Wright-Patterson AFB, DTIC No. AD-892162 (1971).
6. R.F. Harrington, *Time-Harmonic Electromagnetic Fields*, ISBN 07-026745-6, McGraw-Hill, New York (1961).
7. G. Lederer, A transmission line method for the measurement of microwave permittivity and permeability, *Memorandum 4450*, Royal Signals and Radar Establishment (1990).
8. V. Churchill and J.W. Brown, *Complex Variables and Applications*, Fourth Edition, ISBN 0-07-010873-0, McGraw-Hill, New York (1984).
9. A. Rizzi, *Microwave Engineering - Passive Circuits*, ISBN 0-13-586702-9, Prentice-Hall, Englewood Cliffs, NJ (1988).
10. H. Cole, Time-domain spectroscopy of dielectric materials, *IEEE Trans. on Instr. and Meas.*, vol. IM-25, no. 4, pp. 371 - 375 (1976).
11. H. Cole, Evaluation of dielectric behavior by time-domain spectroscopy. I. Dielectric response by real time analysis, II. Complex Permittivity, *J. of Phys. Chem.*, vol. 79, no. 14, pp. 1459 - 1474 (1975).
12. S. Neelakanta, *Handbook of Electromagnetic Materials*, pg. 42, ISBN 0-8493-2500-5, CRC Press, Boca Raton, FL (1995).
13. Knoop, *Infinite Sequences and Series*, pg. 46, Dover Publications, Inc., New York (1956).

MEASUREMENTS OF SHORT-PULSE PROPAGATION THROUGH CONCRETE WALLS

Dr. John F. Aurand

Sandia National Laboratories [1]
High-Power Electromagnetics Department 9323
P.O. Box 5800, MS-1153
Albuquerque, NM 87185-1153

INTRODUCTION

We recently performed a series of experimental measurements of transient electromagnetic (EM) propagation through two different concrete walls. Several different short-duration pulses were used for the incident radiation, with frequency content from VHF to 20 GHz. Both walls were 30 cm thick, with three internal layers of reinforcing steel bars. For this particular set of data, the incident wave polarization was vertical linear only. Corroborating swept-frequency measurements were made with a vector network analyzer.

This paper describes the propagation measurements through the two walls, and the propagation model of a lossy dielectric layer. We also examine the transfer function, dielectric constant, loss tangent, attenuation constant, and time-domain impulse response of these walls. The attenuation increases steadily with frequency, and is a strong function of the moisture content of the concrete. The time-domain pulse attenuation and dispersion are consistent with the lowpass-filtering effect of this attenuation loss vs. frequency.

TIME-DOMAIN PROPAGATION MEASUREMENTS

The transmitter portion of the wideband time-domain measurement system consisted of a very fast low-voltage commercial pulse generator and a wideband TEM-horn antenna with dielectric aperture lens. The receiver consisted of an identical TEM-horn antenna and a wideband sampling oscilloscope. These new antennas are described more fully in the companion paper in this Proceedings, '*A TEM-Horn Antenna with Dielectric Lens for Fast Impulse Response,*' and were expressly developed for this experimental program.

[1] This Work was Supported by the U.S. DOE under Contract DE-AC04-94AL85000.

Ultra-Wideband, Short-Pulse Electromagnetics 3
Edited by Baum *et al.*, Plenum Press, New York, 1997

Considerable effort was spent in optimizing this measurement system to have the fastest transient response possible (thus offering the widest bandwidth of wall characterization).

In order to characterize the pulse attenuation and dispersion of a concrete wall, two different transmission measurements are made in order to form a transfer function representing the lossy dielectric layer under test. One is a free-space reference measurement, characterizing the measurement system response. The other is the 'through' measurement, in which the pulse is actually measured after propagating through the layer of interest. The transfer function is then given by the frequency-domain ratio of the processed received waveforms.

This paper presents data taken with two indoor concrete walls, at different facilities. One wall was located in one of our buildings at Sandia National Labs (Kirtland AFB, New Mexico), and the other wall was located in a building at Wright Laboratory/Armament Directorate (Eglin AFB, Florida). Figure 1 is a photograph of the Wright Lab experimental setup used to capture the measurement system response in a free-space boresight antenna alignment. The antenna aperture separation was 1.10 m (precisely the same as the separation for the wall measurement). The transmit antenna is pointed toward the viewer, with the pulse generator equipment located behind it; the dielectric lens can be seen in the aperture of the antenna. The receive antenna is pointed toward the transmit antenna, in this case with both main-beam boresight directions carefully aligned for maximum time-domain response. The receiving sampling oscillosope can be seen at the left side of the photo, sitting on top of a vector analyzer.

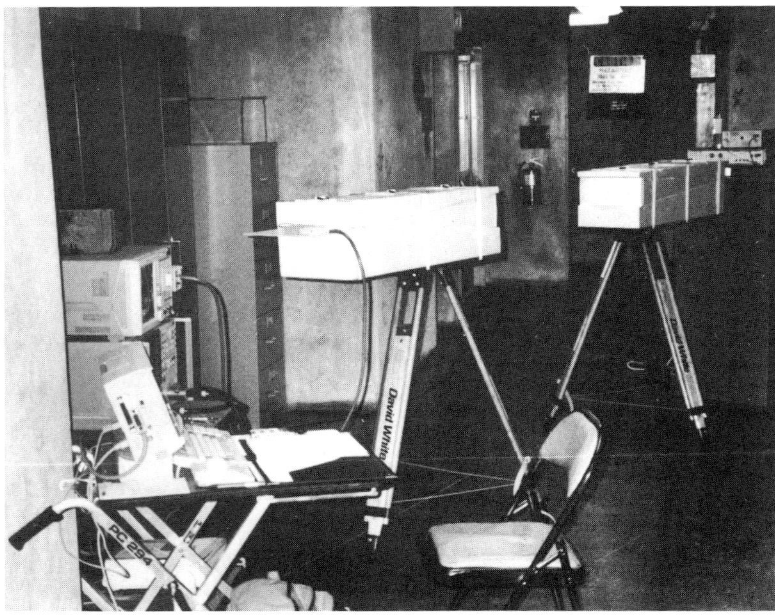

Figure 1. Free-space setup for reference measurement of system, at Wright Lab facility.

For the through measurements, Figures 2 and 3 show the transmitter and receiver setups, respectively. In Figure 2, the transmit antenna is oriented at normal incidence to the wall, with the aperture separated 40 cm from the front surface. In Figure 3, the receive antenna is seen in a similar orientation, on the opposite side of this wall. It is also easier to view the receive equipment in this figure. The antennas were very carefully positioned in order to have a normally-incident propagation line-of-sight and repeatable wall separations (to an accuracy of better than 0.5 cm).

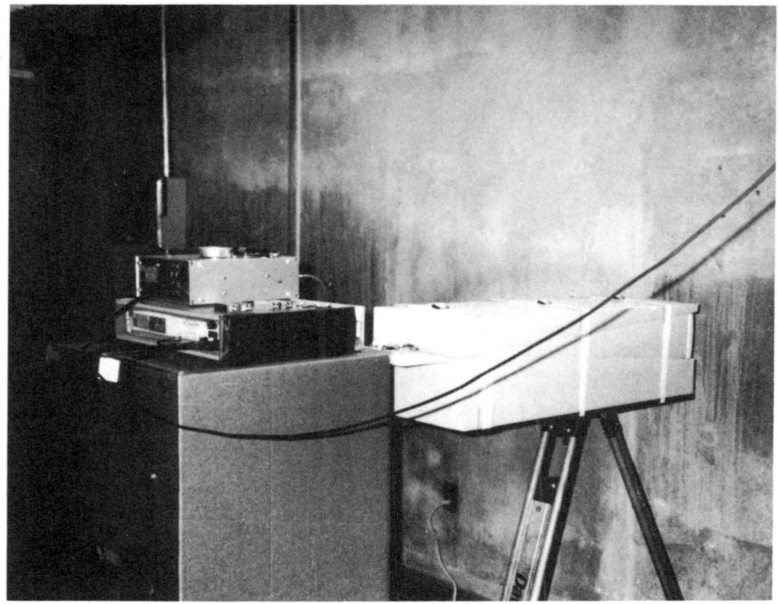

Figure 2. Transmitter setup for wall measurement, at Wright Lab facility.

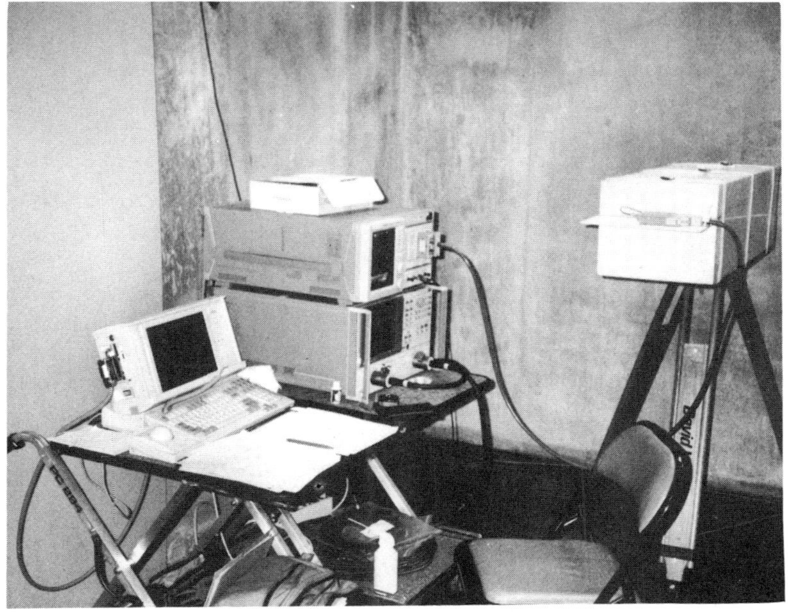

Figure 3. Receiver setup for wall measurement, at Wright Lab facility.

The experimental setup at our facility consisted of the same equipment, but the antennas were supported by custom wood platforms rather than the wood tripods shown in the figures above. These platforms are shown in Figure 6 of the companion paper mentioned above. The aperture separation of the antennas was 90.5 cm.

A variety of time-domain pulse types were used in these measurements in order to determine their transient propagation-loss behavior. These TEM-horn antennas have a clean derivative transmit time-domain response (on boresight), so the radiated EM field is a derivative of the pulse which drives the transmit antenna. The receive response is a time integral of the transmit response, so the antenna produces an output voltage pulse which is a replica of the incident wave. As a result, with step excitation of the transmit antenna, the radiated wave incident on the wall under test is an impulse, and the system has a free-space received pulse which is also an impulse. For the Sandia wall experiments, four impulse EM waves were created, with different pulse durations.

Figure 4 shows one set of data for the Sandia wall, using the fastest impulse excitation. The first and largest impulse is the free-space reference waveform, and the second waveform is that through the wall. The reference impulse has a 50% pulse duration (full-width at half-maximum) of 48 ps, and with a receiving system 10-90% transition duration of about 22 ps (with a 50-GHz sampling head), this means that the radiated field incident on the wall has a pulse duration of about 43 ps. This is the fastest (and widest bandwidth) radiated-wave system we know of for measuring the transient microwave properties of layered dielectric materials. Also note the 1.3-ns time delay due to pulse retardation in the wall. Based on this additional delay, the dielectric constant (relative permittivity) of the wall is 5.2. A further item to note is the lowpass-filtering effect of the through measurement.

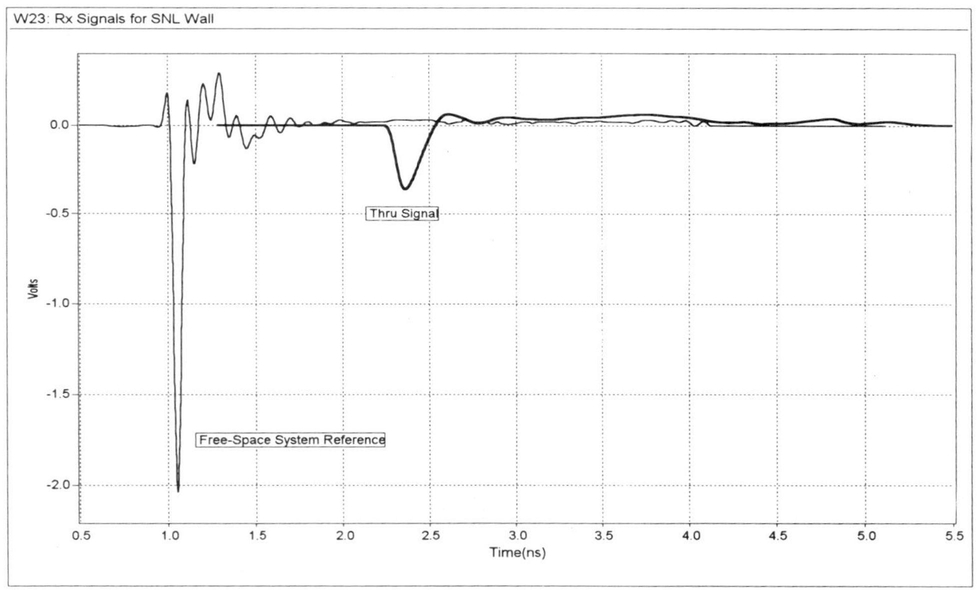

Figure 4. Received waveforms (free-space reference and through-wall), for Sandia Labs wall.

For the Wright Lab wall, five different impulse excitations were used, as well as a bipolar wave excitation. These data are shown in Figure 5, together with the corresponding through waveforms. Note the time delay through the wall (1.15 ns), and how consistent it is for the set of excitation waveforms. This implies that the dielectric constant is relatively constant over the spectral content of the excitation pulses (with a value of 4.5).

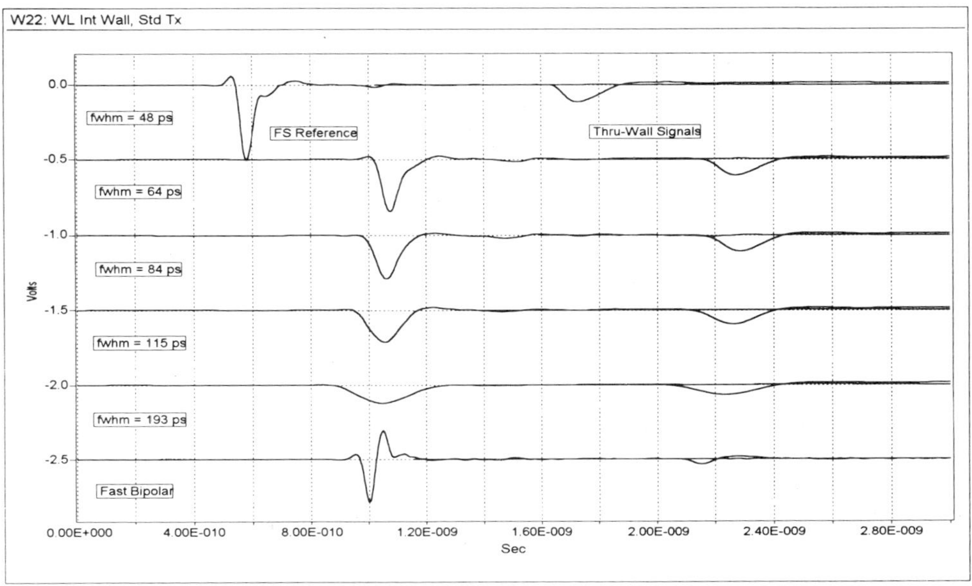

Figure 5. Received waveforms (free-space reference and through-wall), for Wright Lab wall. There are five different impulse excitations and one bipolar excitation.

PROPAGATION MODEL & DATA ANALYSIS

In terms of characterizing the propagation through a wall or other lossy dielectric layer, the most straightforward configuration is a free-space transmission measurement, in which a radiated EM wave is launched by a transmitting antenna, propagates through a material layer of some thickness, and is captured by a receiving system. If we assume plane-wave propagation and a plane-wave layered geometry, then the problem becomes a 1-D model, and the material layer under test can be modeled as a two-port device, with an overall transfer function which relates the output pulse to the input excitation pulse. This transfer function can then be used to unfold or determine the dielectric constant and loss tangent of the material layer.

The single-bounce or partial reflection coefficient at the front surface of a lossy dielectric layer can be modeled as

$$\Gamma(f) \equiv \frac{E_r(f)}{E_i(f)} = \frac{\eta - \eta_0}{\eta + \eta_0} = \frac{1-\sqrt{\varepsilon_r(1-jp_e)}}{1+\sqrt{\varepsilon_r(1-jp_e)}},$$

where $E_i(f)$ and $E_r(f)$ are the Fourier spectra of the incident transient wave and the reflected wave from the front surface. The TEM-wave impedance in the layer is η, with a dielectric constant of ε_r and an effective loss tangent of p_e. The corresponding complex propagation constant is

$$\gamma(f) \equiv \alpha + j\beta = \tfrac{j\omega}{c}\sqrt{\varepsilon_r(1-jp_e)},$$

and the resulting transfer function of the layer, based on a single-pass propagation pulse through it, is

$$H(f) \equiv \frac{E_t(f)}{E_{fs}(f)} = \left(1-\Gamma^2\right) e^{-(\gamma-\gamma_0)\,d}.$$

Here, $E_t(f)$ is the spectrum of the through pulse, and $E_{fs}(f)$ is the spectrum of the free-space reference pulse. Since the same receiving system is used for both waveform measurements, this can also be obtained as the ratio of the received-voltage spectra rather than the wave fields themselves (i.e., it isn't necessary to deconvolve the receiving system out of the measurements).

Figure 6 shows the magnitude of this transfer function for the fastest impulse excitation of each of the walls, with a gradually increasing loss versus frequency. Note that the transfer functions based on all the other pulse excitations were very consistent with this data, but this fastest impulse excitation yielded the cleanest spectral characterization over the widest bandwidth (from UHF to 15 GHz).

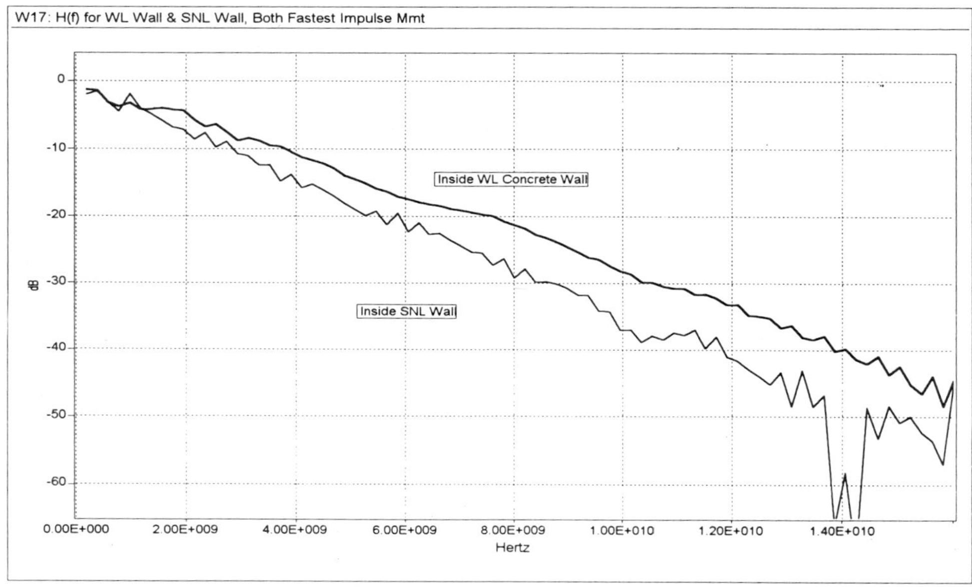

Figure 6. Transfer function for Sandia and Wright Labs' walls, based on fastest impulse excitations.

Then Figure 7 shows the corresponding impulse response of the walls, given by the inverse Fourier transform of the transfer function above. This is the pulse which would emerge from the concrete wall if excited with a perfect delta-function impulse TEM wave. There are several items to note about these waveforms. One is that the Wright wall has lower pulse attenuation and shorter time delay than the Sandia wall. The Wright wall has less attenuation and lower dielectric constant than the Sandia wall; this is evident in the frequency-domain transfer function as well. Note also that the primary impulse response is replica in nature, and that the average 50% pulse duration was 105 ps for the Wright wall and 146 ps for the Sandia wall. Thus, the Sandia wall has more time-domain dispersion or pulse spreading than the Wright wall. The last item to notice is the negative undershoot following the impulse portion of the response; we think that this is partially caused or influenced by the reinforcing bar inside the walls.

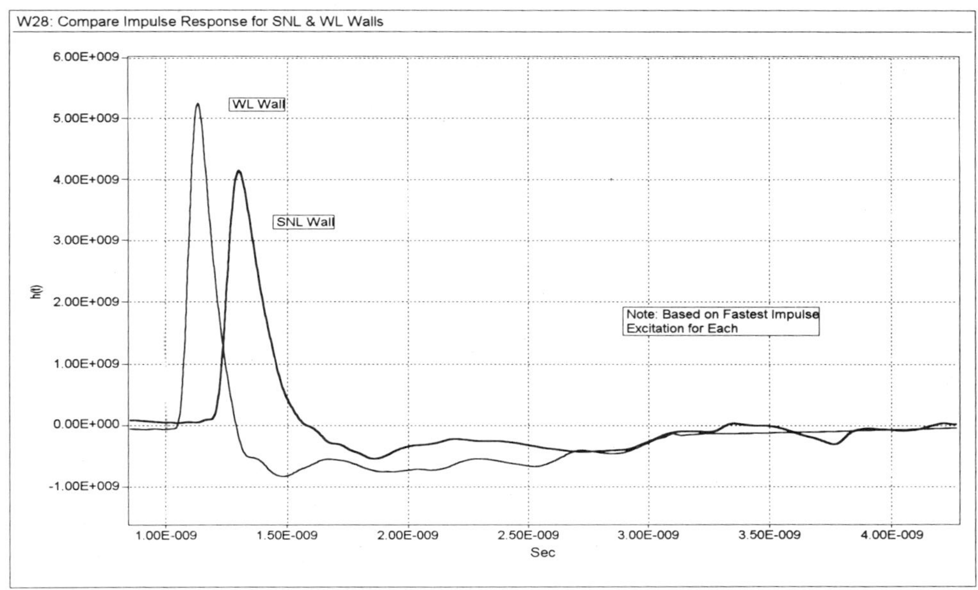

Figure 7. Impulse response of Sandia and Wright Labs' walls, based on inverse Fourier transform of transfer functions shown in Figure 6.

Space in this paper prohibits showing all of the data, but one more useful plot is provided in Figure 8 - this is the resulting attenuation constant $\alpha(f)$ in dB/m for each wall. It represents the volumetric loss of the concrete and re-bar structure, with the pulse reflection losses at each layer interface having been removed.

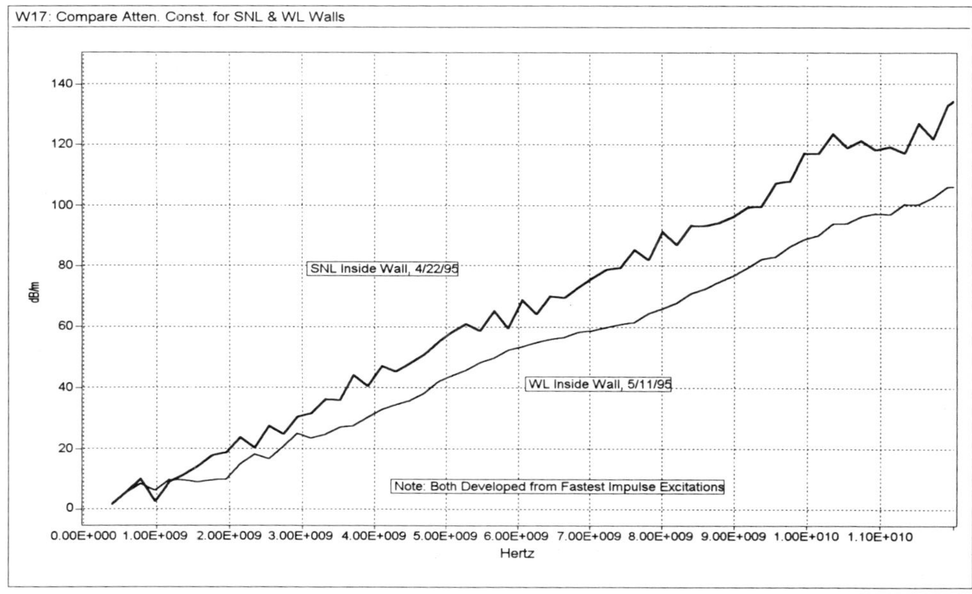

Figure 8. Attenuation constant of Sandia and Wright Labs' walls, based on the transfer functions shown in Figure 6.

The Sandia wall does indeed have higher loss than the Wright wall, with about 20 dB/m at 2 GHz and rising to about 110 dB/m at 10 GHz. The corresponding dielectric constant is fairly flat over this range, consistent with simple time-delay behavior in the impulse response.

The time-domain cross-correlation of the waveform pairs shows similar behavior as the derived impulse response, with the same insertion delays and relative pulse attenuation.

SUMMARY

We performed two series of very careful propagation measurements through two different 30-cm-thick concrete walls, utilizing a variety of transient time-domain pulses. The resulting transfer function and corresponding time-domain impulse response were very consistent with each other, and were additionally validated by frequency-domain vector analyzer measurements of the transfer function. This is the first comprehensive data we have seen for very wideband characterization of actual concrete walls with re-bar construction. The attenuation constant (in dB/m) increases in a fairly linear fashion with frequency, and the dielectric constant is fairly flat over the UHF to 15-GHz band. In addition, the time-domain behavior has been quantified, and will be very useful in time-domain radar studies for ground-penetrating radar, free-space layered-material measurement systems, etc.

ACKNOWLEDGMENTS

Thanks are due to several individuals who assisted in this experimental effort. Joseph Lundstrom helped modify the TEM-horn antennas for best EM performance (here at Sandia). Kwang Min (at Wright Lab) has been a primary supporter of this research, and Marcelious Willis (at Wright Lab) assisted in the measurement effort there.

PROPAGATION OF UWB ELECTROMAGNETIC PULSES THROUGH LOSSY PLASMAS

Steven L. Dvorak, Donald G. Dudley, and Richard W. Ziolkowski

Electromagnetics Laboratory
ECE, Bldg. 104, University of Arizona
Tucson, AZ 85721

INTRODUCTION

An efficient method for the analysis of ultra-wideband (UWB) electromagnetic pulses propagating through dispersive media is indispensable in applications involving UWB radar systems. Closed-form solutions have been obtained for the fields associated with transient sources radiating into lossless plasmas and waveguides (see [1] for historical references). Recently, Dvorak [2] demonstrated that contour integration techniques can be used to analytically evaluate the inverse Fourier transform representation for the potential associated with a continuous-wave transient pulse propagating in a waveguide. Dvorak and Dudley later extended the analysis to the problem of a double-exponential pulse propagating through a simple cold plasma [1].

In this paper, we investigate the propagation of an UWB double-exponential pulse through a homogeneous, lossy plasma which is used to represent a simplified model of the atmosphere. It is well known that an analytical expression exists for the frequency-domain field in a lossy plasma [3]. However, it does not seem possible to derive a closed-form expression for the corresponding impulse response. Thus, the standard procedure for the computation of the transient field involves application of a fast Fourier transform (FFT) to the frequency-domain solution. Unfortunately, due to the long tails in both the time and frequency domains, a large number of sample points are required to compute the transient response. In this paper, we introduce a new extraction technique which dramatically reduces the number of sample points required by the FFT.

Since the high-frequency behavior is similar in the lossless and lossy plasma cases, an analytical frequency-domain expression, which is similar in form to the one encountered for the lossless plasma and encompasses this high-frequency behavior, can be subtracted from the exact expression for the lossy plasma. The extracted term is evaluated analytically. The remaining expression, which can be transformed to the time domain with a FFT, requires only a modest number of sample points. This dramatically improves the numerical efficiency of the approach. Due to space limitations, we only investigate the electric field in this paper. However, the methods

Ultra-Wideband, Short-Pulse Electromagnetics 3
Edited by Baum *et al.*, Plenum Press, New York, 1997

which are developed in this paper can be directly extended to handle the associated transient magnetic field.

FREQUENCY-DOMAIN FIELDS

In this paper, we concentrate on the problem of plane wave propagation through a lossy ionosphere. We utilize a simple model of the ionosphere where it is assumed that the plasma is homogeneous, i.e., the number of free electrons does not depend on the altitude.

We employ a double-exponential pulse excitation at the location $z = 0$, i.e.,

$$E_x(0,t) = A[\exp(-\alpha_1 t) - \exp(-\alpha_2 t)]u(t), \tag{1}$$

where $u(t)$ is the Heaviside unit step function, A is chosen so that the peak amplitude of the pulse is unity, and $\alpha_1, \alpha_2 > 0$. Since the pulse reaches a maximum at

$$t_m = \frac{\ln(\alpha_1/\alpha_2)}{\alpha_1 - \alpha_2}, \tag{2}$$

we find that

$$A = [\exp(-\alpha_1 t_m) - \exp(-\alpha_2 t_m)]^{-1}. \tag{3}$$

If we ignore the earth's magnetic field, then a frequency-domain, x-polarized plane wave propagating in the z-direction can be represented in terms of the following inverse Fourier transform:

$$E_x(z,t) = F(\alpha_1, z, t) - F(\alpha_2, z, t); \quad z \geq 0, \tag{4}$$

where

$$F(\alpha, z, t) = \frac{1}{2\pi} \int_{-\infty}^{\infty} \tilde{F}(\alpha, z, \omega) \exp(j\omega t) d\omega \tag{5}$$

and

$$\tilde{F}(\alpha, z, \omega) = \frac{A \exp\left(-j\frac{z}{c}\sqrt{\omega^2 - \frac{j\omega\omega_p^2}{\nu + j\omega}}\right)}{\alpha + j\omega} \tag{6}$$

is the frequency-domain field for a single-exponential source. In the above expression, ν represents the electron collision frequency and $c = 1/\sqrt{\mu_0 \epsilon_0}$ is the speed of light in a vacuum. The plasma cutoff frequency, ω_p, is related to the physical parameters by

$$\omega_p^2 = \frac{Ne^2}{m\epsilon_0}, \tag{7}$$

where N represents the number of free electrons in a unit volume, m is the mass of an electron, and e denotes the charge. At altitudes above 100 km in the ionosphere, the electron density, plasma cutoff frequency, and electron collision frequency range between $10^8 < N < 10^{12}$ electrons/m^3, $6 \times 10^5 < \omega_p < 6 \times 10^7$ sec^{-1}, and $10 < \nu < 1 \times 10^5$ Hz, respectively [3].

CLOSED-FORM EXPRESSIONS FOR THE LOSSLESS CASE

The electric field in the lossless case can be written as

$$\lim_{\nu \to 0} F(\alpha, z, t) = Ae(\alpha); \quad z \geq 0 \tag{8}$$

where

$$e(\alpha) = \frac{1}{2\pi j} \int_{-\infty}^{\infty} \frac{\exp\left[j\left(\omega t - \frac{z}{c}\sqrt{\omega^2 - \omega_p^2}\right)\right]}{(\omega - j\alpha)} d\omega. \tag{9}$$

As was demonstrated in [2], the inverse Fourier transform in (9) can be evaluated using contour integration techniques. Later, a simpler differential-equation-based technique was developed [1]. It was determined that the canonical integral $e(\alpha)$ can be represented in closed form in terms of complementary incomplete Lipschitz-Hankel integrals (CILHIs) of the first kind:

$$\mathcal{J}e_n(a_\pm, \delta_\pm, \zeta) = \int_{\delta_\pm}^{\zeta} \exp(-a_\pm t) t^n J_n(t)\, dt, \tag{10}$$

where $\delta_\pm = \infty$ when $a_\pm \geq 0$ or $\delta_\pm = -\infty$ when $a_\pm < 0$. The desired closed-form, transient expression in the lossless case is given below:

$$e(\alpha) = \frac{u(t - z/c)}{2}\Big\{G \exp(a_+\zeta) + \frac{1}{\omega_p\sqrt{t^2 - (z/c)^2}}\Big[\Big(\alpha z/c - t\sqrt{\alpha^2 + \omega_p^2}\Big)\exp(a_+\zeta)$$

$$\cdot \mathcal{J}e_0(a_+, \delta_+, \zeta) + \Big(\alpha z/c + t\sqrt{\alpha^2 + \omega_p^2}\Big)\exp(a_-\zeta)\mathcal{J}e_0(a_-, \delta_-, \zeta)\Big]\Big\}, \tag{11}$$

where

$$a_\pm = \frac{-\alpha t \pm (z/c)\sqrt{\alpha^2 + \omega_p^2}}{\omega_p\sqrt{t^2 - (z/c)^2}}, \tag{12}$$

$$\zeta = \omega_p\sqrt{t^2 - (z/c)^2}, \tag{13}$$

and

$$G = 2u\left[\mathrm{Re}(a_+)\mathrm{Re}\left(\alpha z/c - t\sqrt{\alpha^2 + \omega_p^2}\right)\right]. \tag{14}$$

TRANSIENT FIELDS IN A LOSSY PLASMA

Unfortunately, it is not possible to obtain a closed-form electric field expression which is written in terms of readily computable special functions when losses are included. Furthermore, as was demonstrated in [1], a large number of sample points are required to numerically compute the transient fields using a FFT.

As is well known, asymptotic extraction techniques can greatly improve the numerical convergence of integrals and series. For example, application of asymptotic extraction techniques to (5) requires first finding an asymptotic expression, i.e.,

$$\tilde{F}^A(\alpha, z, \omega) = \lim_{\omega \to \infty} \tilde{F}(\alpha, z, \omega), \tag{15}$$

that can be integrated analytically in closed form. This term can then be subtracted from the integrand, thereby yielding an exact expression for the transient field that is in a form which is better suited for numerical computation, i.e.,

$$F(\alpha, z, t) = F^A(\alpha, z, t) + F^F(\alpha, z, t); \quad z \geq 0, \tag{16}$$

where $F^A(\alpha, z, t)$ denotes the inverse Fourier transform of $\tilde{F}^A(\alpha, z, \omega)$ and

$$F^F(\alpha, z, t) = \frac{1}{2\pi} \int_{-\infty}^{\infty} [\tilde{F}(\alpha, z, \omega) - \tilde{F}^A(\alpha, z, \omega)] \exp(j\omega t) d\omega \tag{17}$$

is the analytical expression to be computed numerically using a FFT algorithm. The goal of the asymptotic extraction technique is to subtract off the high-frequency behavior of the signal, thereby resulting in a more confined spectra that can more readily be handled using a FFT. We choose to use the high-frequency (early-time) behavior to develop our asymptotic expression $\tilde{F}^A$. Thus, $\tilde{F} - \tilde{F}^A$ will have a faster rate of decay at high frequencies than $\tilde{F}$. Therefore, the maximum frequency required to obtain accurate results when using a FFT algorithm is lower for (17) than it is for (5) which means fewer sample points are required for (17), as compared with (5). Consequently, the asymptotic extraction technique will dramatically improve the computational efficiency and accuracy of the FFT algorithm.

In addition to modeling the high-frequency behavior, the extraction term (i.e., $\tilde{F}^A$) must possess a real-valued, causal, analytical inverse Fourier transform. The following extraction term has these properties:

$$\tilde{F}^A(\alpha, z, \omega) = \frac{A(C_1 + j\omega)(C_1^* + j\omega)}{(\alpha + j\omega)(C_2 + j\omega)(C_2^* + j\omega)} \exp\left(-j\frac{z}{c}\sqrt{\omega^2 - \omega_p^2}\right); \quad 0 < z < z_{\max}, \tag{18}$$

where $C_1 = R_1 + jX_1$ and $C_2 = R_2 + jX_2$ are complex-valued coefficients, and $*$ denotes the complex conjugate operation. This extraction term is useful only out to a maximum propagation distance $z_{\max}$. An expression for $z_{\max}$ is obtained later. Note that $F^A(\alpha, z, t)$, the Fourier transform of (18), is real-valued since $\tilde{F}^A(\alpha, z, \omega) = [\tilde{F}^A(\alpha, z, -\omega)]^*$, and it satisfies causality provided all of the poles lie in the upper half of the ω-plane [2], i.e., $\mathrm{Re}(C_2) > 0$. In addition, a partial fraction expansion of (18) allows $F^A(\alpha, z, t)$ to be written in terms of the canonical integral (9), which possesses an analytical, closed-form solution (11).

If we employ the identity $e(C_2^*) = [e(C_2)]^*$, then it can be shown that

$$F^A(\alpha, z, t) = A\left\{D_1 e(\alpha) + 2\mathrm{Re}\left[D_2 e(C_2)\right]\right\}; \quad 0 < z < z_{\max}, \tag{19}$$

where

$$D_1 = 1 + \frac{b}{\alpha^2 - 2R_1\alpha + |C_2|^2} \tag{20}$$

$$D_2 = \frac{jb(\alpha - C_2^*)}{2X_2(\alpha^2 - 2R_1\alpha + |C_2|^2)} \tag{21}$$

$$b = \frac{z\omega_p^2\nu}{2c}. \tag{22}$$

The unknown coefficients in (18) are determined by equating $\tilde{F}^A$ to $\tilde{F}$ at high frequencies, i.e., the operation defined by (15). Since there are four unknown real-valued constants, we include all terms up to ω^{-4} in the asymptotic expansions. After equating terms in the asymptotic expansions, we find that

$$R_1 = R_2 = \frac{\nu}{2}, \tag{23}$$

$$X_1 = \sqrt{(\omega_p^2 + b)/2 - 5\nu^2/4}, \tag{24}$$

and

$$X_2 = \sqrt{(\omega_p^2 - b)/2 - 5\nu^2/4}. \tag{25}$$

Since X_1 and X_2 must be real valued, the asymptotic approximation in (18) can only be applied if

$$z < z_{\max} = \frac{c}{\nu}\left[2 - 5\left(\frac{\nu}{\omega_p}\right)^2\right]. \tag{26}$$

Since the asymptotic expressions were equated to fifth order, we denote (18) and (19) as the fifth-order, asymptotic frequency- and time-domain extraction terms, respectively. They can be applied whenever (26) is satisfied; thus the propagation distances over which we can apply the extraction technique is limited by the properties of the asymptotic ionospheric model in use, i.e., (6).

Asymptotic extraction techniques can still be applied when (26) is violated. However, we now employ a third-order asymptotic extraction. If we use (18) with $X_1 = X_2 = 0$, then the first three terms (i.e., up to and including ω^{-2}) in the asymptotic expansions can be equated, thereby yielding expressions for the unknown constants,

$$C_1 = C_2 = \frac{\nu}{2}. \tag{27}$$

Note that these results are valid for all values of z. Therefore, the appropriate third-order, frequency-domain extraction term is

$$\tilde{F}^A(\alpha, z, \omega) = \frac{A}{(\alpha + j\omega)} \exp\left(-j\frac{z}{c}\sqrt{\omega^2 - \omega_p^2}\right). \tag{28}$$

A comparison between (9) and (28) shows that the third-order extraction is equivalent to subtraction of the lossless result. Therefore, the desired transient third-order analytical extraction term is given by

$$F^A(\alpha, z, t) = Ae(\alpha). \tag{29}$$

Unlike the fifth-order extraction term (18) which has limits on its validity (26), the third-order result is valid for all z. Nonetheless, we utilize the fifth-order extraction term when possible since it further reduces the FFT requirements to obtain the desired time-domain results.

NUMERICAL RESULTS

We use the extraction technique to study in an efficient manner how losses affect the propagation of transient electromagnetic pulses in the ionosphere. We only vary the electron collision frequency, ν, and employ a propagation distance of $z = 500$ km. This is the approximate propagation distance from the earth's surface to a satellite. The time history and frequency spectrum for the double-exponential source are shown in Figure 1. The double-exponential pulse parameters are chosen as $\alpha_1 = 1.0 \times 10^7$, $\alpha_2 = 1.0 \times 10^8$, $A = 1.435$. The homogeneous plasma is modeled by a plasma cutoff frequency of $\omega_p = 1.0 \times 10^7$ rad/s.

In Figure 2a, we plot the frequency spectra for the electric field, and the spectra obtained by subtracting off the third- and fifth-order asymptotic results, i.e., $|\tilde{F}^F(\alpha_1, z, \omega) - \tilde{F}^F(\alpha_2, z, \omega)|$ using (28) and (18) respectively. In order to study the impact of loss on the technique, we plot the results for three different electron collision frequencies: $\nu = 1.0 \times 10^3$ Hz, $\nu = 1.0 \times 10^2$ Hz, and $\nu = 1.0 \times 10^1$ Hz.

The sharp drop in the frequency spectra at the plasma cutoff frequency observed in Figure 2a is associated with the high-pass filter properties of the plasma. Reference to (6) shows that losses will exhibit the greatest effect near the plasma cutoff frequency. This behavior is illustrated in Figure 2a where the electric field spectra (i.e., $|\tilde{E}_x| = |\tilde{F}(\alpha_1, z, \omega) - \tilde{F}(\alpha_2, z, \omega)|$) for the three different electron collision frequencies look very similar for $\omega \gg \omega_p$. Since the low-frequency components propagate at a slower velocity than the high-frequency components, losses exhibit the greatest effect on the late-time behavior of the transient waveform.

Figure 2a also indicates that the third- and fifth-order extraction spectra are reduced to a level which is 60 dB down from the peak level at frequencies which are

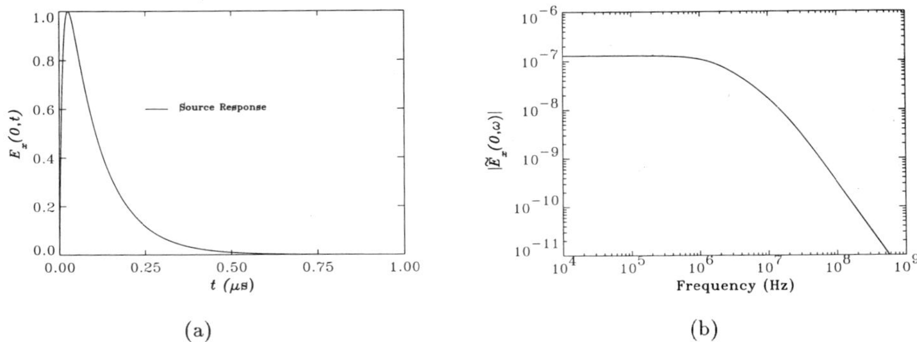

(a) (b)

Figure 1. The double-exponential pulse excitation which is used as the boundary condition at z=0.0 m. The double-exponential pulse parameters are chosen as $\alpha_1 = 1.0 \times 10^7$, $\alpha_2 = 1.0 \times 10^8$, and $A = 1.435$. (a) Source transient response. (b) Source frequency spectrum.

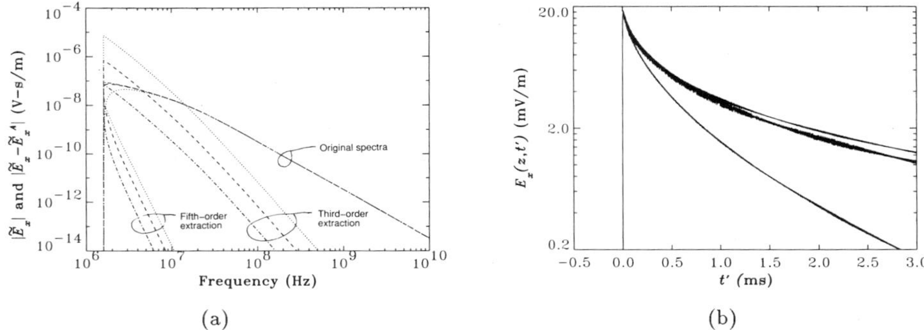

(a) (b)

Figure 2. Plots of the frequency- and time-domain fields associated with a double-exponential pulse that has propagated 500.0 km into a plasma characterized by $\omega_p = 1.0 \times 10^7$ rad/s. The three different line styles denote different electron collision frequencies, i.e., $\nu = 1.0 \times 10^3$ Hz (dotted line), $\nu = 1.0 \times 10^2$ Hz (dashed line), and $\nu = 1.0 \times 10^1$ Hz (dot-dashed line). (a) Frequency spectra for the electric field, and the spectra obtained by subtracting off the three- and five-term asymptotic results in (28) and (18), respectively. (b) Envelopes for the transient electric fields associated with the three different electron collision frequencies.

approximately two and three orders of magnitude smaller than those contained in the exact field spectra, respectively. This indicates that approximately two and three orders of magnitude fewer points are required to resolve the third- and fifth-order extraction spectra as compared with the original field spectra. Thus, the asymptotic extraction technique greatly improves the computational efficiency of the FFT. As expected, the fifth-order technique works better than the third-order one; however, the fifth-order technique is only valid if (26) is satisfied. It is interesting to note that both the third- and fifth-order extraction spectra fall off more rapidly for smaller values of ν. This is due to the fact that the asymptotic extraction term better approximates the true field spectrum when the losses are small.

In order to obtain the desired transient response, a FFT is used to evaluate (17). Since the high-frequency components have already been extracted from (17), the time history F^F is relatively slowly varying. The high-frequency content of the total transient signal is computed via the closed-form analytical solutions for either the third- or fifth-order extracted terms, i.e., (29) or (19), respectively. Since a denser time spacing is required for the FFT data, a cubic spline interpolation is used to obtain F^F at the additional required times. The interpolated FFT results are then added to the results produced by the closed-form analytical expression, i.e., see (29) and (19). The real advantage of this technique is that it allows for nonuniform gridding, which is not possible if a FFT is directly applied to the computation of the transient field. A large number of time steps can be computed at early times where the high-frequencies are important. At late times we only need to compute the field at the time steps naturally associated with the FFT.

Due to the highly oscillatory nature of the transient signal, we only plot the envelopes of the waveforms associated with propagation in a lossy plasma for three different values of ν in Figure 2b. We employed the fifth-order extraction term to compute this data. Note that $z_{\max} = 599,591$ m for $\nu = 1.0 \times 10^3$ Hz and $\omega_p = 1.0 \times 10^7$ rad/s., therefore, the fifth-order expansion is being pushed close to its limit. This figure clearly demonstrates the additional decay associated with the high-loss case. It also shows that the decay associated with losses is most prominent in the late-time (low-frequency) portion of the signal. In order to create Figure 2b, we employed an FFT with a maximum sampling frequency of $f_{\max} = 6 \times 10^6$ Hz and $N_{\text{FFT}} = 524,288$ sample points. These inputs correspond to a $\Delta t = 83.3$ ns time step and a 43.7 ms time history. Even with this large number of points, there are still aliasing errors present in the FFT data, i.e., the high-frequency oscillations which corrupt the smooth envelopes in Figure 2b. In order to obtain the correct high-frequency, early-time behavior, a cubic spline interpolation of the FFT data was used to fill in an additional 40 evenly-space time steps within the first FFT time step Δt, and we logarithmically tapered the number of additional time increments down to eight in the last FFT time interval at 3 ms. Thus, we computed 380,767 points analytically using (19). Note that only 35,800 of the 524,288 points produced by the FFT are included in the plot in Figure 2b. Unfortunately, the large number of points in the FFT is required to reduce the aliasing errors. Based on the results of Figure 2a, we believe that an FFT with approximately N_{FFT}=536.9 $\times 10^6$ points and $f_{\max} = 6 \times 10^9$ Hz would be required to reproduce this data if the asymptotic extraction technique was not employed. The large number of data points required by the FFT makes the $z = 500.0$ km propagation case intractable using a brute-force FFT computation.

Since the early-time portion of the waveform is probably of most interest, we plot its first few cycles in Figure 3. This figure clearly shows that the high-frequency components arrive at early times and the lower-frequency components arrive later. Note that the only FFT point that appears in this waveform is the point at $t' = 0$ s, and the field is zero at that point. Therefore, the closed-form, analytical result in (19) solely provides the early-time information. The FFT contribution is only important at late times.

CONCLUSIONS

In this paper, we developed third- and fifth-order asymptotic extraction tech-

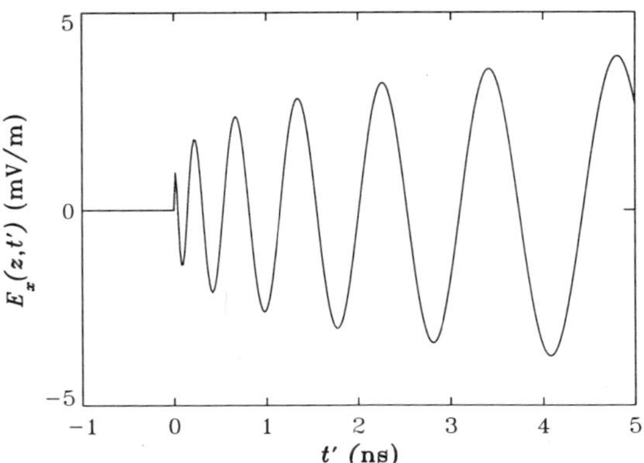

Figure 3. Early-time history for the electric field observed at $z = 500.0$ km for $\nu = 1.0 \times 10^3$ Hz.

niques for the problem of double-exponential pulse propagation through a homogeneous, lossy plasma. The high-frequency behavior of the pulse was subtracted out and the inverse Fourier transform was carried out analytically. The remaining contribution was evaluated numerically via a FFT. We found that the third- and fifth-order asymptotic extraction techniques reduced the number of points required by the FFT by a factor respectively of 100 and 1000. Because of the large reduction in the number of FFT sample points and flexibility of the analytical extracted signal, which allows for the computation of the field at non-uniform time steps, the asymptotic extraction technique was found to greatly improve the computational accuracy and efficiency for this important problem.

The fifth-order technique has a limited range of validity (see (26)) while the third-order technique is valid for all z. Both extraction techniques were found to provide better accuracy for smaller losses. The analytical extracted signal was found to accurately model the early-time portion of the waveform. Therefore, the analytical extracted signal can be considered as an early-time approximation for the transient pulse.

We are currently investigating the application of this technique to other problems. For example, we are looking at the development of asymptotic extraction techniques for pulse propagation through layered plasmas, Lorentz media, and Debye media. Our results from these efforts will be reported elsewhere.

REFERENCES

1. S. L. Dvorak and D. G. Dudley, "Propagation of Ultra-Wideband Electromagnetic Pulses Through Dispersive Media," *IEEE Trans. Electromagn. Compat.*, Vol. 37, No. 2, pp. 192–200, 1995.
2. S. L. Dvorak, "Exact, Closed-Form Expressions for Transient Fields in Homogeneously Filled Waveguides," *IEEE Trans. Microwave Theory Tech.*, Vol. 42, No. 11, pp. 2164–2170, 1994.
3. M. A. Messier, "A Standard Ionosphere for the Study of Electromagnetic Pulse Propagation," *EMP Theoretical Notes,* Note 117, 1971.
4. S. L. Dvorak and E. F. Kuester, "Numerical Computation of the Incomplete Lipschitz-Hankel Integral $Je_0(a, z)$," *J. Comput. Phys.*, Vol. 87, No. 2, pp. 301–327, 1990.

EARLY TIME SIGNATURE ANALYSIS OF DIELECTRIC TARGETS USING UWB RADAR

Shane Cloude[1], Alec Milne[1], Chris Thornhill[2], Graeme Crisp[2]

[1]Applied Electromagnetics
St. Andrews, KY16 9XD, Scotland, UK

[2]DRA Malvern
Great Malvern, WR14 3PS, England, UK

1. INTRODUCTION

In this paper we consider the problem of ultra wide band (UWB) electromagnetic wave scattering by a dielectric cylinder. Our main objective is to study the use of UWB Radar for the detection of non-metallic objects in noise and clutter. UWB Radars utilise a very wide transmitted bandwidth to ensure that the target scatters energy at very long, as well as at very short wavelengths. This provides information about the shape and material structure of the object which can then be utilised for detection and classification.

The scattering of EM waves by dielectric cylinders is a classical boundary problem with a well known convergent series solution[1]. Here we examine the details of the various phenomenological wave contributions to this solution, with a view to using very short pulse Radar to isolate the mechanisms in time, amplitude and polarization.

A key element of the wave solution for dielectric bodies is its increased complexity compared to metallic objects. This arises because of internal wave effects. Consequently, the classical Radar Cross Section (RCS) of dielectric targets varies rapidly with frequency, making it unlikely that narrow band systems could be used for reliable detection. However, by employing a broad band signal and by considering its time domain structure, we shall see that we can establish a relatively simple description of such objects.

Two features are important in establishing our approach to this problem: the first is that we concentrate on the early time response i.e. we consider the scattering response over a time interval which is of the order of the wave transit time through the object. Concentrating on this regime is advantageous for detection because it contains the bulk of the scattered energy. However, due to the presence of a forcing function (the incident wave) we cannot usefully employ a complex pole description of the object in this time window. Instead we consider development of a target template based on the time and amplitude structure of the various wave mechanisms which dominate the early time response.

The second key feature of our approach is that we consider only the case of backscatter. This greatly simplifies the analysis and matches closely the main application developments in Radar system design. By restricting ourselves to wave mechanisms which yield large scattered signals only in the backscattered direction, we shall see that we can characterise generic features of the UWB signature for such objects.

Ultra-Wideband, Short-Pulse Electromagnetics 3
Edited by Baum *et al.*, Plenum Press, New York, 1997

2. UWB SCATTERING BY DIELECTRIC CYLINDERS

We consider the problem of wave backscattering from a homogeneous cylinder of radius a, length L and refractive index n as shown in figure 1.

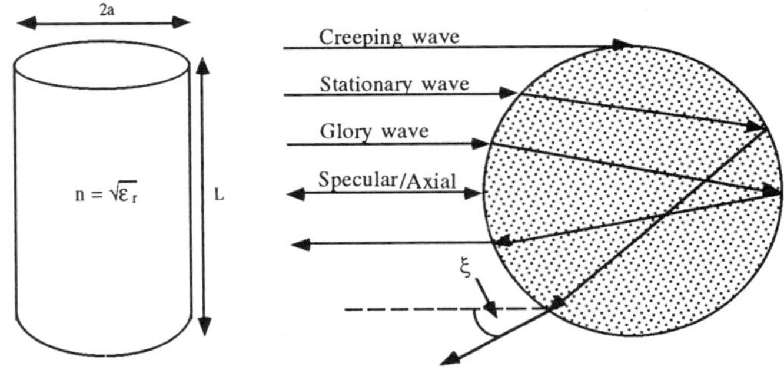

Figure 1: Cylinder geometry and Dominant Backscattering Mechanisms

We consider firstly the case of normal incidence when L is infinite (i.e. the 2-D cylinder problem). Here we can use a convergent series solution together with a 2-D finite difference time domain (FDTD) solver to examine the structure of the solution. We then consider the more general case of a finite cylinder, using a 3-D FDTD solver to generate a Synthetic Aperture Radar (SAR) image. Finally, we compare this solution with experimental results obtained on the DRA ISAR Range at Pershore[2].

In general, there are six important backscattering mechanisms to consider[1] (figure 1). In the time domain, the Radar equation for impulse backscatter from a dielectric cylinder can then be written in the form

$$V_{oc}(t) = \frac{D_r^2 V_{pk} \Gamma(t)}{8\pi^{1.5} c\tau R^2} \qquad \text{- 1)}$$

where R is the range to the target and V_{oc} is the open circuit voltage available at the antenna terminals. We assume the antenna feed pulse is a step of rise time τ and peak amplitude V_{pk}, radiated through an antenna with real aperture width D_r. The function G is related to the RCS of the target and for a dielectric cylinder can be written in the form

$$\Gamma_{\perp,\parallel}(t) = \frac{\partial}{\partial t}\sqrt{\pi a}\{\alpha^{s}_{\perp,\parallel}\delta(t) + \alpha^{a}_{\perp,\parallel}\delta(t-t_a)\,e^{-\beta_a t_a} + \alpha^{g}_{\perp,\parallel}\delta(t-t_g)\,e^{-\beta_g t_g}$$
$$+\,\alpha^{r}_{\perp,\parallel}\delta(t-t_r)\,e^{-\beta_r t_r} + \alpha^{c}_{\perp,\parallel}\delta(t-t_c)\,e^{-\beta_c t_c} + \alpha^{cg}_{\perp,\parallel}\delta(t-t_{cg})\,e^{-\beta_{cg} t_{cg}}\} \qquad \text{-2)}$$

We now consider each term in equation 2 (note that for simplicity we assume the material is loss-free and so all β terms expect β_c and β_{cg} are set equal to zero).

2.1 Specular Waves (α^s)

The specular return from the front of the cylinder is taken as the time reference for the signature. It has an amplitude given by the well known reflection formula

$$\alpha^{s}_{\parallel} = -\alpha^{s}_{\perp} = \frac{1-n}{1+n} \qquad \text{- 3)}$$

2.2 Axial Waves (α^a)

Internal propagation and reflection from the back surface gives rise to a response with amplitude and timing

$$\alpha^a_{\perp,\parallel} = T^{in}_{\perp,\parallel} R^{in}_{\perp,\parallel} T^{out}_{\perp,\parallel} \left(1 - \frac{2}{n}\right)^{-0.5} \qquad t_a = \frac{4na}{c}$$

$$T^{in}_{\perp,\parallel} = \frac{2}{1+n} \quad T^{out}_{\perp,\parallel} = \frac{2n}{1+n} \quad R^{in}_{\parallel} = -R^{in}_{\perp} = \frac{n-1}{n+1} \qquad -4)$$

where the last term in equation 4 breaks down for ε_r=4 due to internal focusing of the wave by the curved cylinder surface.

2.3 Glory Waves (α^g)

Waves which are refracted, reflected and then refracted into the backscatter direction are called glory waves. These are often the largest feature in the signature and are polarization dependent, so we consider their amplitude and timing in some detail as shown in equation 5

$$\alpha^g_{\perp,\parallel} = 2T^{in}_{\perp,\parallel}(\phi)\, R^{in}_{\perp,\parallel}(\psi)\, T^{out}_{\perp,\parallel}(\psi)$$

$$t_g = \frac{2a}{c}\left(2 + \frac{\varepsilon_r}{2}\right) \qquad \psi = \frac{\phi}{2} = \cos^{-1}\left(\frac{n}{2}\right)$$

$$T^{in}_{\parallel}(\Omega) = \frac{\frac{2}{n}\sqrt{1-\frac{\sin^2\Omega}{\varepsilon_r}}}{\cos\Omega + \frac{1}{n}\sqrt{1-\frac{\sin^2\Omega}{\varepsilon_r}}} \qquad T^{in}_{\perp}(\Omega) = \frac{2\cos\Omega}{\cos\Omega + n\sqrt{1-\frac{\sin^2\Omega}{\varepsilon_r}}}$$

$$R^{in}_{\parallel}(\Omega) = \frac{\sqrt{1-\frac{\sin^2\Omega}{\varepsilon_r}} - \frac{\cos\Omega}{n}}{\sqrt{1-\frac{\sin^2\Omega}{\varepsilon_r}} + \frac{\cos\Omega}{n}} \qquad R^{in}_{\perp}(\Omega) = \frac{n\cos\Omega - \sqrt{1-\varepsilon_r\sin^2\Omega}}{n\cos\Omega + \sqrt{1-\varepsilon_r\sin^2\Omega}} \qquad -5)$$

$$T^{out}_{\parallel}(\Omega) = \frac{2n\sqrt{1-\varepsilon_r\sin^2\Omega}}{\cos\Omega + n\sqrt{1-\varepsilon_r\sin^2\Omega}} \qquad T^{out}_{\perp}(\Omega) = \frac{2\cos\Omega}{\cos\Omega + \frac{1}{n}\sqrt{1-\varepsilon_r\sin^2\Omega}}$$

The most important feature to note about this wave is that it only exists for material in the range $2 \le \varepsilon_r \le 4$ but that over this range it is the dominant backscatter wave mechanism.

2.4 Rainbow Waves (α^r)

While rainbow or stationary waves are well known in optics, it is important to realise that significant backscattered energy can still result from these waves (especially for $\varepsilon_r > 2.5$), even though their ray paths are not exactly in the backscatter direction. In quantitative terms, this wave is only important for backscatter if $\xi \le 20^\circ$. The angle ξ is defined in figure 1 and the timing for the rainbow wave is given in equation 6

$$t_r = \frac{a}{c}\left((1-\cos\psi_0) + 4n\cos\psi_1 + (1-\cos\gamma)\right)$$

$$\cos\psi_0 = \sqrt{\frac{\varepsilon_r - 1}{3}} \qquad \cos\psi_1 = 2\sqrt{\frac{\varepsilon_r - 1}{3\varepsilon_r}} \qquad -6)$$

$$\gamma = 4\psi_1 - \psi_0 \qquad \xi = 4\psi_1 - 2\psi_0$$

2.5 Creeping Waves (α^c)

The creeping wave, guided by the curved surface of the cylinder, is a well known backscattering mechanism in UWB Radar and causes backscattered radiation with a timing

$$t_c = \frac{(2+\pi)a}{c} \qquad -7)$$

However, even for the case of a lossless dielectric, this wave maintains an exponential damping due to radiation loss around the surface of the cylinder. Hence, although it is important in the nonpenetrable target case, its effect on the early time signature of low loss targets is generally negligible compared to other wave mechanisms. It is still important however in the mixed glory/creeping wave to be discussed next.

2.6 Creeping/Glory Wave (α^{cg})

While the direct creeping wave contribution is small and the first glory wave only exists for $\varepsilon_r > 2$, there is always the possibility of a hybrid mechanism causing enhanced backscattering, as shown schematically in figure 2.

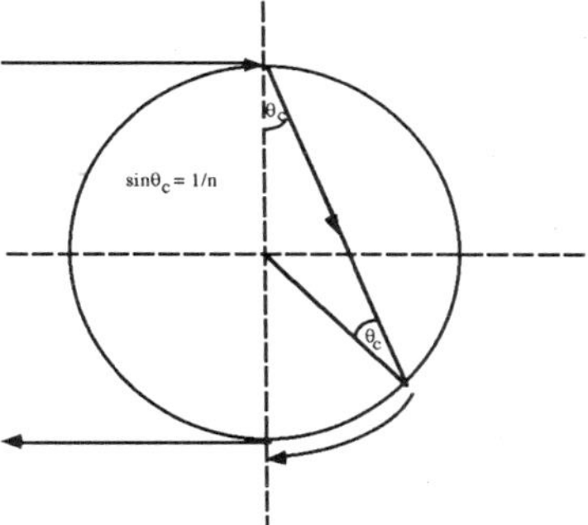

Figure 2: Creeping/Glory Wave Backscatter Mechanism

Here the wave is backscattered by first being coupled into the cylinder at the critical angle before striking the internal surface, again at the critical angle, after which either a second internal reflection takes place (important for $\varepsilon_r < 2$) or a surface wave is launched to propagate along a *portion* of the curved surface before radiating in the backscatter direction (important for $\varepsilon_r > 4$). The timing of this path can be calculated as

$$t_{cg} = \frac{2a}{c}(1 + n\cos\theta_c + \theta_c) \qquad -8)$$

where θ_c is the critical internal angle for dielectric ($\sin\theta_c = n^{-1}$). These two mechanisms become especially important outside of the material range dominated by the glory wave (equation 5).

Figure 3 shows the timings for these wave components as a function of ε_r in the range 2 to 4. Also shown are points derived as the maxima of signatures from FDTD simulations for a dielectric cylinder illuminated with a Gaussian pulse of fwhh = 0.1665 diameters (which matches the experimental results shown in figure 6). We see that across the range the glory/stationary wave combination is dominant.

Figure 4 shows a comparison between theory, computer modelling and experiment for the amplitude ratio of the signal maxima to the front specular return (α^g/α^s).

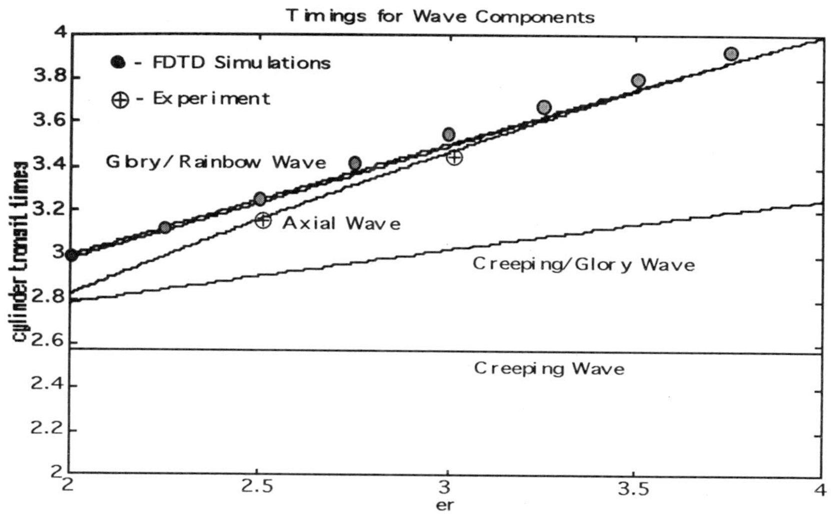

Figure 3: Timings for Scattered Wave Components

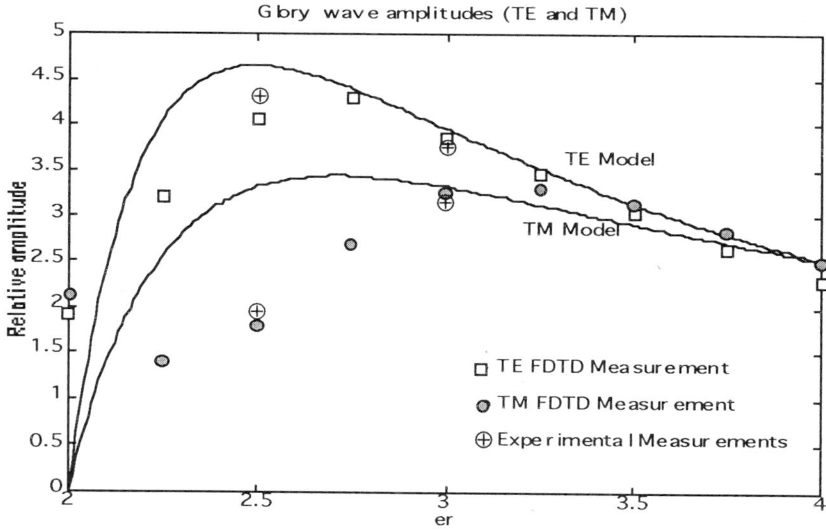

Figure 4: Relative Amplitudes for Glory wave

We see that there is good general agreement for TE but that the TM values are lower than predicted for small values of ε_r. Also note that there is poor agreement for both TE and TM at $\varepsilon_r = 2$. This can be explained as a failure of the simple model of equation 4 (which assumes zero transmission for tangential rays). Also shown in figure 4 are experimental measurements for $\varepsilon_r = 3$ and 2.5. The data for these points was obtained from an ISAR image at DRA Pershore (see figure 6).

3. UWB-SAR IMAGING OF DIELECTRIC CYLINDERS

We now turn to consider the more general case of a finite dielectric cylinder of length L. Figure 5 shows the variation of TE backscatter signature for a dielectric cylinder (where L = 1.2a). The intensity in this image is proportional to signal amplitude and the calculations were performed using a 3-D FDTD solver. We can see that around normal incidence ($\theta = 0^o$), specular waves from the top and bottom surfaces dominate the signature but that as the incidence angle increases, we encounter a wide range where the glory wave dominates the amplitude response. Hence we conclude that the glory wave contribution is still important in the 3-D case, especially at large angles.

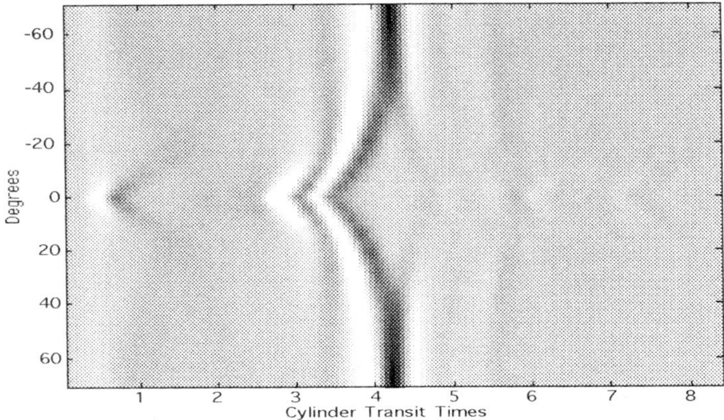

Figure 5: 3-D FDTD Simulation of Cylinder Backscatter vs Angle of Incidence

Such a dielectric cylinder was imaged on the indoor ISAR range operated at DRA Pershore[2]. This range is a time domain short pulse facility transmitting a Gaussian pulse of fwhh = 0.1 ns. The target is located on a straight rail across the aperture and target motion is translated via a k-space Fourier processing method[3] into a high resolution image of the target. Note that an autofocus technique has also been used to focus the image onto the brightest feature in the signature. The experimental configuration used was $\theta = 90^o$ for which this brightest feature is the glory wave.

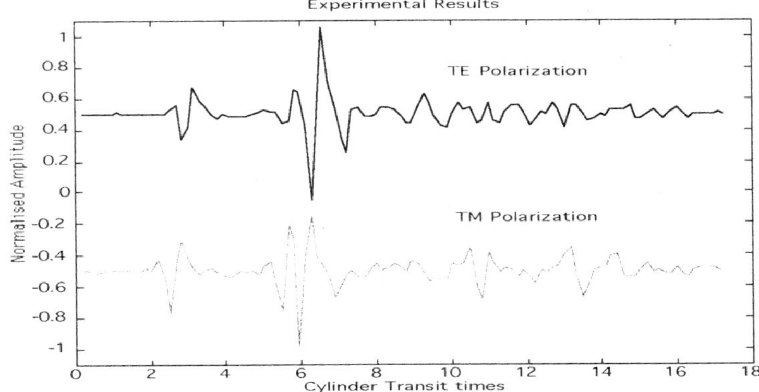

Figure 6: Experimental Data from ISAR Image of Dielectric Cylinder

Figure 6 shows experimental results for TE and TM polarizations. These time series were obtained directly from the autofocused ISAR images. We can see the front specular flash followed some time later by the much larger glory wave contributions. Data from these time series shows good agreement with theory as indicated in figures 3 and 4. We now consider how we might use these observations to design time domain templates for the detection of targets in noise or clutter.

4. DIELECTRIC TARGET DETECTION USING UWB SAR

We now turn to consider how the above observations may be used to establish detection probabilities for dielectric targets in noise and clutter. For the case of detection in noise we note that by using an UWB SAR, we can improve the signal to noise ratio by a gain factor G_S given by

$$G_s = \sqrt{\frac{Rc\tau f_p}{2D_r v}} \qquad -9)$$

which is valid for transmitted pulse widths such that $\frac{c\tau}{D_r} \leq 1$. There are two new important system parameters in equation 9; the platform velocity v and the pulse repetition frequency (PRF) of the source f_p. The ratio of these two is constrained by the sampling theorem so that

$$\frac{f_p}{v} \geq \frac{2}{D_r} \qquad -10)$$

and the PRF has an upper bound limited by the maximum unambiguous range of the Radar $f_p \leq \frac{c}{2R_{max}}$. Within these constraints we can then design a SAR system for CFAR detection using standard Radar signal modelling methods.

If the target response were limited to the front specular return, then G_S would be the only gain factor available for improved Radar detection. However, we have seen that the UWB signature has various wave contributions so that, in simplified form, we may model the target backscatter as a finite impulse response signature as shown in figure 7 (where we have taken only the first three terms of equation 4 i.e. specular, axial and glory waves)

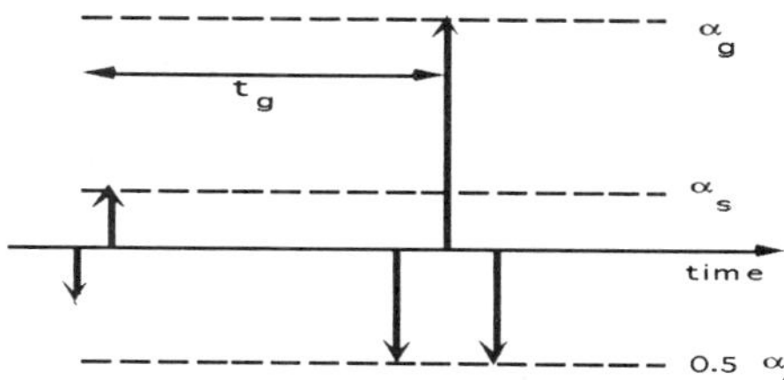

Figure 7: Simplified Template for UWB Cylinder Detection

Here we express the template as a doublet and triplet separated by the glory wave transit time. If we combine templates for both TE and TM polarizations then we obtain a maximum target signature detection gain G_T given by equation 11, where we have taken α_S as the reference for 0dB gain and have assumed that the Radar uses a step fed aperture antenna system.

$$G_T = \sqrt{4 + 1.5\left(\frac{\alpha_{g,\parallel}}{\alpha_s}\right)^2 + 1.5\left(\frac{\alpha_{g\perp}}{\alpha_s}\right)^2} \qquad -11)$$

Figure 8 shows the target gain calculated using equations 3, 5 and 11. Note that the fall off in gain for low values of ε_r will not be expected in practice because of other wave mechanisms becoming important (as observed in figure 4). The final processing gain for the UWB detection problem is given by $G = G_t G_S$. Calculation of these gain terms then enables a quantitative performance analysis of CFAR detection based on polarized broad band signals. Our main conclusion is that the glory wave enhances the detection potential of UWB systems.

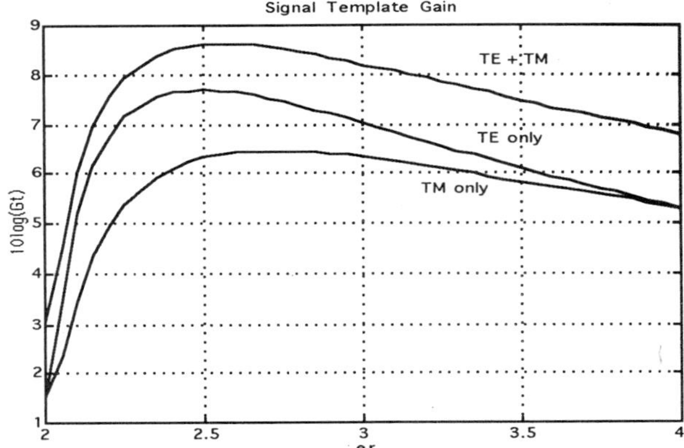

Figure 8: Template Gain for Cylinder Detection using UWB Radar

5. CONCLUSIONS

In this paper we have considered in detail the early time scattering behaviour of dielectric cylindrical targets. We have shown that the backscatter response, although complicated in the frequency domain, is easily characterised by a few parameters in the time domain. We have used a mixture of series solutions, 2-D and 3-D FDTD numerical predictions and experimental measurements made at DRA Pershore to study the relative amplitudes, timings and polarization dependence of the various wave scattering mechanisms involved.

In particular, we have demonstrated the importance of the glory wave in UWB dielectric target backscatter problems and illustrated how it may be used to design a simple FIR template for enhanced detection performance. It is important to bear such gain terms in mind when comparing the performance of UWB Radar with conventional narrow band detection systems. We have shown that through a combination of SAR Imaging and multi-polarization template design, significant processing gains can be achieved for target detection.

Future studies will be addressed at the more difficult problem of achieving sub-clutter visibility using UWB radar techniques.

REFERENCES

[1] "Radar Cross Section Handbook", G T Ruck, D E Barrick, W D Stuart C K Krichbaum, Plenum Press, 1970, Chapter 4

[2] "Analysis of Time Domain Ultra Wide Band Radar Signals", S R Cloude, P D Smith, A Milne, D Parkes, K Trafford, SPIE Volume 1631, Pulse Engineering, January 1992, pp 111-122

[3] "K-Space Imaging Algorithms applied to UWB SAR", S. R. Cloude, P.D. Smith, A. Milne, C. Thornhill, G Crisp, Proceedings of 1st IEEE International Conference on Image Processing (ICIP), Austin, Texas, pp 13-16 November 1994

CONSERVATION OF POWER IN THE GALERKIN APPROXIMATION OF THE ELECTRIC FIELD INTEGRAL EQUATION

Stuart M. Booker

Department of Mathematics
New Jersey Institute of Technology
Newark, NJ 07102-1982

INTRODUCTION

The problem of transient scattering from an arbitrary surface is an extremely important one with many potential applications. In order to solve this problem recourse must generally be made to a numerical method. One such approach which has received considerable attention recently is that of a time-marching algorithm based on the electric field integral equation (EFIE) (Rynne, 1991; Rao and Wilton, 1991). This approach has several advantages over other numerical methods, however, the solution obtained is prone to numerical instability. In order to suppress the onset of this instability Rynne and Smith (1990) introduced a scheme for time-averaging the solution for the current density obtained from the EFIE. These averaging schemes are reminiscent of Crank-Nicholson schemes for a numerical solution of the wave equation. Other considerations have also been found to affect the stability of numerical solutions. Rynne (1985) noted that the use of centred differences to approximate time derivatives tended to suppress the onset of solution instability. Whilst such methods have been shown to suppress (although not eliminate) the numerical instabilities observed in time marching solutions of the EFIE, a clear physical interpretation of these results has not been given.

It can be shown that the time domain EFIE gives rise to a power conservation law which has a clear physical interpretation based on the principle of conservation of energy. For a numerical solution scheme based on the EFIE to be accurate it is necessary (although not sufficient) for the form of this power conservation law to be preserved under the discretisation used. In this paper it is shown that the power conservation law of the EFIE is preserved under a discretisation if: Galerkin's method is used to discretise the spatial part of the problem; a Crank-Nicholson scheme is used to compute the surface current density; centred differences are used to approximate the time derivatives. The approach taken therefore clarifies the physical significance of the results cited above. In addition, a novel numerical scheme is derived for surfaces of finite conductivity which is consistent with the power conservation law of the EFIE. This scheme necessarily involves the solution of a second kind integral equation and possesses desirable stability properties.

FORMULATION OF THE PROBLEM

Consider an arbitrary surface, S, excited by an incident electric field, $\mathbf{E}_{\mathrm{i}}(\mathbf{r},t)$, and surrounded by a lossless, homogeneous medium of constant permittivity, ϵ, and constant permeability, μ. Denote by $c = 1/(\mu\epsilon)^{1/2}$ the speed of light in the surrounding medium. The incident field induces a surface current density, $\mathbf{J}(\mathbf{r},t)$, and surface charge density, $\rho(\mathbf{r},t)$, on S; these in turn give rise to a scattered electric field, $\mathbf{E}_{\mathrm{s}}(\mathbf{r},t)$. Denote by V a volume, bounded by the closed surface Ω, which completely encloses S. Denote by $\mathbf{J}_{\mathrm{v}}(\mathbf{r},t)$, $\mathbf{r} \in V$, a volumetric current density associated with the surface current density $\mathbf{J}(\mathbf{r},t)$, $\mathbf{r} \in S$, such that for an arbitrary vector field $\zeta(\mathbf{r})$, $\mathbf{r} \in V$,

$$\iiint_V \mathbf{J}_{\mathrm{v}}(\mathbf{r},t) \cdot \zeta(\mathbf{r})\, dV \;=\; \iint_S \mathbf{J}(\mathbf{r},t) \cdot \zeta(\mathbf{r})\, dS \;. \tag{1}$$

In order to determine the electric field, $\mathbf{E}_{\mathrm{s}}(\mathbf{r},t)$, scattered by S when it is excited by the incident electric field the usual magnetic vector potential, $\mathbf{A}(\mathbf{r},t)$, and electric scalar potential, $\phi(\mathbf{r},t)$, are introduced. The vector potential satisfies

$$\nabla^2 \mathbf{A}(\mathbf{r},t) - \frac{1}{c^2}\frac{\partial^2 \mathbf{A}}{\partial t^2}(\mathbf{r},t) \;=\; -\mu\, \mathbf{J}_{\mathrm{v}}(\mathbf{r},t) \;, \tag{2}$$

for all points $\mathbf{r} \in V$ and for all times $t \in \mathbf{R}$. This wave equation may be solved using a Green's function approach giving the integral representation

$$\mathbf{A}(\mathbf{r},t) \;=\; \frac{\mu}{4\pi}\iint_S \frac{\mathbf{J}(\mathbf{r}',\tau)}{|\mathbf{r}-\mathbf{r}'|}\, dS' \;, \tag{3}$$

where $\tau = t - |\mathbf{r}-\mathbf{r}'|/c$ is a retarded time. The scalar potential, $\phi(\mathbf{r},t)$, satisfies a similar wave equation and may be represented as a time-retarded surface integral of the surface charge density, $\rho(\mathbf{r},t)$. In addition to the wave equations for the vector and scalar potentials, Maxwell's equations yield an expression for the scattered electric field

$$\mathbf{E}_{\mathrm{s}}(\mathbf{r},t) = -\nabla\phi(\mathbf{r},t) - \frac{\partial \mathbf{A}}{\partial t}(\mathbf{r},t) \;. \tag{4}$$

Denote by $Z(\mathbf{r})$ the surface resistance per square at a point $\mathbf{r} \in S$. Applying to (4) the boundary condition $[\mathbf{E}_{\mathrm{i}}(\mathbf{r},t) + \mathbf{E}_{\mathrm{s}}(\mathbf{r},t)]_{\mathrm{tan}} = Z(\mathbf{r})\, \mathbf{J}(\mathbf{r},t)$, the electric field integral equation (EFIE) for a surface of arbitrary conductivity may be derived

$$Z(\mathbf{r})\, \mathbf{J}(\mathbf{r},t) + \left[\frac{\partial \mathbf{A}}{\partial t}(\mathbf{r},t)\right]_{\mathrm{tan}} = [\mathbf{E}_{\mathrm{i}}(\mathbf{r},t)]_{\mathrm{tan}} - \nabla_{\mathrm{s}}\phi(\mathbf{r},t) \;, \tag{5}$$

where the subscript 'tan' denotes the surface tangential component of a vector and where ∇_{s} denotes the surface gradient operator.

In addition to the EFIE, the continuity equation may be derived relating the surface current and charge densities. Given suitable initial and boundary conditions the EFIE (5) and continuity equation may be solved together with the integral representations for the vector and scalar potentials. A time-marching algorithm may be constructed from this set of equations and a numerical solution obtained (Rynne, 1991).

Before a suitable discretisation is determined for this problem, however, we consider the form of the conservation law which arises from the EFIE. Taking the scalar product of each term of (5) with $\mathbf{J}(\mathbf{r},t)$ and integrating over the surface S, we obtain

$$\iint_S \mathbf{E}_{\mathrm{i}}\cdot\mathbf{J}\, dS = \iint_S Z|\mathbf{J}|^2\, dS + \iint_S \mathbf{J}\cdot\nabla_{\mathrm{s}}\phi\, dS + \iint_S \mathbf{J}\cdot\frac{\partial\mathbf{A}}{\partial t}\, dS\,. \tag{6}$$

This conservation law has a clear physical interpretation. The term on the left hand side of (6) represents the rate at which power couples to the surface, S, from the incident field generating a surface current distribution, $\mathbf{J}(\mathbf{r},t)$. The transfer of power to S has two effects. First, the surface S is heated due to power dissipation by the surface impedance distribution, Z; this power loss is given by the first term on the right hand side of (6). Second, there is a reradiation of power in the form of a scattered electric field, as given by equation (4); the power of this scattered field is given by the final two terms of (6). This explicit representation of the power conservation law is similar to that derived by Amitay and Galindo (1969) for the time-harmonic scattering problem. It is necessary that the form of (6) be preserved under a discretisation of the problem if the resulting numerical scheme is to be accurate. Only then will the energy from the incident electric field which couples to S be conserved and manifested either as a scattered electric field or in the form of a surface heating of S. However, equation (6) is not itself an appropriate necessary condition for the accuracy of the discretisation procedure, as was suggested by Amitay and Galindo (1969). The reason for this is that the final term on the right hand side has a physical interpretation in terms of energy conservation which affects the choice of a suitable discretisation for $\mathbf{J}(\mathbf{r},t)$.

It may be noted that since the magnetic vector potential $\mathbf{A}(\mathbf{r},t)$ satisfies a wave equation it will satisfy a familiar conservation law. Denote by $\hat{n}$ an outward normal on Ω, the surface which bounds V, and by $\hat{n}_1$ and $\hat{n}_2$ outward normal vectors on either side of the surface S. Taking the scalar product of (2) with the time-derivative of $\mathbf{A}(\mathbf{r},t)$ and integrating over V we obtain

$$\begin{aligned}\frac{1}{2\mu}\frac{\partial}{\partial t}\int\int\int_V \frac{|\mathbf{A}_{\mathrm{t}}|^2}{c^2} + |\mathrm{grad}(\mathbf{A})|^2\, dV - \frac{1}{\mu}\int\int_\Omega \frac{\partial\mathbf{A}}{\partial n}\cdot\mathbf{A}_{\mathrm{t}}\, d\Omega \\ +\frac{1}{\mu}\int\int_S\left(\frac{\partial\mathbf{A}}{\partial n_1}+\frac{\partial\mathbf{A}}{\partial n_2}\right)\cdot\mathbf{A}_{\mathrm{t}}\, dS = \int\int_S \mathbf{J}.\mathbf{A}_{\mathrm{t}}\, dS\,,\end{aligned} \tag{7}$$

where the subscript 't' denotes the time derivative, and where the partial derivatives with respect to n, n_1 and n_2 are normal derivatives in the direction of $\hat{n}$, $\hat{n}_1$ and $\hat{n}_2$ respectively. By $\mathrm{grad}(\mathbf{A})$ denote the gradient tensor of the vector potential and by the scalar quantity $|\mathrm{grad}(\mathbf{A})|^2$ the componentwise contraction of $\mathrm{grad}(\mathbf{A})$ with itself. Equation (7) has a clear physical interpretation in terms of energy conservation. The first term on the left hand side of the equation corresponds to the time rate of change of energy stored in the magnetic vector potential field, $\mathbf{A}(\mathbf{r},t)$, in the exterior region V. The second term on the left hand side describes the power loss across the surface Ω due to the flux of vector potential out of V. The final term on the left hand side of the equation is a power reaction against the surface, S, which is necessary to preserve the non-smooth variation in $\mathbf{A}(\mathrm{r},t)$ at S. The right hand side of (7) describes the rate of generation of energy in the form of the vector potential field due to the surface current distribution $\mathbf{J}(\mathbf{r},t)$ on S. Since the term on the right hand side of (7) is the final term of (6), it is necessary that any discretisation of the scattering problem preserves the form of the conservation law for the vector potential (7) if it is to be accurate.

Spatial discretisation of the problem

Approximate S by a grid of triangular patches such that the grid possesses N_c such triangular elements separated by N_e internal edges; denote by T_m, $1 \leq m \leq N_c$, the triangular elements thus defined. Define on this grid a set of scalar basis functions associated with the triangular patches, and a set of vector basis functions, $\mathbf{f}_n(\mathbf{r})$, $1 \leq n \leq N_e$, associated with the internal edges of the grid. An appropriate choice for such basis functions would be those exploited for transient scattering problems by Rynne (1991), and Rao and Wilton (1991).

The surface current density, $\mathbf{J}(\mathbf{r}, t)$, may be approximated in terms of the vector basis functions as

$$\tilde{\mathbf{J}}(\mathbf{r}, t) = \sum_{n=1}^{N_e} I_n(t)\ \mathbf{f}_n(\mathbf{r}) \ , \quad \mathbf{r} \in S \ , \tag{8}$$

and the surface charge density, $\rho(\mathbf{r}, t)$, may be approximated in terms of the scalar basis functions in a similar manner. Using this approximation for $\mathbf{J}(\mathbf{r}, t)$ the magnetic vector potential may be represented in terms of the vector basis functions, via equation (3). Clearly, such a representation is unsuitable for a numerical approach since the expansion functions $I_n(\tau)$ would depend upon the retarded time $\tau = t - |\mathbf{r} - \mathbf{r}'|/c$, which varies continuously over the surface S with the source point $\mathbf{r}'$. The magnetic vector potential is therefore approximated by

$$\tilde{\mathbf{A}}(\mathbf{r}, t) = \sum_{m=1}^{N_c} \sum_{n=1}^{N_e} \frac{\mu}{4\pi} I_n(t - |\mathbf{r} - \mathbf{r}_m|/c) \iint_{T_m} \frac{\mathbf{f}_n(\mathbf{r}')}{|\mathbf{r} - \mathbf{r}'|}\, dS' \ , \quad \mathbf{r} \in V \ , \tag{9}$$

where $\mathbf{r}_m$, $1 \leq m \leq N_c$, is a set of reference points associated with the triangular elements T_m, $1 \leq m \leq N_c$ (for example, the centroids of the triangles). The scalar potential may be approximated in a similar manner, in terms of the representation for the surface charge density.

We may reformulate the original, continuous scattering problem as a semi-discrete problem (continuous in time but discrete in space). The spatial discretisation of the continuity equation and the integral representation for the scalar potential is not considered here; the approach of Rynne (1991) may be adopted for this part of the problem. In order to discretise the EFIE Galerkin's method is applied. Replacing each term in equation (5) by its semi-discrete approximation as defined above, we obtain the following residual error in the semi-discrete approximation of the EFIE:

$$\begin{aligned} \mathbf{R}_E \ &= \ \mathbf{E}_i(\mathbf{r}, t) - Z(\mathbf{r}) \sum_{n=1}^{N_e} I_n(t)\ \mathbf{f}_n(\mathbf{r}) - \nabla_s \tilde{\phi}(\mathbf{r}, t) - \\ & \sum_{m=1}^{N_c} \sum_{n=1}^{N_e} \frac{\mu}{4\pi} \frac{\partial I_n}{\partial t}(t - |\mathbf{r} - \mathbf{r}_m|/c) \iint_{T_m} \frac{\mathbf{f}_n(\mathbf{r}')}{|\mathbf{r} - \mathbf{r}'|}\, dS' \ . \end{aligned} \tag{10}$$

Galerkin's method ensures that this residual error is orthogonal to the set of basis functions in which the unknown current density $\mathbf{J}(\mathbf{r}, t)$, which is to be determined from the equation, is expanded. Hence, we enforce the following condition:

$$\iint_S \mathbf{R}_E \cdot \mathbf{f}_\nu(\mathbf{r})\, dS = 0 \ , \quad 1 \leq \nu \leq N_e \ . \tag{11}$$

This testing procedure gives rise to a system of N_e equations in the unknown expansion functions $I_n(t)$, from which the expansion functions may be determined. However, multiplying (11) by $I_\nu(t)$ and summing over the range $1 \le \nu \le N_e$ we obtain

$$\iint_S \mathbf{E}_i \cdot \tilde{\mathbf{J}} \, dS = \iint_S Z|\tilde{\mathbf{J}}|^2 \, dS + \iint_S \tilde{\mathbf{J}} \cdot \nabla_s \tilde{\phi} \, dS + \iint_S \tilde{\mathbf{J}} \cdot \tilde{\mathbf{A}}_t \, dS \,. \tag{12}$$

This equation is clearly the semi-discrete analogue of the explicit power conservation law (6). Furthermore, it is evident that any discretisation of the problem, based on an expansion for $\mathbf{J}(\mathbf{r}, t)$ with the form of equation (8), will give rise to a discrete conservation law which is analogous to (6). This was the observation of Amitay and Galindo (1969) for the time-harmonic scattering problem. However, the physical significance of the final term of equation (12) may only be realised if the discretisation used is consistent with equation (7), the power conservation law for $\mathbf{A}(\mathbf{r}, t)$. If this is not the case then the final term of (12) will not bear a credible physical interpretation and equation (12) will lose its physical significance as a power conservation law.

Denote by $P : V \mapsto S$ a continuous mapping with the property that $P(\mathbf{r}) = \mathbf{r}$, for all $\mathbf{r} \in S$, and define a volumetric current density by

$$\mathbf{J}_v(\mathbf{r}, t) = \mathbf{J}(P(\mathbf{r}), t) \, \delta(\mathbf{r} - P(\mathbf{r})) \,. \tag{13}$$

This volumetric current density satisfies property (1) and may be approximated in terms of the spatial basis functions $\mathbf{f}_n(\mathbf{r})$, $\mathbf{r} \in S$, via equation (8).

Using equation (13) and the approximation for the magnetic vector potential given in (9), a semi-discrete approximation of the wave equation for $\mathbf{A}(\mathbf{r}, t)$ may be obtained. Taking the scalar product of each term of the resulting equation with $\tilde{\mathbf{A}}_t(\mathbf{r}, t)$ and integrating over the region V, it may be shown that

$$\begin{aligned} \frac{1}{2\mu} \frac{\partial}{\partial t} \int\int\int_V \frac{|\tilde{\mathbf{A}}_t|^2}{c^2} + |\mathrm{grad}(\tilde{\mathbf{A}})|^2 \, dV - \frac{1}{\mu} \int\int_\Omega \frac{\partial \tilde{\mathbf{A}}}{\partial n} \cdot \tilde{\mathbf{A}}_t \, d\Omega \\ + \frac{1}{\mu} \int\int_S \left(\frac{\partial \tilde{\mathbf{A}}}{\partial n_1} + \frac{\partial \tilde{\mathbf{A}}}{\partial n_2} \right) \cdot \tilde{\mathbf{A}}_t \, dS = \int\int_S \tilde{\mathbf{J}}.\tilde{\mathbf{A}}_t \, dS \,. \end{aligned} \tag{14}$$

It is apparent, therefore, that Galerkin's method preserves the form of the conservation law (7) in a semi-discrete approximation of the transient scattering problem.

Temporal discretisation of the problem

In order to implement a full numerical solution of the transient scattering problem the semi-discrete system described above must be discretised in time. The resulting discrete system of equations may then be solved by a time-marching algorithm. It was shown in the last section that the power conservation law implied by the EFIE is preserved by a spatial discretisation of the problem based upon Galerkin's method. It will now be shown that the power conservation law implicit in equations (6) and (7) is preserved by a full discretisation of the problem if such a spatial discretisation is applied, and if the temporal discretisation is based upon a centred difference scheme with the surface current density being approximated by a Crank-Nicholson formula.

Let $t = k\Delta t$, k integer, and denote by ζ^k the value of the function $\zeta(k\Delta t)$; approximate the expansion function of the surface current density by

$$I_n(k\Delta t) \approx \frac{1}{4} \left(I_n^{k+1} + 2I_n^k + I_n^{k-1} \right) \,.$$

Using these approximations and a centred difference scheme to approximate the time derivative in equation (10) the following residual error in our discrete approximation to the EFIE may be obtained

$$\mathbf{R}_E = \mathbf{E}_{\rm i}^{\rm k}(\mathbf{r}) - Z(\mathbf{r}) \sum_{n=1}^{N_{\rm e}} \left(\alpha I_{\rm n}^{\rm k+1} + \beta I_{\rm n}^{\rm k} + \gamma I_{\rm n}^{\rm k-1}\right) \mathbf{f}_{\rm n}(\mathbf{r}) - \nabla_{\rm s}\tilde{\phi}^{\rm k}(\mathbf{r}) - \sum_{m=1}^{N_{\rm c}} \sum_{n=1}^{N_{\rm e}} \frac{\mu}{4\pi} \left(\frac{I_{\rm n}^{[\rm k+1]} - I_{\rm n}^{[\rm k-1]}}{2\Delta t}\right) \iint_{T_{\rm m}} \frac{\mathbf{f}_{\rm n}(\mathbf{r}')}{|\mathbf{r}-\mathbf{r}'|}\, dS' , \tag{15}$$

where $\alpha = \gamma = 1/4$ and $\beta = 1/2$. The superscript '[k′]' denotes a time-retarded expansion coefficient; i.e. $I_{\rm n}^{[\rm k']} = I_{\rm n}^{\kappa}$, where $\kappa\Delta t = k'\Delta t - |\mathbf{r} - \mathbf{r}_{\rm m}|/c$ (note that if $|\mathbf{r} - \mathbf{r}_{\rm m}|/c$ is not an integer multiple of Δt then $I_{\rm n}^{[\rm k']}$ must be linearly interpolated).

Applying Galerkin's method, equation (11), to this residual error gives rise to a system of $N_{\rm e}$ equations in the unknown expansion coefficients, $I_{\rm n}^{k'}$, k' integer. Multiplying each of these equations by $(I_\nu^{\rm k+1} + 2I_\nu^{\rm k} + I_\nu^{\rm k-1})/4$ and summing over the range $1 \le \nu \le N_{\rm e}$ results in the discrete analogue of the explicit conservation law (6). It only remains to show that the form of (7) is preserved under this discretisation.

The magnetic vector potential may be expressed in terms of the expansion coefficients for the surface current density as

$$\tilde{\mathbf{A}}^k(\mathbf{r}) = \sum_{m=1}^{N_{\rm c}} \sum_{n=1}^{N_{\rm e}} \frac{\mu}{4\pi} I_{\rm n}^{[\rm k]} \iint_{T_{\rm m}} \frac{\mathbf{f}_{\rm n}(\mathbf{r}')}{|\mathbf{r}-\mathbf{r}'|}\, dS' , \quad \mathbf{r} \in V . \tag{16}$$

Applying the above approximations to the wave equation for $\mathbf{A}(\mathbf{r}, t)$, equation (2), we obtain

$$\nabla^2 \left(\frac{\tilde{\mathbf{A}}^{\rm k+1} + 2\tilde{\mathbf{A}}^{\rm k} + \tilde{\mathbf{A}}^{\rm k-1}}{4}\right) - \frac{1}{c^2}\left(\frac{\tilde{\mathbf{A}}^{\rm k+1} - 2\tilde{\mathbf{A}}^{\rm k} + \tilde{\mathbf{A}}^{\rm k-1}}{\Delta t^2}\right) = -\mu \left(\frac{\tilde{\mathbf{J}}_{\rm v}^{\rm k+1} + 2\tilde{\mathbf{J}}_{\rm v}^{\rm k} + \tilde{\mathbf{J}}_{\rm v}^{\rm k-1}}{4}\right) , \tag{17}$$

where $\tilde{\mathbf{J}}_{\rm v}^{\rm k'}$, k' integer, is defined in the obvious manner. Taking the scalar product of each term of equation (17) with $(\tilde{\mathbf{A}}^{\rm k+1} - \tilde{\mathbf{A}}^{\rm k-1})/2\Delta t$ and integrating over the volume V, the discrete analogue of equation (7) may be determined. The right hand side of (17) gives rise to precisely the same expression as that which results from the final term of the residual error (15) in the discrete analogue of (6). Hence the discrete analogues of equations (6) and (7) bear the same physical interpretation as does the original conservation law given by (6) and (7).

It is clear, therefore, that a discretisation of the time domain EFIE which is based upon Galerkin's approximation, with time derivatives approximated by centred differences and in which the surface current density is averaged according to a Crank-Nicholson formula, has the following useful property. The discretisation used not only ensures a valid approximation of the electric field integral equation, it also ensures a valid approximation of the power conservation law associated with the EFIE. This means that the numerical scheme which results is not only consistent with an accurate solution for the surface current density on the scatterer, but it also ensures an accurate solution for the various energies which are associated with the vector potential.

It may be noted that whilst the discrete analogue of equation (6) will hold true for different discretisation schemes (for example, one based on forward differences, or without a Crank-Nicholson averaging of the surface current density) no discrete analogue of (7) may be found. Under such conditions the numerical solution scheme which results will not satisfy the physical principle of energy conservation which is implied by the EFIE (5).

NUMERICAL RESULTS

It may be noted that, in general, other choices of α, β and γ (where $\alpha + \beta + \gamma = 1$) may be made in equation (15). The choice $\alpha = \gamma = 0$, $\beta = 1$ requires the solution of a first kind integral equation for the unknown expansion coefficient $I_{\mathrm{n}}^{\mathrm{k}+1}$; such equations are generally ill-conditioned and numerically unstable. Booker et al (to appear) proposed the choice $\alpha = \gamma = 1/2$, $\beta = 0$, which requires the solution of a second kind integral equation, in order to ensure that the resultant solution scheme possesses desirable stability properties. It is apparent from the previous section, however, that the choice $\alpha = \gamma = 1/4$, $\beta = 1/2$ not only requires the solution of a second kind integral equation, for surfaces of finite conductivity, but also ensures that the correct power conservation law is obeyed by the vector potential.

In order to illustrate the effect of using a Crank-Nicholson formula to average the solution for the surface current density and of choosing $\alpha = \gamma = 1/4$, $\beta = 1/2$ in (15) as the basis for a numerical solution, the following scattering problem was considered. Let S be a 2×2 plate of constant surface impedance, Z, lying in the plane $x = 0$. The monostatic backscattered far-field due to the illumination of S by the incident field $\mathbf{E}_{\mathrm{i}}(\mathbf{r}, t) = \exp(x - ct)^2\ \hat{\mathbf{z}}$ was determined by three different solution schemes. Firstly, $\alpha = \gamma = 1/2$, $\beta = 0$ was taken in equation (15) as a basis for the numerical scheme, without averaging of the solution for the surface current density. Secondly, for the same choice of α, β and γ a Crank- Nicholson average of the surface current density was performed. Thirdly, $\alpha = \gamma = 1/4$, $\beta = 1/2$ was taken as a basis for the solution with Crank-Nicholson averaging of the surface current density. All solutions shared the same approximations (e.g. equation (9) was approximated by a simple quadrature after Rynne (1991)) and the same grid design, and solutions were calculated for 1000 time steps. The results of these simulations are summarised in Table 1 for three different values of the surface impedance, Z.

Table 1. Stability properties of solution schemes 2 and 3.

	Scheme 2		**Scheme 3**		**Comparison**
Z ($\Omega/\square$)	Peak field	Error, ϵ_2	Peak field	Error, ϵ_3	ϵ_3/ϵ_2
1.0	8.02×10^{-1}	$1.02 \times 10^{+3}$	8.02×10^{-1}	$5.77 \times 10^{+2}$	5.7×10^{-1}
5.0	7.56×10^{-1}	1.53×10^{-1}	7.55×10^{-1}	5.24×10^{-2}	3.4×10^{-1}
10.0	7.04×10^{-1}	5.03×10^{-6}	7.04×10^{-1}	3.70×10^{-7}	7.3×10^{-2}

It may be noted that all of the solutions obtained from scheme 1 (without averaging of the surface current density) became numerically unstable in the early time and caused a floating-point overflow after approximately 300 time steps. The introduction of Crank-Nicholson averaging (schemes 2 and 3) was sufficient to ensure that a solution was obtained over the full time interval in each simulation; in each case the solution was stable in the early time but became unstable in the late time. Table 1 shows the peak, early time far-field determined by schemes 2 and 3 together with an estimate for the late time error in the solution caused by the numerical instability. This error is taken as the maximum deviation of the numerically computed solution from the actual solution over 1000 time steps. All quantities are measured in arbitrary units. Also recorded is a comparison of schemes 2 and 3 given by the ratio of their respective errors.

As may be seen from Table 1, the results obtained from schemes 2 and 3 are in excellent agreement with each other before the onset of solution instability. As the surface impedance is increased the solutions obtained by schemes 2 and 3 become more stable; this is to be expected since the limit $Z \to 0$ of equation (15) gives rise to a first kind integral equation (Booker et al, to appear). However, it may also be noted that scheme 3 stabilises more rapidly than does scheme 2. It is clear from equation (15) that in the limit $Z \to 0$ all choices of α, β and γ will give rise to the same solution scheme (since the second term of the equation vanishes). As Z increases, however, the differences in solution stability which result from the particular choices of α, β and γ become more pronounced.

These results confirm that a Crank-Nicholson averaging of the surface current density helps to stabilise the EFIE solution scheme, as was demonstrated by Rynne and Smith (1990). They also suggest that by basing the choice of a solution scheme on the criterion discussed above (preserving the relevant conservation laws of the original system of equations) an optimally stable scheme may be obtained for the solution of transient electromagnetic problems via an EFIE approach.

CONCLUSIONS

It has been shown that the power conservation law which arises from the electric field integral equation can be satisfied by a discrete, numerical scheme provided that the scheme results from: a spatial discretisation based on Galerkin's method; a temporal discretisation based on a centred difference approximation of time derivatives, together with a Crank-Nicholson average of the solution obtained for the surface current density. The resulting numerical scheme is not only consistent with an accurate solution for the surface current density but also with the power conservation law associated with EFIE.

This approach clearly demonstrates the physical significance of the time-averaging scheme derived by Rynne and Smith (1990). The approach also suggests a discretisation procedure, for surfaces of finite conductivity, which ensures that the power conservation law associated with the EFIE is obeyed and which gives rise to a novel numerical scheme with desirable stability properties.

REFERENCES

Amitay, N., and Galindo, V., 1969, On energy conservation and the method of moments in scattering problems, *IEEE Trans. Antennas Propagat.* 17:747.

Booker, S. M., Lambert, A. P., and Smith, P. D., To appear, Calculation of surface impedance effects on transient antenna radiation, *Radio Science.*

Rao, S. M., and Wilton, D. R., 1991, Transient scattering by conducting surfaces of arbitrary shape, *IEEE Trans. Antennas Propagat.* 39:56.

Rynne, B. P., 1985, Stability and convergence of time marching methods in scattering problems, *IMA J. Appl. Math.* 35:297.

Rynne, B. P., 1991, Time domain scattering from arbitrary surfaces using the electric field integral equation, *J. Electromag. Waves Applic.* 5:93.

Rynne, B. P., and Smith, P. D., 1990, Stability of time marching algorithms for the electric field integral equation, *J. Electromagn. Waves Applic.* 4:1181.

SCATTERING OF SHORT RADAR PULSES FROM MULTIPLE WIRES AND FROM A CHAFF CLOUD

Herbert Überall[1] and Yanping Guo[2]

[1]Physics Department, Catholic University of America
Washington, DC 20064
[2]Johns Hopkins Applied Physics Laboratory
Laurel, MD 20723

INTRODUCTION

We here present a calculation of the transient, short-pulse response of perfectly conducting wires, and its extension to chaff clouds consisting of multiple, randomly distributed and oriented wires (modeled by up to 12,000 wires of equal length). This calculation is based on the analytic wire cross section formulas of Einarsson[1]. These have been programed by us using computer-generated random-number sets of position and orientation parameters, in order to obtain responses characteristic for the size of the chaff cloud, and its wire distribution in location and orientation.

Previous calculation of the scattering of short pulses by a single wire have been carried out by us[2] by evaluating the radiation of pulses of "traveling waves" that re-radiate from the wire ends as the traveling waves get reflected at the wire ends; in addition we discussed the resonance effects that occur when approximately an integer number of half-wavelengths of the traveling wave span the wire length. The scattering of electromagnetic waves from chaff clouds containing up to 1000 wires has also been considered by us[3-5] for the case of long incident pulses (of 40-cycle sine wave form), with a carrier frequency at the first wire resonance, and it has been shown how the return pulse shape (including the ringing of the resonance) depends on the size and number of wires in the chaff cloud. In addition, the dependence of the echo on incident and response polarizations was considered.

In the present study, we consider the backscattering of short (1 cycle sine) pulses from chaff clouds containing up to 12,000 wires. This is done for the case of vertical linear polarizations of both transmitter and receiver, and the dependence of the echo return shape and spectrum on cloud size and number of wires is obtained, as well as the dependence of the amplitude of the echo signal on the number of wires in the cloud. This is found to have a decreasing (non-linear) character, which, together with the echo shape and spectrum, may permit a possible remote determination of cloud size and wire distribution in the cloud, as well as a discrimination from other (solid) targets.

CALCULATION

Parameters

N = number of wires
τ_o = incident pulse duration, here the period of the assumed one-cycle incident pulse

Ultra-Wideband, Short-Pulse Electromagnetics 3
Edited by Baum *et al.*, Plenum Press, New York, 1997

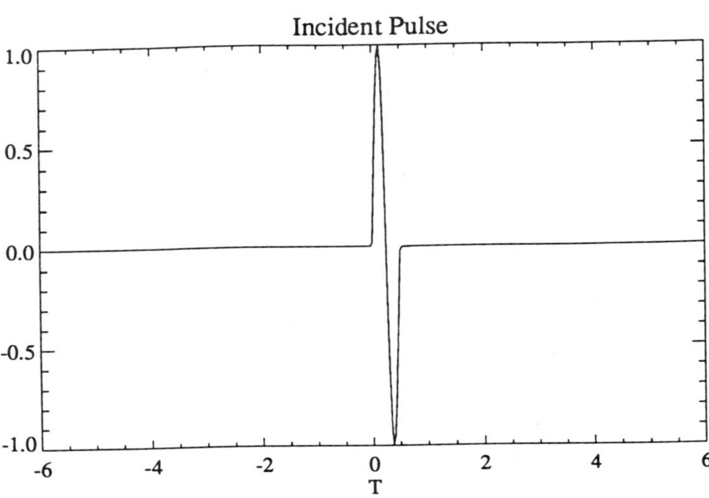

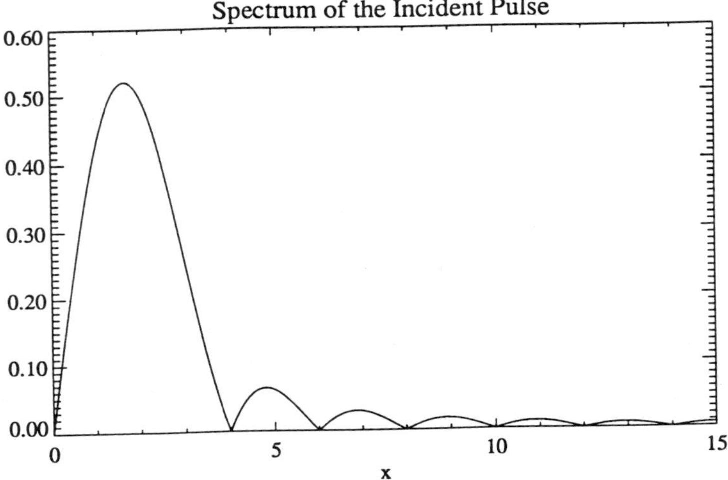

Incident pulse with T_0=0.5.

Fig. 1

form (shown in Fig. 1 with its spectrum)

$T_o = c\tau_o/\ell$ = normalized pulse duration

c = speed of light in air

ℓ = wire length

a = wire radius

λ = wavelength

x $= \ell/\lambda$ = normalized frequency

T = $(ct-r)/\ell$ = normalized time

r = distance of the receiver from the center of the chaff cloud, or the center of the wire for single-wire scattering

Assumptions for the calculation

(a) Backscattering; incidence in the negative x-direction, backscattering in the positive x-direction.

(b) Transmitter and receiver both polarized in the positive z-direction, and both far from the scatterer.

(c) No multiple scattering between wires, since the average wire spacing is assumed twice[6] the incident pulse wavelength ($T_o = 0.5$).

(d) Wires are distributed in space according to a Gaussian distribution function, wire orientations are randomly uniform over the 4π spherical angle.

Scattering formula

$$E_{sc}(T) = \frac{1}{\pi N} Re \int_0^{\infty} \frac{T_{0c}(1-e^{i2\pi NT_{\alpha}x})}{x[1-(T_{0c}x)^2]} \sum_{i=1}^{N} \Big\{ S(\Theta_i,\Theta_{0i}) P(\alpha_{Ti};\alpha_{Ri})$$
$$\times\, e^{-i\frac{2\pi}{l}x[(\sin\theta_T+\sin\theta_R\cos\phi_R)d_{xi}+\sin\theta_R\sin\phi_R d_{yi}+(\cos\theta_T+\cos\theta_R)d_{zi}]} \Big\} e^{-i2\pi xT}\, dx$$

For definitions, see Ref. 5; in particular, $T_{oc} \equiv T_o$, and $S(\Theta_i,\Theta_{oi})$ is the amplitude function for scattering from a single wire given by Einarsson's formula[1].

Scattering from a single wire

In order to gauge the contribution of an individual wire to the chaff formula, we show in Fig. 2 the short-pulse backscattering amplitude of a wire for the following cases:

(a) wire in the xz plane and parallel to the z axis ($\theta = 0°$)
(b) tilted from the z axis by $\theta = 20°$
(c) $\theta = 40°$
(d) $\theta = 60°$
(e) $\theta = 80°$
(f) wire in the yz plane, tilted by $\theta = 30°$.

For the cases (a) and (f) the large first pulse at T = 0 is due to specular reflection. The subsequent pulses are due to the reradiation of back-and-forth traveling pulses when they get reflected by the wire ends. In case (f) the wire is not in the polarization (xz) plane, and the echo amplitude is that projected onto the polarization plane.

Generation of randomly distributed wires

Wire location (d_x, d_y, d_z) and orientation (θ,ϕ) are generated by a random number generator with the following distribution functions:

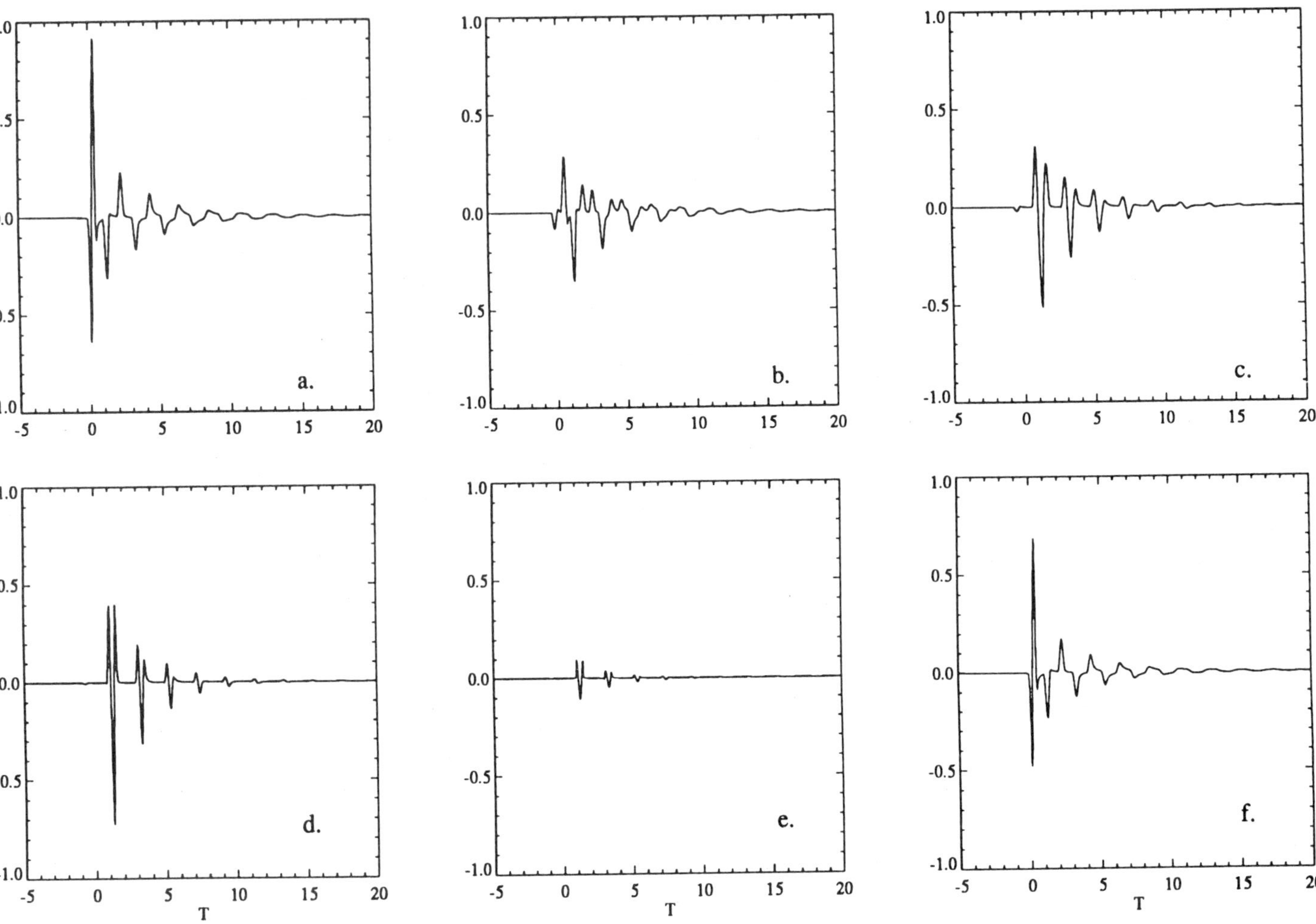

Fig. 2. Backscattering from a single wire with various orientation angles.

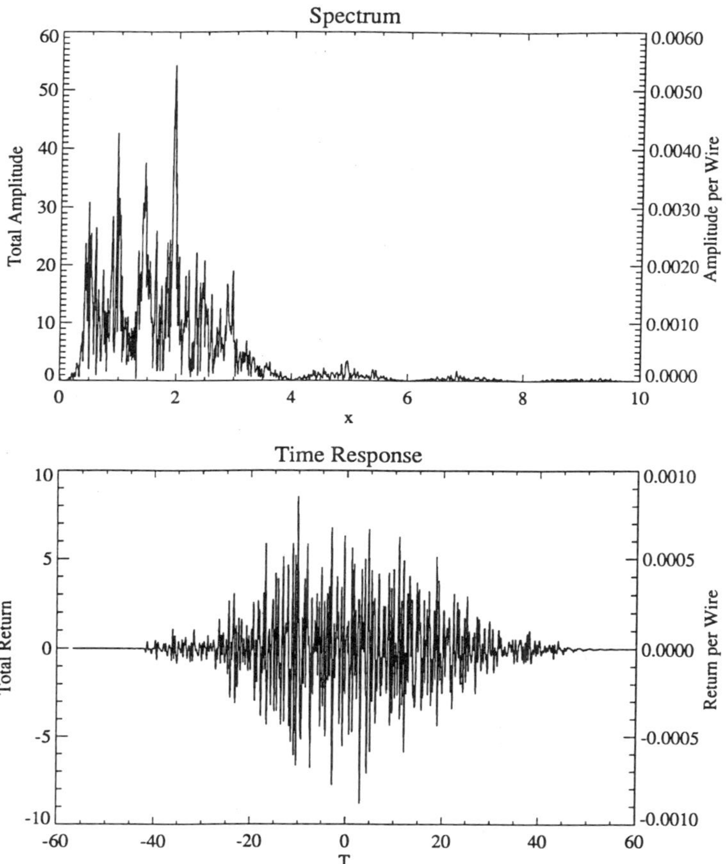

Fig. 3. Backscattering from 10000 wires with $T_0 = 0.5$ and $1/a = 400$.

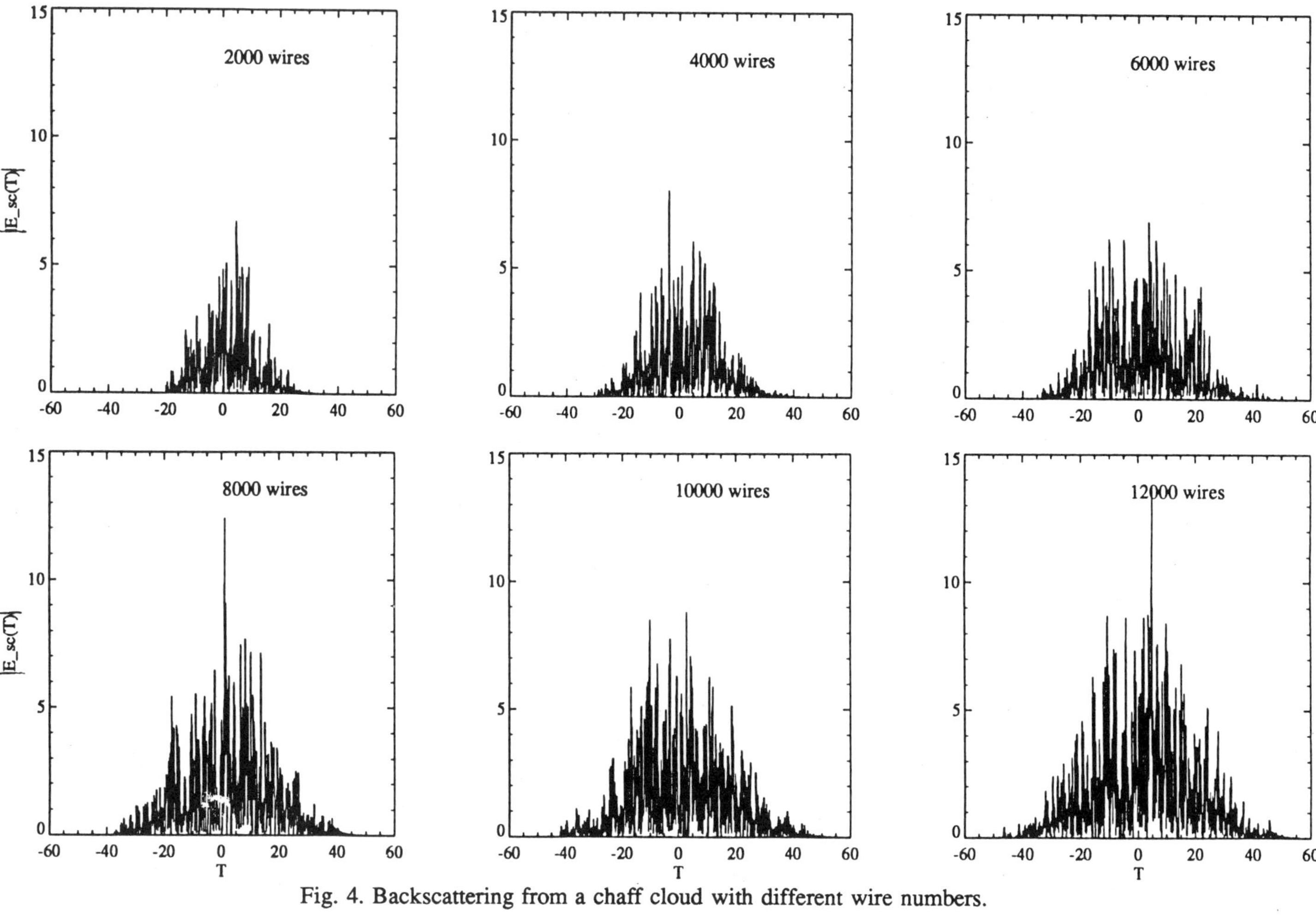

Fig. 4. Backscattering from a chaff cloud with different wire numbers.

Peak Amplitude
15
10
5
0
0
2
4
6
8
10
12
number of wires in thousand
Peak Amplitud per Wire
0.0040
0.0030
0.0020
0.0010
0.0000
0
2
4
6
8
10
12
number of wires in thousand
Average Amplitude
3.0
2.5
2.0
1.5
1.0
0.5
0.0
0
2
4
6
8
10
12
number of wires in thousand
Average Amplitud per Wire
0.0012
0.0010
0.0008
0.0006
0.0004
0.0002
0.0000
0
2
4
6
8
10
12
number of wires in thousand

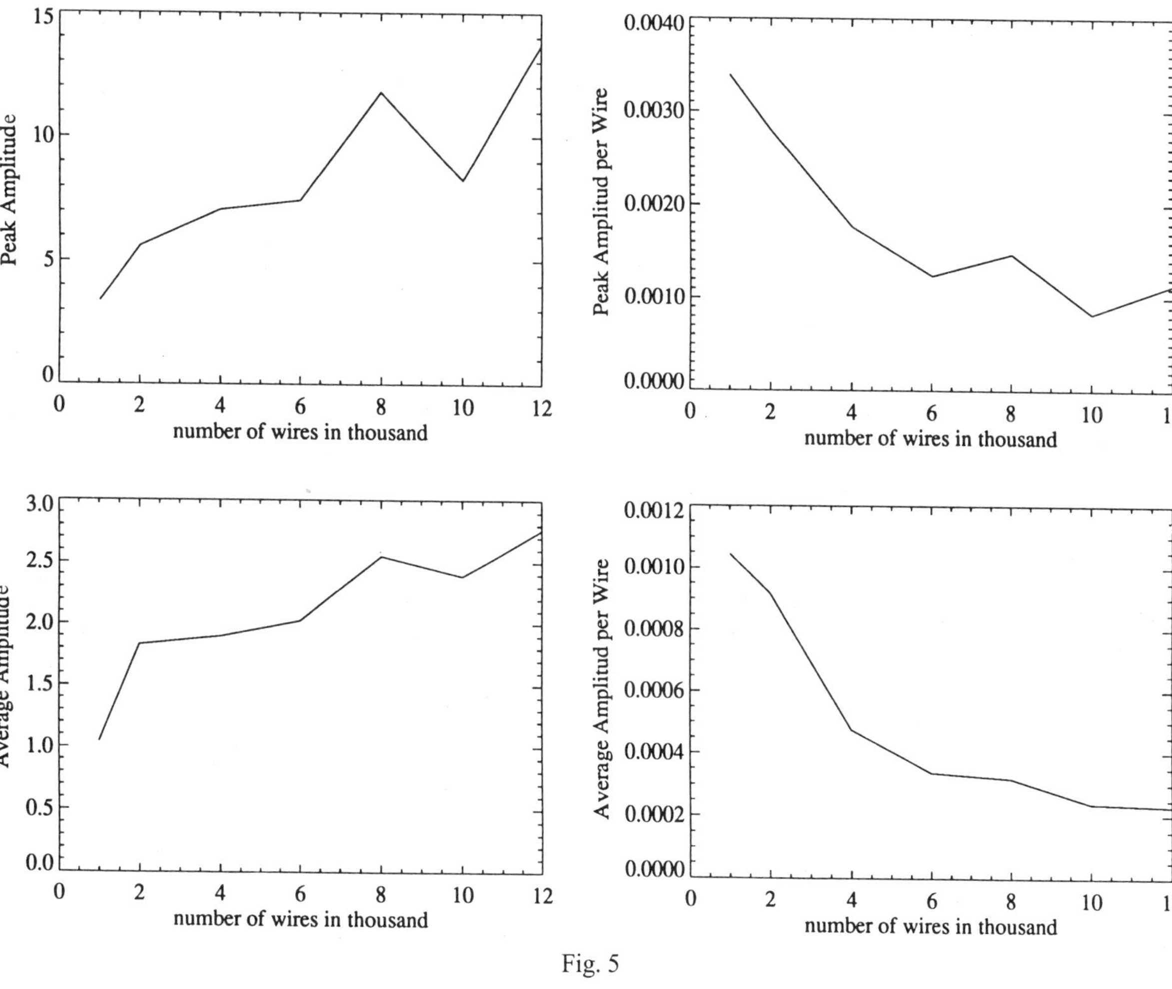

Fig. 5

$$p(d_x) = (2\pi)^{-1/2}\sigma^{-1} \exp(-d_x^2/2\sigma^2)$$
$$p(\theta) = (1/2)\sin\theta$$
$$p(\phi) = 1/2\pi$$

The standard deviation σ is related to the average wire spacing s by

$$s = 1.123(32\pi/3)^{1/3} N^{-1/3}\sigma$$

where N is the total number of wires in the chaff cloud. We here take $s = 2\lambda_o$ where $\lambda_o = c\tau_o = T_o\ell$ is the period of the incident one-cycle waveform.

RESULTS

Two calculations were done for each number of wires chosen, with different random locations and orientations; the results were found close to each other. Figure 3 shows the time response and spectrum for 10,000 wires, and Fig. 4 the time responses for 2,000 up to 12,000 wires. Since the wire spacings were kept constant, the lengths of the return pulses spread with wire number (and cloud size); the shapes start low (return from the rarefied cloud apex), they peak (dense center), and decline again. While the echo pulse shapes appear to consist of random fluctuations, the spectra (Fig. 3 top) clearly show the wire length resonance peaks (fundamental and overtones) which remain at the same x-values for all numbers of wires (not shown). These return pulses (shapes and spectra) are characteristic for extended chaff clouds, and should be quite different from return pulses from a large solid target such as a rocket or an airplane.

Figure 5 shows peak and average return pulse amplitudes, and amplitudes per wire, as a function of the number of wires. The average is taken over a time window corresponding to a standard deviation of the wire location distribution. The averages per wire are found decreasing (non-linear). This and the echo pulse length (and spectrum) may permit the remote sensing of cloud size and wire distribution.

We also calculated return pulses for chaff clouds of 100 wires with different incident pulse lengths (T_o = 0.1, 0.5, 1.0, and 1.5), not shown. Shorter (longer) pulses are found to have returns spread over smaller (larger) time intervals, with corresponding wider (narrower) spectral ranges, as expected.

References

1. O. Einarsson, The wire, "Electromagnetic and Acoustic Scattering by Simple Shapes" J. J. Bowman, T. B. Senior, and P. L. E. Uslenghi, eds. Hemisphere, New York (1987); Acta Polytech. Scan., Electr. Eng. Series, No. 23 (1967).
2. Y. Guo and H. Überall, The resonance effect of a wire in response to transient waves, *J. Electromag. Waves Applic.* **8**:355 (1994).
3. Y. Guo and H. Überall, Bistatic radar scattering by a chaff cloud, *IEEE Trans. Antennas Propag.* **AP-40**: 837 (1992).
4. Y. Guo; "Electromagnetic wave interactions with wires," Ph.D. dissertation, Catholic University of America, Washington, DC., University Microfilms International, Ann Arbor, MI (1992).
5. Y. Guo and H. Überall, Short pulse scattering from wires and chaff clouds, in "Ultra-Wideband, Short-Pulse Electromagnetics 2" (L. Carin and L. Felsen, eds.), Plenum Press, New York (1995).
6. R. G. Wickliff and P. G. Garbacz, The average backscatter cross section of clouds of randomized resonant dipoles, *IEEE Trans. Antennas Propag.* **AP-22**: 503 (1974); Addendum: ibid 842 (1974).

F.D.T.D. METHOD APPLIED TO THE GENERATION AND PROPAGATION OF SHORT PULSE

F. TRISTANT, F. TORRÈS,P. LEVEQUE
Pr. B. JECKO
Institut de Recherche en Communications Optiques et Microondes
CNRS-URA n°356
123, av Albert THOMAS
87 060 LIMOGES CEDEX FRANCE
E-mail : tristant@unilim.fr

D. SERAFIN
Centre d'étude de GRAMAT
46500 GRAMAT FRANCE

C. CRUCIANI[*],P. NOËL[**]
[*] C.T.S.N.-T.I.R.N. - BP 77 - 83 800 TOULON NAVAL
[**] D.C.N. Ing. Sud - BP 30 - 83 800 TOULON NAVAL FRANCE

INTRODUCTION

Directed beam weapons concept has emerged in recent years as a new aggression and it involves new approaches to protect ships. This study is realized in collaboration with the French Navy Technical Department : The DCN/CTSN TIRN in TOULON (FRANCE). So this paper aims to present methods to characterize current navy structures when struck by directed beam weapons.

OUTLINE

We can split the directed Beam Weapons into two types : High Power Microwave (H.P.M) and Ultra Wide Band (U.W.B) with different electromagnetic characteristics. Compared with NEMP, we can notice a frequency shift up and a power elevation. The aim of the study is to assess the coupling of these new aggressions with ships.

In this paper, the Finite Difference in Time Domain method applied to Maxwell's equations to compute the Electromagnetic fields. It involves the whole structure to be modelled with 3D meshes. But with the structure, the number of cells exceeds computers' capacities, because of its large dimensions compared to the wavelength (figure 1 shows a frigate modelled with FDTD code applied to NEMP aggression).

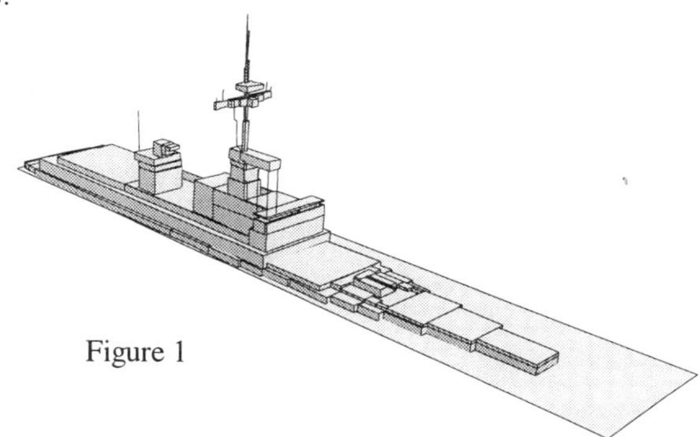

Figure 1

Ultra-Wideband, Short-Pulse Electromagnetics 3
Edited by Baum *et al.*, Plenum Press, New York, 1997

Since we cannot model the whole ship struck by this aggression, we will only study the coupling of the wave onto an alone device (for example an antenna) removed away from the rest of the ship.

Experiments have been conceived to perform such a survey :

-the first at GRAMAT(CEG) with a high power microwave device corresponding to a front door coupling.

-the second in TOULON (DCN/CTSN TIRN,France) with a characterisation bench to test marine panels to a back door coupling.

The first step of our survey is to model in our FDTD code these experiments and to study the characteristics of the field emitted by these devices. Thus we will understand all the emerging phenomena.

Then, we will analyse the coupling mechanism of the wave onto the antennas and new structures to be tested.

The results in this paper concern only the first step.

TWO COUPLING MODES

The main ways of coupling of these directed beam weapons[1] are :

-the front door coupling (Figure 2), which appears when the incident wave is coupled to antennas both within and outside its frequency range.

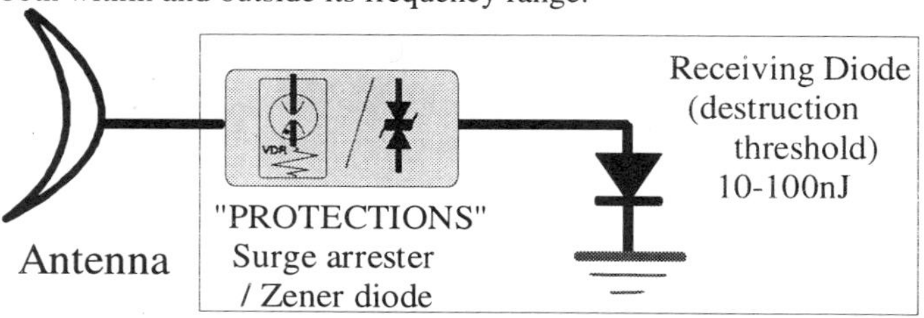

Figure 2

-the back door coupling includes (Figure 3) : penetration across the hull or the walls by conduction, electromagnetic coupling through open and sealed apertures, and indoor coupling by electric coupling with electric cables or their shielding screen.

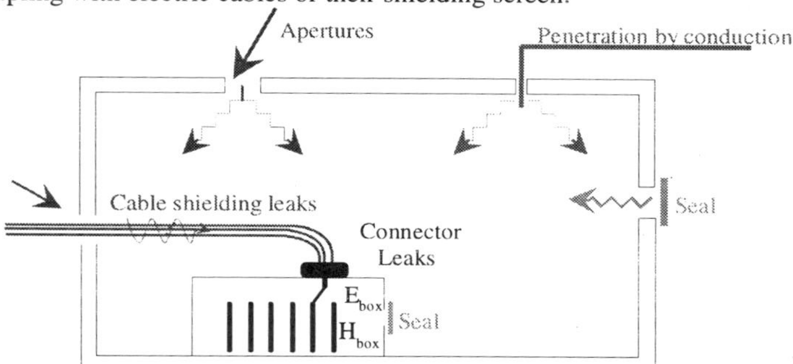

Figure 3

FRONT-DOOR EXPERIMENTATION

This experimentation has been realised at CEG (Centre d'étude de GRAMAT, FRANCE).We can observe on Figure 4 the H.P.M. anecoïc chamber. All the H.P.M. devices exceed 100 MW in peak power and their frequency range is within 1 GHz and 30 GHz in narrow band, and currently higher. Here in this device, the high frequency oscillator works in the 2.1-3.3 GHz band. We extract this wave into two guides, and the adaptation between the rectangular guide and the free space is performed by pyramidal horn antennas.

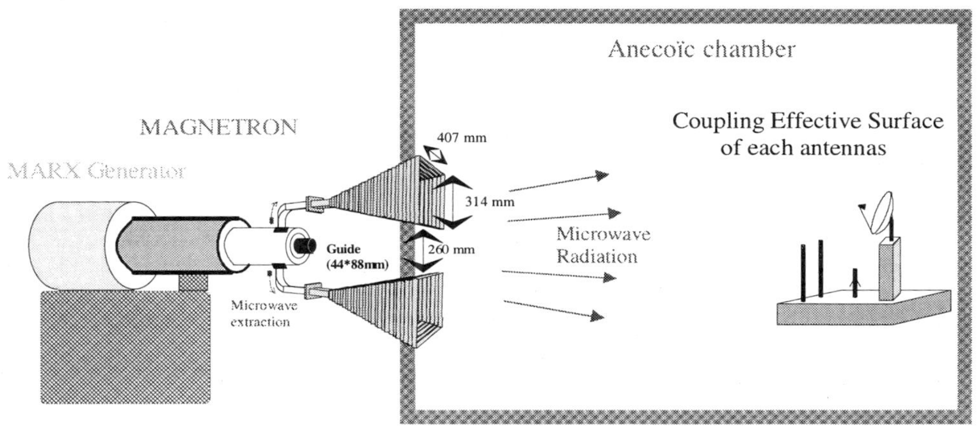

Figure 4

Then, we place in the anecoïc chamber test zone some structures struck by this high frequency and high power pulse. We wish to simulate a real case, that means a distance of about 10 kms between the source and the target ; therefore we must cut down the high level of radiated power and energy densities. It is impossible to modify the power of the source (MARX generator and H.F. oscillator), so the solution is to put a dielectric strip in front of the radiating device to absorb an important part of this power. We wish to get a 10 dB decrease of the field without both signal distortion in the test zone and an important reflection on the strip because we don't want to modify the adaptation of the antennas.

Simulation of this experiment

Antennas are the interface between the H.P.M. source and free space. The two features of H.P.M that stress conventional antenna[2-3-4-5] technology are the high power and short pulse duration. Consequently, H.P.M. antennas have been direct extrapolations of conventional antenna technology, usually in its simplest form, with allowance made for high electric field effects and for the shortness of the pulse.

The radiation field of a horn antenna is not a simple spotlight[6], but has several regions. In the reactive near-field region, the electromagnetic field is not yet fully detached from the antenna (Figure 5). This is followed at $0.62\sqrt{D^3/\lambda}$ by the intermediate radiating near field region where the beam is cylindrical, reaching a peak at roughly $0.2D^2/\lambda$, depending on the antenna's aperture and the wavelength.

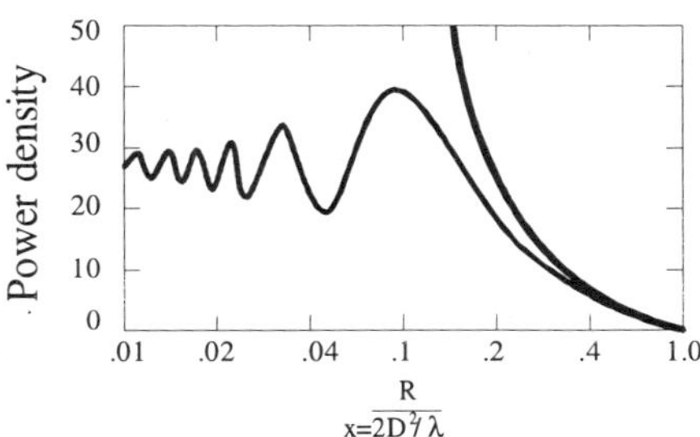

Figure 5.
Intensity of beam from a horn antenna normalised to unit at $0.2D^2/\lambda$.

There are three different radiating zones for a horn antenna (figure 9) :

- Rayleigh's zone
- Fresnel's zone
- Fraunhoffer's zone

: ➾ Far zone $R >> 2D^2/\lambda$
D = max. aperture of antenna(s)

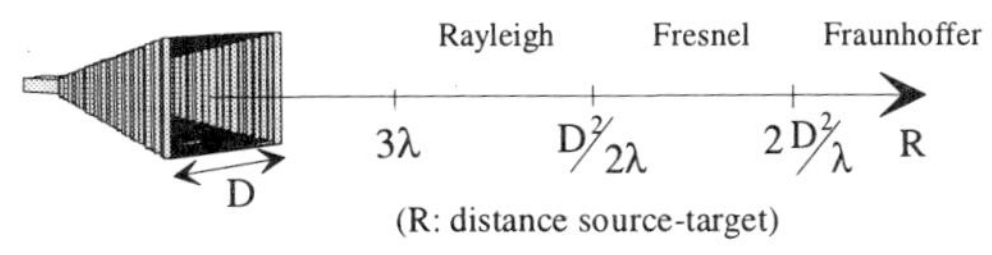

Figure 6

The first work was carried out with pyramidal horns. We simulate both the guide and the horn including perfectly matched layers to simulate free space[7-8]. The guide is excited on its first propagation mode. The mode considered in this case (TE01) is placed in a transverse plane's guide, so that we only have an electric field repartition in $\sin(\pi x/a)$. These static values are modulated by a sinusoid at the frequency "f0", itself modulated by a gaussian covering the bandwidth of the considered mode (Figure 5). This bandwidth will be chosen according to the cutoff frequency (determined by the geometry of the guide) in order to minimise the wave dispersion in the guide. Because perfectly matched layers can be ineffective for near cutoff frequencies waves and evanescent waves[9].

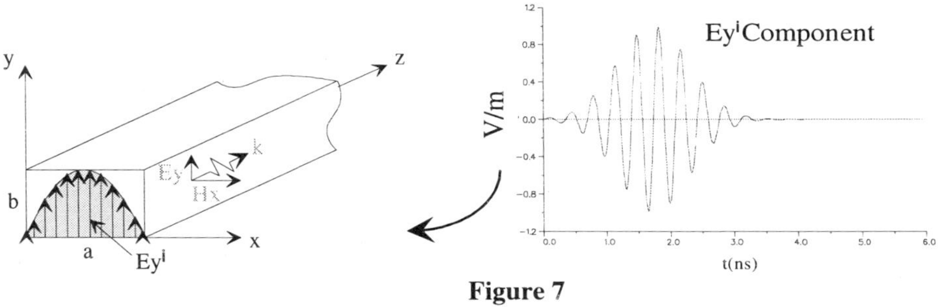

Figure 7

In the second work, FDTD is used to simulate the whole High Power Microwave radiating device. We have added Teflon right against the horn aperture in order to create vacuum inside because breakdown phenomena can appear due to the field's high level existing in these antennas, and of course, the dielectric strip which makes the field level decrease. In our code, we represent this material by Debye model[10].

We must choose the best position and dimension of this strip in front of the radiating device not to add distorting signal in the test zone. And we achieve this study to compute the reflection and transmission coefficients. The power density absorbed by this strip depends on three parameters : the frequency, its geometry and its electromagnetic characteristics (permittivity $\varepsilon(f)$, conductivity $\sigma(f)$). Two strips have been selected to decrease the field of -10dB : the thinner one presents an important reflection coefficient and the second one give a good compromise with an important thickness.

We can notice (on figures 6-7) an important desadaptation of the horn in free space, increased with the TEFLON (figure 6). And it is important to note that both the thicker (LS14) and the thinner (LS24) dielectric strips have a little influence on this desadaptation (figure 7). But, for example, we can observe, at 2.7 GHz that half the power is reflected into the guide, and the generator and the H.F. oscillator running must not be disturbed by this phenomenon.

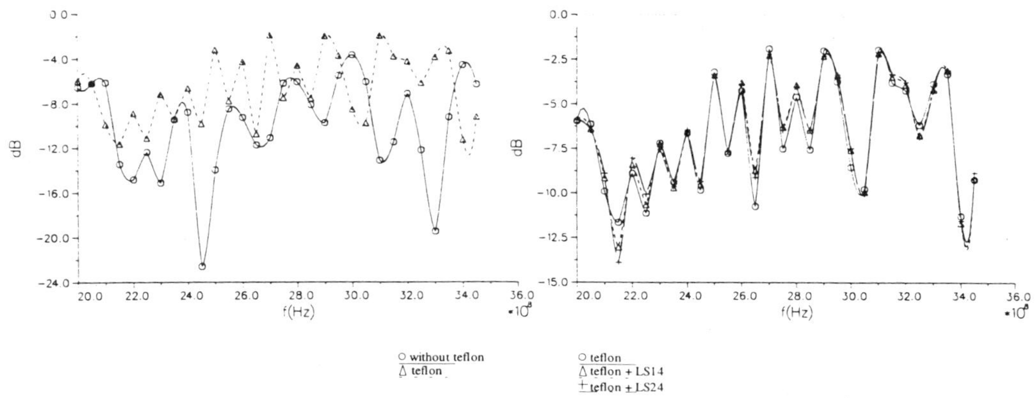

Figures 8-9

On the two following figures (10) (11) , we can observe exactly the foreseeable oscillations, and after 1.5 meters the decrease by 1/R of the electric field Fourier Transform represents the beginning of the far field region.

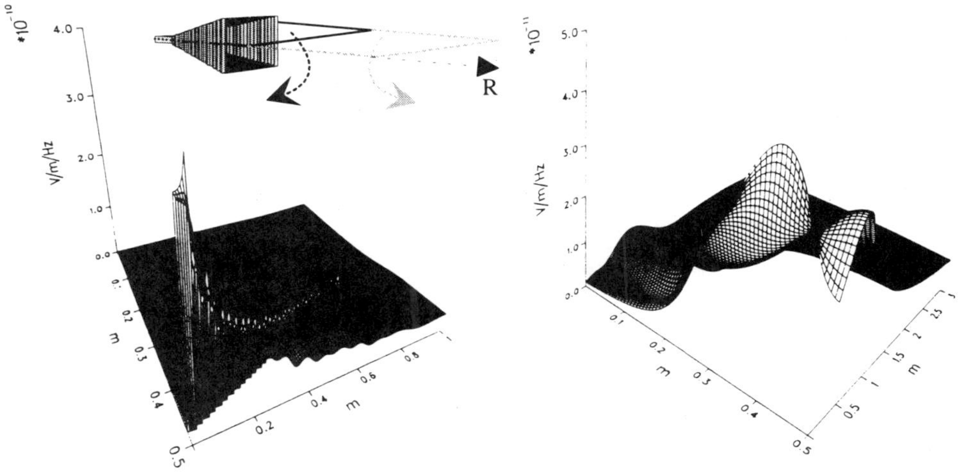

Figures 10-11. Fourier Transform of the electric field at 2.5GHz.
Half symmetry plane for one horn antenna in free space

For each device, it was important to compute the gain and the reflection coefficient S11 at different frequencies. We notice an important desadaptation of the horn which can be explained by the three discontinuities constituted by :

- the guide/horn transition
- horn's aperture
- TEFLON strip.

Due to the diffraction by these discontinuities, the electric and magnetic fields are distorted in the radiating direction. We can calculate the field in all the radiating zones :
The method to assess the fields before the beginning of the Fraunhoffer's zone or far zone consists in computing the whole or the simplified radiating integral from electric and magnetic current sources taken on a Huygens' surface.

The radiating device presented (figure 12) emits the electric field in the Fresnel's zone shown on figure 13.

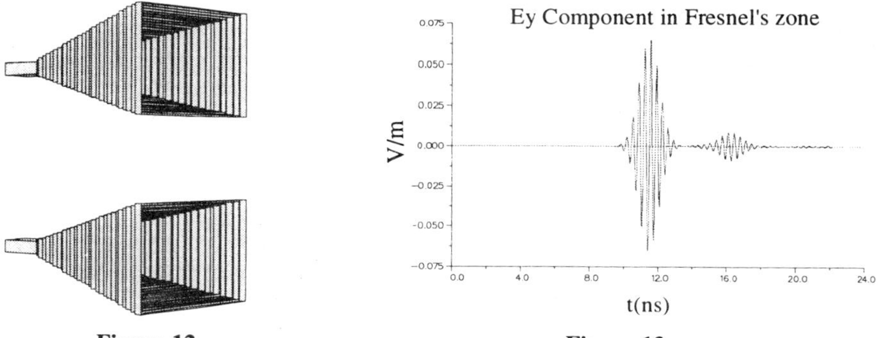

Figure 12 **Figure 13**

Spatial distributions of radiated fields in the time and frequency domains will be presented in different planes up to the test zone to estimate the homogeneity of the field, the ellipticity rate of the wave, and the directivity of the antenna. Some simulations have been performed with several horns, to increase the radiated power and the spot size.

The gain of two horns in the main direction is about 23 dB, the more the frequency increases , the more the central lobe is narrow. Important field variations appear at little angles of the main direction in all the band. We observe exactly the same results (figure 14), at 3 meters from the horns apertures, on the perpendicular plane to propagation direction (holes at 0.65 meter for 2.5 GHz). It was very interesting to illuminate a target with this device because you can strike the target with a controlled field.

Then we obtain (on figure 15) 10dB attenuation in all the frequency range after the dielectric strip (the distance between the dielectric strip and the horn's aperture is 0.5 meter).

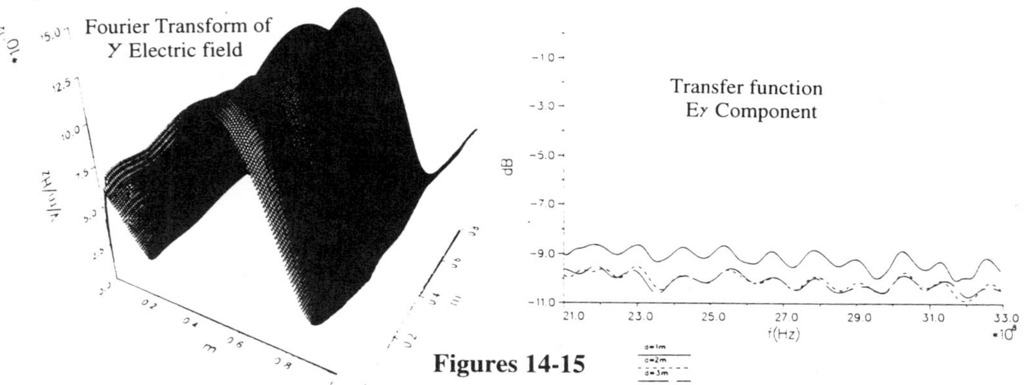

Figures 14-15

STUDY OF RIDGED HORNS (Back door coupling)

In the second experiment, the French Navy Technical Department would like to test the performances of marine panels in the 1-18 GHz band. In this experiment, two facing identical horns antennas are placed at 1 meter from each other and the panels are located between these antennas (figure 16). We would like to assess the radiation of the structures. Because of the symmetry of the antennas, we will only simulate one half.

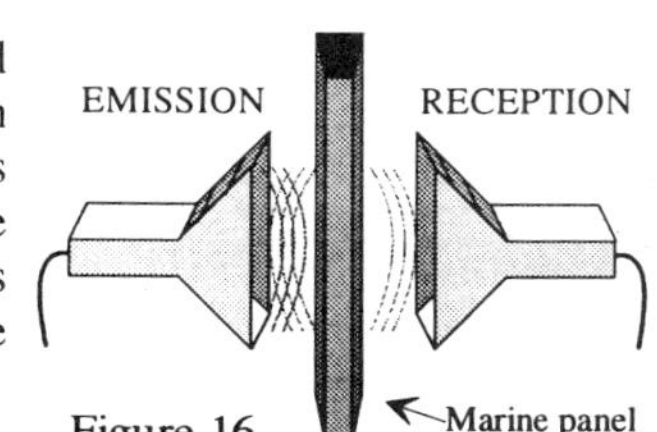

Figure 16

The guide and the exponential decrease of the ridge are quite complex to model with FDTD code. Currently, we are introducing in our study the Contour FDTD Method which involves applying Ampere's and Faraday's laws in cells because Yee's algorithm cannot accurately describe the geometry of the modelled object[11-12-13]. Further results apply to complex antennas like ridged horns will be published soon. We can see the discretisation of this antenna on figure 16 (with one symmetry plane and without contour meshing) and the transient radiated field in near field zone.

The time evolution of electric field is observed in free space at 3 different points (0.2m between two points) on figure 16. We see that the radiated pulse magnitude decreases with modifications when the distance from the antenna increases. At 0.4 meter the electromagnetic fields are in the Fresnel's region (the beginning of the far filed zone is at about 0.9 meter at 5GHz).

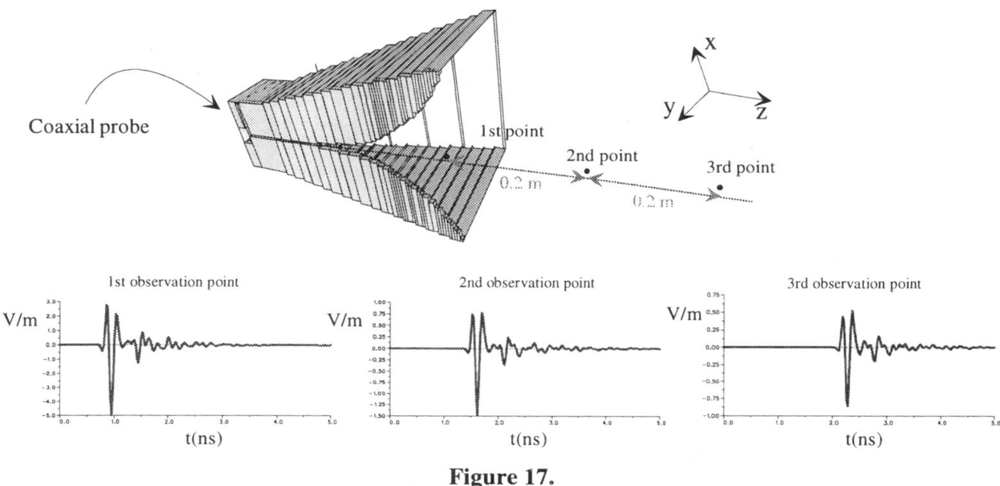

Figure 17.

We observe on figure 17 the Fourier Transform of X Electric field component for two perpendicular planes at 5 GHz. These representations show the propagation of waves in this complex antenna.

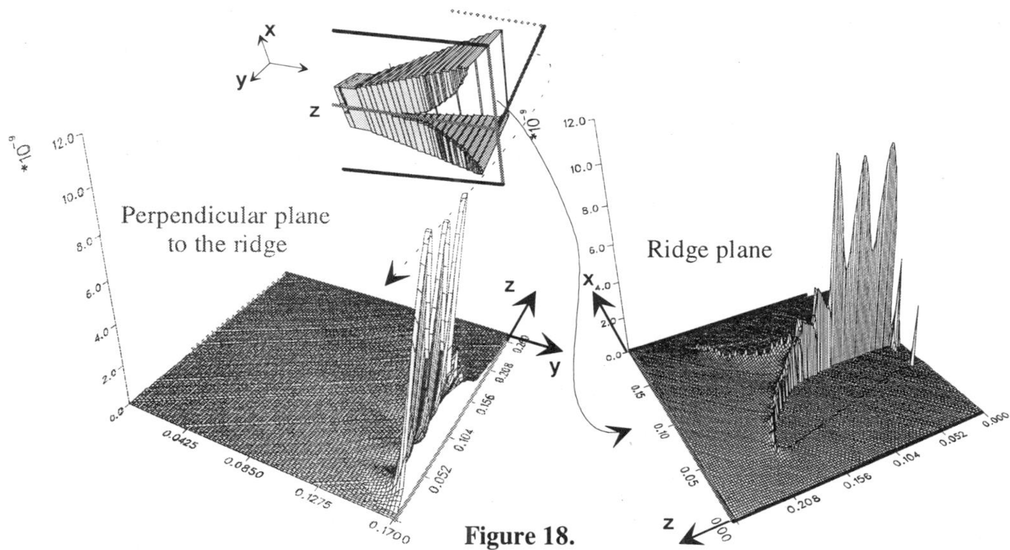

Figure 18.

We place new generation panels to test their performances, to assess the back door coupling, and to quantify the penetration by conduction if a high frequency pulse strikes these walls.

CONCLUSION

As a conclusion, radiating devices with one or several horns can be now simulated with the FDTD algorithm. This method allows to model open space, complex structures and all the propagation and radiation modes are taken into account. This survey enables us to know the behaviour of each three dimensional radiating structure, with different feedings, in a wide frequency range, and moreover in one run. To know the validity domain of these simulations, we have compared theoretical and experimental reflection coefficient and gain of each devices. However, it was impossible to compare transient results because the experiment has been performed in CW mode at discrete frequencies.

REFERENCES

1 P. MORIN, Publication "Les armes microondes".

2 J.B. Keller, "Diffraction by aperture ," J.Appl Phys., vol. 28 pp 426-44, Apr. 1957.

3 J.B. Keller, "Geometric theory of diffraction," J.Opt. Soc. Am.,vol. 52, pp. 116-130, Feb. 1962.

4 J. Huang, Y Rahmat-Samii, and K. Woo, "A GTD study of pyramidal horns for offset reflector antenna applications, " IEEE Trans. Antennas Propagat., vol AP-31, pp.305-309, Mar. 1983.

5 R.C. Menendez and S.W. Lee, "Analysis of rectangular horn antennas via uniform asymptotic theory, "IEEE Trans. Antennas Propagat., vol AP-30, pp.241-250, Mar. 1982.

6 James Benford, John Swegle, High Power Microwaves, 1992

7 J.P. Berenger, "A perfectly matched layers for free-space simulation in finite difference computer codes", presented at Euro. Electromagnetics 94, Bordeaux, France, May 30-June 4, 1994.

8 J.P. Berenger, "A perfectly matched layers for the absorption of electromagnetic waves, "Journal of Comput. Physics, 114, pp. 185-200, oct. 1995

9 Z Wu, J Fang, "Numerical Implementation and Performance of Perfectly Matched Layer Boundary Condition for Waveguides Structures", IEEE Transactions Micro. Theory and Techniques, vol. 43, No.-12, pp 2676-2683, December 1995.

10 Ph.Leveque,A.Reineix, B. Jecko, "Modelling of dielectric losses in microstrip patch antennas : application of FDTD method", Electronics letters, vol.28,n°6, Mars 1992, pp. 539-541.

11 Daniel S. Katz, Allen Taflove, "F.D.T.D. Analysis of Electromagnetic Wave Radiation from Systems Containing Horn Antennas.", I.E.E.E. Transaction Antennas and Propagation, vol. 39, No.8, pp 1203-1212, August 1991.

12 Thomas G. Jurgens, Allen Taflove,"Finite-Difference Time-Domain Modeling of Curved Surfaces.", I.E.E.E. Transaction Antennas and Propagation, vol. 40, No.4, pp 357-366, April 1992.

13 Jiayuan Fang, Jishi Ren, A Locally Conformed Finite-Difference Time-Domain Algorithm of Moleling Arbitrary Shape Planar Metal Strips.", I.E.E.E. Transaction on Microwave Theory and Techniques, vol. 41, No. 5, pp 830 837, May 1993.

SHORT PULSE SCATTERING MEASUREMENTS ON CONDUCTING CYLINDRICAL CAVITIES

Marc Piette[1], David Perrot [2]

[1] Royal Military Academy / Department Telecommunications
Avenue de la Renaissance 30, B-1000 Brussels, Belgium
[2] Ecole Spéciale Militaire de Saint-Cyr Coëtquidan
F-56381 Guer Cedex, France

INTRODUCTION

Since the founding paper of Crispin and Maffett [1] and that of Witt and Price [2] in the sixties, many authors have studied the plane wave scattering of a hollow circular cylinder with a Perfect Electric Conductor-termination, seen as a simplified model of a jet intake. A more realistic model of a jet intake has been later proposed by Moll and Seekamp [3], who designed the termination of the cylinder in the form of a multirowed bladed structure in order to represent the first stages of the jet engine. By regarding the duct as an excited waveguide and applying appropriate techniques that take into account the various modes propagating in the cavity, they predicted the monostatic RCS pattern as a function of the aspect angle for different termination depths. Their results show in particular that a short circuit termination for the cavity is generally a worst case in comparison with the idealized bladed structure with regard to the monostatic RCS. It is worth noting that these results refer only to the echo from the interior of the intake because the contribution from the exterior surfaces is not taken into account in the model.

The case of large hollow circular cylinders with a PEC- termination has been worked out by Huang [4], who give simple formula for the backscattered RCS. The field reradiated from the cavity is estimated by the Kirchhoff approximation and the reciprocity theorem while the scattering from the exterior surface is approximated by the GTD, in conjunction with the equivalent current method. Nevertheless, the simple closed form solution obtained neglects the multiple scattering between the open and closed ends of the cavity.

The attenuation properties of normal modes propagating in an overmoded circular waveguide coated with a lossy material have been calculated exactly by C.S. Lee, S.W. Lee and Chuang [5] who suggested that the RCS of a jet engine inlet can be reduced by coating its interior walls with a lossy material. They demonstrated this in a next paper [6], where the RCS of a PEC-terminated circular waveguide of radius a is calculated with and without coating. Some authors [7] have also studied by high frequency techniques the case of an open-ended waveguide, especially the reflection of the normal propagating modes at the open end of the waveguide.

In the early nineties the first analysis of the time-domain response of cavities seems to be due to Moghaddar and Walton [8] who have described the scattering from circular waveguide cavities in terms of Time-Frequency Distribution. This method is particularly suitable to identify the propagating modes and cutoff frequencies and to untangle the mixed scattering phenomena due to the frequency dispersive character of the interior of the cavity. Some measurements carried out on various targets with an impulse UWB system have been recently reported [9], in particular on cavities at different aspect angles.

The present paper reports recent measurements of short pulse scattering on cylindrical cavities carried out *directly in the time-domain* with the Transient Scattering Range of the Royal Military Academy (RMA) Brussels. The monostatic responses of a PEC-terminated cylinder are measured under gaussian excitation with pulsewidths of 90 ps and 320 ps and their waveforms are analysed. One of the measured responses is also compared with the transient response of the cavity calculated by FDTD. Because of the high spatial resolution obtained with pulse durations as short as 90 ps, the various physical phenomena involved in the scattering of a cavity like edge diffraction, bottom reflection and creeping waves can be separed in their time occurence and compared in their relative amplitude.

THE TRANSIENT SCATTERING RANGE AT THE RMA

In the framework of its research project "Ultra-wideband Radar Signature of Complex Targets", the Royal Military Academy Brussels has developed an indoor Transient Scattering Range for measuring *directly in the time-domain* the monostatic and bistatic short pulse scattering of 3D-targets [10].

The system consists at the transmitting side of a 2 meter long monocone antenna on square ground plane (3m x 3m) coaxially base-driven by a fast step function generator and, at the receiving side, of a broadband e-sensor connected to a 20 GHz digitizing oscilloscope trough a set of ultrabroadband amplifiers. The target under test is put on the ground plane and illuminated with the transient TEM wave radiated by the monocone antenna in the vertical polarisation.

In so far the target cross section is kept small (< 20 cm x 20 cm) the spherical wave impinging the target can be considered as a plane wave locally. On the other hand the length of the monocone and the size of the ground plane have been taken large enough so that the travelling time of the current launched on it from the monocone apex is large with regard to the pulse length. So during the time the current pulse needs to travel up to the cone top, the conical antenna looks just like an infinitely long antenna and its instantaneous bandwidth is virtually infinite. By keeping the feeding coaxial aperture of the cone small with regard to the shortest wavelength of interest to limit the radiation from it and by choosing a small value for the cone flare angle (< 8°) which makes the higher order modes negligible with regard to the dominant TEM mode, the field radiated by the antenna is in any fieldpoint (even at short distance from the apex) the replica of the stepfunction voltage applied to its apex. There is of course some reflections of the current pulse at the top of the monocone and at the edge of the ground plane but they occur much later than the time the impulse response of the target needs to reach the sensor.

Two generators can be used to feed the antenna. Both use an avalanche transistor and generate a very pure waveform with distortions less than 2%. They yield step voltages of respectively 50V/350 ps and 10V/90 ps. Because the sensor used is an e-sensor (short monopole on ground plane) the output of which is proportional to the time derivative of the incident electric field, the stepfunction radiated by the antenna is in fact seen by the oscilloscope as a gaussian impulse with a duration of 320 ps or 90 ps (Fig.1). The latter confers to the scattering range a *radial resolution of 1,4 cm* on the target and an instantaneous bandwidth ranging from a few kHz to 3,5 GHz. The waveform displayed on the sampling oscilloscope (*hp* 54121T) consists of 501 samples digitized on 14 bits. Up to 2048 periods of the signal can be averaged to enhance the signal to noise ratio. After calibration with various conducting and dielectric canonical targets like spheres and flat plates [11], the Transient Scattering Range is now used to measure the short pulse response of non canonical bodies.

CAVITIES USED IN THE MEASUREMENTS

Three different cavities of length L and diameter 2a have been designed for the measurements . The first one has a square cross section of 5 cm on a side, a length of 24 cm and a wall thickness of 3 mm. The second cavity has the same length, wall thickness and aperture area but with a circular cross section. The third cavity has a square cross section of 10 cm on a side, a length of 20 cm and the wall thickness is 1 mm. The cavities exhibit a physical length of about 5 pulse lengths with regard to the 320 ps excitation pulse and of

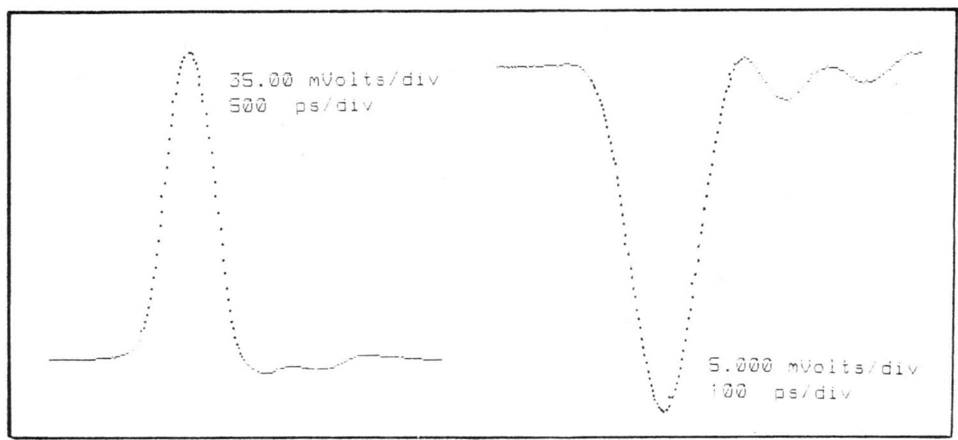

Figure 1 The excitation pulses used to illuminate the target a) the positive 320 ps pulse b) the negative 90 ps pulse

about 18 pulse lengths with regard to the 90 ps pulse. In this last case the cross dimensions of the cavities are large with regard to the 1,4 cm spatial resolution of the excitation pulse. The direction of the incident field is refered to the symmetry axis of the cavity by the aspect angle θ. Because the target is lying on the ground plane of the Scattering Range, this angle is measured in the horizontal plane. For all the measurements the distance from the cone apex to the center point of the cavity entrance is 100 cm and the distance from this point to the sensor is 60 cm.

RESULTS OF THE MEASUREMENTS

The results shown in Fig.2 are obtained with the 320 ps excitation pulse at normal incidence ($\theta = 0°$). Fig.2a shows the responses of the small rectangular cavity (2a=5 cm) with and without backplate. From the identity of the two traces one can infer that no energy is coupled into the cavity since the removal of the bottom seems to have no influence on the backscattering. This also means that the short response observed (duration less than 1 ns) only consists of the diffraction on the aperture, which is electrically too small to enable any coupling of the incident wave into the waveguide formed by the cavity. For a square cross section waveguide of 5 cm on a side, the cutoff frequency of the first modes (TE_{1O} and TE_{O1}) given by c/2a (where c is the speed velocity) is indeed 3 GHz, while the 3 dB bandwidth of the incident pulse spectrum is only 1 GHz. The backscattering of the large cavity (2a=10 cm) on the contrary is a 4 ns long response, as shown in Fig.2b. Its peak value is about 4 times stronger than that observed with the small cavity, according to the 4:1 ratio of their cross sections : at the higher frequencies contained in the early time response, the RCS of an aperture is indeed proportional to the squared aperture area, as pointed out by Crispin and Maffett [1], so the backscattered fieldstrength is simply proportional to the cross section area. But the response of the large cavity differs from that of the small one by two features : the total duration, which is much longer (> 4 ns) and the fact that the late time response (after 1,4 ns) is clearly influenced by the removal of the backplate. These two facts indicate that besides the diffraction of the incident field at the entrance of the cavity, some energy is coupled into it, is reflected on the backplate and travels back to the aperture where it is diffracted after some propagation delay in the waveguide. For a square cross section waveguide of 10 cm on a side, the cutoff frequency of the first modes is 1,5 GHz. If one keeps in mind that the incident gaussian pulse still contains a significant fraction of its total energy beyond the 3 dB cutoff (1 GHz), it follows that the dominant mode can be well excited at the aperture by the vertically polarised TEM wave. It is worth to note that for the bottomless cavity (lower trace in Fig.2b), which is

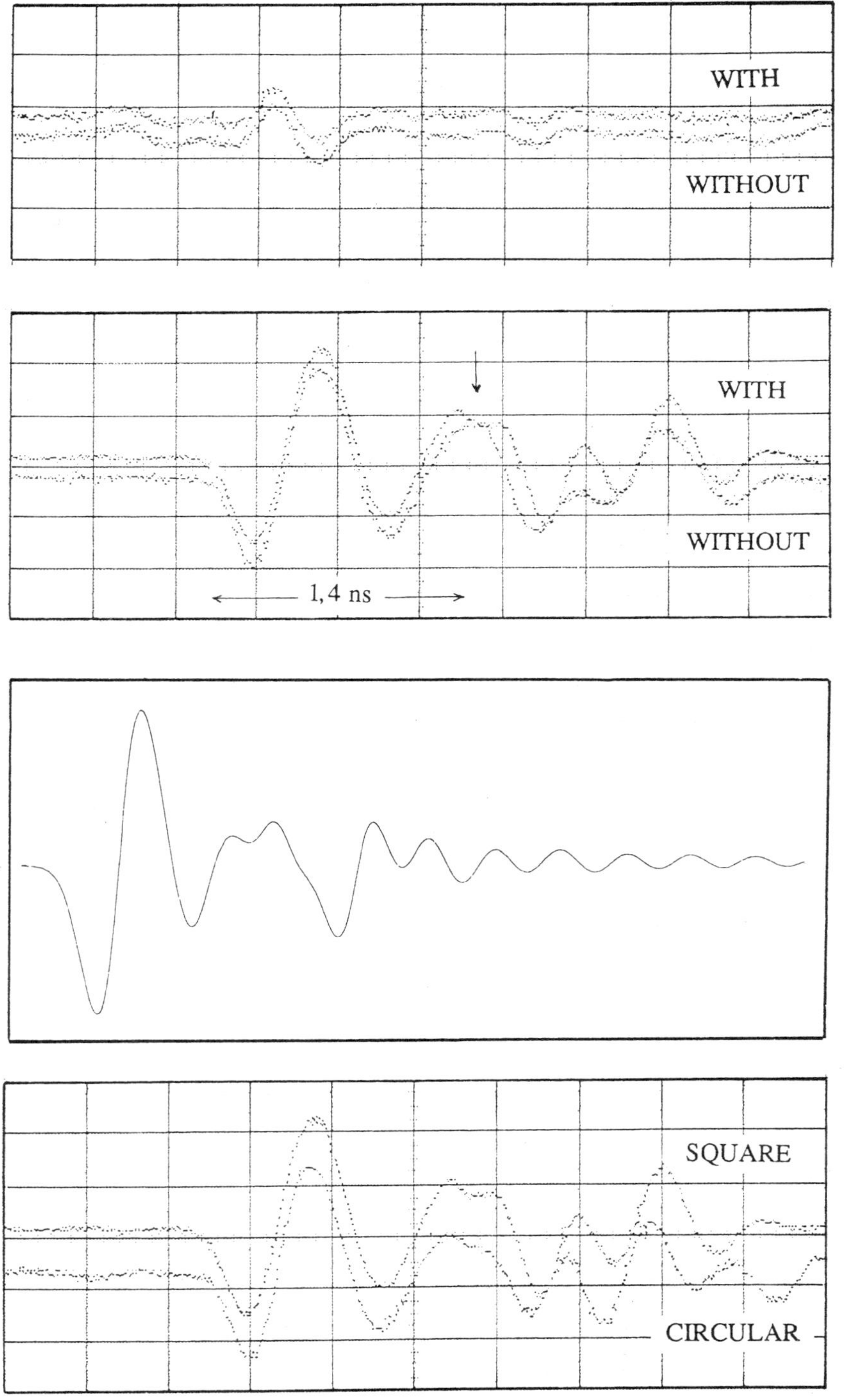

Figure 2 On-axis backscattering of the cavities to the 320 ps pulse (500 ps/div) : (a) square cavity (2a=5 cm) (b) square cavity (2a=10 cm) (c) FDTD response of the large square cavity (d) large square and circular cavities

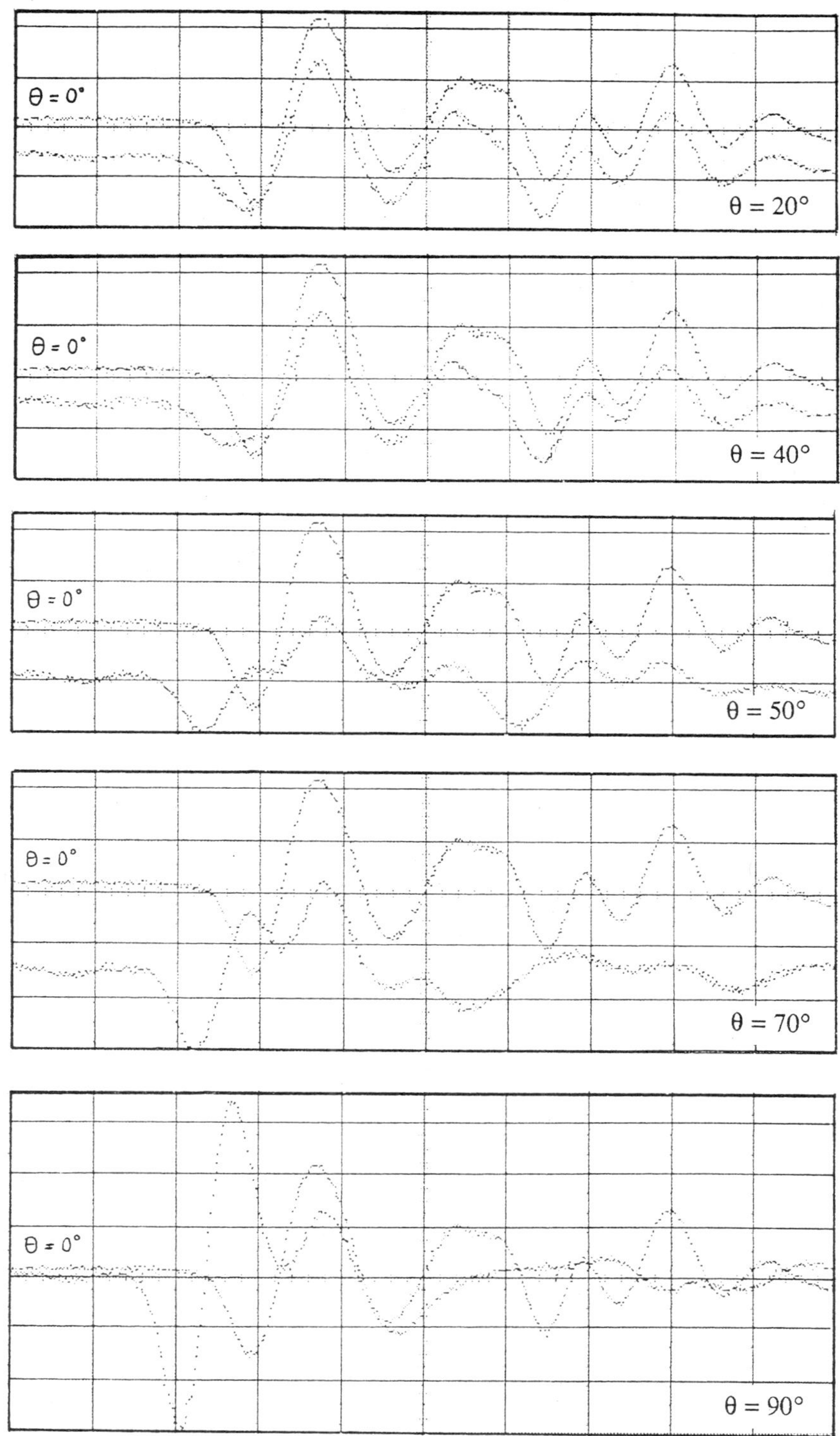

Figure 3 Off-axis backscattering of the large square cavity to the 320 ps pulse (500 ps/div)

nothing else than an open-ended waveguide, the response doesn't vanish after 1,4 ns. That confirms that even for an open-ended waveguide some energy is reflected back at the open end, as predicted by Chuang and Pathak [7]. For comparison with these measurments, Fig.2c shows the response of the PEC-terminated square cavity (2a=10 cm) to the 320 ps pulse at normal incidence calculated by FDTD. The agreement with the measured response (upper trace in Fig.2b) is good, which augments the confidence in the results. This calculated waveform shows also the very late time response of the cavity that cannot be directly measured with the Scattering Range because of its limited clear time window. The ringing phenomenon in the cavity is clearly visible. If one compares now the response of the square cross section cavity with that of a circular one of the same aperture area (Fig.2d), it turns out that the diffraction at the entrance is independent of the exact shape of the aperture, at least for this excitation, while the late time response keeps the same form as for the square cavity but comes a bit sooner. This time shift of the late time response is due to the fact that for two cavities having the same aperture area A - one with a square cross section ($A=4a^2$) and the other with a circular one ($A=\pi a^2$) - the group velocity of the TE_{11} mode in the circular waveguide is larger than that of the TE_{10} mode in the square waveguide.

The backscattering of the large square cross section cavity (2a=10 cm) has also been measured as a function of the aspect angle θ of the cavity (Fig.3). For $\theta < 45°$, the impulse response is quite similar to the on-axis response (upper trace), apart from a slight lead of the first negative peak. This is due to the fact that by rotating the cavity in the horizontal plane with an angle θ, one of the vertical edges of the aperture comes closer to the sensor so that the scattering from it reaches the sensor a little sooner than for the on-axis incidence. Fig.3a and 3b show that the backscattering of the cavity is quite constant over an angular sector as wide as 2 x 45°= 90°, which corroborates in some extent the measurements reported by Madonna [9]. For aspect angles larger than 45°, the backscattered waveform begins to differ significantly from the on-axis response : on one hand the early time response becomes larger because the return from the cavity sidewall adds to the rim diffraction; on the other hand the late time response becomes smaller since less energy can penetratre into the waveguide at off-axis incidence.

A second set of measurements has been carried out with the 90 ps pulse excitation, which offers a better spatial resolution (1,4 cm). The amplitude of the waveforms shown in Fig.4 are smaller than for the 320 ps pulse because the fast generator yields a step voltage of only 10 V and the physical size of the sensor has to be reduced by a factor 3,5 to give it a sufficient bandwidth. Fig.4a shows the on-axis response of the large cavity (2a=10 cm) with and without the backplate. The first small peak is the rim diffraction and the second peak much larger than the first one is due to the reflection on the backplate of the cavity. It is worth to note that under the 90 ps excitation the nose-on response from the backplate is larger than that of the rim diffraction in contrast with what is observed under the 320 ps pulse excitation. This is in good agreement with the prediction of Lee and Lee [6] who stated that at axial incidence, the RCS from the rim diffraction becomes less significant as a/λ becomes larger, whereas the interior irradiation becomes predominant. To be absolutely sure that the second peak is well due to the return from the interior of the cavity and not to a travelling wave creeping along the exterior wall, the aperture has been closed with a PEC-plate (Fig.4b). Only the rim diffraction is still visible, which confirms the previous assumption. Note that the response of the closed cavity is larger than that of the open cavity ! This agrees with the common assertion that the high frequency nose-on return from a cavity is bounded by the return from a flat plate whose area is the same as that of the aperture [1,2].

The influence of the aspect angle on the response to the 90 ps pulse excitation is shown in Fig.4c. The comparison of the early time responses between on the first three traces (θ = 0°, 10°, 20°) brings to the fore the progressive separation in time of the diffractions coming from the vertical edges of the aperture. The time delay between them increases logically with θ as can be seen on the next traces (θ = 40°, 60°). The slight negative peak following the first two positive ones accounts for the double diffraction on the aperture edges. It is more difficult to identify the origin of the train of positive peaks visible on the late time response, because of the dispersive character of the propagation in the waveguide. From these measurements it appears nevertheless that the amplitude of the rim diffraction is independent of the aspect angle. This is logical because a vertical edge impinged by a vertically polarized incident wave coming in the horizontal plane diffracts rays with the

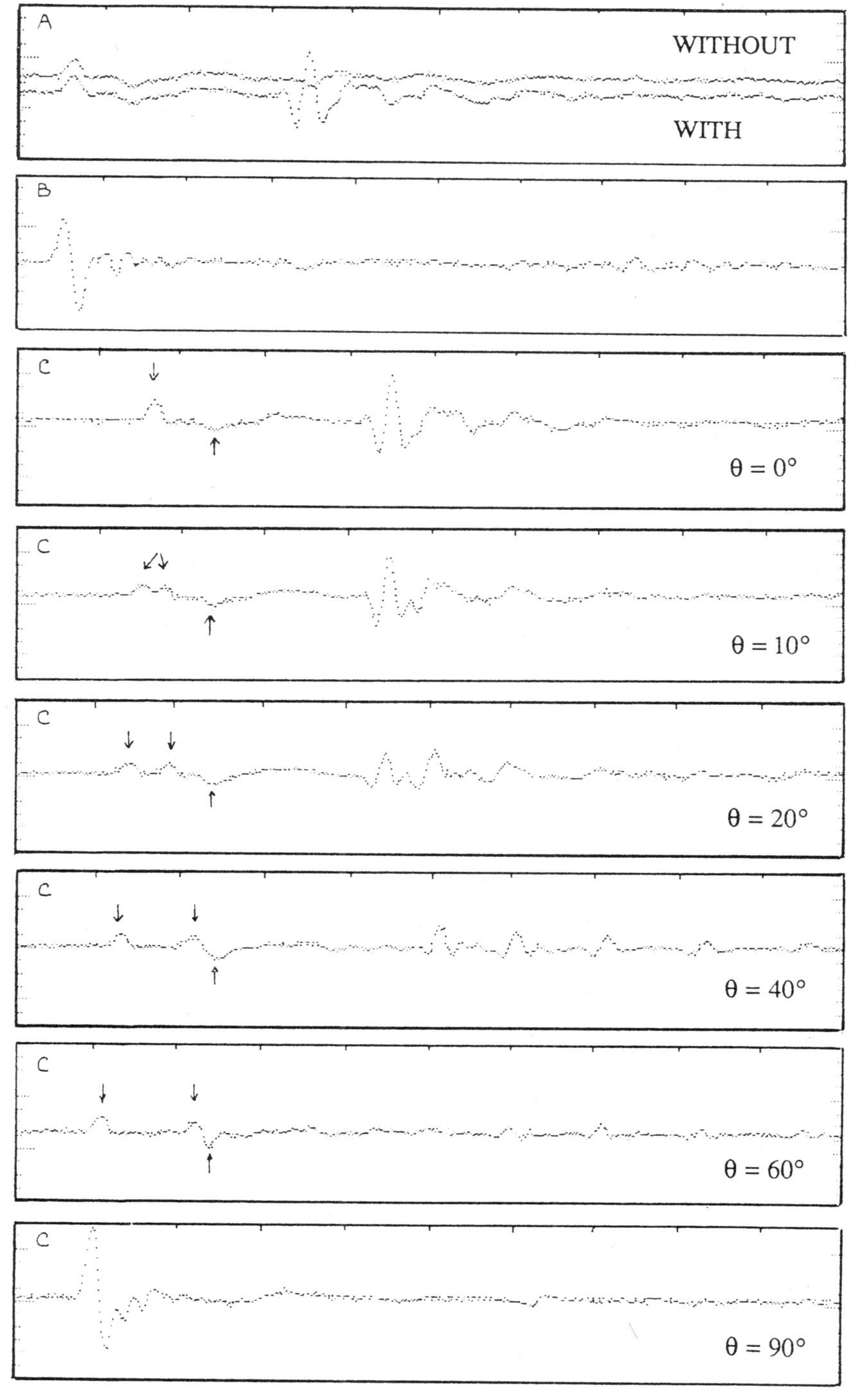

Figure 4 Backscattering of the large square cavity to the 90 ps pulse (500 ps/div) : (a) On-axis response with and without bottom (b) On-axis response with closed aperture (c) Off-axis responses

same fieldstrength in all directions of the horizontal plane. The amplitude of the late time peaks on the contrary decreases for larger values of the aspect angle because less energy can penetrate the cavity at off-axis incidence. Besides, the broadside backscattering of the cavity depicted on the last trace ($\theta = 90°$) only consists of the strong return from the waveguide sidewall without any late time response from the interior.

CONCLUSION

Re-entrant cavities on modern jet aircrafts like the engine intakes, the exhaust ducts and the cockpit are usually considered as main contributors to the global RCS of the target and as wide FOV scattering centers. Though the scattering of a cavity has been studied theoretically by many authors, most of the past works were particularly focused on the frequency domain, whereas few experimental data are available in the time-domain.

After having reviewed some previous papers dealing with the scattering of cavities, the authors have reported recent measurements on cylindrical cavities carried out *directly in the time-domain* with the Transient Scattering Range of the Royal Military Academy Brussels. The monostatic impulse responses of a conducting cylinder open at one end and shorted at the other end have been measured under gaussian pulse excitation with pulsewidths of 90 ps and 320 ps. One of the measured responses has also been compared with the transient response calculated by FDTD. In contrast with the theoretical approaches where some effects are sometimes neglected (multiple scattering between the open end and the closed end, return from the exterior walls), these measurements take into account all the physical phenomena involved in the scattering of the cavity. They have lead to conclusions that agree with previous studies and give also an original insight into the various mechanisms contributing to the global RCS of a cavity because of the high spatial resolution of the 90 ps pulse (1,4 cm), in particular the relative amplitude of the scattering phenomena and their dependence or independence on the many factors like the aspect angle, the aperture area, the aperture shape, and the pulse lenght.

REFERENCES

1. J. Crispin Jr. and A. Maffett, Radar cross-section estimation for simple shapes, *Proc.IEEE,* 53 : 833-848 (1965).
2. H. Witt and E. Price, Scattering from hollow conducting cylinders, *Proc.IEE,* 115: 94-99 (1968).
3. J. Moll and R. Seecamp, Calculation of radar reflecting properties of jet engine intakes using a waveguide model, *IEEE Trans. on Aerospace and Electronic Systems* , 6-5 : 675-683 (1970).
4. C. Huang, Simple formula for the RCS of a finite hollow circular cylinder, *Electronics Letters* 19 : 854-856 (1983).
5. C. Lee, S. Lee and S. Chuang, Normal modes in an overmoded circular waveguide coated with lossy material, *IEEE Trans. on Microwave Theory and Techniques* 34:773-785 (1986).
6. C. Lee and S. Lee, RCS of a coated circular waveguide terminated by a perfect conductor, *IEEE Trans. on Antennas and Propagation* 35 : 391-398 (1987).
7. C. Chuang and P. Pathak, Ray analysis of modal reflection for three-dimensional open-ended waveguides, *IEEE Trans. on Antennas and Propagation* , 37 : 339-346 (1989).
8. A. Moghaddar and E. Walton, Time-frequency distribution analysis of scattering from waveguide cavities, *IEEE Trans. on Antennas and Propagation* , 41 : 677-1993 (1993).
9. R. Madonna, P. Scheno, G. Vilardi, C. Hum and J. Scannepieco, Scattering, resonance, creeping wave, travelling wave and all that : UWB measurements of various targets, in *Ultra-wideband Short-Pulse Electromagnetics*, H.Bertoni, L. Carin, L. Felsen, Plenum Press, New-York: 83-91(1993).
10. M. Piette, Banc de mesure en régime transitoire de la signature radar d'objets tridimensionnels conception, développement et validation, Thèse de doctorat, Université Catholique de Louvain (1995).
11. M. Piette, E. Schweicher, A. Vander Vorst, Calibration of an impulse radar scattering range with conducting and dielectric canonical targets*1995 IEEE AP-S Symposium and URSI Radio Science Meeting*, Newport Beach, URSI Proceedings: 268.

RCS DETERMINATION FROM LOCALIZED SHORT-PULSE SCATTERING MEASUREMENTS: THEORY AND EXPERIMENT

Morris P. Kesler, James G. Maloney, Eric J. Kuster,
Paul G. Friederich and Brian L. Shirley

Signature Technology Laboratory
Georgia Tech Research Institute
Atlanta, GA 30332-0824

INTRODUCTION

The Radar Cross Section (RCS) of a target has become an important metric for the characterization of electromagnetic performance. As a result, much effort has gone into developing techniques to accurately measure RCS. By its definition, RCS is a plane-wave concept, i.e., it is determined by the far-field scattering of an object when illuminated by a plane-wave. Most RCS measurement techniques involve illuminating the object under test with an approximation to a plane-wave (one exception is the near-field scanning technique). Specialized facilities, such as outdoor and compact ranges, are currently used to measure RCS; these facilities are generally very large in terms of the electromagnetic wavelength and are often located at remote sites. In this paper, we present a new RCS measurement technique that can be applied in much smaller spaces, and is potentially transportable.

At the last Ultra-WideBand meeting, we presented results of short-pulse, localized scattering measurements [1]. A localized scattering measurement is defined as one in which the transmitter and receiver are located in the near-field of the target, typically less than a target diameter away. Thus, non-planar fields are incident upon the target and the scattered fields are measured in a region where they are also non-planar. Detailed comparisons between the measurements and model predictions (FDTD and MoM) were presented which showed excellent agreement for complex, laboratory-scale targets.

We have since developed a technique to determine RCS from a series of localized measurements. The technique is related to the near-field scanning approach in that we perform scattering measurements at an array of transmit and receive positions [2,3]. The differences lie in how the resulting scattering data are used to determine RCS. In the traditional near-field approach, the data for the scattered field are processed as if they were available over an infinite surface (in the case of the planar scan). Our approach takes into account the finite size of the scan plane to optimize the result. This considerably reduces

Ultra-Wideband, Short-Pulse Electromagnetics 3
Edited by Baum *et al.*, Plenum Press, New York, 1997

the size of the scan plane. Moreover, our approach can be applied to any type of scan surface.

The RCS measurement technique is illustrated in Figure 1, which shows a single transmit/receive pair of $\hat{y}$-oriented antennas located in a scan plane in front of the test volume. By using a series of transmit positions in the scan plane, a planar wavefront is synthesized within the test volume. The field scattered from the target is likewise sampled at a series of positions in the scan plane and combined to obtain the far-field scattering.

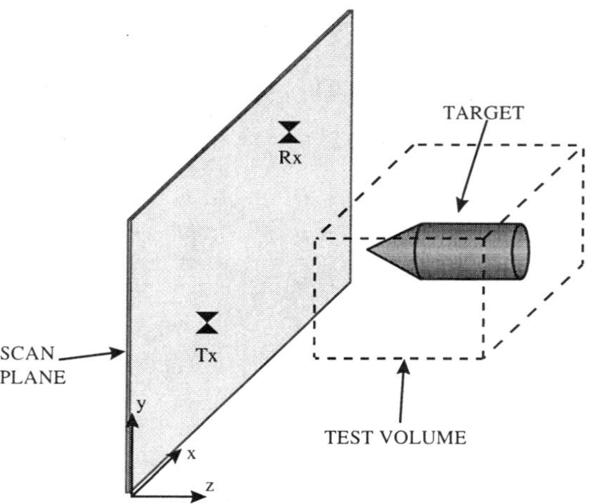

Figure 1. Sketch of the RCS measurement technique.

The process can be expressed in terms of a series of localized measurements as

$$kR\frac{\hat{y}\cdot\mathbf{E}^{FF}}{E_0}=\sum_{Tx}\sum_{Rx}W_{Tx}W_{Rx}S_{SCAT}(Tx,Rx) \tag{1}$$

where W_{Tx} and W_{Rx} are weighting factors for the plane wave synthesis and far-field extraction, respectively, and E_0 is the amplitude of the synthesized field in the test volume. The calibrated localized measurements are contained in $S_{SCAT}(Rx,Tx)$, which is given by

$$S_{SCAT}(Rx,Tx)=\frac{V_{LM}(Rx,Tx)-V_{CS}(Rx,Tx)}{V_{CS}(Rx,Tx)}\hat{y}\cdot\mathbf{S}_{Dipole}(Tx,Rx), \tag{2}$$

where V_{CS} is the voltage measured by the receiver with no target present (a clearsite measurement), V_{LM} is the voltage measured with the target present, and $\mathbf{S}_{Dipole}$ represents dipole propagation from the transmit to receive position [1]. The RCS is given by

$$\sigma=\frac{4\pi}{k^2}\left|kR\frac{\hat{y}\cdot\mathbf{E}^{FF}}{E_0}\right|^2. \tag{3}$$

In the following sections, we discuss the synthesis problem for transmission and reception, the measurement system we have developed, and representative RCS measurements showing the effectiveness of the technique.

PLANE-WAVE SYNTHESIS FOR TRANSMISSION

The transmission problem can be viewed as determining how to create a planar wavefront within the test volume from the coherent addition of multiple local sources (a synthesized array). The weighting factors, W_{Tx}, which are used to combine the sources are defined by

$$\sum_{Tx} W_{Tx} \mathbf{S}_{Dipole}(Tx, r) \approx \hat{y} e^{-j\bar{k}\cdot\bar{r}} \tag{4}$$

for every $\bar{r}$ in the desired test volume. The direct approach to finding the best set of weights is to predict the field generated by the array throughout the test volume, compare this with the desired plane wave field and then minimize the difference by adjusting the weights. However, computing the field throughout a volume requires a great deal of computer resources. A uniqueness theorem due to Hill reduces the problem to computing the field only over the surface of the test volume [4].

Hill showed that if the vector quantity,

$$\mathbf{F}(\mathbf{r}) = \hat{n}(\mathbf{r}) \times (\mathbf{E}(\mathbf{r}) - \eta_0 \mathbf{H}(\mathbf{r})) \tag{5}$$

is zero on the surface of the test volume, and there are no sources within the test volume, then the **E** and **H** fields are zero within the test volume. Here $\hat{n}$ is the unit vector normal to the surface of the test volume. To apply this theorem to our problem, we want the difference between **F** from the array and **F** for the desired plane wave to be zero. The field from the array will then be the desired plane wave. The computational procedure is to sample **F** over the surface of the test volume, form a least squares solution matrix, and then invert the matrix to find the weights. This is accomplished using the constrained source norm approach [4,5].

For an n x n array with element spacing s, the center frequency (where the element spacing is $\lambda_0/2$) is given by $f_0 = c/2s$. In general, it is possible to synthesize good, planar fields over the test volume at frequencies from approximately $f_L \approx c/(n-1)s$ to $f_H \approx c/s$. At the low end, the overall size of the array is the limiting factor, while the upper frequency is restricted by the spacing between elements. These limits are only approximate and depend, of course, on the size of the test volume.

The synthesis procedure is performed in the frequency domain, and the weights obtained are a function of the size and element spacing for the array, and size and location of the test volume (all in wavelengths). The synthesis procedure can also be applied to a broadband pulsed excitation; however, the element weights must then be determined separately for each frequency in the pulse. The coefficients vary because the antenna patterns change with frequency, and because the element spacing and test volume size, although fixed in absolute units, are also functions of frequency. This means, in general, that each element will have its own frequency dependent excitation. Equivalently, in the time domain, each element will be excited with a different waveform.

To illustrate the optimized array using broadband excitation, we performed a simulation of a 7 x 7 array using our FDTD code. The optimized element excitations were computed at 400 frequencies between $f_0/200$ and $2f_0$. The test volume used in the optimization was a cubical region $2\lambda_0$ on a side, centered $2\lambda_0$ in front of the array. The direction of propagation for the plane wave was chosen to be normal to the array. In order to provide band limiting, the computed frequency dependent excitations for each element were multiplied by the frequency spectrum of a differentiated Gaussian pulse (spectral peak at $0.4f_0$). The resulting product spectra were transformed back into the time-domain and used as the excitations for each of the elements in an FDTD simulation. Examples of the temporal waveforms which resulted from this process are shown in Figure 2 for three elements: the lower corner (1,1) element (left plot), the (2,2) element (middle plot) and the center (4,4) element (right plot). There are considerable differences in these waveforms, and notice that the elements near the edges are excited more strongly than the center element.

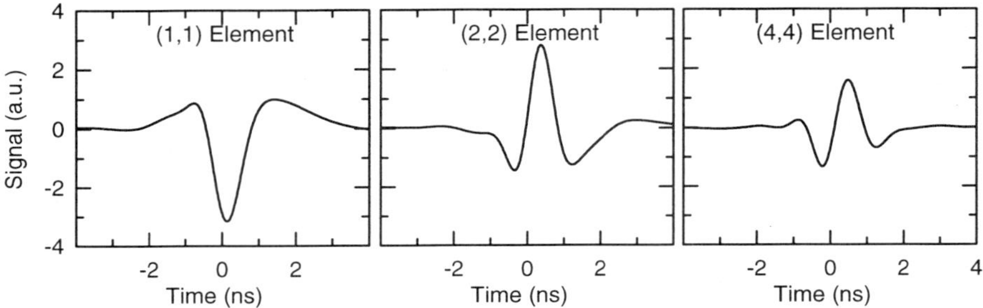

Figure 2. Examples of the temporal waveforms needed to create the pulsed plane wave.

When these excitations are applied to the 49 array elements, the radiated fields are as shown in Figure 3. The two plots represent the magnitude of the field in two orthogonal planes at one instant in time. If the dipoles are oriented in the $\hat{y}$-direction, located on the z = 0 plane and centered about the z-axis (see Figure 1), then the left plot is for the x = 0 plane, and the right plot is for the z = $2\lambda_0$ plane. For comparison, the same two cuts for an array with uniform excitation by the same differentiated Gaussian pulse are shown in Figure 4. The field for the optimized array is noticeably more planar than that for the uniform array.

SELECTION OF FAR-FIELD SCATTERING

The second part of the measurement problem was determining the desired far-field component of the scattered field from the local field measurements. The received signals must be combined with appropriate weights, W_{Rx}, to extract the far-field result:

$$kR\frac{\mathbf{E}^{FF}}{E_0} = \sum_{Rx} W_{Rx} \frac{\mathbf{E}_{Loc}(Rx)}{E_0}. \tag{6}$$

where $\mathbf{E}_{Loc}(Rx)$ is the field measured at the receiver positions. (The equivalent problem for a time-domain signal is to find the appropriate impulse response to apply to each received waveform, which in general will be different for each element in the array.) We have investigated three techniques for solving equation (6). The first technique was a plane wave method. In this approach the scattered field is expanded in a plane wave basis so that the desired bistatic scattering in a particular direction is just the amplitude of the appropriate plane wave in the expansion. The second technique was a spherical mode expansion. In this technique the scattered electric field is expanded in vector spherical modes, and a set of weighting coefficients is sought for predicting the far-field scattering independent of the target. The third technique was to simply use the weights for transmission also for reception. This choice is based on the principle of reciprocity.

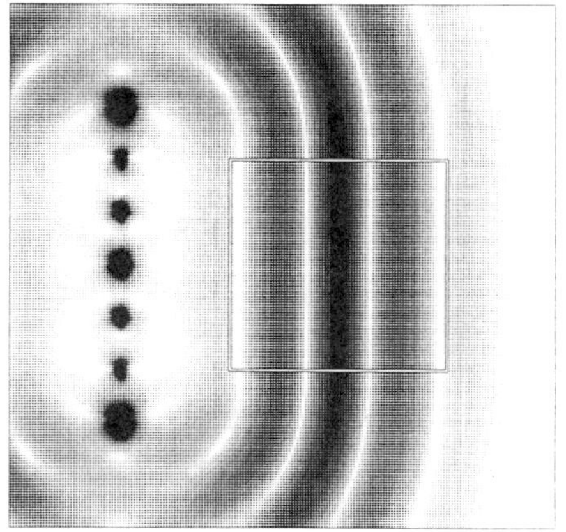
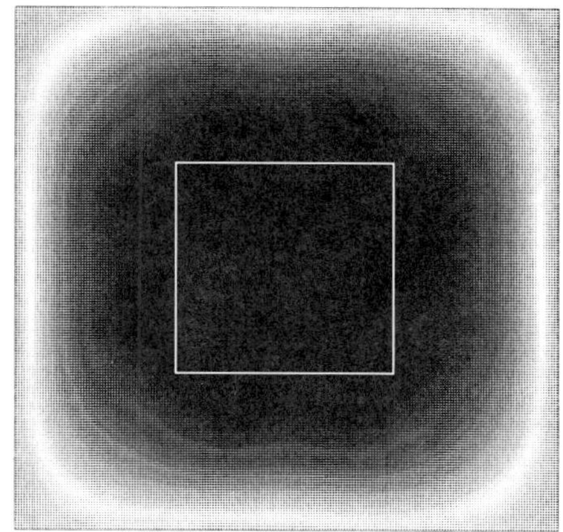

Figure 3. Snapshot of the fields radiated from the 7 x 7 array excited with the optimized waveforms.

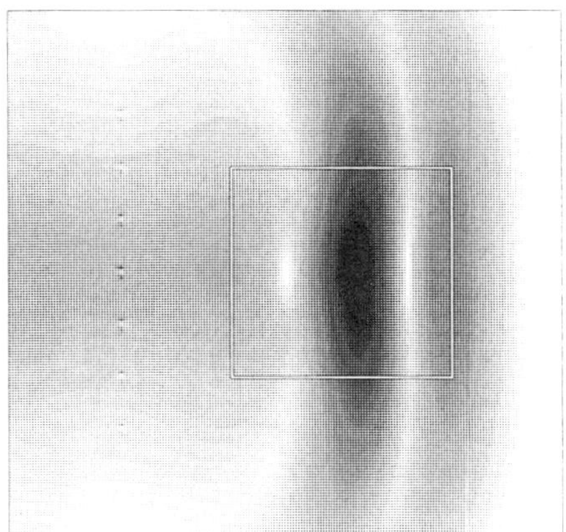
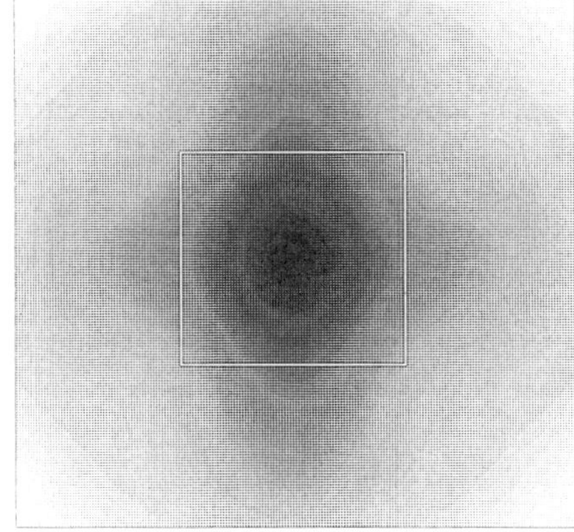

Figure 4. Snapshot of the fields radiated from the 7 x 7 array uniformly excited with differentiated Gaussian pulses.

A simple example will serve to illustrate the reciprocity concept. Consider an array of antennas which have been excited with a set of weighting coefficients such that the array radiates a plane wave over a region. To drive the array, a unit amplitude signal on a single coaxial line is split and sent to the array elements with the appropriate amplitudes and phases consistent with the weighting coefficients. Now place a second antenna in the plane wave region to receive the field. As the second antenna is rotated, the received voltage will be proportional to the pattern of the second antenna, since it is being illuminated with a plane wave. Reciprocity states that if the problem is reversed, i.e. the second antenna is driven with the unit amplitude signal and the array is used in the receive mode with the same weighting coefficients, then the same voltage will be measured as the second antenna is rotated. This means that the signal measured by the array will also be proportional to the pattern of the second antenna. Thus the array has selected the appropriate far-field component of the field incident on it.

We investigated each of these techniques and found that using the coefficients for transmission also on reception gave results as good or better than the other two techniques. Because of this, all of the RCS results contained in this paper were processed using transmit coefficients to select the far-field component.

One other point about the reception process is important. To eliminate the need for measurements in which the transmitter and receiver are co-located, we perform the synthesis procedure for reception with that particular measurement excluded. Thus, there is actually a set of receive coefficients for each transmitter position. Equation (1) now becomes

$$kR\frac{\hat{y}\cdot\mathbf{E}^{FF}}{E_0} = \sum_{Tx}\sum_{Rx\neq Tx} W_{Tx}W_{Rx}(Tx)S_{SCAT}(Tx,Rx). \tag{7}$$

MEASUREMENT SYSTEM

There are several options for implementing this technique for RCS measurements. Perhaps the simplest one conceptually is a direct implementation in which a physical array is used to create the plane wave. Unfortunately, this approach requires considerable hardware, since each array element would need individual amplitude and phase control (frequency dependent). The time-domain approach would require individual short-pulse waveform generators for each element. A second option is to again construct a physical array, but use only one element at a time for transmitting, and one element for receiving. The full array response can be synthesized by combining the individual measurements with the appropriate weighting factors. This reduces the hardware requirements somewhat, but an RF switching network is required.

The third option is to synthesize the array response using only two antenna elements, one for transmitting and one for receiving. The elements are sequentially moved to each of the array element locations, and scattering measurements are performed for each combination of antenna positions. This approach has the simplest RF hardware requirements, and it provides greater flexibility than the other approaches, because the array size and element spacing can be easily varied. The cost comes in the time for data collection, which now includes the time to accurately position the two antennas.

We have constructed a laboratory-scale measurement system using the scanned element approach. A schematic of the system is shown in Figure 5. Two, computer-controlled, x-y scanners are used to position the antennas at the desired locations, and a network analyzer is used to make the scattering measurement. For each set of antenna positions, scattering measurements are performed as a function of frequency. (We chose a frequency domain approach because of available hardware; a time-domain approach can also easily be used.) Time-gating is used to eliminate scattering from outside of the region of interest. Bi-conical antennas were used because of their bandwidth, their simple dipole patterns, and because we had used them extensively in earlier localized scattering measurements. This system is capable of scan planes up to 7 feet on a side, allowing objects up to approximately 4 feet in length to be measured. The scanners are elevated so that the bottom of the scan plane is at least 4 feet from the ground. We recently added absorber on the floor. However, we have also obtained good results without it.

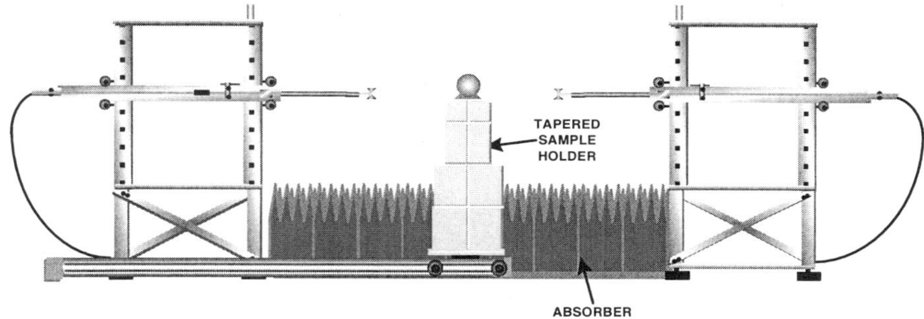

Figure 5. Schematic of the laboratory RCS measurement system.

Computer control of the scanning and measurement system is required because of the large number of measurements involved. For example, a 7 x 7 array of positions involves 1176 individual scattering measurements (we use reciprocity to reduce the number of measurements). For our system, this sequence of measurements takes between 7 and 10 hours depending on the element spacing, number of frequencies measured, and averaging time. For the largest size scan planes, approximately half of this time is spent moving the antennas. Faster scanners and RF equipment would substantially reduce the time required for the measurements.

MEASUREMENT RESULTS

We have performed numerous RCS measurements using our scanned measurement system. This section contains representative results that illustrate the capabilities of the system. The data were taken at a variety of element spacings, but all used a 7 x 7 array of measurement positions. No calibration target is used: the RCS data are derived from measurements with the target present and a clearsite measurement with no target present. To present the data in a common format, the frequency axis has been normalized to the "center" frequency for the scan, $f_0 = c/2s$, defined to be the frequency for which the element spacing s is one-half wavelength (making the overall scan plane $3\lambda_0$ by $3\lambda_0$). The RCS has been normalized to $(\lambda_0)^2$, and the object dimensions have been normalized to λ_0.

The first two cases are for no absorber present. Figure 6 shows the normalized RCS of a slotted conducting cylinder oriented parallel to the incident electric field. The cylinder had a diameter of $0.75\lambda_0$, a length of $0.75\lambda_0$, and a slot ($0.25\lambda_0$ wide and $0.25\lambda_0$ deep) around the center. For the measurement, the cylinder was located $2\lambda_0$ in front of the scan plane. In the figure the solid points represent measured data while the solid curve shows the RCS predicted with our FDTD code. Overall, there is good agreement between the measured and predicted data. This particular target illustrates the large dynamic range possible and the typical bandwidth of 4 to 1 obtained from a 7 x 7 array.

A dielectric coated conducting cylinder was selected as the second target. This cylinder was $0.75\lambda_0$ long, $0.375\lambda_0$ in diameter, with a $0.125\lambda_0$ thick coating. The coating material was a low-loss dielectric with a relative permittivity around 10 (Stycast manufactured by Emerson and Cuming). The cylinder was also located $2\lambda_0$ in front of the array, oriented parallel to the incident electric field. Figure 7 shows the results of the measurement compared to an FDTD prediction. In this case, the agreement between the measured and predicted data is also good. Notice the agreement at the null near the frequency $1.2f_0$.

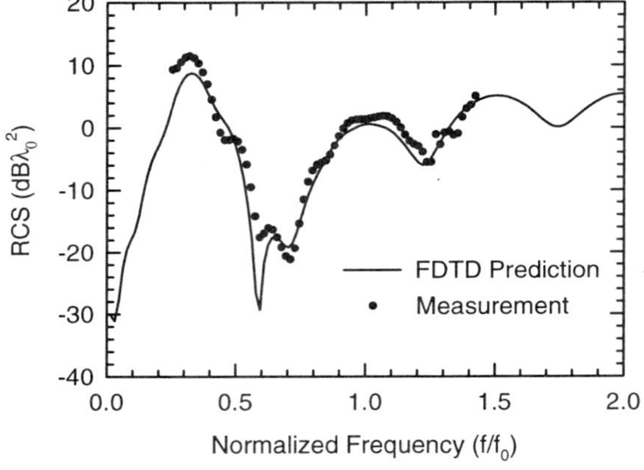

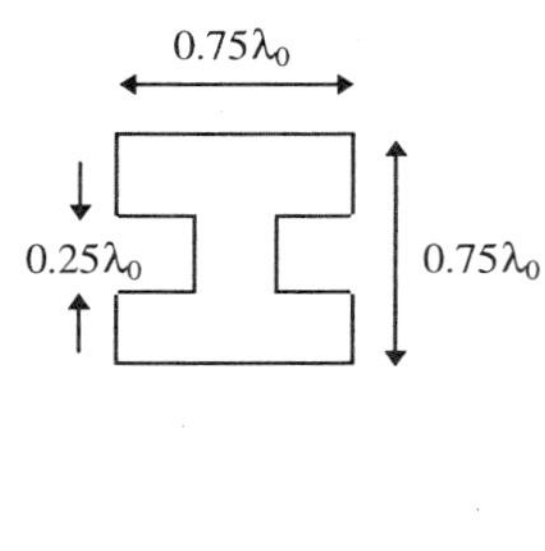

Figure 6. RCS of a $0.75\lambda_0$ long and $0.75\lambda_0$ diameter slotted cylinder.

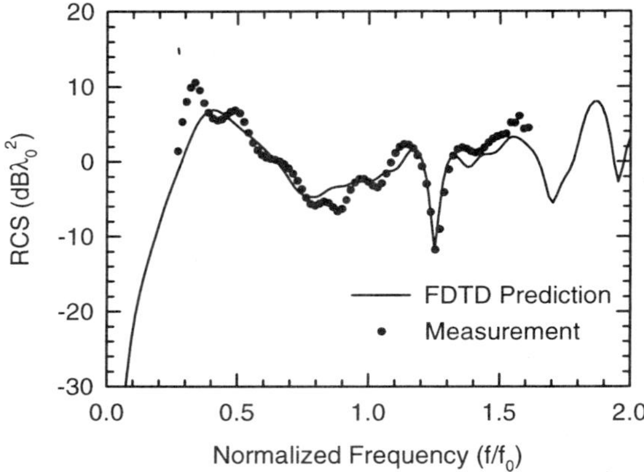

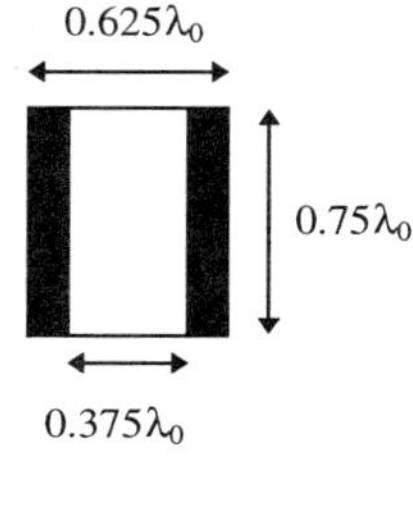

Figure 7. RCS of a dielectric coated cylinder.

For these two data sets, the upper frequency was limited by the antennas used in the measurement. At the upper frequencies, the pattern of the bi-conical antennas began to deviate from that of a simple, short dipole antenna. Since our plane-wave synthesis procedure assumed this type of pattern, the processing did not produce the correct result. However, the synthesis procedure is general and other antenna types could be used as long as they are fully characterized in the near-field.

The remaining data were taken with the absorber present (as indicated in Figure 5) and with other changes in the RF system to increase the signal-to-noise ratio. Figure 8 shows the RCS measured for two different conducting spheres. The data on the left are for a sphere with a diameter of $\lambda_0/4$, while those on the right are for a sphere with a diameter of $\lambda_0/2$. The solid curves are the RCS calculated using a Mie series solution. In this case the target was located λ_0 in front of the scan plane. Notice the excellent agreement over a wide range of frequencies. The upper frequency limit (in normalized units) is significantly higher for these data than the earlier data. This is because the element spacing was 3 times larger, and so the actual upper frequency is still below the point where the antenna patterns deviate from the assumed dipole shape. However, the larger overall scan plane improves the low-frequency performance giving a bandwidth of around 6 to 1.

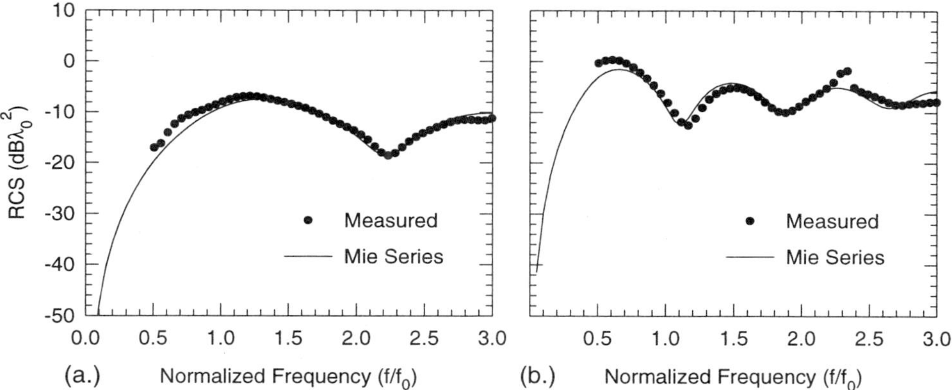

Figure 8. RCS of two different spheres. (a.) Sphere diameter $\lambda_0/4$, (b.) Sphere diameter $\lambda_0/2$.

Next we will show that one set of measured data can be used to obtain the RCS at multiple angles. A square conducting plate ($\lambda_0/2$ on a side) was used as the target, which was located λ_0 in front of and oriented parallel to the scan plane. The plot on the left in Figure 9 shows the data processed to give monostatic RCS at normal incidence, while that on the right shows the same data processed for monostatic RCS at a 20 degree angle (with respect to the normal). The results agree very well with the predicted curves at normal incidence, and reasonably well with those for 20 degrees. As the angle is increased, it becomes more difficult to synthesize the desired plane with the test volume directly in front of the scan plane. Depending on the size of the test volume, we have been able to get good results for angles up to 30 degrees. Bistatic RCS can also be obtained by processing for different transmit and receive angles (a bistatic angle of up to 60 degrees).

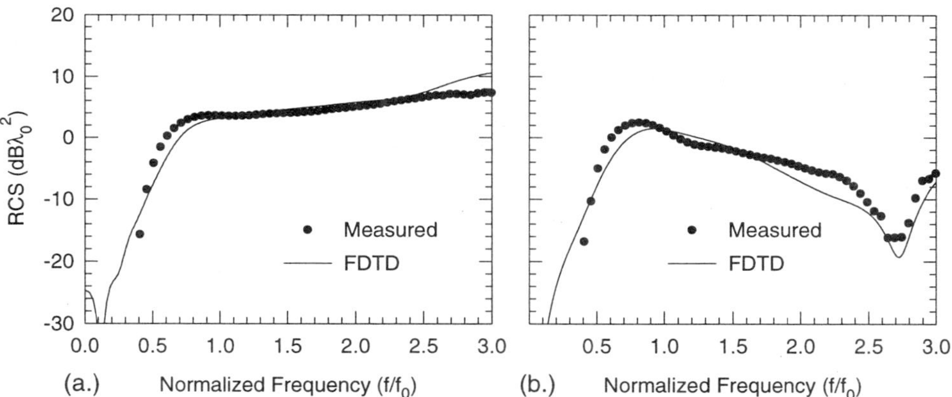

Figure 9. RCS data for a flat plate $\lambda_0/2$ on a side. (a.) Data processed for normal incidence, (b.) Data processed for 20 degrees off normal.

The final two data sets illustrate some of the limits of the current system. Figure 10(a.) shows the RCS of a small pointed cylinder ($\lambda_0/8$ in diameter and approximately $3\lambda_0/8$ in total length with a cone angle of 53 degrees) looking towards the conical end. This plot shows good agreement even for relatively low RCS levels, although the results at the low-frequency end are beginning to deviate substantially because of a poor signal-to-noise ratio. Figure 10 (b.) shows the results for a conducting cylinder whose length is more than half size of the scan plane. (The cylinder was oriented with the end face parallel to the scan plane.) Even with such a large target, the measured RCS agrees with the calculated results to within a few dB.

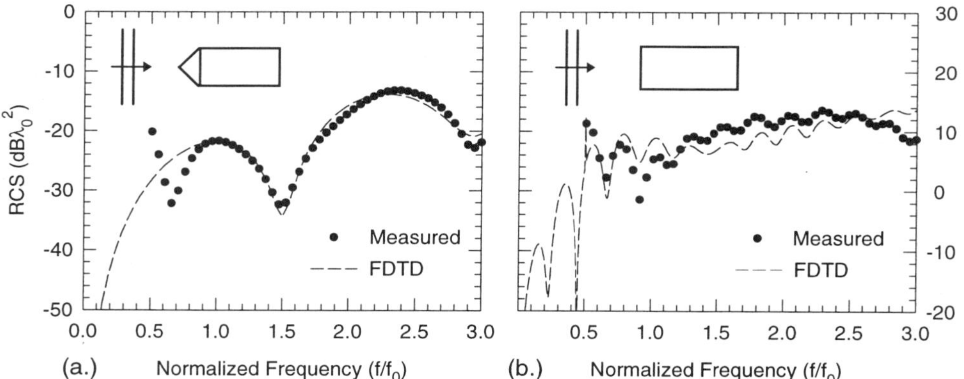

Figure 10. (a.) RCS of a small pointed conducting cylinder (see text for dimensions). (b.) RCS of a large conducting cylinder ($0.75\lambda_0$ diameter by $2\lambda_0$ long).

CONCLUSIONS

We have described the application of localized measurements to the problem of measuring RCS. The technique developed involves performing a series of local scattering measurements, then combining the results in an optimized fashion to obtain RCS. We have developed a laboratory-scale measurement system, and with this we have demonstrated that accurate RCS measurements can be performed in a relatively small space.

The measurements made to date have demonstrated that a scan surface approximately two times the size of the target is sufficient to obtain accurate RCS. This is considerably smaller than for the traditional near-field technique where the scan plane can be many times the size of the target.

REFERENCES

1. J. G. Maloney, M. P. Kesler and E. J. Kuster, "Localized short-pulse scattering from coated cylindrical objects: experimental measurements and numerical models," in: *Ultra-Wideband Short-Pulse Electromagnetics 2*, L. Carin and L. B. Felsen, ed., Plenum Press, New York, (1995).
2. M. A. Dinallo, "Extension of Plane-Wave Scattering-Matrix Theory of Antenna-Antenna Interactions to Three Antennas: A Near-Field Radar Cross Section Concept", in Proceedings of the Antenna Applications Symposium, Urbana, Illinois, pp. 665-686, Sept. 1984.
3. B. J. Cown and C. E. Ryan, Jr., "Near-Field Scattering Measurements for Determining Complex Target RCS", IEEE Trans. Antennas Propagat., Vol. 37, pp. 576-585, May 1989.
4. D. A. Hill, "Theory of Near-Field Phased Arrays for Electromagnetic Susceptibility Testing", National Bureau of Standards Technical Note 1072, February 1984.
5. J. R. Mautz and R. F. Harrington, "Computational Methods for Antenna Pattern Synthesis", IEEE Transactions on Antennas and Propagation, Vol. 23, pp 507-512, July 1975.

FEATURE EXTRACTION FROM ELECTROMAGNETIC BACKSCATTERED DATA USING JOINT TIME-FREQUENCY PROCESSING

L. C. Trintinalia and H. Ling

Department of Electrical and Computer Engineering
The University of Texas at Austin
Austin, TX 78712-1084

INTRODUCTION

In radar cross section applications, it is well known that the electromagnetic backscattering from a complex target can be approximately modeled as if it is emanating from a discrete set of points on the target called scattering centers. The scattering center model, while only an approximation, is conceptually simple and provides a sparse abstraction of the actual target for numerous radar applications. For instance, by storing the strength and position of the scattering centers, the 1D range profile and 2D inverse synthetic aperture radar (ISAR) imagery of the target can be easily reconstructed in real time, alleviating the need for the storage of large data sets. Therefore, the extraction of the scattering center model of a target from electromagnetic signature data is an important problem[1].

Most real targets, however, are not exactly amenable to such scattering center, or point scatterer, description. One example is the resonant or dispersive scattering phenomena due to sub-skinline features such as inlet ducts, cockpits and antenna windows on an aircraft. These scattering mechanisms appear in an ISAR image as blurred clouds that extend down-range and often do not correspond to the spatial features on the target. This type of scattering occurs most prominently when the characteristic wavelength of the radar is on the order of the resonant dimension of the sub-skinline feature. Another example of scattering data which deviate from a perfect point scatterer model is when a scattering feature shows strong dependence on aspect angle. This issue of limited visibility of scattering centers is in fact rather commonplace in realistic targets due to self occlusion and shadowing from other parts of the target. It is especially prominent for multiple-bounce returns.

Joint time-frequency analysis is a well-established technique in signal processing[2,3] and has been successfully applied to electromagnetic scattering data recently[4-10]. In this paper, we present two applications of the joint time-frequency technique to extract target features from electromagnetic backscattered data in the presence of non-point scattering behaviors. In the first part, we address the issue of extracting resonant mechanisms in signature data.

We present a new processing technique, the joint time-frequency ISAR, to decompose an ISAR image into an enhanced ISAR image containing only scattering centers and a frequency-aspect plot to display target resonance features and cutoff phenomena. In the second part of this paper, we present the application of joint time-frequency processing technique to the aspect dimension of the backscattered data (in this context, it is perhaps more appropriate to use the term joint (aspect)-(cross range) as oppose to joint time-frequency). This implementation allows the parameterization of the backscattered data as a summation of point scatterers whose visibility is aspect dependent. It should be, therefore, a very compact way to store the backscattered data collected over a large angular sector. Once such representations are available, it allows fast reconstruction of the ISAR image at any observation angle.

JOINT TIME-FREQUENCY ISAR

We shall describe the conceptual idea behind the joint time-frequency ISAR algorithm for extracting resonant and dispersive mechanisms from the ISAR image. The joint time-frequency processing is first applied to the range (or time) axis of the conventional (range)-(cross range) ISAR image to gain an additional frequency dimension. The result is a three-dimensional (range)-(cross range)-(frequency) matrix, with each (range)-(cross range) slice of this matrix representing an ISAR image at a particular frequency. Consequently, by examining how the ISAR image varies with frequency, we can distinguish the frequency-independent scattering mechanisms from the frequency-dependent ones. In the actual implementation of the joint time-frequency ISAR, the choice of the joint time-frequency processing engine is critical. Our choice is the adaptive Gaussian representation[11], a signal representation scheme that uses adaptive normalized Gaussian as basis functions (a similar algorithm called matching pursuit was developed independently by Mallat and Zhang at around the same time[12]). The objective of this method is to expand a signal $s(t)$ in terms of normalized Gaussian functions $h_p(t)$ with an adjustable standard deviation σ_p and a time-frequency center (tp, fp):

$$s(t) = \sum_{p=1}^{\infty} B_p h_p(t) \tag{1a}$$

where

$$h_p(t) = \left(\pi\sigma_p^2\right)^{-0.25} \exp\left[-\frac{\left(t-t_p\right)^2}{2\sigma_p^2}\right] \exp\left(j2\pi f_p t\right) \tag{1b}$$

The coefficients B_p are found one at a time by an iterative search procedure. One begins at the stage $p=1$ by choosing the parameters σ_p, t_p and f_p such that $h_p(t)$ is most "similar" to $s(t)$, i.e., for which the inner product between $s(t)$ and $h_p(t)$ is the largest. The next B_p is found using the same procedure after the orthogonal projection of $s(t)$ onto $h_p(t)$ has been removed from the signal. This procedure is iterated to generate as many coefficients as needed to accurately represent the original signal. The major difficulty in implementing this algorithm is the determination of the optimal elementary function at each stage. In our implementation we used the same guidelines as those given by Qian and Chen[11]. We start with a large σ_p and scan the data in frequency and time for a peak. Then we divide σ_p by two and find the new peak. We continue this procedure until the standard deviation is small enough and then select the highest peak and extract the residual. It should be pointed out that the fast Fourier transform is used during the search procedure to obtain the coefficients for all the frequency centers at once, speeding up the search that would otherwise be very time consuming.

The adaptive Gaussian representation has two distinct advantages over conventional time-frequency techniques such as the short-time Fourier transform. First, it is a parameterization procedure that results in very high time-frequency resolution. More importantly for our application, the adaptive spectrogram allows us to automatically distinguish the frequency-dependent events from the frequency-independent ones through the extent of the basis functions. From expression (1b) it can be seen that scattering centers, i.e., signals with very narrow length in time, will be well represented by basis functions with very small σ_p. Frequency resonances, on the other hand, will be better depicted by large σ_p. Therefore, if we reconstruct the ISAR image using only those Gaussian bases with small variances, a much "cleaner" image can be obtained showing only the scattering centers. The remaining mechanisms, i.e., those related to the large variance Gaussians, are more meaningful to view in a dual frequency-aspect display, where resonances and other frequency-dependent mechanisms can be better identified.

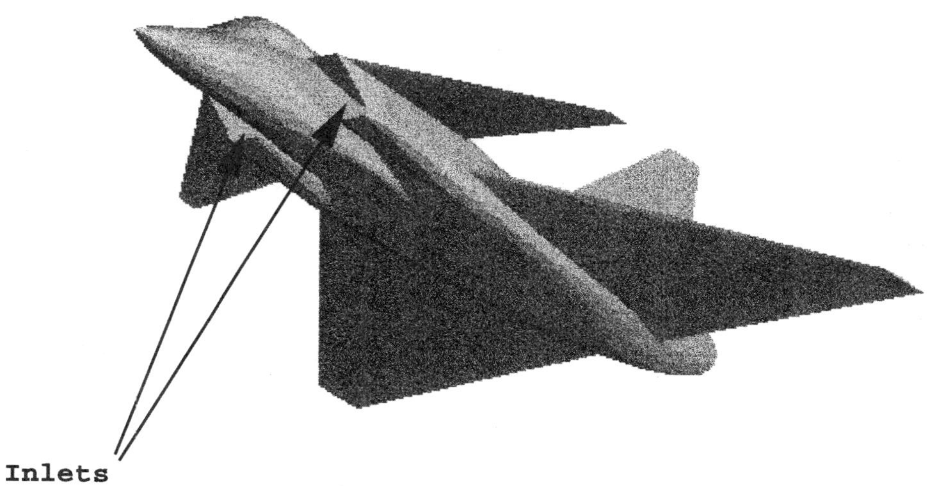

Figure 1. VFY-218 airplane model.

Results

The algorithm is demonstrated using the chamber measurement data of a 1:30 scale model Lockheed VFY-218 airplane provided by the EMCC (Electromagnetic Code Consortium)[13]. The airplane has two long engine inlet ducts, as shown in Figure 1, which are rectangular at the open ends but merge together into one circular section before reaching a single-compressor face. As we can see in the conventional ISAR image of Figure

2 for the horizontal polarization at 20° near nose-on, the large cloud over the right wing is the inlet return. Figure 3 shows the enhanced ISAR image of Figure 2, obtained by applying the above joint time-frequency ISAR algorithm (622 basis functions were used to represent the whole ISAR image) and keeping only the small variance Gaussians. We see that only the scattering center part of the original signal remains in the image, as expected. Notice that the strong return due to engine inlet has been removed, but the scattering returns from the wing tips remain. Figure 4 shows the frequency-aspect display of the high variance Gaussians. A number of equispaced vertical lines can be seen between 10.5 and 13.5 GHz. Given the dimension of the rectangular inlet opening (approximately 1.5 cm × 2.5 cm), we estimate that these frequencies correspond approximately to the cutoff frequencies of the higher order modes in the waveguide-like inlet. We have also run our algorithm on the simulated scattering data, generated using a numerical Maxwell's solver, for a plate-waveguide configuration to verify this claim. The frequency resonances displayed indeed correspond very closely to the theoretical cutoff frequencies of the waveguide.

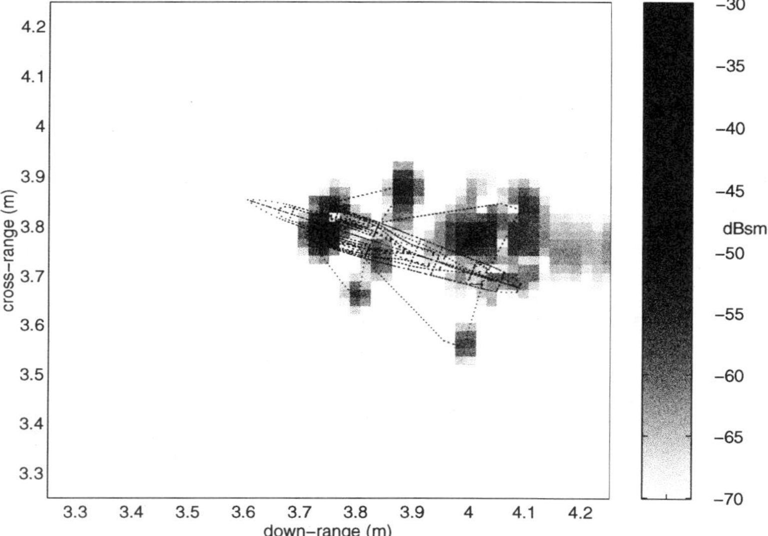

Figure 2. Standard ISAR image of the VFY-218 model, obtained for f=8-16 GHz and θ=0°-40°, horizontal polarization.

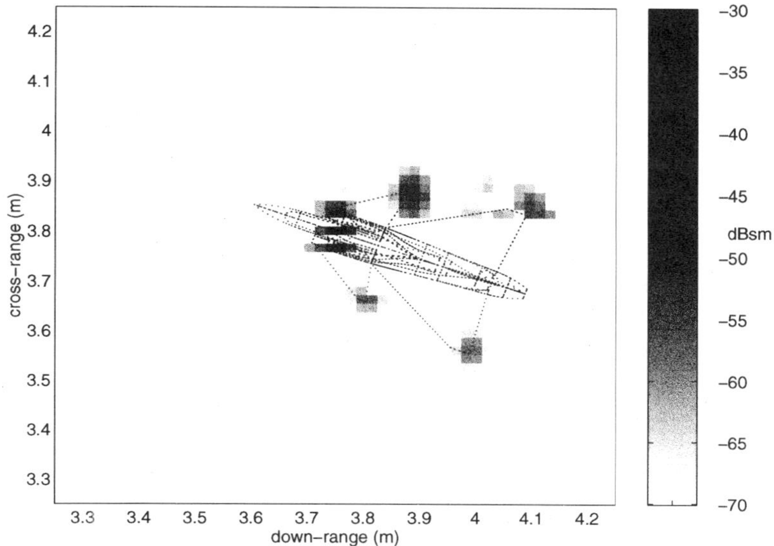

Figure 3. Enhanced ISAR image of the VFY-218 model, obtained by applying the adaptive Gaussian representation to the ISAR image of Figure 2, and keeping only the small variance terms.

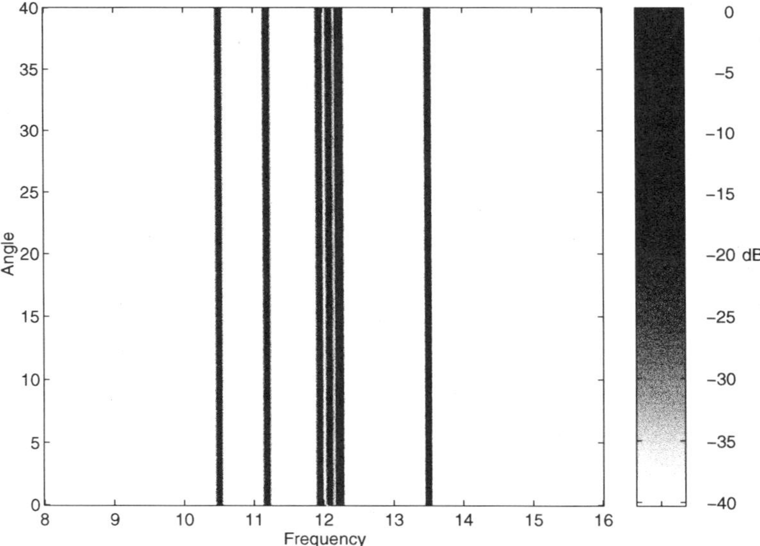

Figure 4. Frequency aspect display of the VFY-218 model, obtained by applying the adaptive Gaussian representation to the ISAR image of Figure 2, and keeping only the large variance terms.

EXTRACTING SCATTERING CENTER VISIBILITY USING ADAPTIVE GAUSSIAN REPRESENTATION

We will now present a second application of the joint time-frequency processing concept to handle the issue of scattering center visibility. We assume that the target is composed of point scattering centers only, but that each scattering center is visible only over certain ranges of observation angles. Therefore data collected over a broad angular window could be better represented by a summation of point-scatterers multiplied by their respective visibility function:

$$S(f,\theta)=\sum_n A_n(\theta)\exp\left[-2j\left(k_x x_n + k_y y_n\right)\right] \qquad (2)$$
$$k_x = 2\pi f\cos\theta/c \quad k_y = 2\pi f\sin\theta/c$$

If we assume that this angular dependence can be well represented by Gaussian basis functions with arbitrary variances (like those used in equation (1)), we can apply a modified version of our joint time-frequency ISAR to obtain the representation showed in (2) by rewriting it as:

$$S(f,\theta)=\sum_n A_n \exp\left[-\frac{(\theta-\theta_n)^2}{2\sigma_n^2}\right]\exp\left[-2j\left(k_x x_n + k_y y_n\right)\right] \qquad (3)$$

To obtain such representation we use a search algorithm similar to the one described earlier with two major differences. The first difference is that the search algorithm must be applied in the two-dimensional k_x-k_y plane (or θ-f plane) rather than just over a one dimensional time or frequency axis as in the previous section. The second is that now 4 parameters (θ_n, σ_n, x_n, y_n) rather than 3 need to be determined. The search for θ_n and σ_n is performed by a zoom-in algorithm as before, and at each stage a two-dimensional fast Fourier transform is used to determine the values for x_n and y_n that gives us the highest coefficient A_n. After obtaining the expansion given by (3) one can easily reconstruct the ISAR image of the target at any observation angle θ_0.

Results

We applied our algorithm to backscattered data of a generic complex target, simulated using XPATCH[14]. The simulation was carried out for frequencies from 9.75 up to 10.25 GHz over a 360° angular span. Using the procedure described earlier, we extracted 500 terms to represent the total backscattered of this target. The energy in the remaining residue was only 4% of the total backscattered energy.

Figure 5(a) shows the original ISAR image of this target obtained for an angular window of 3°. Figure 5(b) shows the image reconstructed using (4), and we can see that we obtained a good agreement between those images. The agreement for other observation angles (not shown here) is also satisfactory. Other types of basis functions that could provide faster convergence are currently being investigated.

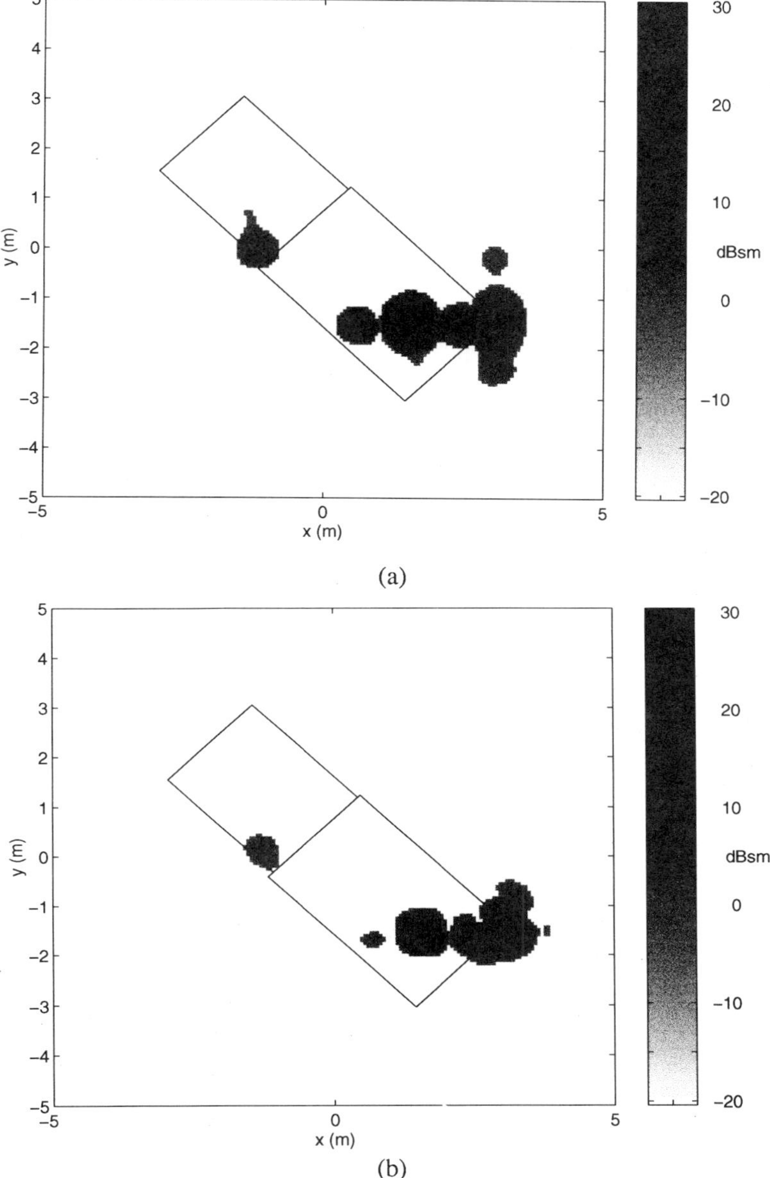

Figure 5. (a) Original ISAR image of a generic target (shown in outline), obtained for an angular window of 3°, and (b) the image reconstructed using equation (4).

CONCLUSIONS

In this paper we presented two applications of the joint time-frequency technique to extract target features from electromagnetic backscattered data in the presence of non-point scattering behaviors. In the first application, we developed a joint time-frequency ISAR algorithm to process data from complex targets containing not only scattering centers but also other frequency-dependent scattering mechanisms. The adaptive joint time-frequency ISAR algorithm allows the enhancement of the ISAR image by eliminating non-point scatterer signals, thus leading to a much cleaner ISAR image. It also provides information on the extracted frequency-dependent mechanisms such as resonances and frequency dispersions. This is accomplished without any loss in resolution. In the second application, we extended the same technique to the aspect axis of the backscattered data. This alternative implementation allowed the parameterization of the backscattered data as a summation of point scatterers multiplied by an aspect dependent visibility function. This expansion is a very compact way to store the backscattered data collected over large angular sectors. Once such representations are available, it enables fast reconstruction of the ISAR image at any observation angle.

ACKNOWLEDGMENTS

This work was sponsored by the Joint Services Electronics Program under Contract No. AFOSR F49620-95-C-0045. The United States Government is authorized to reproduce and distribute reprints for governmental purposes notwithstanding any copyright notation hereon.

L. C. Trintinalia is on leave from Department of Electronic Engineering of Escola Politécnica da Universidade de São Paulo, Brazil and is also sponsored by CNPq.

REFERENCES

1. R. Bhalla and H. Ling, "3D scattering center extraction using the shooting and bouncing ray technique," to appear in *IEEE Trans. Antennas Propagat.*, 1996.
2. L. Cohen, *Time-Frequency Analysis*, Prentice Hall, Englewood Cliffs, NJ, 1995.
3. S. Qian and D. Chen, *Introduction to Joint Time-Frequency Analysis - Methods and Applications.* Prentice Hall, Englewood Cliffs, NJ, 1996.
4. K. F. Casey, D. G. Dudley, and M. R. Portnoff, "Radiation and dispersion effects from frequency-modulated (FM) sources," *Electromagnetics*, vol. 10, pp. 349-376, 1990.
5. A. Moghaddar and E. K. Walton, "Time-frequency-distribution analysis of scattering from waveguide cavities," *IEEE Trans. Antennas Propagat.*, vol. 41, pp. 677-680, May 1993.
6. H. Kim and H. Ling, "Wavelet analysis of radar echo from finite-size targets," *IEEE Trans. Antennas Propagat.*, vol. AP-41, pp. 200-207, Feb. 1993.
7. H. Ling, J. Moore, D. Bouche and V. Saavedra, "Time-frequency analysis of backscattered data from a coated strip with a gap," *IEEE Trans. Antennas Propagat.*, vol. 41, pp. 1147-1150, Aug. 1993.
8. L. Carin and L. B. Felsen, "Wave-oriented data processing for frequency- and time-domain scattering by nonuniform truncated arrays," *IEEE Antennas Propagat. Mag.*, vol. 36, pp. 29-43, June 1994.
9. J. Moore and H. Ling, "Super-resolved time-frequency analysis of wideband backscattered data," *IEEE Trans. Antennas Propagat.*, vol. 43, pp. 623-626, June 1995.
10. L. C. Trintinalia and H. Ling, "Interpretation of scattering phenomenology in slotted waveguide structures via time-frequency processing," *IEEE Trans. Antennas Propagat.*, vol. 43, pp. 1253-1261, Nov. 1995.
11. S. Qian and D. Chen, "Signal representation using adaptive normalized Gaussian functions," *Signal Processing*, vol. 36, no. 1, pp. 1-11, Mar. 1994.
12. S. G. Mallat and Z. Zhang, "Matching pursuits with time-frequency dictionaries," *IEEE Trans. Signal Processing*, vol. 41, pp. 3397-3415, Dec. 1993.
13. H. T. G. Wang, M. L. Sanders and A. Woo, "Radar cross section measurement data of the VFY 218 configuration," *Tech. Rept. NAWCWPNS TM-7621*, Naval Air Warfare Center, China Lake, CA, Jan. 1994.
14. D. J. Andersh, M. Hazlett, S. W. Lee, D. D. Reeves, D. P. Sullivan and Y. Chu, "Xpatch: A high-frequency electromagnetic scattering prediction code and environment for complex three-dimensional objects," *IEEE Antennas Propagat. Mag.*, vol. 36, pp. 65-69, Feb. 1994.

CLASSIFICATION OF BURIED TARGETS USING TIME-FREQUENCY SIGNATURES EXTRACTED BY A GROUND PENETRATING RADAR

H. C. Strifors,[1] A. Gustafsson,[2] S. Abrahamson,[2] and G. C. Gaunaurd[3]

[1]National Defense Research Establishment (FOA 6)
S-17290 Stockholm, Sweden
[2]National Defense Research Establishment (FOA 3)
P. O. Box 1165, S-58111 Linköping, Sweden
[3]Naval Surface Warfare Center (Code 684)
White Oak, Silver Spring, MD 20903-5640

INTRODUCTION

It is gaining acceptance by the radar signature community that effective target recognition requires more advanced signal analysis methods than those available for the frequency or time domains, separately. In recent years, the various distributions in the time-frequency domain of the Wigner-type, or Cohen class[1–3] have proven their feasibility of extracting features in the cross-section of returned echoes and be more informative by their ability to display the time evolution of these features. As we demonstrated earlier, resonance features generated in the time-frequency domain from an echo backscattered by a target in free space,[4–6] or buried underground,[7] when a pulse from an impulse radar is incident on it, could be used to identify the target. A precondition in these time-frequency analyses is that the broad-band pulses emitted by the impulse radar do not exhibit distorting anomalies in their spectra. These anomalies could detrimentally affect the extracted resonance features in the time-frequency distribution of the pulses backscattered by any target.

The effective use of an impulse radar as a ground penetrating probe requires that it cannot only detect buried targets but that it is also capable of identifying or classifying detected targets. In these situations the antenna and the target are located in different half-spaces having different electrical properties, and the extracted echoes will be distorted by multiple scattering between the target and the surface of the ground and also by scattering from discontinuities in the ground itself. Analyzing returned echoes from targets buried at selected depths we found[7] that a pseudo-Wigner distribution (PWD) is capable of generating the dominant features of a target that could be used to identify it at different depths. On the other hand, the presence of clutter and multiple scattering obscuring the echoes, corrupted the frequency-domain target signatures altogether at the larger depths.

The Wigner distribution (WD) has a number of properties[2,8,9] that need not be repeated here, and the basic difficulty in applications is the occurrence of cross-terms interference that is due to the fact that it is a bilinear functional of the signal. However, using various smoothing techniques such as windows some trade-off is possible between the de-

Ultra-Wideband, Short-Pulse Electromagnetics 3
Edited by Baum *et al.*, Plenum Press, New York, 1997

sirable time-frequency concentration of the features and the undesirable distortion caused by the cross-terms in the then resulting pseudo-Wigner distribution (PWD). Sufficient smoothing is often conveniently achieved by time-windowing the signal before evaluating the WD.

Fuzzy C-means (FCM) is a data clustering technique where each data point in a cluster belongs to each subcluster specified by a degree of membership that depends on its relative distance to the centers of the respective subclusters. To use this technique we first convert each PWD to a point cluster representation. Then, using the FCM each PWD used as a template for a target signature can be further reduced to a set of cluster centers. A straightforward classification algorithm is then applied to validation data taken from an additional set of returned echoes. Here, the same targets are used but they are buried at different depths, illuminated with the GPR antennas at slightly different positions, or both. Class membership of a target is then decided using a simple metric. The results of our investigation serve to assess the possibility of classifying subsurface targets using a GPR.

TARGET RESPONSE IN THE TIME-FREQUENCY DOMAIN

A novel approach to processing signals returned from a target when it is illuminated by a pulse emitted from an ultra-wideband radar is to analyze them in the joint time-frequency domain. This approach exhibits the time-evolution of the target resonances in a more informative way than in the time or frequency domains, separately. It is thus more advantageous for target identification purposes. The general class (or, the Cohen class) of bilinear distributions[1,2] is defined by:

$$C_{\tilde{f}}(t,\omega;\Phi)=\frac{1}{2\pi}\int_{-\infty}^{+\infty}\int_{-\infty}^{+\infty}\int_{-\infty}^{+\infty}\Phi(\tau,\xi)\,e^{i(\xi t-\omega\tau-\xi u)}\,\tilde{f}(u+\frac{\tau}{2})\,\tilde{f}^*(u-\frac{\tau}{2})\,du\,d\tau\,d\xi, \tag{1}$$

where different choices of the kernel function $\Phi(t,\omega)$ yield the numerous distributions in current use. For example, if $\Phi(t,\omega)\equiv 1$, the Wigner distribution[8] (WD) follows:

$$\mathrm{WD}_{\tilde{f}}(t,\omega)=\int_{-\infty}^{+\infty}\tilde{f}(t+\frac{\tau}{2})\,\tilde{f}^*(t-\frac{\tau}{2})\,e^{-i\omega\tau}\,d\tau. \tag{2}$$

In Eqs. (1) and (2), we could use the analytic function[2,8] $\tilde{f}$, which is related to the given signal f by:

$$\tilde{f}(t)=f(t)-\frac{i}{\pi}\,\mathrm{p.v.}\int_{-\infty}^{+\infty}\frac{f(\tau)}{t-\tau}\,d\tau, \tag{3}$$

where p.v. indicates the Cauchy principal value of the integral. The analytic signal has a one-sided Fourier transform, i.e., the Fourier transform of $\tilde{f}$ is zero for negative frequencies. Using the analytic signal when computing any of the bilinear time-frequency distributions (having a period of π rather than the 2π of the Fourier transform) eliminates the need of using a sampling rate that is twice the Nyquist rate.

Introducing a window-function gives the pseudo-Wigner distribution (PWD) in the form used here:

$$\mathrm{PWD}_{\tilde{f}}(t,\omega)=\int_{-\infty}^{+\infty}\tilde{f}(t+\frac{\tau}{2})\,\tilde{f}^*(t-\frac{\tau}{2})\,w_f(\frac{\tau}{2})\,w_f^*(-\frac{\tau}{2})\,e^{-i\omega\tau}\,d\tau, \tag{4}$$

where we select a Gaussian window of the form: $w_f(t)=\exp(-\alpha t^2)$. The parameter α controls the width of the window. While the cross-terms interference between different portions of a noise-free signal backscattered from a target produces desirable signature features in the PWD, cross-terms associated with added noise give rise to undesirable "artifacts" that tend to obscure the target's signature. However, selecting an appropriate value of α, the cross-terms interference can be suppressed to a desirable degree. For numerical calculations we use a discrete-time version of the PWD of f in the form:

$$\mathrm{PWD}_f(l,k)=2\sum_{n=0}^{N-1} f(l+n)\,f^*(l-n)\,w_f(n)\,w_f^*(-n)\,e^{-i(4\pi/N)kn}, \tag{5}$$

where l,k=0,1,2,...N-1 represent time and frequency, respectively.

Any PWD calculated using Eq. (5) is then known in a rectangular meshgrid defined by intervals of the integer coordinates l and k. In each mesh the value of the PWD is constant and we represent it by a small number of points being randomly distributed within the mesh and with the number being proportional to the function value. In this way we get a point cluster representation of the PWD that can be used in conjunction with fuzzy classification model based on cluster estimation.[10,11] A typical cluster representation of a PWD is, in our applications, typically composed of about a thousand points. Using the "fuzzy C-means" (FCM) clustering algorithm[12,13] we can considerably reduce the number of characterizing cluster points and get a more feasible representation of a target's signature consisting of a few tens of points, called cluster centers.

The main idea is that any set of points that are distributed over a bounded plane surface can be represented by a (much) smaller number of cluster centers, which in a fuzzy or indistinctly meaning are located as close as possible to all of the given points. This could be achieved by first introducing a cost function that is defined to be the sum of each point's squared distance to the most adjacent cluster center, and then adjusting the position of each cluster center so as to minimize the cost function. This procedure is based on the assumption that all points that are closest to a given cluster center form a subcluster where these points have unit membership. If now this cost function is modified by the introduction of a slightly different membership function that instead associates a degree of membership μ_{ik} with each point k and each cluster center i, such that $\sum_{i=1}^{c}\mu_{ik}=1$ for each k, we arrive at the idea of the fuzzy C-means algorithm. The FCM algorithm is an iterative optimization algorithm that minimizes the cost function J:

$$J=\sum_{k=1}^{n}\sum_{i=1}^{c}\mu_{ik}^{m}\left\|x_k-\mathrm{v}_i\right\|^2, \tag{6}$$

where n is the (total) number of data points x_k in the cluster, c is the number of cluster centers v_i, and m is a real number larger than 1 (typically, m = 2).

The degree of membership μ_{ik} is defined by

$$\mu_{ik} = \frac{1}{\sum_{j=1}^{c} \left(\frac{\|x_k - v_i\|}{\|x_k - v_j\|} \right)^{2/(m-1)}}. \tag{7}$$

Starting with a desired number of clusters c, and an initial guess for each cluster center $v_i, i = 1,2,\ldots,c$, the FCM algorithm will converge to a solution for the cluster centers that represent either a local minimum or a saddle point of the cost function.[11] The exponent m in the cost function J establishes the relative emphasis that is placed on the membership of the cluster points.

EXPERIMENTAL SETUP AND IMPULSE RADAR MEASUREMENT

In the experiments, an impulse radar system plays the role of a ground penetrating radar (GPR), and the target is buried in dry sand contained in an indoor sandbox with length and width of 5 m and depth of 2 m. The sandbox is located at the Linköping site of the National Defense Research Establishment (FOA 3). The impulse radar system transmits pulses of length 0.3 ns with a pulse repetition frequency of 250 kHz and a peak power of 50 W. The antenna unit contains two broad-band dipoles (transmitting and receiving) that are perpendicular to each other (to suppress the strong specular return from the ground surface). The radar system operates in the frequency band of 0.2-2 GHz. Data are sampled and preprocessed using a digitizing signal analyzer that has a signal level resolution of 14 bits, and the equivalent-time sampling rate is chosen to be 20 GS/s (gigasamples per second). The analog bandwidth of the signal analyzer system does not impose any further restriction on the total bandwidth of the complete radar and signal analyzer system. To suppress noise, the time-series are each composed of 10 ensemble averaged, or time-overlaid, sampled waveforms.

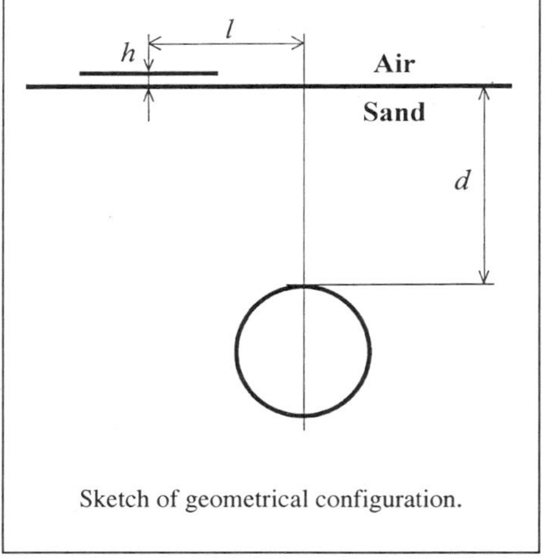

Sketch of geometrical configuration.

The GPR antenna is horizontally mounted on a computer controlled positioning system above the ground surface. We use here an antenna location at a height $h = 2$ cm above the ground surface; see sketch. Also when using crossed dipole antennas and the target is a metal sphere or a short metal cylinder with plane end surfaces and with the symmetry axis in the vertical direction, stronger echoes can be obtained if the center of the antenna is located at a small horizontal distance from the vertical symmetry axis of the target. We use here a horizontal distance of $l = 20$cm. This means that the echoes received from the target are not strictly monostatic returns. The diameter of the metal spheres used as targets in the experiments is either 10 cm (4") or 30 cm (12") and the metal cylinder has diameter 10 cm (4") and height 5 cm (2"). We have found that there exists a strong coupling between the transmitting and receiving antenna elements within this particular antenna unit. This coupling can be reduced by first performing a measurement with the antenna unit elevated (in this case) to a height of about 70 cm above the ground surface (without any target buried in the sand). We then subtract that recorded time-series from each one of the time-series obtained with a target present.

AUTOMATIC CLASSIFICATION OF BURIED TARGETS

We compute the time-frequency signature as given by the PWD of the above three metal targets when they are buried at the depth 5 cm and are illuminated by the impulse radar. For each PWD the cluster representation is then computed and the FCM algorithm is used to get a set of cluster centers. We select the value $m = 2$ for the exponent in the membership function and $c = 40$ for the number of cluster centers. Figure 1 displays the cluster representations of the PWD for the three metal targets, viz., the two spheres of diameter 10 cm and 30 cm and the short cylinder of diameter 10 cm and height 5 cm. These plots also display the cluster centers that result from the application of the FCM algorithm. We found that using a smaller value of the exponent m the cluster centers tend to be more outspread over the area covered by the cluster points. Conversely, a larger value will tend to concentrate the cluster centers at the central portion of the PWD cluster where the density of cluster points is large.

The obtained sets of cluster centers are then used as templates for classification using backscattered signals when any of the above targets is illuminated by the impulse radar. In these cases the targets are buried at the larger depth of $h = 15$ cm, while the location of the antenna unit is the same. Figure 2 displays the results of one attempt at classifying a target at the depth 15 cm using the three templates. (Note that the time scale is different from Fig. 1.) The system automatically record the information of the particular target and the number of inputted waveform, in this case Wfm 326, which establishes the actual antenna location. Each one of the sets of cluster centers, or templates, is then made to slide along the time axis in small steps together with a time window of width 4 ns denoted by the horizontal lines that appear in front of and behind the cluster centers in Fig. 2. At each location of the template the relative cost function $j = J/n$ is computed, where n is the number of data points contained within the time window. (The data points that occur at a time of about 1 ns is clutter contribution from a possibly unevenly packed sand layer covering the target.) The system keeps track of the smallest value of j and the location where it occurs. This procedure is repeated for each template and the optimum location of the template is displayed in each plot in Fig. 2, and the corresponding value of the cost function j is recorded together with the name of the template used. We note that the smallest value of the cost function occurs in the upper left plot, which means that the target is classified as the metal sphere of diameter 10 cm, which indeed it is.

In the next classification situation we have selected as target the larger metal sphere (diameter 30 cm) buried at depth 15 cm. The optimum localization of each template is displayed overlaid on the cluster plot of the target in Fig. 3. The cost function j has the smallest value when the template is the large sphere, which means that also this target is correctly classified. Finally, Fig. 4 displays the result of the classification when the target is the short metal cylinder buried at depth 15 cm. The classification is apparently correct, but we note that there is a relatively small difference between the values of the cost function for the correct template and the template of the large sphere. In fact, when the target is buried at larger depths or the antennas are moved horizontally, the classification becomes less reliable. Most often it is the cylindrical target that is wrongly classified.

We have noted that the classification performance is sensitive to how well one can eliminate the effect of antenna coupling, which generates a strong signal. In some cases it seems that the coupling signal ought to be time shifted only a fraction of the time step used in the sampling of the signals. That would require a signal processing that uses re-sampling at a higher sampling rate, which will be implemented soon into the algorithms we are now applying.

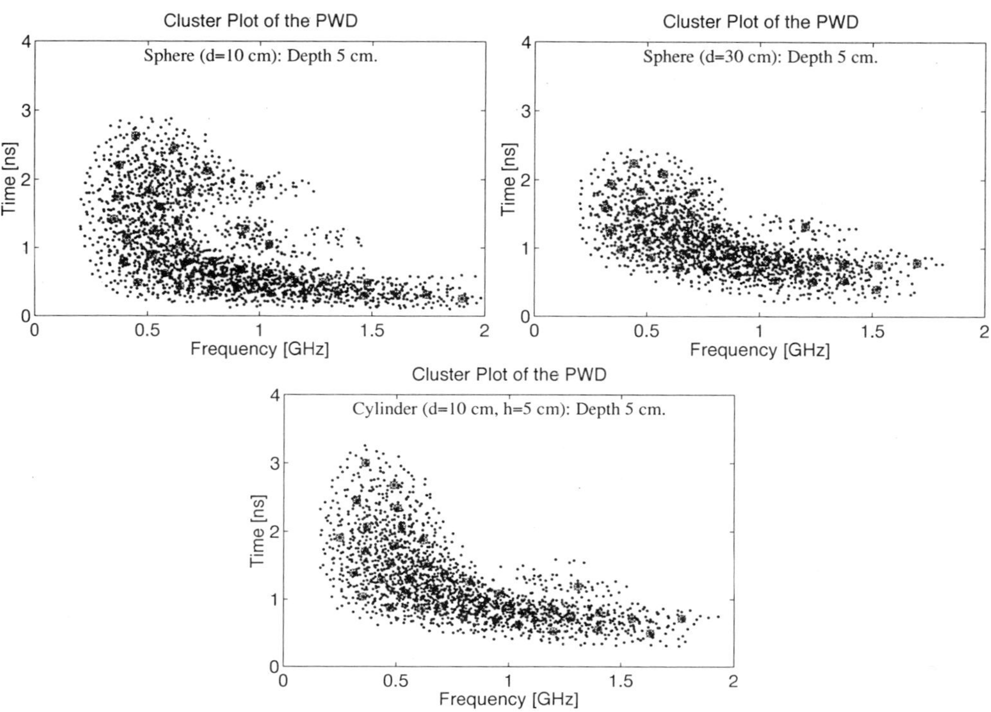

Figure 1. Cluster plots of the PWDs and the template cluster centers of the three targets.

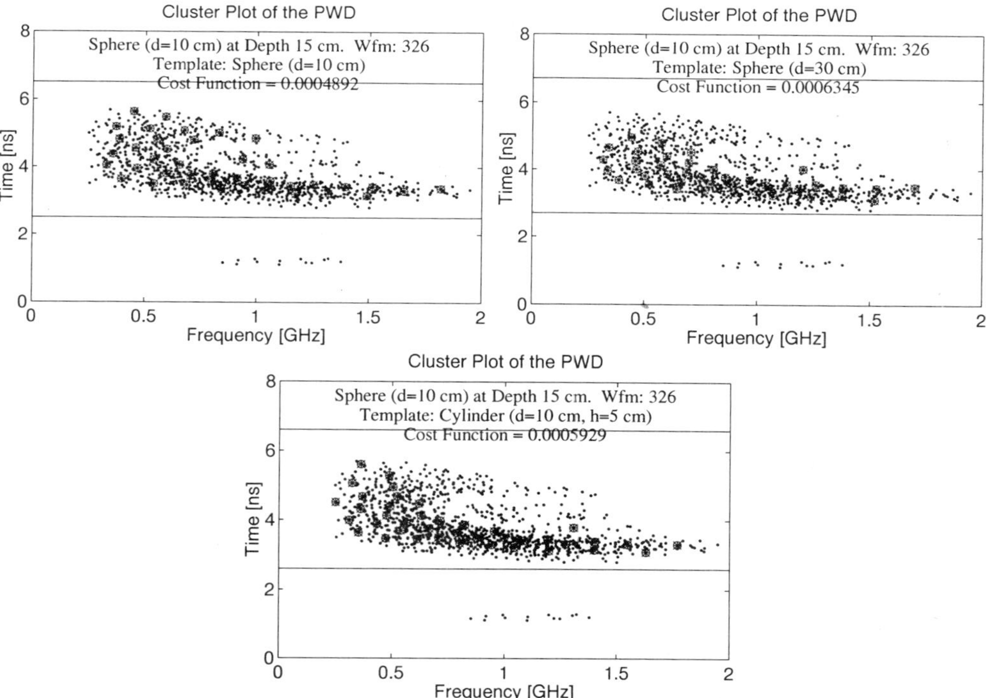

Figure 2. Cluster plots of the PWDs of the smaller spherical target at a burying depth of 15 cm together with each one of the three templates at optimum location.

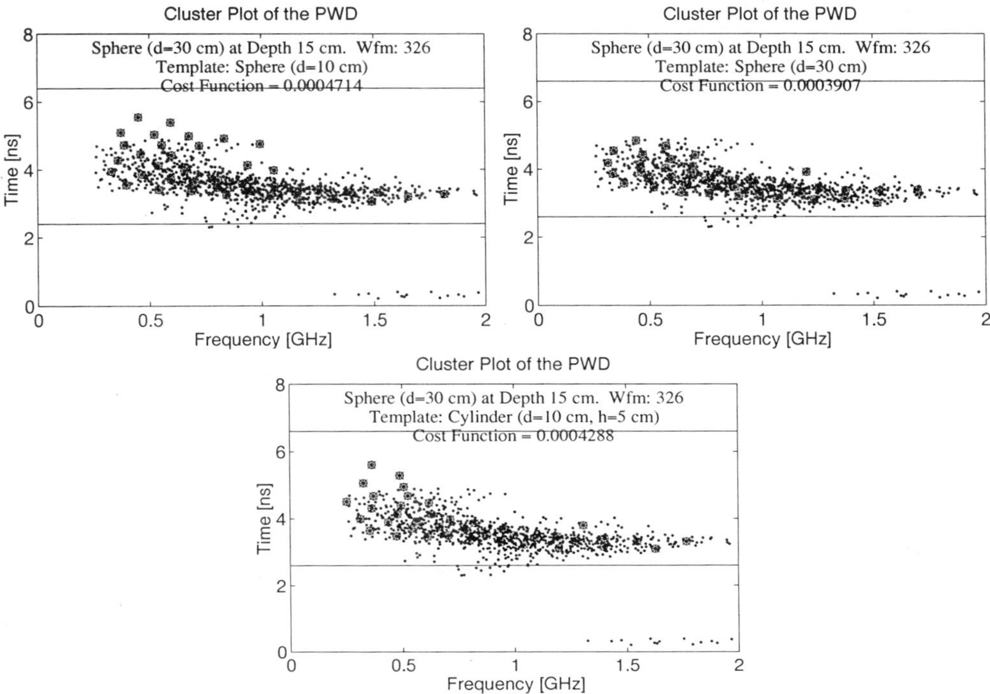

Figure 3. PWDs of the larger spherical target together with each one of the three templates at optimum location.

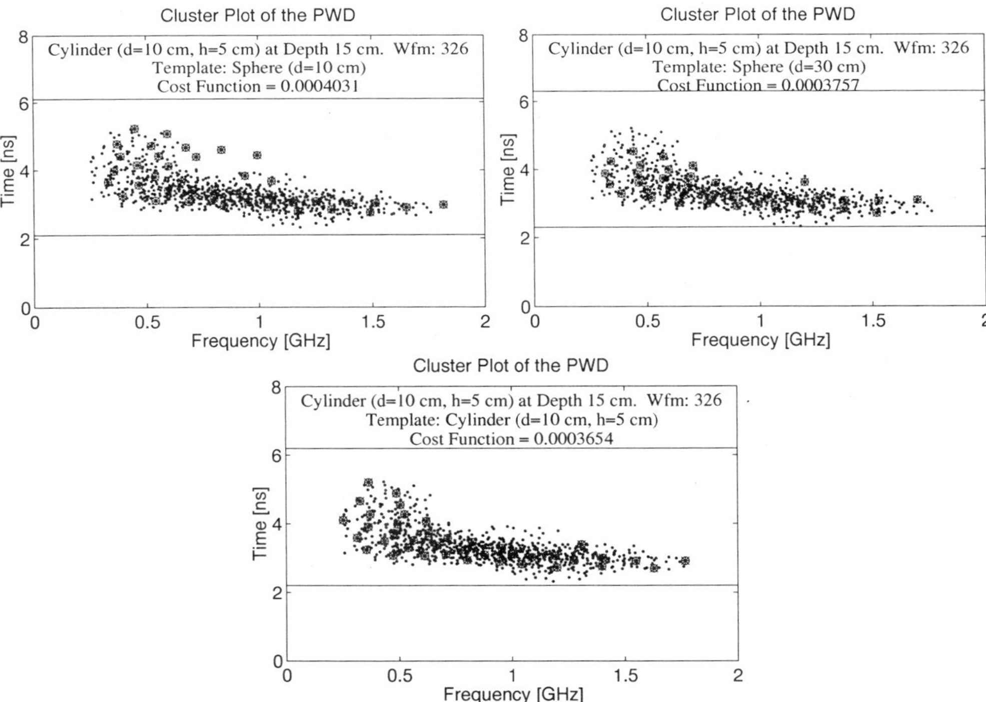

Figure 4. PWDs of the cylindrical target together with each one of the three templates.

DISCUSSION

We have studied the transient interaction resulting when a short, broad-band pulse generated by an impulse radar is incident on an underground target. We have used the impulse radar system as a ground penetrating radar (GPR) for a few targets buried in the (dry) sand contained in an indoor sandbox. The targets were metal spheres of two different sizes (viz., diameter 10 cm or 30 cm) and a short metal cylinder (viz., of diameter 10 cm and height 5 cm). We have examined the target information that is available in the combined time-frequency domain using the pseudo-Wigner distribution (PWD). The PWD, usually being presented as a surface plot, was transformed into a point cluster representation. From such representations templates for each considered target were constructed by reducing the point cluster representation to a set of cluster centers using the fuzzy C-means (FCM) algorithm. These templates then furnished the basis for a straightforward method of automatic target classification.

Time-frequency distributions like the PWD seem to be well suited to handle situations in which the target response is corrupted by clutter and multiple scattering effects, which are both present when underground targets are considered. The method for automatic target classification that is used in this work, and which utilizes the PWD, has an intuitive attractiveness that could make it interesting for comparison with other, less computation demanding, methods. It remains, however, to be seen how well it will perform when tested using larger data sets.

ACKNOWLEDGMENTS

The authors gratefully acknowledge the support of the Independent Research Boards of their respective Institutions. The work was supported in part by the Defense Materiel Administration (FMV), Sweden and by the Swedish Rescue Service Agency.

REFERENCES

1. L. Cohen, "Time-frequency distribution—A review," *Proc. IEEE*, Vol. 77, 941–981, July 1989.
2. L. Cohen, *Time-Frequency Analysis*, Prentice Hall, Englewood Cliffs, NJ, 1995.
3. B. Boashash, Ed., *Time-Frequency Signal Analysis*, Longman Cheshire, Melbourne, Australia, 1992.
4. S. Abrahamsson, B. Brusmark, H. C. Strifors, and G. C. Gaunaurd, "Extraction of target signature features in the combined time-frequency domain by means of impulse radar," in *Automatic Object Recognition II*, F. A. Sadjadi, Ed., Proc. SPIE Vol. 1700, 1992, 102–113.
5. H. C. Strifors, G. C. Gaunaurd, B. Brusmark, and S. Abrahamson, "Transient interactions of an EM pulse with a dielectric spherical shell," *IEEE Trans. Antennas Propagat.*, Vol. 42, 453–462, Apr. 1994.
6. H. C. Strifors, B. Brusmark, S. Abrahamson, and G. C. Gaunaurd, "Aspect dependence of impulse radar extracted signature features in the combined time-frequency domain," in *Automatic Object Recognition IV*, Sadjadi, F. A., Ed., Proc. SPIE Vol. 2234, 1994, 2–13.
7. H. C. Strifors, B. Brusmark, S. Abrahamson, and G. C. Gaunaurd, "Signature features in time-frequency of simple targets extracted by ground penetrating radar," in *Automatic Object Recognition V*, Sadjadi, F. A., Ed., Proc. SPIE Vol. 2485, 1995, 175–186.
8. T. A. C. M. Claasen and W. F. G. Mecklenbräuker, "The Wigner distribution — A tool for time-frequency signal analysis," *Philips J. Res.*, Vol. 35, Part I, 217–250; Part II, 276–300; Part III, 372–389; 1980.
9. L. Cohen, "Quantization problem and variational principle in the phase-space formulation of quantum mechanics," *J. Math. Phys.*, Vol. 17, 1863–1366, 1976.
10. J. C. Bezdek, *Pattern Recognition with Fuzzy Objective Function Algorithms*, Plenum Press, New York, 1981.
11. S. L. Chiu, "Fuzzy model identification based on cluster estimation," Journal of Intelligent and Fuzzy Systems, Vol. 2, 267–278, 1994.
12. J. Dunn, "A fuzzy relative of the ISODATA process and its use in detecting compact, well separated clusters," *J. Cybernetics*, Vol. 3, 32–57, 1974.
13. J. Bezdek, "Cluster validity with fuzzy sets," *J. Cybernetics*, Vol. 3, 58–71, 1974.

SHORT-PULSE RADAR VIA ELECTROMAGNETIC WAVELETS *

Gerald Kaiser

Department of Mathematical Sciences
University of Massachusetts Lowell
Lowell, MA 01854, USA
Email: gkaiser@cs.uml.edu

INTRODUCTION

The new theory of physical wavelets makes it possible to perform radar and sonar analysis directly in the space-time domain, based on fundamental principles underlying the emission, reflection, and reception of electromagnetic and acoustic waves[1,2]. Being independent of the Fourier transform and even of the usual (affine) wavelet transform, this formalism is therefore equally well suited for ultrawideband or short-pulse radar as for narrowband or continuous-wave radar. However, Fourier analysis does have a natural place in this theory and can be used easily when spectral questions are of interest.

A transmitting antenna following an arbitrary (possibly accelerating or nonlinear) space-time trajectory $\alpha(t)$ emits a physical (acoustic or electromagnetic) wavelet which is propagated in space by the appropriate Green function. This defines an *emission operator* E_α which, acting on any time signal $\psi(t)$, gives the emitted space-time wave $(E_\alpha\psi)(\mathbf{r},t)$. The *reception operator* R_α is dual to E_α, measuring any incident space-time wave $F(\mathbf{r},t)$ along the given antenna trajectory $\alpha(t)$ to produce the received time signal $(R_\alpha F)(t)$. Reflection is modeled as reception followed by re-emission, i.e., by the operator $E_\alpha R_\alpha$ transforming any incident space-time wave to the reflected space-time wave. Let the receiving antenna follow another arbitrary space-time trajectory $\gamma(t)$ (possibly different from the trajectory $\alpha(t)$ of the transmitter), let the target follow a third arbitrary space-time trajectory $\beta_T(t)$, and let the transmitted and received signals be $\psi(t)$ and $\chi(t)$, respectively. The objective is to estimate the target trajectory $\beta_T(t)$ from a knowledge of $\alpha(t), \gamma(t), \psi(t)$ and $\chi(t)$. This is achieved by maximizing the modulus of the the *normalized ambiguity functional* $\tilde{\chi}_N(\beta)$, obtained by matching the *actual* return $\chi(t)$ with the *computed* return due to a *trial trajectory* $\beta(t)$. When the radar is monostatic and the target is assumed to move uniformly in the radial direction, then $\tilde{\chi}_N(\beta)$ reduces to the usual *wideband ambiguity function,* which is just the ordinary *time-scale (wavelet) transform* of $\chi(t)$ with $\psi(t)$ as the basic wavelet. In the narrowband approximation, it

* Supported by AFOSR grant F4960-95-1-0062 and ONR grant N00014-96-1-0584.

reduces further to the usual time-frequency ambiguity function, which is a windowed Fourier transform of the *video signal* of the return. This shows that our "physical wavelet analysis" is a generalization of the usual ("mathematical") wavelet analysis, which is in turn a generalization of time-frequency analysis. In particular, our analysis applies equally well to *bistatic wideband radar*, where the Doppler effect can no longer be represented simply by scaling and hence the usual affine wavelet analysis breaks down.

In Reference 2, the transmitting and receiving antennas and the target were all assumed to be *points*, so that the trajectories $\alpha(t), \beta(t)$, and $\gamma(t)$ fully describe their motions. Consequenctly, the emitted and reflected waves are omnidirectional and hence are not very useful in practice. In this paper we generalize the above model to include *extended* antennas and targets, follwing an idea introduced by Heyman and Felsen[3]. The associated radar beams are much more useful since they have a measure of directivity.

EXTENDED PHYSICAL WAVELETS AND AMBIGUITY FUNCTIONALS

Suppose we are given an antenna located at the space point $\mathbf{x}$, emitting the response to an impulse at time t. Ignoring polarization for simplicity, the resulting wave at the observation point $\mathbf{x}'$ at time t' can be represented by a solution of the scalar wave equation, which we write as

$$K(\mathbf{x}', t' \,|\, \mathbf{x}, t),$$

with a source distribution appropriate to the antenna. If the antenna is very small (essentially a point) and omnidirectional, then $K(\mathbf{x}', t' \,|\, \mathbf{x}, t)$ is well approximated by the retarded Green function

$$G(\mathbf{x}' - \mathbf{x}, t' - t) = \frac{\delta(t' - t - |\mathbf{x}' - \mathbf{x}|/c)}{4\pi|\mathbf{x}' - \mathbf{x}|} \,,$$

where c is the speed of light. Following Heyman and Felsen[3], a simple model can be formulated for an *extended* antenna by allowing *complex* antenna space-time coordinates $\mathbf{x} \rightarrow \mathbf{z} = \mathbf{x} + i\mathbf{y}$ and $t \rightarrow u = t + is$ and taking $K(\mathbf{x}', t' \,|\, \mathbf{z}, u)$ to be an analytic extension of the above retarded Green function.* Heyman and Felsen showed that for $|\mathbf{y}| < cs$, $K(\mathbf{x}', t' \,|\, \mathbf{z}, u)$ can be interpreted as a *pulsed beam field* emitted by a circular disk of radius $|\mathbf{y}|$ in the direction of $\mathbf{y}$. Thus $\mathbf{y}$ is a convenient "handle" by which the radius and orientation of the antenna can be controlled, without having to construct a messy model for the antenna involving a continuous distribution of point sources. More complicated extended antennas and arrays can be modeled by a distribution (continuous or discrete) of such complex source points.

To keep the notation uncluttered, we combine the space and time coordinates into a single symbol:

$$x' \equiv (\mathbf{x}', t') \in \mathbf{R}^4, \qquad z \equiv (\mathbf{z}, u) = x + iy \in \mathbf{C}^4,$$

* Note that our convention differs from that of Reference 3 in that our positive-frequency time-harmonic waves vary as $e^{i\omega t}$ rather than $e^{-i\omega t}$. For this reason, the analytic continuation of the retarded Green function (using the analytic signal of the delta function) is to the upper-half time plane ($s > 0$) rather than the lower-half time plane. This is consistent with the convention used in Reference 1.

where

$$x = (\mathbf{x}, t) \quad \text{and} \quad y = (\mathbf{y}, s).$$

The condition $|\mathbf{y}| < cs$ means that the imaginary space-time four-vector y belongs to the *future cone*, so that z actually belongs to the complex *future tube*[1,4]

$$z \in \mathcal{T}_+ \equiv \{x + iy \in \mathbf{C}^4 : x \in \mathbf{R}^4 \text{ and } y = (\mathbf{y}, s) \text{ with } |\mathbf{y}| < cs\},$$

which is a four-dimensional generalization of the upper-half complex time plane.

Suppose now that our antenna executes an arbitrary motion, including possible rotations and accelerations. Using the complex source coordinates, this can be parameterized as

$$z = \alpha(t) = x(t) + iy(t), \quad \text{where } x(t) = (\mathbf{x}(t), t) \in \mathbf{R}^4 \text{ and } y(t) = (\mathbf{y}(t), s). \tag{1}$$

It is reasonable (but mathematically unnecessary) to assume that the radius of the antenna remains constant during the motion, so that $|\mathbf{y}(t)| = |\mathbf{y}(0)| \equiv R < cs$, although the direction of $\mathbf{y}(t)$ may vary to allow tracking, scanning, etc. While executing this motion, the antenna is fed an input time signal $\psi(t)$. Then the output beam is

$$\Psi_\alpha(x') = \int_{-\infty}^{\infty} dt\; K(x' \,|\, \alpha(t))\, \psi(t).$$

For reasons explained in Reference 2, we call $\Psi_\alpha(x')$ the *extended physical wavelet* generated by $\psi(t)$ along the antenna motion $\alpha(t)$. Given $\alpha(t)$, we define the *emission operator* E_α as the operator transforming the time signal $\psi(t)$ to the space-time wave $\Psi_\alpha(x')$, i.e.,

$$(E_\alpha \psi)(x') \equiv \int_{-\infty}^{\infty} dt\; K(x' \,|\, \alpha(t))\, \psi(t). \tag{2}$$

Thus E_α takes a function of one variable (the input signal) to a function of four variables (the output beam). On the other hand, if the antenna is used as a receiver, it converts space-time waves into time signals. Again, assume that the complex antenna motion $\alpha(t)$ is given as in (1). Then the simplest model for the received signal due to an incident wave $F(x')$ is

$$(R_\alpha F)(t) = g_\alpha\, F(\alpha(t)), \tag{3}$$

where g_α is a "gain factor." Thus R_α simply measures the field along the complex trajectory $\alpha(t)$. More complicated receivers can be formulated which measure derivatives of F along $\alpha(t)$. (In the full electromagnetic formalism, for example, R_α could measure the induced current rather than the field.) Since $\alpha(t)$ is complex, the "evaluation" of the field $F(x')$ at $x' = \alpha(t)$ must be defined in (3). For this we use the *analytic-signal transform* of F, which extends F to complex space-time[1,4]:

$$F(x + iy) \equiv \frac{1}{\pi i} \int_{-\infty}^{\infty} \frac{d\tau}{\tau - i}\; F(x + \tau y).$$

When $y = (\mathbf{0}, s)$ with $s > 0$, $F(x + iy)$ reduces to the usual Gabor analytic signal $F(\mathbf{x}, t + is)$ corresponding to $F(\mathbf{x}, t)$, with $\mathbf{x}$ regarded as an external parameter; this function is analytic in the upper-half complex time plane. It is further shown in Reference 1 that if $F(x')$ is any solution of the homogeneous wave equation (or Klein-Gordon equation[4]), then $F(x + iy)$ is analytic in the future tube $\mathcal{T}_+$ (i.e., $|\mathbf{y}| < cs$). The reception operator then evaluates the analytic-signal transform $F(x + iy)$ in its region of analyticity.

With emission and reception modeled by (2) and (3), we are almost ready to formulate a general radar problem. The only missing element is a model for *reflection.* In the spirit of regarding a scattered electromagnetic wave as being emitted by the current induced on the scatterer by the incident wave, we propose the following model: Suppose we are given an oriented circular "target" disk executing a motion described by a complex space-time trajectory $\alpha(t) = x(t) + iy(t)$ as in (1). Again, we interpret the imaginary position vector $\mathbf{y}(t)$ as defining the radius and orientation of the disk. (To say that the disk is "oriented" means that its two sides are not equivalent; for example, one side could be reflective while the other side is not. Then every unit vector $\hat{\mathbf{y}} = \mathbf{y}/|\mathbf{y}|$ corresponds to a unique orientation of the disk. This is useful if, for example, we approximate a complicated target by patching together disks of various sizes and orientations, as in Section 10.2 of Reference 1; their non-reflecting sides should then be oriented towards the interior.) A given space-time wave $F(x')$ will now be assumed to be reflected from the disk as follows: First the disk acts as a *receiver,* then as a *transmitter.* Thus the reflected wave is

$$F_{\text{refl}}(x') = (E_\alpha R_\alpha F)(x') = g_\alpha \int_{-\infty}^{\infty} dt\ K(x' \,|\, \alpha(t))\, F(\alpha(t)).$$

Note that in the present context, the original "gain factor" g_α is re-interpreted as a *reflection coefficient.* When a complicated target is patched together from circular targets of various radii and orientations, the reflection coefficient becomes a function defined over the target surface as desired.

The ambiguity functional formalism developed in Reference 2 generalizes easily and naturally to the present setting of extended physical wavelets. Given the outgoing time signal ψ and the motions α, β, and γ of the transmitter, target, and receiver (all complex), our model for the time signal received at γ is

$$\begin{aligned}\psi_\beta(t'') &= (R_\gamma E_\beta R_\beta E_\alpha \psi)(t'') \\ &= g_\gamma g_\beta \iint dt'\, dt\ K(\gamma(t'') \,|\, \beta(t'))\, K(\beta(t') \,|\, \alpha(t))\, \psi(t).\end{aligned} \tag{4}$$

Of course, the received signal depends functionally on all three trajectories α, β, γ, as is evident from the right-hand side of (4). But to simplify the notation, we have suppressed the dependence on the *known* trajectories α and γ and displayed only the dependence on the target trajectory β. To estimate the *actual* target trajectory $\beta_T(t)$, we compute $\psi_\beta(t)$ for a *trial trajectory* $\beta(t)$ and match the result with the actual return $\chi(t)$ by taking the inner product of the two time signals. We denote the result by $\tilde{\chi}(\beta)$, which we call the *ambiguity functional* of the return:

$$\begin{aligned}\tilde{\chi}(\beta) \equiv \langle\, \chi, \psi_\beta \,\rangle &\equiv \int_{-\infty}^{\infty} dt''\ \chi(t'')\, \psi_\beta(t'') \\ &= \iiint dt''\, dt'\, dt\ \chi(t'')\, K(\gamma(t'') \,|\, \beta(t'))\, K(\beta(t') \,|\, \alpha(t))\, \psi(t).\end{aligned} \tag{5}$$

(We assume that $\psi(t)$ and $\chi(t)$ are real; if they are complex, then $\chi(t)$ should be replaced by its complex conjugate in (5).) Assuming that $\psi_\beta(t)$ and $\chi(t)$ have finite energies $\|\psi_\beta\|^2$ and $\|\chi\|^2$, the Schwarz inequality implies that

$$\begin{aligned}|\tilde{\chi}(\beta)\,| &= |\langle\, \chi, \psi_\beta \,\rangle| \le \|\chi\|\, \|\psi_\beta\| \\ |\tilde{\chi}(\beta)| &= \|\chi\|\, \|\psi_\beta\| \Longleftrightarrow \chi(t) = C\psi_\beta(t).\end{aligned} \tag{6}$$

Therefore, to estimate the true target trajectory $\beta_T(t)$, we need to maximize the *normalized* ambiguity functional

$$\tilde{\chi}_N(\beta) \equiv \frac{\tilde{\chi}(\beta)}{\|\psi_\beta\|}.$$

By (6),

$$|\tilde{\chi}_N(\beta)| \leq \|\chi\| \quad \text{and} \quad |\tilde{\chi}_N(\beta)| = \|\chi\| \Longleftrightarrow \chi(t) = C\psi_\beta(t).$$

Equivalently, we can minimize the *error functional* defined by

$$\mathcal{E}(\beta) \equiv 1 - \frac{|\chi(\beta)|}{\|\chi\| \, \|\psi_\beta\|},$$

since the Schwarz inequality states that

$$0 \leq \mathcal{E}(\beta) \leq 1 \quad \text{and} \quad \mathcal{E}(\beta) = 0 \Longleftrightarrow \chi(t) = C\psi_\beta(t).$$

Thus $|\tilde{\chi}_N(\beta)|$ and $\mathcal{E}(\beta)$ attain their maximum and minimum values, respectively, only when the trial return is indistinguishable from the actual return. Of course, this does not guarantee that the trial trajectory $\beta(t)$ coincides with the actual target trajectory $\beta_T(t)$, since the return does not, in general, uniquely determine the target trajectory. That is, *the functionals $\tilde{\chi}_N(\beta)$ and $\mathcal{E}(\beta)$ are generally not one-to-one.* The class of all trajectories β such that $\tilde{\chi}_N(\beta) = \tilde{\chi}_N(\beta_T)$ or, equivalently, $\mathcal{E}(\beta) = \mathcal{E}(\beta_T)$, represents the inherent *ambiguity* of the radar problem. A problem of obvious importance is to find outgoing signals $\psi(t)$ which minimize this ambiguity class.

We have assumed above that the return is due to a reflection from a *single* target. If N distinct targets are involved, then we can approximate the return as a superposition

$$\psi_{\beta_1,\beta_2,\ldots,\beta_N} \approx \psi_{\beta_1} + \cdots + \psi_{\beta_N}\,. \tag{7}$$

As noted, (7) is an approximation because it ignores *multiple reflections.* Although these can often be ignored, they can also cause resonances (ringing), hence must sometimes be taken into account. This can be easily done, in principle. For example, the signal received by the doubly-reflecting path $\alpha \to \beta_m \to \beta_n \to \gamma$ is

$$\psi_{\beta_m,\beta_n} = R_\gamma E_{\beta_n} R_{\beta_n} E_{\beta_m} R_{\beta_m} E_\alpha \psi\,,$$

which can be immediately converted to a triple integral by using the definitions (2) and (3). Sums of contributions from various "trial" scattering paths may then be matched with the actual return, defining a generalized ambiguity functional

$$\tilde{\chi}(\beta_1, \beta_2, \ldots, \beta_N) \equiv \langle\, \chi, \psi_{\beta_1,\beta_2,\ldots,\beta_N} \,\rangle,$$

and the Schwarz inequality may be used as in the case of a single path to optimize the match.

This method is reminiscent of *Feynman diagrams*[5], where fundamental processes are represented by multiple integrals with corresponding intuitive diagrams. Because the physics is built into the formalism from the beginning through the Green functions, our model can handle such complications in a conceptually straightforward (if computationally nontrivial) way. The resemblance to Feynman diagrams is no coincidence, and the present formalism may be modified to include quantum (photonic) aspects of radar simply by using *Feynman propagators* in place of the retarded Green functions.

REFERENCES

1. G. Kaiser, *A Friendly Guide to Wavelets,* Birkhäuser, Boston, 1994.

2. G. Kaiser, Physical wavelets and radar, *IEEE Antennas and Propagation Magazine*, February, 1996.

3. E. Heyman and L.B. Felsen, Complex-source pulsed-beam fields, *Journal of the Optical Society of America A* **6**, 806–817.

4. G. Kaiser, *Quantum Physics, Relativity, and Complex Spacetime: Towards a New Synthesis,* North-Holland, Amsterdam, 1990. Second edition to be published by Birkhäuser, Boston.

5. R.P. Feynman, *QED: The Strange Theory of Light and Matter*, Princeton University Press, 1985.

THE E-PULSE TECHNIQUE FOR DISPERSIVE SCATTERERS

S. Primak,[1] J. LoVetri,[2] Z. Damjanschitz,[2] and S. Kashyap[3]

[1]Department of Electrical and Computer Engineering
Ben-Gurion University of the Negev
POB 653, Beer-Sheva, 84105, Israel
[2]Department of Electrical Engineering
The University of Western Ontario,
London, Ontario, Canada N6A 5B9
[3]Department of National Defence
Defence Research Establishment Ottawa,
3701 Carling Ave., Ottawa, Ontario, Canada K1A 0K2

INTRODUCTION

The E-pulse radar target discrimination scheme, employed in the frequency domain to extract aspect dependent information about targets was recently presented by Rothwell *et al.*[1] This approach assumes that the scattering response is approximated by the model of point scatterers, and consequently can be represented in frequency domain as a sum of complex exponents. Further investigation[2-5] has shown that the scattering mechanism is more complicated and is better modelled by an exponential expansion with polynomial coefficients. This dispersive property of the scattering response is found in many real situations and motivates the adaptation of E-pulse based discrimination schemes to this type of target. Here we present procedures, both for the construction of the E-pulses and the extraction of the scattering features from the measured data. This algorithm will be applied to angle discrimination of an open-ended rectangular cavity in which the scattering response was obtained via frequency domain measurements ant to a fin structure in which the scattering response was obtained directly in the time domain using the FDTD technique.

MODELLING OF EARLY-TIME TRANSIENT SCATTERING

Let us assume that a very short interrogating pulse is incident on a radar target and its scattered transient response $r(t)$ is measured. A simple model for the early-time portion of this response, $r_E(t)$, was suggested by Altes [6] in the form

$$r_E(t) = \sum_{m=1}^{M} g_m(t - T_m)u(t - T_m). \tag{1}$$

Ultra-Wideband, Short-Pulse Electromagnetics 3
Edited by Baum *et al.*, Plenum Press, New York, 1997

Here $g_m(t - T_m)$ is the impulse response of the m-th point scattering centre, originating at time T_m, $u(t)$ is the unit step function, and M is the number of point scattering centres considered in the model. The corresponding frequency domain response is written as

$$R_E(\omega) = \sum_{m=1}^{M} G_m(\omega) \exp[-j\omega T_m] \tag{2}$$

where $G_m(\omega)$ is the Fourier transform of $g_m(t)$. The assumption, that $G_m(\omega)$ can be represented as a sum of real exponentials

$$G_m(\omega) = \sum_{k=1}^{K_m} b_{m,k} \exp[-\alpha_{m,k}\omega] \tag{3}$$

where $b_{m,k}$ are complex amplitudes, allowed Rothwell *et al.*[1] to represent the early-time scattering response in the form

$$R_E(\omega) = \sum_i B_i \exp[-\tau_i \omega] \tag{4}$$

where $\tau_i = \alpha_i - jT_i$, and to apply their E-pulse target discrimination scheme in the frequency domain.

It has been found in recent investigations[2-5], that the model described by (3) is not a good approximation for a wide class of the scattering problems. A more accurate model for the frequency domain response is given by assuming a form

$$G_m(\omega) = \sum_{n=0}^{N_m} a_{mn} \omega^n \tag{5}$$

and replacing the exponential in (2) with the more general form $\exp[\tau_m \omega]$. This form of the frequency domain response reflects the dispersive properties of the scattering centres[2-4]. The construction of E-pulses for this model of the early time response will now be considered.

E-PULSES FOR DISPERSIVE SCATTERERS

It can be easily seen that the E-pulse construction procedure suggested by Rothwell[1] cannot be applied, without modification, to the construction of E-pulses for an early time response model in the form of (2) with frequency domain responses modeled by (5). In fact, for this case, we would obtain an underdetermined system of linear equations. Here we suggest another approach to construct the desired E-pulses.

Consider first the backscattered response for only one scattering centre. Thus we have

$$R_{E_1}(\omega) = \sum_{k=0}^{K_1} a_{1k} \omega^k \exp[\tau_1 \omega] \tag{6}$$

where τ_1 can be any complex number with non-positive real part. Now consider the following set of differential operators:

$$L_{\tau_1,0} = \left(\frac{d}{d\omega} - \tau_1\right), \qquad L_{\tau_1,k} = L_{\tau_1,0}^{k+1} = \left(\frac{d}{d\omega} - \tau_1\right)^{k+1}. \tag{7}$$

Applying (7), with $k = 0$, to the function $y_{K_1}(\omega) = \omega^{K_1}\exp[\tau_1\omega]$ we find that

$$L_{\tau_1,0}[y_0(\omega)] = 0, \qquad L_{\tau_1,0}\left[y_{K_1}(\omega)\right] = K_1\omega^{K_1-1}\exp[\tau_1\omega] \tag{8}$$

and consequently

$$L_{\tau_1,K_1}\left[y_{K_1}(\omega)\right] \equiv L_{\tau_1,0}^{K_1+1}\left[y_{K_1}(\omega)\right] = 0. \tag{9}$$

Thus, the operator L_{τ_1,K_1} annihilates all terms in (6) and being a linear operator, L_{τ_1,K_1} does not change the spectral content of the response $R_E(\omega)$; it removes only the response of one scattering centre. Consequently, the differential operator

$$L = \prod_{m=1}^{M} L_{\tau_m,K_m} \tag{10}$$

annihilates the entire response (2). The composite operator (10) can be considered a manifestation of the E-pulse approach in continuous time.

For practical applications, a discrete-time signal processing scheme is required. The discrete-time annihilation scheme corresponding to the continuous composite operator of (10) can be easily derived for the model of equations (2) and (5). Let Z denote the frequency domain shift operator which shifts a frequency domain function by an amount $\Delta\omega$, that is

$$Zx(\omega) = x(\omega - \Delta\omega). \tag{11}$$

Then, introducing the operator

$$L_{\tau_1,p}^{\Delta\omega} = (e^{\Delta\omega\tau}Z - 1)^{p+1} \tag{12}$$

we see that it annihilates the frequency series

$$x(n) = (n\Delta\omega)^i\exp[n\Delta\omega\tau_1],\ 0 \le i \le p \tag{13}$$

which can be considered the discrete analog of the terms in the summation of (6). Being also a linear operator, $L_{\tau_1,p}$ preserves the spectral content of the frequency response (2), removing components generated by a single scattering centre, located a time-distance τ from the measurement point. Consequently, the operator

$$L^{\Delta\omega} = \prod_{m=1}^{M} L_{\tau_m,K_m}^{\Delta\omega} \tag{14}$$

completely annihilates $R_E(\omega)$.

The operator $L^{\Delta\omega}$ can be considered as an FIR digital filter of length

$$K = \sum_{m=1}^{M} K_m, \tag{15}$$

that is,

$$L^{\Delta\omega} = \prod_{m=1}^{M} (e^{\Delta\omega\tau_m} Z - 1)^{K_m} = \sum_{i=0}^{K} c_i Z^i, \tag{16}$$

where the coefficients c_i can be obtained after expanding out the product in (16) and collecting all terms containing the corresponding term Z^i. The set of discrete coefficients c_i represent the E-pulse in the discrete frequency domain. Applying the operator $L^{\Delta\omega}$ to the response $R_E(\omega)$ is equivalent to convolving this frequency domain E-pulse with the response. Let us note, that in contrast to the technique described by Rothwell[1], we do not need to solve a system of linear equations to obtain the desired E-pulse, and consequently, this approach is more numerically stable.

CONSTRUCTION OF THE MODEL FROM MEASURED DATA

Although the construction of the E-pulses for the model in the form of (2) and (5) is easily obtained using the above method, extraction of the parameters for this model from the measured response cannot be done using the standard Prony's method or similar techniques directly. For this purpose we follow the ideas presented by Moore and Ling[2] as well as Carin *et al.*[3] We divide the entire frequency domain into R intervals $[\omega_{r-1}, \omega_r]$, $r = 1, \ldots, R$. For each such interval of the frequency response we extract the features using the matrix pencil or ESPRIT method[8]. Thus, on each r-th interval, we obtain the approximation for (3) in the form

$$R_E(\omega) = \sum_{m=1}^{M} g_{rm} \exp[\omega\tau_m] \quad, \omega \in [\omega_{r-1}, \omega_r]. \tag{17}$$

Then, the series of coefficients $\{g_{rm}\}$, $r = 1, \ldots, R$ can be approximated by polynomials in a mean square sense which gives us the desired result

$$G_m(\omega) = \sum_{k=0}^{K_m} a_{mk}\omega^k. \tag{18}$$

In this way the model for the frequency response model in the form given by (2) and (5) can be extracted from the measured data.

EXPERIMENTAL DATA AND NUMERICAL SIMULATION

As an example we construct a family of E-pulses for angle discrimination of an open-ended rectangular cavity. The geometry of the target and the configuration of the measurement apparatus are shown in Figure 1. The measurements were performed in a 6m×6m×6m cubic anechoic chamber. The measurement set-up has a range of about 1.25 m as shown in Figure 1. A set of two Dalmo Victor dual-polarized quadridged horns (model

A6100) were used: one to illuminate the target and the other to receive the scattered field. This introduces a bistatic angle of about 8°, hence an error in the RCS. The error incurred, however, was assessed using the Numerical Electromagnetic Code (NEC)[9] by comparing the monostatic and the bistatic RCS on a wide variety of metallic and dielectric targets, and was found to be negligible for small metallic targets. A frequency range of 2-18 GHz was used. Time-gating, with a gate span of several nanoseconds, was used to isolate the target response.

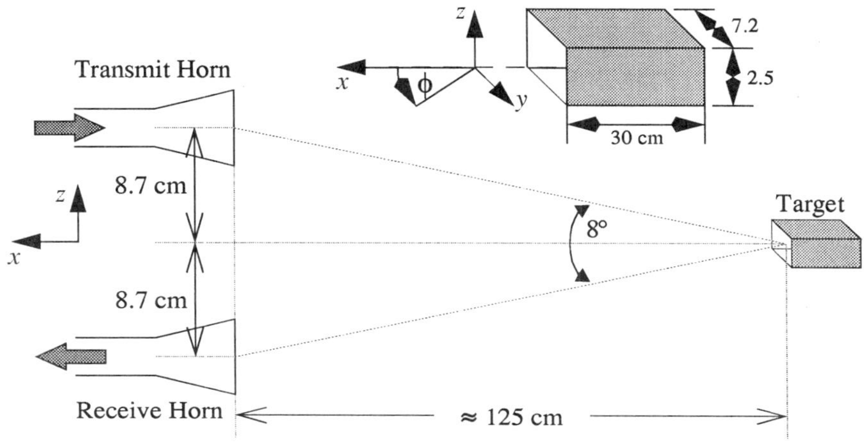

Figure 1. Geometry of the target and relationship to transmit and receive horns.

An example of the raw frequency domain data obtained from measurements is shown in Figure 2 (solid curve). The scattering centers and the order of the approximation for the corresponding models in the form of (2) and (5) for some typical angles are given in Table 1. Figure 2 also shows good agreement between the reconstructed frequency response (dotted curve) and the measured one. Corresponding E-pulses and E-pulse discriminating numbers (EDN)[1] are given in Figure 3 and Table 2 respectively. It can be seen that E-pulses built for dispersive scatterers are better in discriminating the aspect angle of the target.

Table 1. Scattering center and order of approximation for cavity and fin structure.

	Angle ϕ	(Scattering Center (μs), Order of scattering center K_m)
c	-90	(1.6, 1)
a	-60	(1.7, 1) (14.0 ,0) (19.4, 0) (21 ,0)
v i	-45	(3.0, 0) (11.6, 0) (13.5, 0) (15.8, 0) (18.0, 0) (20.7, 0) (22.8, 0) (25.6, 0)
t	-30	(9.14, 1) (10.65, 0)
y	0	(5.9, 1) (6.6, 0)
fin	25	(0.46, 0) (0.58, 3) (1.23, 0) (2.80, 4)

We also used a custom FDTD code[10] to obtain the scattering response of the fin structure given in Figure 4. The calculated and reconstructed frequency responses are shown in Figure 5. The EDN for the dispersive as well as the traditional E-pulse schemes are given in Table 2. The dispersive E-pulse which was obtained using the method described in this paper is shown in Figure 6.

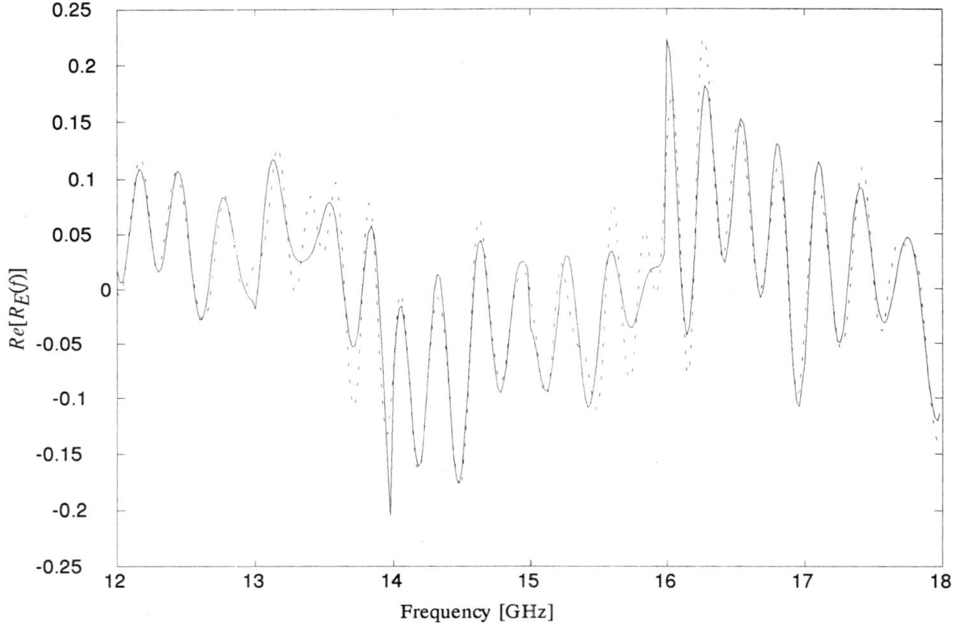

Figure 2. Measured (solid line) and reconstructed (dotted line) of cavity frequency response at angle ϕ = -60.

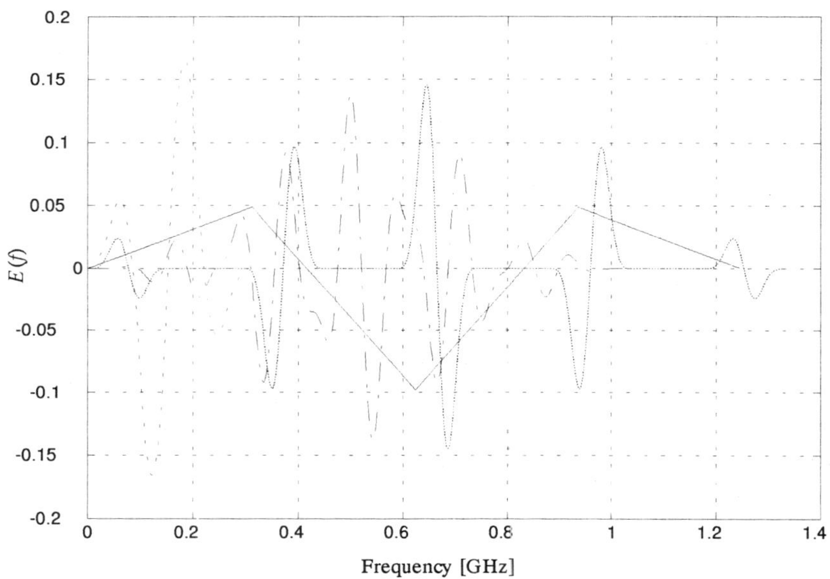

Figure 3. E-pulse for cavity target angle ϕ = -90, -60, -45, -30, 0.

Table 2. EDN for rectangular cavity (left - dispersive, right traditional)

	-90		-60		-45		-30		0	
-90	1.25	5	4.7	2.6	7.8	7.8	3.8	6.5	2	8.4
-60	1.5	5.6	0.7	2.4	1.9	1.9	1.9	4.3	1.1	2.8
-45	1.4	0.96	4.8	2.1	0.2	0.2	1.8	0.58	0.8	3.3
-30	4.8	7.5	1.1	7.3	1.5	1.5	0.4	0.7	2.4	1.6
0	3.7	1.0	1.9	0.6	3.7	3.7	5.6	4.5	0.18	0.19

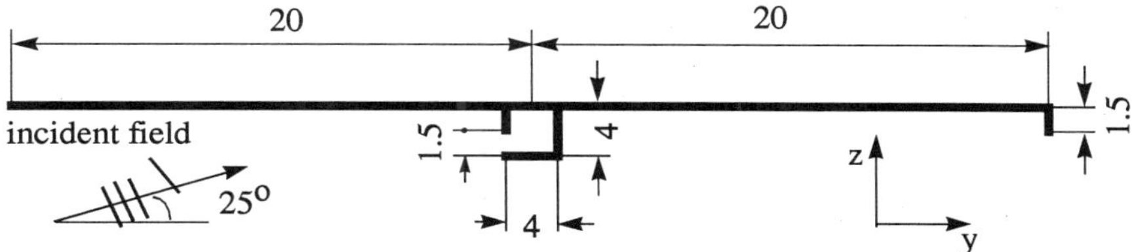

Figure 4. Fin structure geometry.

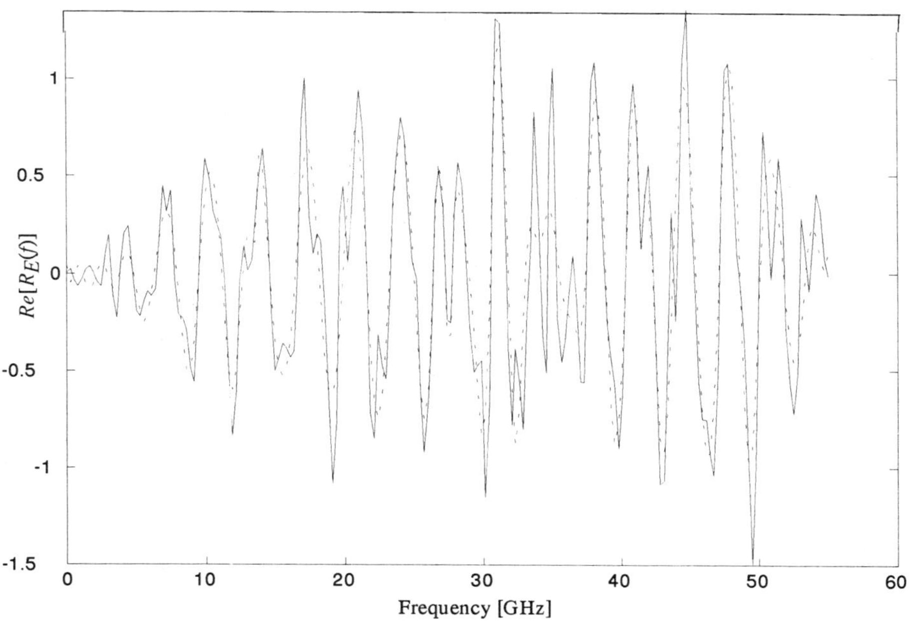

Figure 5. Measured (solid line) and reconstructed (dotted line) frequency response of the fin structure.

CONCLUSIONS

In this paper we have investigated the construction of frequency domain E-pulses for dispersive scatterers. It was pointed out that traditional schemes fail in this case. To extract the features from the measured data for rectangular open-ended cavity the modified Prony or ESPRIT algorithm can be used. A number of E-pulses for angle discrimination of open-ended cavity is given.

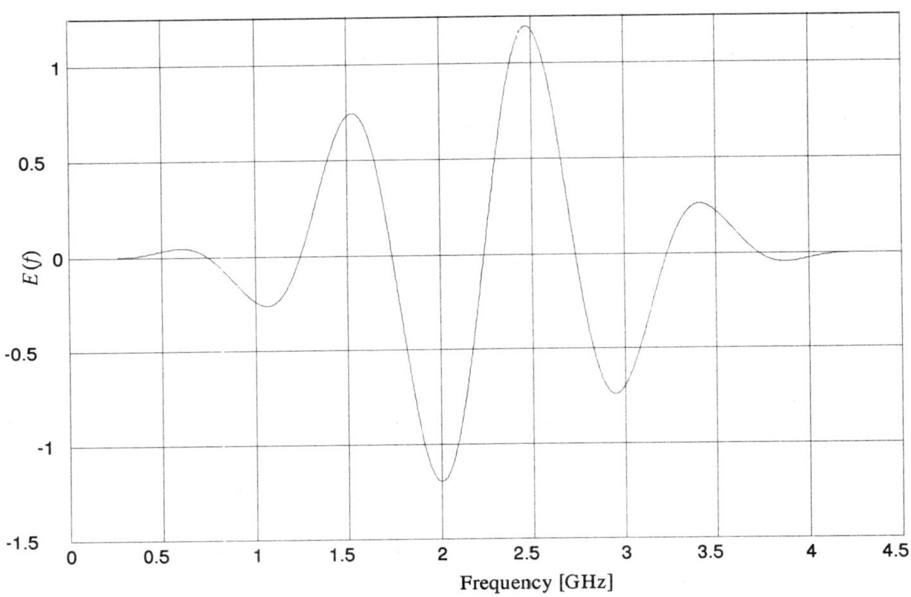

Figure 6. E-pulse for the fin structure.

REFERENCES

1. E. J. Rothwell, K.M. Chen, D.P. Nyquist, P. Ilavarasan, J. E. Ross, R. Bebermeyer, and Q. Li, A general E-Pulse scheme arising from the dual early-time/late-time behavior of radar scatteres, IEEE Trans. on Ant. and Prop., vol. 42, no. 9, pp. 1336-1341, September (1994).

2. J. Moore, and H. Ling, Super-resolved time-frequency analysis of wideband backscattered data, *IEEE Trans. on Ant and Prop*, Vol. 42, No. 9, pp. 1336-1341, September (1994)

3. J. Moore, and H. Ling, Super-resolved time-frequency processing of surface-wave mechanism contained in wideband radar echo, *Microwave and Optical Technology Letters*, Vol. 9, No. 5, pp. 237-240, August (1995).

4. L. Carin, L.B. Felsen, D. Kralj, S.U. Pillai, and W.C. Lee, Dispersive modes in the time domain: analysis and time-frequency representation, *IEEE Microwave and Guided Wave Letters*, Vol. 4, No. 1, pp. 23-25, January (1994).

5. Q. Li, E.J. Rothwell, K.-M. Chen, and D.P. Nyquist, Scattering Center Analysis of Radar Targets Using Fitting Scheme and Genetic Algorithm, *IEEE Trans. on Ant and Prop*, Vol. 44, No. 2, pp. 198-207, Februar (1996).

6. R.A. Altes, Sonar for generalized target description and its similarity to animal echolocation system, *J. Acoust. Soc. Amer.*, Vol. 59, pp. 97-105, January (1976).

7. R. Roy, A. Paulraj, and T. Kailath, ESPRIT - a subspace approach to estimation of parameters of cisoids in noise, *IEEE Trans. on Acous., Speech and Sig. Proc.*, Vol. 34, No. 5, pp. 1340-1342, October (1986).

8. G.J. Burke, and A.J. Poggio, Numerical Electrical Code, *Technical Document 116*, Naval Electronic Systems Command, Naval Ocean Systems Center, San Diego, California, 18 July (1977).

9. K. S. Kunz, and R.J. Luebbers, The Finite Difference Time Domain Method for Electromagnetics, CRC Press, London, 1993.

SPECTRAL CORRELATION OF WIDEBAND TARGET RESONANCES

Vincent Sabio

Microwave Sensors Branch
U.S. Army Research Laboratory
Adelphi, MD 20783-1197

INTRODUCTION

Recognition of target resonances in ultra-wideband (UWB) radar imagery has been a topic of investigation for several years, with a common method of resonance extraction being the singularity expansion method (SEM)—a contemporary adaptation of Prony's method. SEM requires high signal-to-noise (S/N) ratios—about 15 to 20 dB[1]—and performs poorly in the presence of noise and multipath effects. In this paper, I demonstrate the application of spectrally matched filters, employing the discrete cosine transform (DCT), to achieve an efficient means of target resonance recognition.

BACKGROUND

The radar-imaging system used in this investigation is the Army Research Laboratory (ARL) UWB synthetic-aperture radar (SAR) instrumentation system. The ARL UWB SAR operates across a 1-GHz-wide band, from 50 to 1050 MHz. A BASS (Bulk Avalanche Semiconductor Switch) is used as the transmitter; it drives a TEM (transverse electromagnetic) horn antenna that effectively differentiates the 1-ns transmit pulse to form the doublet shown in figure 1a. The spectral distribution of the pulse is shown in figure 1b;

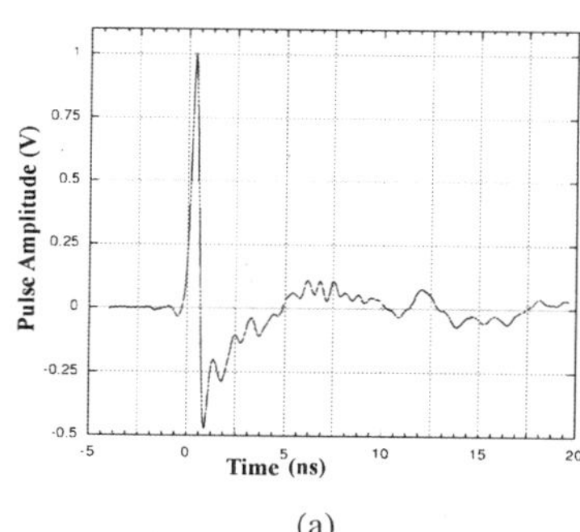

(a)

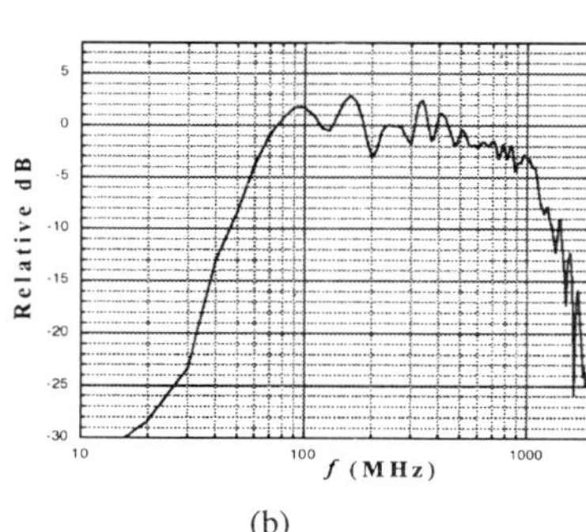

(b)

Figure 1. UWB transmitted pulse: (a) time-domain waveform; (b) spectrum.

Ultra-Wideband, Short-Pulse Electromagnetics 3
Edited by Baum *et al.*, Plenum Press, New York, 1997

the low-end roll-off is largely attributed to the frequency response of the antenna, and the high-end roll-off is primarily limited by the data-acquisition system. The SAR traverses a 104-m laser-leveled track on the roof of a four-story building on the laboratory campus. The target area extends from 112 to 267 m in range, and is primarily populated by deciduous trees and smaller flora, providing a suitable testbed for evaluation of the SAR's foliage-penetration capabilities.

The UWB SAR is fully polarized, with the SAR polarization planes inclined 45° to the radar slant plane; since the radar looks almost directly to the north, the transmit and receive planes are defined by the direction (east or west) of the upward-pointing E-field vector. The polarization-channel nomenclature follows the standard transmit-receive format; thus, an east-transmit, west-receive channel is denoted EW. (This is in contrast to the more familiar horizontal/vertical orientations of most radars, for which the polarization channels are designated HH, HV, etc.)

The results of the canonical-target studies have been widely reported,[2,3,4] so this paper will focus primarily on the complex-target investigations.

The complex target employed in this analysis was a standard commercial utility cargo vehicle (CUCV), shown in figure 2; this target measures 5.13 by 1.98 by 1.78 m. Six UWB SAR images were used in this investigation, comprising 15 occurrences of CUCV targets at various aspect angles, plus four CUCVs embedded in dense summer foliage.

Figure 2. Commercial utility cargo vehicle (CUCV).

TARGET RESONANCE EFFECTS

The response of a resonant scatterer to an incident wideband pulse generally comprises two temporally distinct parts, referred to as the early-time (driven) response and the late-time (resonant) response. The early-time response is the echo of the incident pulse, caused by local currents being driven on the surface of the object; alone, it does not convey a great deal of information about the scatterer. (ARL has recently been studying unique aspect-angle dependencies of early-time UWB target signatures for the purpose of target detection.[5]) The late-time response is a ringdown of the natural frequencies of the target excited by the incident pulse. These natural frequencies are a function of the electrical dimensions of the object, which are generally unique to each target.

The late-time resonance phenomenon is best illustrated through a canonical example; for relevance, consider a dipole target. The spatio-temporal distribution of current along a thin-wire dipole of length L is described by

$$i(x,t) = I_0\, e^{-at} \sin\left(\frac{2\pi x}{\lambda}\right) \sin\left(\frac{2\pi ct}{\lambda}\right), \tag{1}$$

where I_0 is the value of the current at a current antinode.[6,7] Boundary conditions require $i(x,t) = 0$ at $x = 0$ and $x = L$, which lead to the condition $2\pi L/\lambda = k\pi$, $k = 1, 2, \ldots$. Solving equation (1) at the midpoint of the dipole ($x = L/2$), substituting ω for $2\pi c/\lambda$ in (1), and using the relation $L/\lambda = k/2$, we obtain

$$i(L/2,t) = \frac{1}{n^2} e^{-a_k t} \sin\left(\frac{\pi k}{2}\right) \sin(\omega t), \qquad k = 1, 2, \ldots . \tag{2}$$

Clearly, $i(L/2,t) = 0$ for k even; these are the cases in which the current distribution is antisymmetrical along the dipole. Meaningful solutions to equation (2) exist for k odd; these are the fundamental ($k = 1$) and higher ($k = 3, 5, \ldots$) harmonic resonances of the dipole.

SPECTRAL FILTERING

The proposed resonance-based recognition approach involves the use of "spectrally matched" filters. A time-domain "synthetic ringdown" is constructed from either a theoretical model of the target or empirical UWB target data. The synthetic ringdown is projected onto a frequency-space transform basis (such as a Fourier or wavelet basis), creating a set of spectral coefficients, referred to as the "spectral template." An image chip (subset of the image data) is then analyzed by the same set of basis functions, creating a similar set of spectral coefficients. These two sets of coefficients are individually vectorized, and a simple correlation coefficient is generated from the two coefficient vectors. The correlation coefficient for two vectors $\mathbf{x}$ and $\mathbf{y}$ is computed as

$$\rho = \frac{E(\mathbf{xy}) - E(\mathbf{x})E(\mathbf{y})}{\sqrt{V(\mathbf{x})V(\mathbf{y})}}, \tag{3}$$

where $E(\mathbf{x})$ is the expectation and $V(\mathbf{x})$ is the variance of $\mathbf{x}$. The correlation coefficient is a measure of the degree of linearity between the vectors $\mathbf{x}$ and $\mathbf{y}$—higher correlation values (i.e., closer to unity) indicate greater linearity. Thus, the correlation value measures the degree of symmetry—in the frequency domain—between the spectral template and the transformed image data. This allows a target-declaration threshold to be set—if the threshold is exceeded, a target is declared to have been recognized; otherwise, no target is declared.

Clearly, the target-declaration threshold selection is somewhat arbitrary: Higher threshold values yield fewer false alarms, but carry a correspondingly greater probability of "missing" a target; lower thresholds yield more false alarms with lower probabilities of miss. Similar to other ATR systems, a point on the receiver-operating characteristic (ROC) curve must be selected, representing an acceptable tradeoff between false-alarm rate and probability of recognition. Since the spectral correlator is destined to be a single piece of the overall ATR architecture for the UWB SAR processor, a high probability of recognition (and correspondingly high false-alarm rate) was selected as the operating point. This translates to a relatively low target-declaration threshold. Recent studies have focused on reducing the false-alarm rate; the results of these studies are discussed later in this paper.

This frequency-domain target-recognition process—the "spectral correlation method"—is shown in figure 3. It is critical that the same transform basis be employed in the spectral correlator as was used in the creation of the spectral template, although selection of the specific basis is very application dependent. Earlier studies had employed various Fourier and wavelet bases, with the wavelets demonstrating superior recognition performance.[3] However, subsequent studies employing the discrete cosine transform indicated that the DCT outperformed both the wavelet and complex-Fourier bases in terms of recognition performance and false-alarm rate.[4] Thus, the CUCV analyses employed only the DCT, with a 64-point analysis window.

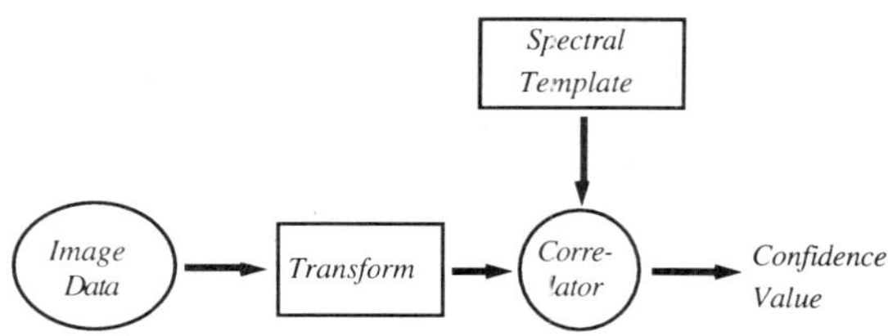

Figure 3. Spectral correlation schematic.

The spectral correlator has demonstrated relatively high noise immunity. Since only a very small number of subbands actually transform ringdown components, most subbands contain only noise. The subbands of interest contain signal plus noise, but—since additive white Gaussian noise spreads evenly across all subbands—the S/N ratio within those subbands is increased substantially (especially when compared to equivalent spatially-matched filters or time-domain correlators). Thus, the spectral correlation method exhibits good performance even in relatively noisy environments, as are inherent in nonsynthetic data.

TARGET MODELING

The key to successfully implementing the spectral correlator is developing a high-quality synthetic ringdown, from which a correspondingly high-quality spectral template can be created. Since there exists no model that will sufficiently predict the CUCV's complex radar cross section, it was not possible to use a modeling approach to construct the synthetic ringdown. Instead, a template was created by selecting range-line cuts (range profiles) through the resonant portions of CUCV signatures in two different images; profiles from one image are shown in figures 4a-c. In both images, the CUCV was oriented broadside, with the right side of the vehicle to the SAR aperture. The range-line cuts were taken through the front portion of the vehicle, where the resonant signature appeared to be the cleanest. To help cancel noise, range-line cuts from two different images were averaged to form the synthetic ringdown (Figure 4d). Examination of the synthetic ringdown shows close agreement with the raw range profile in figure 4c, suggesting that the signature at that point on the vehicle is reasonably consistent from image to image.

TARGET CUEING

Processing speed and efficiency—always considerations in the design of an ATR system—motivated the introduction of a target cueing stage. The time required to process a single 8-mega-pixel image on an otherwise unloaded Sun Sparc 1 was well over 100 hours, which simply was not practical. Porting the software from PV-Wave to i860-based DSP cards brought the processing time down to less than six minutes. To improve performance further, I developed a simple target cuer that looks for the large driven-response signal at the "start" (i.e., near-range portion) of the target signature; this early-time response is clearly observed in figure 4b. Setting a threshold on the minimum excursion necessary to cue the recognizer can improve performance by a factor of 2 to 5. (Actual performance improvement depends upon the threshold selected—higher thresholds yield better performance, but run the risk of missing targets.) Note that the target cuer performs a "simple" compare with the selected threshold; it does not compare target-window amplitude with a reference or clutter window, so it is not a constant false-alarm-rate (CFAR) detector. This is desirable for finding foliage-embedded targets, where the target-window-to-clutter-window contrast is generally quite low.

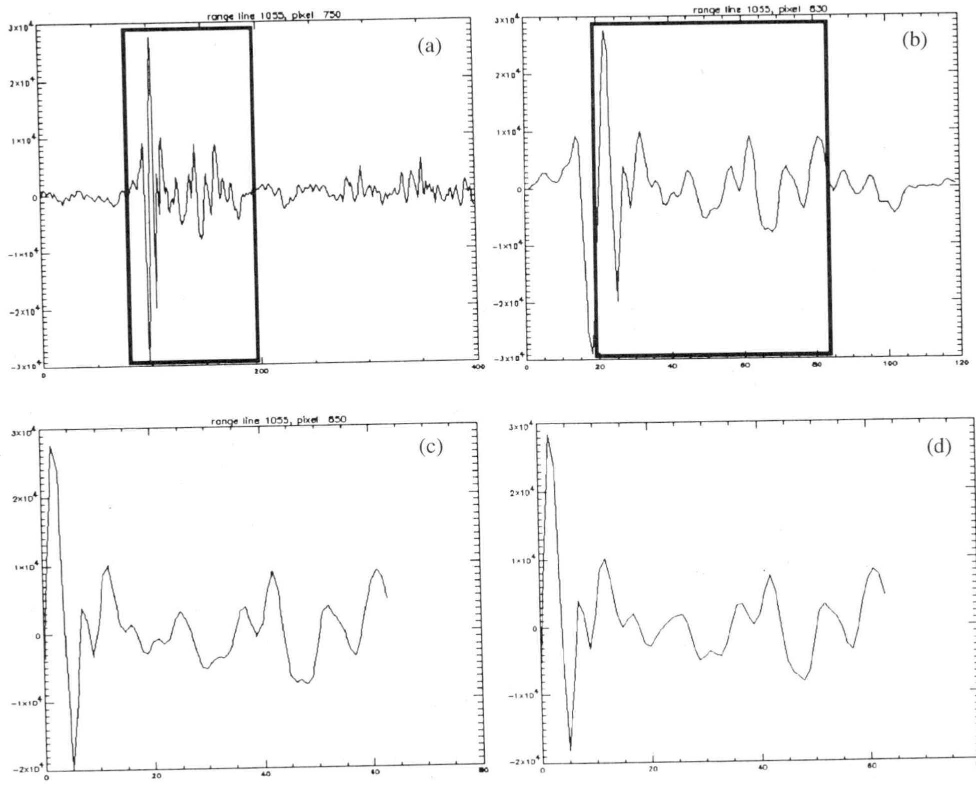

Figure 4. CUCV signatures: (a) target plus surrounding clutter; (b) driven and resonant responses; (c) resonant portion of signature; (d) synthetic ringdown.

TARGETS IN THE CLEAR

Seven different images containing a total of 15 CUCVs in the clear were analyzed; all images were west-west polarized. Target aspect angles are shown in Table 1; 0° aspect represents the passenger side of the vehicle facing the SAR aperture, and angles increase counter-clockwise (thus, 270° aspect is the front of the vehicle oriented toward the

Table 1. CUCV aspect angles and recognition performance.

Aspect angle (degrees)	Number of of occurrences	Number recognized
0	4	4
40	1	1
70	1	1
90	2	2
180	1	1
225	1	0
270	2	2
295	1	1
315	2	2

aperture). Additionally, the single target at 180° aspect had a fiberglas "cap" over the truck bed. Each image contained at least one CUCV, plus natural clutter and non-CUCV "confuser" targets, such as high-mobility multi-purpose wheeled vehicles (HMMWVs), Bradley fighting vehicles, M60s, civilian trucks, canonical targets (dipoles, corner reflectors, dihedrals, etc.) and the occasional lamp post.

Note in Table 1 that a single target failed to be recognized (at 225° aspect), representing a recognition rate of 93 percent *for a single template*. Each target generated an average of 11 recognitions, representing a total of 159 recognitions for the 14 targets recognized.

TARGETS IN FOLIAGE

In addition to the 15 targets in the clear, I analyzed four more targets embedded from 30 m to 47 m in dense foliage. These target signatures were heavily corrupted by foliage clutter. Figure 5a shows a range profile of a foliage-embedded CUCV at 10° aspect; the specular flash from the broadside of the vehicle is within the boxed area. This range profile is taken through roughly the same point as the broadside (0° aspect) CUCV signature shown in figure 4. For comparison, the range profile of the CUCV in the clear is shown in figure 5b. Note that—in addition to the substantial clutter corruption—the signature in clutter is attenuated by about 6 dB compared to the signature in the clear.

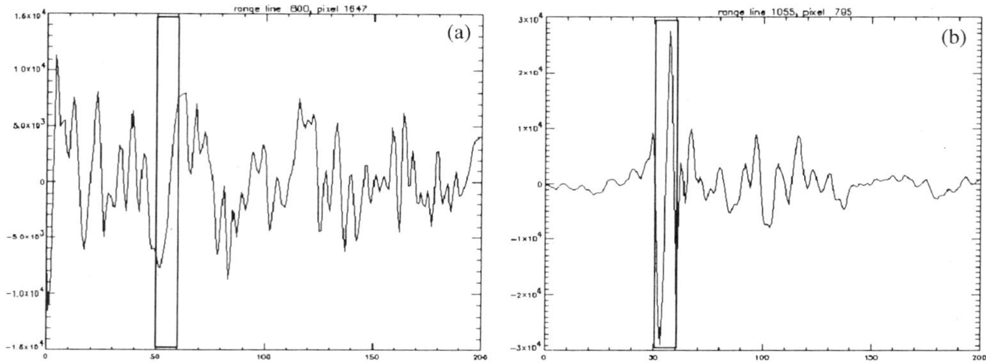

Figure 5. CUCV signatures (driven responses are shown in boxes): (a) in foliage clutter; (b) in the clear.

Despite the severe clutter-induced degradation of the signature, two of the four foliage-obscured targets were recognized using the same template and thresholds that were used in the foregoing analysis. The results are summarized in Table 2. Recognition rates appear to decrease with depth into foliage; this is to be expected, since the low grazing angle of the rooftop SAR requires the full depth of foliage to be penetrated to reach the target—and similarly for the scattered field return, thus yielding a substantial two-way loss in signal strength.

Table 2. Foliage-embedded CUCV aspect angles and recognition rates.

Aspect angle (degrees)	Minimum depth into foliage (m)	Number of recognitions on each target
270	30	5
10	44	2
185	46	0
180	47	0

FALSE -ALARM REDUCTION

Unprocessed false alarms averaged about 550 per image; for each 8-mega-pixel image, this is about one false alarm per 15,000 pixels. Since recognitions are typically "clumped," implementation of a target-clustering algorithm substantially reduced the false-alarm rate. Resonances in UWB imagery extend primarily downrange, but there is also an azimuthal "spread" to the resonance. The azimuthal angle subtended by each component of the ringdown is a function of frequency; higher-frequency modes subtend smaller angles than the broader-beam low-frequency modes.[8]

Target declarations on real targets averaged about 10 per target in the clear; for false alarms, this figure dropped to about 3 per false "target." Thus, target-clustering could reduce the false-alarm rate by about 67% without impacting the recognition rate. Moreover, setting a threshold on the number of recognitions required within a cluster for a target to be declared could provide a further reduction in false-alarm rate (but with a corresponding impact on recognition performance).

As shown by Carin and Pillai,[8] low-mode resonances can span more than 40 range lines in our rooftop UWB SAR imagery; while this would motivate a cluster window with a large azimuthal extent, I employed a more conservative 15-range-line clustering window. In the range dimension, high-Q target ringdown components have extended for several hundred range "pixels" (resolution bins) before decaying into clutter;[8] the lower-Q targets investigated in this paper generally ring for about 200 pixels, but the useful range for resonance-based recognition is about 100 pixels, thus yielding the 100-pixel clustering window.

To simplify the clustering task, a simple rectangular clustering window was used, spanning 15 range lines by 100 range pixels. Multiple targets within this window were "collapsed" into a single target; as shown in Table 3, this reduced the average number of false alarms per image from 550 to 180, representing a 67-percent decrease in false-alarm rate. Additionally, imposing a threshold of two targets within the sliding window reduced the average number of false alarms per image to 120, representing an overall 78-percent decrease in false-alarm rate. Increasing the threshold to three targets reduced the average false alarms to 74 per image, an 86-percent decrease.

Table 3. Target clustering, thresholding, and associated probabilities of recognition and false-alarm rates.

Clustering and threshold level	Average false alarms per image	Probability of recognition (in the clear)	Probability of recognition (in foliage)
No clustering	550	0.93	0.50
Clustering, no threshold	180	0.93	0.50
Clustering, threshold = 2	120	0.80	0.50
Clustering, threshold = 3	74	0.73	0.25

Of course, the thresholding operation can potentially impact the recognition rate, as shown in Table 3; of fourteen targets successfully recognized in the clear with no threshold, two were missed when the threshold was established at two targets within the clustering window, and a third target was missed when the threshold was increased to three. Since only one or two hits on a target would be considered a "weak" recognition in the resonance-based case, the tradeoff of P_R for reduction in false-alarm rate is quite reasonable. Although there are too few targets in the foliage case to infer any conclusive performance statistics, the impact of thresholding on the foliage-embedded target is provided in Table 3 for comparison with the "in-the-clear" case.

CONCLUSIONS

One is cautioned against drawing too many conclusions from the limited data presented here; true statistical significance can be achieved only with a substantially larger data set (statistically-significant results can be acquired from the recently-collected BoomSAR data[9]). Nevertheless, the recognition rates presented in this paper are quite high for targets in the clear, and are unprecedented for any algorithm employing a single template and recognizing targets at virtually all aspect angles. The recognition rate presented for foliage-embedded targets is similarly unprecedented for targets embedded more than 30 m into dense foliage. Clustering has demonstrated a substantial reduction false-alarm rate with no impact on recognition performance; imposing a target-declaration threshold on the clustering window has shown an even greater decrease in false-alarm rate, with relatively minor impact on recognition performance. Additionally, optimizing the spectral templates should reduce false alarms even more, and should improve recognition performance for foliage-embedded targets.

The resonance-based recognition results presented in this paper have never before been shown for complex targets in the field—either for targets in the clear or for targets obscured by foliage. Moreover, no single-template algorithm has ever demonstrated such high recognition rates on a complex target. These results suggest that resonance phenomena can be effectively employed as bases for target discrimination.

ACKNOWLEDGMENTS

Though there are many people both within and outside ARL who have contributed to the results reported in this paper, I would like to take this opportunity to acknowledge the contributions of—and dedicate this paper to—the late Dr. Joseph Sattler of the Army Research Laboratory. Dr. Sattler's technical guidance and persistence contributed greatly to the progress and success of the target-recognition methodology presented in this paper.

REFERENCES

1. M. Morgan, "Singularity Expansion Representations of Fields and Currents in Transient Scattering," *IEEE Trans. Antennas Propag.*, **AP-32**, No. 5, pp. 466–467, May 1984.

2 V. Sabio, "Target Recognition in Ultra-Wideband SAR Imagery," Army Research Laboratory, ARL-TR-378, August 1994.

3 V. Sabio and R. Chellappa, "Efficient Method of Target Recognition Based On Spectral Correlation of Wideband Resonance Effects," *SPIE Algorithms for Synthetic Aperture Radar Imagery*, Vol. 2230, pp. 328-335, April 1994.

4 V. Sabio, "An Efficient Method of Target Resonance Recognition Using Spectrally Matched Filters," *Proc. of the ATR Science and Technology Conference*, November 1994.

5 R. Kapoor and N. Nandhakumar, "Multiaperture Ultrawideband SAR Processing with Polarimetric Diversity," *SPIE Algorithms for Synthetic Aperture Radar Imagery II*, Vol. 2487, pp. 26-37, April 1995.

6 H. F. Harmuth, *Nonsinusoidal Waves for Radar and Radio Communication*, pp. 114–115, Academic Press, New York, NY, 1981.

7 S. Silver, *Microwave Antenna Theory and Design*, pp. 65–99, McGraw-Hill Book Company, New York, NY, 1949.

8 L. Carin and S. U. Pillai, "Ultra-Wideband/Short-Pulse Electromagnetics and Signal Processing," Polytechnic University Dept. of Electrical Engineering Technical Report, produced under ARO Contract DAAH04-93-02-0010.

9 M. Ressler, L. Happ, L. Nguyen, T. Ton, M. Bennett, "The Army Research Laboratory Ultra-Wideband Testbed Radars," IEEE International Radar Conference, May 1995.

ROBUST TARGET IDENTIFICATION USING A GENERALIZED LIKELIHOOD RATIO TEST

Jon E. Mooney, Zhi Ding, and Lloyd Riggs

Department of Electrical Engineering
200 Broun Hall
Auburn University, AL 36849

INTRODUCTION

The concept of deciding among a set of alternatives (or hypotheses) based upon the observation of a set of random variables has been a topic studied by statisticians for many years. This concept, known as hypothesis testing, provides a mathematically solid foundation to perform target identification. Target identification with known signatures can be easily formulated using Bayes hypothesis testing. However, a significant challenge lies in the need to accurately discriminate among known targets with only partial knowledge of target signatures. The lack of complete target signature knowledge results from the unknown orientation of the target and the dependency of the target signature on the target's orientation. For practical purposes, it is important to derive efficient and reliable schemes to accurately identify the target without a priori knowledge concerning the target's orientation.

Previous works by Rothwell[1,2], et al. introduced an E-pulse filter approach for multiple target discrimination using only the knowledge of the poles of a target's impulse response. This method, though effective, represents only one particular utilization of the prior knowledge of a target's poles. In this paper, we present a mathematically rigorous formulation of generalized hypothesis testing to perform target identification. In addition to the mathematical development, numerous results are provided demonstrating the effectiveness of the generalized likelihood ratio test. These results, which are shown as percent correct identification versus signal-to-noise ratio, contrast the performance of the GLRT to the E-pulse filter technique.

PROBLEM FORMULATION

The concept of using a target's poles to perform target ID is based on the singularity expansion method (SEM) representation of the transient scattered field returned from a target which has been illuminated by an "impulsive" (wide bandwidth) radar pulse. The SEM, as formulated by Baum[3] in 1971 , represents electromagnetic in-

Ultra-Wideband, Short-Pulse Electromagnetics 3
Edited by Baum *et al.*, Plenum Press, New York, 1997

teraction or scattering in terms of simple pole terms (or singularities) in the complex frequency plane, or correspondingly, as simple damped sinusoids in the time domain. For example, the SEM representation of the late-time-scattered field impulse response of a conducting object may be written as

$$r(t) = \sum_{i=1}^{N} a_i \, e^{(s_i \, t)}, \qquad t > T_L \tag{1}$$

where the complex amplitude coefficient (coupling coefficient) of the ith mode, a_i, depends on the orientation of the target with respect to the radar (aspect dependent parameters). The pole term s_i is aspect-independent and represents the frequency and damping constant of the ith mode. Note that the summation is over poles and **not** over conjugate pole pairs. Thus, only N modes are assumed excited by the incident field waveform. Late time, denoted by $t > T_L$, is defined as the time period after the incident pulse has passed over the target, so that subsequent radiation is associated with the target's free natural resonances[1] . Equation (1) is constructed using what is referred to as a class I coupling coefficient[4] . An SEM representation for the scattered field employing a class II coupling coefficient may also be constructed and enjoys the advantage of greater accuracy than the class I form in early time, $0 < t < T_L$, albeit at the expense of greater complexity[5] .

Assuming a target exists and its from a family of M possible candidates, then the return from the k-th target in the presence of noise can be written as

$$y(t) = \sum_{i=1}^{N} a_i^{(k)} b_i^{(k)}(t) \; + \; n(t) \qquad t > T_L, \quad 1 \leq k \leq M \tag{2}$$

where

$$b_i^{(k)}(t) \; = \; e^{s_i^{(k)} t}$$

and $n(t)$ is additive white Gaussian noise with zero mean and variance σ^2. For convenience, we denote the various signals in (2) by their uniform samples at the interval T_s:

$$\mathbf{y} \equiv \begin{bmatrix} y(T_L) \\ y(T_L + T_s) \\ y(T_L + 2T_s) \\ \vdots \end{bmatrix} ; \qquad \mathbf{b}_i^k \equiv \begin{bmatrix} b_i^{(k)}(T_L) \\ b_i^{(k)}(T_L + T_s) \\ b_i^{(k)}(T_L + 2T_s) \\ \vdots \end{bmatrix} ; \qquad \mathbf{n} \equiv \begin{bmatrix} n(T_L) \\ n(T_L + T_s) \\ n(T_L + 2T_s) \\ \vdots \end{bmatrix} .$$

Thus, the return signal vector $\mathbf{y}$ becomes

$$H_k : \qquad \mathbf{y} = \mathbf{B}_k \mathbf{a}_k \; + \; \mathbf{n} \tag{3}$$

where

$$\mathbf{a}_k \equiv \begin{bmatrix} a_1^k \\ a_2^k \\ a_3^k \\ \vdots \\ a_N^k \end{bmatrix}$$

and

$$\mathbf{B}_k = \begin{bmatrix} \mathbf{b}_1^k & \mathbf{b}_2^k & \cdots & \mathbf{b}_N^k \end{bmatrix} .$$

For the analysis presented here, $\mathbf{a}_k$ is an unknown parameter vector in the identification of target k. The only known parameters are the poles (or $\mathbf{B}_k$), and the measured return $\mathbf{y}$.

Generalized Hypothesis Testing

Having developed a model structure, it is now possible to apply hypothesis testing to generate a generalized likelihood ratio test (GLRT). Because the noise has been characterized as being white and Gaussian, the probability density function for $\mathbf{n}$ can be expressed as

$$p(\mathbf{n}) = \frac{1}{(2\pi)^{q/2}\sigma^q} \exp\left(-\frac{1}{2}\mathbf{n}^H\mathbf{n}\right) \tag{4}$$

where q represents the total number of samples and H denotes the Hermitian operator. If we know that target k is present, then the probability of getting $\mathbf{y}$ is simply the probability the noise would make up the difference in (4). Thus, we can write the conditional probability density function of $\mathbf{y}$ given target k as

$$p(\mathbf{y} \mid \text{target } k) = \frac{1}{(2\pi)^{q/2}\sigma^q} \exp\left(-\frac{1}{2\sigma^2}(\mathbf{y} - \mathbf{B}_k\mathbf{a}_k)^H(\mathbf{y} - \mathbf{B}_k\mathbf{a}_k)\right). \tag{5}$$

The conditional probability density function in (5) is a function of the unknown parameter $\mathbf{a}_k$ and is often referred to as the likelihood function.

Without loss of generality, a Bayes criterion can be used to develop a likelihood ratio test (LRT)[6] to decide between target 1 and target 2 . The LRT is written in terms of the likelihood functions as

$$\text{LRT}: \quad \frac{p(\mathbf{y} \mid \text{target } 1)}{p(\mathbf{y} \mid \text{target } 2)} \underset{H_2}{\overset{H_1}{\gtrless}} \gamma \tag{6}$$

where γ is the threshold. The threshold is a function of the prior probabilities and the cost. If we assume that all targets are equally probable and when uniform cost (zero for a correct decision and one for an incorrect decision) is assumed, then $\gamma = 1$. If the left side of the LRT is greater than γ, then we say hypothesis H_1 is true, or equivalently, target 1 is present. Similarly, if the ratio of the two likelihood functions is less than γ then target 2 is present.

In order to use the LRT in (6), the likelihood functions must be evaluated. Unfortunately, since the parameters $\mathbf{a}_1$ and $\mathbf{a}_2$ are unknown, it is not possible to use the LRT. One method of remedying this situation is to estimate $\mathbf{a}_1$ assuming target 1 is present, and then estimate $\mathbf{a}_2$ assuming target 2 is present. These estimates are then used in the LRT as if they were correct. If the maximum likelihood estimates are used, then the resulting LRT is referred to as a generalized likelihood ratio test (GLRT)[6] . The GLRT can be written in a similar form to the LRT as

$$\text{GLRT}: \quad \frac{\max_{\mathbf{a}_1} \; p(\mathbf{y} \mid \text{target } 1)}{\max_{\mathbf{a}_2} \; p(\mathbf{y} \mid \text{target } 2)} \underset{H_2}{\overset{H_1}{\gtrless}} \gamma \tag{7}$$

The maximum likelihood (ML) estimate $\hat{\mathbf{a}}_1$ is that value of $\mathbf{a}_1$ that maximizes the likelihood function $p(\mathbf{y} \mid \text{target } 1)$. Once an ML estimate has been obtained, then $p(\mathbf{y} \mid \text{target } 1)$ is evaluated using $\hat{\mathbf{a}}_1$. A similar procedure is used to obtain the denominator of (7).

To obtain an ML estimate, the likelihood function $p(\mathbf{y} \mid \text{target } k)$ must be maximized under the assumption that target k is present. Maximizing the expression

$$\exp\left(-\frac{1}{2\sigma^2}(\mathbf{y} - \mathbf{B}_k\mathbf{a}_k)^H(\mathbf{y} - \mathbf{B}_k\mathbf{a}_k)\right)$$

is equivalent to minimizing $||\mathbf{y} - \mathbf{B}_k\mathbf{a}_k||$. A least squares solution to $\mathbf{y} = \mathbf{B}_k\mathbf{a}_k$ yields

$$\hat{\mathbf{a}}_{k,\,\max} = (\mathbf{B}_k^H\mathbf{B}_k)^{-1}\mathbf{B}_k^H\mathbf{y}. \tag{8}$$

Substituting the least squares solution into the GLRT for the simple two target case (with $\gamma = 1$) yields after some manipulation the decision rule

$$\mathbf{y}^H\mathbf{B}_1(\mathbf{B}_1^H\mathbf{B}_1)^{-1}\mathbf{B}_1^H\mathbf{y} \underset{H_2}{\overset{H_1}{\gtrless}} \mathbf{y}^H\mathbf{B}_2(\mathbf{B}_2^H\mathbf{B}_2)^{-1}\mathbf{B}_2^H\mathbf{y}. \tag{9}$$

If we maintain the conditions of equal prior probabilities and uniform cost, then for multiple hypothesis testing, the above decision rule can be generalized for M target discrimination as

$$\text{decide } \{y(t)\} = \text{target } k \text{ if } \mathbf{y}^H\mathbf{B}_k(\mathbf{B}_k^H\mathbf{B}_k)^{-1}\mathbf{B}_k^H\mathbf{y} \text{ is maximum.} \tag{10}$$

SIMULATION RESULTS

To demonstrate the effectiveness of the GLRT as a function of signal-to-noise ratio (SNR), several simulations were conducted using various combinations of the four targets shown in Figure 1. Target A is a simple 1 meter (m) long thin cylinder lying along the x-axis and centered at the origin. Target B is a swept wing aircraft model. This example was chosen for its obvious relevance to target ID. The fuselage of the aircraft lies along the x-axis with forward and aft sections of 1/3 m and 2/3 m, respectively. The wings are swept back 45° from the normal to the fuselage and are 1/2 m in length. Target C is a symmetric tripole whose arms are each a length of 1/2 m. Target D is also a swept wing aircraft model similar to Target B. The only distinguishing feature between the two is the angle at which the wings are swept back. The wings on target D are swept back 60° from the normal to the fuselage. Also shown in Figure 1 is the orientation of the incident field E_ϕ relative to each target.

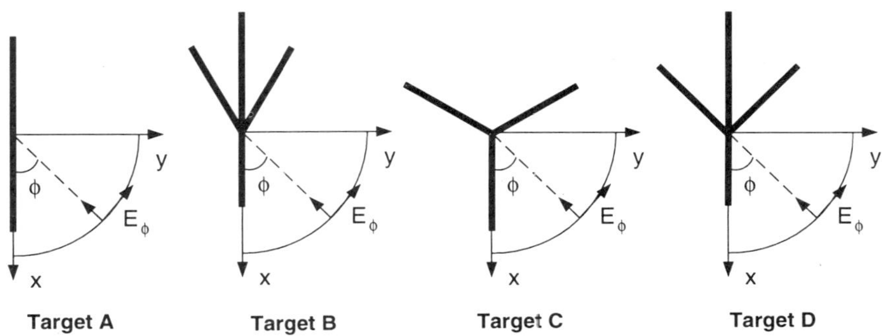

Figure 1: Targets A, B, C, and D used in the simulations to demonstrate the performance of the GLRT.

The first simulation involves targets A, B, and C. In this simulation, a computer randomly selects one of these three targets to be present. Recall that in the development of the GLRT detector, we assumed that each target has an equal probability of being present. Thus, for this specific simulation, each target has a probability of 1/3 of being present. This condition is enforced in the selection of targets A, B, and C. Once a target has been selected, random white Gaussian noise is added to a signature ($r(t)$, see (1)) of the selected target. The value of the noise power σ^2 is adjusted accordingly for a specified SNR. The corrupted return is then given to the GLRT detector (10) which yields a decision as to which target is present. To verify this decision, it is compared to the target selected at the beginning of the experiment. This process is repeated 1000 times at each specified value of SNR. For the purposes of this experiment, the SNR values where chosen to range from -30 dB to 40 dB. The results of this experiment are shown in Figure 2 for two different target orientations. In order to provide an upper bound on the performance of the GLRT, the performance of a LRT is also given. The results for both the GLRT and LRT were obtained simultaneously. Note that at very low SNR values, the confidence level of the GLRT is very low. At these values, the detector correctly identifies the target only 33% of the time which corresponds to the prior information. However, as the SNR increase, the performance of the detector increases as would be expected. At SNR values above 15 dB, the GLRT detector is correctly identifying the target 100% of the time for both target orientations.

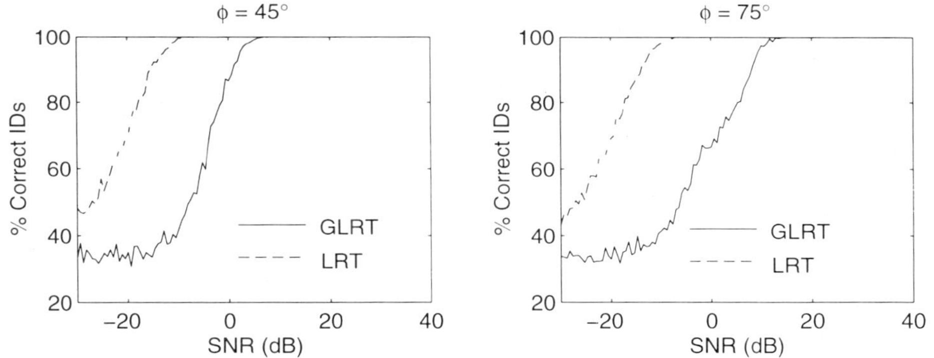

Figure 2: The performance of the GLRT and LRT as a function of SNR for different aspect angles using targets A, B, and C.

As a way of showing the sensitivity of the GLRT, another simulation is performed using only targets B and D. These two targets are chosen because of their similar features. The conditions for this simulation are the same as in the previous simulation. A thousand trials are performed at each SNR value, and the SNR values range from -30 dB to 40 dB. Furthermore, since only two targets are involved, each target has a 1/2 probability of being present. Figure 3 shows the results of this simulation for two different aspect angles. Both the performance of the GLRT and LRT are shown. Again, at very low SNR values, the confidence level of the GLRT is low. At these low values, the best the GLRT can do is identify the target correctly 50% of the time. As the SNR increases, the performance of the GLRT increases greatly. When the aspect angle ϕ is 30°, the GLRT identifies the correct target in every trial at SNR values above 10 dB. At $\phi = 60°$, a slightly higher resolution is required for the GLRT to obtain the same performance level. In this case, the GLRT identifies the correct target 100% of the time at SNR values above 17 dB.

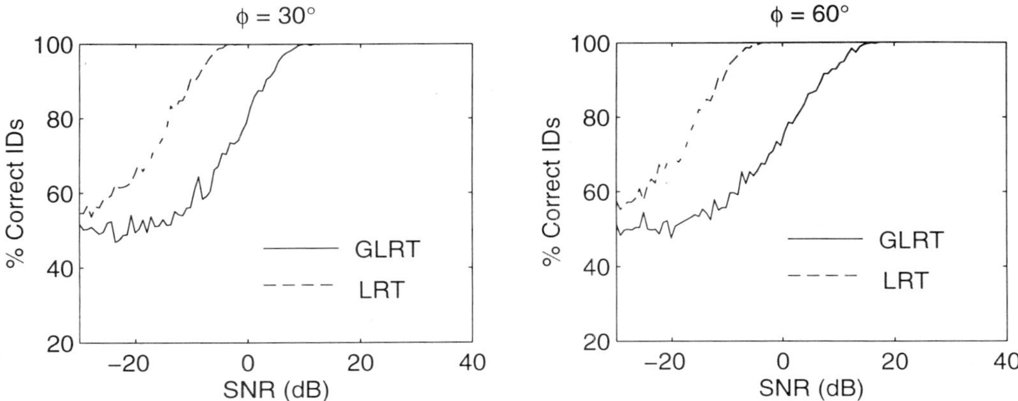

Figure 3: The performance of the GLRT and LRT as a function of SNR for different aspect angles using two similar targets (B,D).

The final simulation compares the performance of the GLRT detector to the E-pulse method. In this simulation, targets A, C, and D were used. An E-pulse filter was created for each target so that if the correct return was matched to its E-pulse filter, then a zero late time response would result. For example, if the return of target A (with no noise) is passed through the E-pulse filter for target A, then a zero late-time response would result. Passing this same response through the E-pulse filter for either target C or D would result in a non-zero response.

In order to provide a basis of comparison between the E-pulse technique and the GLRT, the scheme of using an energy ratio[7] at the output of each E-pulse filter is adopted. For example, the energy ratio to be computed at the output of E-pulse filter A is defined as

$$E_A = \frac{\int_{T_{LES_A}}^{T_{LEE_A}} c^2(t)\, dt}{\int_0^{T_{e_A}} e_A^2(t) dt}. \tag{11}$$

The parameter $c(t)$ represents the convolution of the E-pulse $e_A(t)$ with the received return. If $c(t)$ is the correct target (target A), then ideally the energy ratio would be zero. The time T_{LES_A} is defined as

$$T_{LES_A} = T_{e_A} + 2T_{L_A} \tag{12}$$

and represents the "earliest time at which the unknown target convolution is certain to be a unique series of natural modes."[7] The time T_{e_A} is the duration of the E-pulse for target A, and T_{LEE_A} is the end time of the energy ratio. In general, the time T_{LEE} is selected so that the window length $T_{LEE} - T_{LES}$ is the same for each ratio. A correct identification is determined by the minimum energy ratio at the output of the E-pulse of the unknown target. For example, if the energy ratio at the output of the E-pulse filter for target A is the smallest, then target A is selected to be the correct target. This scheme is illustrated in Figure 4.

As in the other simulations, this simulation consists of performing a thousand trials at each SNR value. The SNR values span the range from -30 dB to 50 dB. At each trial, the randomly generated target is corrupted with white Gaussian noise having zero mean and noise power σ^2. The corrupted signal is then passed through the GLRT detector as well as the three E-pulse filters. The output waveforms from the E-pulse

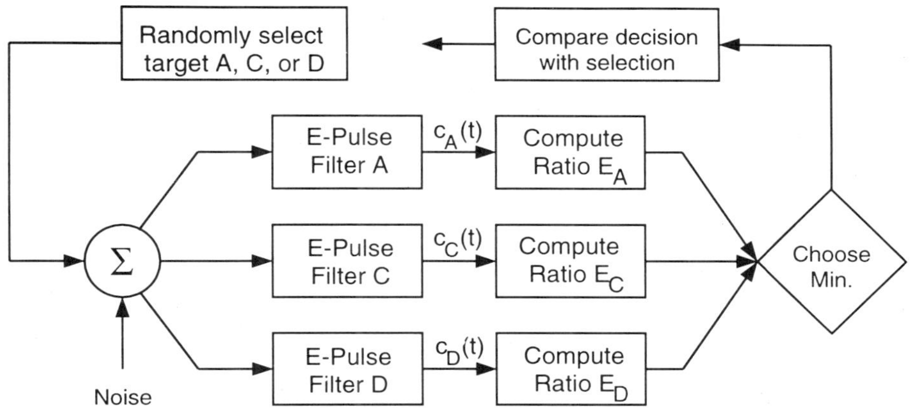

Figure 4: The scheme used to compare the performance of the E-pulse method to the GLRT using targets A,C, and D.

filters are used to compute the energy ratios E_A, E_C, and E_D. The ratio having the minimum value is concluded to be the target. This decision as well as the decision of the GLRT are then compared with the target selected at the beginning of the trial to verify the accuracy of each method.

Figure 5 shows the results of this simulation for two different aspect angles. The performance of the GLRT and the E-pulse technique are plotted as percent correct identifications per 1000 trials versus signal-to-noise ratio. The difference in performance between the GLRT and the E-pulse technique is profound. At SNR values above 0 dB there is roughly a 25 dB difference in SNR for the same level of performance. For the case when $\phi = 30°$, the GLRT begins to correctly identify the target in every trial at an SNR of approximately 12 dB. This same level of performance does not occur with the E-pulse method until the SNR reaches approximately 35 dB. The results are similar for the case when $\phi = 45°$. In this case, the GLRT achieves flawless performance at an SNR of approximately 17 dB. This kind of performance is not observed with the E-pulse technique until the SNR reaches a value of 43 dB.

CONCLUSIONS

In this paper, we have used well established mathematical models and rigorous statistical analysis to develop a simple but reliable method to perform target identification. Beginning with an SEM representation of the scattered field, we have developed a detector based on a generalized likelihood ratio test (GLRT) that is capable of identifying a specific target out of a family of M candidates. The GLRT assumes only a knowledge of a target's natural resonances thereby making the method aspect independent.

Numerous numerical results were presented demonstrating the effectiveness of the GLRT in the presence of random noise. These results showed the ability of the GLRT to identify the correct at low SNR values. Furthermore, the GLRT was compared to E-pulse technique. In the simulations we performed, the GLRT out performed the E-pulse method by a considerable margin.

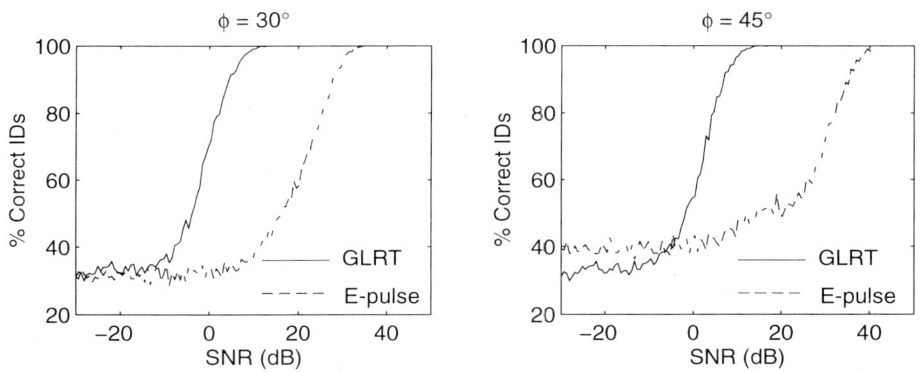

Figure 5: The performance of the GLRT compared to the E-pulse technique using targets A, C, and D.

REFERENCES

1. E.J. Rothwell and D.P. Nyquist, "Radar target discrimination using the extinction-pulse technique," *IEEE Trans. Antennas Propagation*, 33:929–936 (1985).
2. E.J. Rothwell, K.M Chen and D.P. Nyquist, "Frequency domain E-pulse synthesis and target discrimination," *IEEE Trans. Antennas Propagation*, 35:426–434 (1987).
3. C.E. Baum, The Singularity Expansion Method, in: *Transient Electromagnetic Fields* Springer-Verlag (1976).
4. C.E. Baum, "Representation of surface current density and far scattering in EEM and SEM with entire functions," Interaction Note 486, Phillips Laboratory, Kirtland AFB (1992).
5. M. Richards, "SEM representation of the early and late time fields scattered from wire targets," *IEEE Trans. Antennas Propagation*, 42 (1995).
6. H.L. Van Trees, *Detection, Estimation, and Modulation Theory Part I.* John Wiley & Sons (1967).
7. C.E. Baum, E.J. Rothwell, K.M. Chen, and D.P. Nyquist, "The singularity expansion method and its application to target identification," *IEEE Proceedings*, 79:1481–1492 (1991).

NEW METHODS OF DESIGNING OPTIMUM BROAD-BAND RADAR SIGNALS

Ovarlez Jean-Philippe and Dulost Jacques

Office National d'Études et de Recherches Aérospatiales
DES/SR, BP72, F92322 Châtillon Cedex, France
Email : ovarlez@onera.fr

1. INTRODUCTION

In radar or sonar, estimating the parameters such as the velocity or the position of a target is often a delicate problem. Let $z(t)$ be the transmitted and analytic signal with a constant propagation velocity c. The echo $x(t)$ reflected from a target moving with velocity v can be expressed as :

$$x(t,\theta) = A_0 T_\theta z(t)\, e^{i\phi} + b(t) \tag{1}$$

where T_θ is a transformation acting on the signal $z(t)$ with a vector θ of unknown parameters (time delay, Doppler shift, Doppler compression, etc..), and A_0 is the amplitude, ϕ a phase change and $b(t)$ a zero-mean white gaussian noise with σ^2 variance. When the probability density of the parameters A_0 and ϕ is unknown, the Maximum Likelihood ratio Λ to maximize, according to the Maximum Likelihood estimation theory, is given by the square modulus of the cross-ambiguity function :

$$\Lambda(\theta,\theta_0) = \frac{1}{2\sigma^2}\,\frac{\left|\displaystyle\int_{-\infty}^{+\infty} x(t,\theta)\, T_{\theta_0}^{*} z(t)\, dt\right|^2}{\displaystyle\int_{-\infty}^{+\infty} |T_{\theta_0} z(t)|^2\, dt} \tag{2}$$

The efficiency of an estimate $\hat{\theta}$ is generally measured by its variance $\mathrm{var}(\hat{\theta})$. For an unbiased estimate ($E(\hat{\theta}) = \theta$), this variance has a lower value given by the Cramer Rao Bounds (CRB) [1]. The CRB are obtained by inverting the Fisher Information Matrix (FIM) defined as :

$$J_{i,j} = \left(-E\left[\frac{\partial^2 \Lambda}{\partial\theta_i \partial\theta_j}\right]\right)_{i,j} \tag{3}$$

where θ_i denotes each component of the vector θ.

1.1 The Narrow-Band Case

Under Woodward's conditions [2], the Doppler effect can be approximated by a shift in frequency of the signal $z(t)$. Hence, the received signal $x(t,\theta)$ can be put in the form :

$$x(t,\theta) = A_0\, z(t-\tau)\, e^{2i\pi\nu t}\, e^{i\phi} + b(t) \tag{4}$$

where $\nu = 2vf_0/c$ is the Doppler frequency shift and τ the time delay (radial position $c\tau/2$, f_0 center

Ultra-Wideband, Short-Pulse Electromagnetics 3
Edited by Baum *et al.*, Plenum Press, New York, 1997

frequency). In this case the FIM (3) can be easily calculated and leads to :

$$J = \frac{4\pi^2 A_0^2}{\sigma^2} \begin{pmatrix} \sigma_t^2 & f_0 t_0 - m \\ f_0 t_0 - m & \sigma_f^2 \end{pmatrix} \tag{5}$$

where the first order moments f_0 and t_0 represent the mean frequency and the mean epoch, and the second order moments σ_f and σ_t represent the bandwidth and the duration of the signal. The parameter m is the modulation index of the signal. Each lower bound of the variance of estimates is obtained by inverting the matrix (5). These well known results prove that the best signal in radar (good range and velocity resolutions) is characterized by a high time-bandwidth product.

1.2 The Broad-Band Case

In that case, the problem of estimating a velocity does not consist in estimating a Doppler shift but a Doppler compression factor. Thus, the echo $x(t)$ can be put in the form :

$$x(t) = A_0\, z(a_0^{-1}t - b_0)\, e^{i\phi} + b(t) \tag{6}$$

The statistic to maximize is given by the square modulus of the broad-band cross-ambiguity function which is rewritten in the frequency domain :

$$\Lambda = \frac{a}{2\sigma^2} \left| \int_0^{+\infty} X(f)\, Z^*(af)\, e^{2i\pi abf}\, df \right|^2 \tag{7}$$

where the parameters $a = (c+v)/(c-v)$ and b represent the Doppler compression and the time delay parameters to estimate. The direct calculation of the FIM (3) by classical methods is not very easy and its coefficients do not lead to a simple interpretation as in the narrow-band case [3]. In the next section and using the Mellin transform [4, 5] already used in Broad-Band signal analysis [6, 7, 9, 10], the FIM computation is easily performed and leads to a perfect physical interpretation of its coefficients in the time-frequency half plane.

2. THE FISHER INFORMATION MATRIX IN BROAD-BAND CASE

2.1 The Mellin Transform

The Mellin transform which plays an important part in the computation and the physical interpretation of the FIM's coefficients has been well defined in [4] and acts on the analytic signal $Z(f)$ in frequency by :

$$M^{\xi}[Z](\beta) = \int_0^{+\infty} Z(f)\, e^{2i\pi\xi f}\, f^{2i\pi\beta + r}\, df \tag{8}$$

This transform can be interpreted as the coefficient of the decomposition of the signal onto a hyperbolic signals basis with a group delay law given in the time-frequency half plane by the equation $t = \xi + \beta/f$ with the invariant scalar product given by :

$$\int_0^{+\infty} Z_1(f) Z_2^*(f) f^{2r+1} df = \int_{-\infty}^{+\infty} M^{\xi}[Z_1](\beta) M^{\xi *}[Z_2](\beta) d\beta \tag{9}$$

The dual Mellin variable β therefore characterizes the coefficient of an hyperbola in the time-frequency half plane. The parameter r is free but is chosen here equal to $-1/2$ to preserve the classical scalar product. The study of the tomographic construction of the unitary affine time-frequency distribution $P_0(t, f)$ [11] has shown that a signal localized in the time-frequency half plane has a Mellin transform support bounded in Mellin space (cf. figure1). The connection between the P_0 distribution :

$$P_0(t, f) = f \int_{-\infty}^{+\infty} \frac{u}{2\sinh u/2}\, Z\left(\frac{u f e^{-u/2}}{2\sinh u/2}\right) Z^*\left(\frac{u f e^{u/2}}{2\sinh u/2}\right) e^{-2i\pi f t u}\, du \tag{10}$$

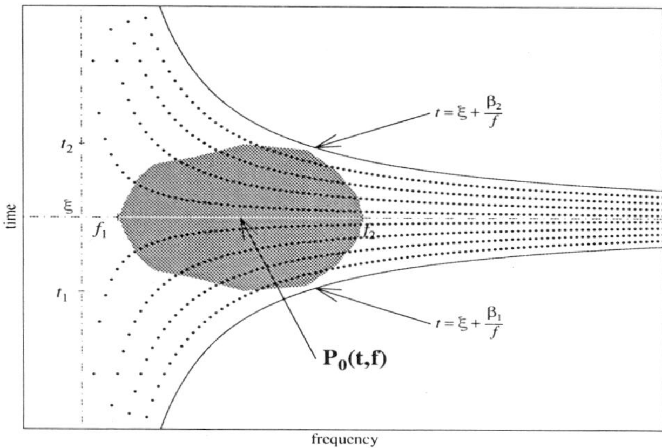

Figure 1: Localization in the time-frequency half plane of a signal $Z(f)$ having a time-frequency energy distribution $P_0(t,f)$. The two hyperbolas defined by equations $t = \xi + \beta_1/f$ and $t = \xi + \beta_2/f$ delimit the support $[\beta_1, \beta_2]$ of its Mellin transform.

and the Mellin transform is nothing but a hyperbolic Radon transform :

$$\int_{-\infty}^{+\infty} dt \int_0^{+\infty} P_0(t,f)\, \delta(t - \xi - \beta/f)\, f^{-1}\, df = \left| M^{\xi}[Z](\beta) \right|^2 \tag{11}$$

Using an a priori knowledge of the localization of the signal in the time-frequency half plane (bandwidth, relative bandwidth, duration), it is now possible to perfectly determine the spread $\sigma_\beta = \beta_2 - \beta_1$ of the signal in the Mellin space (cf. figure 1). In the following, the ξ parameter will be chosen equal to zero and the transform will be noted $M[Z](\beta)$. The main property of the Mellin transform is the property of scale invariance :

$$\begin{array}{ccc} Z(f) & \longrightarrow & Z'(f) = \sqrt{a}\, Z(af) \\ \downarrow & & \downarrow \\ M[Z](\beta) & \longrightarrow & M[Z'](\beta) = a^{-2i\pi\beta} M[Z](\beta) \end{array} \tag{12}$$

which is useful when rewriting (7) :

$$\Lambda = \frac{1}{2\sigma^2} \left| \int_{-\infty}^{+\infty} M[X](\beta)\, M^*[Z_b](\beta)\, a^{2i\pi\beta}\, d\beta \right|^2 \tag{13}$$

with $Z_b(f) = Z(f) \exp(2i\pi bf)$. Another important property of the Mellin transform, useful for computation of the FIM coefficients, is the diagonalization of the operator B defined by :

$$BZ(f) = -\frac{1}{2i\pi} \left(f \frac{d}{df} + \frac{1}{2} \right) Z(f) \tag{14}$$

which is transformed as $M[BZ](\beta) = \beta M[Z](\beta)$

2.2 Broad-Band Expression of the Fisher Information Matrix

The Fisher Information Matrix has the form [8] :

$$J = \frac{4\pi^2 A_0^2}{\sigma^2} \begin{pmatrix} \sigma_\beta^2 & f_0\beta_0 - M \\ f_0\beta_0 - M & \sigma_f^2 \end{pmatrix} \tag{15}$$

where the parameters σ_f and f_0 define the bandwidth and the mean frequency of the signal and where *the parameters β_0, σ_β are given by :*

$$\beta_0 = \int_{-\infty}^{+\infty} \beta \, |M[Z](\beta)|^2 \, d\beta \qquad \sigma_\beta^2 = \int_{-\infty}^{+\infty} (\beta - \beta_0)^2 \, |M[Z](\beta)|^2 \, d\beta \tag{16}$$

The first and second order moments can be viewed respectively as the mean β and the spread of the signal Z in Mellin space. The broad-band modulation index M defined by :

$$M = -\frac{1}{2\pi} Im \int_0^{+\infty} f^2 \, \frac{dZ}{df} \, Z^*(f) \, df \tag{17}$$

plays the same role for the hyperbolic signals as the narrow-band modulation index m for the chirp signals. Finally, the ratio A_0^2/σ^2 is the Signal-to-Noise Ratio. The proof of this result can be found in annexe.

To estimate the quality of the compression and delay parameters, the FIM must be inverted. Each term of the inverse matrix J^{-1} gives the variance lower bound of each estimate. As the estimates are unbiased and efficient (high SNR), the CRB are reached and we obtain the following important new results :

- The variance of the time delay estimate $\hat{b}$ is given by :

$$\text{var}(\hat{b}) = \frac{\sigma^2}{4\pi^2 A_0^2} \frac{\sigma_\beta^2}{\sigma_f^2 \sigma_\beta^2 - (M - \beta_0 f_0)^2} \geq \frac{\sigma^2}{4\pi^2 A_0^2} \frac{1}{\sigma_f^2} \tag{18}$$

- The variance of the compression estimate $\hat{a}$ is given by:

$$\text{var}(\hat{a}) = \frac{\sigma^2}{4\pi^2 A_0^2} \frac{\sigma_f^2}{\sigma_f^2 \sigma_\beta^2 - (M - \beta_0 f_0)^2} \geq \frac{\sigma^2}{4\pi^2 A_0^2} \frac{1}{\sigma_\beta^2} \tag{19}$$

- The variance of the velocity estimate $\hat{v}$ is given by : $\text{var}(\hat{v}) = \dfrac{c^2}{4} \text{var}(\hat{a})$

- The covariance of the cross-estimates is given by :

$$\text{cov}(\hat{a}, \hat{b}) = \frac{\sigma^2}{4\pi^2 A_0^2} \frac{M - \beta_0 f_0}{\sigma_f^2 \sigma_\beta^2 - (M - \beta_0 f_0)^2} \tag{20}$$

The first result (18) shows that the time delay (or range) resolution is always related to the inverse of the signal spread in frequency as in the narrow-band case. The result (19) is very important because it proves that the compression (or velocity) resolution depends only on the inverse of the signal spread in the Mellin space instead of the signal duration as in the narrow-band case. As an example, let us consider the so-called Doppler invariant signals as hyperbolic signals (cf. figure 2) which are characterized by a no spread in Mellin space ($\sigma_\beta = 0$) : this kind of signals does not lead to a good velocity resolution (well known result). The figure 3 shows, on the contrary, that a signal with very short duration can have a no negligible velocity resolution. These two extreme examples prove the difference with the results classicaly obtained in the narrow-band case.

Under Woodward's assumptions (narrow-band case, $v/c \ll 1$), the hyperbolas which delimit the signal in the time-frequency half plane, may be replaced by straight lines parallel to the frequency axis. In that case, the parameter $\beta, \beta_0, \sigma_\beta, M$ and $a = (c+v)/(c-v)$ can be respectively approximated by $f_0 t, f_0 t_0, f_0 \sigma_t, f_0 m$ and $1 + 2v/c$. Substituting these approximations in (19), we obtain the classical narrow-band results :

$$E\left[(v - \hat{v})^2\right] = \frac{c^2 \sigma^2}{16\pi^2 A_0^2 f_0^2} \frac{\sigma_f^2}{\sigma_t^2 \sigma_f^2 - (m - f_0 t_0)^2} \tag{21}$$

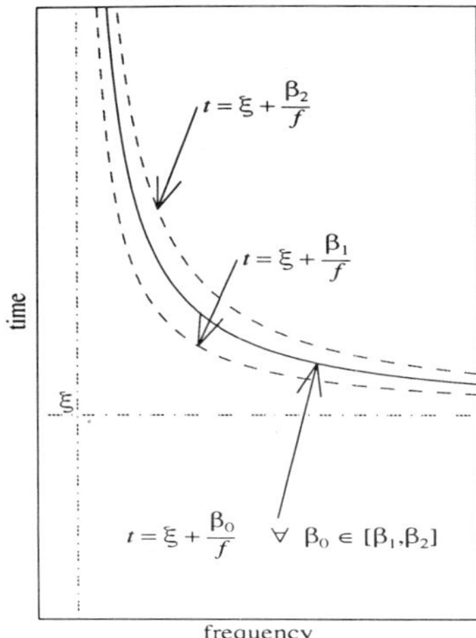

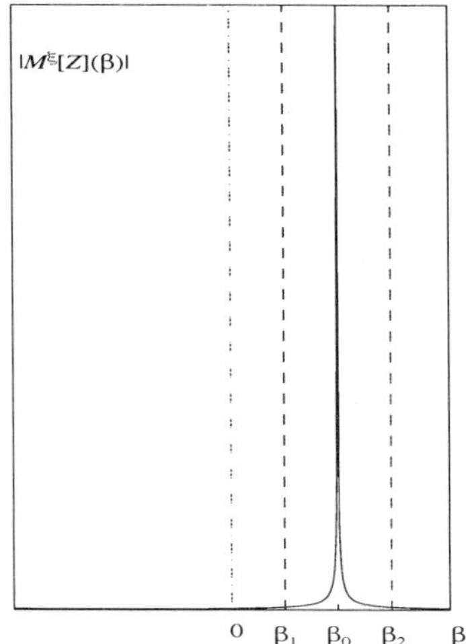

Figure 2: Localization in the time-frequency half plane of a hyperbolic signal $Z(f)$ labeled by its parameter β_0 and its Mellin transform. Any pair of hyperbolas with equation $t = \xi + \beta_1/f$ and $t = \xi + \beta_2/f$ (with $\beta_1 < \beta_0 < \beta_2$) can delimit the signal. Such a signal, although it has a infinite duration, has a zero-spread in Mellin space and therefore no velocity resolution.

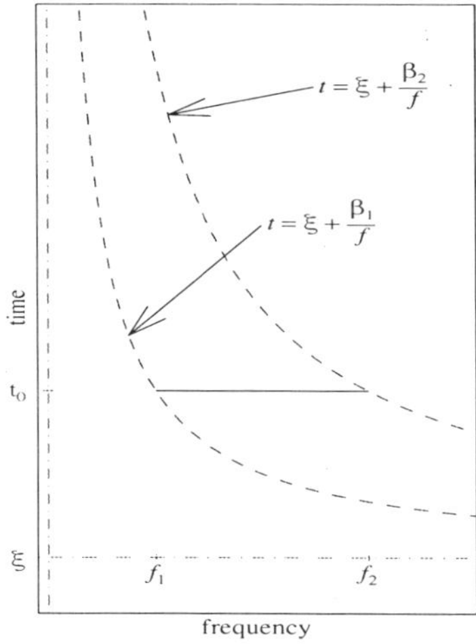

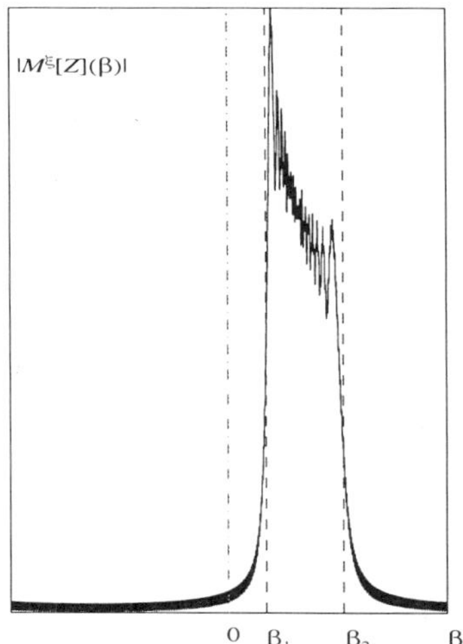

Figure 3: Localization in the time-frequency half plane of a short pulse centered around $t = t_0$ with a bandwidth $B = f_2 - f_1$ around $f_0 = (f_1 + f_2)/2$ and its Mellin transform. Such a signal, although of very short duration, has a spread $\sigma_\beta = (f_2 - f_1)t_0$ in Mellin space and therefore a finite velocity resolution.

3. HIGH $B\Delta_\beta$ SIGNALS SYNTHESIS METHODS

The two methods which will be presented are useful when looking for signals which minimize the Cramer-Rao lower bounds. The first method is devoted to the construction of optimal broad-band signals with given autocorrelation functions in velocity and delay spaces. The second one determines a phase law which allows the signal to reach the desired spreads in the Mellin and frequency spaces.

3.1 The Stationary Phase Method

The Stationary Phase Principle method already used for designing high time bandwidth product signals [12] is applied here but is extended to the Mellin and frequency spaces. The main idea is to construct high $B\Delta_\beta$ product signals (asymptotics signals) in the same way. The inverse Mellin transform is defined by :

$$Z(f) = e^{-2i\pi\xi f}\, f^{-1/2} \int_{-\infty}^{+\infty} M^{\xi}[Z](\beta)\, f^{-2i\pi\beta}\, d\beta \tag{22}$$

Following the stationary phase principle method and applying it on (22), we have :

$$Z(f) = f^{-1/2}\, e^{-2i\pi\xi f} \sqrt{\frac{2\pi}{|\phi''(\lambda)|}} \left|M^{\xi}[Z](\lambda)\right| e^{i(\phi(\lambda)-2\pi\lambda\log f \pm\pi/4)} \tag{23}$$

where we note $M^{\xi}[Z](\beta) = \left|M^{\xi}[Z](\beta)\right| \exp\left(i\phi(\beta)\right)$ and where λ is the stationary point defined by the following equation :

$$\frac{d}{d\beta}\left[\phi(\beta) - 2\pi\beta\log f\right]_{\beta=\lambda} = 0 \tag{24}$$

or, if we note ϕ'^{-1} the reciprocal function of ϕ', defined by $\lambda = \phi'^{-1}\,(2\pi\log f)$.

The spectrum phase law has the form $\Psi(f) = -2\pi\xi f + \phi(\lambda) - 2\pi\lambda\log f \pm \pi/4$ and is thus defined by its group delay ($T(f) = -\dfrac{1}{2\pi}\dfrac{d\Psi}{df}$) given by $T(f) = \xi + \frac{\lambda}{f}$.

Acting on the shape of $|Z(f)|$ and $|M^{\xi}[Z](\beta)|$ by choosing the distance autocorrelation function $F(b)$ and velocity autocorrelation function $G(a)$ defined according to :

$$|Z(f)|^2 = \int_{-\infty}^{+\infty} F(b)\, e^{-2i\pi bf}\, db \qquad \left|M^{\xi}[Z](\beta)\right|^2 = \int_{0}^{+\infty} G(a)\, a^{-2i\pi\beta-1}\, da \tag{25}$$

we thus define the phase law $\phi(\lambda)$ given by the differential equation :

$$\phi''(\lambda) = 2\pi\, \frac{\left|M^{\xi}[Z](\lambda)\right|^2}{f\, |Z(f)|^2} \tag{26}$$

Choosing $\psi(\lambda) = \dfrac{1}{2\pi}\phi'(\lambda) = \log f$, the last equation can be integrated with respect to λ and leads to :

$$\int_{0}^{\exp\psi(\lambda)} |Z(f)|^2\, df = \int_{-\infty}^{\lambda} \left|M^{\xi}[Z](\beta)\right|^2\, d\beta \tag{27}$$

By choosing a given λ, it is now possible by (27) to find $\psi(\lambda)$ and to determine the phase law $\phi(\beta)$ by :

$$\phi(\beta) = 2\pi \int_{-\infty}^{\beta} \psi(u)\, du \tag{28}$$

3.2 Construction of Optimal Signals

Consider a monochromatic and analytic signal given by its equation $Z(f) = \delta(f - f_0)$. This signal has a Mellin transform given by $M^{\xi}[Z](\beta) = f_0^{2i\pi\beta-1/2}\, e^{2i\pi\xi f_0}$. We can therefore perfectly determine the frequency law of the signal $Z(f)$ as the function of the Mellin variable :

$$f_0 = \exp\left(\frac{1}{2\pi}\frac{d\phi}{d\beta}\right) \tag{29}$$

where $\phi(\beta)$ is the phase of the Mellin transform of Z. Extending this relation, we obtain the expression of the frequency in terms of the β variable :

$$f(\beta) = \exp\left(\frac{1}{2\pi}\frac{d\phi(\beta)}{d\beta}\right) \tag{30}$$

Given a frequency law $f(\beta)$ in Mellin space, we can obtain by solving (30) the derivative of the Mellin phase and hence the expression of the signal in Mellin space $M^{\xi}[Z](\beta) = e^{i\phi(\beta)}$. This procedure is the analogous construction of a signal from time to frequency space using the definition of the instantaneous frequency. It only ensures that the signal will have, at one and the same time, a given bandwidth and spread in Mellin space but does not ensure the sidelobes quality of the two autocorrelation functions in range and velocity spaces.

4. CONCLUSION

The analytical expression of the Cramer Rao bounds for velocity estimation in the broad-band case has been established using the Mellin transform. The most impressive result concerns the velocity resolution of active radar (and particularly sonar) which is not related to the inverse of the signal duration as in narrow-band case but to the inverse of the spread of the signal in Mellin space. This spread has a direct geometrical interpretation in the time-frequency half plane and can be easily estimated when duration, bandwidth and relative bandwidth are known. Thanks to this interpretation, two interesting procedures have been proposed to construct optimal broad-band signals which minimize the Cramer-Rao lower bounds.

ANNEXE : Proof of the Proposition

The main idea for FIM derivation is to compute the FIM coefficients from the statistics Λ (13) rewritten in Mellin space rather than a direct computation. To simplify the demonstration, all the partial derivatives of the statistic Λ with respect to parameters a and b will be evaluated at the point $\mathrm{O}(a = a_0 = 1, b = b_0 = 0)$. If we note $A(a, b)$, the classical cross-ambiguity function rewritten in Mellin space :

$$A(a,b) = \int_{-\infty}^{+\infty} M[X](\beta)\, M^*[Z_b](\beta)\, a^{2i\pi\beta}\, d\beta \tag{31}$$

all the partial derivatives of A with respect to parameters a and b and evaluated at the point $\mathrm{O}(a = 1, b = 0)$, when using the property of unitarity of the Mellin transform (9), lead to :

$$\frac{\partial A}{\partial a} = 2i\pi \int_{-\infty}^{+\infty} \beta\, M[X](\beta)\, M^*[Z](\beta)\, d\beta \tag{32}$$

$$\frac{\partial^2 A}{\partial a^2} = 2i\pi \int_{-\infty}^{+\infty} \beta(2i\pi\beta - 1)\, M[X](\beta)\, M^*[Z](\beta)\, d\beta \tag{33}$$

$$\frac{\partial A}{\partial b} = 2i\pi \int_{-\infty}^{+\infty} M[X](\beta)\, M^*[fZ(f)](\beta)\, d\beta = 2i\pi \int_{0}^{+\infty} f\, X(f)\, Z^*(f)\, df \tag{34}$$

$$\frac{\partial^2 A}{\partial a \partial b} = -4\pi^2 \int_{-\infty}^{+\infty} \beta\, M[X](\beta)\, M^*[fZ(f)](\beta)\, d\beta \tag{35}$$

$$\frac{\partial^2 A}{\partial b^2} = -4\pi^2 \int_{0}^{+\infty} f^2\, X(f)\, Z^*(f)\, df \tag{36}$$

The first coefficient J_{11} of the FIM takes the form :

$$J_{11} = -\frac{1}{\sigma^2} E\left[Re\left(A^* \frac{\partial^2 A}{\partial a^2} + \left|\frac{\partial A}{\partial a}\right|^2\right)\right] \tag{37}$$

The noise $b(t)$ is a zero mean white gaussian noise. If we note $C(\beta_1 - \beta_2)$ the covariance of the Mellin transform of X, it is easy to show that :

$$C(\beta_1 - \beta_2) = E\left[M[X](\beta_1) M^*[X](\beta_2)\right] = A_0^2\, M[Z](\beta_1) M^*[Z](\beta_2) + \sigma^2\, \delta(\beta_1 - \beta_2) \tag{38}$$

while the covariance $C(f_1 - f_2)$ of the Fourier transform of the signal is given by :

$$C(f_1 - f_2) = E\left[X(f_1) X^*(f_2)\right] = A_0^2\, Z(f_1)\, Z^*(f_2) + \sigma^2\, \delta(f_1 - f_2) \tag{39}$$

Using relations (32), (33) and (38), J_{11} can be rewritten :

$$J_{11} = -\frac{1}{\sigma^2} Re \int_{-\infty}^{+\infty} \int_{-\infty}^{+\infty} C(\beta_1 - \beta_2) \left[2i\pi\beta_1(2i\pi\beta_1 - 1) + 4\pi^2\beta_1\beta_2\right] d\beta_1\, d\beta_2 \tag{40}$$

and we obtain the expression of the J_{11} coefficient :

$$J_{11} = \frac{4\pi^2 A_0^2}{\sigma^2} \int_{-\infty}^{+\infty} (\beta - \beta_0)^2 \left|M[Z](\beta)\right|^2 d\beta = \frac{4\pi^2 A_0^2}{\sigma^2} \sigma_\beta^2 \tag{41}$$

The computation of the J_{22} coefficient of the FIM is easily performed in the same way using relations (34) and (36) and leads to :

$$J_{22} = \frac{4\pi^2}{\sigma^2} Re \int_{0}^{+\infty} \int_{0}^{+\infty} C(f_1 - f_2)\, Z^*(f_1)\, Z(f_2)\, (f_1^2 - f_1 f_2)\, df_1\, df_2 \tag{42}$$

Substituting in (42) the covariance given by (39), we obtain the J_{22} coefficient proposed in (15). Finally, the last symmetrical coefficient J_{12} or J_{21} computation is given by :

$$J_{12} = -\frac{1}{\sigma^2} E\left[Re\left(X^* \frac{\partial^2 X}{\partial a\, \partial b} + \frac{\partial X}{\partial a}\frac{\partial X^*}{\partial b}\right)\right] \tag{43}$$

which, using relations (32), (34) and (35) and substituting the covariance given in (38), leads to :

$$\begin{aligned} J_{12} \;=\; & -\frac{4\pi^2 A_0^2}{\sigma^2} Re\left[\int_{-\infty}^{+\infty} \beta_1 M[Z](\beta_1)\, M^*[fZ(f)](\beta_1)\, d\beta_1 \right. \\ & \left. - \int_{-\infty}^{+\infty} M[Z](\beta_1)\, M^*[fZ(f)](\beta_1)\, d\beta_1 \int_{-\infty}^{+\infty} \beta_2 \left|M[Z](\beta_2)\right|^2 d\beta_2\right] \end{aligned} \tag{44}$$

This last expression can be easily transformed in the frequency domain using the unitarity property (9) of the Mellin transform and using the operator B defined by equation (14). With the definition of the broad-band modulation index M defined by (17), the coefficient J_{12} is finally derived.

References

[1] H.L. Van Trees, *Detection, Estimation and Modulation Theory*, Part I, II and III, John Wiley and Sons, New York, 1971

[2] P.M. Woodward, *Probability and Information Theory with Applications to Radar*, Pergamon Press, New York, 1953

[3] Qu Jin, Kon Max Wong and Zhi-Quan (Tom) Luo, The estimation of time delay and Doppler stretch of wideband signals, *IEEE Trans. on ASSP*, Vol. 43, No. 4, April 1995

[4] J. Bertrand P. Bertrand and J.P. Ovarlez, The Mellin transform, in : *The Transforms an Applications Handbook*, ed. A. D. Poularikas, CRC Press Inc., 1995, chapter 12

[5] J.P. Ovarlez, La Transformation de Mellin: un Outil pour l'Analyse des Signaux à Large-Bande, *Thesis University Paris 6*, Paris, April 1992.

[6] J. Bertrand P. Bertrand and J.P. Ovarlez, The wavelet approach in radar imaging and its physical interpretation, in : *Progress in Wavelet Analysis ans Applications*, ed. Y Meyer and S Roques, Editions Frontières, 1993

[7] J. Bertrand P. Bertrand and J.P. Ovarlez, Frequency directivity scanning in laboratory radar imaging, *International Journal of Imaging Systems and Technology*, Vol. 5, pp. 39-51, 1994

[8] J.P. Ovarlez, Cramer-Rao bound computation for velocity estimation in the broad-band case using the Mellin transform, *Proc. IEEE-ICASSP*, Minneapolis, MN, USA, 1993

[9] J. Bertrand P. Bertrand and J.P. Ovarlez, Discrete Mellin transform for signal analysis, *Proc. IEEE-ICASSP*, Albuquerque, NM, USA,1990.

[10] J.P. Ovarlez J. Bertrand and P. Bertrand, Computation of affine time-frequency distributions using the fast Mellin transform, *Proc. IEEE-ICASSP*, San Francisco, CA, USA, 1992

[11] J. Bertrand and P. Bertrand, Affine time-frequency distributions, in : *Time Frequency Signal Analysis - Methods and Applications*, ed. B. Boashash, Longman-Cheshire, Australia, 1992, chapter 5

[12] A. Papoulis, *Signal Analysis*, McGraw Hill, New York, 1977

ULTRA-WIDEBAND RADAR DETECTION IN WHITE NOISE

M. Steiner, K. Gerlach, and F.C. Lin

Naval Research Laboratory
Washington DC, 20375-5320

INTRODUCTION

The purpose of this paper is to examine in detail aspects of UWB detection in white Gaussian noise. Specifically, we assume a range-spread target using a point scattering model with a single polarization. We also assume a single pulse has been transmitted as opposed to multiple pulses. Hence we are not considering the problem of compensating for range walk or cancelling clutter with multiple pulses.

There have been several papers published previously in this area which examined different aspects of the detection problem. These papers usually make different assumptions regarding either the target or the noise model which results in different detection forms. Ref. [1] examines several detector forms for detection of range spread targets. These techniques and others were further examined in [2] for applicability to UWB. In [3] the target is also considered spread over many range cells. Two detectors are considered: 1 out of M detection logic and a noncoherent integration detector which noncoherently integrates across the range cells. It was shown that when few scatterers are present, the 1 out of M performance dominates the noncoherent integrator whereas when there are many scatterers, the noncoherent integrator gives the best performance. Similar results were obtained by Rose in [4]. In [5], [6] Farina et al examine more specific techniques for UWB detection and identification. In regards to detection, they derive a likelihood ratio test for the detector under the assumption that the target is a Gaussian distributed stochastic time sequence with zero mean and a known covariance matrix. The likelihood ratio test is a quadratic form with respect to a received signal vector which we represent here by $\mathbf{x}$. That is, the test takes the form $\mathbf{x}^\dagger Q\mathbf{x}$ which is compared with a threshold. The matrix Q is a function of the covariance matrix of the target and on the interference. In [7] a detector form is derived under the generalized likelihood ratio test (GLRT) assumption. They assume that the target consists of a sum of point scatterers with unknown amplitudes. A maximum likelihood estimate is used to estimate the scatterer amplitudes. The positions of the scatterers are assumed known in the derivation although simulations are presented that show performance when the scatterer locations are not known exactly. The form of their GLRT is also a quadratic form, with matrix Q that is different from that shown in [5]. A major difference in assumptions between those made in [7] and in [5] is that the target is modeled by a stochastic model in the latter paper and is assumed unknown in the former.

Another problem related to target detection that has been examined in the literature is the optimization of the bandwidth. In this problem, the question arises as to whether or not the bandwidth can be optimized for a given target. This question appears to have been first addressed in [8] by Nitzberg. Nitzberg shows that the optimal bandwidth occurs approximately when the scatterers first begin to be resolved.

An area which does not appear to be fully addressed in these papers is the issue of in-phase and quadrature (I/Q) sampling versus real sampling. That is, what are the theoretical underpinnings of the I/Q representation and what is the tradeoff in performance between I/Q and real processing. This issue is examined in depth in this paper in Section III. We will examine the model assumed in the narrowband derivation and see that it does not necessarily hold in the UWB area. We will show however, for a Gaussian modulated waveform, that the difference appears to be mainly a tradeoff of sampling rates and that by properly designing the match filters, nearly equivalent performance can be obtained when the real sampling rate is twice the I/Q rate, even as the waveform approaches having only a single cycle.

We also compare the performance of the GLRT for cases when the target scatterer amplitude and positions are known exactly versus when they are unknown. Additionally, an optimal BW is found similar to Nitzberg for the GLRT.

PRELIMINARIES

In this section we define the modeling of the target and noise and also the design of the transmitted waveform. Additionally, various definitions, including bandwidth and resolution, are given. We also show how the model is normalized so that results presented can be applied for any center frequency of the transmitted waveform.

There are a number of ways to describe and model a target. In this paper the target is modeled by a sum of point scatterers,

$$s(t) = \sum_{j=1}^{n} a_j g(t - \tau_j) \tag{1}$$

where the a_j are assumed to be Gaussian distributed and $g(t)$ is the received waveform. The τ_j will be generated by a Poisson random process with mean time $2dsep/c$. The scatterers are separated by a distance that is modeled using a Poisson process with the parameter *dsep* equal to the mean distance between scatterers. When the probability of detection (hereafter referred to as performance) for a given false alarm probability is considered the results are found by averaging over many realizations of targets generated by a Poisson process specified by *dsep*.

There are numerous potential waveforms $g(t)$ that can be assumed. We assume a Gaussian transmitted waveform that modulates a sinusoidal carrier given by,

$$g(t) = \left(\frac{1}{\pi T^2}\right)^{\frac{1}{4}} e^{-\frac{t^2}{2T^2}} \sin(2\pi f_c t). \tag{2}$$

Note that $g(t)$ is parameterized by a constant proportional to the duration T and a carrier frequency f_c. We choose to use $\sin(2\pi f_c t)$ as opposed to a $\cos(2\pi f_c t)$ since the former, when used in (2) never has a DC component, whereas the latter can have a significant DC component (which does not propagate) for UWB as T becomes small. For arbitrary waveforms, the carrier frequency becomes more difficult to define, but we will only use the Gaussian waveform in this paper.

The bandwidth of this waveform can be defined in various ways. The RMS bandwidth β for a baseband waveform $f(t)$ is defined by

$$\beta^2 = \frac{\int_{-\infty}^{\infty} f^2 |S(f)|^2 df}{\int_{-\infty}^{\infty} |S(f)|^2 df}. \tag{3}$$

For the Gaussian waveform at baseband ($f_c = 0$), β can be derived as $\frac{1}{2\sqrt{2}\pi T}$. The effective RMS bandwidth, is another bandwidth representation given by 2β [9, p. 299] for signals that are not at baseband. Throughout the remainder of this paper when we refer to RMS bandwidth it is the effective RMS BW given by $B_{RMS} = \frac{1}{\sqrt{2}\pi T}$.

Another definition often used is the 3 dB bandwidth, B_{3dB}. The 3 dB bandwidth is given by twice the frequency span from the center frequency to the 3 dB point of the power spectral density of the waveform. For the Gaussian waveform, the power spectral density is proportional to $e^{-(2\pi T f)^2}$ which yields $B_{3dB} = \frac{2\sqrt{\log 2}}{2\pi T}$. Note that B_{RMS} and B_{3dB} differ only by a constant, hence results that are derived assuming B_{RMS} (which we will assume throughout) can easily be converted if one prefers B_{3dB}.

The resolution of a waveform can also be defined in a variety of ways. In this paper, we adopt the definition given in [10, p. 101]. Here one considers the 3 dB power width of the output response from a matched filter. The output response is also the autocorrelation function of the waveform, $A(\tau)$, and for the Gaussian waveform is proportional to, $e^{-\frac{1}{4}\frac{\tau^2}{T^2}}$. When this is set equal to $\frac{1}{\sqrt{2}}$ the result is that $\tau = T\sqrt{2\log 2}$ and hence the resolution, $\Delta = 2\tau = 2T\sqrt{2\log 2}$. Note that the bandwidth resolution product is a constant as expected.

Another useful parameter that can help to distinguish UWB waveforms as opposed to narrowband (NB) waveforms is a measure of the number of cycles in the transmitted waveform. We define nc as the number of cycles of the carrier frequency in a resolution cell. Note that $nc = f_c\Delta$ is a function of both bandwidth and the center frequency. Hence, if the center frequency is doubled, the bandwidth must be doubled to keep nc the same. Note that nc is similar in notion to the percent bandwidth or fractional bandwidth which is the bandwidth normalized by the carrier frequency.

Now, so far we have discussed the center frequency, bandwidth, resolution, and nc characteristics of the transmitted waveform and also the mean distance between target scatterers. It would be useful to normalize these parameters in a meaningful way to reduce the number of parameters considered while keeping the results as broad as possible. To this end we define the normalized separation, $nsep$ as the ratio of $dsep$ (the mean distance between scatterers) to the resolution cell in meters ($2c/\Delta$) where c is the speed of light in m/s. This parameter is a direct measure indicating when the scatterers have been resolved. As $nsep$ increases, the returns from scatterers interfere less. Another simplifying parameter is the ratio of f_c to sampling rate (f_s) which we define as $fcsr$. The sampling rate f_s is the rate at which discrete samples are taken of the RF waveform. Throughout this paper we will assume this to be a constant (note this rate is different then the rate at which the matched filter output is sampled, which we will vary later). Now, writing Eqn. (2) in terms of the above parameters we arrive at the discrete representation,

$$g(k) = (\frac{1}{\pi T^2})^{\frac{1}{4}} e^{(\frac{-k\ fcsr\ nsep\ c}{2\ f_c\ dsep})^2} \sin(2\pi k\ fcsr), \tag{4}$$

where we have substituted $t = fcsr\ k$, $\Delta = 2\tau = 2T\sqrt{2\log 2}$, and $nsep = dsep/(2c/\Delta)$ into Eqn. (2). Now, it can be seen that the percent bandwidth or nc parameter can be kept the same if f_c is increased by a constant C via $f_c' = Cf_c$ and additionally the resolution is reduced by $\Delta' = \Delta/C$ (or the bandwidth is increased by C). If additionally the mean scatterer separation distances are reduced via $dsep' = dsep/C$ it can be seen from (4) that $g(k)' = g(k)$. Hence the discrete representation is unchanged. Note furthermore that $nsep' = nsep$ does not change. Hence, any performance results derived for a given f_c are also valid for a new f_c' provided the percent bandwidth (or nc) is the same and the mean scatterer separation distance $dsep$ is scaled by f_c/f_c'. Therefore, in the simulations which follow, we will fix f_c. Observe from Eqn. (4) that when f_c is fixed (and $f_c sr$), the transmitted waveform can be parameterized by only the two parameters, $nsep$ and $dsep$, significantly simplifying the problem representation.

GLRT FORMULATION

Before the GLRT formulation is discussed, it is desirable to consider the optimal detector (in terms of maximizing the probability of detection P_d for a fixed probability of false alarm P_{fa}). Assuming n scatterers and additive white Gaussian noise, it can be shown that the likelihood ratio test (LRT) statistic Λ is given by,

$$\begin{aligned} \Lambda &= \int \cdots \int e^{\left(\frac{2}{N_0}\int x(t)s(t)dt - \frac{1}{N_0}\int s^2(t)dt\right)} f(\tau_1, \cdots, \tau_n) d\tau_1 \cdots d\tau_n \\ &= \int \cdots \int e^{\left(\frac{2}{N_0}\int x(t)\sum_i g(t-\tau_i)dt - \frac{1}{N_0}\int (\sum_i g(t-\tau_i))^2 dt\right)} f(\tau_1, \cdots, \tau_n) d\tau_1 \cdots d\tau_n, \end{aligned} \quad (5)$$

where $x(t)$ is the received signal. The optimal LRT formulation requires the specification of an *a priori* distribution $f(\tau_1, \cdots, \tau_n)$ on the position of the scatterers. Such information is generally not known, and even if a model is assumed, the LR detector is typically difficult to implement since it requires averaging over many possible outcomes.

Due to the difficulty in implementing the LRT, the GLRT formulation is often used which first makes estimates of the unknown quantities. A comparison of a LRT and GLRT was given in [11] for a slightly different radar problem. Here it was shown that the GLRT performance is only slightly degraded from the LRT.

Summary of GLRT

The GLRT for the UWB detection problem has been formulated in [7] assuming real sampling and using maximum likelihood estimation of the scatterer amplitudes. We will summarize these results here for the single pulse in white Gaussian noise case.

When the τ_j are assumed known in Eqn. (1), the waveform vector is defined as

$$G(t) = [g(t-\tau_1) \;\; g(t-\tau_2) \;\; \cdots \;\; g(t-\tau_J)], \quad (6)$$

where $J = n$, the number of target scatterers. Note that $G(t)$ is a vector with J entries. The i th entry consists of a time waveform that represents the return from the i th scatterer. We will see later that the final detector form performs match filtering by multiplying each time signal entry of $G(t)$ by the received signal. This is done for each entry to form a J dimensional vector $\mathbf{x}$.

When the τ_j are unknown, the waveform vector is chosen such the waveforms are offset by J (not necessarily equal to n) equally spaced intervals of length τ,

$$G(t) = [g(t-\tau) \;\; g(t-2\tau) \;\; \cdots \;\; g(t-J\tau)]. \quad (7)$$

The purpose of using equally spaced intervals when the scattering positions are unknown is to cover all the possibilities of the scatterer positions as best as possible.

When the τ_j are unknown, we will use a sampling parameter $sp = 1/(\tau B_{RMS})$ as a means of quantifying the rate offset τ in $G(t)$. Note that if $sp = 1$, the sample rate

$1/\tau$ is the same as B_{RMS}. If $sp = 2$, the sampling rate has increased to twice the I/Q Nyquist rate.

Assuming the thermal noise $n(t)$ has covariance that is given by

$$\mathrm{E}(n(t)n(t+\tau)) = \frac{N_0}{2}\delta_{\mathrm{I}}(t-\tau)$$

, where δ_{I} denotes the impulse function, the GLRT processor is given as follows. First, the received time signal $x(t)$ is matched filtered against $G(t)$

$$\mathbf{x} = \frac{2}{N_0}\int G(t)^T x(t)dt. \tag{8}$$

Note that each waveform component of $G(t)$ is individually matched filtered against $x(t)$ to form the components of $\mathbf{x}$. For example, for the case where $G(t)$ is given as in Eqn. 6, the i th component of $\mathbf{x}$ is,

$$x_i = \frac{2}{N_0}\int g(t-\tau_i)^T x(t)dt.$$

Note that the values in x_i represent the match filter outputs for each scatterer.

Next, a cross coupling matrix C is defined as,

$$C = \int G(t)^T G(t)dt. \tag{9}$$

Lastly, a detection is made if the quantity

$$\frac{\mathbf{x}^T C^{-1}\mathbf{x}}{2} \tag{10}$$

exceeds a threshold η_0.

Observe that under both hypotheses $\mathbf{x}$ is a Gaussian correlated vector with unknown amplitudes. The statistic $\mathbf{x}^T C^{-1}\mathbf{x}/2$ results when making a maximum-likelihood estimate of the amplitudes and when decorrelating $\mathbf{x}$. It is easy to prove that if $C^{-1} = A^T A$, then $\mathbf{x}' = \mathbf{x}A$ is an i.i.d. (independent and identically distributed) Gaussian vector.

Also observe that for the case when the scatterers are resolved in Eqn. (6), then C is proportional to the identity matrix, and the detector reduces to a noncoherent integrator of the scatterer matched filter outputs. That is, it is proportional to the sum of the squares of the x_i.

One can also derive a GLRT using baseband samples although for UWB there is not a significant complexity difference between processing the RF samples as done here and the baseband signal, due to the high fractional bandwidth.

I/Q modifications to GLRT

Now, the previous UWB GLRT does not assume the use of an I/Q representation. It is desirable to determine what modifications are necessary and also if an I/Q representation is useful.

Before proceeding, we examine the narrowband signal model and the assumptions made. There are a number of narrowband signal models that can be examined in the literature, but the derivation in [12, p. 299] is fairly in-depth. There is one key assumption made in narrowband detector derivations. This is that the signal can be represented as

$$g(t) = a(t)\sin(\omega_0 t + \theta) \tag{11}$$

such that $a(t)$ is a baseband signal with bandwidth much less than ω_0.

This assumption leads to the modeling of θ as uniformly distributed. The variable θ is assumed to be uniform due to the uncertainty in when the signal is received. Since $a(t)$ is slowly varying by the assumption, there are generally many cycles in a pulse and hence θ appears to be approximately uniformly distributed. Note that the uniform phase model is an approximate model of the unknown time of return, and hence not absolutely true, as is often misconstrued.

Now, for UWB detection, the uniform phase assumption is not fully valid, and hence modeling the return signal simply as having an angle θ that is uniformly distributed is generally not correct. In fact, the signal return from a single point scatterer simply can best be modeled (not accounting for propagation or scattering effects) as $g(t-\delta)$ where δ is an unknown time delay. This leads to the GLRT shown above, whereby the real waveform matrix $G(t)$ is designed to match to many time offsets. However, there is no claim to optimality. In fact, when the scatterer locations are not known, there will generally still be a mismatch between the selected time offsets in $G(t)$ and those actually received. It would therefore be useful to determine to what degree I/Q representations can compensate for such time alignment mismatches in UWB detection for the position unknown case.

Proceeding with the narrowband derivation, the cosine in (11) can be expanded as,

$$g(t) = g_I \cos\theta - g_Q \sin\theta \tag{12}$$

where $g_I = a(t)\cos w_0 t$ and $g_Q = a(t)\sin w_0 t$ are the RF I/Q samples [12, p. 299]. Now, for the narrowband model whereby θ is uniform, Eqn. (12) is often used to show that the optimal detector is found by match filtering to g_I and g_Q and summing the squares of the outputs.

The conventional detector therefore follows from the standard derivation [12, p. 299] which assumes a narrowband signal. For the GLRT with an UWB signal, the waveform matrix $G(t)$ positions match filters that are spaced at intervals τ determined by the sampling parameter, *sp*. However, a target will generally not align exactly with the matched filters and hence a loss occurs. However, this loss can be reduced by the use of I/Q samples. The amount of the loss is clearly a function of *sp* and whether or not I/Q samples are used. In order to quantify whether or not I/Q representations are useful, we will examine three schemes for generating I/Q.

The first scheme we will examine, incorporates the NB I/Q representation into the GLRT detector. To see heuristically if the NB representation is useful for UWB signals, examine Fig. 1.

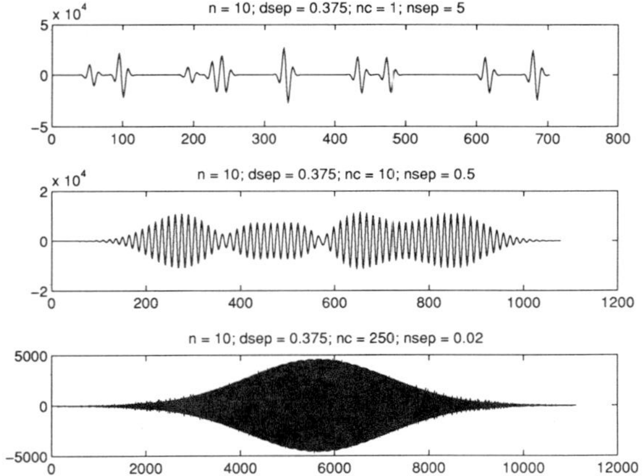

Fig. 1. Envelope of an I/Q (NB representation) signal. $nc = .5$, $nsep = 1$, $dsep = .375$

The NB I and Q, and envelope of the NB representation of I and Q found from (12) are plotted for a signal with $nc = .5$. Note that the envelope is significantly more broader than the I or Q separately. This I/Q representation can be incorporated into the GLRT by modifying Eqn. (7) to include the complex term, i.e.,

$$G(t) = [g_I(t-\tau)+ig_Q(t-\tau) \;\; g_I(t-2\tau)+ig_Q(t-2\tau) \cdots \;\; g_I(t-J\tau_J)+ig_Q(t-J\tau_J)]. \quad (13)$$

The detector is $\mathbf{x}^\dagger C^{-1}\mathbf{x}/2$ where † is conjugate transpose. Note that this is valid as long as G_I and G_Q are orthogonal, which is true for the Gaussian waveform integrated over a long enough window. If G_I and G_Q are not orthogonal (they are not guaranteed to be), a solution is to double the length of $G(t)$ and append the quadrature signals, i.e. $G(t) = [G_I(t) \;\; G_Q(t)]$, where $G_I(t)$ is as in (7) and $G_Q(t)$ is the quadrature representation of $G(t)$. This is valid since the detector form performs prewhiting between all components. Note that for orthogonal I and Q the quadratic detector is identical (adds the squares of the I and Q) for both cases above.

A second scheme is based on the Taylor series expansion of a function,

$$g(t-\tau) = g(t) + \alpha\dot{g}(t) + \cdots. \quad (14)$$

To a first order approximation, the signal with unknown time delay can be expanded in terms of an unshifted version and its derivative. Since the $\dot{g}(t)$ and $g(t)$ are orthogonal, we can use these as basis functions in the GLRT derivation. In this case the GLRT is modified by changing Eqn. (7) to,

$$G(t) = [g(t-\tau)+ic\dot{g}(t-\tau) \;\; g(t-2\tau)+ic\dot{g}(t-2\tau) \cdots \;\; g(t-J\tau_J)+ic\dot{g}(t-J\tau_J)], \quad (15)$$

where c is a normalizing constant equal to $||g(t)||/||\dot{g}(t)||$ where $||g(t)||^2$ is $\int_{-\infty}^{\infty} g(t)^2 dt$.

The third scheme is based on analytic signal representation, by the use of Hilbert transforms which exists for all signals, including UWB signals. Theoretically, a signal and its Hilbert transform can be written in complex notation, i.e. $z(t) = g(t) + i\tilde{g}(t)$ where $\tilde{g}(t)$ is the Hilbert transform of $g(t)$. One benefit of using complex notation is that the sampling rate is half of what is required for the real signal. It is shown in [12, p. 60] that the envelope of $z(t)$ and the envelope of the conventional I/Q are nearly the same for narrowband signals. However, for UWB signals there is a difference between the two representations. It is therefore desirable to determine if there is a significant difference in detection performance when considering $z(t)$ as opposed to the narrowband representation. The derivation of the GLRT detector form for analytic signals is as follows. First, an analytic signal is constructed from the received signal $x(t)$ as $x_c(t) = x(t) + i\tilde{x}(t)$. Now, under hypothesis H_1 the analytic signal is $s(t) + i\tilde{s}(t) + n(t) + i\tilde{n}(t)$ while under hypothesis H_0 it is $n(t) + i\tilde{n}(t)$. Hence this derivation is the same as that for the real GLRT, except that complex signals are used. The result is that in Eqn. (8) $x(t) = x_c(t)$ is matched against

$$G(t) = [g(t-\tau)+i\tilde{g}(t-\tau) \;\; g(t-2\tau)+i\tilde{g}(t-2\tau) \cdots \;\; g(t-J\tau_J)+i\tilde{g}(t-J\tau_J)], \quad (16)$$

and the detector is $\mathbf{x}^\dagger C^{-1}\mathbf{x}/2$.

PERFORMANCE OF GLRT

We now examine the performance of the GLRT for several cases. We begin with the case where the location of the target scatterers are known exactly (position known) so that $G(t)$ is assumed as in Eqn. (6). After that we will examine the position unknown case and the performance of the I/Q schemes discussed in the previous subsection.

Before beginning, we give a few details specific to the simulation. For each SNR case, there were 1000 Monte-Carlo trials. Each trial consists of a RF received signal with new scatterers that are chosen in time via a Poisson random variable with average distance *dsep* and with amplitudes that are Rayleigh distributed.

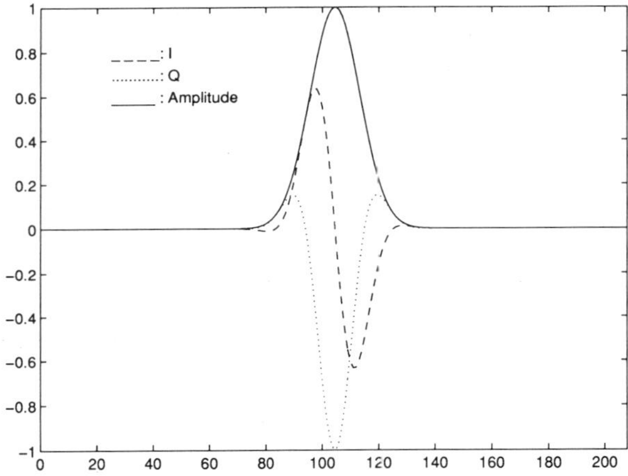

Fig. 2. Examples of simulated received signal: Top figure, $n = 10$, $dsep = .375$, $nc = 1$, $nsep = 5$; Middle figure, $n = 10$, $dsep = .375$, $nc = 10$, $nsep = .5$; Bottom figure, $n = 10$, $dsep = .375$, $nc = 250$, $nsep = .02$

Some examples of received signals with various parameters *dsep*, *nsep*, and *nc* are shown in Fig. 2.

Position Known

For the position known case, we use the GLRT with known scatterer positions or time delays seen in Eqn. (6). Shown in Fig. 3 is the performance of the GLRT for three cases of *nsep*.

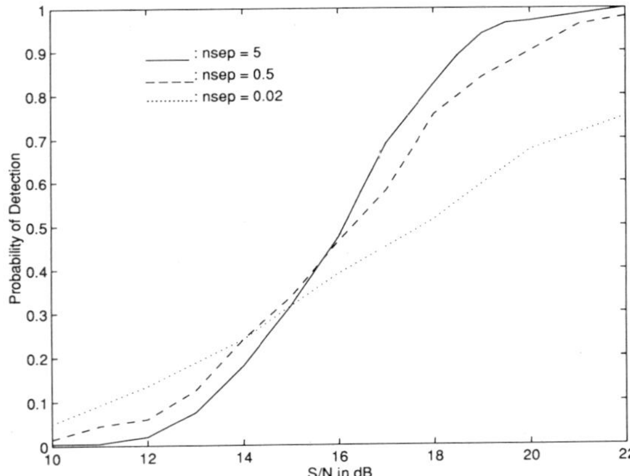

Fig. 3. Position known performance: $n = 10$, $dsep = .375$, $nsep = 5, .5, .02$

It is seen that the best performance occurs when *nsep* is large or when the scatterers are fully resolved. This observation was made for a number of other cases not presented here where *nc* was varied. The case when the position is known requires knowledge of scatterer positions which is rarely available. Hence we examine the position unknown case.

Position Unknown

We will begin the position unknown simulations using the real GLRT and later consider the usefulness of the various I/Q methods discussed previously.

The major addition to the parameters of the position unknown case is the specification of the sampling parameter *sp*. Note that too small a sampling parameter will result in a mismatch with the received signal. However, too large a sampling parameter will result in too many filters causing a collapsing loss. In order to determine the best sampling parameter, we simulated the performance over different cases of *nsep* and *dsep*. One such case is shown in Fig. 4. We found that $sp = 4$ results in fairly robust performance although for certain *nsep* and *dsep* better performance can be had by optimizing *sp*.

In Fig. 5 the performance is shown with $sp = 4$, for the same cases of *nsep* and *dsep* as the position known of Fig 3.

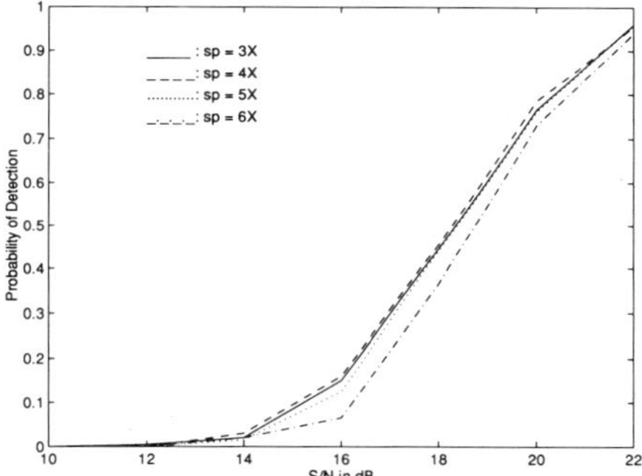

Fig. 4. Effect of sampling rate - real sampling, $sp = 3, 4, 5, 6$, $n = 10$, $nc = .5$, $nsep = 2$, $dsep = .075$

Note the performance as a function of *nsep* is significantly different from the performance of the position known case. In the position known case, the best performance was found for large *nsep*. For the position unknown case large *nsep* requires a large number of degrees of freedom, J, so that many filters exist that do not contain signal. Such cases result in a collapsing loss. Note that the best performance as a function of *nsep* is a function of SNR.

For NB or low percent bandwidth waveforms, we found a technique that can significantly improve the performance of the GLRT when $sp > 1$ or when there is oversampling. This technique consists of increasing the phase of the signals in $g(t)$ by 90 degrees relative to the carrier frequency. That is, the new $G(t)$ becomes

$$G(t) = \mathrm{Re}[e^{i\Delta_1} g(t - \tau) \quad e^{i\Delta_2} g(t - 2\tau) \quad \cdots \quad e^{i\Delta_J} g(t - J\tau)], \tag{17}$$

where $\Delta_k = k\tau\omega_c + k\pi/2$. The reason for including the offset variable Δ is as follows. Suppose that $\mathbf{x}$ is zero in some component. This can occur for NB signals (sinusoids) if, for some m, $g(t - m\tau)$ and the received signal are mismatched in phase by 90 degrees. In such a case we would like to guarantee that at least one of the filters in adjacent components is matched properly. Note that Eqn. (7) does not guarantee this, in fact it is possible that nearly all the components could be small in magnitude if the phases are matched improperly. Hence, if a component of $\mathbf{x}$ is zero, at least one of the adjacent components of $\mathbf{x}$ must peak in amplitude since it will match in phase with the received signal.

The performance of this technique is shown in Fig. 6 for the same *nsep* and *dsep* parameters of Fig. 5. Note there is a significant improvement in the NB case, whereas the UWB case is not as much affected. However, note that changing the phase of the UWB signal can result in a change of the character of the signal. Hence, we recommend for the real GLRT using (17) for NB signals and (7) for UWB signals.

We now present a comparison of the I/Q detection schemes proposed earlier. The sampling parameter is $sp = 2$ for the I/Q methods and $sp = 4$ for the real detector for comparison purposes. The real detector is the original GLRT which uses Eqn. (7). We use $nc = .5$, $nsep = 4$, and $dsep = .15$ which amounts to 1/2 cycle in a resolution cell, and mean scatterer separation that is 4 times as great as the resolution cell width. In Fig. 7 the performance is shown. Note that all the methods are nearly identical. This shows that the conventional I/Q detector can still be used for UWB detection purposes without substantial loss.

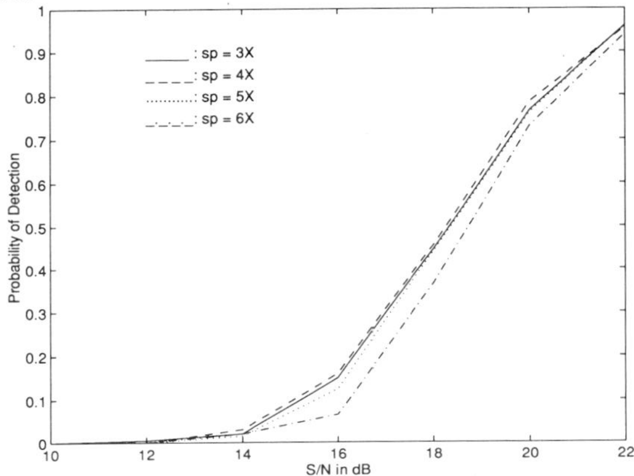

Fig. 5. Position unknown performance-no phase offset in $G(t)$: $n = 10$, $dsep = .375$, $nsep = 5, .5, .02$

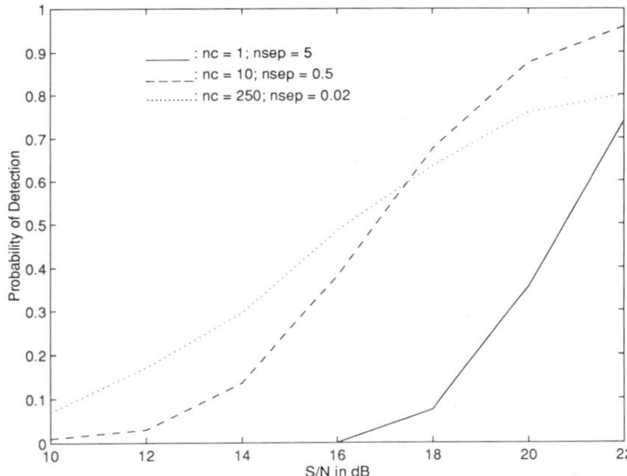

Fig. 6. Position unknown performance-phase offset in $G(t)$: $n = 10$, $dsep = .375$, $nsep = 5, .5, .02$

On the other hand, for the same case as above except $nc = 20$, we show in Fig. 8 that the performance of the different I/Q representations are comparable although the real detector is slightly lagging. This is reasonable to expect since the NB I/Q and Hilbert representations are nearly same for NB signals and the I/Q representation is the optimal detector for such cases. Note that the real detector is degraded without the phase offset.

Overall, we believe (although we have not simulated every case) the 4X oversampled real detector, conventional I/Q detector, Hilbert transform I/Q representation, or derivative I/Q representation, should be used for UWB signals (considered here) with $nc < 1$ and the conventional I/Q can be used with little loss for $nc > 1$.

We now examine the detection performance as a function of *nsep*. The question of interest is what is the optimal bandwidth to transmit to maximize detection performance. This question was examined initially in [8] (in a simplified manner) which concluded that the optimal bandwidth occurs once the scatterers begin to separate. We will use the NB representation here since in the optimizing region of interest, $nc > 1$.

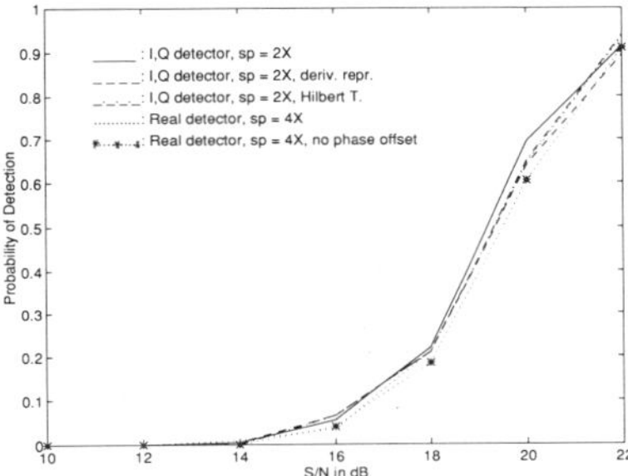

Fig. 7. IQ Comparison: NB, derivative, and Hilbert transform representations, also real detector with and without phase offset, $n = 10$, $nc = .5$, $nsep = 4$, $dsep = .15$

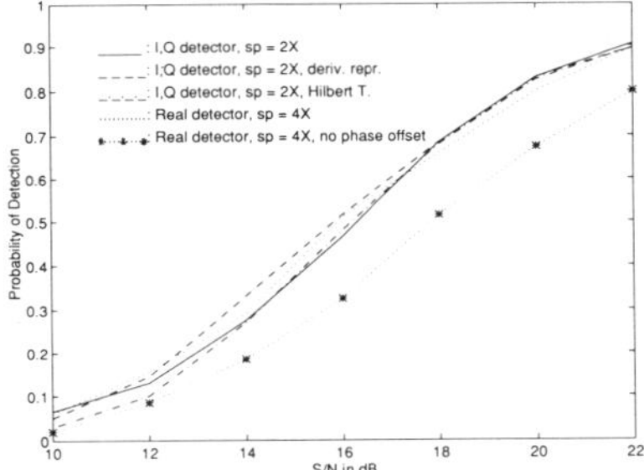

Fig. 8. IQ Comparison: NB, derivative, and Hilbert transform representations, also real detector with and without phase offset, $n = 10$, $nc = 20$, $nsep = 4$, $dsep = .15$

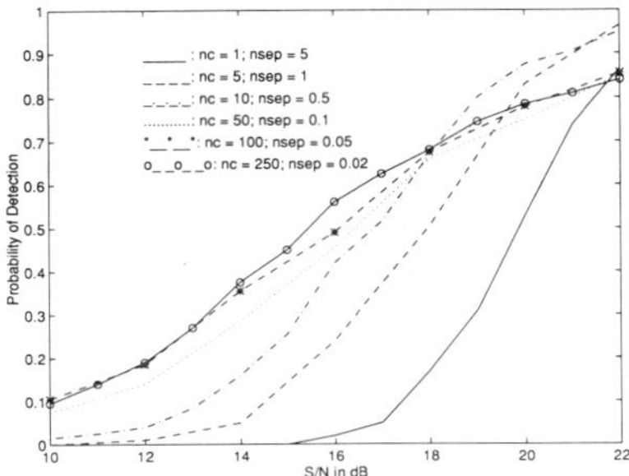

Fig. 9. Effect of bandwidth with real detector (sp=4) position unknown: *nsep* ranges from .02 to 5, *nc* ranges from 1 to 250, $dsep = .375$

In Fig. 9 we show the performance for $dsep = .375$ for various *nsep*. Note that at high SNR, we observe that there appears to be an optimal bandwidth which occurs in the range of *nsep* between .5 and 1 for SNR between 20 and 22. However, at low SNR the NB appears to dominate. Hence we confirmed the findings in [8] for high SNR, but not for medium to low SNR. The heuristic explanation is that the NB scatterers interfere causing significant fluctuations in the matched filter output. For UWB, the scatterers are separated so they do not interfere as much and there is less spreading in the distribution of the matched filter. This favors the UWB at high SNR since the NB scatterers will occasionally cancel themselves. It favors the NB output at low SNR since the NB scatterers will occasionally add in phase which significantly increases the output energy of the matched filter and provides detections.

CONCLUSIONS

The detection of UWB radar signals in white Gaussian noise has been considered. We have seen that the NB signal model, which imposes a random phase on the received carrier, is not valid for representing UWB signals. When the received signal is properly represented as in Eqn. (1), the optimal detector is difficult to implement due to the stochastic nature of the unknown time delays.

Several suboptimal detectors have been examined. The GLRT was simulated for both position known and unknown cases. It was found that the UWB out-performed the NB waveform for the position known cases. However, for medium to high SNR an optimal intermediate BW exists for the position unknown case. This optimal intermediate BW occurs when the radar resolution is on the same order as the mean scatterer separation. For low SNR, the opposite was true such that the NB signal dominates, at least for the cases we investigated.

We also investigated the role that I/Q representations play in UWB detection. As mentioned above, the NB signal model is not valid for UWB signals. Since the I/Q detector follows directly from the NB signal model, it is questionable as to the detection performance when using I/Q as opposed to real sampling. However, we showed through simulation that there does not appear to be any significant difference (<.5 dB), when comparing real sampling with I/Q sampling when the I/Q is sampled at half the rate of the real samples, and the rate is sufficiently above Nyquist to avoid collapsing losses. In fact, we examined several methods for generating the I/Q, and all performed satisfactory.

REFERENCES

[1] H. L. Van Trees, *Detection, Estimation, and Modulation theory*, vol. 3. John Wiley & Sons, 1971.

[2] M. J. Steiner, "On the detection of ultrawideband radar signals," report 92-9517, Naval Research Laboratory, September 1992.

[3] P. K. Hughes II, "A high-resolution radar detection strategy," *IEEE Transactions on Aerospace and Electronic Systems*, vol. AES-19, pp. 663–667, September 1983.

[4] G. C. Rose, "A look at automatic detection algorithms for a x-band radar using a high resolution search mode," Report TSC-W75-65/kks, Technology Service Corporation, October 1987.

[5] A. Farina and A. Russo, "Radar detection of correlated targets in clutter," *IEEE Transactions on Aerospace and Electronic Systems*, vol. AES-22, pp. 513–532, September 1986.

[6] A. Farina and F. A. Studer, "Detection with high resolution radar: Advanced topics and potential applications," *Chinese J. of Systems Engineering and Electronics*, vol. 3, no. 1, pp. 32–34, 1992.

[7] H. Wang and L. Cai, "A localized adaptive MTD processor," *IEEE Transactions on Aerospace and Electronic Systems*, vol. AES-27, pp. 532–540, May 1991.

[8] R. Nitzberg, "Effect of a few dominant specular reflectors target model upon target detection," *IEEE Transactions on Aerospace and Electronic Systems*, vol. AES-14, pp. 670–673, July 1978.

[9] C. E. Cook and M. Bernfeld, *Radar Signals, An Introduction to Theory and Applications*. New York: Academic Press, 1967.

[10] A. W. Rihaczek, *Principles of High-Resolution Radar*. New York: McGraw-Hill, Inc., 1969.

[11] L. E. Brennan, I. S. Reed, and W. Sollfrey, "A comparison of average-likelihood and maximum-likelihood ratio tests for detecting radar targets of unknown doppler frequency," *IEEE Transactions on Inform. Theory*, vol. 14, pp. 104–110, January 1968.

[12] J. V. DiFranco and W. L. Rubin, *Radar Detection*. Massachusetts: Artech House, 1980.

ERROR CORRECTION IN TRANSIENT ELECTROMAGNETIC FIELD MEASUREMENTS USING DECONVOLUTION TECHNIQUES

Jian-Zhong Bao, Jonathan C. Lee, Michael E. Belt, David D. Cox, Satnam P. Mathur, and Shin-Tsu Lu

McKesson BioServices and U.S. Army Medical Research Detachment
Brooks Air Force Base, Texas 78235

INTRODUCTION

It is difficult to make an accurate transient measurement on electromagnetic pulses in the pico-second domain because of the limitations of measurement components. In this paper, we present a two-step deconvolution routine to compensate for the measurement distortions in our short electromagnetic pulse (EMP) exposure facility for studying biological effects. As depicted in Figure 1, the facility mainly consists of a pulse generator and a GTEM cell. The measurement system includes a Tektronix SCD 5000 transient digitizer (4.5 GHz bandwidth), connection cables, and two Asymptotic Conical Dipole (ACD) D-dot (dD/dt) sensors: ACD-1(A), which is mounted on the top ground wall of the cell and utilized for real time monitoring during exposures, and ACD-1(R), which is used to map the field on the bottom ground wall of the cell where the specimens are placed.

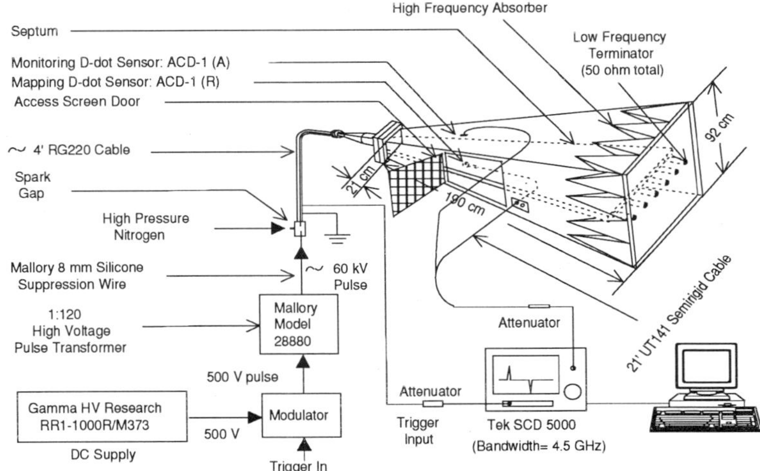

Figure 1: Short electromagnetic pulse (EMP) exposure facility and data acquisition system.

Because of the low-pass nature of the connection cables and the limited bandwidth of the SCD 5000, the measured signal is a distorted output of a D-dot sensor. An empirical transfer function of the cable-digitizer system is evaluated using a reference impulse generated by a Pico Second Pulse Lab (PSPL) 4050B step generator with a 5210 Impulse Forming Network (IFN) and characterized with a Tektronix CSA 803 communication signal analyzer with a SD 30 sampling head (40 GHz bandwidth). The reference impulse is injected into the connection cable at the D-dot sensor end and measured with the SCD 5000 while the cable is kept in the same position as for making D-dot measurements to ensure an in-place calibration.

Due to its right-angle structures, the ACD-1(R) gives a different output from that of ACD-1(A) to the same pulse, especially to the fast leading edge although the sensing elements for both sensors are the same. The right-angle bends in ACD-1(R) cause reflections of D-dot signals in the sensor. To correct the errors due to the reflections, we have developed a semi-empirical procedure: the impulse response of ACD-1(R) D-dot sensor is assumed as a summation of δ-function and its parameters are determined with a reference measurement using ACD-1(A) and an optimization procedure with Levenberg-Marguardt algorithm [1].

ITERATIVE CAUSAL DECONVOLUTION

By assuming that the measurement system is linear and time-invariant (LTI), we can have

$$u(t) = v(t) * h(t) + n(t) = \int_{-\infty}^{+\infty} v(\tau)h(t-\tau)d\tau + n(t), \quad (1)$$

where $u(t)$ is the measured signal, $v(t)$ is the true signal before any degradations, $h(t)$ is the system impulse response, $n(t)$ is the additive noise, and $*$ denotes convolution, which smears fast changing features in $v(t)$. Here we have an inverse problem: finding $v(t)$ from $u(t)$, $h(t)$, and $n(t)$, i.e., deconvolution, which is an ill-posed problem mathematically: small changes in $u(t)$ can be mapped into large changes in $v(t)$. This is such a serious problem that an effective noise-control procedure has to be implemented because none of the measurements are noise-free. Performing Fourier transform on Eq. (1), we get

$$U(f) = V(f)H(f) + N(f), \quad (2)$$

where upper case letters are Fourier transforms of the corresponding lower case letters, respectively, $H(f)$ is the system transfer function, and f is the frequency. Now $v(t)$ can be solved by performing inverse-Fourier transform:

$$v(t) = F^{-1}\{\frac{U(f) - N(f)}{H(f)}\}, \quad (3)$$

where F^{-1} stands for inverse Fourier transform. If we had an exact knowledge about $U(f)$, $N(f)$, and $H(f)$ in the entire frequency range, and if the measurement system, as assumed, were LTI, $v(t)$ can be recovered exactly. Unfortunately, none of the above information can be obtained exactly in practice because of the ever presence of noise and error. In general, the noise spectra $N(f)$ cannot be separated from the measured signal spectra $U(f)$ unless there are other information or assumptions available. So Eq. (3) can only be applied approximately with $U(f) - N(f) \approx U(f)$. To minimize the error due to $N(f)$ in the least square sense, we applied a Wiener filter[1]: $\Phi(f) = 1- \mid N(f) \mid^2 / \mid U(f) \mid^2$. If we assume that the power spectra of noise $\mid N(f) \mid^2$ with a D-dot signal excitation is the same as that after the excitation, we can obtain it from a measurement in a time window of the same size that is remote after the excitation. In this study, all the waveforms were sampled with 1024 points in a 10 ns window, which gives a maximum frequency of 51.15 GHz, and are the average of 200 waveforms to remove random noises and increase the signal-to-noise ratio. All the raw data was pre-treated before applying the Fast Fourier Transform (FFT). The pre-treatment includes subtracting baseline offset, converting attenuator factor, zero-padding to avoid aliases and to make u periodically causal, and data-windowing to avoid sharp changes at the edges. Since the signal is over sampled, a low pass filter, $L(f)$, is mandatory. With $\Phi(f)$ and $L(f)$, Eq. (3) can be rewritten as

$$v(t) \approx F^{-1}\{\tilde{V}(f)\} = F^{-1}\{\frac{U(f)\Phi(f)L(f)}{H(f)}\}, \tag{4}$$

where the cutoff edge of $L(f)$ is a half Hann window [1]. Another problem is that, in general, directly performing F^{-1} in Eq. (4) does not guarantee a causal $v(t)$ from band-limited frequency domain data, and the absence of causality may cause errors although an acausal inverse-transform generally gives satisfactory results. To enforce causality on $v(t)$, the real ($\tilde{V}_R$) and imaginary ($\tilde{V}_I$) part of $\tilde{V}(f)$ in Eq. (4) must follow the Hilbert transform[2]:

$$\tilde{V}(k) = \frac{1}{N}\sum_{m=0}^{N-1} \tilde{V}_R(m)Y_N(k-m), \quad \text{where} \quad Y_N(k) = \begin{cases} N, & k=0, \\ -j2\cot(\pi k/2), & k=\text{odd}, \\ 0, & k=\text{even}, \end{cases} \tag{5}$$

where $j = \sqrt{-1}$ and N is the total number of data point including zero-padding,

$$j\tilde{V}_I(k) = \frac{1}{N}\sum_{m=0}^{N-1} \tilde{V}_R(m)Z_N(k-m), \qquad \text{and} \tag{6}$$

$$\tilde{V}_R(k) = \frac{1}{N}\sum_{m=0}^{N-1} j\tilde{V}_I(m)Z_N(k-m) + v(0) + (-1)^k v(N/2), \tag{7}$$

where $Z_N(k) = Y_N(k) - N\delta(k)$. In the time domain, above relations can be represented as:

$$v(n) = v_e(n)y_N(n), \qquad \text{and} \qquad v_o(n) = v_e(n)z_N(n), \tag{8}$$

where $y_N = \text{IFFT}\{Y_N\}/N$, $z_N = \text{IFFT}\{Z_N\}/N$, $v_e = \text{IFFT}\{\tilde{V}_R\}/N$, and $v_o = \text{IFFT}\{j\tilde{V}_I\}/N$, where IFFT stands for inverse FFT, v_e and v_o are even and odd parts of $v = v_e + v_o$, respectively. v is referred periodically causal if $v(n) = 0$ for $n \geq N/2$.

Eqs. (5) to (7) suggest that a causal v can be restored completely from the real part or almost from the imaginary part. In practice, since $\tilde{V}_R$ and $\tilde{V}_I$ may not follow the Hilbert transform, the causal v recovered from $\tilde{V}_R$ alone will not be consistent with that restored from $\tilde{V}_I$. To overcome this disagreement, we have improved an algorithm initially developed by Sarkar et al.[3] to extract the causal time domain sequence in an iterative manner. Our algorithm shows a faster and more stable convergence. In Eqs. (4) and (12), F^{-1} means the following iterative procedure:

1. Calculate an even and an odd time domain sequence with $v_e = \text{IFFT}\{\tilde{V}_R\}/N$ and $v_o = \text{IFFT}\{j\tilde{V}_I\}/N$, respectively, and calculate $S = \sum_{i=N/2}^{N-1}[v_e(i) + v_o(i)]^2$.

2. Create a new odd time domain sequence as
$v_{o,new}(0:N-1) = \{0, [v_o(1:N/2-1) + v_e(1:N/2-1)]/2,$
$0, [v_o(N/2+1:N-1) - v_e(N/2+1:N-1)]/2\}.$

3. Calculate a new imaginary part of frequency domain data with $\tilde{V}_{I,new} = \text{FFT}\{v_{o,new}\}$.

4. Create an average imaginary part of frequency domain data as
$\tilde{V}_{I,avg}(0:N/2) = \{0, [\tilde{V}_I(1:N/2-1) + \tilde{V}_{I,new}(1:N/2-1)]/2, 0\}$, and
$\tilde{V}_{I,avg}(N/2+1:N-1) = -\tilde{V}_{I,avg}(N/2-1:1)$, where $\tilde{V}_{I,avg}(0:N-1)$ has been forced to follow the symmetric property that ensures a real inverse transform.

5. Calculate another new odd time domain sequence with $v_{o,avg} = \text{IFFT}\{j\tilde{V}_{I,avg}\}/N$.

6. Create a new even sequence as
$v_{e,new} = \{v_e(0), [v_e(1 : N/2-1) + v_{o,avg}(1 : N/2-1)]/2,$
$v_e(N/2), [v_e(N/2+1 : N-1) - v_{o,avg}(N/2+1 : N-1)]/2\}.$

7. Calculate a new real part of frequency domain data with $\tilde{V}_{R,new} = \text{FFT}\{v_{e,new}\}$.

8. Create an average real part of frequency domain data as
$\tilde{V}_{R,avg}(0 : N/2) = \{[\tilde{V}_R(0 : N/2) + \tilde{V}_{R,new}(0 : N/2)]/2\}$, and
$\tilde{V}_{R,avg}(N/2+1 : N-1) = \tilde{V}_{R,avg}(N/2-1 : 1)$, where $\tilde{V}_{R,avg}(0 : N-1)$ has been forced to follow the symmetric property that ensures a real inverse transform.

9. Calculate another even part of time domain sequence with $v_e = \text{IFFT}\{\tilde{V}_{R,avg}\}/N$.

10. Calculate a better causal sequence with $v = v_e + v_o$ and $S_{new} = \sum_{i=N/2}^{N-1}[v(i)]^2$. A smaller S_{new} implies a better periodically causal v.

11. Calculate $\mid S_{new} - S \mid / S_{new}$. If it is less than a sufficient resolution, say 10^{-5}, stop the iteration and output a causal time domain sequence v; otherwise, let $\tilde{V}_R = \tilde{V}_{R,new}$, $\tilde{V}_I = \tilde{V}_{I,new}$, $v_e = v_{e,new}$, $v_o = v_{o,new}$, and $S = S_{new}$, and go to step 2.

COMPENSATION FOR CONNECTION CABLE AND SCD 5000

A PSPL 4050B step generator with a 5210 IFN was utilized to generate an impulse with a magnitude of 2.8 V and a pulse-width of 58 ps as the reference. Since the output impulses from the IFN are identical, they can be characterized by a CSA 803 with a SD30 sampling head. The characterized pico-second impulse was injected into the connecting cable between the D-dot sensor and the SCD 5000 at the D-dot sensor end. Figure 2 shows the reference impulse characterized by the CSA 803, the impulse measured by the SCD 5000, and the respective Fourier spectra. There is a significant difference between the reference and the measured impulses in both time and frequency domain. The magnitudes of the FFT spectra suggest that the reference impulse is much stronger than measured impulse in a wide frequency range (DC-20 GHz), which gives us the physical basis that we can numerically expand the bandwidth of the measurement hardware to a wider frequency range.

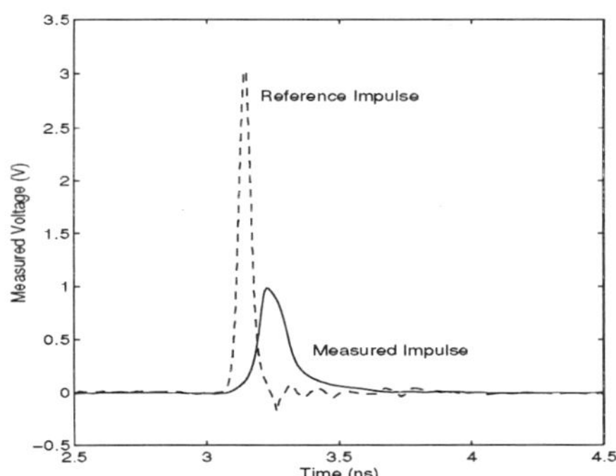

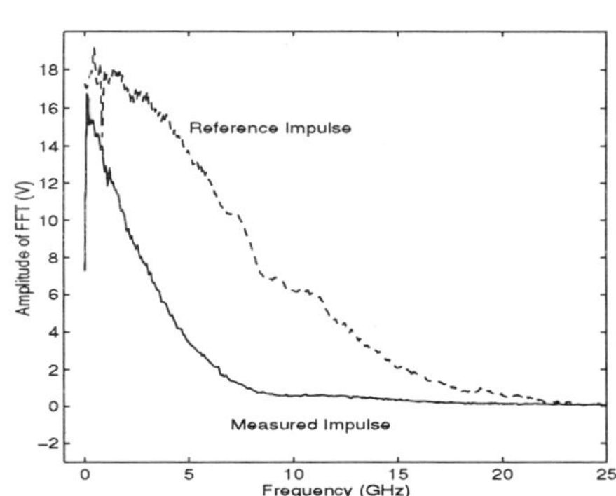

Figure 2: Comparison between the reference impulse characterized using a CSA 803 with a SD30 sampling head and the measured impulse with SCD 5000 in time and frequency domain.

Rearranging Eq. (1) to a form for the evaluation of system transfer function, we have:

$$x_{out}(t) + n_{out}(t) = [x_{in}(t) + n_{in}(t)] * h_{cd}(t), \tag{9}$$

where $x_{out}(t)+n_{out}(t)$ is the impulse measured by the SCD 5000, in which $x_{out}(t)$ is the desired signal we want to obtain and $n_{out}(t)$ is the additive noise, $x_{in}(t)+n_{in}(t)$ is the characterized impulse by the CSA 803, in which $x_{in}(t)$ is the real signal and $n_{in}(t)$ is the noise, and $h_{cd}(t)$ is the impulse response of the cable-digitizer system to be obtained from the measurements. By taking Fourier transform on Eq. (9), we can get the frequency domain form:

$$H_{cd}(f) = \frac{X_{out}(f) + N_{out}(f)}{X_{in}(f) + N_{in}(f)} = H(f) \cdot \frac{1 + N_{out}(f)/X_{out}(f)}{1 + N_{in}(f)/X_{in}(f)}, \tag{10}$$

where the upper case letters stand for the Fourier spectra of the corresponding lower case letters, respectively, and $H(f) = X_{out}(f)/X_{in}(f)$ is the true transfer function. The fraction term behind $H(f)$ is the noise contribution, which is close to one at the frequencies at which the signal-to-noise ratios of measured data are very large. At the frequencies where $X_{in}(f)$ is close to zero, $X_{out}(f)$ should be close to zero, too. Consequently, $H_{cd}(f)$ is just noise. Since $H(f)$ can only be obtained approximately in practice, an exact deconvolution is inherently impossible[4].

H_{cd} was utilized in Eq. (4) as the transfer function, H, to compensate the effect of the connection cable and the SCD 5000 for the measurements obtained with a mounted ACD-1(A) on the top ground wall. Figure 3 shows the measured and cable-digitizer compensated D-dot, E field, and energy density spectra of corresponding E field at a pulse repetition frequency of 60 Hz. Clearly, the compensated data give a faster rise time and a higher magnitude than those directly measured. The comparison of their pulse parameters is listed in Table 1.

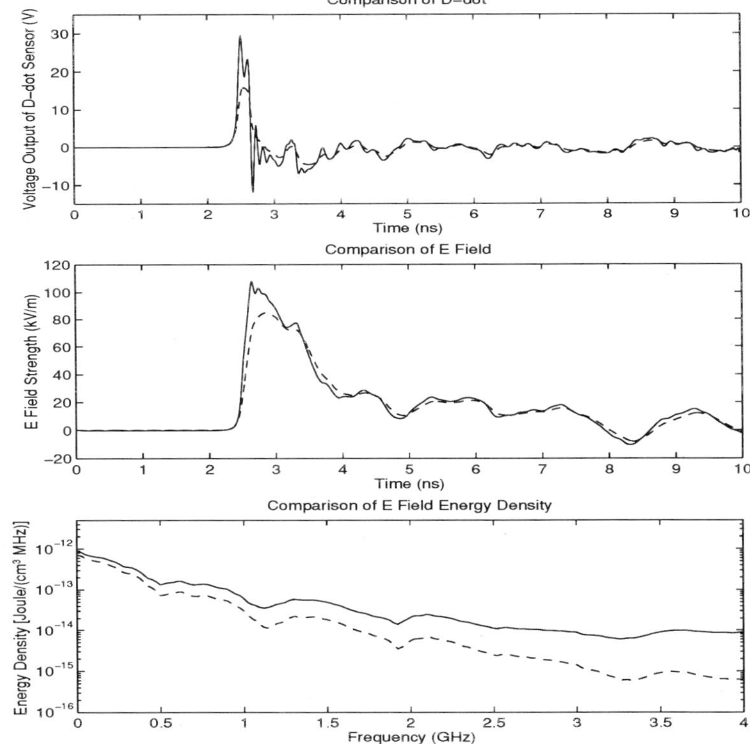

Figure 3: Comparison between cable-digitizer compensated (solid lines) and uncompensated (dashed lines) D-dot, E field, and energy density spectra of E field. The measurement was done with an ACD-1(A) D-dot sensor at a pulse repetition frequency of 60 Hz.

Table 1. Comparison of pulse parameters between the uncompensated and the cable-digitizer compensated E filed presented in Figure 3.

	rise time (ps)	magnitude (kV/m)	pulse width (ns)
uncompensated E field	234	84.8	1.11
compensated E field	166	107.6	0.98

COMPENSATION FOR REFLECTIONS IN D-DOT SENSOR ACD-1(R)

The sensing elements ("eggs") in both ACD-1(R) and ACD-1(A) are the same while they are connected to the respective SMA connectors differently. In ACD-1(A), the sensing element and the SMA connector are joined in the axial direction directly while, in ACD-1(R), they are connected by a coaxial line through three right-angle bends in the radial direction, as drawn in Figure 4.

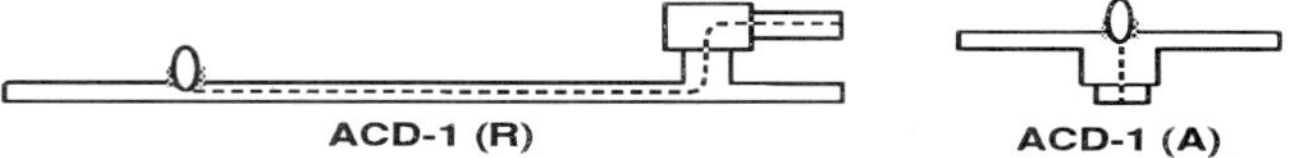

Figure 4: Schematic diagrams of ACD-1(R) and ACD-1(A) D-dot sensors.

The right-angle bends of ACD-1(R) cause reflections of D-dot signals in the sensor. The reflections result in a magnitude reduction and a "shoulder" on the falling edge of the first spike, as shown in Figure 5(a). Although there are only minor influences on its slow variation part, there is a clear effect on the fast leading edge of the pulse, as shown in Figure 5(b), which is important for estimating the rise time.

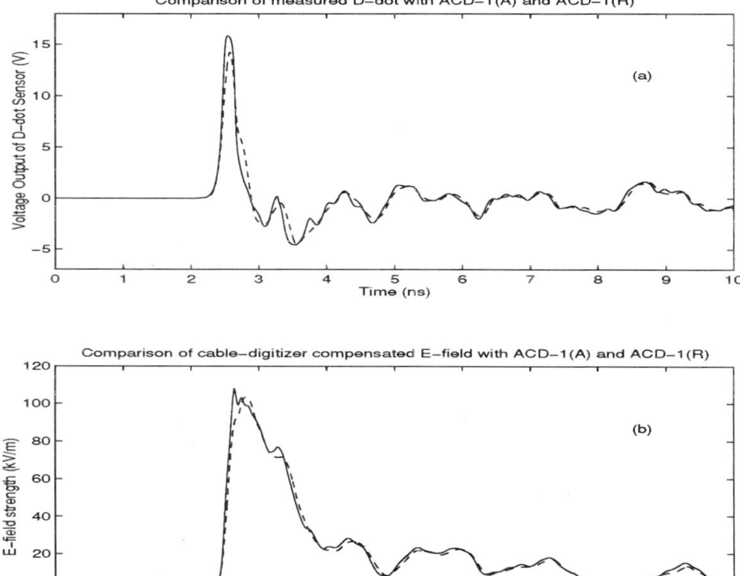

Figure 5: Comparison of (a) measured D-dot with ACD-1(A) (solid lines) and ACD-1(R) (dashed lines) at a symmetric pair of locations on the top and the bottom ground walls, respectively, and (b) their corresponding cable-digitizer compensated E field wave forms.

The output of ACD-1(R) is the transmission of the original signal plus scaled and delayed replicas of the original signal, therefore its impulse response can be modeled as

$$h(t) = a_0\delta(t) + a_1\delta(t - \tau_1) + a_2\delta(t - \tau_2), \tag{11}$$

where $a_k < 1, (k = 0, 1, 2)$ are the scaled coefficients, and $\tau_k > 0, (k = 1, 2)$ are the delay times. We did not take the effects of the sensing element into account because it has a bandwidth ≥ 10 GHz [5], and ignored the low pass nature of the coaxial line and the SMA connector. The first δ-function in Eq. (11) is for the transmitted D-dot signal through the right-angle bends, and rest of them are for the reflections. From Figure 5(a), it seems that only the first reflection has an obvious effect on the E field, so we only include two reflections: one for the right-angle band at the joint between the "egg" and the coaxial line, and another for a bend at SMA connector. For a given set of parameters, $\mathbf{P} = \{a_0, a_1, \tau_1, \cdots\}$, the distortion due to the reflections in ACD-1(R) and the cable-digitizer limitations can be compensated together with

$$w_{rec}(t, \mathbf{P}) = F^{-1}\{\frac{U(f)\Phi(f)L(f)}{H_{cd}(f)H_{dd}(f, \mathbf{P})}\}, \tag{12}$$

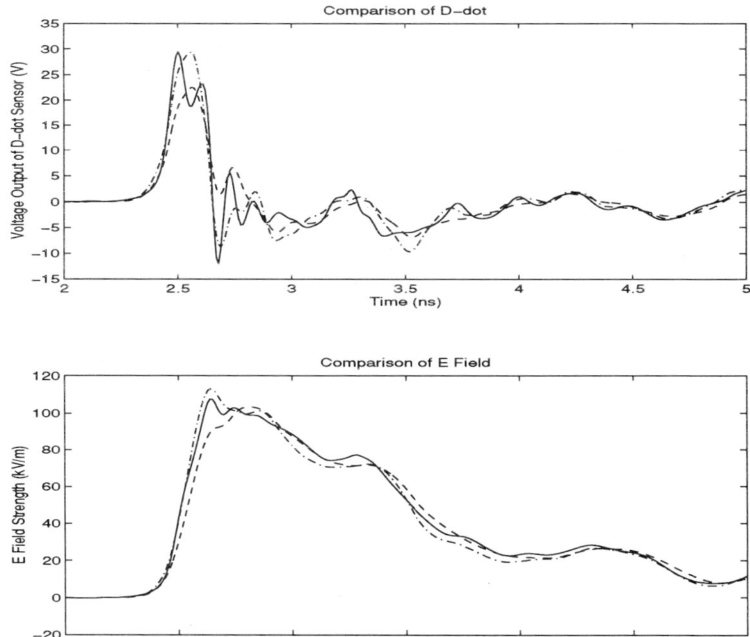

Figure 6: D-dot and E field measured using an ACD-1(R) compensated with Eq. (4) (dashed line) and Eq. (12) with the optimized parameters (dash-dot line), and those measured using an ACD-1(A) compensated with Eq. (4) (solid line).

where $H_{dd}(f, \mathbf{P})$ is the transfer function given by

$$H_{dd}(f, \mathbf{P}) = a_0 + a_1 e^{-j\omega\tau_1} + a_2 e^{-j\omega\tau_2}, \tag{13}$$

where ω is the angular frequency. For a different $\mathbf{P}$, we get a different w_{rec}. To determine $\mathbf{P}$, we have developed an optimization procedure which uses the Levenberg-Marguardt algorithm [1] to minimize an objective function (S): the parameters are adjusted in time domain while the derivatives are calculated in frequency domain and then transformed back to time domain [6]. S is defined as

$$S(\mathbf{P}) = \sum_{n=0}^{N/2-1} W(t_n)[E_{ref}(t_n) - E_{rec}(t_n, \mathbf{P})]^2, \tag{14}$$

where W is the weighting factor, $E_{ref}(t_n) = c\sum_{k=0}^{n} v_{ref}(t_k)$, used as a reference, is the E field obtained from the cable-digitizer compensated D-dot measured using the mounted ACD-1(A), $E_{rec}(t_n, \mathbf{P}) = c\sum_{k=0}^{n} w_{rec}(t_k, \mathbf{P})$ is the E field after cable-digitizer *and* D-dot sensor compensation for a measurement using an ACD-1(R) at a specific location on the bottom ground wall that is symmetric to the ACD-1(A) about center septum, as indicated in Figure 1. c is a converting coefficient. By the symmetry of the cell, E_{rec} and E_{ref} should be very close to each other, and the observed difference is mainly due to the reflections in ACD-1(R). It has been found that utilizing the wave forms of E field in the optimization procedure gives a better recovery to its leading edge than that obtained with D-dot wave forms. Figure 6 shows the results for the best fitted parameters: $a_0 = .72$, $a_1 = .28$, $\tau_1 = 81.2$ ps, $a_2 = .01$, and $\tau_2 = 510$ ps, from which we can see the reflection at the SMA connector is very minor. Although the semi-empirical method requires more computation, it is of following advantages: (i) more clear physics, (ii) easier portability for transform function due to a small number of parameters, and (iii) wider valid frequency range because of using an analytical model.

In summary, we have presented an error compensation routine for the electromagnetic pulse measurements in the pico second domain, which includes an algorithm of an iterative causal deconvolution and evaluations of an empirical transfer function for a cable-digitizer system and a semi-empirical transfer function for an ACD-1(R) D-dot sensor.

ACKNOWLEDGMENTS

This work is supported by the U.S. Army Medical Research and Materiel Command under contract DAMD17-94-C-4069 awarded to McKesson BioServices. The views, opinions and/or findings contained in this report are those of the authors and should not be construed as an official Department of Army position, policy or decision unless so designated by other documentation.

REFERENCES

[1] W. H. Press, B. P. Flannery, S. A. Teukolsky, and W. T. Vetterling. *Numerical Recipes in C.* Cambridge University Press, Cambridge, UK, 2 edition, 1992.

[2] A. V. Oppenheim and R. W. Schafer. *Discrete-Time Signal Processing.* Prentice Hall, New Jersey, 1989.

[3] T. Sarkar, H. Wang, R. Adve, and M. Moturi. Evaluation of a causal time domain response from bandlimited frequency domain data. In H. L. Bertoni, L. Carin, and L. B. Felsen, editors, *Ultra-Wideband, Short-Pulse Electromagnetics*, pages 501–515, Plenum Press, New York, 1993.

[4] N. S. Nahman. Software correction of measured pulse data. In J. E. Thompson and L. H. Luessen, editors, *Fast Electrical and Optical Measurements. Volume I*, pages 351–417, Martinus Nijhoff Publishers, Boston, 1986.

[5] *Specification of ACD D-dot Sensor (Data Sheet 1119).* EG&G Washington Analytical Services Center, Inc., Albuquerque, New Mexico, 1987.

[6] J.-Z. Bao. Time domain optimization with frequency domain derivatives. To be published.

Ultrawide Band Sources and Antennas: Present Technology, Future Challenges

W.D. Prather, C.E. Baum, F.J. Agee, J.P. O'Loughlin,
D.W. Scholfield, J.W. Burger, J. Hull, J.S.H. Schoenberg, R. Copeland

Phillips Laboratory
Kirtland AFB, New Mexico 87117

INTRODUCTION

Ultrawide Band (UWB) sources and antennas are of interest for a variety of potential applications that range from transient radar systems to high power jammers and communications systems. As a result, research efforts into sources and antennas for the production of high power UWB energy has made significant progress in the past 6 to 8 years. The development of high pressure gas switching technology has made steady progress in recent years as has the solid state technology. Recent research in high voltage insulating oil has made significant strides, and developments in stacked Blumlein technology are opening up new possibilities for compact, high power UWB devices. Significant progress has also been made in UWB transient antennas and lenses, due in large part to the solid technical foundation which was laid down in the early 90's by C.E. Baum and others [1, 2, 3].

GAS-SWITCHED SOURCES

H-Series Sources: The H-Series of devices produced at Phillips Laboratory uses high pressure hydrogen switches to produce extremely powerful and very compact UWB sources. The H-device work began in 1991 with the construction of the H-1, and has continued to the present development of the H-3 and H-4 pulsers. The first pulser, H-1 was designed for a 300 kV output in a 2" diameter, 50 Ω coaxial line. The output power of this early device into a resistive load was about 275 MW and successfully demonstrated the concept of a high pressure (1400 psi) hydrogen-switched source. The H-2, completed in late 1992, represented a substantial improvement over the earlier version and was capable of generating 3.9 GW into a resistive load. It produced a 500 kV pulse in a 4" diameter, 50 Ω line. The pulse has a risetime of about 250 picoseconds (ps) and a total pulse length of 1.5 to 2 nanoseconds

Figure 1. H-2 with PGC and TEM Horn

(ns). The H-3, which is currently under development, is designed to generate 1 MV in a 6" diameter 8-ohm coaxial line tapering to a 40Ω line which equates to 40 GW of peak power into a resistive load. It produces a risetime of 150 ps and has a pulse duration of only 300 ps. In order to achieve the very fast risetime, the H-3 makes use of a unique Nanosecond Transmission Line Charging Technique and a multi-channel switch in its final stage [4]. The next pulser in the series, the H-4, is still undergoing final design. It is essentially a larger version of the H-3 and will therefore derive much of its final design from the ongoing series of developmental tests of the H-3. The goals for the H-4 are to produce 2 MV into an 8" diameter, 6 Ω line which tapers to a 25Ω line. This equates to 100 GW of peak power into a resistive load. The waveform will have a risetime of 130 ps and a duration (FWHM) of 250 ps or greater.

The IRA: The Impulse Radiating Antenna (IRA) shown in Figure 2 also utilizes a high pressure hydrogen switch (1400 psi) to produce an extremely powerful UWB pulse. With a charge of only ± 60kV, the system puts out a unipolar transient which was measured to be 4.6 kV/m at 300m. The pulse has a risetime of 85 ps and a duration of over 1 ns resulting in a 3 dB bandwidth of 35 MHz to 4 GHz [5,6]. The upper 3 dB point is governed by the voltage and risetime of the gas switch and the optical correction afforded by a very precise oil and polyethylene lens at the apex. The lower frequency point is a function of the diameter of the antenna aperture and the **p x m** compensation network at the base of each arm which allows wavelengths longer than the diameter of the antenna to be radiated [7]. When the switch fires, the transient is guided by a 200 Ω, four-wire TEM transmission line to a 4m diameter parabolic reflector. The resulting 3 dB beam width of the IRA is ±1°, producing a very efficient long distance UWB source.

Figure 2. The IRA

Technology Challenges with the Gas-Switched Sources

Choice of Gases: The main disadvantage which we have found with hydrogen is that it is reactive. Brass electrodes in the presence of H_2 and a spark channel produce copper hydride powder which coats the inside of the switch housing and eventually causes them to fail. Even stainless steel electrodes have been found to be reactive to the high pressure hydrogen in these types of switches. Thus, for hydrogen gas switches, we have found that tungsten works the best.

Repetition Rates: Repetition rates of 1 to 2 kHz have been achieved with the H-2 and H-3 devices. Higher rates are desirable, but we find that the rep rate is one of the limitations with these high voltage devices. The pulsers tend to work quite well when fired at single shot or at very low rep rates, but when fired at higher rep rates for very long, local heating can change the characteristics of the gas and the shape of the waveform.

Insulators and Windows: Typically, the material of choice for dielectric standoffs in the H-series of pulsers has been Lexan™ , a brand of polycarbonate manufactured by

Dupont. Lexan has good dielectric strength, is easy to machine, and is strong enough so that it will not flow when placed under high pressure. It also has a dielectric constant which is close to that of transformer oil, so it is optically compatible. Where it is used at an interface with the gas, we try to use a Brewster Angle window in order to minimize reflections and improve efficiency. The disadvantage of Lexan is that it has a rather high loss tangent at frequencies above 1 GHz. There are other materials which have a smaller loss tangent, but they typically don't have the mechanical strength which we find in Lexan. For example, high density polypropylene, polyethylene, and Teflon all have desirable electrical qualities, but are not suitable for use in high pressure interfaces. It would be desirable to find another material to use in the final stages of the H- machines where we are trading off extremely fast risetimes, high voltages, and high gas pressures, but so far we have not settled on an alternative. We are investigating a boron nitride ceramic which might prove useful. It has a much lower loss tangent than Lexan, but it is a ceramic, and therefore limited in the mechanical stresses it will take.

SOLID STATE SOURCES

BASS™ Technology: The technology which is used in the GEM series of pulsers is the Bulk Avalanche Solid State (BASS™) switches manufactured by Power Spectra Inc. These are GaAs semiconductor switches which operate around 10 kV and are capable of extremely short risetimes. When they are illuminated by laser light of the right wavelength, they become conductive and fire in avalanche mode. They require a very small amount of light, easily that which can be supplied by a laser diode and carried by a fiber-optic link. This affords a considerable amount of flexibility to the designer in the placement of the switches with respect to the control system, and also permits the design of arrays of individually controlled elements.

GEM-Series Sources: The GEM series of sources which was produced for Phillips Laboratory by Power Spectra, Inc. uses this technology [8]. The GEM I is a 2 x 4 array of antenna elements as shown in Figure 3, and the GEM II is an 8 x 12 array as seen in Figure 4. Each element is individually controlled (switched) by its own laser diode and fiber-optic link. Thus, this design can be steered over about ±30° by switching the elements in the correct sequence. It also allows one to vary the radiated waveform/spectrum by switching the elements in different sequences and with different

Figure 3. GEM I

Figure 4. GEM II

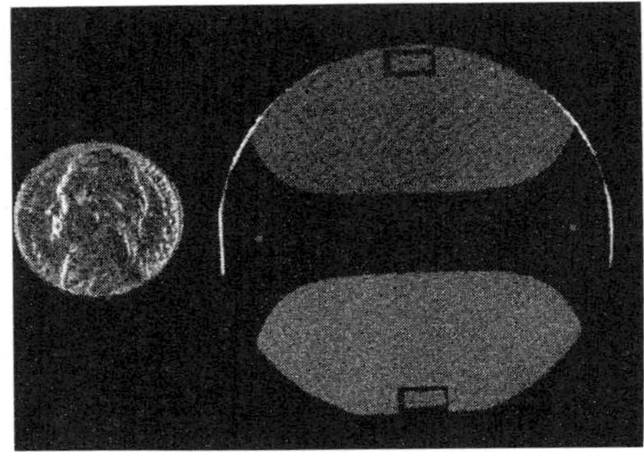

Figure 5. Lateral PCSS Switch

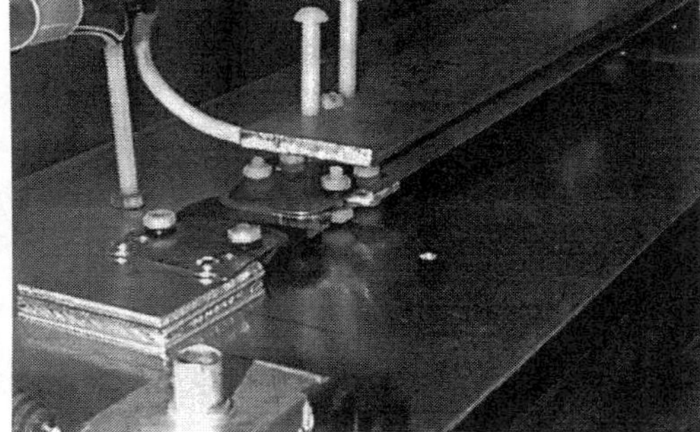

Figure 6. Optically Switched Blumlein

timing. For example, by switching the columns in a fast sequence, one can create a short burst of very fast rep rate pulses or a swept rep rate of pulses. The trade-off, of course, is that in doing so, the radiated energy is less than if the whole array were switched at once.

Lateral PCSS Technology: Some of these limitations are being overcome by the current research on Lateral Photo Conducting Solid State (PCSS) technology. The lateral switches also use GaAs wafers, but they are designed such that the current path is along the surface of the wafer rather than through it. These switches, which are illustrated in Figure 5, can operate at much higher voltages and currents than the bulk avalanche technology. The ones currently being used by Phillips Laboratory operate in the 50 kV to 130 kV range which greatly increases the radiated power. At the present time, these type of switches are being incorporated into switched transmission lines and Blumleins which are in turn connected to wideband TEM horns. Since these systems are also optically switched, they may be used singly or in arrays as was done in the GEM sources. However, our objective is to use the higher voltage lateral switch technology and a voltage multiplying device such as the stacked Blumlein to significantly increase the radiated power of the source. Also, the use of wideband TEM horns in the array instead of the band-limited horns currently used in the GEM series will increase the low frequency radiated energy.

Technology Challenges with the Solid State Sources: The limitations seen in the solid-state technology sources, both the BASS™ and the Lateral PCSS are the maximum voltage and current which they will carry versus the size and, therefore, risetime of the devices. Their reliability (lifetime) decreases as one increases the voltage, current, pulse length, rep rate, etc. Local heating is the problem, and in these solid state devices, the heating is usually fatal. The solid state devices do not degrade gracefully like the gas devices nor can they be recovered. Once the GaAs is damaged, it must be replaced. We find that solid state devices operate with good reliability at around half of their maximum rated stress.

Transmission Line Technology: In order to build more powerful, efficient solid state sources, we must find ways to increase the voltage output of the sources without sacrificing the reliability. The method which is currently being investigated is the Optically Switched Transmission Line (OSTL). This basically consists of combining a high voltage Lateral PCSS switch with some type of charged transmission line, not a new idea, but one which is being given a novel twist by the requirements for UWB transients. At present, three types of OSTL are being investigated. The first is a single parallel plate transmission line with a Lateral PCSS switch at one end. This will produce a step function pulse at the load or antenna interface. The second type is a parallel plate line with a resonant stub placed part way down the line which generates a bipolar pulse. The third type is a three-conductor Blumlein as shown in Figure 6. This is a very effective way of doubling the voltage with the use of only one switch [9]. Such a configuration when combined with the Lateral switches offers great promise for increasing the power output of solid state sources.

MODE CONVERTORS

The coaxial geometries employed by these devices make scaling calculations and machining more simple, but at the same time, create a problem by delivering the energy to the antenna in an awkward manner. In order to be able to radiate the UWB energy from the H-series sources, it must be converted from a coaxial mode to a TEM mode. The most common example of a mode convertor or transition between a coaxial geometry and a TEM horn is commonly called a "zipper." This is a common device which has been used for many years, and the mode convertors being designed and used at Phillips Lab today are variations on this basic theme. The challenge is to design a device which will pass the short risetimes and hold off the extremely high voltages which these machines put out. One version of this device which has been successfully used with both the H-2 and H-3 machines is called the Point Geometry Convertor or PGC which is shown in Figure 7. In designing this device, Voss Scientific and Fiore Industries overcame the risetime limitation by making the zipper short with respect to the times and wavelengths of interest. The result, of course, was that the electric field stress in the device was larger than could typically be used with standard transformer oil. To overcome this limitation, they passed the oil through a super-fine filter which removed all particulates larger than 1 micron. They discovered that this oil could then hold off much higher stresses than before. For the very short UWB pulses, tests have shown that it is possible to reach stresses of as high as 15 MV/cm without breakdown (single shot). At moderate rep rates, this number reduces to about 7 MV/cm. Even so, this is a remarkable development and is what makes the present PGC design possible [10].

Figure 7. The PGC Balun

Another version of the zipper is being designed and built by Farr Research and EG&G [11]. Their design does not try to keep the convertor short, but relies on a smooth impedance change and a carefully controlled geometry which keeps the electric field very uniform throughout the zipper. Low voltage tests have shown that this design will transmit risetimes on the order of 50 ps, and it is being designed to operate with the full 2 MV output of the H-4 pulser. This design is making excellent progress and will be ready for high voltage tests before the end of 1996.

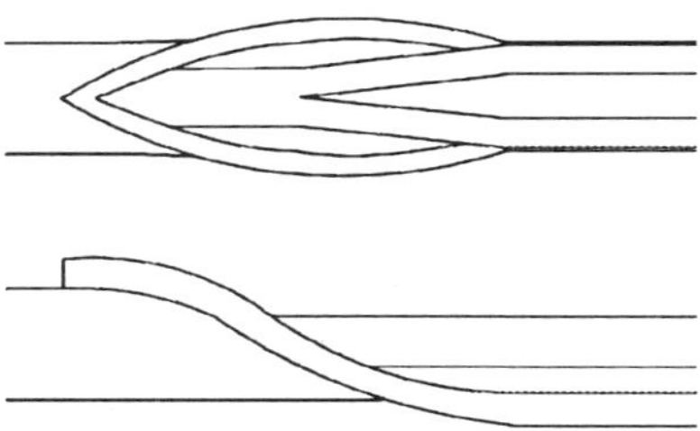

Figure 8. High Voltage Zipper

LENSES: For the very fast risetimes with which we are dealing, it is quite often necessary to include optics in the design of the antennas and mode convertors. Since we are dealing with high voltages, this is most often done with the same materials used for insulation. Combinations of Lexan, polyethylene, and transformer oil work well together since they have about the same relative dielectric constant. A huge amount of progress has been made in the last 5 years in the research into ultra wideband lenses by Stone and Baum. Their exemplary work has made possible many of the UWB antenna designs which we have today [12]. In addition, the Phillips Laboratory is now investigating the feasibility of developing artificial dielectrics which will work with UWB transients. The objectives are to develop dielectrics for lenses which are very light weight and which can include graded dielectric constants for special applications [13, 14].

ANTENNA TECHNOLOGIES: For radiating UWB transients, we need antennas which are specifically designed and optimized to that task. As a result, there has been a considerable amount of research into fast transient antennas in the past five years, and now the design principles for UWB antennas are much better understood. There are many articles and technical reports available now and more than a few very impressive antenna designs available.

As demonstrated by Baum, Farr, and Buchenauer [15], frequency dispersion is most easily overcome by a conical TEM horn or a biconical antenna, frequency independent structures which produces spherical wave fronts. The spatial dispersion caused by the non-planarity of the wavefront is corrected by the use of a lens or a reflector to straighten out the wavefront.

Originally demonstrated in the laboratory by Buchenauer [15], both of these concepts have recently brought new standards of performance to the antenna world. A TEM horn with a polyethylene lens from Sandia Labs has been measured to emit a pulse with a FWHM of 23 ps [16]. Also, the team of Farr and Frost have recently built and demonstrated a 9" diameter reflector IRA with a risetime of 25 ps, an improvement over the 18" IRA which they had previously shown to radiate a pulse with a FWHM of 35 ps [17, 18]. This corresponds to upper frequency limits of around 20 GHz. Truly remarkable achievements for both teams.

These kinds of risetimes/bandwidths, of course, have only been achieved at relatively low voltages. However, the higher voltage machines which operate in the 100kV to 1 MV range are now also achieving very fast risetimes. The IRA, which operates at only 120 kV has achieved a radiated risetime of 85 ps, and the H-3 pulser which operates in the 500 kV to 1 MV range has a risetime of around 150 ps.

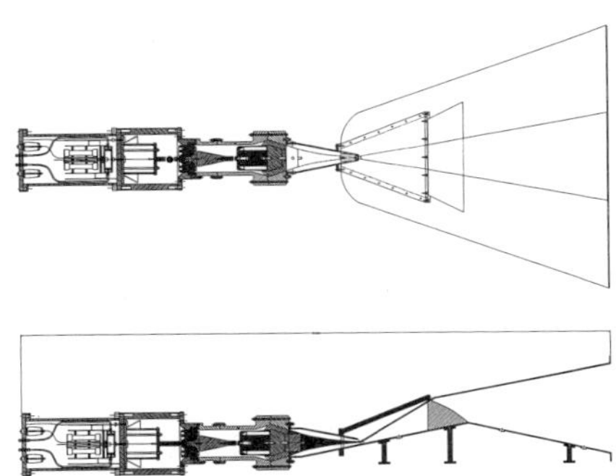

Figure 9. Half-TEM Horn with Brewster Angle Lens

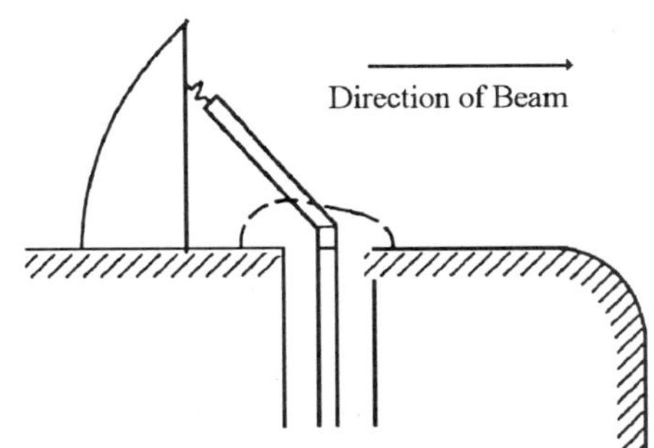

Figure 10. The Half-IRA

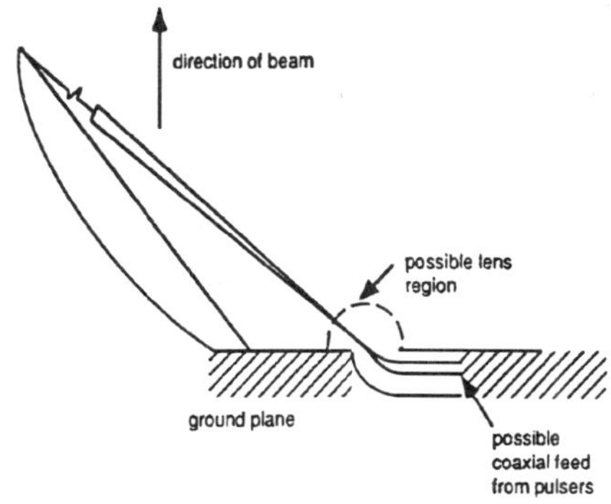

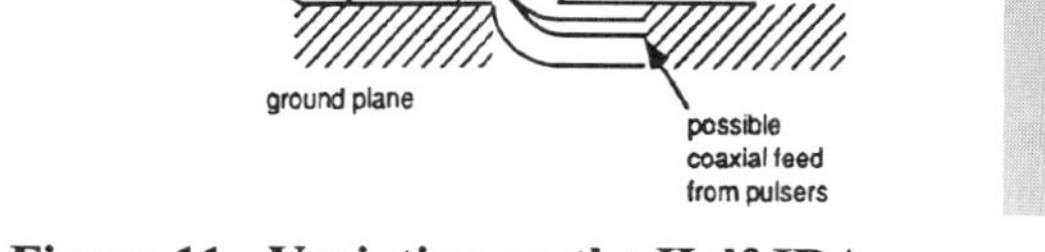

Figure 11. Variation on the Half-IRA

Figure 12. Compact Lensed Horn

Wideband TEM Horn Arrays versus Band-limited Horn Arrays: The horns presently used in the GEM sources are mounted in enclosed waveguides. As a result, their low frequency performance is limited by the cut-off wavelength of the guides. Also, these waveguide-mounted horns are not additive in frequency. That is, we do not get the frequency response of a larger aperture when we stack them up into a large array. The spectral content of the large GEM II array is exactly the same as the spectrum of each single element. In some applications, this may be desirable, but in others, we may need a larger (lower frequency) aperture. Thus, in order to increase the low frequency bandwidth of such an array, Dr. Baum has proposed the use of TEM horns [19, 20]. The TEM horns are additive in both voltage and frequency so that an array of small antennas can have the same radiated field and low frequency performance (effective aperture size) as a larger antenna. These can be designed to have a single polarization, dual polarization, or a rotating polarization. The antenna array concept also includes a **p x m** loop which increases the low frequency radiated energy for the size of the antenna aperture.

Half-IRA and Half-TEM Horn: The output of the H-series sources is a coaxial transmission line and is therefore not appropriate for balanced antennas such as the IRA or a TEM horn. Therefore, it was necessary to design antennas which were unbalanced.

The basic idea was to simply take half of a balanced antenna over a symmetry plane (ground plane). This has resulted in two designs which are currently being manufactured, a Half-TEM Horn [21, 22, 23] and a Half-IRA [24, 25]. Both of these are fed from below the ground plane by the coaxial output as shown in Figures 9 and 10.

Variations on a Theme: In some of his latest notes on UWB antennas, Dr. Baum has developed a number of variations on these antenna design concepts. As can be seen in Figures 11, and 12, these designs can be varied to provide for different pulser configurations, antenna mounting geometry, and offsets to change the angle of propagation [26, 27, 28].

SUMMARY

In the last 5 years, there has been steady advancement in UWB sources and related technology. The research has reached for higher voltages and shorter risetimes until we are now pushing the limits of our knowledge in material properties. In particular, the dielectric breakdown properties of insulating materials at these extremely high voltages and extremely fast risetimes is being challenged. Likewise, with the fast risetimes, we are now seeing the high frequency loss tangents of some materials show up where they were not a problem before.

The high voltage properties of insulators and the breakdown properties of gases and oils have not been systematically investigated and documented for the very fast risetimes we are dealing with.. Such a program has recently begun at the Phillips Laboratory and at various universities through the sponsorship of AFOSR. Hopefully, this program can lay the foundations for future research in this area by the UWB community at large.

Also, as a result of the research in UWB transient antennas, several new concepts are emerging which are beginning to have an effect in other areas of the electromagnetics world. The design of non-dispersive designs has made possible antennas which can be used over an extremely wide range of frequencies, simultaneously if desired. These multi-band antennas offer up the tantalizing possibility for replacing several individual antennas with one multi-function antenna, thus reducing the space and weight required by the antenna farms currently used on many systems. Designers of future spacecraft, aircraft, and Navy ships are taking a hard look at the advantages which may be had from these unique designs. Likewise, the p x m concept where one can increase the low frequency radiation from a small antenna offers possibilities for reducing antenna size in flight vehicles and other applications.

REFERENCES

1. C.E. Baum, "General Properties of Antennas," Sensor and Simulation Note 330, Phillips Laboratory, July 1991.
2. C.E. Baum and E.G. Farr, "Extending the Definitions of Antenna Gain and Radiation Pattern Into the Time Domain," Sensor and Simulation Note 350, November 1992.
3. C.E. Baum and A.P. Stone, *Transient Lens Synthesis: Differential Geometry in Electromagnetic Theory,* Hemisphere Publishing Corp, 1991.
4. J.P. O'Loughlin, "A Nanosecond Transmission Line Charging Technique," U.S. Patent 5,444,308.
5. D.V. Giri, W.D. Prather, et al, "A Reflector Antenna for Radiating Impulse-Like Waveforms," Sensor and Simulation Note 382, Phillips Laboratory, July 1995.
6. C. Courtney, W. Prather, et al, "Measurement and Characterization of the Impulse Radiating Antenna," Prototype IRA Memo 5, September 1995.
7. D.V. Giri and S.Y. Chu, "On the Low-Frequency Electric Dipole Moment of Impulse Radiating Antennas (IRAs)," Sensor and Simulation Note 346, 5 October 1992.
8. J. Oicles, et al, "High Power Impulse Generators for UWB Applications," *Ultra Wideband/Short-Pulse Electromagnetics 2,* Plenum Press, New York, 1994.
9. F. Davanloo, et al., "High Power, Sub-nanosecond Rising Waveforms Created by the Stacked Blumlein Pulsers," *Ultra Wideband/Short-Pulse Electromagnetics 3*, this volume.
10. J. Wells, et al., "A Device for Radiating High Power RF Fields from Coaxial Structures," *Ultra Wideband/Short-Pulse Electromagnetics 3*, this volume.
11. E.G. Farr, Gary D. Sower, and C.J. Buchenauer, "Design Considerations for Ultra-Wideband, High Voltage Baluns," Sensor and Simulation Note 371, October 1994.
12. A.P. Stone, C.E. Baum, and J.J. Sadler, "A Prolate Spheroidal Uniform Isotropic Dielectric Lens Feeding A Circular Coax," *Electromagnetics*, v15, Taylor & Francis, 1995.
13. J.S.H. Schoenberg and C.J. Buchenauer, "Artificial Dielectrics for Ultra-Wideband Application," *Ultra Wideband/Short-Pulse Electromagnetics 3*, this volume.
14. C.E. Baum, "Dielectric Jackets as Lenses and Application to Generalized Coaxes and Bends in Coaxial Cables," Sensor and Simulation Note 394, March 1996.

15. E.G. Farr, C.E. Baum and C.J Buchenauer, "Impulse Radiating Antennas, Part 2," *Ultra Wideband/Short-Pulse Electromagnetics 2,* Plenum Press, New York, 1994.
16. J.F. Aurand, "A TEM-Horn Antenna with Dielectric Lens for Fast Impulse Response," *Ultra Wideband/Short-Pulse Electromagnetics 3*, this volume.
17. E.G. Farr and C.A. Frost, "Compact Ultra-Short Pulse Fuzing Antenna Design and Measurements," Sensor and Simulation Note 380, June 1995.
18. C.E. Baum, "Self-Complementary Array Antennas," Sensor and Simulation Note 374, October 1994.
19. C.E. Baum, "Timed Arrays for Radiating Impulse-Like Transient Fields," Sensor and Simulation Note 361, July 1993.
20. C.E. Baum, "Low-Frequency-Compensated TEM Horn," Sensor and Simulation Note 377, January 1995.
21. C.E. Baum, "Brewster Angle Interface Between Flat-Plate Conical Transmission Lines," Sensor and Simulation Note 389, November 1995.
22. M.H. Vogel, "Design of the Low-Frequency Compensation of an Extreme-Bandwidth TEM Horn and Lens IRA," Sensor and Simulation Note 391, February 1996.
23. E.G. Farr and G.D. Sower, "Design Principles of Half Impulse Radiating Antennas," Sensor and Simulation Note 390, December 1995.
24. L.M. Atchley, G.D. Sower, and E.G. Farr, "Scale Model Ground Plane Measurements of a Half IRA," Prototype IRA Memo 6, January 1996.
25. C.E. Baum, "Variations on the Impulse-Radiating-Antenna Theme," Sensor and Simulation Note 378, February 1995.
26. C.E. Baum and E.G. Farr, "Impulse Radiating Antennas With Two Refracting or Reflecting Surfaces, Sensor and Simulation Note 379, May 1995.
27. C.E. Baum, "Steerable Lens Surface for Use with the IRA Class of Antennas," Sensor and Simulation Note 387, September 1995.
28. E.G. Farr and C.A. Frost, "Development of a Reflector IRA and a Solid Dielectric Lens IRA, Part I: Design, Predictions, and Construction," Sensor and Simulation Note 396, April 1996.

A DEVICE FOR RADIATING HIGH POWER RF FIELDS FROM A COAXIAL SOURCE

Jimmy Wells,[1] Clifton Courtney,[3] Tracey Bowen, [2] David Eckhardt,[3] Norman Keator,[3] Carl Noggle,[1] Donald Voss,[3] Gary Watt,[1] Harvey Wigelsworth,[1]

[1]Fiore Industries Inc.
1009 Bradbury Drive SE
PO Box 9243
Albuquerque, NM 87119-9243

[2] USAF Phillips Laboratory
3550 Aberdeen Ave.
Kirtland AFB, New Mexico 87117

[3]Voss Scientific
416 Washington SE
Albuquerque, NM 87106

INTRODUCTION

Many high power RF sources are designed in a coaxial geometry. This is done for a variety of reasons. The cylindrical geometry is convenient for containing extremely high gas pressures, the parts are inexpensive to machine, the impedance is relatively easy to control, and the design can be easily scaled up or down. However, the coaxial geometry poses a problem when one wants to feed the energy to an antenna. Whereas a coaxial geometry might be ideal for some applications, it is not conducive to radiation. In order to effectively radiate the energy, we want to put it into a TEM mode. This requires a mode convertor. In this case, it requires a mode convertor which has extremely wide bandwidth and the ability to withstand stresses as high as 1 MV without flashing over.

Ultra-Wideband, Short-Pulse Electromagnetics 3
Edited by Baum *et al.*, Plenum Press, New York, 1997

BASIC DESIGN

Attempting to change the output geometry to a planar geometry usually results in flash-over in the balun region which terminates the high voltage pulse rather than driving the antenna. An example of this is the standard zipper balun in air. As a solution to this problem we have developed and refined the design of an oil-insulated Point Geometry Converter (PGC) which abruptly transitions from a coaxial source output to a parallel plate geometry for feeding a TEM antenna.

Properties of Insulating Oil

Our experience on the Air Force Phillips Lab's Phoenix UWB source suggests that highly filtered transformer oil is capable of withstanding extremely high field stress levels for a very short duration without breaking down. Our basic assumption was that if the maximum field stress can be kept below the breakdown level by a factor of about four, then breakdowns could be avoided. The problem then became one of accurately predicting this field stress level. We chose to predict the breakdown voltage using formulas empirically derived by I.D. Smith, Thomas H. Martin, and J. C. Martin [1-3]. Some of the formulas were originally developed from gas breakdown data, but has been shown to be scaleable to other insulating materials. Existing oil breakdown data from several sources, including Phoenix, were plotted for comparison with gas data. Comparison of the oil breakdown data to the gas data revealed practically identical slopes. Based upon this data a PGC section was developed that has a maximum electric field stress of about 1 MV/cm, a factor of four to seven below the electric field stress necessary to initiate breakdowns in the oil for this pulse duration. Using Thomas Martin's equation scaled for the oil data (1), we calculated the field intensity necessary for initiating breakdowns in transformer oil with a two-nanosecond voltage pulse width to be about 5 MV/cm.

$$\rho\tau = 5000\,(\mathbf{E}/\rho)^{-3.44} \qquad (1)$$

Where the density ρ is in gm/cc, the electric field intensity $\mathbf{E}$ is in kV/cm, and the time for which the voltage is above 80% of the breakdown value τ is in seconds. As mentioned previously the highest field stress in the PGC transition region is about 1 MV/cm and thus no carbon should be formed. Since no carbon is formed, there is no need for filtering of the insulating oil. The transition region consists of a short tapered section of 56 Ω coaxial line It maintains a constant impedance while reducing the inner conductor diameter to 0.328 inches and an outer diameter of 1.312 inches. The total length of this tapered section is about 4.2 inches as shown in Figure 1.

From this smaller coaxial diameter we complete a transition to parallel plate geometry in a length of less than one inch, while continuing to use the transformer oil for insulation. The total transition section is only 15 inches in length. This length includes the coaxial taper and a section for expanding the parallel plate width to match that of the TEM antenna as shown in Figure 1. The design should minimize degradation of the fast rise time pulse as it propagates toward the antenna.

MEASUREMENT INSTRUMENTATION

All the components for the modification were fabricated in our own machine shop and fitted to the machine in our anechoic chamber for radiating. Additional diagnostics (B-dots

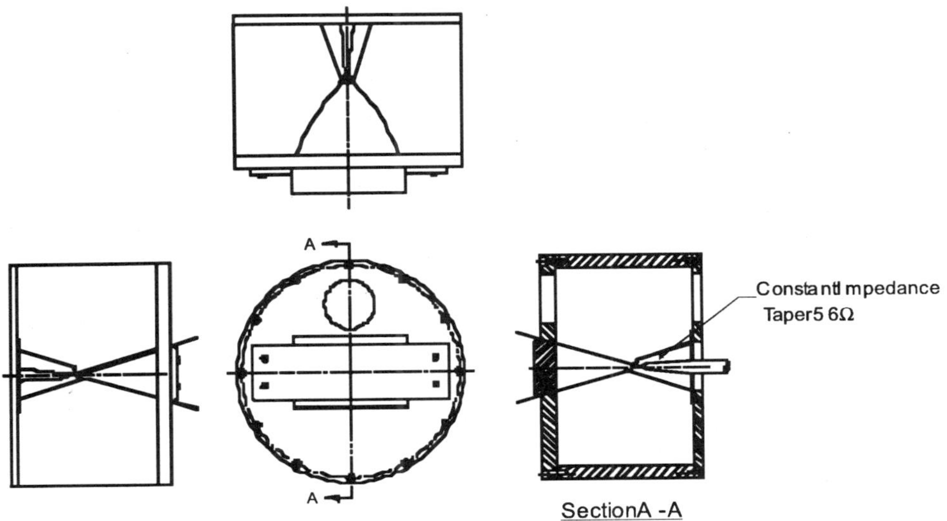

Figure 1. Point Geometry Convertor

and self-integrating D-dots at each location) were installed to allow measurement of both electric and magnetic fields at the post peaking switch location and thus accurately determine the power level. Since voltage and current values will change with impedance variations along the length of the source, obtaining power levels by measuring either of these parameters alone may be inaccurate and there is no means to cross-check the results.

These diagnostics allow us to determine the power along the length of the machine after the switch and thus determine the efficiency of the PGC. Calibration of these sensors was done using a 50 Ω parallel plate transmission line and a pulse generator capable of producing a 40 volt pulse with an 80 picosecond risetime. When fabricated correctly, the self-integrating D-dot sensors have excellent band width and reproduce the transmission line pulse very well.

HIGH VOLTAGE TESTS

Tests were conducted with the H-2+ and H-3 UWB sources to measure the performance of the PGC and to compare its behavior with that of a standard zipper balun with air insulation.

The basic premise used for the characterization was that if there is no flash-over in the antenna feed, then we should find a linear relationship between charge voltage on the primary capacitor bank and the resulting radiated field inside the test chamber. When flash-over begins to occur, we would expect to see flattening or rollover of this curve.

RESULTS

Figure 2 is a plot of the primary charge voltage versus the radiated electric field intensity generated by the source at an axial location five meters from the antenna with the unmodified source using a "zipper" balun to feed the antenna. Note the roll-over in the curve and the increase in field spread at the higher charge voltages indicating the onset of

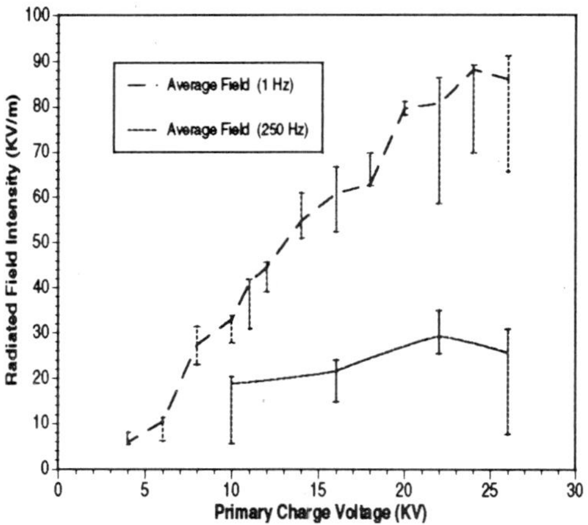

Figure 2. Radiated Fields with Zipper.

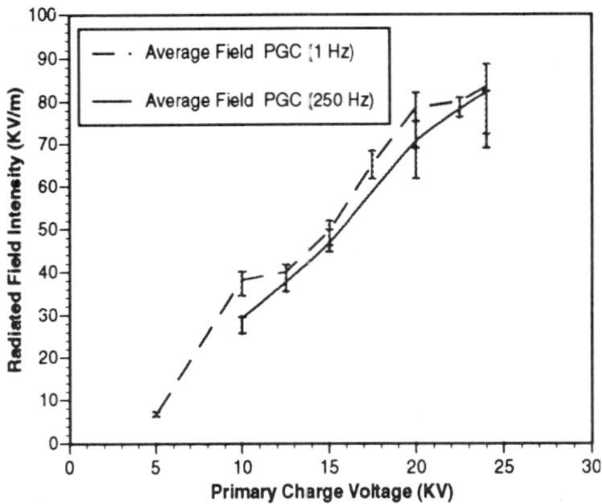

Figure 3. Radiated Fields with PGC.

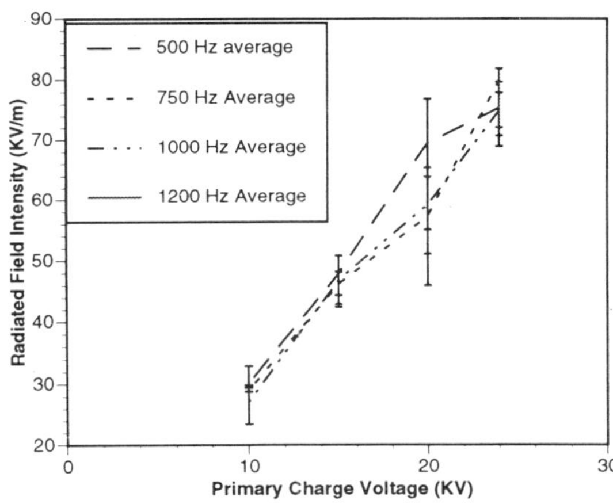

Figure 4. Repetition rate data for H-2+ with the PGC hardware installed.

flash-over in the antenna feed section. This rollover point will change as a function of the repetition rate, occurring at lower charge voltages for higher repetition rates. Thus, these field values represent the best region of operation and highest fields possible since they were generated at one hertz. Shots were also taken at a repetition rate of 250 Hz which show that radiated field levels are substantially reduced even at this moderate repetition rate. These reduced field intensities and increased spread of the data indicate that corona and/or flash-over is occurring in the balun. The effect becomes worse as the repetition rate is increased. (See Figure 2.)

When rollover does begin to occur on this source however, one other possible explanation could be that the secondary windings of the high voltage transformer are losing energy due to corona discharge inside the pressure vessel. The onset of corona discharge from the secondary windings will have the same effect as flash-over in the antenna feed section on the radiated fields. As a precaution against the corona problem, we also recorded measurements of the power level at the post-peaking switch location as an indication of whether the problem was in the transformer or the antenna feed section, since the corona will reduce the power at this location while flashing in the antenna feed will not.

The radiated field tests were then repeated with the oil insulated PGC installed in place of the zipper balun. The results are shown in Figure 3. Note the absence of any indication of rollover in the data and the decreased spread of the radiated field levels at the higher charge voltages. This indicates that flash-over is no longer occurring in the antenna feed region as a result of the added transformer oil insulation.

Radiated field versus charging voltage curves were also generated at higher repetition rates of 500, 750, and 1000 Hz. These results are presented in Figure 4. Note the absence of any rollover, even at the highest repetition rates. One data point was collected at 1200 Hz repetition rate and 24 kV primary charge voltage. At this repetition rate we obtained an average radiated field intensity level of 71.5 kV/m, indicating no rollover at this repetition rate. At repetition rates of 1000 Hz and above we saw a slight decrease in the secondary output voltage at the highest charge voltages. This indicates that energy is being lost to corona discharges from the transformer secondary windings, as mentioned earlier. The primary switch worked very well at a repetition rate of 1200 Hz (the highest repetition rate attempted), however, the maximum repetition rate has not yet been determined. The diagnostics indicate a power level just after the peaking switch of over 2 GW with a line voltage of 360 kV and a line current of over 6 kA. The electric and magnetic field sensors agree with one another to within 10% which is encouraging. The power level at a point just after the PGC section measured 1.85 GW which gives an efficiency of 87% through this section. Note however that this value is near the 10% variation in readings between sensors.

The spectrum for typical output pulses was essentially unchanged from the zipper balun to the PGC tests. The radiated field has content out to about 3 GHz which is consistent with the measured rise time of about 250 picoseconds. The rise time for the pulse at the input to the PGC is the same as that of the radiated pulse indicating that the PGC is capable of faster rise times than H-2+ can provide. Recent tests with the new H3 HPM source fitted with similar PGC hardware have revealed radiated risetimes of 185 picoseconds.

CONCLUSION

A device for converting coaxial to planar geometry, and one which is capable of operating at very high voltages has long been needed for radiating high electric field strengths. The PGC represents a major break through in this area. Radiated TEM field levels from machines with coaxial outputs, such as the H-series sources, have now produced greater radiated field intensities at higher repetition rates than was previously possible using other structures.

REFERENCES

1. I.D. Smith, "Breakdown of Transformer Oil," Dielectric Strength Note 12, AFWL, Nov 1966.

2. J.C. Martin, "Pressure Dependency of the Pulse Breakdown of Gases," Dielectric Strength Note 15, Air Force Weapons Laboratory, 1967.

3. T.H. Martin, "An Empirical Formula for Gas Switch Breakdown Delay," 7th Pulse Power Conference, 1989.

4. T.H. Martin, "Gaseous Breakdown Processes That Are Important for Pulsed Power Switching," 8th Pulsed Power Conference, San Diego, 1990.

5. T.H. Martin, "Macroscopic Gas Breakdown Relationship," Proc. Gaseous Dielectrics VI, 1990.

6. J. Wells, et al., "Anamalous Sub-nanosecond Pulse Breakdown Strength of Transformer Oil," *Ultra Wideband/Short-Pulse Electromagnetics 3*, this volume.

HIGH VOLTAGE UWB HORN ANTENNAS.

P.D. Smith† and C.J. Brooker‡.

†Applied Electromagnetics, 5A Alexandra Place, St Andrews, Fife, UK
‡Defence Research Agency, Fort Halstead, Sevenoaks, Kent, UK

INTRODUCTION

Antenna structures capable of supporting (essentially) transverse electromagnetic (TEM) spherical waves are characterised by their frequency independence over a wide band width, making them ideal candidates for transient field radiation. In particular the triangular plate configuration of the TEM horn is of interest for a variety of directive wide-band applications including ultrawideband radar pulse transmission [1] and feeds for paraboloidal reflector antenna systems [2]. The basic design is founded upon idealised, infinitely long, conical antennas which have been studied by several authors including [3] and [4]. However in producing practical antennas which optimise the desired characteristics of the radiated pulse train, several modifications are required to accommodate the presence of high strength dielectric media preventing breakdown at high voltage and high pulse repetition frequency operation. In particular the dielectric/air boundary must be shaped to minimise reflection of energy at the interface.

The pressurised gas spark gap is widely used in high power switching operations. There is recent interest in developing these switches for use in very fast (< 200ps risetime), high pulse repetition rate (>1 kHz) and high voltage pulse generators[5,6]. High pressure spark gaps are versatile, lightweight, inexpensive and robust. This paper describes a low inductance gas sharpening gap for fast risetimes, a high pulse repetition frequency (PRF) modulator and a matched antenna to radiate the transient pulses.

SPARK GAP

For optimal performance it is desirable that the spark gap should be integral with the antenna in order to minimise losses and preserve the fast risetime. A parallel plate (or stripline) arrangement, rather than coaxial configuration, was used for the spark gap because the transition to a TEM horn antenna is more straightforward. The effect of single point switching dictated a maximum width to the transmission line plates at the gap location, and

Ultra-Wideband, Short-Pulse Electromagnetics 3
Edited by Baum *et al.*, Plenum Press, New York, 1997

the production of higher order modes in the waveguide was avoided by restricting the dimensions.

The sharpening gap was incorporated into a 50 ohm stripline with conductor width 32mm separated by 10mm of polymethylmethacrylate (Perspex) dielectric. The strips were manufactured from 3mm thick brass strip with rounded edges and the electrodes were shaped to ensure that breakdown occurred in the centre. The electrode separation was 0.5mm and the estimated inductance of the gap housing was 2nH. The modulator was used to pulse charge a 150mm long section of this line (the pulse forming line or PFL). This section then discharged through the sharpening gap to the antenna.

The sealed gap was filled with nitrogen at up to 300psig, without gas flow, for high PRF operation. The burst duration was up to five seconds. The gap was also operated at low PRF (about 1 Hz) in nitrogen using a Marx generator to extend the measurement range to higher voltages and pressures (90 kV breakdown voltage and 600psig). The coaxial Marx generator was constructed by Veradyne Corporation and had a 2ns risetime to over 100 kV.

ANTENNA

For any TEM horn antenna, the characteristic impedance is related to the plate separation angle and the plate apex half-angle, as well as the ambient dielectric medium (which is assumed to be homogeneous in our case). A conformal mapping calculation of the characteristic impedance of idealised infinitely long horns is given in [3]; studies of the effect of truncation to a finite length as well as profiling or flaring of the horn arms are given in [4]. These studies provided the required angles defining the TEM horn both in the dielectric filled section adjacent to the source as well as in the air-filled section radiating into free space.

The complete unit comprises a spark gap incorporated in a stripline matched to a TEM horn, both encased in dielectric; a transition section to be described shortly provides a match to an air-filled TEM horn (of finite length) which radiates into free space. A design also closely related to this has recently been given in [7].

The TEM horn antenna was designed to fit directly onto the output of the spark gap with minimal discontinuity in impedance. This design was intended for free space operation from a balanced 50 ohm stripline feed to permit radiation of transient pulses with risetimes of less than 150 ps. A feature of this antenna was the use of Brewster Angle matching techniques to ensure a constant 50 ohm transition across the solid dielectric/air interface [8,9].

Given a pair of homogeneous dielectric materials the Brewster Angle is that angle at which a plane wave incident on a planar dielectric interface with the magnetic field in a direction parallel to the plane of the interface undergoes no reflection. For a wave passing from medium 2 of higher refractive index and permittivity ε_2 to medium 1 of lower refractive index and permittivity ε_1 the angles ψ_1 and ψ_2 are given by

$$\tan\psi_2 = \sqrt{\frac{\varepsilon_2}{\varepsilon_1}} \quad \text{and} \quad \tan\psi_1 = \sqrt{\frac{\varepsilon_1}{\varepsilon_2}}, \quad \text{where} \quad \psi_1 + \psi_2 = \pi/2 .$$

The Perspex to air transition angles are $\psi_2 = 58.7°$ and $\psi_1 = 31.3°$. In order to use these principles to match the impedances of two TEM horns which are located on either side of the media interface and meet along it consider firstly two striplines, of equal width w, in media 1 and 2 of heights h_1 and h_2 respectively. Then

$$\frac{h_1}{h_2} = \left(\frac{\sin \psi_1}{\sin \psi_2}\right) = \tan \psi_1 = \sqrt{\left(\frac{\varepsilon_1}{\varepsilon_2}\right)}$$

This implies that the impedances of the striplines, given approximately by $Z_0(h_1 / w)$ and $Z_0\sqrt{(\varepsilon_1 / \varepsilon_2)}(h_2 / w)$ when h_1, $h_2 << w$ and $Z_0 = 120\pi$ ohms, are matched across the interface (ignoring any effects arising from the finite width of the stripline). The striplines can be replaced by TEM horn elements provided their elevation and taper is shallow to ensure that the wavefront arriving at the interface is nearly planar; small deviations from planarity will produce small reflections. There must be a slight flaring of the plates through the transition section in order to maintain 50 ohms on either side of the interface. Since the flare angles do not create major discontinuities within this region, mismatch is expected to be at most 10%.

The antenna and spark gap are shown in figure 1. The antenna comprised two 50 ohm horns, one 35cm long enclosed in dielectric (Perspex) and the other in air (free space) with length 30 cm, together with a transition section. The impedance in the Perspex section was 50 ohms; the air filled section was initially a constant 50 ohms but was later modified to taper exponentially from 50 ohms to approximately 115 ohms at the aperture in order to achieve a better match to free space. The relative dielectric constant of Perspex was taken to equal 2.7, giving a height to width ratio of 3.2 for a 50 ohm structure. The Perspex filled TEM horn had 8° angular separation between the upper and lower arms (i.e. 4° elevation of each arm to the horizontal), and 25.5° apex angle. The air filled TEM horn was 8° angular separation and 46° virtual apex angle. Plate separation, h_1, at the transition exit into air was chosen as 3.7 cm.

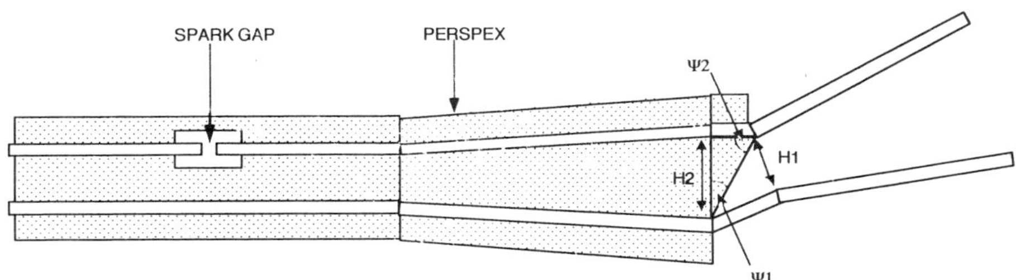

Figure 1. Diagram of Spark Gap and Antenna.

THE MODULATOR

The high PRF modulator included a 25 kV, 0.25A high voltage power supply, inductive/resistive charge circuit, an EEV 33 kV CX1159 thyratron and associated circuitry. The load for the modulator was a 1 ns long 50 ohm PFL, of capacitance 20pF and inductance 50nH, incorporating the sharpening gap. The final element of the modulator was a Blumlein line of 50 ohm output impedance connected to the PFL by 50 ohm coaxial cabling. The Blumlein comprised two 1 metre long sections of 25 ohm, 50 kV cable giving a total capacitance for the Blumlein of 350pF and a pulse length of 11 ns full width half maximum (FWHM). A hydrogen filled spark gap formed the Blumlein switch with an inductance of 20nH and risetime of 1.75 ns into the 25 ohm cable.

The Blumlein was charged through a transformer to step up the voltage from the 20 kV or less available from the thyratron to the output voltage required and the pulse transformer had a turns ratio of 5:1 with a 15 turn primary and 80 turn secondary. Due to the imperfect coupling in the transformer, however, the actual voltage gain in the circuit was 3.3:1. The primary of the transformer was connected between five Murata 2nF capacitors connected in parallel and ground. The other side of the capacitors was connected to the anode of the thyratron and the HV power supply lead.

DIAGNOSTICS

Tektronix 7250 and SCD5000 digitisers (single shot bandwidths 6GHz and 4.5GHz resp.) recorded spark gap waveforms. Picosecond Pulse Laboratories 5056 delay lines with 50 ps risetime were used with these digitisers together with low loss connecting cable (Sucoflex by Suhner). Most waveforms at high PRF were acquired with the SCD5000 digitiser capturing up to 16 consecutive traces at ten waveforms per second or once every 100 shots at 1 kHz.

Radiated far field measurements were made at a distance of 5 metres from the antenna aperture with a small conical D-dot sensor mounted in the centre of a 500mm diameter ground plane. The charging voltage waveform was measured using a capacitance divider built into the stripline and the output voltage from the spark gap was measured using small D-dot probes also built into the stripline and antenna throat sections. The D-dot probe output was integrated numerically. Our waveforms are uncorrected for diagnostic system response.

SPARK GAP RESULTS

The gap breakdown voltages at varying pressures of nitrogen using DC voltage, the modulator operated at 10 Hz and 1 kHz and the Marx are shown in figure 2. The values from the modulator are the average of ten shots captured at a rate of ten per second during a one second burst. The increased breakdown voltage under pulsed conditions relative to the DC level can be seen and was dependent on the delay time taken as the interval between the moment the DC breakdown field was reached and actual breakdown. The average delay time was 1.4 ns with an RMS jitter of 120 ps at 1 kHz PRF and 300psig. The slope of breakdown voltage against pressure for the DC case, the Marx and the modulator was very similar and showed a nearly linear dependence of breakdown voltage and pressure up to 1.8 MV/cm at 600psig when charged by the Marx.

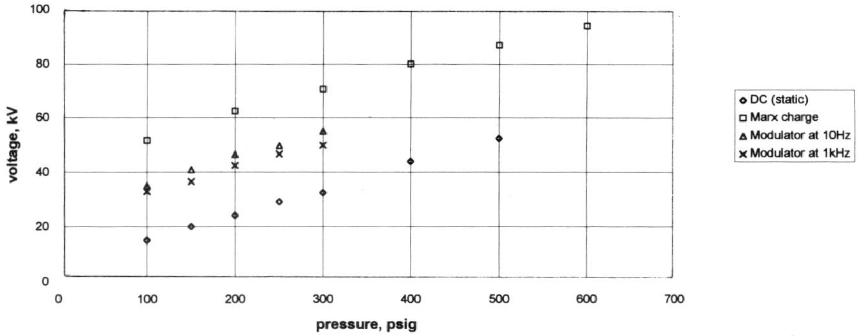

Figure 2. Breakdown Data for Nitrogen.

Figures 3 and 4 compare pulse charging waveforms, including the first, captured at a rate of ten per second during a one second burst at 10 Hz and 1kHz for nitrogen at 300psig respectively. The shot to shot repeatability in breakdown voltage (the minimum voltage in the figures) was similar when the PRF was increased from 10 Hz to 1kHz. The breakdown voltage of nitrogen was lower at 1 kHz than 10 Hz by a maximum of 9.4% at 300psig. Standard deviation of breakdown voltage at 1kHz PRF and 10 Hz PRF was 4%.

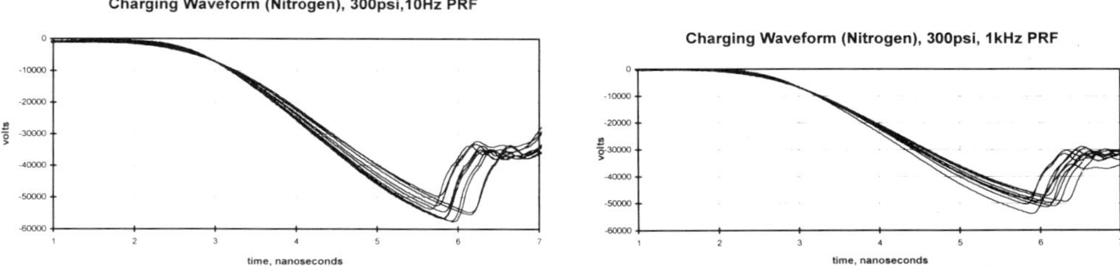

Figure 3. Nitrogen Breakdown at 10 Hz. Figure 4. Nitrogen Breakdown at 1 kHz.

Figure 5 displays the waveform output by the sharpening gap operated with nitrogen at 500 psig using the Marx as charge. The risetime was 130 ps. With the modulator as charge, a similar waveform was obtained, but having negative polarity, and a falltime of 150 ps at 300 psig and 25 kV output. At atmospheric pressure the falltime reduced to 130 ps at a lower output voltage of 15 kV.

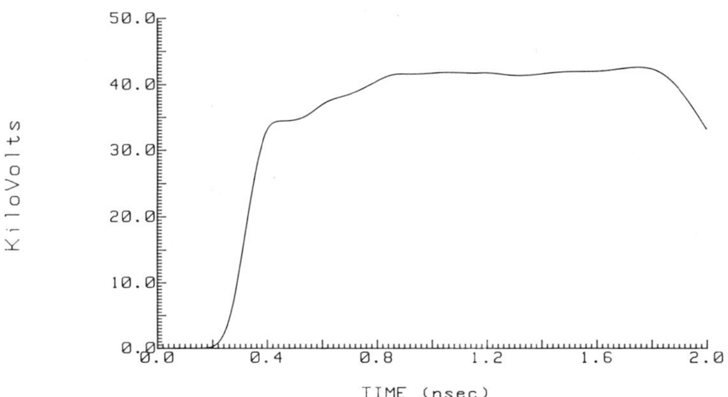

Figure 5. Spark Gap Output Waveform (500psig Nitrogen).

IMPEDANCE MEASUREMENTS

The impedance of the horn measured by time domain reflectometry (TDR) is shown in figure 6, effected by an abrupt coaxial to stripline connection of a Tektronix 11801B to the horn via a 0.5 metre length of coaxial cable. The stripline feed was decoupled from the horn for this measurement. The 50 ohm coaxial cable is visible up to the 10 ns time point, where the sharp discontinuity in the mating section of coax to horn is visible. The double transit time of the Perspex filled and air filled horns was about 6 ns. The TDR measurements indicated that over this length, corresponding to the 6 ns trace following the discontinuity, impedance varied between 46.5 and 50.5 ohms. Beyond the end of the horn, impedance rapidly increased.

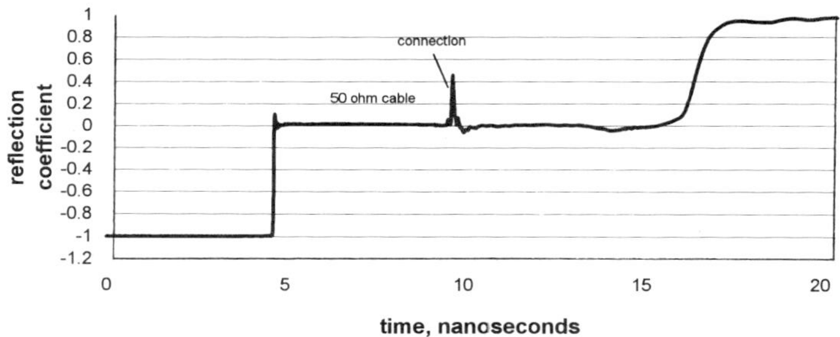

Figure 6. Impedance Measurement of Antenna using Time Domain Reflectometry.

In figure 7 is shown a 140 ps falltime pulse in the Perspex section 15cm prior to the transition obtained with the spark gap at atmospheric pressure. Perspex is not a perfect dielectric, so that some slowing of the falltime was expected, due to the non-zero loss tangent. Figure 8 shows a similar pulse in the air section after it has passed through the interface. The falltime of this pulse was 150 ps. The transient waveform generated by the source propagated successfully in TEM mode to the radiating aperture, with at most minor loss of power or slowing of wavefront falltime. It provides a nearly constant impedance transition from source via stripline to dielectric filled horn through the shaped dielectric-air interface.

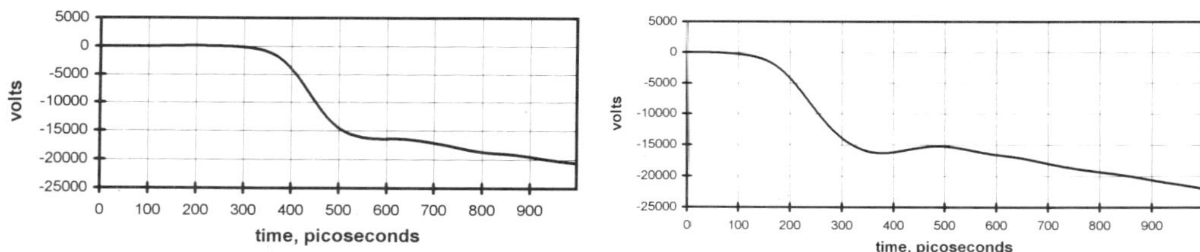

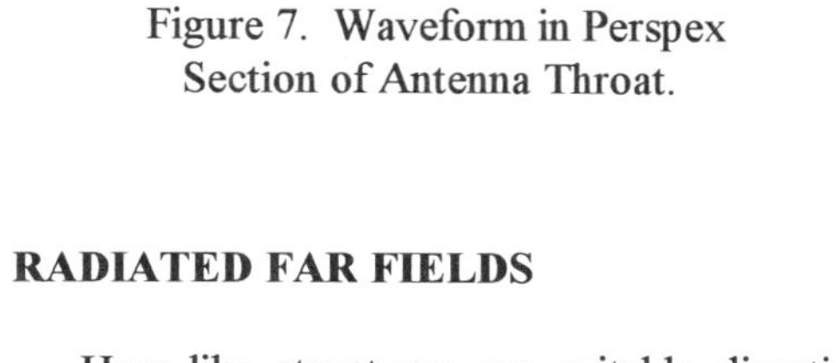

Figure 7. Waveform in Perspex Section of Antenna Throat.

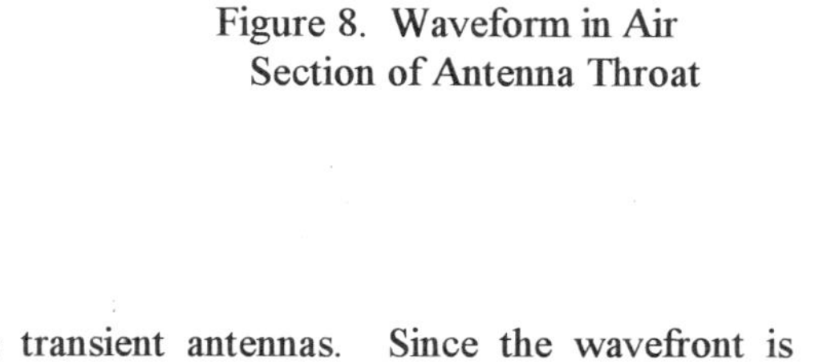

Figure 8. Waveform in Air Section of Antenna Throat

RADIATED FAR FIELDS

Horn-like structures are suitable directive transient antennas. Since the wavefront is nearly planar at the mouth of the horn, the radiated far field pattern will in early time be proportional to the *derivative* of the driving voltage (as well as inversely proportional to distance). Subsequently the reflected current waves on the horn generate later time features of the radiated field.

A simple model of the horn can be used to demonstrate why the derivative waveform is radiated. The designs are chosen to ensure, as nearly as possible, that the electric field $E^{ap}(q,t)$ impressed at the mouth of the horn, which to a first approximation, can be taken to be a rectangular aperture A (say of width w and height h), is *uniform*, and therefore equals $V(t)/h$. At a point P in the farfield, with distance R from the aperture, and at an angle θ to the axial direction of the horn in the equatorial plane, the electric field equals

$$\frac{1}{4\pi Rc}\int_A \frac{\partial E^{ap}}{\partial t}\left(q,t-\frac{|P-q|}{c}\right)dS_q,$$

where the point q ranges over the surface A. On axis ($\theta = 0$) this equals

$$\frac{hw}{4\pi Rc}\frac{\partial E^{ap}}{\partial t}=\frac{w}{4\pi Rc}\frac{\partial V(t-R/c)}{\partial t}.$$

A successful horn design must therefore ensure that (i) the aperture is matched to the source, (ii) the wavefront does not significantly deviate from being planar at the aperture, and (iii) the reflection at the termination is minimised. The first condition is guaranteed by the high frequency independent impedance of these TEM horns. The second puts a constraint on the maximal useful length of the horn for given apex angle 2ψ. The final condition is addressed by modifying the end shape of the horn, or by various types of loading. If the source waveform is modelled as proportional to the error function $\frac{2}{\sqrt{\pi}}\int_0^{t/t_0}\exp(-u^2)du$ (plus a DC offset), for a suitable time constant t_0, the 10-90% falltime should agree (to within 10%) with the FWHM of the radiated (derivative) waveform. Figure 9 shows the radiated pulse at 5 metres from the aperture. The applied pulse had 15 kV amplitude. The radiated pulse had a peak field strength of 3.8 kV/m and 150 ps FWHM, in good agreement with prediction.

Surface loading or resistive termination of the antenna demonstrably suppresses later time reflections. In particular two 100Ω resistors connecting the vertices of the aperture to ground provide in effect a matched 50Ω load for the low frequency components for which the antenna should be regarded more nearly as a transmission line, rather than a radiating structure. This arrangement has the advantage of providing a simply implemented termination and does not reduce the peak radiated field strength as does a continuous surface loading (such as the Wu-King profile) [10]. Alternative terminations have also been considered [11].

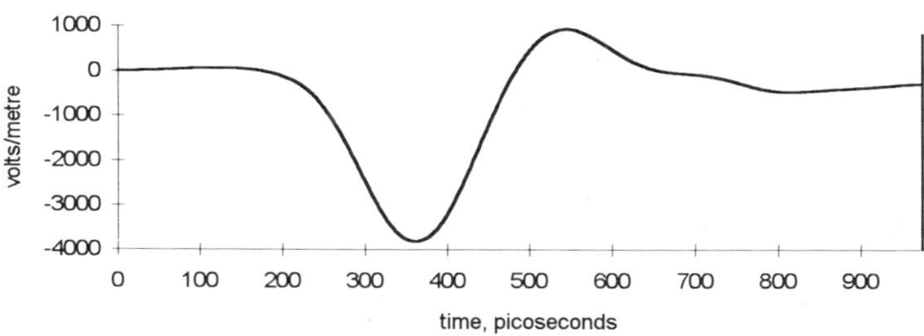

Figure 9. Radiated Far-Field.

DISCUSSION

Nitrogen showed good recovery characteristics when operated at 1 kHz PRF; it recovered to over 90% of its voltage hold-off compared to that at 10 Hz PRF. In order to achieve the highest over-voltages the time for the charge voltage to reach its peak should be equal to, or less than, the delay time. Substantial improvements in applied voltage and output risetime should be achieved if the charging voltage risetime were to be reduced to below 1 ns.

Operation at higher PRF and higher voltages may require greater plate separation in the air section. Alternatively, this section might be enclosed in pressurised sulphur hexafluoride for greater breakdown strength. Faster risetimes will require a dielectric with lower loss than Perspex (e.g. PTFE or a liquid dielectric, suitably contained); the size of the horn must also be reduced in order to radiate the derivative of the spark gap waveform.

CONCLUSIONS

A pulse generator and antenna utilising a high pressure gas sharpening gap is described. The sealed spark gaps operate at over 25 kV output at 1 kHz pulse repetition frequency with falltimes of below 150 ps using nitrogen at up to 300psig without gas flow. The antenna is directive, preserves the fast risetime of the source, radiates a transient waveform very nearly the derivative of the applied voltage and is able to withstand a 25 kV voltage at 1 kHz PRF. The antenna represents a constant impedance transition from source to radiating aperture and the transient waveform generated by the source propagates successfully in TEM mode to the radiating aperture with at most minor loss of power or slowing of wavefront risetime. Upon using a small Marx generator to pulse charge the gas sharpening gap at 1 Hz PRF, its operating range was extended to 45 kV output at higher pressures (600 psig), whilst preserving the fast risetime.

REFERENCES

1. D.M. Parkes, M.F. Lewis, R.L.S. Devine, K. Trafford and D. Richardson, Practical Measurements using Ultrawideband Radar. Ultrawideband Radar, SPIE Vol. 1631, pp 232-242 (1992).
2. E.G. Farr and C.E. Baum, Prepulse associated with the TEM feed of an Impulse Radiating Antenna. Sensor and Simulation Note 337 (1992).
3. A.P. Lambert, S.M. Brooker and P.D. Smith, Calculation of the Characteristic Impedance of TEM Horn Antennas using the Conformal Mapping Approach. IEEE Trans. Antennas Propagat. AP-43, 47-53 (1995).
4. F.C. Yang and K.S.H. Lee, Impedance of a two-conical-plate transmission line. Sensor and Simulation Note 221 (1976).
5. C.A. Frost, T.H. Martin, P.E. Patterson, L.F. Rinehart, G.J. Rohwein, L.D. Roose, J.F. Aurand, M.T. Buttram, Ultrafast gas switching experiments, Proc. 9th IEEE Pulsed Power Conference, pp 491-494 (1993).
6. W.R. Cravey, E.K. Freytag, D.A. Goerz, P. Poulsen, P.A. Pincosy, Picosecond High Pressure Gas Switch Experiment, *Proc. 9th IEEE Pulsed Power Conference*, pp483-486 (1993).
7. Baum C.E., Brewster-angle interface between flat-plate conical transmission lines. *Sensor and Simulation Note* **389** (1995).
8. C.E. Baum and A.P. Stone, *Transient Lens Synthesis: Differential Geometry in Electromagnetic Theory*, Hemisphere Publishing Corporation (1991).
9. C.J. Brooker and P.D. Smith, Ultra Wide Band Antenna, *UK Patent Application 9410274*, (1994).
10. Booker S.M., Lambert A.P. and Smith P.D., A calculation of surface impedance effects on transient antenna radiation. Proc. International Symposium on Electromagnetic Theory (URSI), pp 480-482 (1995).
11. Baum C.E., Low frequency compensated TEM horn. *Sensor and Simulation Note* **377** (1995).

ANTENNAS AND ELECTRIC FIELD SENSORS FOR ULTRA-WIDEBAND TRANSIENT TIME-DOMAIN MEASUREMENTS: APPLICATIONS AND METHODS

C. Jerald Buchenauer,[1,2] J. Scott Tyo,[1] and Jon S. H. Schoenberg[1]

[1]Phillips Laboratory / WSQW
Kirtland AFB, NM 87117

[2]Los Alamos National Laboratory, NIS-9
Los Alamos, NM 87545

ABSTRACT

Many time-domain electromagnetic measurements require sensors that generate accurate signals proportional to the incident electric field for some finite clear time, after which the response may be of little interest, except for a possible frequency-domain requirement on the damping of resonances. In a review of earlier work,[1] examples of such devices are given that combine more conventional antennas with open transmission lines. In designs that can have highly directional properties, antenna effective height h_{eff}, risetime t_r, and clear time t_c may be chosen independently. Current work focuses on extending the parameter range of these sensors to greater sensitivity and shorter risetimes, where sensor performance becomes limited by the effects of skin and dielectric loss and dispersion. These limitations are largely overcome through the use of guided-wave optics in sensor designs.

INTRODUCTION

Alternatives to the popular B-dot and D-dot time-differentiating electromagnetic field sensors are of interest for measuring low-amplitude or nonrepetitive radiated-field signatures having ultrawide bandwidths and moderately long durations. A sensor with desirable properties might generate a signal that accurately replicates the incident electric field and preserves its time integral for some finite clear time t_c. This may be achieved with a transmission-line E-field sensor illustrated in Fig. 1. The transmission-line electrode is assumed to be highly conducting, thin compared with other system dimensions, and everywhere perpendicular to the incident electric field. After the step-function electromagnetic wave passes, the electrode is raised to an electrical potential hE. No net surface currents flow on the electrode, and no appreciable scattering of the incident wave occurs until, at $t = 0$, the wave passes the feed point, where a yet unspecified electrical structure transports a signal current to the ground-plane output port. The prompt early-time response of the signal current depends critically upon the design of this transmission-line to feed-point connection and upon the fields in the

Ultra-Wideband, Short-Pulse Electromagnetics 3
Edited by Baum *et al.*, Plenum Press, New York, 1997

immediate proximity that are incident upon this structure. The late-time ($t_r \ll t \leq t_c$) sensor response is insensitive to the incident-wave propagation direction in the plane and is determined by the properties of the open transmission line and its ability to propagate without reflection an outward traveling wave. The measured signal current, which drives this outgoing wave, persists unaltered so long as no reflection returns to the feed point. If the stripline is of uniform impedance Z, the signal amplitude will be

$$V_0(t) = h_{\text{eff}} E(t), \quad \text{where} \quad h_{\text{eff}} = \frac{Z_0}{Z_0 + Z} h. \tag{1}$$

The signal V_0 is delivered to the feed impedance Z_0 for a clear time equal to twice the electrical length of the line $t_c = 2L/c$, where c is the speed of light. For matched impedances, $Z = Z_0$, the signal is $V_0 = hE/2$ during the clear time and rapidly approaches zero thereafter.

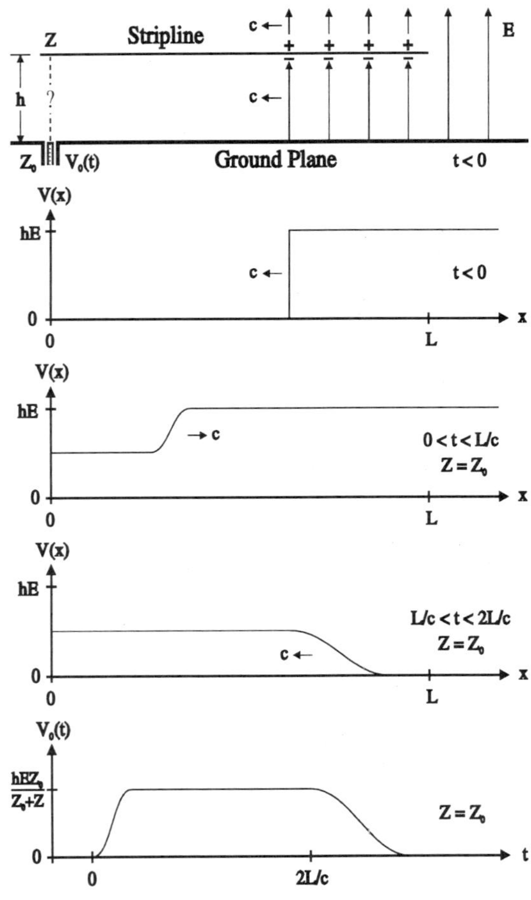

Figure 1. Behavior of a transmission-line E-field sensor in the presence of a step-function incident field. The signal current is sustained by an outgoing wave on the open transmission line for a time $2L/c$.

TEM-HORN TRANSMISSION-LINE E-FIELD SENSORS

Sensors with highly directional properties that allow sensitivity, risetime, and clear time to be chosen independently are shown in Figs. 2 through 5. The unbalanced ground-plane-mounted TEM-horn transmission-line E-field sensor of Fig. 2 consists of two flat-plate conical electrodes joined at their intersection line: the first with its vertex at the feed point; and the second, if it were extended, with its vertex at the source point. For plane-wave measurements, the second conical plate becomes a parallel-plate stripline, as shown in Fig. 3 for a balanced sensor to be used for free-field measurements. The TEM-horn bistripline E-field sensor of Fig. 4 divides the transmission-line section into two striplines that are directed parallel the incident H-field. These TEM-horn transmission-line sensors produce accurate results when properly configured for specific rf environments. The TEM-horn transmission-line sensor response is sensitive to longitudinal field gradients and insensitive to transverse field gradients,[1] but this sensor can be configured for specific field gradients[2,3] as shown in Fig. 2. With a very long forward-looking transmission line, the risetime of the incoming electromagnetic wave may degrade because of dissipation in the electrodes due to skin resistance. This does not occur with the TEM-horn bistripline sensor with its transmission lines oriented transverse to the propagation direction. The bistripline sensor response is insensitive to longitudinal field gradients and first-order transverse field gradients, but it is sensitive to higher order transverse field gradients, the effects of which can be eliminated in some cases by constructing the stripline in an arc centered at the source point.[3]

The results of a sensor risetime scaling study were presented in an earlier work.[1] The risetimes of TEM-horn transmission-line sensors were expressed in terms of the difference between the flight times of signals traveling along the direct path S and an indirect path S' shown in Fig. 2.

$$t_r = t_{rise}\,(10 - 90\%) = \frac{0.8\,[S'\,(S, d, b, x) - S]}{c}. \tag{2}$$

The quantity $\eta = \frac{x}{a}$ was used as a scaling parameter because it was known from theory to be bounded as $\frac{a}{b}$ approaches zero. The scaling results may be expressed as

$$\eta = \frac{x}{a} = 0.65 + 0.1\,\frac{a}{b} \pm 0.1, \quad \text{for } 1.0 \leq \frac{a}{b} \leq 2.5, \text{ and } \theta \leq 30^\circ, \tag{3}$$

which covers a useful range of TEM-horn transmission-line-sensor impedances from 50 to 100 ohms for unbalanced sensors or 100 to 200 ohms for balanced sensors. For

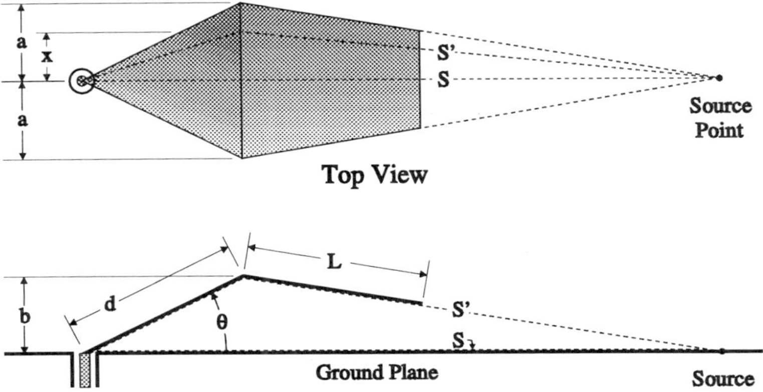

Figure 2. Unbalanced TEM-horn tapered-transmission-line E-field sensor for spherical wave measurements on a ground plane, with dimensional parameters defined.

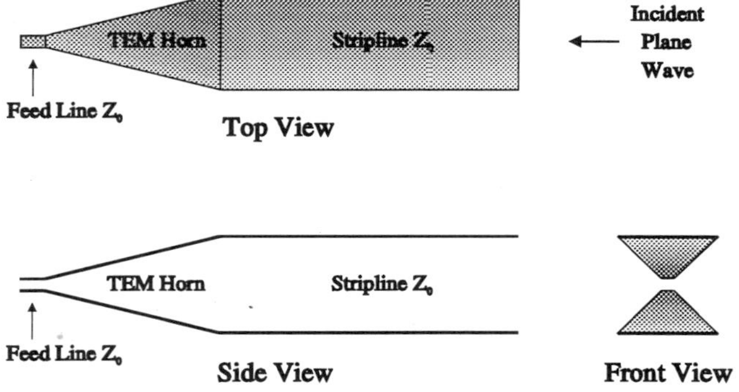

Figure 3. Balanced TEM-horn transmission-line E-field sensor for free-field plane-wave measurements.

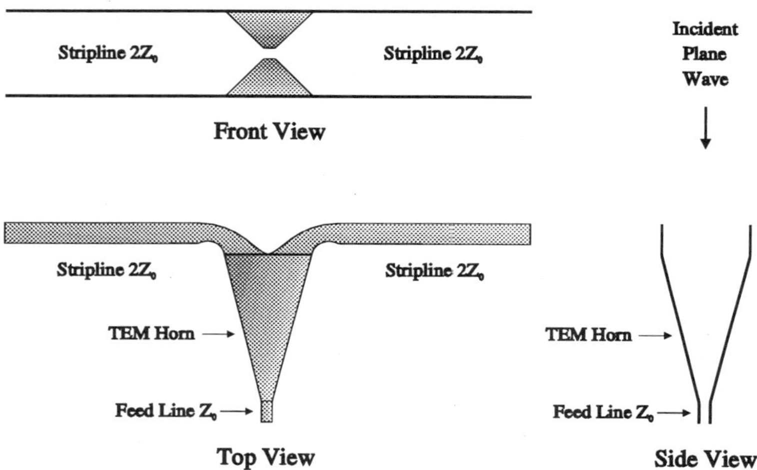

Figure 4. Balanced TEM-horn bistripline-line E-field sensor for free-field plane-wave measurements.

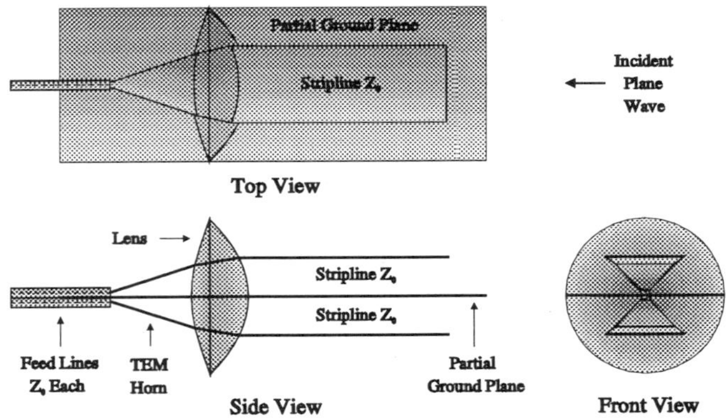

Figure 5. Balanced TEM-horn transmission-line E-field sensor with lens for free-field plane-wave measurements. The electrodes pass through the lens along the paths of classical light rays, not exactly as shown. The partial ground plane is a special option.

the simplified case of a plane wave ($S \rightarrow \infty$) incident upon a small-angle TEM-horn transmission-line sensor ($a \ll d$ and $b \ll d$), Eq. (2) may be approximated as

$$t_r = t_{rise}\,(10 - 90\%) \approx \frac{0.8\,(b^2 + \eta^2 a^2)}{2dc}. \tag{4}$$

One might conclude from Eqs. (2) and (4) that sensor sensitivity and risetime may be chosen independently so long as η remains bounded. However, for a fixed risetime, the horn length must increase as the square of the sensitivity or effective height. In addition to practical limits on size, very long horns may exhibit undesirable off-boresight responses as well as degraded risetime performance due to skin and dielectric loss and dispersion. It was observed with horns of length $d \approx 1$ m and heights b or h of a few centimeters that risetimes began to exceed the scaling-equation predictions. The excess risetime was assumed to be due to conductor skin loss. To minimize the effects of skin loss, the horn length must be kept short. For very short risetimes, this can be done only by introducing optical elements into the design, as shown in Fig. 5.

CONDUCTOR SKIN-LOSS EFFECTS

Linear Transmission Lines

The time-domain response of a coaxial transmission line with skin resistance has been derived by Wigington and Nahman.[4] For time intervals much smaller than the magnetic diffusion time ($\mu\sigma\Delta^2 \gg t \geq 0$), the impulse response is given by

$$h\,(t) = \sqrt{\frac{t_s}{\pi}}\, t^{-\frac{3}{2}}\, e^{-\frac{t_s}{t}}, \tag{5}$$

and the unit step response is

$$g\,(t) = \int_0^t h\,(t')\,dt' = \mathrm{Cerf}\left[\sqrt{\frac{t_s}{t}}\right] = 1 - \frac{2}{\sqrt{\pi}} \int_0^{\sqrt{t_s/t}} dx\, e^{-x^2}, \tag{6}$$

where $t = 0$ corresponds to the high-frequency group arrival time, and

$$t_s = \frac{\mu}{\sigma}\left[\frac{l}{4\,Z_0\,W_{dis}}\right]^2. \tag{7}$$

Δ is the smaller of the conductor separation distance, the inner-conductor radius, or the outer-conductor thickness, Z_0 is the transmission line impedance, l is its length, σ and μ are the electrical conductivity and magnetic permeability of the conductors, and $\frac{1}{W_{dis}} = \frac{1}{2\pi r_1} + \frac{1}{2\pi r_2}$ for a coaxial transmission line with conductor radii r_1 and r_2. These expression may be generalized to transmission lines of arbitrary geometry by defining W_{dis} as the effective dissipation width for the transmission line; i.e. if the transmission-line current were uniformly distributed over a width of conductor W_{dis}, the dissipation would be the same as that of the actual transmission line. For an arbitrary two-conductor transmission line

$$\frac{1}{W_{dis}} = \sum_{k=1}^{2} \frac{1}{W_{dis[k]}}, \quad \text{with } W_{dis[k]} = \frac{\left[\oint ds_k\, J_k\,(s_k)\right]^2}{\oint ds_k\, J_k^2\,(s_k)}, \tag{8}$$

where $J_k(s_k)$ is the surface current distribution on the k'th conductor, and the closed path integrals are taken around the conductors over paths perpendicular to the currents.

Several commonly-used risetimes of the step response function due to skin resistance are

$$\begin{aligned}\tau_{10-90\%} &= 125.9\,t_s,\ \tau_{20-80\%} = 29.9\,t_s,\ \tau_{0-50\%} = 4.396\,t_s,\\ \tau_{1/2} &= 1.80\,t_s,\ \text{and}\ \tau_{dir} = [\max h(t)]^{-1} = 4.325\,t_s.\end{aligned} \tag{9}$$

$\tau_{10-90\%}$, $\tau_{20-80\%}$, and $\tau_{0-50\%}$ are based upon the time intervals between the indicated relative amplitudes of the step response $g(t)$. $\tau_{1/2}$ is the full width at half maximum of the impulse response $h(t)$, and τ_{dir} is the derivative risetime derived from the maximum slope of $g(t)$. There are large differences between differently defined risetimes, which are all proportional to the *square* of the cable length to width ratio: $t_s \propto \left(\frac{l}{W}\right)^2$.

Conical Transmission Lines

The TEM horns in the antenna / sensors described thus far consist of tapered transmission lines in which the cross-sectional dimensions depend upon an axial coordinate z. To obtain a value of t_s for tapered lines, Eq. (7) must be modified by integrating over the length of the line $l = z_2 - z_1$, i.e.

$$t_s = \frac{\mu}{\sigma}\left[\frac{1}{4Z_0}\sum_{k=1}^{2}\int_{z_1}^{z_2}\frac{dz}{W_{dis[k]}(z)}\right]^2 = \frac{\mu}{\sigma}\left[\frac{l}{4Z_0}\sum_{k=1}^{2}\left\langle W_{dis[k]}^{-1}\right\rangle\right]^2. \tag{10}$$

The sum is over both electrodes, and $\langle W_{dis}^{-1}\rangle$ is defined as a mean inverse dissipation width. In transmission lines with precisely conical symmetry, all cross-sectional dimensions such as the physical width $W(z)$ are proportional to the axial coordinate z. Consequently, $W_{dis}(z) \propto W(z) \propto z$, and the integration over z yields

$$l\left\langle W_{dis}^{-1}\right\rangle = \frac{1}{Y}\left[\frac{z_2 - z_1}{W(z_2) - W(z_1)}\right]\ln\left[\frac{W(z_2)}{W(z_1)}\right],\ \text{with}\ Y(z) = \frac{W_{dis}(z)}{W(z)}. \tag{11}$$

$W(z_1)$ and $W(z_2)$ are the physical widths at the ends of the line, $l = z_2 - z_1$ is the line length, and $Y(z)$ in conical symmetry is a geometry-dependant constant independent of z.

In practice TEM-horn electrodes are rarely made with exactly conical symmetry. While the physical widths $W(z)$ may be proportional to z, the electrode thicknesses are generally constant. This leads to a $Y(z)$ or ratio of W_{dis} to W that is z dependent. This problem will be treated for flat-plate transmission lines in the next section. Our tentative findings suggest the following empirical relationship for thin conical flat plates of uniform thickness:

$$[Y(z)]^{-\beta} = \left[\frac{W_{dis}(z)}{W(z)}\right]^{-\beta} \propto \ln(\alpha z),\ \text{and}\ W_{dis}(z) \propto \frac{W(z)}{[\ln(\alpha z)]^{\frac{1}{\beta}}} \propto \frac{z}{[\ln(\alpha z)]^{\frac{1}{\beta}}}, \tag{12}$$

where α and β are constants with $1 \le \beta \le 2$, and $\beta \to 1$ for low impedance lines and $\beta \to 2$ for high impedance lines. Using the above expression in Eq. (10), we obtain

$$l\left\langle W_{dis}^{-1}\right\rangle = \left[\frac{\beta}{1+\beta}\right]\left[\frac{[Y(z_2)]^{-(1+\beta)} - [Y(z_1)]^{-(1+\beta)}}{[Y(z_2)]^{-\beta} - [Y(z_1)]^{-\beta}}\right]\left[\frac{z_2 - z_1}{W(z_2) - W(z_1)}\right]\ln\left[\frac{W(z_2)}{W(z_1)}\right]. \tag{13}$$

The transient responses of conical lines will be the same as those of linear lines with the same values of t_s. The above expressions are strictly valid only within the transmission-line approximation, which requires that the shortest radian wavelength be larger than the transverse dimensions of the transmission line. A large-aperture TEM horn will violate this condition near its exit aperture, where the effects of edge dissipation cannot propagate across the entire wave front. However, for a conical transmission line with a large width ratio, most of the dissipation occurs at the smaller end of the line. Thus, the errors introduced by applying Eqs. (10) through (13) to large TEM horns should not be large, but these expression must be applied with caution. The effects of edge dissipation on aperture field distributions and gain of focused aperture antennas remains a subject for future study.

Anomalous Edge Losses

In transmission lines and antennas consisting of flat plates, the currents tend to concentrate at the plate edges. These concentrated currents lead to enhanced losses and longer risetimes. For a thin flat conducting plate of width $2a$ that is far from other electrodes, the surface current density is approximately given by[5]

$$J(x) \simeq \frac{J(0)}{\sqrt{1-\left(\frac{x}{a}\right)^2}}, \tag{14}$$

where x is the distance from the center of the plate. The denominator integral in Eq. (8) does not converge in the thin-plate approximation, and

$$W_{dis} = \frac{\left[\oint ds\ J(s)\right]^2}{\oint ds\ J^2(s)} \to 0. \tag{15}$$

Therefore, the dissipation must depend on thickness even when the thickness is much larger than the skin depth and much smaller that the width.

A numerical analysis was performed to determine the actual current distributions and dissipation widths for thin flat-plate striplines. The electric field of the TEM mode on a two-conductor structure satisfies the two-dimensional Laplace equation. Using a finite-element-method (FEM) code available with the Matlab (ver. 4.2c) Partial Differential Equation Toolbox (ver. 1.0), the two-dimensional Laplace equation can be solved numerically for arbitrary conductor cross sections that cannot be solved analytically. The magnitude of the current density is proportional to the magnitude of the E-field for the TEM mode, so $E(s)$ can be substituted for $J(s)$ in Eq. (8) for W_{dis}.

To account for edge effects in a thin flat plate, the sharp edge is replaced by a semicircle with a diameter equal to the plate thickness. The effects of thickness are first examined by analyzing a modified coaxial transmission line consisting of a cylindrical outer conductor and a flat-plate inner conductor shown in Fig. 6a. The dissipation width of the inner conductor is determined as a function of inner-conductor thickness using Eq. (8).

To establish the accuracy of the FEM code, computations were performed on a pair of confocal ellipses. For a confocal elliptical geometry, W_{dis} can be determined analytically in terms of complete elliptic integrals of the first kind.[6] For inner conductor thicknesses in the range of those studied in this investigation, the FEM code was in agreement with the analytic solutions to within 1%. A finer FEM mesh is needed to achieve the proscribed accuracy as the thickness of the conductor is decreased.

Figure 7 shows the results of the FEM calculations on the modified coaxial transmission line. W_{dis} is plotted as a function of the inner-conductor thickness-to-width

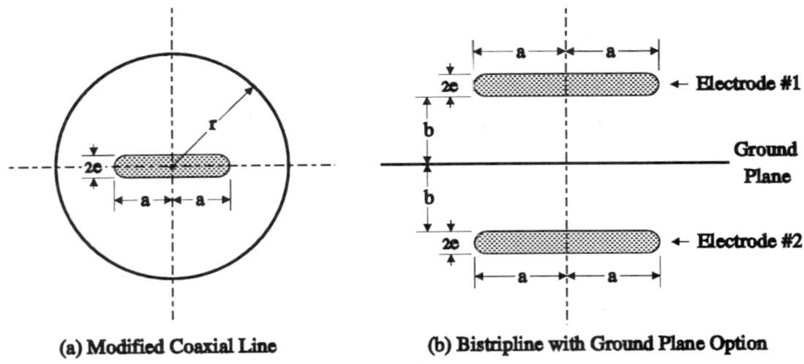

Figure 6. Transmission-line geometries and dimensional parameters used in the FEM numerical dissipation-width computations.

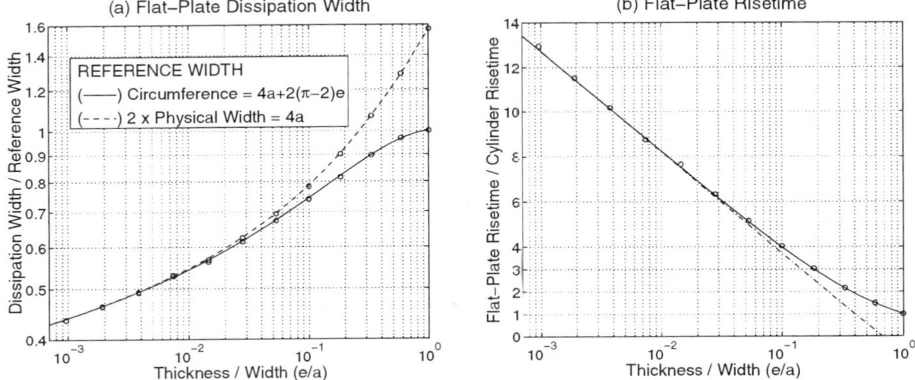

Figure 7. Dissipation widths (a) and risetimes (b) for the modified coaxial transmission line of Fig. 6a. The data points represent the FEM computed values, which closely fit the curves given by the empirical releatioship: $\ln\left[1+\frac{2a}{3e}\right] W_{dis}^2\left(\frac{e}{a}\right) = \ln\left[\frac{5}{3}\right] W_{dis}^2(1)$, where the dissipation width for a circular cylinder ($\frac{e}{a}=1$) is $W_{dis}(1) = 2\pi a$. Plot (b) shows the risetimes due to dissipation in the center electrode. In this log-linear plot, the dashed curve is an asymptotic straight-line fit given by $\tau'_{rise}\left(\frac{e}{a}\right) = \ln\left[\frac{2a}{3e}\right] \tau_{rise}(1)$.

ratio in Fig. 7a. The ratio of the inner diameter of the outer conductor to the width of the inner conductor was held fixed at $\frac{r}{a} = 2$. A thickness-to-width ratio $\frac{e}{a}$ of one corresponds to a cylindrical coaxial transmission line. Figure 7b shows a plot of the normalized risetime attributed to skin dissipation on the inner conductor only. The results indicate that a thickness-to-width ratio of 1 % results in a factor of eight increase in risetime relative to that for a cylindrical coaxial transmission line.

Next, the parallel-plate transmission-line of Fig. 6 b was analyzed. The unbounded geometry of this structure required special techniques in order to apply the FEM code, which requires a bounded two-dimensional structure. One quadrant of the transmission line was analyzed with a circular outer boundary. By using an iterative procedure, the potential of this circular outer boundary was determined. The procedure involved choosing a functional form for the electrostatic potential on the outer boundary, solving for the charge distribution on the electrode, and updating the potential on the outer boundary. The iterative procedure was determined to converge rapidly to its final value. The accuracy of the iterative procedure was verified by comparing calculated values of the line impedance to theoretical values, which are available for zero-thickness

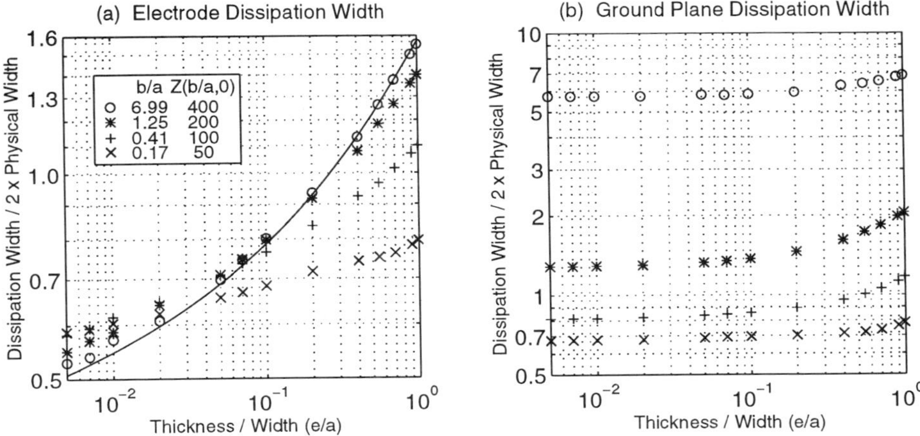

Figure 8. Dissipation widths as a function of electrode thickness for the electrodes and ground plane of the stripline in Fig. 6b. The data points represent the FEM computed values. The data points for a single electrode are shown in plot (a). The solid curve is the same as the dashed curve in Fig. 7a for the modified coaxial line. The balanced stripline impedance $Z\left(\frac{b}{a}, \frac{e}{a}=0\right)$is the theoretical value at zero thickness. The ground plane single-surface dissipation width relative to twice the stripline width $4a$ is shown in plot (b).

conductors.[7] The nature of the FEM solution allowed for calculation of W_{dis} for the transmission-line electrodes, as well as for a ground plane. For the geometries studied, W_{dis} was smaller for the electrodes than the ground plane as expected, but the effects of a ground plane were not negligible. The results of the FEM analysis for parallel-plate lines are presented in Fig. 8. Several different geometrical configurations were analyzed, and the results indicate that the dissipation width does decrease significantly with decreasing plate thickness.

Dissipation-width data in Fig. 8a for $\frac{e}{a} \lesssim \frac{1}{10}$ fits the scaling relationship expressed in the previous section by Eq. (12). Therefore, Eq. (13) may be used in computing $l\left\langle W_{dis}^{-1}\right\rangle$ for thin flat conical plates of uniform thickness. The data in Fig. 8b for ground planes are nearly constant in the same parameter range. Thus, dissipation in a ground plane near a thin flat conical plate of uniform thickness will be identical to that of a perfectly conical electrode, and Eq. (11) may be used to compute $l\left\langle W_{dis}^{-1}\right\rangle$ for these surfaces. Further examination of Fig. 8 reveals that the dissipation width of an electrode is larger than the actual physical width of the plate for the geometrical configurations studied $\left(\frac{1}{6} \leq \frac{b}{a} \leq 7, \frac{1}{200} \leq \frac{e}{a} \leq 1\right)$. If one were to guess at what the dissipation width should be without taking edge effects into account, one might guess that the current is flowing approximately evenly over the inside face of the electrode. In fact, the current is not distributed in this manner except in the very low impedance limit. However, the aggregate effect of the actual current distribution with the enhanced edge losses yields a dissipation width that is close to this naively approximated value for thin plates.

DIELECTRIC LOSS EFFECTS

Unlike the formalism that describes the transient response due to skin loss, there are no convenient universal formulas that describe the time-domain behavior of all dielectrics over wide parameter ranges. For frequencies below 1 GHz, risetimes greater than 1 ns, and long transmission lines, qualified universal dielectric responses have been identified and characterized in the time domain.[8–10] For the purpose of developing the widest-bandwidth low-dispersion antennas possible, our interest lies in cable lengths and lens thicknesses of less than 1 m, risetimes much less than 1 ns, and frequencies above 1 GHz, where broad resonances and impurity effects complicate the temporal response. In this parameter range, the exact form of the transient dielectric response will be very material dependent. Despite these difficulties, there are some generalizations that can be made at this stage of our ongoing work.

During the time scale of interest, dielectric loss is a bulk phenomenon. In characterizing transmission lines, TEM horns, and optical elements, one is not constrained to operate within the transmission-line approximation. However, at late times or in coarsely inhomogeneous materials, lateral surface polarization may appear, and, as with the skin effect, this is not a homogenous process. Also, transverse differences in absorption and reflection and some subsequent mode conversion can occur in large variable-thickness optical elements.

Two classes of dielectrics are of interest: low-loss polymers and synthetics. The popular plastics used in microwave work are polyethylene, polystyrene, and "Teflon" polytetrafluoroethylene . These are extremely low loss materials in pure form, with loss tangents no larger than a few times 10^{-4} in the frequency range from below 100 MHz to at least 25 GHz.[14,15] These losses are so small that the risetime limitations due to dielectric loss in short transmission lines or thin lenses will be determined only by absorption at much higher frequencies, where data is scarce, or by impurities and molecular damage in the polymers. Absorption of water by foams made from these polymers and even by solid polystyrene causes strong increases in absorption at audio frequencies as well as frequencies above 1 GHz[15]. Liquid water has a strong broad absorption band near 25 GHz.

Synthetic dielectrics have the potential to reduce the weight of large optical elements. Such materials might consist of suspended metal particles or dispersed oriented filaments of high-dielectric-constant material. Absorption and dispersion will undoubtedly be the major factors limiting the use of such materials in time-domain sensors and antennas. Characterizing their transient responses will be the subject of future work.

An important consideration in design optimization is the fact that dielectric transient responses will not generally scale with length or thickness in same way as the transient response due to skin loss. Nahman[8,9] has treated the problem of a frequency-dependent attenuation constant $\gamma \propto \omega^m$, where $0 < m < 1$. He derived a relationship for the risetime that scales with cable length as $t_r \propto l^{1/m}$. Data on many dielectrics[10–13] at frequencies below 1 GHz suggest a value of $m \lesssim 1$, which gives risetimes that increase only slightly faster than cable lengths: $t_r \gtrsim t \propto l$. The Debye dipole relaxation model does not work well to explain dielectric losses in most solids, but it does approximate the response of dilute solutions of polar molecules such as water. When strongly absorbing, such a mixture has an impulse response that approaches a Gaussian function, whose width squared is proportional to cable length. From this model one gets a risetime that scales as $t_r \propto \sqrt{l}$. Consequently, risetimes due to dielectric loss are likely to increase more slowly with cable length than they would with skin loss alone, where $t_r \propto l^2$. It is therefore possible to see a situation in which risetimes of short cables are dominated by dielectric losses, while those in equivalent longer cables are dominated by skin losses.

At this time it is appropriate to comment on the applicability of the quadrature method of combining the risetimes of series filter elements, which computes the net risetime to be the square route of the sum of the squares of the component risetimes. We see that this method would be accurate only for the rarely applicable Debye model.

Indeed, *the quadrature method should not be used* in computing risetimes due to skin or dielectric loss in most cases.

The question remains, What will be the transient dielectric response of low-loss materials well under one meter thick? We speculate that the initial step response may be very fast, limited only by weak absorption in the infrared spectrum and by impurities or molecular defects. This initial rise may occur in a few picoseconds, but it will be followed by a smaller slower rise that stretches out over periods of nanoseconds. This slower rise is derived from the lower-frequency collective mechanisms described earlier.[11–13] In future work we will attempt to fill in the missing details that will facilitate a basis for engineering design and optimization.

FOCUSED-APERTURE TEM-HORN TRANSMISSION-LINE SENSORS

To achieve maximum sensitivity and minimal risetime, a focusing lens must be used to reduce the horn length. The subsequent reduction in conductor skin losses must be balanced against any increased dielectric losses in the lens to achieve the greatest ratio of effective height to risetime. A schematic illustration of a lens used in a replicating E-field sensor was shown in Fig. 5. The electrodes pass through the lens along the paths of classical light rays. If the lens is lossless and sufficiently large, the sensor will behave as desired: it will produce a signal that precisely replicates the E-field incident from the boresight direction for the specified clear time. How large the lens should be and how it should be truncated to achieve the desired result, is a nontrivial question.

Figures 9 and 10 illustrate the problems caused by lenses with finite apertures. By blocking regions of the aperture with net positive or negative vertical field components,

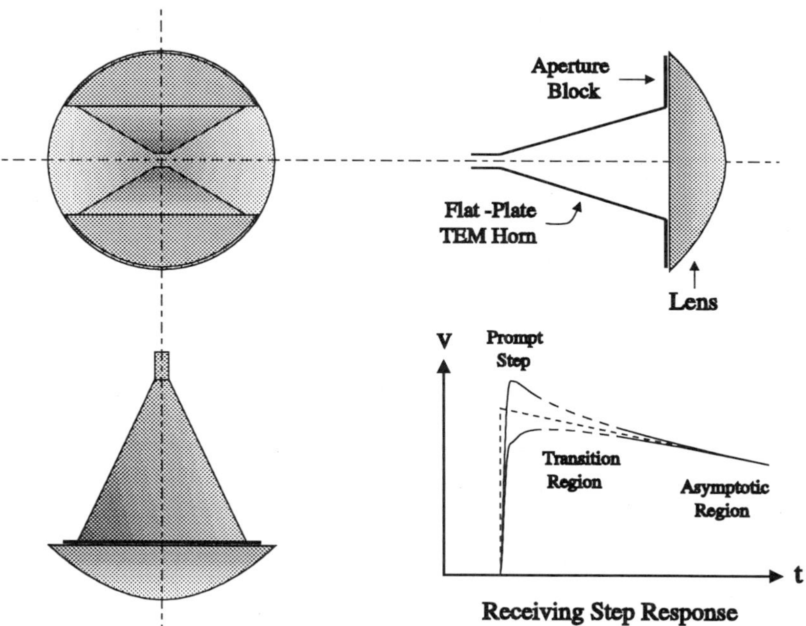

Figure 9. TEM-horn E-field sensor with lens and optional aperture block. The relative amplitude of the early- and late-time responses depends critically upon lens size and aperture construction. The intermediate-time response is is unknown (Transition Region).

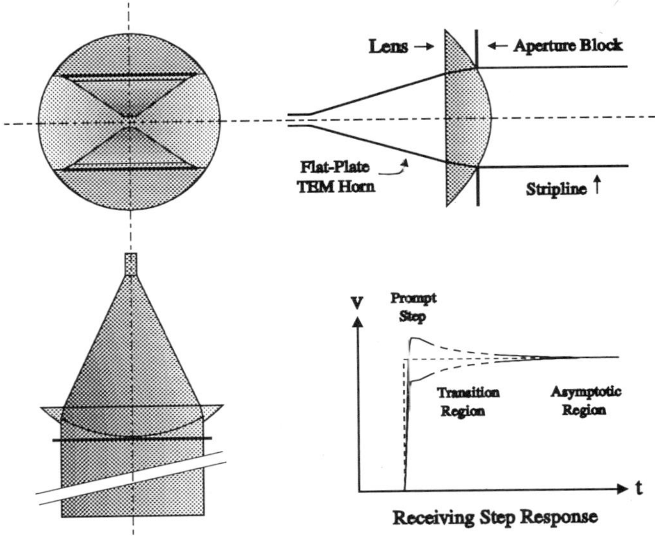

Figure 10. TEM-horn transmission-line E-field sensor with lens and optional aperture block. The relative amplitude of the early- and late-time responses depends critically upon lens size and aperture construction. The intermediate-time response is unknown (Transition Region).

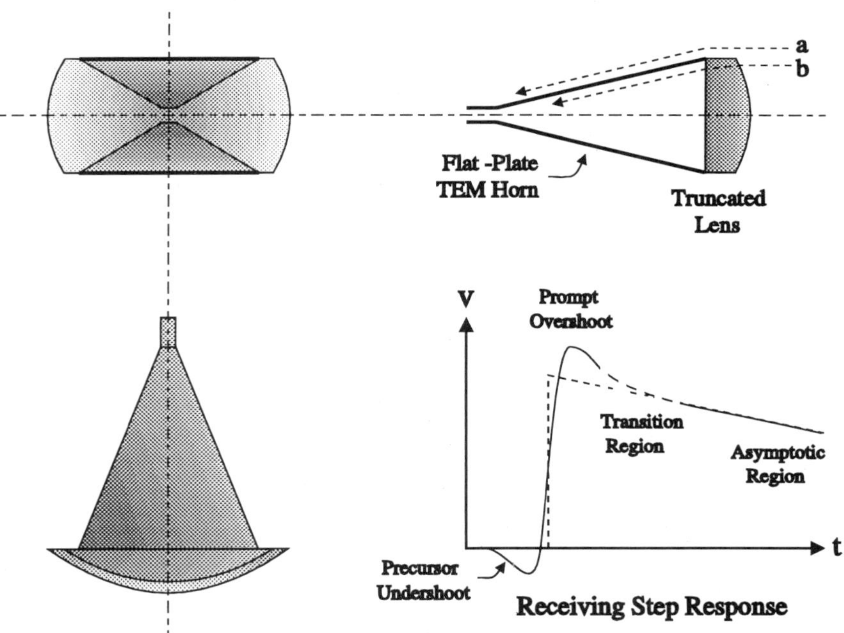

Figure 11. TEM-horn E-field sensor with truncated lens. The lens truncation and aperture shape result in a negative precursor signal followed by a prompt overshoot. The negative precursor signal traversing path (a) arrives earlier than the positive signal traveling through the lens along path (b). The intermediate-time response is unknown (Transition Region).

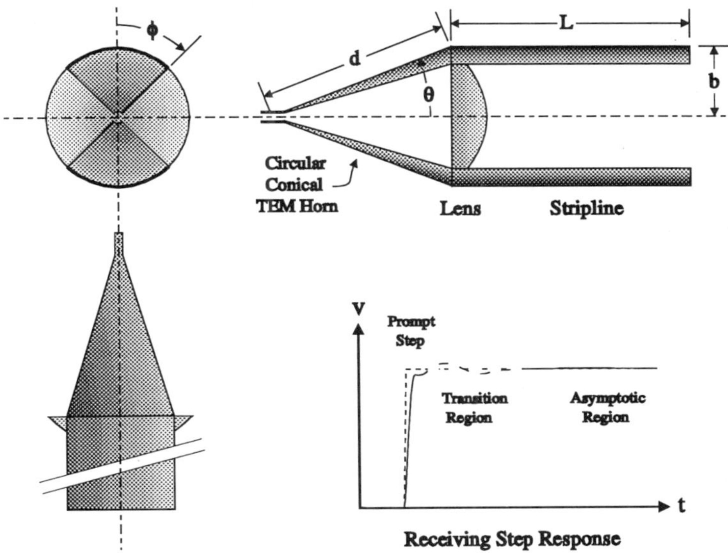

Figure 12. Circular conical TEM-horn transmission-line E-field sensor with circular lens terminating at the horn edge. The relative amplitude of the early- and late-time responses depends only upon reflection and absorption losses in the lens. The intermediate-time response has not been treated theoretically (Transition Region).

the prompt response of the sensor can be made to overshoot or undershoot the late-time asymptotic signal. The prompt response can theoretically be infinite in the physically unrealizable case of an infinitely large focused-aperture antenna with an appropriately chosen aperture block.[16] Field energy falling outside the circular lens boundary is defocused and effectively blocked at early times, contributing to the sensor signal only through edge diffraction at late times. An exception to this generalization is shown in Fig. 11, where a truncated lens is used. Here, a primary signal that traverses path (b) through the lens is preceded by an earlier signal that traverses path (a) by diffracting around the truncated portion of the lens. The result is a negative precursor signal. In most other cases, the early- and late-time responses of these devices are understood and can be predicted from theory. However, there is no theory to cover the intermediate time scale when the early- and late-time signals must join together somehow. Here, the step response will depend very critically upon the exact details of the aperture structure.

Figure 12 shows one antenna / sensor design with special properties that might solve some or all of the problems associated with finite apertures. It is a circular conical TEM-horn transmission-line sensor consisting of two 90°-section circular-conical electrodes joined to two cylindrical electrodes attached at the edge of a circular aspheric lens. This is a self-reciprocal antenna structure, which has the property that fields outside the circular aperture contribute nothing to its prompt response.[17] The natural aperture choice is therefore the circle of symmetry. There will be no prepulse or overshoot, and the prompt and late time responses will differ only by the amount of reflection and absorption losses in the lens. This design uses the lightest lens and is also the most efficient two-electrode antenna structure for a circular aperture, with an effective height h_{eff} of 0.85 times the radius.[18] The nature of the transitional response was unknown until very recently.

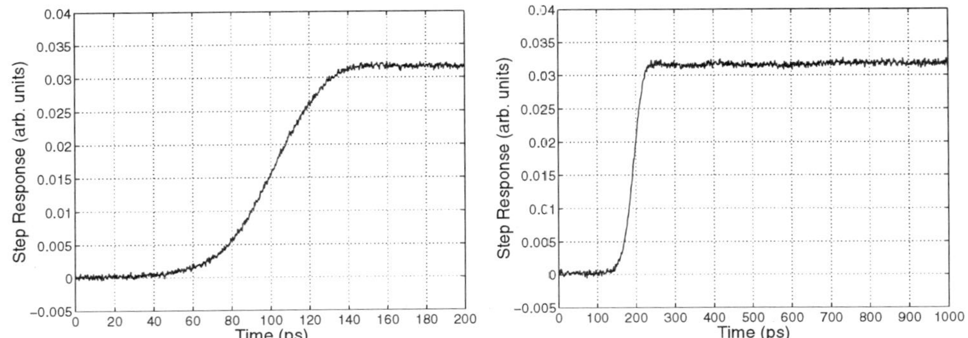

Figure 13. Preliminary receiving step-response data on boresight for the circular conical TEM-horn transmission-line E-field sensor with lens of Fig. 12. Antenna dimensions in cm were $b = 30$, $d = 79$, and $L = 61$, with $\phi = 45°$, and $\theta = 22.7°$. The measured risetime is nearly identical to that of the measurement system at 50 ps. Signal aberrations are minimal and are largely attributable to the measurement apparatus.

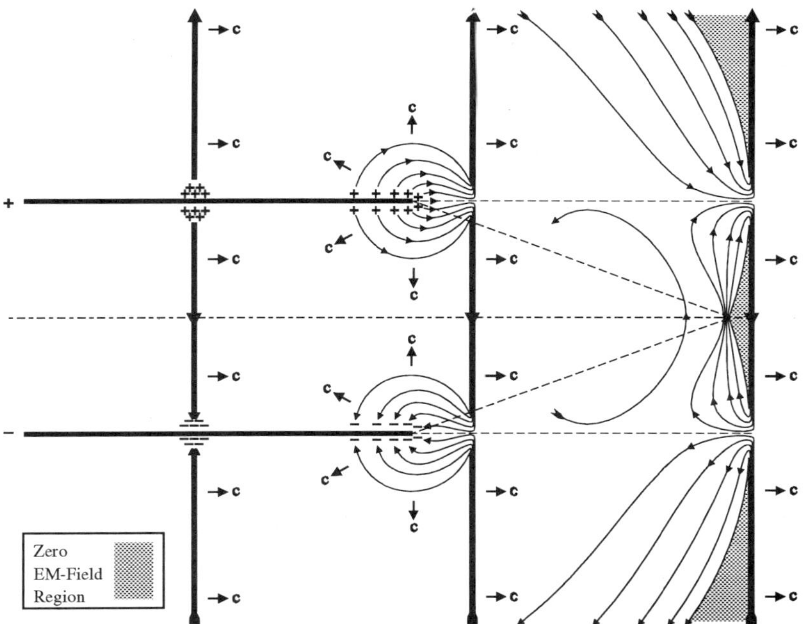

Figure 14. Conceptual field-line map of an infinitely-long circular conical TEM horn driven by three successive current impulses. A negative precursor field is seen just outside of the radiating collum.

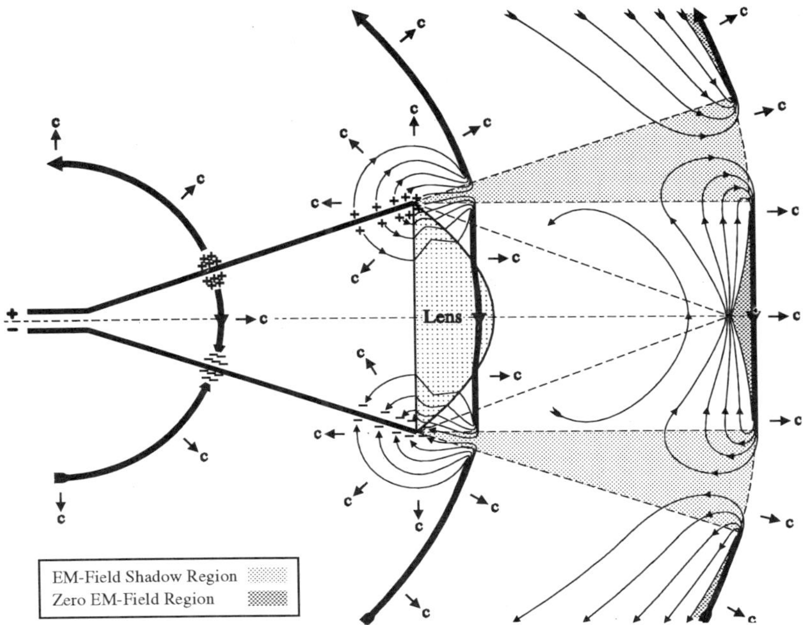

Figure 15. Conceptual field-line map of a circular conical TEM horn with lens driven by three successive current impulses. A shadow region exists between the aperture collum and the conical volume inclosing the horn electrodes.

Preliminary measurements of a half-model of the sensor of Fig. 12 were performed on our new twelve-meter-long ground-plane antenna range. The plano-convex polyethylene lens had a 30-cm radius and a 72-cm focal length. The 61-cm-long linear transmission-line extension provided a 4 ns clear time. The receiving step response shown in Fig. 13 has no prepulse and minimal post-step aberrations. The measured risetime is virtually the same as the system measurement risetime of 50 ps. The transition between early- and late-time responses appears to be perfectly smooth, and the sensor risetime is unresolvable with $\tau_{10-90\%} \ll 50\,\text{ps}$. This particular antenna / sensor appears to have exceptional properties.

Off-boresight responses are strongly influenced by details of aperture construction. From the reciprocity principal we know that the transmission response is proportional to the time derivative of the receiving response for any antenna. Because the transmission response is more easily visualized in this case, we will examine the transmission impulse response. In the infinite aperture limit, the radiated fields are identical to those from a very long transmission line, as shown schematically in Fig. 14. Here the fields from an impulse reverse direction above and below the aperture collum, as the fields are reversed on the outside of the transmission line. For the conical horn with lens shown in Fig. 15, a shadow region exists between the aperture collum and the conical volume inclosing the horn electrodes. Here the fields can penetrate only at late times by diffraction, and a negative response is delayed and excluded from a conical region with at least half the horn throat angle. Figure 16 shows some old off-boresight transmission impulse-response data for the conical TEM-horn with lens less transmission-line extension measured on an earlier version of our antenna range. The horn throat angle was 22.7°, and no evidence of a negative precursor exists in the E-plane data for off-boresight angles out to the observation limit of about 8°.

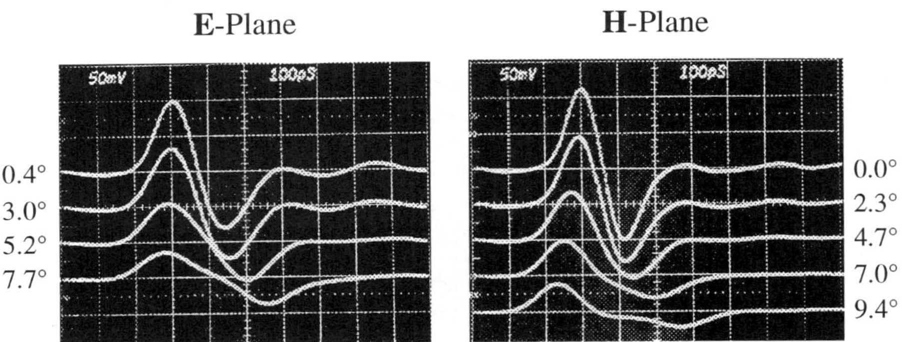

Figure 16. Off-boresight transmission impulse respoinse of a circular conical TEM horn with lens at $r = 5\,\text{m}$. Measurement bandwidths were 2.0 GHz for the E-plane data and 2.4 GHz for the H-plane data. Antenna dimensions were the same as that of Fig. 13 except for $L = 0$.

CONCLUSIONS

Numerous E-field sensors have been described that generate signals that accurately replicate an incident electric field for some finite clear time. A substantial increase in sensitivity and decrease in risetime was achieved, without loss of precision or introduction of spurious responses, through the use of guided wave optics. These sensors may also be used as transmitting antennas to radiate, for a specified duration and direction, a field that is precisely proportional to the time derivative of the applied voltage. Formulas were derived to aid in design optimization. Future work will attempt to achieve similar performance in antennas with different aperture geometries. Attempts are being made to develop lighter synthetic dielectric lenses with minimal loss and dispersion.

ACKNOWLEDGMENTS

The authors express their appreciation to Norris S. Nahman, Carl E. Baum, and James R. Andrews for helpful discussions and suggestions.

REFERENCES

1. C. J. Buchenauer and J. R. Marek, "Antennas and Electric Field Sensors for Time-Domain Measurements: An Experimental Investigation," in: *Ultra-Wideband Short-Pulse Electromagnetics 2*, Plenum Press, New York, 1995.
2. E. G. Farr and C. J. Buchenauer, "Experimental Validation of IRA Models," Sensor and Simulation Note 364, Jan. 1994.
3. C. J. Buchenauer and J. R. Marek, "Hybrid Antenna-Sources for Radiating High-Power Impulsive Fields," in: *Intense Microwave Pulses II*, H.E.Brandt Editor, Proc. SPIE 2557, pp. 209-213, 1995.
4. R. L. Wigington and N. S. Nahman, "Transient Analysis of Coaxial Cables Considering Skin Effect," Proc. IRE, vol. 45, pp. 166-174, Feb. 1957.
5. D. V. Giri and C. E. Baum, "Equivalent Displacement for a High-Voltage Rollup on the Edge of a Conduction Sheet," Sensor and Simulation Note 294, Oct. 1986.
6. C. E. Baum and J. S. Tyo, "Transient Skin Effects in Cables," Measurement Notes # 47, Aug. 1996.
7. T. L. Brown and K. D. Granzow, "A Parameter Study of Two-Parallel-Plate Transmission-Line Simulators of EMP Sensor and Simulation Note XXI," Sensor and Simulation Note LII, Apr. 1968.
8. N. S. Nahman, "A Discussion on the Transient Analysis of Coaxial Cables Considering High-Frequency Losses," IRE Trans. Circuit Theory, pp. 144-152, Jun. 1962.

9. N. S. Nahman, "A Note on the Transition (rise) Time Versus Line Length in Coaxial Cables," IEEE Trans. Circuit Theory, vol. 20, pp. 165-167, Mar. 1973.
10. H. Curtins and A. V. Shah, "Pulse Behavior of Transmission Lines with Dielectric Losses," IEEE Trans. Circuits and Systems, vol. 8, Aug. 1985.
11. E. A. Konscher, "Physical Basis of Dielectric Loss," Nature, vol. 253, pp. 717-719, Feb. 1975.
12. E. A. Konscher, "The 'Universal' Dielectric Response," Nature, vol. 267, pp. 673-679, Jun. 1977.
13. R. M. Hill, "Characterisation of Dielectric Loss in Solids and Liguids," Nature, vol. 275, pp. 96-99, Sep. 1978.
14. A. von Hippel and W. B. Westphal, "Tables of Dielectric Materials," MIT Cambridge Laboratory for Insulation Research, Apr. 1957.
15. A. von Hippel, *Dielectric Materials and Applications*, Artech House, Boston, 1994.
16. C. E. Baum, "Aperture Efficiencies for IRA's," Sensor and Simulation Note 328, Jun. 1991.
17. E. G. Farr, and C. E. Baum, "Radiation from Self-Reciprocal Apertures," in: *Electromagnetic Symmetry*, C. E. Baum and H. N. Kritikos, Editors, Taylor and Francis, Washington, D.C., 1995.
18. E. G. Farr, "Optimization of the Feed Impedance of Impulse Radiating Antennas Part II: TEM Horn and Lens IRA's," Sensor and Simulation Note 384, Nov. 1995.

DENSE MEDIA PENETRATING RADAR

Kwang Min and Marcelious Willis, Jr.

Air Force Wright Laboratory, Armament Directorate
Munitions Division, Fuzes Branch
101 W. Eglin Blvd., Suite 219
Eglin AFB, FL 32542-6810

INTRODUCTION

A prototype of compact ultra-wideband radar capable of penetrating dense media has been built and tested through layers of concrete walls and layers of sand. Objects behind walls or buried in the sand are detected with this device. The following discussion briefly describes the device, the experimental results obtained, the signal processing currently utilized, and future plans. The ultimate goal of the device is for media identification and media interface detection.

The Air Force has been interested in developing an ultra-wideband, short pulse radar, which will penetrate through dense media and simultaneously detect objects behind or buried in them. We opted to utilize an extremely short duration impulse source for the radar. This choice was opted in favor of the finer resolution and smaller antenna size. Returns from the interfaces of strata of material provide object detection and media layer ranging and detection.

A schematic of our radar is shown in Figure 1. An extremely short pulse is generated by a source is transmitted through a transmitting antenna, interacts with dense material and objects ahead, and radar returns are captured with the receiving antenna and sampled with a sampler. Following appropriate signal processing, pertinent information will be obtained.

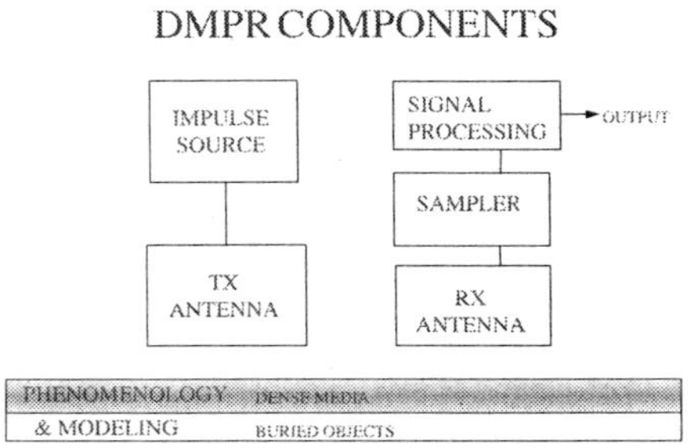

DMPR COMPONENTS

IMPULSE SOURCE	SIGNAL PROCESSING → OUTPUT
	SAMPLER
TX ANTENNA	RX ANTENNA
PHENOMENOLOGY & MODELING	DENSE MEDIA BURIED OBJECTS

Figure 1. Radar Schematic

Ultra-Wideband, Short-Pulse Electromagnetics 3
Edited by Baum *et al.*, Plenum Press, New York, 1997

HARDWARE DESCRIPTION

ANTENNAS

We utilized a reflector impulse radiating antenna (IRA) for the transmission and a transverse electromagnetic (TEM) horn antenna for the reception. Detailed descriptions of these antennas are given in reference 1. Simple characteristics of the above antennas are shown below:

Reflector Impulse Radiating Antenna

A reflector impulse radiating antenna (IRA) was developed under a Small Business Innovative Research (SBIR) contract. This antenna is 46 cm in diameter, (see Figure 2). The IRA is currently being further developed to reduce its size to approximately 23 cm in diameter, (See Figure 3). During further studies, a lens was put on the 23 cm diameter antenna. We then found that the antenna's gain pattern, now being more focused, increased by a factor of five times that of the antenna without the lens. (See Figures 3 & 4).

Figure 2. 46 cm Diameter Reflector IRA

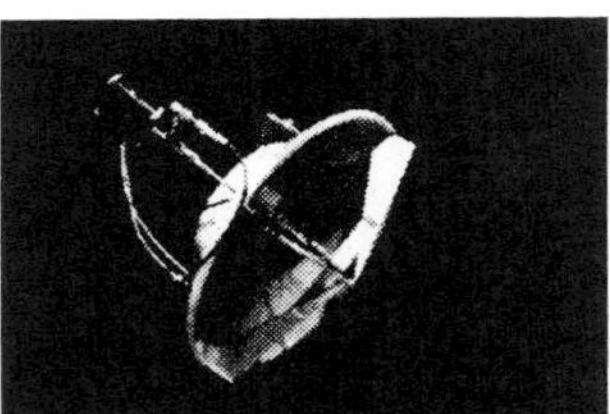

Figure 3. 23 cm Diameter Reflector IRA

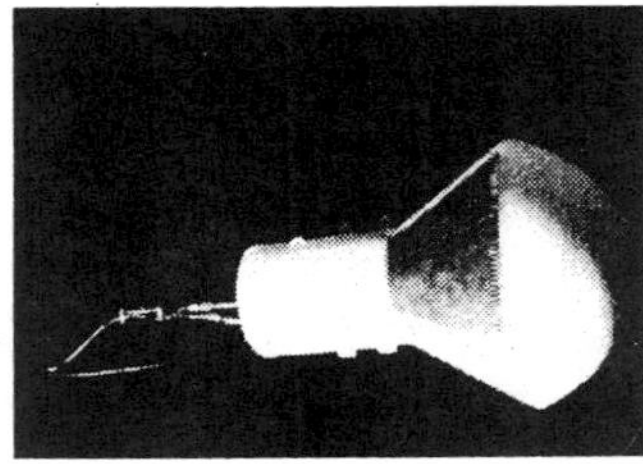

Figure 4. 23 cm Diameter IRA with Lens

Transverse Electromagnetic Horn Antenna

Radar measurements were made using the 76 cm transverse electromagnetic (TEM) horn, (See Figure 5), and the IRAs as transmitter/receiver (Tx/Rx) pairs with the IRA transmitting and the TEM horn receiving. To reduce mutual coupling and scattering from surrounding objects, a sheet of Tuff-R (foil-on-two-sides) was inserted between the Tx/Rx antennas. Two different radar target configurations were used: 1) a target (18 cm wide by 45 cm high by 0.3 cm thick aluminum plate) was suspended in air at a distance of approximately 96 cm) the same target was buried in sand at a distance of approximately 91 cm from the outside wall of the test apparatus (sandbox). For both cases, the target was in the same vertical position and on the bore sight to the Tx/Rx antennas. The target was easily seen when suspended in air; however, background subtraction was required to detect the target when it was buried in the sand.

SOURCES

The various sources that were used, either commercial-off-the-shelf (COTS) or developed through SBIR contracts were an ultra-fast pulser, developed by Pulse Power Physics through an SBIR contract, a BASS-01X, BASS-01X-S, and a BASS-02X all COTS of Power Spectra, Inc. and the PSPL 4015C, a COTS pulse generator of Picosecond Pulse Labs. Of these sources, we obtained the best performance using the sources developed by Pulse Power Physic (ultra-fast pulser) and Picosecond Pulse Labs (PSPL 4015C). When using the above sources, we found that the BASS-01X and the BASS-01X-S had too much jitter for our data acquisition system. The BASS-02X works well, however, for our application we need a very fast risetime and the BASS-02X's risetime was not quite as fast as what we desire and the pulse duration is longer than desired. For our application, shorter pulse durations and faster risetimes produced better empirical results. The PSPL 4015C works well also, but the output voltage is not as high as desired. The ultra-fast pulser developed by Pulse Power Physics has shown the best performance to date for our application. The ultra-fast pulser's risetime is approximately ninety picoseconds, the falltime is twenty-three hundred picoseconds, the pulse jitter is less than twenty picoseconds RMS, the output voltage is 2.5 kV, the pulse repetition (PRF) is 500Hz, the volume is 12 cubic centimeters (not including the power converters), and the mass is less than 0.23 kg. This pulser is the fastest of the sources we have used in our experiments. Although the output voltage is adequate and stable for our testing, we desire a slightly higher output voltage to increase the resolvable depth of penetration by our instrumentation. The ultra-fast pulser is being further developed to increase its present output voltage from 2.5 kV to 10 kV. The risetime is presently less than 100 picsoseconds and through further development, the risetime will be shortened to less than 70 picoseconds. The falltime will be greater than one

Figure 5. 76 cm TEM Horn Antenna

nanosecond and the PRF greater than 1 kHz. The goal is to keep the mass and volume down to a minimum as in the present source development. The pulser's output waveform (See Figure 6)and the pulser's D.C. voltage power supply (See Figure 7) are shown below:

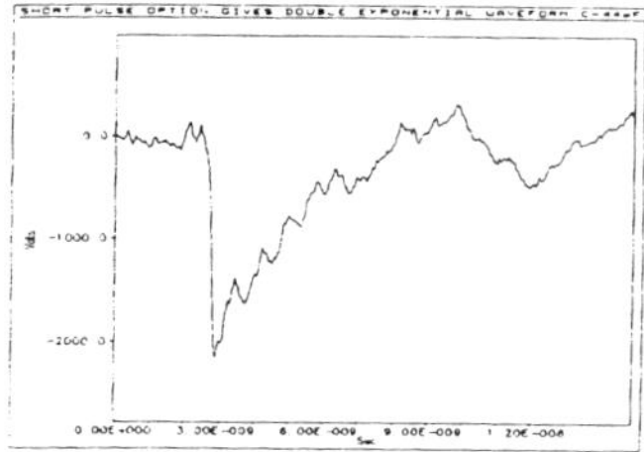

Figure 6. Ultra-Fast Pulser Output

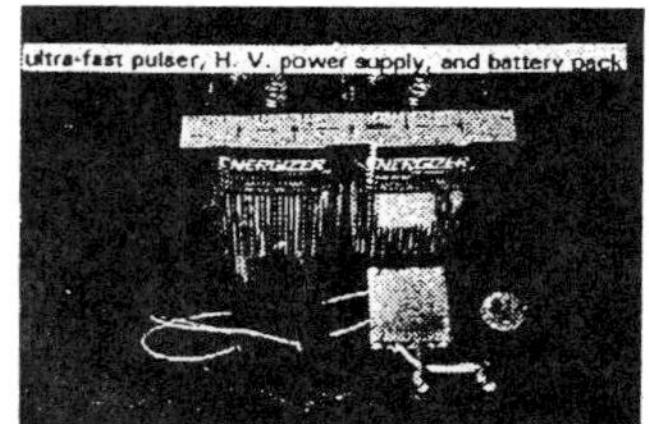

Figure 7. Ultra-Fast Pulser, H. V. Power Supply & Battery Pack

The D.C. power supply makes the pulser portable and it facilitates use in field testing. The power supply consists of two 6 volt batteries connected in series to give a total of 12 volts which is needed to power the pulser.

SAMPLER

We utilized existing commercially available samplers. They are Tektronix SCD5000 Transient Waveform Digitizer and Communications Signal Analyzer CSA803A. A Small Business Innovative Research (SBIR) program aimed at development of an inexpensive and compact fast sampler has been initiated.

BISTATIC RADAR TEST AT RANGE A-22 ('CASTLE')

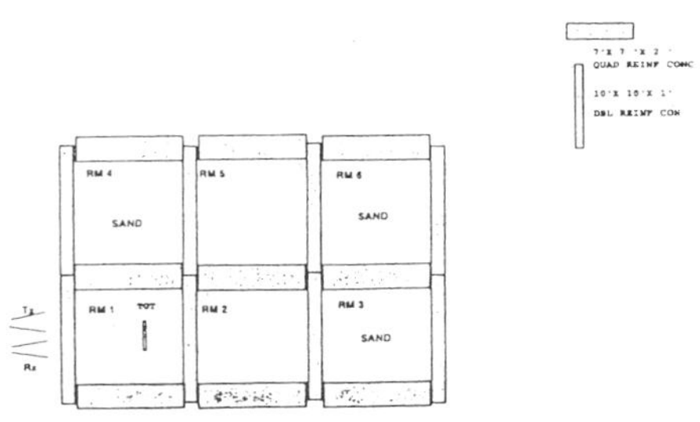

Figure 8. Range A-22, Eglin AFB, FL

BISTATIC RADAR TEST AT VITRO FACILITY (SAND BOX)

BISTATIC RADAR TEST AT VITRO FACILITY (BOMB BAYS)

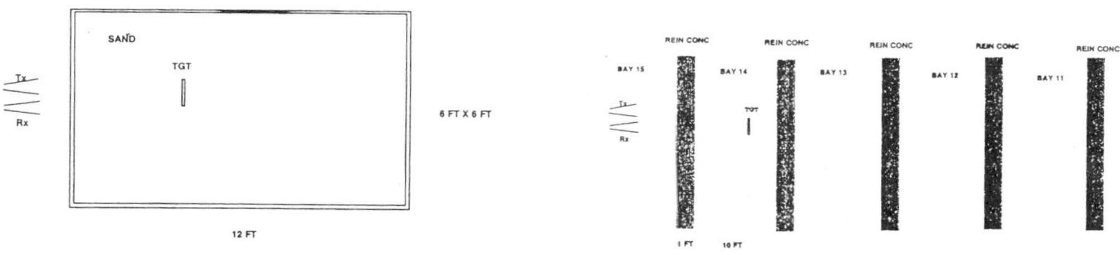

Figure 9. Sand Transmission Test Setup

Figure 10. Bomb Bays

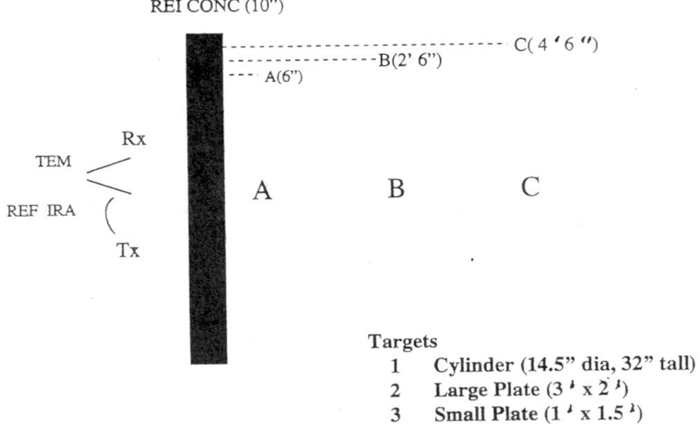

Figure 11. WSMR Elevator Shaft Room Test Setup

Experimental Setups

Figures 8 through 11 illustrate our experimental facilities used for the radar testing. They are as follows: (1) Range A-22 (Figure 8), (2) Sand Transmission test setup (Figure 9), (3) Bomb Bay (Figure 10), and (4) WSMR (Figure 11).

Figure 11 illustrates the experimental setup we utilized at the White Sands Missile Range (WSMR). A reflector IRA (an SBIR Phase I deliverable) antenna was used to transmit a short pulse using a Pulse Power Physics source (a SBIR Phase I deliverable). A TEM horn (See Figure 5), we procured from E-Systems was used to receive the returns. Targets were placed inside and elevator shaft room with thick steel door. Reinforced concrete wall is 25 cm thick. Targets used were (1) 37 cm (dia) x 81 cm (tall) cylinder, (2) a 91 cm x 61 cm metal plate (larger), and (3) a 30 cm x 46 cm metal plate (smaller). Each of them were placed at three different locations A (15 cm from the wall), B (75 cm from the wall), and C (137 cm from the wall).

Figure 12 illustrates the raw radar returns on the left, its interim processed results in the center, and the final processed results on the right. There are four groups of traces shown in this figure. The first group shows the radar returns without the target, the second group shows radar returns from a target at position A, the third shows radar returns from a target at position B, and the last group shows radar returns from a target at position C. Notice that a clear signature of the target (a metal plate) appears

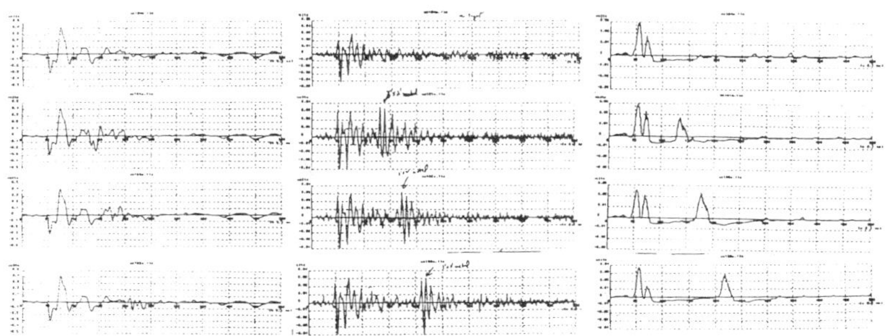

Figure 12. Target Signatures from 91 cm x 61 cm metal plate

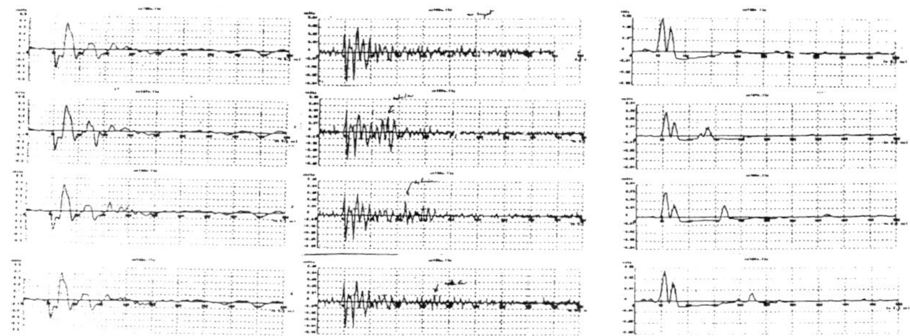

Figure 13. Target Signatures from metal cylinder

in the second, third, and fourth groups of signatures illustrated on the right hand side, corresponding to the target 15 cm, 75 cm, and 137 cm away from the wall, respectively. It is encouraging to see nothing appears on the top illustration except the signature corresponding to the concrete wall when there is no target.

Figure 13 illustrates the same for a cylinder when it was used as the target and with the same test setup. It appears to have a different signature corresponding to a different target. It suggests that there is a possibility of identifying the target type.

Final stage processing to perform the ranging is about 70% complete. We are currently working on target and media identification algorithm developments using the interim processing results.

SIGNAL PROCESSING

Signal processing techniques for this effort are under development in-house. This processing will enable detection and ranging of a target behind dense media such as a reinforced concrete wall buried within wet or dry sand. In our field testing, we used the CSA803A Communications Signal Analyzer to difference two measured signals, first a return from a concrete wall with no target behind it or from an empty room with no target in it and the second, a return from a concrete wall with a scatterer or target behind the wall or within a room. Figure 14 shows the graphs from the CSA803A using subtraction to get the target signatures. Note that in the final difference trace the target signal shifted in time. This is due to the target being moved to a different location; for every 15 cm that the target was moved farther away from the wall, the signal return was delayed in time by one nanosecond which is in line with the speed of light (3×10^8 meters/second) round-trip delay.

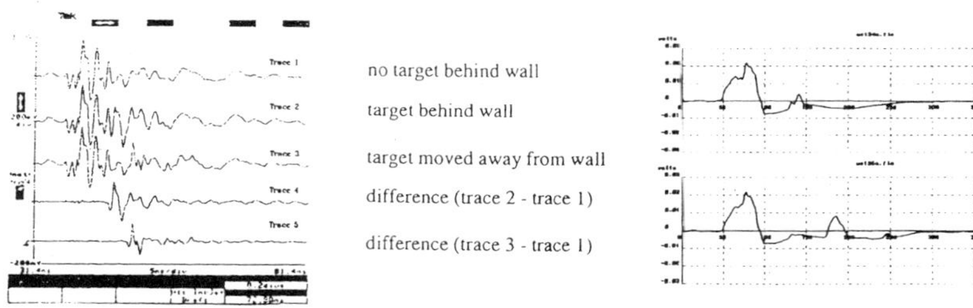

Figure 14. Returns measured on CSA803A

Figure 15. Final Processed Data

Signal Processing Algorithms

We have developed in-house algorithms that allow us to detect a target without taking the difference of measured signals with a target and without a target. The raw data signal is taken through various processing iterations that enhances the target signatures. Both small and large targets can be detected and their signatures displayed. Algorithm developments will continue to be further enhanced as our ultra-wideband technology development work continues. Figure 15 shows the results of data that has been processed using the algorithms developed.

CURRENT STATUS

Several recent Air Force programs have successfully developed components relevant to the short pulse unipolar emission technique for short pulse technology development. These components include a pulse source for producing the high peak power signal and antennas for radiating the proper waveform toward the media of interest. Preliminary data taken includes attenuation and propagation speed of these pulses through several media. A most recent experiment showed that is possible to detect a human being and a 60 cm x 120 cm wooden board behind a reinforced concrete wall. This shows that the target does not have to be a good RF scatterer to be detected.

FUTURE PLANS

Future plans are to continue developing this technology through programs and tests that will enhance the capabilities of current munitions. An improved impulse source is currently being developed under an SBIR contract and a fast-sampler SBIR program is scheduled to start sometime in late May or early June 1996. A compact sampler is needed to facilitate field testing by making it easier to setup tests in different locations and therefore increase the types of target structures that can be tested. We will continue to develop the essential parts of the radar system and integrate them into a compact UWB radar system. Through further development of this technology, we hope to identify the target media, detect the target's location, detect buried mines, and optimize media interface identification.

CONCLUSION

The compact UWB radar system is a state-of-the-art radar system which uses extremely short risetime impulses to penetrate dense media. The radar returns from the material interfaces are processed for target detection. The essential parts of the system are the source, the antennas, the sampler, the digitizer, and the processing algorithms. Based on our positive results, we have decided to continue to investigate generators for extremely fast risetime pulses and antennas to radiate extremely short duration pulses. There is a need for excellent antenna isolation, low loss cabling, computer controlled instrumentation, and data collection. A tactical application will probably require a one-shot (non repetitive) sampler to meet the real time munitions timing requirements for signal processing. These parts are all being developed and/or refined and will be integrated into a compact radar system to meet the requirements needed to accomplish Air Force goals.

REFERENCES

1. M. Willis, Jr., R. Orgusaar, K. Min. *UWB Radar Technology Development,* AeroSense 96 Proceedings paper # 2747-01, April 1996.
2. C. A. Frost. *Radar Fuze Sensor for Detecting Buried Structures* Report #: WL-TR-96-7008, Wright Laboratory Armament Directorate, Eglin AFB FL 32542-6810 (1995).
3. C. A. Frost. Compact Solid -State Ultra-Fast Sources for Impulse Radar, in: *Ultra-Wideband, Short-Pulse Electromagnetics 3,* A. Stone, C. Baum, and L. Carin, eds., Publisher, Plenum Press, New York (1996).
4. E. G. Farr and C.E. Baum. Impulse Radiating Antennas, Part III, *Ultra-Wideband, Short-Pulse Electromagnetics 3,* A. Stone, C. Baum, and L. Carin, eds., Publisher, Plenum Press, New York (1996).
5. E. G. Farr and C. A. Frost. Design and Predictions of Two Types of Impulse Radiating Antennas, *Ultra-Wideband, Short-Pulse Electromagnetics 3,* A. Stone, C. Baum, and L. Carin, eds., Publisher, Plenum Press, New York (1996).

FIRST ACHIEVEMENT OF PUMP AND PROBE EXPERIMENTS INVOLVING AN OPTOELECTRONIC GIGAHERTZ ULTRASHORT PULSE GENERATOR FOR MEASUREMENTS OF TRANSIENT PROPERTIES IN MATERIALS

by Jean-François ELOY*, Nicolas BREUIL, Vincent GERBE and Jean Hugues TROMBERT.

CEA/DAM/ Centre d'Etudes Scientifiques et Techniques d'Aquitaine, BP2, F33114 Le Barp.
e-mail: eloy@geocub.greco-prog.fr

INTRODUCTION

In recent years, we have seen a growing interest for the field of short and ultrawideband electromagnetic pulses[1,2]. To this end, experimental setups based on optoelectronical technologies have been developed [3].

Picosecond optoelectronic pulse generation and sampling have proved to be an effective method for characterizing the properties of high-speed electronic devices[4], antenna structures[5,6], and materials[7,8]. We recently described an optoelectronic measurement system[9,10] which was developed to measure the transient dielectric properties of materials over a wide frequency range in the microwave region (10-100 GHz). Our experiment uses a particular arrangement of the well-known time-equivalent sampling technique, involving a fast optoelectronic gate pulse generator and a sampling detector (see Fig. 1). In order to perform reliable measurements with these optoelectronic devices, a first check on the characterization of laser features such as average power, wavelength of radiation, pulse time duration and noise level followed by lock-in electronic optimization (signal/noise ratio) has been achieved[11]. Data acquisition and analysis are accomplished using a PC computer, LabView for windows software, and National Instrument GPIB board. A computer code then manages translation stage displacement, data acquisition, and a preliminary numerical treatment in an appropriate timing sequence. A brief description of the optoelectronic measurement technique is provided, but we emphasize in this work the first experimental results and measurements.

OPTOELECTRONIC DEVICES AND SAMPLING TIME MEASUREMENT TECHNIQUE

Both the pulse generator and sampling detector are optoelectronic microdevices consisting of metallic coplanar striplines photolithographically processed on a radiation-damaged silicon on sapphire (rd:SOS) material. In order to compare the potential advantages, including the ability to transmit wide band electrical signals, two types of microdevice structure with a feed gap of 10 μm have been processed and tested (see Fig. 2b and 2c) in a particular optical arrangement (see Fig. 2a) : Grischkowsky dipoles and coplanar stripline ETCS antennas (Exponentially Tapered Coplanar Striplines) with an end aperture of 1.25 and 2.5 mm.

A Kerr lens mode-locked Ti:sapphire laser[12] is used for photoexcitation of the semiconductor. The full-width-at-half-maximum (FWHM) of the laser pulse is approximately 100 fs with 10nJ pulse energy. The wavelength of the laser light is close to 800 nm with a 76 MHz pulse repetition rate. The short time duration incident laser pulse (the pump laser beam) focused between the biased pulser line induces a fast conductivity change in the semiconducting material which causes an electrical pulse to be radiated both directly in free space (Fig.2a), and simultaneously launched onto the transmitting line to be finally radiated by the antenna structure (Fig.1). The free-space electromagnetic pulse is then detected by a reciprocal receiving optoelectronic microdevice. The entire waveform can be measured by inserting a variable delay line between the pump and the probe laser beams. This technique thus provides a cross-correlation measurement between

Ultra-Wideband, Short-Pulse Electromagnetics 3
Edited by Baum *et al.*, Plenum Press, New York, 1997

the pulser and the sampler time-response.

We tested successively Grischkowsky dipole transmitters (see front-view, Fig. 2b) and ETCS coplanar stripline (see front-view Fig. 2c) in a particular optical arrangement (see cross-view, Fig. 2a) .

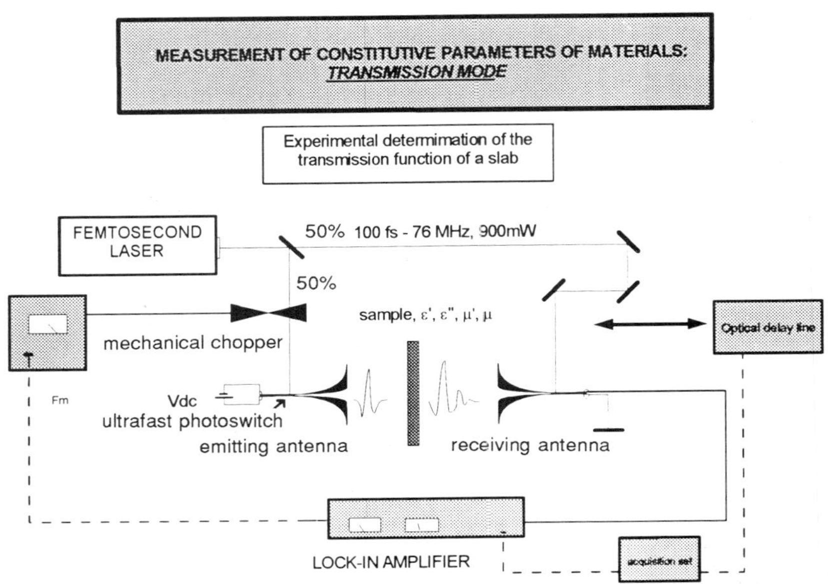

Figure 1. Optical arrangement and measurement system with ETCS antennas.

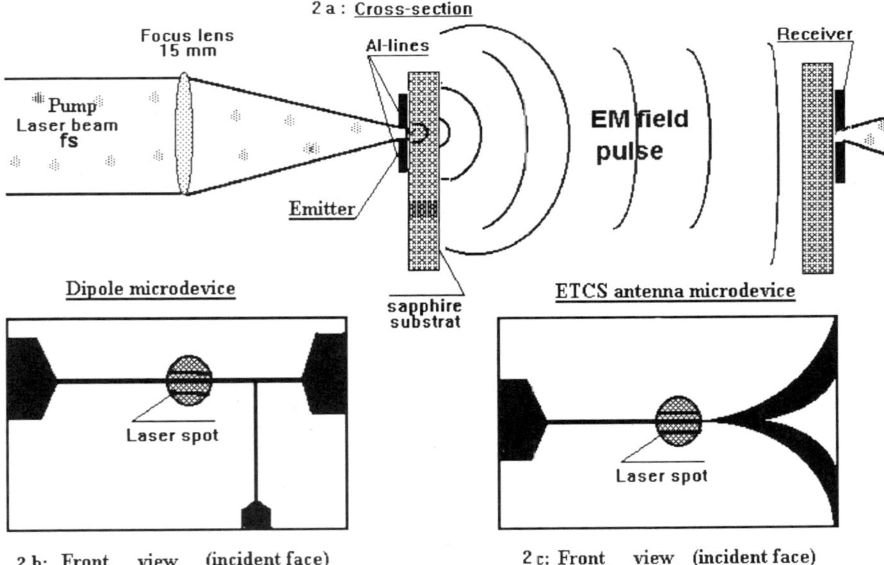

Figure 2 . Particular optical arrangement (2a) to study the direct transmitted emission of laser spot through the both dipole structure (2b) and ETCS transmission (2c) lines.

The pump and probe laser beams were impinging on the optoelectronic devices (transmitter line biased with a 5V dc voltage) as presented in the cross-section view Fig. 2a, and the EM field emission transmitted through the substrats recorded after 3 cm of propagation in the direction of the laser axis. We observed quite similar time profiles of electromagnetic pulses as shown Fig. 3. Considering the rise time of the electromagnetic signal emitted by the dipole structure, we can assume the lifetime of the carriers to be about 3-4 ps for our photoconducting rd-SOS material.

With the above configuration, we can compare the different electromagnetic emission of ETCS coplanar stripline and dipole transmitters which are directly radiated from the laser spot of interaction with the semi-conducting material in the gap zone. The peak to peak amplitudes of signal delivered and detected using biased pulser line of dipoles is greater by 26 % than the one emitted directly by ETCS antenna biased line. We have in this manner directly studied the on-off switching process of the photoconducting rd-SOS material. We note that the comparison of each fast Fourier Transform (FFT) of measured EM pulses shows strong differences in the lower frequency contributions of the dipolar antenna emission (Fig. 4).

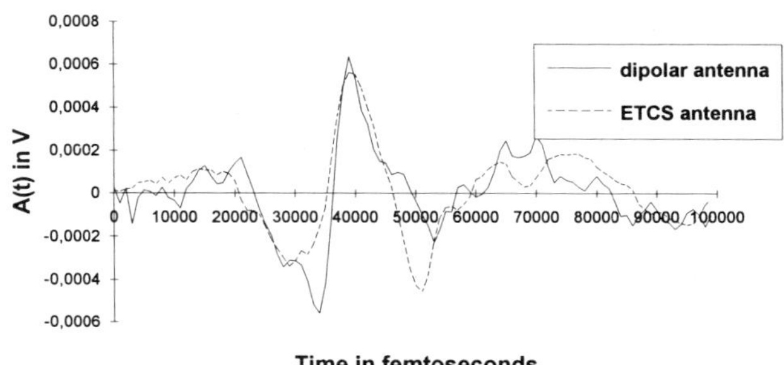

Figure 3 . Comparison of time profiles of pulses generated respectively by an ETCR antenna and a dipole emitter after 3 cm of propagation.

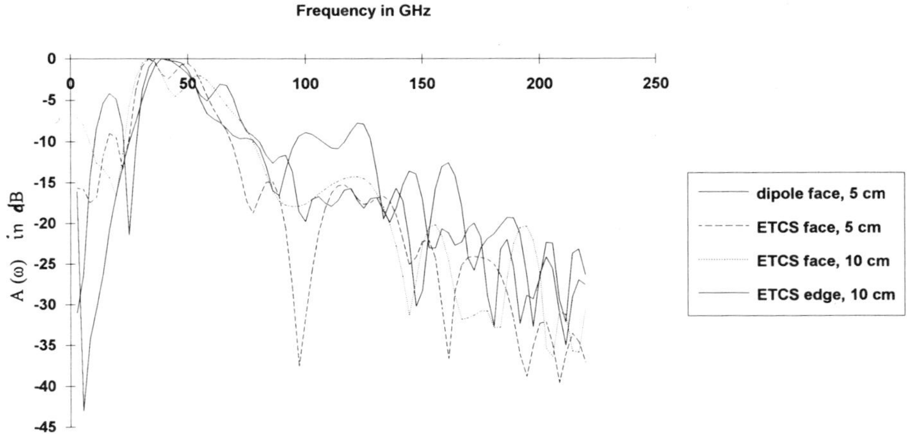

Figure 4 . Comparison of (FT) pulse spectra generated respectively by an ETCR antenna and a dipole emitter after 5 and 10 cm of propagation.

OPERATIVE CONDITIONS OF MATERIALS MEASUREMENTS

In order to implement the measurements of transient constitutive properties of materials, it was necessary to control the spatial and temporal distribution of the radiated wave front using the ETCS coplanar stripline antennas microdevices. For a transmitting ETCS antenna line biased with a 5 V dc voltage, we have recorded the following raw data presented in Fig. 5, for two optoelectronic micro-devices previously separated by a 5 cm distance in the optical configuration shown in Fig. 1. The operative conditions consist of a 2 kHz mechanical chopper used as a modulation frequency reference to a L.I.A (Lock In Amplifier) for an output filter time constant set to 1 s. A current preamplifier provides a 10^8 gain on the weak picoampere signal.

The transient electromagnetic radiation in free space is assumed to be strongly polarized with the E field parallel to the plane of the transmission lines. Thanks to 450 points acquired during 40 mn (consistant with an 1s time constant), the long time window of Fig. 5 allows us to observe secondary peaks caused by the reflection from the end of the pulser line. The spectral resolution is close to 7 GHz. With the associated 145 ps time window, the power spectrum of the pulse is then presented in the inset of Fig. 5. This trace shows a peak centered on 50 GHz with 10 - 100 GHz spectral bandwidth.

For the characterization of sample slabs, the two optoelectronic micro-devices were separated by 10 cm and deliver a collected and measured time-resolved signal presented in Fig. 6. A peak-to-peak amplitude of about 1 mV and a 5 ps rising time are estimated. This time acquisition is too long to avoid slow laser fluctuations and so forth. Moreover, the time window is too short to determine the spectrum resolution limit of the setup (limit imposed by the pulser line length). The time acquisition is reduced by a factor of 2 (45 mn) by increasing the step length (130 fs to 400 fs) and thereby decreasing the number of points (1000 to 450). The detailed comparison of signal traces presented Fig. 5 and Fig. 6 illustrates the difference between 130 fs and 400 fs step of measurement in operating conditions.

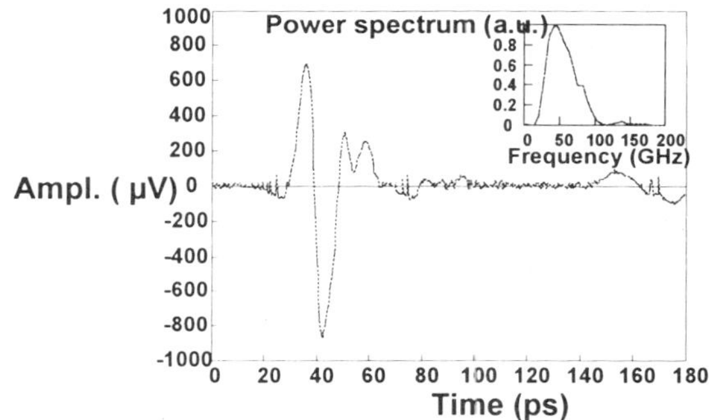

Figure 5. Radiated electromagnetic pulse for ETCS antenna microdevices separated by : 5 cm with 130 fs step length.

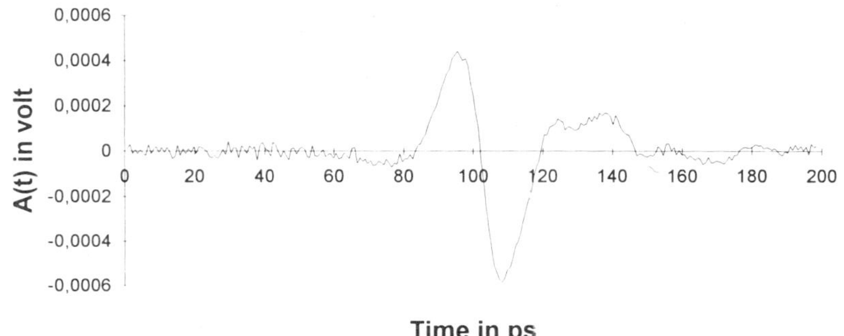

Figure 6 . Sampled signal of time-resolved electric field amplitude with 400 fs step length.

In order to determine the dielectric properties of materials, the space-time characterizations of the electromagnetic field generated by this kind of ETCS antenna structure[13] is first required.

ANTENNA PATTERN CHARACTERIZATION

The spatial distribution of the amplitude pattern resulting from our first attempt to measure the spatial dependence of peak-to-peak amplitudes for a transmitting ETCS antenna biased with a 5 V dc voltage, is plotted Fig. 7. These measurement have been achieved by removing the receiver microdevice holder step by step (each 2.5 cm) respectively along Oz axis (that of the transmitter line) and Oy axis (orthogonally to the transmitter line plane) . This graph should be considered in the context of the incertainties of the optimal adjustement of the laser spot position between the biased pulser line of the receiver microdevice and the 10% of reproducibility of laser energy (between each period of measurement) . In order to reduce this apparatus parameter, we check on the accurate position of laser spot by looking at the maximum of photovoltaic current in the unbiased line of the receiver until we obtain a reproductible value of this current.

Statistically, the spatial distribution of the peak-to-peak amplitude pattern can be expressed as a mathematical function of space parameters, z and y :

$$F(z, y) = 151 + 538 \exp[-(y/60)^2] + {}^1/_{z^{0.95}} \left(6928 + 147053 \exp[-(y/60)^2] \right) \quad (1)$$

From these spatial distribution measurements, it appears that the pulse exhibits approximately a $1/z^{0.95}$ decay rate dependence, with z the distance from the transmitter.

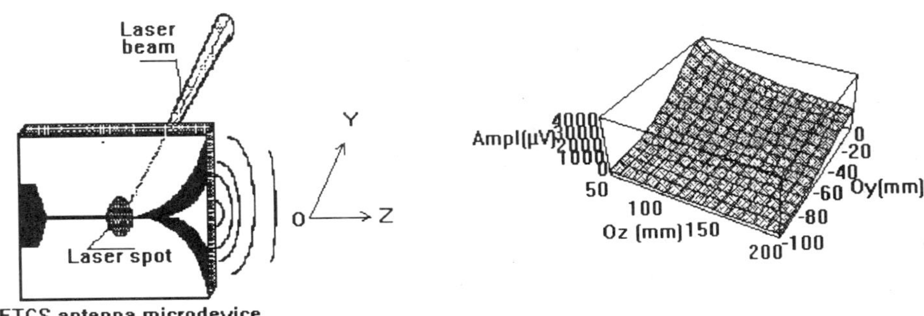

Figure 7 . Characterization of the ETCS antenna spatial distribution of max. peak to peak amplitudes.

Our previously published results[1] of the computed field amplitude decay of a wave train over a finite distance for a picosecond pulse shape revealed an asymptotical behavior of the amplitude decay dependence of $1/z^{0.9}$ close to a classical inversed dependence of length.

DATA MEASUREMENTS OF MATERIAL SAMPLE

In order to validate this method of transient electromagnetic properties determination, we have selected four kinds of standard material having the following determined constitutive properties :

- fused silica with 10 mm of thickness and announced values[14,15] for refractive index , n_r , and imaginary permittivity , ε_i , in the frequency range from 5 GHz to 150 GHz :

$$1.9558 < n_r < 1.9562 \text{ and } 2.5 * 10^{-3} < \varepsilon_i < 10*10^{-3}$$

- nylon with 10 mm of thickness and announced values[16] for refractive index , n_r , and imaginary permittivity , ε_i , in the frequency range from 5 GHz to 150 GHz :

$$1.728 < n_r < 1.734 \text{ and } 2.5 * 10^{-2} < \varepsilon_i < 4.5*10^{-3}$$

- alumina with 7.5 mm of thickness and announced values[16] for refractive index , n_r , and loss coefficient, tanδ, in the frequency range from 5 GHz to 150 GHz :

$$3.067 < n_r < 3.0645 \text{ and } 10 * 10^{-4} < \tan\delta < 25*10^{-4}$$

- teflon (**PTFE**) with 5 mm of thickness and announced values[17] for refractive index , n_r , and imaginary permittivity , ε_i , in the frequency range from 60 GHz to 300 GHz :

$$1.4386 < n_r < 1.4389 \text{ and } 1.1 * 10^{-3} < \varepsilon_i < 1.5*10^{-3}$$

As an example, Fig. 8 shows an electromagnetic signal collected and measured after 10 cm of propagation and transmission through the nylon sample slab .

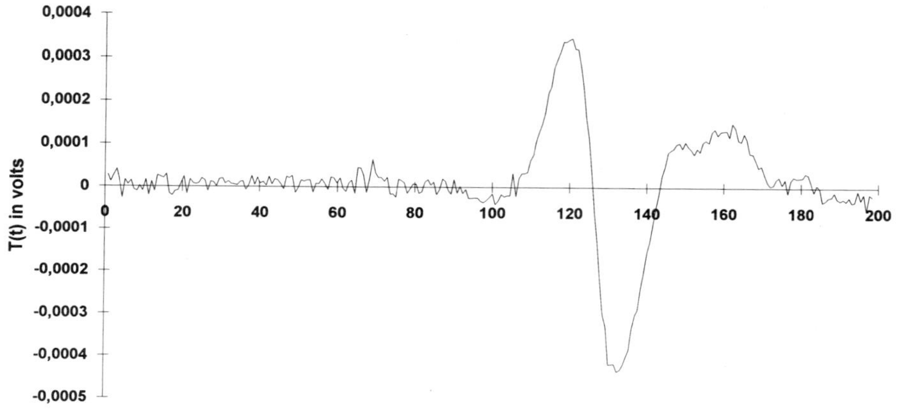

Figure 8 : Sampled signal of time resolved electromagnetic pulse after propagation and transmission through Nylon slab with 'e'= 10 mm of thickness.

Using a previously described time-resolved spectroscopy method[18,7], the experimental approach for determining the complex refractive index n(ω) for a dielectric, homogeneous and isotrope slab of thickness, 'e', we are required to evaluate the spectral transmission coefficients T(ω) by a comparative method. With this aim, it is necessary to record successively the following two generated transient EM pulses :
-1) $A_i(t)$: detected signal by the system after a free space propagation for a determined distance separating the two ETCS antennas which gives the time resolved signal for **reference** as shown Fig. 6.
-2) $A_T(t)$: detected signal by the system through the slab under test for the same distance separating the two ETCS antennas (Fig. 8) .

CALCULATION METHOD OF CONSTITUTIVE PROPERTIES

In the transmission configuration, the transient electromagnetic behavior of a slab subject to an ultrawideband EM pulse is obtained by a time convolution of the two detected signals, $A_i(t)$ and $A_T(t)$. The amplitude and phase spectra are calculated by numerical Fourier Transform (FFT): $A_T(\omega)$=FFT($A_T(t)$) and $A_i(\omega)$=FFT($A_i(t)$). Hence :

$$n(\omega) = n_{re}(\omega) + i.n_{im}(\omega)$$

$$n_{re}(\omega) = 1 + Arg\left[\frac{A_T(\omega)}{A_i(\omega)}\right].\frac{e}{c.\omega}$$

$$n_{im}(\omega) = Ln\left[Abs\left[\frac{A_T(\omega)}{A_i(\omega)}\right]\right].\frac{e}{c.\omega}$$

In order to apply this method, we systematically used a FFT algorithm to convert the time resolved signals to the frequency domain by applying an arbitrary broadening of time window of measurements in order to increase the frequency resolution.

The first results obtained by this method of measurement and calculation are presented graphically and successively in Fig. 9,10,11and 12 corresponding respectively to the four selected standard materials.

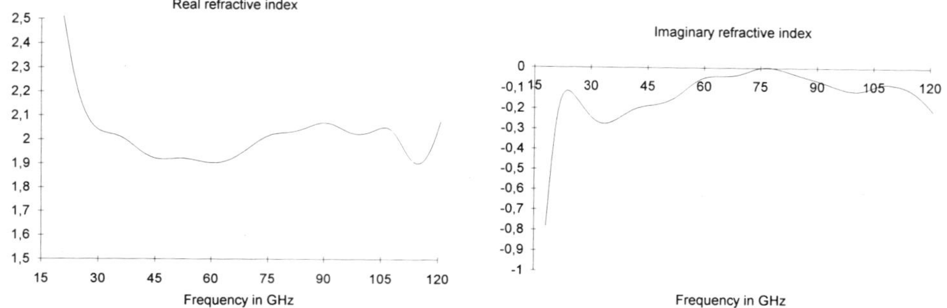

Figure 9 : Real and imaginary parts of the refractive index of **fused silica** slab with 'e' = 10 mm of thickness.

The presented results of calculations reveal a quite good agreement for the real part of refractive index with the announced values for the fused silica sample. A weak difference is however observed for the imaginary part between our results and the published data[14,15].

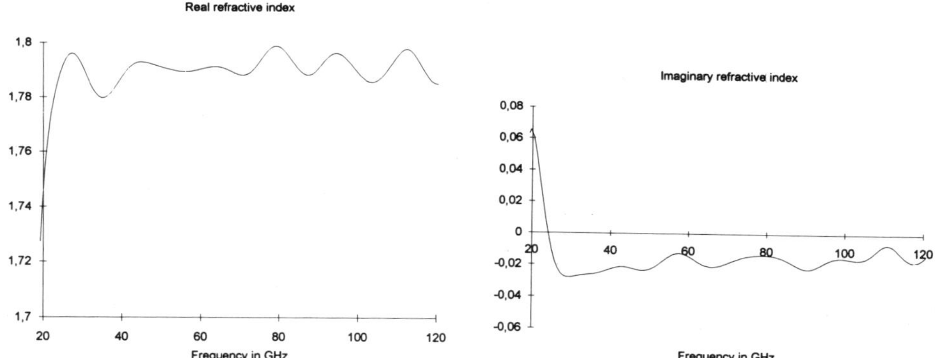

Figure 10. Real and imaginary parts of the refractive index of a **Nylon** slab sample with 'e'=10 mm thickness

For the nylon sample, the calculated results of real part of refractive index are slightly higher than the announced values but close for the imaginary part[16].

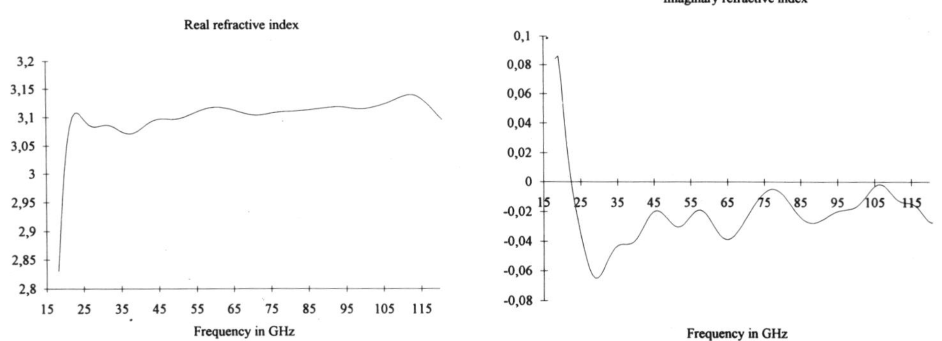

Figure 11. Real and imaginary parts of the refractive index of an **Alumina** slab with 'e' = 7.5 mm thickness.

The calculated results of both real and imaginary parts of refractive index of the alumina sample are slightly higher than the announced values[16].

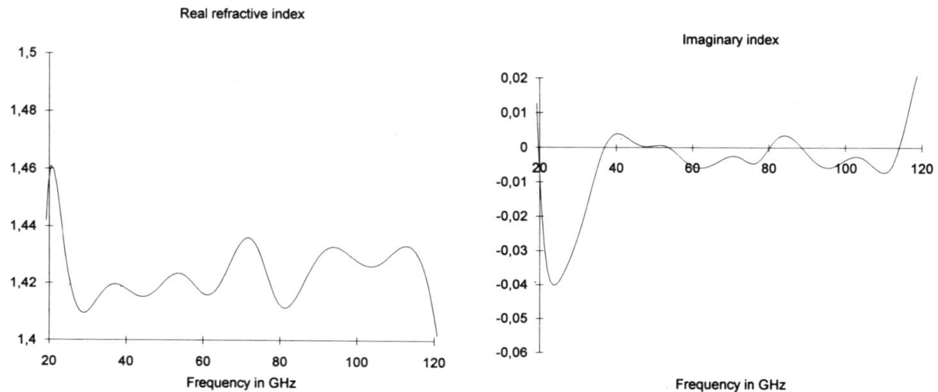

Figure 12 . Real and imaginary parts of the refractive index of a teflon (**PTFE**) slab with 'e'=5 mm thickness

For the teflon sample, the calculated results of real part of refractive index are slightly lower than the announced values but are close for the imaginary one[17]. In order to avoid the uncertainties due to the reflexion problem and consequently to improve the accuracy of these electromagnetic properties characterizations, we intend to apply the method of double measurements by using two thicknesses of each material sample.

CONCLUSIONS

We have tested a pump-probe optoelectronic experimental setup and presented some results of ultrawide band electromagnetic signal measurements. Acquisition time of this method has been reduced to 40 mn without significant noise enhancement, but may be even further reduced.

The spatial distribution of ETCS antenna radiation pattern has been achieved and then compared with the direct electromagnetic emission of the laser spot impinging on the same gap of photoconducting material rd-SOS. The lifetime of the photocarriers can be estimated to be close to 3-4 ps.

Using a comparative method to evaluate the spectral transmission coefficients according to previously recorded time-resolved signals, we measured the microwave properties characterizations of four different standard materials inserted between two ETCS antennas optoelectronic microdevices. With 5 to 10 % accuracy in the 30 to 110 GHz frequency range, the results of both measurements and associated calculations are relatively close to the published values.

REFERENCES

1. J-F. Eloy, F. Moriamez, Spectral Analysis of E-M ultra-short Pulses at coherence limit. Modeling. SPIE Proc, *Intense Microwave and Particle Beams III*, pp.298-309, (1992).
2. H. Wilhelmsson, J.H. Trombert, J-F. Eloy, Dispersive and Dissipative Medium Response to an ultrashort pulse : a Green's Function Approach. *Physica Scripta*, **52**, 102 ,(1995).
3. D.H. Auston, in *Picosecond Optoelectronic Devices*, C.H. Lee, Ed. New York: Academic Press, 1984, pp. 73-116.
4. H. J. Cheng, J. F. Whitaker, T. M. Weller, L. P. B. Katehi, "Terahertz-bandwidth of coplanar transmission lines on low permittivity substrate," *IEEE Trans. Microwave Theory Tech.*, **MTT-42**, (12),pp. 2399-2406, 1994.
5. G. Arjavalingam, Y. Pastol, J. M. Halbout, W. M. Robertson, "Optoelectronically-pulsed antennas : characterization and applications," *IEEE Antennas and Propagation. Magazine*, pp. 6-11, Feb. 1991.
6. M. M. Gitin, F. W. Wise, G. Arjavalingam, Y. Pastol and R. C. Compton, "Broad-band characterization of millimeter-wave Log-periodic antennas by photoconductive sampling," *IEEE Trans. Antennas and Propagation.*, **42**, (3),pp. 335-338, 1994.
7. D. Grischkowsky, S. Keiding, M. Van Exter, C. Fattinger, Far-infrared time-domain spectroscopy with terahertz beams of dielectrics and semiconductors. *J. Opt. Soc. Am B*, **7**, (10),pp.2006-2015, Oct. 1990.
8. W. M. Robertson, G. Arjavalingam, S. L. Shinde, "Microwave dielectric measurements of zirconia-alumina ceramic composites: a test of the Clausius-Mossotti mixture equations," *J. Appl. Phys.*, **70**,(12), pp. 7646-7650, 1991.
9. J-F. Eloy, J-H. Trombert, H. Wilhelmsson, Proc. Journées Maxwell'95, June 6-9th, 1995, edt J-F Eloy, CEA/DAM/CESTA, Le Barp, France (1995).
10. J-F. Eloy, C. Fos, V. Gerbe, J-H. Trombert, H. Wilhelmsson, Journées d'Etudes IEE/SEE "Antennes imprimées", Sept. 25-26th 1995, Lille (France).
11. J-F. Eloy, C. Fos, V. Gerbe, J-H. Trombert, Pump and Probe Experimental Setup with Optoelectronically Pulsed Antenna for Giga/terahertz Spectroscopy, *IEEE Trans. Instrum. Measurement*, to publish in 1996.
12. Coherent Inc., Model Mira 900™ femtosecond Ti:sapphire mode-locked laser.
13. E. G. Farr, C. E. Baum and C.J. Buchenauer, **"Impulse Radiating Antennas, Part II"**, Ultra-Wideband, Short-Pulse Electromagnetics, edt. L. Carin and L.B. Felsen, Plenum Press, New-York, 1995.
14. M.N. Afsar, *IEEE Trans. Microwave Theory Tech.*, **MTT-32**, (12), Dec. 1984.
15. C.D. Capps, R.A. Falk, S.G. Ferrier and T.R. Majoch, *IEEE Trans Microwave Theory Tech.*, **MTT-40**, (1), Jan. 1992.
16. Y. Pastol, G. Arjavalingam, J.-M. Halbout, G.V. Kopcsay, *Electron. Lett.*, **25**, (8), April 1989.
17. M.N. Afsar, *IEEE Trans. Instrum. Measurement*, **IM 36**, (2), June 1987.
18. G. Arjavalingam, Y. Pastol, J. M. Halbout, G. V. Kopcsay, "Broad-Band Microwave Measurements with Transient Radiation from Optoelectronically Pulsed Antennas" *IEEE Trans Microwave Theory Tech.*, **MTT-38**, (5), pp. 615-621, May 1990.

TARGET DETECTION AND IMAGING USING A STEPPED-FREQUENCY ULTRA-WIDEBAND RADAR

E.J. Rothwell, K.M. Chen, D.P. Nyquist,
A. Norman, G. Wallinga and Y. Dai

Department of Electrical Engineering
Michigan State University
East Lansing, MI 48824

INTRODUCTION

Interest in ultra-wideband radar systems for target detection, identification and imaging arises from both the clutter suppression capability of the radar and its potentially high resolution. Several time-domain radar systems have been tested[1,2], but difficulties in the generation, radiation and reception of high-energy pulses, and the potential for interference with existing communications systems, make the construction of a time-domain system problematic. The Naval Ocean Systems Center is developing an alternative radar system, which will have several hundred narrow band channels within a 1 GHz bandwidth centered in C or X-band. The data gathered by this system can be used to synthesize equivalent time-domain target responses for detection or identification.

This paper will consider two applications for a stepped-frequency, ultra-wideband radar. First, the detection of objects located above a disturbed sea surface will be considered. When the clutter signal is large in comparison with the target return, detection using conventional radar is difficult. A time domain radar provides temporal separation of the dominant scattering events, allowing the clutter signal coming from periodic swells to be reduced, and the target signal to be extracted. Second, the imaging of radar targets by a radar system which uses convenient ultra-wide, but band-limited, signals is examined. A bistatic imaging scheme is developed which can be used in a simple physical-optics mode, or can be rigorously implemented to allow for non-specular contributions to the scattered field.

TARGET DETECTION IN A SEA CLUTTER ENVIRONMENT USING BAND-LIMITED WAVEFORMS

The detection of radar targets in a sea clutter environment becomes quite difficult when the signal returned by the target is small compared to the clutter signal. By using an UWB radar system, the approximately periodic nature of the sea clutter signal can be used to create a "clutter reducing transmit waveform" (CRTW) which, when transmitted

Ultra-Wideband, Short-Pulse Electromagnetics 3
Edited by Baum *et al.*, Plenum Press, New York, 1997

or used for post-processing, reduces the background clutter while maintaining the strength of the target signal. This technique, earlier reported for baseband implementation[3], is applied here in the case of band-limited signals.

Because the surface profile of a disturbed sea is approximately periodic over a finite spatial interval, the time-domain scattered field response of the sea is also approximately periodic. This allows the sea clutter signal to be modeled as a finite series of exponentials

$$r(t) = \sum_{n=-N}^{N} A_n e^{Q_n t} \qquad 0<t<T_R \tag{1}$$

As in the E-pulse technique[4], a CRTW $e(t)$ is a waveform of finite duration T_E which, when convolved with the sea clutter signal produces a null result

$$c(t) = e(t)*r(t) = \int_0^{T_R} r(t)e(t-t')dt' = 0 \qquad T_E<t<T_R \tag{2}$$

Thus, if a CRTW is radiated in the presence of sea clutter, there will be no returned signal in the interval $T_E<t<T_R$. The condition for creating the CRTW is[4]

$$E(s{=}Q_n) = 0 \qquad 1\le n\le N \tag{3}$$

where $E(s)$ is the Laplace spectrum of $e(t)$.

To test the potential of using a CRTW on sea clutter signals, the theoretical pulse responses of two finite-length, perfectly conducting, sea surface models with surface profiles shown in Figure 1 were computed. The moment method was used to compute the frequency-domain response in the band 9-14 GHz, with a 3 dB bandwidth of 1.2 GHz (about 11% of the center frequency of 11 GHz), and the time-domain signals were computed using the inverse FFT. Figure 2 shows the magnitudes of the frequency responses. The sizes of the sea models were chosen to allow comparison with measurements taken within an anechoic chamber, and thus do not match the dimensions of actual sea surfaces.

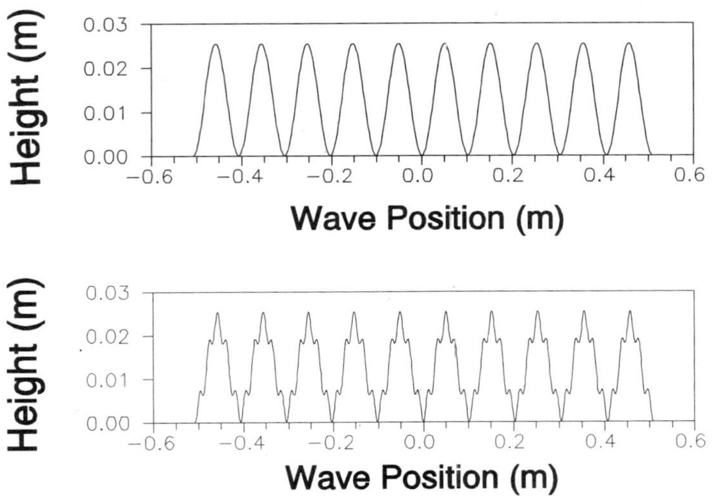

Figure 1. Sea surface profiles generated from a single sinusoid and a double sinusoid.

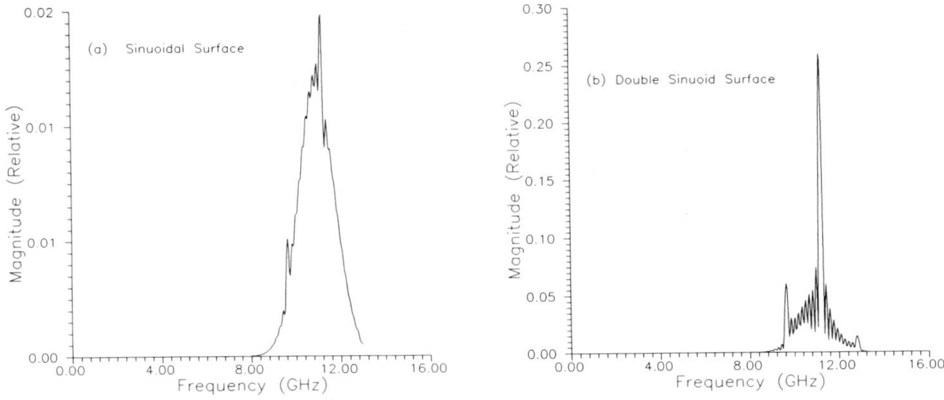

Figure 2. Computed magnitude spectra of fields scattered by (a) single sinusoid and (b) double sinusoid sea surface model.

After the time-domain signals were computed, the response of a five inch long missile target was added. The missile amplitude was scaled to produce a prechosen target-to-clutter ratio (T/C) defined as the ratio of maximum target signal strength to maximum clutter signal strength (excluding the signal produced by the leading edge of the finite-sized surface). Figure 3 shows the superposition of the missile and clutter signals for the two surfaces, while Figure 4 shows the CRTWs created for the clutter signals using (3). It is obviously difficult to discern the presence of the target amongst the band-limited clutter signal. Figure 5 shows the convolution of the CRTWs with the clutter/target combination. The presence of the target is clearly enhanced after the application of the CRTW, allowing the missile target to be more readily detected.

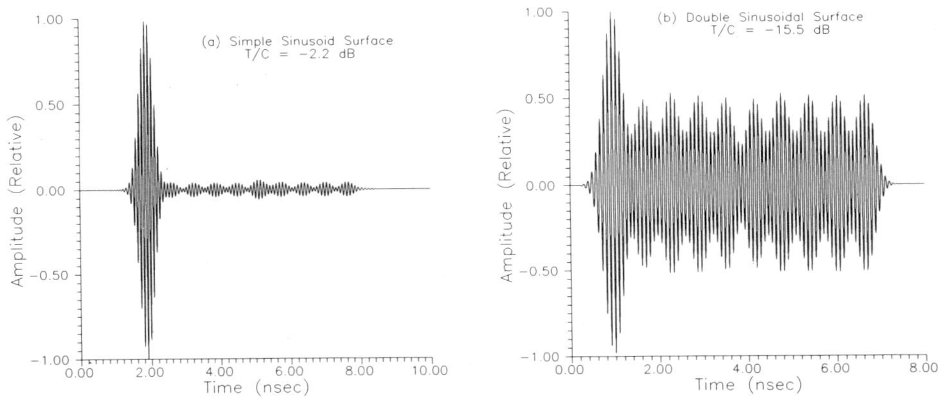

Figure 3. Time-domain scattered field responses for missile above (a) single sinusoid surface and (b) double sinusoid surface.

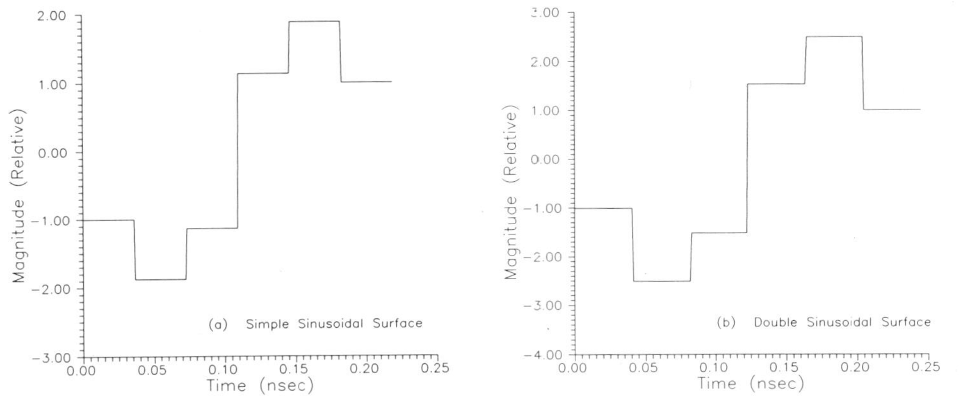

Figure 4. CRTWs constructed for (a) single sinusoid surface and (b) double sinusoid surface.

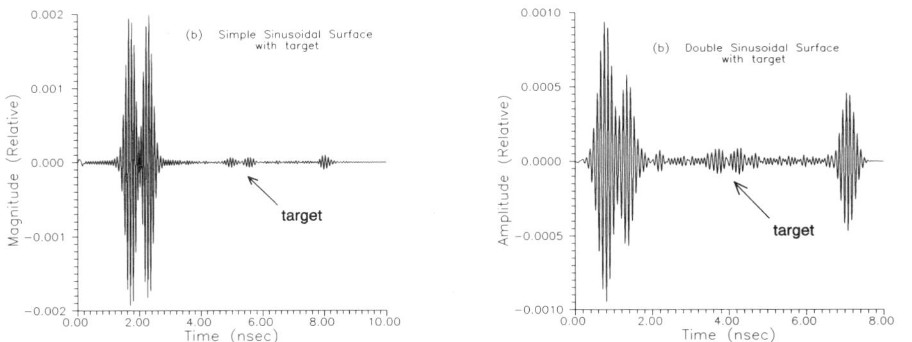

Figure 5. Convolutions of CRTW with missile response immersed in clutter for (a) single sinusoid surface and (b) double sinusoid surface.

TARGET IMAGING USING BAND-LIMITED, ULTRA-WIDEBAND SIGNALS

Time-domain inverse scattering identities have been developed by several researchers[5,6]. Applying physical optics principles, the scatter geometry can be related to the impulse, step and ramp scattered field responses. Unfortunately, each of these responses is very wideband, with dominant frequency content at the low end of the spectrum. More rigorous techniques have been developed[7], but these still require inconvenient interrogation waveforms. This paper proposes a rigorous identity based on a signal with spectral content limited to a portion of the EM spectrum. Thus, it is ideal for use with an ultra-wideband stepped-frequency radar.

Consider a perfectly conducting object illuminated by a transient plane wave. The time domain current induced on the object must obey the space-time integral equation[7]

$$\vec{J}_s(\vec{r},t) = 2\hat{n}\times\vec{h}^i(\vec{r},t) - \frac{1}{2\pi}\int_s \hat{n}'\times\left\{\left[\frac{1}{R^2}+\frac{1}{R}\frac{\partial}{\partial\tau}\right]\vec{J}_s(\vec{r}',\tau)\times\hat{a}_R\right\}ds' \tag{4}$$

Here $\vec{h}^i(\vec{r},t)$ is the incident magnetic field, $\hat{n}$ is the unit vector normal to the scatterer surface, $\vec{r}$ is the position vector, t is normalized time in meters, and $R = |\vec{r}-\vec{r}'|$, $\hat{a}_R = (\vec{r}-\vec{r}')/R$, $\tau = t-R$. If the scatterer current is known, the far-zone field produced by the scatterer can be found from

$$\vec{h}^s(\vec{r}_s,t) = -\frac{1}{4\pi r}\int_s \hat{r}_s \times \frac{\partial \vec{J}(\vec{r}',\tau')}{\partial \tau'} ds', \quad \vec{r}_s = r_s \hat{r}_s \tag{5}$$

Substitution of (4) into (5) gives the following relationship between incident and scattered fields

$$\int_V \frac{\partial^2 h^i(\vec{r}',\tau)}{\partial \tau^2} dv' = \frac{\pi r_s}{(\vec{K}\cdot\hat{r})\cos(\beta/2)} H^s(\vec{r}_s,t) \tag{6}$$

Here , $h^i(\vec{r},t)=\hat{h}^i\cdot\vec{h}^i(\vec{r},t)$, $h^s(\vec{r},t)=\hat{h}^i\cdot\vec{h}^s(\vec{r},t)$, $\vec{K}=\hat{r}_s-(\hat{r}_s\cdot\hat{h}^i)\hat{h}^i$, and β is the bistatic angle. The total magnetic field is $H^s(\vec{r}_s,t)=h^s(\vec{r}_s,t)-h_2^s(\vec{r}_s,t)+h_c^s(\vec{r}_s,t)+h_{c2}^s(\vec{r}_s,t)$, which is the sum of the scattered field, the scattered field under complimentary illumination (shadow zone), and the "correction fields"

$$h_c^s(\vec{r}_s,t) = \frac{1}{4\pi r_s}\left[\int_s \hat{r}_s \times \frac{\partial \vec{J}_c(\vec{r}',\tau')}{\partial \tau'} ds'\right]\cdot\hat{h}^i \tag{7}$$

produced by the correction current $\vec{J}_c(\vec{r},t) = \vec{J}(\vec{r},t) - 2\hat{n}\times\vec{h}^i(\vec{r},t)$ and the correction current due to complimentary illumination.

Now, consider the case of a band-limited incident field waveform. If $\vec{h}^i(\vec{r},t)$ has a time dependence given by the sine-modulated-exponential-pulse (SMEP) waveform

$$h^i(t) = \sin(\omega_c t)e^{-\alpha t}U(t) \tag{8}$$

then (6) can be written as

$$\int_{\infty}^{-\infty}\!\!\int\!\!\int \gamma(\vec{r}')\delta(t_s-\hat{r}\cdot\vec{r}')dv' = \tag{9}$$
$$= \frac{2\pi r_s|\cos(\beta/2)|}{\vec{K}\cdot\hat{r}\cos(\beta/2)\omega_c}\left[H^s(\vec{r}_s,t) + 2\alpha\int_0^t H^s(\vec{r}_s,t)dt + (\omega_c^2+\alpha^2)\int_0^t\!\!\int_0^t H^s(\vec{r}_s,t)dtdt\right]$$

where $t_s = (r_i+r_s-t)/[2\cos(\beta/2)]$, and $\gamma(\vec{r}')$ is the characteristic function of the scatterer

$$\gamma(\vec{r}') = \begin{cases} 1 & \vec{r}'\in V \\ 0 & \vec{r}'\notin V \end{cases} \tag{10}$$

It is seen that the left-hand side of (9) is the cross-sectional area of the scatterer along the $\hat{r}$ direction. Mathematically, this is the radon transform[8] of $\gamma(\vec{r})$, and thus the characteristic function can be found by taking the three-dimensional inverse radon transform of (9). If the two-dimensional inverse transform is taken instead, the thickness function Γ for the target is obtained. Computing the transform in the x-y plane gives

$$\Gamma(\vec{\rho}') = \frac{-2\rho_s|\cos(\beta/2)|}{(\vec{K}\cdot\hat{\rho})\pi\omega_c}\int_{\beta/2}^{\pi+\beta/2}\int_{-\infty}^{\infty}\left[\left(2\alpha+\frac{\partial}{\partial t}\right)[h^s(\vec{\rho}',t)-h_2^s(\vec{\rho}',t)+h_c^s(\vec{\rho}',t)+h_{c2}^s(\vec{\rho}',t)] + \right.$$
$$\left. +(\omega_c^2+\alpha^2)\int_0^t[h^s(\vec{\rho}',t)-h_2^s(\vec{\rho}',t)+h_c^s(\vec{\rho}',t)+h_{c2}^s(\vec{\rho}',t)]dt\right]\frac{dt\,d\phi}{t-\rho_{is}+2\rho'\cos(\beta/2)\cos(\phi'-\phi^i+\beta/2)} \tag{11}$$

where $\vec{\rho}' = x'\hat{x} + y'\hat{y} = \rho'\cos\phi'\hat{x} + \rho'\sin\phi'\hat{y}$, and $\rho_{is} = \rho_i + \rho_s$.

Equation (11) is a rigorous two-dimensional imaging identity for the two-dimensional bistatic case. It requires the knowledge of not only the field scattered by the object (assumed to be measured) but also the correction fields. If the object is not known a priori, these correction fields may be unavailable. Neglecting them leads to a physical optics scattering identity similar to the monostatic, baseband identity described previously by the authors[9].

To test the significance of the physical optics approximation, consider the case of a perfectly conducting sphere. The exact current induced on the sphere due to the SMEP incident field waveform was computed using the marching-on-in-time method. From this, the scattered field and correction field waveforms were constructed and the information substituted into (11). The resulting thickness function is shown in Figure 6, and is found to very accurately reproduce the sphere geometry. In comparison, if the exact scattered field is substituted into (11), but the correction fields are ignored, the thickness function showed in Figure 7 results. It is seen that the resulting error is fairly small. Although a more significant difference is expected for convex scatterers in which shadowing may become an important factor, it is anticipated that the PO approximation should provide a reasonable approximation to the thickness function.

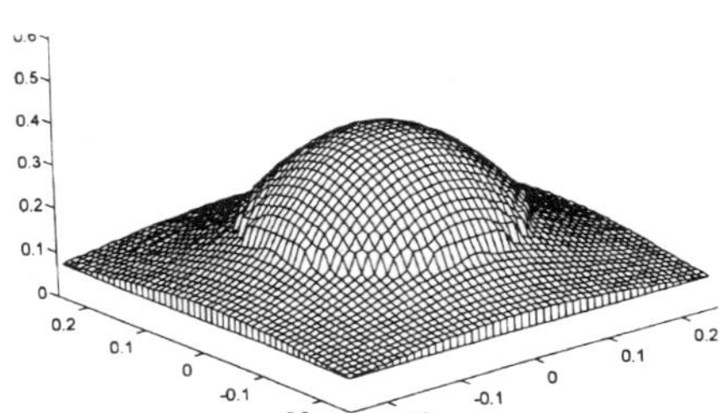

Figure 6. Thickness function for a conducting sphere computed using time-domain imaging identity. Correction fields included.

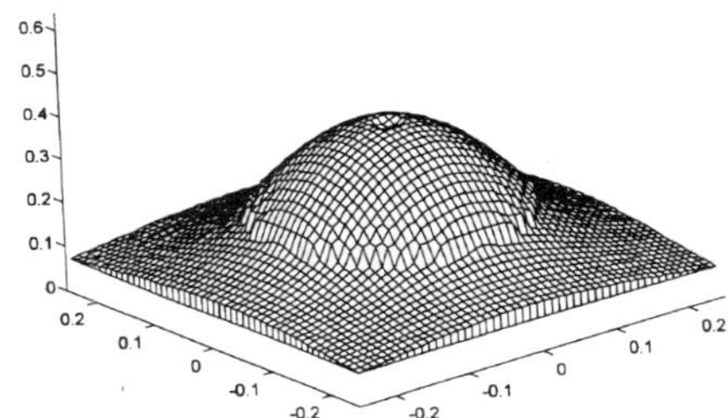

Figure 7. Thickness function for a conducting sphere computed using time-domain imaging identity. Correction fields ignored.

A simulation of the imaging technique has been carried out by measuring the scattered field response of a 1:72 scale model B-52. The response was measured at 200 aspect angles from 0° to 180° (with a bistatic angle $\beta = 10^o$) in the frequency band 4-16 GHz using an HP-8720 network analyzer. Each response was then windowed using the spectrum of two SMEP waveforms with different values of α (as defined in (8)), and transformed into the time domain using the FFT. Figure 8 shows the spectrum of the two SMEP waveforms. The first has a 3 dB bandwidth of 2.4 GHz (24% of the center frequency of 9 GHz), while the second has a bandwidth of 1.2 GHz (12%). Since the spectra were abruptly truncated, the time-domain waveforms, shown in Figure 9, are not precise SMEP waveforms, but only approximately so. Using the time derivative of this data in the imaging identity (11), and ignoring the correction terms, results in the images shown in Figures 10 and 11. Using the derivative allows the edge of the aircraft to be more clearly delineated. It is seen that as the bandwidth is decreased, resolution is diminished, even though the imaging identity is precise for the SMEP waveform. Considerable enhancement is possible by using peak detection on the scattered field waveforms. It is also seen that shadowed portions of the aircraft, such as the trailing edges of the wings, are not visible due to the limitations of the physical optics approximation.

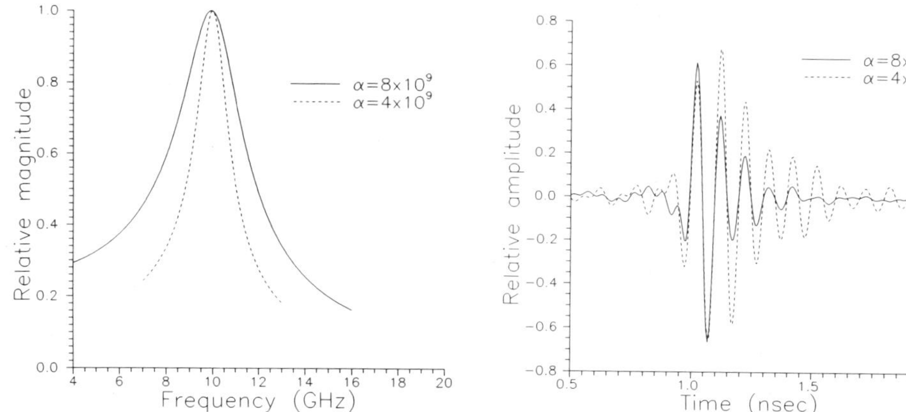

Figure 8. Magnitude spectra of SMEP waveforms with $f_c=10$ GHz.

Figure 9. SMEP waveforms with $f_c=10$ GHz.

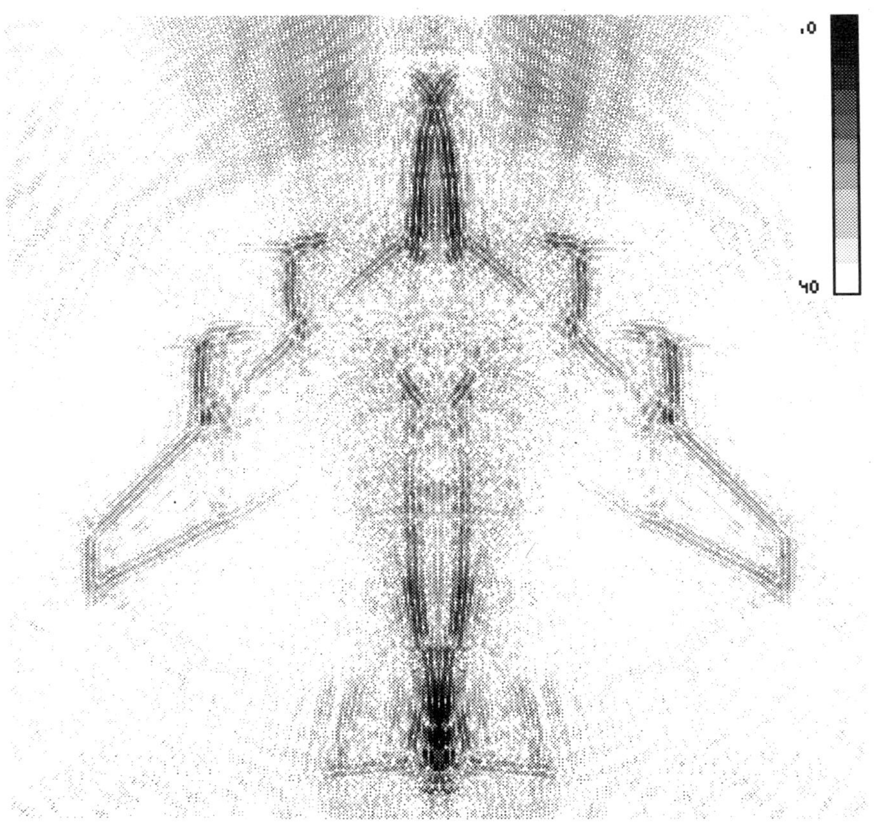

Figure 10. Image of 1:72 scale B-52 aircraft model constructed from measured SMEP response. $f_c=10$ GHz, $\alpha=8\text{x}10^9$.

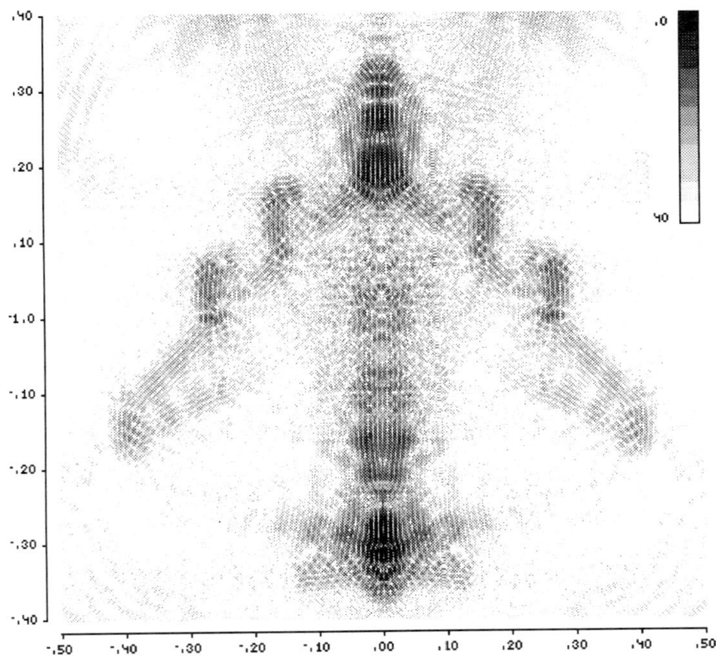

Figure 11. Image of 1:72 scale B-52 aircraft model constructed from measured SMEP response. $f_c=10$ GHz, $\alpha=4\times10^9$.

Acknowledgements

This work was supported by the Office of Naval Research under Grant N00014-93-1-1272.

REFERENCES

1. C. Phillips, P. Johnson, K. Garner, G. Smith, A. Shek, R.C. Chou, and S. Leong, Ultra-high-resolution radar development and test, in: *Ultra-Wideband, Short-Pulse Electromagnetics 2*, Lawrence Carin and Leopold B. Felsen, ed., Plenum Press, New York (1995).
2. P. Hansen, M. Sletten and K. Scheff, Ultrawideband, impulse driven X-band Clutter Measurement Radar, *1995 URSI Radio Science Meeting Digest*, p. 270.
3. K.M. Chen, E. Rothwell, D.P. Nyquist, J. Ross, P. Ilavarasan, R. Bebermeyer, Q. Li, C.Y. Tsai and A. Norman, Radar identification and detection using ultra-wideband/short-pulse radars, in: *Ultra-Wideband, Short-Pulse Electromagnetics 2*, Lawrence Carin and Leopold B. Felsen, ed., Plenum Press, New York (1995).
4. E. Rothwell, K.M. Chen, D.P. Nyquist, P. Ilavarasan, J.E. Ross, R. Bebermeyer, and Q. Li, A general E-pulse scheme arising from the dual early-time/late-time behavior of radar scatters, *IEEE Trans. Ant. Propagat.*, 42:1336 (1994).
5. J. D. Young, Radar imaging from ramp response signatures, *IEEE Trans. Ant. Propagat.*, AP-24:276 (1976).
6. N. N. Bojarski, A survey of the physical optics inverse scattering identity, *IEEE Trans. Ant. Propagat.*, AP-30:980 (1982).
7. C. L. Bennett, Time domain inverse scattering, *IEEE Trans. Ant. propagat.*, AP-29:213 (1981).
8. S. R. Deans. *The Radon Transform and Some of its Applications,* John Wiley & Sons, 1983.
9. E.J. Rothwell, K.M. Chen, D.P. Nyquist, and J.E. Ross, Time-domain imaging of airborne targets using ultra-wideband or short-pulse radar, *IEEE Trans. Ant. Propagat.,* 43:327 (1995).

POLARIMETRY IN ULTRAWIDEBAND INTERFEROMETRIC SENSING AND IMAGING

Wolfgang-Martin Boerner* and James Salvatore Verdi

NAWC-AD-PAX, Bldg. 2187, Rooms 3122, MS-3
Attn: James S. Verdi, Manager, P3-SAR Program, Code 45.552
48110 Shaw Road, PATUXENT RIVER, MD 20670-5304
T/F: +[1](301)342-0048/0121

* Otherwise with:

UIC-EECS/CSN, M/C 154, 900 W. Taylor St., SEL(607)W-4210,
CHICAGO, IL / USA 60607-7018, T&F: +[1](312)996-5480

Abstract: **'WISIP: Wideband (μHz - PHz) Interferometric Sensing and Imaging Polarimetry'** has become an important, indispensable tool in wide area military battlespace surveillance and global environmental stress change monitoring of the terrestrial and planetary covers. It enables dynamic, real-time optimal feature extraction of significant characteristics of desirable targets and/or target sections with simultaneous suppression of undesirable background clutter and propagation path speckle at hitherto unknown clarity and never before achieved quality. **'WISIP'** may be adopted to the **D**etection, **R**ecognition and **I**dentification (**DRI**) of any stationary, moving or vibrating target or distributed scatterer segments versus arbitrary stationary, dynamically changing and/or moving geo-physical/ecological environments, provided the instantaneous 2×2 phasor (Jones/Sinclair) and 4×4 power density (Mueller/Kennaugh) matrices for forward-propagation/backward-scattering, respectively, can be measured with sufficient accuracy. For example, the **DRI** of stealthy, dynamically moving and/or camouflaged stationary objects occluded deeply into heterogeneous stationary and/or dynamically moving inhomogeneous volumetric scatter environments such as precipitation scatter, the ocean sea/lake surface boundary layers, the littoral coastal surf zones, pack-ice and snow or vegetative canopies, dry sands and soils, etc., can now be successfully realized. A comprehensive overview is presented on how these modern high resolution/precision, complete polarimetric coregistered signature sensing and imaging techniques, complemented by full integration of novel navigational electronic tools, such as **DGPS**, will advance electromagnetic vector wave sensing and imaging towards the limits of physical realizability. Various examples utilizing most recent image data take sets of the **NAWC/ERIM-P3-UWB-TOPIF'E-CATI/LTBL-POLSAR** and **NASA-JPL-AIRSAR** airborne, the **NASA/DARA/DASI-SIR-C/X-SAR** shuttle, and the **ESA ERS-1/2**, **JERS** and **RADARSAT** satellite imaging systems will be presented for demonstrating the utility of **WISIP**.

INTRODUCTION

A succinct overview of the pertinent wideband (μHz - PHz) polarimetric theory, metrology and systems calibration, sensor design and device technology as well as of vector signal and tensor image processing is presented covering natural and/or

Ultra-Wideband, Short-Pulse Electromagnetics 3
Edited by Baum *et al.*, Plenum Press, New York, 1997

anthropogenic scatterer scenarios. A comprehensive assessment on how the resulting **'Huynen polarization fork (HPF)', the 'polarimetric entropy coefficients (PEC: H/A/α), the 'optimal polarimetric contrast enhancement (OPCE)'** and the optimal **'polarimetric matched signal/image filters (PMSF/PMIF)'** may be implemented effectively in modern high resolution, interferometric precision, complete polarimetric, co-registered multi-altitudinal/multi-platform signature fusion, sensing and imaging technology is provided. Such **U**ltra**W**ide**B**and **(UWB) M**ulti**B**and **(MB) TOP**ographic, **I**nter**F**erometric digital-**E**levation **(TOPIF'E) POL**arimetric (scattering matrix) **S**ynthetic **A**perture **R**adar **(POL-SAR)** imaging techniques, complemented by full integration of novel navigational electronic tools such as **G**lobal satellite electronic **P**ositioning **S**ystems **(GPS)** and **I**ntegrated inertial **N**avigation **U**nits **(INU)** and by **C**ross-**A**long-**T**rack **I**nstantaneous-inflight **(CATI)** as well as **L**ong-**T**emporal-**B**ase-**L**ine **(LTBL)** repeat-track/orbital overflight image overlay interferometry, will advance electromagnetic vector wave sensing and imaging towards the limits of physical realizability in that in addition to fixed target sections, the **D**etection, **R**ecognition and **I**dentification **(DRI)** of minutest instantaneous-to-long-term environmental scene changes have now become feasible. In addition, **WISIP** will play a major role in developing combined **UWB-TOPIF'E-POLSAR ↔ MTI** (**M**oving **T**arget **I**ndication) imaging systems for precise instantaneous to long-term tracking, detection, recognition and identification of stealthy stationary to rapidly moving objects occluded deeply into volumetric/surface scattering scenarios.

Whereas in this review paper, an exhaustive list of most pertinent references is provided, in the accompanying lecture series, various up-to-date novel, most recently processed image data takes will be presented parallel to introducing the underlying **UWB/MB-TOPIF'E-CATI-LTBL-POLSAR** imaging theory, metrology and calibration, polarimetric sensor design and device technology plus **PMSF/PMIF** optimal vector signal/tensor image processing and 2/3-D display. The associated **WISIP** lecture series includes timely mission-oriented geo-ecological examples obtained for:

I. Polarization in Nature: (1.1) Atmospheric Optical Polarimetry: the Arago sphere of sky-polarization, (1.2) Marine Optical Polarimetry: Waterman's refracted Arago-sphere of underwater sky-polarization; (1.3) Optical Sensory Bionic Polarimetry: the Haidinger brush of the human eye, e-polarization-vector navigation of insects/ants/.../fish/marine creatures; (1.4) Wideband Optical Multi-spectral Sensing and Camouflaging Polarimetry: IR-OPT-UV; (1.5) Low Frequency (ULF/ELF) magnetometric sensory bionic polarimetry of vertebrates: co-seismogenic signature detection versus polarimetric geomagnetic field gradient navigation of migrating birds; (1.6) The quest and need for establishing **"Polarimetric Sensory Bionic Research Laboratories"** straight adjacent to **Centers for Environmental Stress Change Remote Sensing** [1,4,8].

II. Basic to Advanced Theory of Optical-to-Radar Polarimetry: Great care is taken in separating the **optical forward propagation** (2×2 Jones/4×4 Mueller matrix) **polarimetry** from the **microwave backward scattering** (2×2 Sinclair/4×4 Kennaugh matrix) **radar polarimetry** by developing the following concepts:

IIa. Basic Polarimetric Radar Theory: Definition of Polarization State Operators and Polarimetric Matrices: The basic formulation and mathematical representation of polarization vectors and of the set of four distinct scattering matrices, their inter-relations, and their transformations between different polarization bases is considered [1-8]. Great care is taken in formulating the appropriate transmission (anti-monostatic) versus backscattering (monostatic) coordinate systems for treating the two distinct vector wave medium interaction cases -- in order of complexity -- expressed in terms of a set of four distinct matrices [5-8]. In each case, these include: (i) the 2×2 complex phasor (coherent) Jones transmission [T] versus Sinclair scattering [S] matrices; (ii) the associated 2×2 complex coherent power density transmission [F] = $[T]^{\dagger}[T]$ versus the Graves [G] = $[S]^{\dagger}[S]$ complex coherent power scattering matrices (with † denoting the Hermitian conjugate); (iii) the 4×4 real power density Mueller [M] propagation versus Kennaugh [K] scattering matrices (of which an optical, but non-identical alternate is the 4×4 Stokes reflection matrix); and (iv) the 3×3 (symmetric: monostatic reciprocal) or 4×4 (asymmetric: general bistatic and/or non-reciprocal) Polarimetric Covariance Transmission [ϒ] versus Scattering [Σ] matrices [6-8]. It is then shown how each set of four unique polarimetric matrices can be strictly related to one another via a matrix tryptic by utilization of the coherence matrix and the SU (2,3,4) Lie and Lorentz transformation groups [4,5], where use is made of Cloude's group-theoretic expansion [9,10] of the covariance matrices for the optical [21] and radar[25] cases, respectively. Also, the alternative four dimensional polarization sphere and four dimensional polarization ellipsoidal representations of Zhivotovsky [11] and Czyż [12,13], respectively are assessed.

II.b Distinct Polarimetric Matrix Optimization Approaches, the Huynen Polarization Fork for the Coherent and Partially Coherent Cases, and Target Matrix Decomposition: Although considerable progress was made in advancing the Kennaugh radar target characteristic polarization theory **(KRPT)** for the determination of the Optimal (characteristic) Polarization States and Huynen's Polarization Fork Concept **(HPF)** for unique association of these characteristic polarization states on the Poincaré sphere; no fully transparent theory separating the forward scattering (propagation) from the backscattering (monostatic and bistatic) cases and/or its interactive relations was developed until recently. Unfortunately, still today, these distinct wave-scatterer interaction cases are wildly confused in the literature [6,14]. However, with the recent advances made by Horn and Hong [15] in analyzing '**similarity**' versus '**consimilarity**' eigenvalue/vector problems [16], we are now equipped to resolve the fine points (pitfalls) of the coherent radar transformation phase in the formulation of the proper transformation matrices which differ distinctly for the propagation (similarity) and the backscattering (consimilarity) cases and can be determined uniquely by inclusion of energy conservation principles [6,14]. Using these two distinct matrix sets and the associated similarity versus consimilarity problems, the resulting five pairs [17] of optimal polarization states **(KRPT)** [18] are determined together with the complementing set of two Huynen polarization forks **(HPFs)** [19], where specific reference is made to the recent interactive forward propagation versus backscattering formulation along an idealized lossless, reciprocal polarimetric propagation two-port in terms of generalized transmission/reflection formulation of the 4×4 complex cascading propagation matrices [20] which serves to demon-strate that a clear distinction of the forward propagation (optical: Jones/Mueller) versus back-reflection (radar: Sinclair/Kennaugh) matrix cases [14] must be made in all cases [6].

For the partially (polarized) coherent cases, Huynen first proposed the concept of target matrix decomposition, i.e., the separation of the 4×4 power density matrices into an **'average target matrix [H]'**, plus a **'noise residue matrix [N]'**, in strict extension of the **'Chandrasekhar-Kraus'** principle of decomposing the received Stokes vector of a partially polarized wave into a fully polarized and a totally unpolarized component. This simplified canonical target matrix decomposition approach of Huynen [8,10,19] is compared with various other matrix decomposition approaches of Barnes, Krogager and especially Cloude [6,7,21,22]; and the associated hitherto unresolved uniqueness question is addressed [22].

IIc. The Polarimetric Entropy Concept and Polarimetric Contrast Optimization Procedure: Based on the complete (unique) propagation (transmission) versus (back) scattering (reflection) matrix formulation of optical versus radar polarimetry for both the coherent and partially coherent cases, in a next step the **polarimetric entropy concept (PEC)**, first conceived and formulated by Cloude [6,7,10,19], and the optimal contrast procedures for optimally separating desired (target) versus undesired (clutter) signatures (merit factor ratios), first conceived and treated by Kozlov [23] and the Russian polarimetrists [24], may now be uniquely defined and introduced.

The **Polarimetric Entropy Concept (PEC)** was first derived by Cloude from the **'polarimetric covariance [Σ] matrix'** properly defined in order to satisfy energy conservation principles [9,10,21,22]. Specifically, two distinct polarimetric entropy coefficients **(PEC)** are deerived from the engenvalues $\lambda_i\{[\Sigma^{3,4}]\}$ of the covariance matrix $[\Sigma^{3,4}]$: the **'polarimetric entropy H'**, and the **'polarimetric anisotropy A'**; and another **polarimetric phase entropy coefficient α** is derived from the associated eigenvectors (3,4) associated with the 3/4-dimensional covariance matrices for the symmetric reciprocal (3-dim.) and general bistatic (4-dim.) cases, respectively, where $0 \leq H \leq 1$. For null-entropy (H = 0) the scatter scenario is fully polarized, whereas for unit-entropy (H = 1) one deals with polarimetric white noise; and similar interpretations on the 'polarimetric scatterer randomness' exist for the A and coefficients as analyzed by Cloude and Pottier in [25,26], by Krogager and Holm in [27,28], by Jong-Sen Lee et al. in [29,30], and by Jakob J. van Zyl in [31,32]. In fact, **Cloude's polarimetric entropy coefficients PEC(H/A/α)**, derived from a group-theoretic approach [9,10], **is adding a very essential new tool to polarimetry**, in that it is, in general, now possible to determine unequivocally those image scenes which possess polarimetric-dependent properties, and those that don't via the **PEC-H** and **PEC-A** plus the specific principal target (e.g. sphere, bi/tri-hedral, cone-tip, etc.) characteristic features in terms of the **PEC-α** polarimetric entropy identifiers, plus its polarimetric feature -- characteristic transition states [26,32].

Optimal Polarimetric Contrast Enhancement (OPCE): In addition to defining those image sections for which polarization utilization becomes critical with the introduction of the **OPCE** coefficients, as defined first by Kozlov and collaborators, and more recently re-developed and generalized by Boerner together with Tanaka [33] and Mott [34], it is now possible to achieve optimal fine-tuned, post-processing contrast enhancement by the optimization of either the 2×2 coherent phasor Jones/Sinclair, the 2×2 coherent complex power Graves, and/or the 4×4 incoherent power density Mueller/Kennaugh matrices **but not of the covariance matrices**, the latter primarily serving the purpose of establishing polarimetric distributions [29] and entropy identifiers [21]. Here, it should be noted that similar to the coherent Huynen polarization fork **(HPF)** concept [17,19], it is possible to introduce another set of canonical polarimetric merit factors, **the OPCE-ratios**, which become essential tools in developing the **'Polarimetric Matched Vector Signal Filter (PMSF)'** and the **'Polarimetric Matched Tensor Image Filter (PMIF)'** algorithms [35,36].

IId. The Polarimetric Matched Signal/Image Filter (PMSF/PMIF) Algorithms in Wideband Polarimetric Sensing and Imaging Interferometry: With the complete unique formulation of the polarimetric scattering matrices, the polarization fork (**HPF**), the polarimetric entropy coefficients (**PEC: H/A/α**) and the optimal polarimetric contrast enhancement (**OPCE**) concept, it is now possible to establish uniquely the optimal **Polarimetric Matched Vector-Signal Filter (PMSF)** and the **Polarimetric Matched Tensor-Image Filter (PMIF)** concept as shown in [35,37], respectively. In addition, spatial Fourier transform analysis of image sections enables **reduction of image noise and speckle by removing high spatial frequency components. Most** importantly, the PMSF/PMIF algorithms are formulated so as to **enable the optimal**

application of standard computer-assisted digital image processing techniques which are, however, not further discussed here.

IIe Polarimetric Enhancement and Optimization of Interferometric Image Coherence:
Instead, specific attention is paid to most recent accomplishments in polarimetric image over-lay interferometry for both **CATI** (Cross-Along-Track-Inflight: single platform) and **LTBL** (repeat-track/orbit: Long-Temporal-Base-Line multiple platform) polarimetric image overlay interferometry for both altitudinal (height: span-invariant) and transverse (latitudinal and longitudinal [31,32]: complete scattering matrix) co-registered volumetric and surface stress-change analyses. This enormous scientific feat was made possible by implementation of novel **navigational electronic tools** such as Differential and Single Platform **GPS** (Global satellite Positioning System), by **PINS** (Precision Integrated inertial Navigation System), and various further advancing modes of **AMCS** (Automated Motion Compensation System) for airborne, shuttle- and satellite-borne imaging platforms. These and similarly more advanced **DGPS-PINS/ AMCS** techniques and **real-time on-board POL-SAR image processors** have made possible for the first time "**real-time image fusion**" of image data takes collected with separate multi-altitudinal, multi-spectral **UWB-TOPIF'E-CATI/LTBL-POL-SAR** imaging platforms in radar polarimetry; i.e., **TOP**ographic **I**nter**F**reometric (**TOPIF**) digital **E**levation relief mapping (**TOPIF'E**) has now become feasible with decimeter altitudinal and meter lati/longi-tudinal resolutions [38]. Thus, **WISIP** will bring about a complete change and overhaul of, and entirely new digital technologies in cartography and relief map production soon also becoming essential tools of modern navigation [38-48].

However, whereas the underlying '**Ultrawideband Poalrimetric Radar Theory**' for both the coherent and partially coherent cases is rather well developed, in '**POL-SAR Image Overlay Interferometry**' hitherto little attention was paid to fully exhaust the high resolution fine-structure information content of "**dual interferometric sets of complete (pixel-by-pixel) scattering matrix image information**" [49]. Although certainly truly remarkable progress is being made in applying various novel co-registration scemes of repeat-track/orbit images such as by Giles Peltzer, Paul Rosen et al, JPL [50,51,52] or by Richard Bamler, Kostas P. Papathanassiou and João R. Moeira, DLR [53,54], etc.; or by applying various existing yet highly upgraded statistical methods for improving "**interferometric coherence**" (i.e., the reduction of fuzziness of interferograms and speckle reduction, such as by Jong-Sen Lee and Dale Schuler, NRL-RSD [29,30,55,56,57]), integration of complete polarimetric approaches to optimizing interferometric coherence are still lacking. However, major research investigations are currently being pursued in order to overcome this last hurdle in perfecting WISIP-technology [49].

In concluding this sub-section on introducing the most recent accomplishments of **WISIP**, i.e., the demonstration of multi-altitudinal **UWB-TOPIF'E-CATI/LTBL-POL-SAR** image data take fusion, here we add that the combination of **POL-TOPIF'-SAR** with **POL-MTI-RAR** (**POL**arimetric **M**oving **T**arget **I**ndication **R**eal **A**perture **R**adar) has also been achieved in rudimentary modes [38-57] and further rapid advances of these techniques are to be expected soon, as is being discussed in the recent exhaustive survey of missions and sensors by ***Herbert J. Kramer, On Observations of the Earth and Its Environments*** [43].

III APPLICATIONS OF WISIP TECHNOLOGY

he wide ranging applications of WISIP-technology in air/space-borne remote sensing are slowly but steadily being accepted as major, indispensable tools in wide area military battlespace surveillance and local-to-global environmental stress change monitoring. In the complementing lectures series special attention will be paid to:

IIIa Detection, Recognition and Identification (DRI) of Environmental Stress Changes of either Natural and/or Anthropogenic Origin: Various specific applications are demonstrated dealing with the Detection, Recognition and Identification (DRI) of environmental stress changes of either natural and/or anthropogenic origin such as [38-57] of: (i) wetland versus flood plain delineation and the stress/pressure build-up along dams, dikes, and levies as well as deformations of the river beds, coastal shorelines and dunes, and surrounding wetlands during major flash flood and storm events [36,40]; (ii) the DRI of acid rain and acid snow on boreal forests and permafrost tundra environments [49]; (iii) altitudinal height and transverse skewing surface deformations during an entire tectonic stress change episode long before, precisely at, and long after the stress release (earthquake) has occurred [31,32,,39/40,44-57]; (iv) for sea-quakes, it will include tsunami-mapping and sub-ocean surface ocean-bottom fracture zone delineation [44-46]; (v) of geo-ecologic stress changes caused by natural and/or anthropogenic secondary/primary source mechanisms [8,48]; (vi) subtle strategic changes in battlespace scenarios [39,40]; and/or (vii) the camouflaged construction of hidden bunkers, arms caches and the overnight deployment of minefields, etc. [38-47]. Various illustrative polarimetric interferometric images, such as applications of most recent **UWB-TOPIF'E-CATI/LTBL-POL-SAR ↔ POL-MTI-TOP-RAR** [40-43,49,57] image data takes will be presented for the purpose of demonstrating the general applicability of the **WISIP** principles introduced in this state-of-the-art-overview.

IIIb. Principal Applications of Extrawideband Radar Polarimetry in Military Battlespace Surveillance and Environmental Stress Change Monitoring by utilizing: (3i) co-esismogenic/vocanologic 3-axis **ULF/ELF** electro/magneto-metric signature analyses in low frequency polarimetry; (3ii) Beyond the Horizon POL-RAD Imaging and Inverse Scatterometry in **HF/VHF** Polarimetry; (3iii) Polarimetric Radar Meteorology in Ground-based SHF (1 - 30GHz) Doppler Radar Polarimetry; (3iv) Air/Space-borne Ultrawideband (10MHz - 100GHz) **TOP**ographic **I**nter**F**erometric digital **E**levation (**TOPIF'E**) **POL**arimetric **S**ynthetic **A**perture **R**adar (**POL-SAR**) imaging, implementing both **CATI** (**C**ross-**A**long **T**rack **I**nflight) and Repeat-track/orbit **LTBL** (**L**ong-**T**emporal **B**ase**L**ine) interferometry utilizing the airborne **NASA-JPL AIRSAR** and the **NAWC-P3 Quadband TOPIF'E-POLSAR platforms**, the spaceborne shuttle **SIR-C/X-SAR TOPIF'E-POLSAR** and the satellite **ERS-1/2** and **JERS-1 TOPIF'E-POLSAR systems**; (3v) recently advanced **UWB/MB-TOPIF'E-POLSAR** imaging platforms such as the **FOA-CARABAS (10MHz - 100MHz), the ERIM TOPIF'E-POLSAR, the ERIM-DCS-TOPIF'E-POLSAR** imagers; as well as (3vi) **Polarimetric multi-spectral CCD Stokes vector NIR-OPT-NUV and UV spaceborne imagers** [4,8,43]. Specifically, the development of complete **POL**arimetric **LI**ght **D**etection **A**nd **R**anging (**POL-LIDAR**) in atmospheric and oceanographic remote sensing as well as of **POL**arimetric **LA**ser **D**etection **A**nd **R**anging (**POL-LADAR**) in eyesafe long-range space target DRI operated within the extended optical region (NIR-VIS-NUV), deserve to be mentioned here in that these eyesafe environmental remote sensing techniques will indeed play a major role in future environmental stress change monitoring and wide area battlespace surveillance of the boreal and austral polar regions.

The lecture series is concluded with a succinct critical analysis on still unresolved problems encountered in perfecting **'WISIP'** as well as with a preview on anticipated near-future **UWB/MB-TOPIF'E-CATI-LTBL-POLSAR-MTI** imaging systems advances and how with its extrawideband implementation in wide area environmental stress change monitoring and military battlespace surveillance mankind may come a step closer in fully realizing the "**irreversible paradigm conversion from military nationalist toward environmental global defense**" in better monitoring our fragile terrestrial and planetary hydro/bio-spheres [58,59].

CONCLUSION

A comprehensive overview of **Wideband Interferometric Sensing and Imaging Polarimetry** is presented together with a well structured identification of various crucial unresolved problems. Based on these meticulous, diligent analyses of radar polarimetry, very clear methods of solution (**Ansätze**) are provided. First, basic polarimetric radar theory and metrology needs to be perfected and some of the last hurdles must be removed as proposed. In a second step, various vector electromagnetic radar inverse scattering theories of more complicated shapes need to be solved in order to further perfect **PMSF/PMIF** algorithms by simultaneous advancement of the **PEC(H/A/α), OPCEC/OPIFE** concepts. In a third step, it is proposed to rapidly develop spread-spectrum improved **DGPS**-supported **CATI/LTBL-MB/UWB-TOPIF'E-POL-SAR Image Interferometry** which has become feasible for repeat-orbit shuttle/satellite operations and can be resolved also for airborne repreat-track overflights in the nearer future [48]. Because of the tremendous impact **WISIP** has on further prefecting **Day/Night All-Weather High Resolution Wide Area Surveillance of the Terrestrial and Planetary Covers**, more funding support for all R&D teams involved in these timely efforts is requested nationally, internationally, and worldwide [48,43,58, 59].

ACKNOWLEDGEMENTS

This research was supported under various US NAVY-ASEE-SFRP engagements with the Naval Air Surveillance (Code 45) and Advanced Electronic Navigation (NRaD Code 30) Departments, the P3-MB/UWB-POL-SAR program (Code 45.512) and by various complementing research study contracts. The partial research travel interaction support for attending the AMEREM'96 May 25-27, in Albuquerque, NM was received from the US Navy ONR-OE and of the NAWC-AD-WAR Code 45.552, and it is sincerely acknowledged. Also, the invitation of the AMEREM'96 TPC for presenting this State-of-the-Art Review with complementing lectures is sincerely acknowledged.

REFERENCES

[1] **W-M. Boerner, et al., eds., *Direct and Inverse Methods in Radar Polarimetry*, Proc. NATO-ARW-DIMRP'88, Bad Windsheim, FR Germany,m Sept. 18-24, 1988, NATO ASI Series C: Math & Phys. Scie., Vol. C-350, Kluwer Acad. Publ. Co., Dordrecht, Holland: 1,938 pages) 1992 Feb. 15.**

[2] H. Mott, *Antennas for Radar and Communications: A Polarimetric Approach*, John Wiley & Sons, New York, NY, 1992.

[3] **H. Mott and W-M. Boerner, eds., *Radar Polarimetry*, SPIE'92 Int's Symposium, San Diego, CA, 1992 July 20-25, Radar Polarimetry Conference, SPIE, Vol. 1748, 1993.**

[4] **W-M. Boerner, Polarimetry in Wideband Interferometric Sensing and Imaging of Terrestrial and Planetary Environments (Invited Keynote Address, Vol. 1, pp. 1-38) in J. Saillard, E. Pottier, S.R. Cloude, (eds.), Proceedings of the Third International Workshop on Radar Polarimetry, JIPR-3, 1995 March 2-23, IRESTE, U. Nantes, La Chantrerie, Bretagne, France, Vol. 1&2.**

[5] W-M. Boerner, C-L. Liu and X. Zhang, Comparison of the Optimization Procedures for the 2×2 Sinclair and the 4×4 Mueller Matrices in Coherent Polarimetry and Its Application to Radar Target Versus Background Clutter Discrimination in Microwave Sensing and Imaging, Int'l Journal on Advances in Remote Sensing (IJARS), (EARSeL) Boulogne-Billancourt, France, Vol. 2(1-1), pp. 55-82, 1993.

[6] E. Lüneberg, Principles of Radar Polarimetry, IEICE Transactions on Electronics (Special Issue on Electromagnetic Theory), Vol. E78C, No.10, pp.1339-1345, Oct. 1995.

[7] W-M. Boerner, Invited Review: Introduction to Radar Polarimetry -- with Assessments of the Historical Development and of the Current State-of-the-Art, 76 pages, in P.J. Moser, ed., *Electromagnetic Wave Interactions* - Reviews presented during Herbert Überall Sixty-fifth Birthday Emeritation Lecture Series at the Catholic University of America, 1995 June 03, World Scientific Publishing Co., PTE, Ltd, Farrer Road, Singapore 9128, 1996.

[8] **W-M. Boerner, Wideband Interferometric Sensing and Imaging Polarimetry, 86 pages, in A.W. Sáenz and P.-P. Delsanto, eds., *New Perspectives on Problems in Classical and Quantum Physics*, A Festschrift Buch in honor of Professor Herbert Überall, Gordon and Breach, 1996.**

[9] S.R. Cloude, Group Theory and Polarization Algebra, Optik, Vol. 75, No.1, pp. 26-26, Jan. 1996.

[10] S.R. Cloude, Lie Groups in Electromagnetic Propagation and Scattering with Applications to Radar Polarimetry, pp. 91-142, in C.E. Baum and H.N. Kritikos, *Electromagnetic Symmetries*, Taylor & Francis, Bristol, PA, 1995.

[11] L.A. Zhivotovskiy, The Polarization Sphere Modification (to four-dimension) for the Representation of Partially Polarized Electromagnetic Waves, Radiotechnica i Electronica, Vol. 30, No.8, pp. 1497-1504, 1985.

[12] Z.H. Czyż, Polarization Properties of Non-Symmetrical Matrices: A Geometrical Interpretation, D.Sc. Thesis, Warszaw Technical University, Warszaw, Poland, 1986.

[13] Z.H. Czyż, Scattering and Cascading Matrices of the Lossless Reciprocal Polarimetric Two-Port in Their General Similarity Versus Consimilarity Formulations, Proc. SPIE'96, Innsbruck, Tyrol, Austria (1996 July 8-12).

[14] E. Lüneberg and W-M. Boerner, The Backscatter Operator in Radar Polarimetry, Its Con-eigen-value/vector and Con-similarity Representations, and Its Applications, in print, AEÜ, Vol.50 (1996), (PIERS'94 ESA, Noordwijk, NL, 1994 July 11-15).

[15] R.A. Horn and C.A. Johnson, *Matrix Analysis* (Similarity versus Consimilarity), Cambridge Unviersity Press, New York, NY, 1985 (also see: Y-P. Hong and R.A. Horn, A Canonical Form for Matrices under Consimilarity, Linear Algebra and its Applications, **102**, (1989) 143-168; ibid, **73**, (1986) 213-226; **11**, (1975) 189-218; Y-P. Hong, 'Consimilarity: Theory and Applications', Ph.D. dissertation, The Johns Hopkins University, Baltimore, MD/USA, 1985).

[16] R.A. Horn and C.A. Johnson, *Topics in Matrix Analysis*, (similarity versus consimilarity transformations), Cambridge University Press, Cambridge, MD, 1991.

[17] A-Q. Xi and W-M. Boerner, Determination of the Characteristic Polarization States of the Target Scattering Matrix [S(AB)] for the Coherent Monostatic and Reciprocal Propagation Space, J. Opt. Soc. Amer., Part 1A, Optics & Image Science, Series 2, Vol.9, No.3, pp. 437-455, March 1992.

[18] E.M. Kennaugh,Polarization Properties of Radar Reflectors, M.Sc. Tehsis, Dept. of Electrical Engineering, The Ohio State University, Columbus, OH 43212, 1952 (also see: D.L. Moffatt and R.J. Grabacz, Research Studies on the Poalrization Properties of Radar Targets, by Prof. Edward M. Kennaugh, The Ohio Stae University, ElectroScience Laboratory, 1420 Kinnaer Road, Columbus, OH 43212, July 1984, Vols. 1&2).

[19] J.R. Huynen, Phenomenological Theory of Radar Targets, Ph.D. Dissertation, Technical University Delft, the Netherlands, 1970 (revised: 1987/available from author; also see: ibid, Chapter 11 in *Electromagnetic Scattering*, P.L.E. Uslenghi, ed., Academic Press, New York, 1978).

[20] Z.H. Czyż, Scattering and Cascading Matrices of the Lossless Reciprocal Polarimetric (Transmission Reflection) Two-Port in Their General Forms, PIERS'96, July 8-12, Innsbruck, Austria.

[21] S.R. Cloude and E. Pottier, Concept of Polarization Entropy in Optical Scattering, SPIE, OPT-ENG., Vol.34, No.6, pp. 1599-1610, June 1995.

[22] S.R. Cloude, Uniqueness of Target Decomposition Theorems in Radar Polarimetry, pp. 267-296, in Proc. NATO-ARW-DIMRP'88, W-M. Boerner et al, edsl, Part 1, Kluwer Academic Publ., Dordrecht, NL, 1992.

[23] A.I. Kozlov, Radar Contrast of Two Objects, Izvestiya Vuz., Radioelektronika, Vol. 22, No.7, July 1979, pp. 63-67.

[24] D.B. Kanareykin, N.F. Pavlov,and U.A. Potekhin, *The Polarization of Radar Signals,*Moscos: Sovyetskoye Radio, Chap. 1-10 (in Russian), 1966, (English Translation of Chaps. 10-12: Radar Poalrization Effects, CCM Inf.Copr., G. Collier and McMillan, 900 Third Ave., New York, NY 10023).

[25] S.R. Cloude and E. Pottier, A Review of Target Decomposition Theorems Based on Group-Theoretic Concepts in Radar Polarimetry, IEEE Trans. GRS, Vol. GE-34, No.2, March 1996, pp.498-518.

[26] S.R. Cloude and E. Pottier, An Entropy/α-parameter Based Classification Scheme for Land/Sea Applications for the Interpretation of POL-SAR Data Takes, IEEE Trans. GRS, Vol. GE-35, No.1, Jan. 1997, pp..

[27] E. Krogager, Aspects of Polarimetric Radar Imaging, D.Tech.Sci. Thesis, Technical University of Denmark, Lyngby, DK, 1993 (also see conference/symposium papers in PIERS, IGARSS, NATO-ARW/AGARD Proceedings; e.g., E. Krogager and W-M. Boerner, On the importance of utilizing Polarimetric Information in Radar Imaging and Classification, AGARD SPP Symposium on "Remote Sensing: A Valuable Source of Information", Toulouse, France, 1996 April 22-25, AGARD Conference Porceedings 582, pp.17.1 to 17.13, Oct. 1966.).

[28] E. Krogager and Z.H. Czyż, Properties of the Sphere, Diplane, Helix Decomposition of the Complex Sinclair Matrix in Radar Polarimetry, Proceedings JIPR-3, 1995 March 21-23, IRESTE-Nantes, France.

[29] J-S. Lee, M.R. Grunes, R. Kwok, Classification of Multi-look Polairmetric SAR Imagery Based on Complex Wishart Distribution, Int'l Journal Remote Sensing,Vol.15, No.11, pp. 2299-2311, 1994.

[30] J-S. Lee, K.W. Hoppel,S.A. Mango, A.R. Miller, Intensity and Phase Statistics of Multi-look and Interferometric SAR Imagery, IEEE Trans. GRS, Vol.32, No.5, pp. 1017-1028, Sept. 1994.

[31] J.J. van Zyl, Unsupervised Classification of Scattering Behavior Using Radar Polarimetry Data, IEEE Trans. GRS, Vol. GE-27, pp. 36-45, 1989 (also see: ibid, Application of Cloude's Target Decomposition Theorem to Polarimetric Imaging Radar Data Takes, in Radar Polarimetry, Proc. SPIE, 1748, pp. 184-212, 1992).

[32] H.A. Zebker and J.J. van Zyl, Imaging Radar Polarimetry: A Review, Proceedings of the IEEE, Vol.79, pp. 1583-1606, 1991.

[33] H. Mott, M. Tanaka and W-M. Boerner, Optimal Polarimetric Contrast Enhancement of the Coherent and Partial Coherent Radar Scattering Matrices and its Application to POL-SAR Image Analysis, Proc. PIERS'95, 1995 July 24-28, University of Washington, Seattle, WA, p.892, 1995.

[34] H. Mott, W-M. Boerner, M.M. Tanaka and Y. Yamaguchi, Determination of the Optimal Polarimetric Contrast Enhancement (OPCE) Coefficient from the 2×2 Coherent Sinclair [S] and the Partially Coherent 4×4 Kennaugh [K Matrices and its Interpretation in Terms of the Polarimetric Entropy Coefficients (PEC: H/A/α) in POL-SAR Image Analysis, VI.NASA-JPL-AESW, 1996 March (04)06pm - 08, Theodore von Karmann Auditorium, Pasadena, CA.

[35] W-M. Boerner, M. Walther and A. Segal, The Concept of the Polarimetric Matched Signal and Image Filters: Application to Radar Target Versus Clutter Optimal Discrimination in Microwave Imaging and Sensing, Int'l Journal on Advances in Remote Sensing (IJARS), (ERSeL), Boulogne-Billancourt, France, Vol.2, No.1-1, pp.219-252, Jan. 1993.

[36] J.S. Verdi, S. Krasznay, F. Ilseman, J.G. Teti and W-M. Boerner, Application of the Polarimetric Matched Image Filter to the Assessment of SAR Data from the Mississippi Flood Region, IEEE-IGARSS'94, Session: POL-II, Metrology, Calibration and Analysis, 1994 Aug. 8-12, CAL-TECH/JPL, Pasadena, CA, 1994.

[37] W-M. Boerner, E. Lüneberg and Y. Yamaguchi, Optimization of the Mueller [M] and Kennaugh [K] Power Density and Covariance [Σ] Matrices for Analyzing Incoherent Rough Surface Scatter, IEEE-IGARSS'94, Session: POL-II, Metrology, Calibration and Analysis, 1994 Aug.8-12, CAL-TECH/JPL, Pasadena, CA, 1994.

[38] W. Keydel, Session Organizer, IGARSS'96 Sessions 1/2, The World's Airborne SAR Facilities, Session 1: Conventional (CW) Frequencies (P/L/C/X-band) SAR-Systems and Respective Applications; Session 2: Extraordinary & Ultrawideband Frequency (ULF/UHF/S/Ku/W-bands) UWB: 10MHz-100MHz, 100MHz-200MHz, 200MHz-900MHz, 900MHz-1000MHz) SAR-Systems; IGARSS'96 May 26-30, Lincoln, NE.

[39] Y-J. Kim and J.J. van Zyl, NASA-CALTECH/JPL Airborne (P/L/C-band) TOPPOL-SAR System, CALTECH-JPL, Radar Engineering Section, MS 300-243, 4800 Oak Grove Drive, Pasadena, CA 91108-0899, T/F: +[1](818)354-9500/**393-5258**.

[40] D.R. Sheen and R. Rawson, ERIM (NAWC-AD-WAR) P3 UWB and C/X-band (TOPIF'E) POLSAR System, ERIM Radar Laboratory, 3300 Plymouth Road, P.O. # 13-4001, ANN ARBOR, MI / USA 48113-4001, T/F: +[1](313)994-1200 x2414/ **994-1808**. G.F. Adams and N.L. Vanden Berg, ERIM (US-ACE/TOP-LAB) LJ X-band TOPIF'E POL-SAR System. 3300 Plymouth Road, P.O. # 13-4001, ANN ARBOR, MI / USA 48113-4001, T/F: +[1](313)994-1200 x2625/**994-1808**. D.C. Ager and J.W. Burns, ERIM (USAF/WL) CV-580 (Multiband Data Collection System) DCS, 3300 Plymouth Road, P.O. # 13-4001, ANN ARBOR, MI / USA 48113-4001, T/F: +[1](313)994-1200 x2407/ **994-1808**.

[41] D. Held and L.H. Kosowsky, The NORDEN APG-76 TOP-MTI-SAR System, Westinghouse-Norden Systems, 10 Norden Place, NORWALK, CT/USA 06856-5300, T/F: +[1](203)852-7890/**7423**.

[42] B.C. Walker, SANDIA POL-SAR Testbed (DHC-6 Twin Otter STARLOS: 15/10/ 35GHz + 125-950MHz POL-TOPIF'E-SAR; P-3A AMPS 15GHZ ALL(TOPIF'E-POL)SAR; OC-135 SAROS X-band AN/APD-12-LORAL POL-SAR) Systems, Sandia National Laboratories, Bldg. 891, Rm. 4435, P.O. Box 5800, ALBUQUERQUE, NM / USA 87185-0529, T/F: +[1](505)844-1261/**0858**.

[43] **H.J. Kramer, *Observation of the Earth and Its Environment: Survey of Missions and Sensors*, 3rd Edition, New York, NY, Springer Verlag, May 1996 (2nd Edition: ISBN: 3-540-57858-7/0-387-57858-7, May 1994).**

[44] D. Evans and J.J. van Zyl, The NASA/DARA/DASI SIR-C/X-SAR Syhuttle Imaging Radar System, NASA/CALTECH/JPL Earth Sciences Division (180-703), Radar Engineering Section (300-227), 4800 Oak Grove Drive, PASADENA, CA/USA 91109-8099, T/F: +[1](818)393-1492/1365/**5285**.

[45] N. Jensen and B. Arbesser-Rastburg, The ESA-ERS-1/2 (multi-spectral European Remote Sensing Satellite Systems 1 & 2), ESA-ESTEC/XEB, Keplerlaan 1, NL-2200 AG NOORDWIJK, The NETHERLANDS, T/F: +[31](71)565-4541/**4999**.

[46] M. Shimada, The NASDA-JERS (Japanese Earth Resources Satellite) System, Fourth Observation Center, Roppongi First Blvd., 1-9-9 Roppongi, Minato-ku Tokyo 106, Japan, T/F: +[81](3)3224-7056/**7052**.

[47] F. Ahern, The CCRS-RADARSAT, Canadian Center for Remote Sensing, CCRS Data Acquisition Division, 601 Booth St., Ottawa, ONT/CAN K1A-0E8, T/F: +[1](613)947-1295/**1385**.

[48] M. Hayakawa and Y. Fujinawa, Electromagnetic Phenomena Related to Earthquake Prediction, Tokyo: Terra Scientific Publ. Co., 1994 (Proc. Int'l Workshop on 'ibid', University of Electro-Communications, Chofu-shi/Tokyo, Japan, 1993 Sept. 6-8).

[49] W-M. Boerner, J-S. Lee, D-L. Schuler and T-L. Ainsworth, Polarimetric Enhancement and Optimization of Interferometric Image Coherence in Repeat-Orbit (POL)-SIR-C/X-SAR Image Overlay Interferometry, SPIE Annual Meeting, 1997 July 28 - Aug. 02, San Diego Convention Center, Polarimetry and Spectrometry: WISIP Workshop (97 July 28-30), SPIE-Proceedings No..

[50] G. Peltzer and P.A. Rosen, Surface Displacement of the 1993 May 17 Eureka Valley, California Earthquake Observed by SAR Interferometry, Science, Vol. 268, 2 June 1995, pp.1333-.

[51] G. Peltzer, P.A. Rosen, F. Rogez, K. Hudnut, Post-seismic Rebound in Fault Step-Overs Caused by Pore Fluid Flow, Science, Vol.273, 30 Aug. 1966, pp. 1202-.

[52] P.A. Rosen, S. Hensley, H.A. Zebker, F.H. Webb, and E.J. Fielding, Surface Deformation and Coherence Measurements of Kilauea Volcano, Hawaii from SIR-C Radar Interferometry, Journal Geophysical Research, Vol. 101, No.E10, pp.23.109-125, 1996 Oct. 25.

[53] R. Lanari, G. Fomaro, D. Riccio, M. Migliaccio, K.P. Papathanassiou, J.R. Moreira, M. Schwäbisch, L. Dutra, G. Puglisi, G. Franceschetti and M. Coltelli, Generation of Digital Elevation Models by Using SIR-C/X-SAR Multi-frequency Two-Pass Interferometry: The ETNA Volcanologic Case Study, IEEE Trans. Geoscience and Remote Sensing, Vol.34(5), 1996 Sept., pp. 1097-1114.

[54] R. Bamler, N. Adam, G.W. Davidson and D. Just, Noise-induced Slope Distortion in 2D-Phase Unwrapping by Linear Estimates with Applications to SAR-Interferometry, IEEE Trans. Geoscience and Remote Sensing, Vol.35(3), 1997 May (in print).

[55] J-S. Lee, D-L. Schuler and M-R. Grünes, Statistical Analysis of Segmentation of Multi-lock SAR Imagery Using Polarimetric AIRSAR and SIR-C/X-SAR Date, Proc. IGARSS'95, Vol. II, pp. 1422-1424.

[56] D-L. Schuler and J-S. Lee, A Polarimetric Microwave Technique to Improve the Measurement of Directional Ocean Wave Spectra, Ins. J. Remote Sensing, (Technical Note), Vol.16(2), 1995, 199-215.

[57] D-L. Schuler, J-S. Lee and G. DeGrande, Measurement of Topography Using Polarimetric SAR Images, IEEE Trans. GRS, Vol.34(5), 1996 Sept., pp. 1266-1277.

[58] W-M. Boerner and J.B. Cole, (Invited), FROM MILITARY TO PLANETARY ENVIRONMENTAL DEFENSE:The Challenge of the next Century, and a Viable New Role of the US Military in an "ENVIRONMENTAL PLANETARY DEFENSE INITIATIVE" on a Global Scale, Proc. NSIA-DEFENSE INDUSTRY AND THE ENVIRONMENTAL AGENDA-SYMPOSIUM, 1991 Oct. 9-10, Sheraton Premier Hotel at Tyson Corner, Vienna, VA, pp. 314-330, Nov. 1991, (available from Dr. D. Brent Pope, Ed., NSIA 1025 Connecticut Ave., NW, Washington, DC 20036-5405).

[59] W-M. Boerner and J.B. Cole, From Natioal Military Towards Planetary Environmental Defense: A New Role for World Militaries in an International Environmental Defense Initiative, Parts 1&2, IEEE Journal on Society and Technology, No.3, July 1993, (14 pages), in print. (also in reduced form in H. Mott and W-M. Boerner, eds., Radar Polarimetry, SPIE Proc., Vol. 1748, pp. 12-22).

POLARIZATION PROCESSING FOR UWB RADAR

Shane R Cloude,

Applied Electromagnetics
St. Andrews, KY16 9XD, Scotland, UK

1. INTRODUCTION

Radar Polarimetry is now an established discipline within the microwave remote sensing community [1] and several multi-frequency polarimetric synthetic aperture radars (POLSARs) are currently operational [2]. In these systems the complete 2 x 2 complex scattering matrix of the target is imaged in range and azimuth and by analysing the relative amplitude and phase of terms in this matrix, classification and parametric inversion is possible. Usually such systems operate at high resolution but narrow fractional bandwidth in one or more of the frequency bands L(λ = 25 cm) C (λ = 6 cm) or X (λ = 3 cm). Such methods have already been applied in a wide range of remote sensing problems, from surface scattering studies [3] to woody biomass estimation [4].

In the context of recent developments in Radar systems operating with ultra wide band (UWB) signals, it is of timely interest to investigate the possible exploitation of wave polarization information over very wide instantaneous bandwidths, for the purposes of further improved target classification and parameter extraction. Since an UWB POLSAR promises instantaneous measurement of vector target scattering over several decades of wavelength scale, the potential clearly exists for practical inversion of scattered field data. In this paper we address this problem and suggest a formalism which may be used to develop a general theory of UWB polarimetry.

The problem, as with all multi-parameter Radar studies, is the complexity of the received signal. One of the key challenges of all such studies is to consider methods for compressing the data while preserving key target information and maintaining some kind of invariant (i.e. polarization basis independent) description of the scattering process. While such a description has been developed for narrow band polarimetry, no such theory yet exists for the UWB case, although initial developments have recently appeared in the literature [5].

2. GENERAL THEORY OF UWB POLARIMETRY

The natural wave basis to use for a description of UWB polarized signals is Cartesian x-y (horizontal and vertical linear polarizations). In this basis we have a physical representation of any broad band plane wave as an electric field locus in the xy plane. For example, figure 1 shows the UWB time signature of a dihedral scatterer illuminated at normal incidence with a plane wave of $+45^{0}$ linear polarization and Gaussian time history. The signature was predicted using a 2-D FDTD simulator augmented with Huygens surface and near-to-far field modules [8].

Ultra-Wideband, Short-Pulse Electromagnetics 3
Edited by Baum *et al.*, Plenum Press, New York, 1997

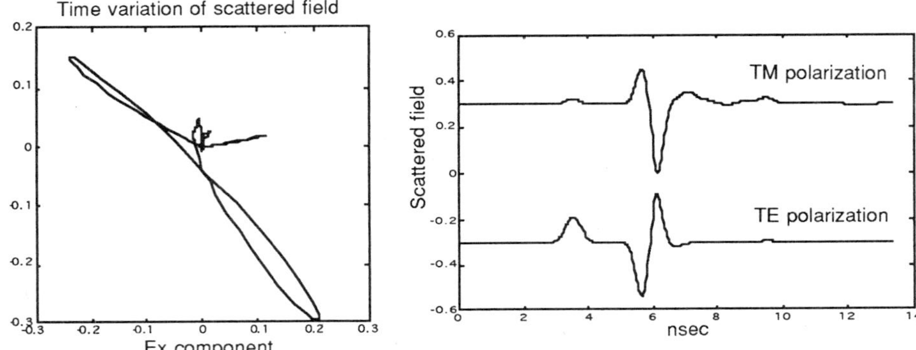

Figure 1: UWB Polarization Response of Dihedral Reflector

From the high frequency asymptotic behaviour, we expect the scattered wave to be polarized at -45^{0} linear (for an incident wave with +45^{0} polarization). However, we can see that the time history involves some departure from this behaviour. We clearly cannot use a simple narrow band polarization ellipse description of such waveforms and so must consider a more general spectral approach. The objective is to obtain an invariant geometrical description of arbitrary trajectories in the xy plane and to use these geometrical descriptors to classify targets. We shall see that we can define the spectral entropy of trajectories such as figure 1 and that this scalar property is of great importance in the UWB polarimetric classification problem.

We summarise in figure 2 the general processing methodology. We begin with a time/space representation of the broad band signal (as in figure 1). We then seek to represent this signal in a complex space where we can apply unitary matrix transformations to simulate a change in polarization state. Finally we transform back to a real space where the new signal represents a time series obtained at the terminals of an elliptically polarized antenna. This filtering will change the structure of the UWB time signal and the geometry of the trajectory in x-y space and so we must seek the best type of filtering to bring out the invariant features of the UWB signal i.e. those which do not depend on a choice of unitary matrix in figure 2.

We develop our approach in three stages: firstly we review the theory of spectral representation of UWB polarized signals, as first outlined by Ulander [6]. This is the first stage of the process as it allows us to construct a complex space in a well defined manner.

We then consider how to represent target scattering in this space and in particular we consider the effects of a change in antenna polarization base i.e. how to apply the unitary matrix transformations of figure 2. Finally we consider the concept of k-space entropy which will permit us to characterise in a basis invariant manner the polarimetric variability of the signal in k-space. This can be used to highlight those frequency bands where the response is most sensitive to wave polarization and can also be used to parametrically classify different types of scatterer by combining the k-space entropy concept with conventional k-space holographic imaging.

3. SPECTRAL REPRESENTATION OF POLARIZED WAVES

The spectral representation associates with the E-field locus in the x-y plane a complex signal z such that

$$z(t) = x(t) + i\,y(t) \qquad i = \sqrt{-1} \qquad -1)$$

The Fourier Transform of z(t) then provides a complex mapping of the x oriented real signal components into positive frequencies and the y oriented components into negative frequencies. In this scheme a harmonic plane polarized wave is then represented in the frequency domain as shown in equation 2

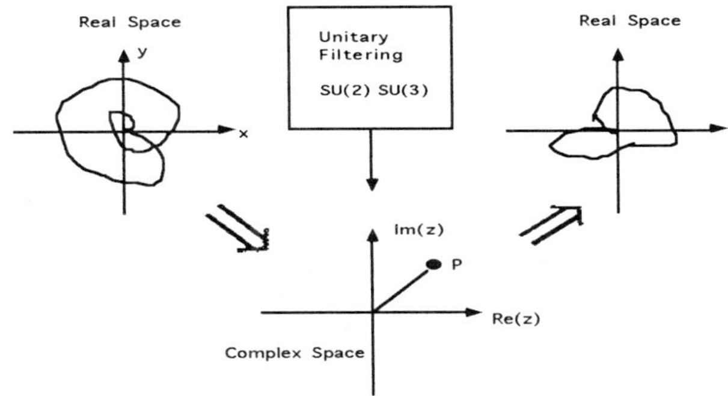

Figure 2: Schematic Flow Diagram of UWB Polarimetric Processing

$$E(\omega) = E_+ \exp(i\phi_+)\delta(\omega - \omega_0) + E_- \exp(i\phi_-)\delta(\omega + \omega_0) \qquad -2)$$

This distributional approach permits us to immediately extend the representation into wide band signals with bandwidth B (figure 3) so that in general we construct a *complex* spectrum $E(\omega)$ from the two channels of time domain data as

$$E(\omega) = E_x(\omega) + iE_y(\omega) \qquad -3)$$

It follows from equation 2 that such a representation is a decomposition of the time domain signal into left and right handed circularly polarized components. In this way the circular polarization basis has a more fundamental role to play in UWB signal analysis than for narrow band Radar. We can of course change the antenna polarization to any orthogonal elliptical state pq. This we can do by invoking a unitary complex transformation of the complex spectral vector so that we obtain a new pair of positive frequency components given by equation 4

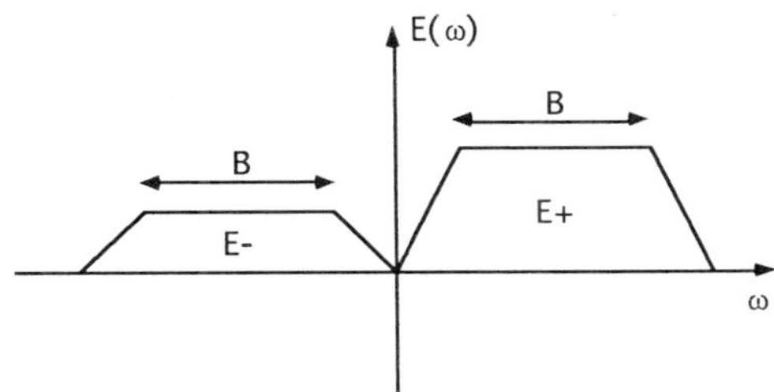

Figure 3: Complex Spectrum of Polarized UWB signals

$$\begin{bmatrix} E_p^+ \\ E_q^+ \end{bmatrix} = \begin{bmatrix} U_p^+ & U_p^- \\ U_q^+ & U_q^- \end{bmatrix} \begin{bmatrix} E_s^+ \\ E_s^- * \end{bmatrix} \qquad \begin{matrix} U_p^+ = e^{i\theta}\cos\mu & U_p^- = e^{-i\theta}\sin\mu \\ U_q^+ = -i\,e^{i\theta}\sin\mu & U_q^- = i\,e^{-i\theta}\cos\mu \end{matrix} \qquad -4)$$

where the parameters μ and θ are spherical angles locating the new polarization base from the pole of the Poincaré sphere. Note the conjugation of the second element of the spectral vector. This arises because of the mapping from left and right circular polarization states into positive and negative frequencies and is an important feature of spectral developments of wide band polarimetry. Also note that the left hand side does not contain negative frequency components. These are easily obtained by using the conjugate symmetry of real signals ($E^- =$

conj(E^+)). Employing an inverse Fourier Transform, we can then generate $e_{pq}(t)$, the real time domain signal obtained at the terminals of p and q polarized antennas.

For example, rotation of an antenna pair about the line of sight by θ is represented in the spectral domain by the unitary matrix operation ($\mu = \pi/4$ in equation 4)

$$\begin{bmatrix} U_p^+ & U_p^- \\ U_q^+ & U_q^- \end{bmatrix} = \frac{1}{\sqrt{2}} \begin{bmatrix} e^{i\theta} & e^{-i\theta} \\ -i\,e^{i\theta} & i\,e^{-i\theta} \end{bmatrix} \qquad -5)$$

Figure 4 shows an example, the result of using $\theta = \pi/4$ in equation 5 on the data in figure 1 i.e. the effect of rotating antenna polarization to match the high frequency - 45° orientation of the field scattered from the dihedral. We see that the UWB signatures show different features and that, as expected, the $+45^o$ channel is much smaller than the - 45° (although not zero). Such a combination might be used for example to remove double bounce clutter scattering from UWB imagery of forests.

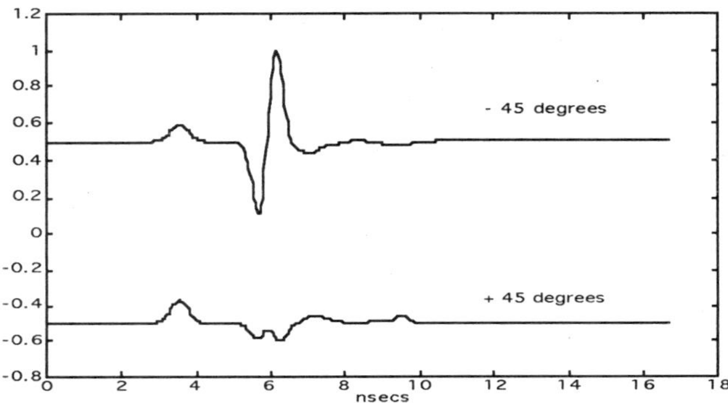

Figure 4: UWB signals at terminals of ± 45° polarized antennas (Dihedral Target)

In summary, by employing a spectral description of UWB polarized waves, we can model the effects of changes in antenna polarization using a unitary matrix transformation (equation 4). This spectral description has some differences to the conventional narrow band representation, most notably the conjugate operation required to relate negative frequencies to positive. There are two main approaches to the exploitation of such unitary transformations of broad band data :

- Null Polarimetry, where the unitary transforms are used to provide a broad band matched filter to enhance or suppress signals from the target (see [1] for details)
- Target Scattering studies, where emphasis is placed on characterising the broad band physics of scattering mechanisms in the object. Such studies must by nature be independent of wave polarization state and so must involve a matrix descriptor of the target which can be made invariant to unitary change of base transformations.

In this paper we concentrate on the second approach, the development of which requires construction of a spectral scattering matrix, to which we now turn.

4. SPECTRAL REPRESENTATION OF TARGET SCATTERING

The target itself can be represented by a 2 x 2 spectral scattering matrix S relating the positive and negative frequencies of the incident and scattered transient waves, such that we can define the scattered field spectral vector E_s in terms of the incident vector E_i as equation 6

$$\begin{bmatrix} E_S^+ \\ E_S^- * \end{bmatrix} = \begin{bmatrix} S^{++} & S^{+-} \\ S^{-+} & S^{--} \end{bmatrix} \begin{bmatrix} E_i^+ \\ E_i^- * \end{bmatrix} = \frac{1}{2} \begin{bmatrix} (S_{xx} - S_{yy} + 2iS_{xy})_{\omega+} & (S_{xx} + S_{yy})_{\omega+} \\ (S_{xx} + S_{yy})_{\omega-} * & (S_{xx} - S_{yy} + 2iS_{xy})_{\omega-} * \end{bmatrix} \begin{bmatrix} E_i^+ \\ E_i^- * \end{bmatrix} \quad -6)$$

where the elements of the scattering matrix are shown explicitly as functions of the linear x and y scattering coefficients. Note that all of the signals S_{xx}, S_{xy} and S_{yy} show conjugate symmetry, so for backscatter problems we always have $S^{-+} = S^{+-}$ but $S^{++} = S^{--}$ only if $S_{hv} = 0$. i.e. the spectral matrix, like the narrow band scattering matrix, is complex symmetric for backscatter problems with 3 independent elements.

Equations 4 and 6 represent the basic spectral formulation of Polarimetric UWB Radar. In concise notational form, the general matrix equation of polarimetry in the spectral representation for backscatter co-ordinates is then

$$\begin{bmatrix} E_p^+ \\ E_q^+ \end{bmatrix}_{receive} = [U_2]\,[S]\,[U_2]^T \begin{bmatrix} E_p^+ \\ E_q^+ \end{bmatrix}_{transmit} \quad -7)$$

Note that the spectral scattering matrix in equation 6 may be decomposed (just like its narrow band counterpart) using the Pauli matrices and target decomposition [7] then applied to generate an infinite set of spectral vectors using an 8 parameter 3 x 3 complex unitary matrix U_3 such that by definition we have

$$t(\omega)' = [U_3]\, t(\omega) \qquad t(\omega) = \begin{bmatrix} S^{++} & S^{--} & S^{+-} \end{bmatrix}^T \quad -8)$$

Physically such transformations represent different choices for an expansion of the spectral matrix in terms of physical scattering mechanisms. The representation of equation 8 allows us to consider just the properties of the scatterer, independent of the incident wave polarization.

Having outlined the basic formulation of UWB Polarimetry, we see that there is flexibility over the choice of unitary antenna polarization basis and over unitary change of target basis for the spectral scattering matrix. Both can be considered forms of complex filtering and the final stage of our analysis is to devise a basis invariant approach to such UWB Polarimetric processing. In the next section we show how the concept of spectral entropy may be defined as an invariant descriptor of UWB polarized signals and used to classify different types of scatterer.

5. SPECTRAL INVARIANTS: K-SPACE ENTROPY

The spectral entropy H is a basis invariant target parameter obtained from the eigenvalues λ_i of a spectral density matrix $J(\omega)$ defined as an integral over frequency of the outer product of spectral vectors (equation 9)

$$J(\Delta\omega) = \int_0^{\Delta\omega} t(\omega)\, t^{*T}(\omega)\, d\omega = \begin{bmatrix} J_{11} & J_{12} & J_{13} \\ J_{12}^* & J_{22} & J_{23} \\ J_{13}^* & J_{23}^* & J_{33} \end{bmatrix} \qquad J' = [U_3]\,[J]\,[U_3]^{*T} \quad -9)$$

H is then defined in the classical von Neumann sense as

$$H = \sum_{i=1}^{n} -P_i \log_n P_i \qquad P_i = \frac{\lambda_i}{\sum_{j=1}^{n} \lambda_j} \quad -10)$$

where n = 3 for backscatter problems so that $0 \leq H \leq 1$. If the spectral entropy is zero then the *polarimetric* properties of the target do not change with frequency, whereas if the entropy is unity then we have rapid spectral variation in polarimetric properties over the integration bandwidth $\Delta\omega$ in equation 9.

The second key basis invariant target parameter obtained from J is the average spectral vector $\bar{t}$, obtained as a weighted sum of eigenvectors of J. This vector can be parameterised as shown in equation 11 [7, 9]

$$\bar{t} = \begin{bmatrix} \cos\bar{\alpha} & \sin\bar{\alpha}\cos\bar{\beta}e^{i\bar{\delta}} & \sin\bar{\alpha}\sin\bar{\beta}e^{i\bar{\gamma}} \end{bmatrix}^T \qquad -11)$$

where, for example, the α-parameter is obtained from the normalised eigenvalues as shown in equation 12. Here α_1 is obtained from the eigenvector corresponding to λ_1 etc. (using the parameterisation shown in equation 11)

$$\bar{\alpha} = P_1\alpha_1 + P_2\alpha_2 + P_3\alpha_3 \qquad -12)$$

The angle α has a range of 90° and corresponds to a smooth change of behaviour from symmetric scattering where HH = VV ($\alpha = 0^o$) through scattering mechanisms where HH $\neq$ VV, encompassing dipole scattering (HH or VV = 0 and $\alpha = 45^o$)) and moving into double bounce scattering mechanisms between dielectric surfaces (as arises in foliage penetration applications), finally reaching dihedral scatter from metallic surfaces at $\alpha = 90°$. In this way we see that α can be used to distinguish several scattering mechanisms .

If we combine the information in α with that in H, we obtain a 2-dimensional H-$\bar{\alpha}$ classification space as shown in figure 5.

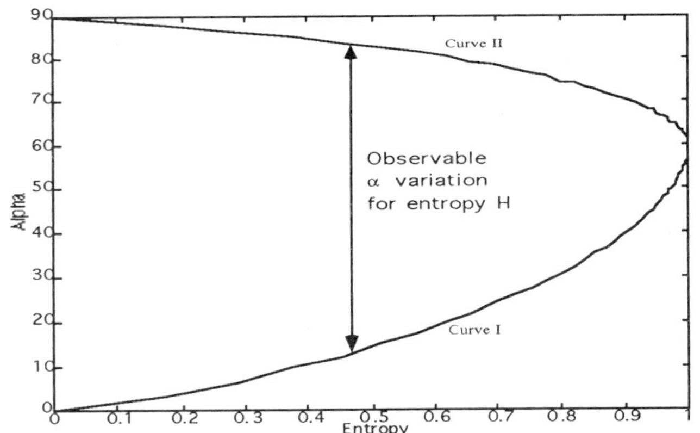

Figure 5: Feasible Region of H-α space lies between curves I and II

All possible UWB scattering mechanisms can be represented in this space. However, because of the averaging inherent in equation 11, not all regions of this space are equally populated. For example, when H = 1 there is only one possible value for alpha ($\alpha = 60°$). This reflects our increasing inability to distinguish between scattering mechanisms as the entropy increases. On the other hand, at H =0 we have access to the full range of possible α values and can distinguish many different types of polarimetric behaviour.

We can quantify the variation in this feasible region of figure 5 by considering the two curves shown as I and II. For each value of entropy H, we can identify a possible variation of α lying between curves I and II. These curves are determined by the H-α variation for a spectral coherency matrix with degenerate minor eigenvalues with amplitude m ($0 \leq m \leq 1$) of the form shown in equation 13

$$[J]_{I} = \begin{bmatrix} 1 & 0 & 0 \\ 0 & m & 0 \\ 0 & 0 & m \end{bmatrix} \quad [J]_{II} = \begin{bmatrix} m & 0 & 0 \\ 0 & 1 & 0 \\ 0 & 0 & m \end{bmatrix} \quad 0 \leq m \leq 1 \qquad -13)$$

6. FDTD SIMULATIONS OF UWB POLARIMETRIC SIGNATURES

We now apply the above procedure to FDTD simulations of UWB scattering from canonical structures [8]. We consider backscatter from two targets; a metallic dihedral corner and a metal strip, as shown in figure 6. The problem is solved using an 2-D FDTD package employing a Gaussian incident field for both TE and TM Polarizations. Because of the symmetry of the object, there is no linear cross polarization and hence the corresponding spectral matrix has only two independent elements. The spectral matrix is calculated using the TE (VV) and TM (HH) time signatures and a numerical FFT to obtain the positive and negative frequency components required to evaluate the elements of S.

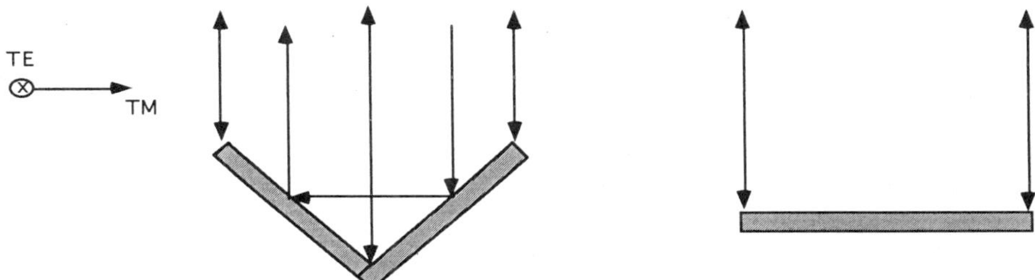

Figure 6: 2-D Dihedral Corner and Metallic Strip Scattering

In an UWB sense the corner scatterer shows three main features: edge scattering from the front, specular multiple internal scattering and contributions due to current flow over the surface of the dihedral. All of these features are visible in the time series shown in figure 1. Each of these mechanisms has its own polarization dependence, and each depends on frequency. We see that even for a simple object such as this, the UWB response is complicated. However, figure 7 shows the entropy and α parameter as a function of integration bandwidth for the dihedral. We see that despite the underlying complexity, the scatterer tends towards a fixed point, namely $(H, \alpha) = (0.3, 78^{o})$.

The strip target on the other hand shows much less polarimetric variation with frequency and so has a much lower spectral entropy. The FDTD simulations yield a point $(0.04, 6^{o})$ in the H-α space. This point is different from that for the corner and hence we can see how target classification becomes possible using a representation of UWB targets in H-α space.

7. CONCLUSIONS

In this paper we have outlined a general theory of geometrical processing for UWB polarized signals. We have shown that the spectral description of such signals leads to a matrix description of target scattering which relies on a mapping of the positive and negative frequency components of the complex spectrum for wave and target. This mapping can be used to define a spectral density matrix, the eigenvalues of which can be used to effect a basis invariant description of the polarimetric variability of the target signature. In this way we can compress the data obtained using an UWB polarimetric radar into a representation in a 2-D space of target points. In particular we have defined the concept of spectral entropy and showed how it may be employed for target classification studies.

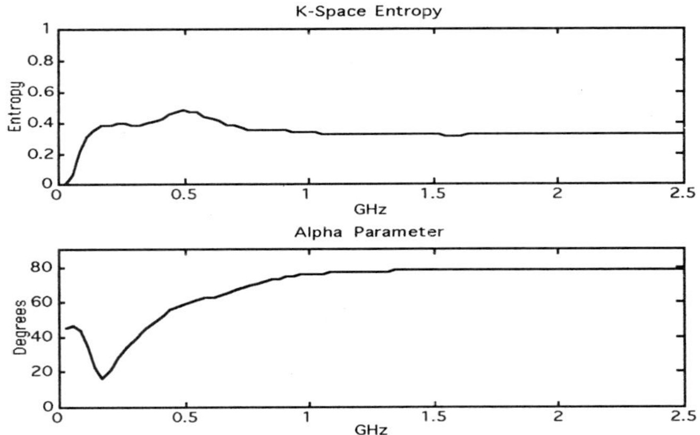

Figure 7 : Entropy and α parameter for UWB Dihedral response

8. REFERENCES

[1] W M Boerner, "Polarimetry in ... Sensing and Imaging of Terrestrial and Planetary Environments", Invited Keynote Address, Proceedings of 3rd International Workshop on Radar Polarimetry (JIPR '95), IRESTE, University of Nantes, France, April 1995, pp 1-38

[2] H J Kramer "Observation of the Earth and its Environment", Springer Verlag, Berlin, ISBN 3-540-60933-4, 1996

[3] P Dubois, J van Zyl, T Engman "An Empirical Soil Moisture Estimation Algorithm using Imaging Radar", Proceedings of IGARSS '94, pp 1573-1575, Pasadena, August 1994

[4] T Le Toan, A Beaudoin, J Riom, D Guyon "Relating Forest Biomass to SAR data", IEEE Trans GRS Vol. 30, March 1992, pp 403-411

[5] A Nelander " Analysis of Wide-band Polarimetric Radar", Proceedings of 3rd International Workshop on Radar Polarimetry (JIPR '95), IRESTE, University of Nantes, France, April 1995, pp 89-98

[6] S. R. Cloude, P.D. Smith, A. Milne, C. Thornhill, G Crisp, "K-Space Imaging Algorithms applied to UWB SAR", Proceedings of 1st IEEE International Conference on Image Processing (ICIP), Austin, Texas, 13-16 November 1994, pp 491-495

[7] S. R. Cloude, E. Pottier, "A Review of Target Decomposition Theorems in Radar Polarimetry", IEEE Transactions on Geoscience and Remote Sensing, Vol. 34 No. 2 March 1996, pp 498-518

[8] S R Cloude, A.M Milne, P D Smith ,"Time Domain Modelling: Integral Equations, Finite Difference and Experimental Results " IEE Proceedings of International Conference on Computational Electromagnetics, Publication No. 350, London, November 1991, pp 241-244

[9] S.R. Cloude, E. Pottier, "An Entropy Based Classification Scheme for Land Applications of Polarimetric SAR", Proceedings of International Symposium on Retrieval of Bio-and Geophysical Parameters from SAR data for land applications, CNES, Toulouse, France, October 10-13, 1995

IMPLEMENTATION OF THE OPTIMAL POLARIZATION CONTRAST ENHANCEMENT CONCEPT IN ULTRAWIDEBAND (MULTISPECTRAL) POL-SAR IMAGE ANALYSIS

Harold Mott[1] and Wolfgang-M. Boerner[2]

[1]Electrical Engineering Department
The University of Alabama
Box 870286
Tuscaloosa, Alabama 35487

[2]University of Illinois at Chicago
UIC-EECS
851 S. Morgan Street SEO-1120
Chicago, IL 60607-7053

INTRODUCTION

It is often desirable to select antenna polarizations that maximize the contrast in received powers from two classes of scatterers. A terrain-mapping radar may use polarization to maximize the contrast between forested areas and farmland. Alternatively, it may be desirable to choose polarizations to maximize received power or target-clutter ratio.

RECEIVED POWER

Received power can be written as

$$W = \frac{Z_0^2 I^2}{256 \pi R_a \lambda^2 r_1^2 r_2^2} \boldsymbol{G}_A^{rT} \boldsymbol{K} \boldsymbol{G}_A^t \tag{1}$$

where

$$\boldsymbol{G}_A = \begin{bmatrix} |h_x|^2 + |h_y|^2 \\ |h_x|^2 - |h_y|^2 \\ 2\operatorname{Re}(h_x^* h_y) \\ 2\operatorname{Im}(h_x^* h_y) \end{bmatrix} \tag{2}$$

is the Stokes vector of an antenna. The superscripts refer to transmitter and receiver respectively.

The real matrix **K** is often called the Mueller matrix of the target. As used here it comes from the target Sinclair matrix in coordinates at the transmitter and receiver with z axes pointing to the target, unlike Mueller matrix coordinates. Then **K** will be called here the "Kennaugh matrix."

For the transmitter, **G** is a Stokes vector in coordinates with z axis in the direction of wave travel. For the receiver, **G** is in coordinates with the z axis pointing from receiver to target. For this reason we use "Kennaugh vector" to denote the scattered wave and receiving antenna [1].

The received power may also be written as

$$W = \frac{Z_0^2 I^2}{128\pi R_a \lambda^2 r_1^2 r_2^2} \boldsymbol{H}^T \boldsymbol{C} \boldsymbol{H}^* \tag{3}$$

where

$$\boldsymbol{C} = \langle \boldsymbol{X}\boldsymbol{X}^{T*} \rangle \tag{4}$$

is the covariance matrix of the target. The angle brackets denote an ensemble average.

For bistatic scattering, vector **X** representing the target is defined as

$$\boldsymbol{X} = \begin{bmatrix} S_{xx} \\ S_{xy} \\ S_{yx} \\ S_{yy} \end{bmatrix} \tag{5}$$

and the Kronecker product

$$\boldsymbol{H} = \boldsymbol{h}_r \times \boldsymbol{h}_t = \begin{bmatrix} h_{rx}h_{tx} \\ h_{rx}h_{ty} \\ h_{ry}h_{tx} \\ h_{ry}h_{ty} \end{bmatrix} \tag{6}$$

is the joint antenna vector. For backscattering the target vector can be simplified to

$$\boldsymbol{X} = \begin{bmatrix} S_{xx} \\ \sqrt{2}\, S_{xy} \\ S_{yy} \end{bmatrix} \tag{7}$$

and the joint antenna vector is redefined to be

$$\boldsymbol{H} = \begin{bmatrix} h_{rx}h_{tx} \\ \frac{1}{\sqrt{2}}(h_{rx}h_{ty} + h_{ry}h_{tx}) \\ h_{ry}h_{ty} \end{bmatrix} \tag{8}$$

The power density of the scattered wave from a target is

$$\mathcal{P}^s = \frac{Z_0 I^2}{32\pi\lambda^2 r^4}\boldsymbol{h}_t^{T*}\sigma\boldsymbol{h}_t \tag{9}$$

where the ensemble average of Kronecker products of the Sinclair matrix,

$$\sigma = \langle \boldsymbol{S}^{T*}\boldsymbol{S}\rangle \tag{10}$$

is the Graves matrix [1].

CLUTTER MATRICES

Requirements for an unpolarized clutter matrix are that the scattered wave is unpolarized and received power is independent of transmitter and receiver polarizations.

The bistatic covariance clutter matrix is the 4 x 4 identity matrix and the Graves clutter matrix is the 2 x 2 identity matrix.

A matrix cannot be found that will make

$$\boldsymbol{H}^T\boldsymbol{C}\boldsymbol{H}^* = \text{constant} \tag{11}$$

for arbitrary polarizations of transmitter and receiver, where

$$\boldsymbol{H} = \begin{bmatrix} h_{rx}h_{tx} \\ \frac{1}{\sqrt{2}}(h_{rx}h_{ty} + h_{ry}h_{tx}) \\ h_{ry}h_{ty} \end{bmatrix} \tag{12}$$

Then no 3 x 3 covariance matrix can in general represent unpolarized scattering.

CONTRAST WITH THE 3 X 3 COVARIANCE MATRIX

The ratio of received powers from two scatterers is

$$R_{ab} = \frac{\boldsymbol{H}^T\boldsymbol{C}_{a3}\boldsymbol{H}^*}{\boldsymbol{H}^T\boldsymbol{C}_{b3}\boldsymbol{H}^*} \tag{13}$$

It can be maximized by choosing the components of **H** with the use of a Lagrange multiplier method [2] and leads to

$$(\boldsymbol{C}_{a3} - \lambda\boldsymbol{C}_{b3})\boldsymbol{H}^* = \boldsymbol{0} \tag{14}$$

This equation has the form of the generalized eigenvalue problem,

$$(\boldsymbol{C}_{a3} - \lambda\boldsymbol{C}_{b3})\boldsymbol{X} = \boldsymbol{0} \tag{15}$$

with eigenvalues λ and eigenvectors **X**. An eigenvector will cause R_{ab} to be an extremum. Since the elements of **H** are independent, $\mathbf{H}^*$ can be set equal to the eigenvector **X**. After $\mathbf{H}^*$ is found, transmitting and receiving antennas can be found.

BISTATIC CONTRAST WITH THE COVARIANCE MATRIX

The contrast between two scatterers in the bistatic case can be maximized by maximizing

$$R_{ab} = \frac{\boldsymbol{H}^T \boldsymbol{C}_{a4} \boldsymbol{H}^*}{\boldsymbol{H}^T \boldsymbol{C}_{b4} \boldsymbol{H}^*} \tag{16}$$

The elements of **H** are not independent. Eigenvectors found with a Lagrange multiplier approach have four independent elements and cannot be related to **H**.

If

$$R = \frac{\boldsymbol{X}^{T*} \boldsymbol{C}_{a4} \boldsymbol{X}}{\boldsymbol{X}^{T*} \boldsymbol{C}_{b4} \boldsymbol{X}} \tag{17}$$

is maximized, **H** can be set approximately to $\mathbf{X}^*$. Form 3-element vectors from the 4-element relationship, thus

$$\begin{bmatrix} 1 \\ P_r + P_t \\ P_r P_t \end{bmatrix} \approx \begin{bmatrix} 1 \\ X_2^*/X_1^* + X_3^*/X_1^* \\ X_4^*/X_1^* \end{bmatrix} \approx \begin{bmatrix} 1 \\ H_2/H_1 + H_3/H_1 \\ H_4/H_1 \end{bmatrix} \tag{18}$$

This vector equation can be solved to give approximate values of the antenna polarization ratios.

SIGNAL-CLUTTER RATIO WITH COVARIANCE MATRIX

Unpolarized clutter can be represented by the 4 x 4 identity matrix, so the ratio to be maximized is

$$R = \frac{\boldsymbol{H}^T \boldsymbol{C}_4 \boldsymbol{H}^*}{\boldsymbol{H}^T \boldsymbol{I} \boldsymbol{H}^*} \tag{19}$$

From eigenvector **X** corresponding to the largest eigenvalue of $\mathbf{C}_4$, an approximate joint antenna vector **H** is found. The approximation can be improved numerically.

The 3 x 3 covariance matrix cannot be used in this way if one of the targets is unpolarized clutter, since a 3 x 3 matrix for unpolarized clutter does not exist.

OPTIMUM POLARIZATION WITH NOISE OR UNPOLARIZED CLUTTER

The effects of noise and unpolarized clutter can be ameliorated if antennas are selected to maximize received power from a target.

The target's Graves matrix is used to select a transmitting polarization to maximize power density at the receiver. From the equation for power density and the constraint that the transmitter effective length have unit magnitude, the power density ratio to be maximized is

$$R_{an} = \frac{\boldsymbol{h}_t^{T*} \sigma \boldsymbol{h}_t}{\boldsymbol{h}_t^{T*} \boldsymbol{I} \boldsymbol{h}_t} \tag{20}$$

This leads to the standard eigenvalue problem,

$$(\sigma - \lambda \boldsymbol{I})\boldsymbol{X} = \mathbf{0} \tag{21}$$

Eigenvector **X** can be equated to the effective length $\mathbf{h}_t$ of the transmitter. From $\mathbf{h}_t$ the Stokes vector of the transmitter is formed. From it and the target Kennaugh matrix the Kennaugh vector of the scattered wave is found by

$$\boldsymbol{G}^s = \frac{Z_0^2\, I^2}{16\,\pi\,\lambda\, r_1^2\, r_2^2}\, \boldsymbol{K}\boldsymbol{G}_A^t \tag{22}$$

The Kennaugh vector of the scattered wave is separated into polarized and unpolarized parts, thus

$$\boldsymbol{G}^s = \boldsymbol{G}^{s\,(1)} + \boldsymbol{G}^{s\,(2)} = \begin{bmatrix} (1 - R)\, G_0^{\,s} \\ 0 \\ 0 \\ 0 \end{bmatrix} + \begin{bmatrix} R\, G_0^{\,s} \\ G_1^{\,s} \\ G_2^{\,s} \\ G_3^{\,s} \end{bmatrix} \tag{23}$$

where R is the degree of polarization of the scattered wave,

$$R = \frac{\sqrt{\left(G_1^{\,s}\right)^2 + \left(G_2^{\,s}\right)^2 + \left(G_3^{\,s}\right)^2}}{G_0^{\,s}} \tag{24}$$

Received power is the product of the wave Kennaugh vector with the Kennaugh vector of the receiving antenna, thus

$$\begin{gathered} W = \boldsymbol{G}_A^{r\,T}\boldsymbol{G}^s = \boldsymbol{G}_A^{r\,T}\boldsymbol{G}^{s\,(1)} + \boldsymbol{G}_A^{r\,T}\boldsymbol{G}^{s\,(2)} \\ \boldsymbol{G}_A^{r\,T} = \begin{bmatrix} 1 & G_{A1}^r & G_{A2}^r & G_{A3}^r \end{bmatrix} \end{gathered} \tag{25}$$

where the first element of the antenna Stokes vector makes the antenna effective length equal to one.

Power from the wave will be maximum if the antenna Stokes vector is proportional to the Kennaugh vector of the polarized part of the wave, thus

$$\boldsymbol{G}_A^r = \text{constant} \times \boldsymbol{G}^{s\,(2)} \tag{26}$$

The receiving antenna effective length is found from this vector.

CONTRAST WITH GRAVES AND KENNAUGH MATRICES

The contrast in received power from two targets can be approximately maximized in four ways:

A. Target b is ignored and power from target a is maximized as described in the preceding section.

B. The transmitter is found as in A. The Kennaugh matrix of target b is used to find the scattered wave from b. The wave is separated into unpolarized and polarized parts, and the Stokes vector of the receiving antenna is selected to make the power received from the polarized part zero.

Received polarized power is zero if

$$\boldsymbol{G}_A^r = \text{constant} \times \begin{bmatrix} G_0^{s(2)} \\ -G_1^{s(2)} \\ -G_2^{s(2)} \\ -G_3^{s(2)} \end{bmatrix} \tag{27}$$

C. Transmitter polarization is selected to be an eigenvector of the generalized eigenvalue equation of the Graves matrices. This maximizes

$$R_{ab} = \frac{\boldsymbol{h}_t^{T*} \sigma_a \boldsymbol{h}_t}{\boldsymbol{h}_t^{T*} \sigma_b \boldsymbol{h}_t} \tag{28}$$

Receiver polarization is found as in A.

D. Transmitter polarization is found as in C and receiver polarization as in B.

EXAMPLES

Bistatic covariance matrices of two targets were generated and other target matrices were generated from them. Element choices were arbitrary with appropriate constraints.

Covariance matrices for scatterers a and b are

$$\boldsymbol{C}_{a4} = \begin{bmatrix} 5.0 & 0.2+j0.1 & 0.2+j0.06 & 0.3+j0.1 \\ 0.2-j0.1 & 1.0 & 0.9-j0.2 & 0.1-j0.03 \\ 0.2-j0.06 & 0.9+j0.2 & 1.2 & 0.1-j0.01 \\ 0.3-j0.1 & 0.1+j0.03 & 0.1+j0.01 & 3.0 \end{bmatrix}$$

$$\boldsymbol{C}_{b4} = \begin{bmatrix} 2.0 & 0.1-j0.02 & 0.09-j0.03 & 0.8-j0.1 \\ 0.1+j0.02 & 0.5 & 0.45-j0.08 & 0.2+j0.1 \\ 0.09+j0.03 & 0.45+j0.08 & 0.7 & 0.16+j0.12 \\ 0.8+j0.1 & 0.2-j0.1 & 0.16-j0.12 & 4.0 \end{bmatrix} \tag{29}$$

Contrast with the 3 x 3 Covariance Matrix

The eigenvalues for the generalized eigenvalue equation of the matrices of targets a and b are 2.7398, 2.0461, and 0.7356. Antenna polarization ratios, tilt and ellipticity angles, and rotation senses corresponding to the largest eigenvalue are:

$P_r = 0.5037 + j0.0614$ $\tau = 26.8°$ $\epsilon = 2.8°$ RS = L
$P_t = -0.4490 + j0.1232$ $\tau = -24.5°$ $\epsilon = 5.8°$ RS = L

The ratio of received powers is equal to the largest eigenvalue.

Contrast with the Bistatic Covariance Matrix

The largest eigenvalue of the bistatic covariance matrices is 2.7399, and the approximate antenna polarization ratios are:

$P_r = 0.5125 - j0.0340$	$\tau = 27.2°$	$\epsilon = -1.5°$	RS = R
$P_t = -0.4589 + j0.0366$	$\tau = -24.7°$	$\epsilon = 1.7°$	RS = L

The power ratio with these approximate valuses is 2.6776. If the ratio is maximized numerically, the power ratio can be increased to 2.6922. Both are smaller than the eigenvalue, because of a constraint on the joint antenna vector.

Signal-Clutter Optimization with Graves and Kennaugh Matrices

The largest eigenvalue of the Graves matrix of target a is 6.0464. Antenna polarizations for maximum signal-noise or maximum signal-clutter ratio are:

$P_r = 0.1043 + j0.0366$	$\tau = 6.0°$	$\epsilon = 2.1°$	RS = L
$P_t = 0.1466 + j0.0342$	$\tau = 8.4°$	$\epsilon = 1.9°$	RS = L

The signal-noise or signal-clutter ratio is 5.0135. This ratio compares favorably to the value of 5.0362 arrived at by a numerical maximization.

Contrast with Graves and Kennaugh Matrices

The power ratios that result from the four solution methods described previously are:

A. R = 2.440
B. R = 1.692
C. R = 2.504
D. R = 2.070

Method C is the best solution, differing from the exact power by 0.4 dB. It may not be best for all target pairs.

REFERENCES

1. H. Mott, Antennas for Radar and Communications: A Polarimetric Approach, Wiley, New York, 1992.

2. J. A. Kong, et al, "Classification of Earth Terrain Using Polarimetric Synthetic Aperture Radar Images," Polarimetric Remote Sensing, PIER 3, J. A. Kong, ed., Elsevier, New York, 1990.

POLARIZATION STRUCTURE OF ULTRA-WIDE-BAND RADAR SIGNALS.

V.A. Sarytchev and G.B.Katchalova

"Leninez Holding" Company
212 Moskovsky Pr.,
Sankt-Petersburg 196066 RUSSIA

At first we give the definitions of principal conceptions such as the signal temporal structure and the polarization structure. We define the "structure" as the combination of whole system connection, corresponding to the object presentation and being necessary for investigation of its dynamic and development. The signal temporal structure corresponds to the procedure of realization (i.e. the dynamic) and to the characteristics of realizations group (i.e. development). We will associate the temporal structure with the signal dynamic only and leave in the side a signal group properties at least for pulse, where the uncertainty can be only parameteric.

The introduction of polarization structure is based on assumption that a signal can have components which are orthogonal on some basis. This orthogonality has not depending on the temporal structure of components and it is preserved under so call polarization transforms associated with the inclination angle α and the ellipticity angle β of transform unit vectors. Can we define the polarization structure of pulse signals, especially when fixed components have in a principle different temporal structures, e.g., triangular and Gaussian pulses, or monochromatic and pulse signals? This problem becomes more important for ultra-wide-band signals because their temporal structure is hardly defined.

Consider classical case in radar polarimetry when signal (viz. ultra-wide-band signal) is presented by means of the two polarization (real-valued) components with the $E_1(t)$ and $E_2(t)$ temporal structures respectively. They have the one argument and this means that signal is considered at fixed point and that all effects associated with propagation are not considered.

Introduce the set of components pairs. These pairs are connected in following a way:

$$E_2(t)=a_1E_1(t)+a_2(t)\Gamma(E_1(t)), \tag{1}$$

where a_1 and a_2 are real number, $\Gamma(f(x))$ - the Hilbert transform of function f(x) defined as the principal value integral

$$\Gamma(f(x))= \pi^{-1}\, \mathrm{PV}\int_{+\infty}^{-\infty} f(x)(x-y)^{-1}dx\ .$$

Ultra-Wideband, Short-Pulse Electromagnetics 3
Edited by Baum *et al.*, Plenum Press, New York, 1997

In this case we will speak that $E_1(t)$ and $E_2(t)$ components are matched on the temporal structures. Because of $E_1(t)$ and $E_2(t)$ are ultra-wide-band signals we can consider them as real-valued functions. However we can postulate their amplitudes:

$$| E_1(t)| = \sqrt{E_1^2(t) + \Gamma^2(E_1(t))}\ , \tag{2a}$$

$$| E_2(t)| = \sqrt{E_2^2(t) + \Gamma^2(E_2(t))}\ , \tag{2b}$$

phases:

$$\varphi_1(t) = \cos^{-1}(E_1(t)/| E_1(t)|), \tag{3a}$$

$$\varphi_2(t) = \cos^{-1}(E_2(t)/| E_2(t)|), \tag{3b}$$

and frequencies:

$$\omega_1(t)=(E_1(t)\frac{d}{dt}\Gamma(E_1(t))-\Gamma(E_1(t))\frac{d}{dt}E_1(t))/| E_1(t)|^2, \tag{4a}$$

$$\omega_2(t)=(E_2(t)\frac{d}{dt}\Gamma(E_2(t))-\Gamma(E_2(t))\frac{d}{dt}E_2(t))/| E_2(t)|^2, \tag{4b}$$

It is easy convinced that for matched on temporal structure components

$$\omega_1(t)=\omega_2(t).$$

Matched on temporal structure components have evident polarization structures. To be convinced in this one must compose the analytic signal from these components:

$$E(t)=E_1(t)+iE_2(t), \tag{5}$$

that admits following presentation:

$$E(t)= e^{i\beta}(E_1(t) \cos \alpha + i\, \Gamma(E_1) \sin \alpha). \tag{6}$$

The spectrum of this signal has following appearance:

$$E(\omega)= e^{i\beta}E_1(\omega)(\cos \alpha + \text{sgn}\, \omega \sin \alpha). \tag{7}$$

Signals orthogonal to (6) or (7) in meaning of the scalar production

$$((E^{(1)}(t),E^{(2)}(t))=\int_{+\infty}^{-\infty} E^{(1)*}(t)E^{(2)}(t)dt,$$

(where * is the complex conjugation) are described as

$$E^{\perp}(t)=(-E_1(t) \sin \alpha + i\, \Gamma(E_1(t)) \cos \alpha)\, e^{i\beta}, \tag{8}$$

$$E^{\perp}(\omega)=E_1(t)\, e^{i\beta}(\cos \alpha\ \text{sgn}\, \omega - \sin \alpha). \tag{9}$$

It is seen from (7) and (8) that spectral components of matched on temporal structures signal have identical polarization states, described by means of α and β parameters. The orthogonality of (6) and (8) or (7) and (9) follows from polarization one of every spectral component.

The envelope, the phase and the frequency of (6) can be determined in agreement with expression (1)-(4):

$$|E(t)| = E_1(t)\sqrt{(1+tg^2\beta)(1+tg^2\alpha\, tg^2\varphi_c(t))}, \tag{10}$$

$$\varphi(t)=tg^{-1}((tg\,\beta + tg\,\alpha\, tg\,\varphi_c(t))/(1- tg\,\alpha\, tg\,\beta\, tg\,\varphi_c(t))), \tag{11}$$

$$\omega=\omega_c(t)(tg\,\alpha + tg^2\varphi_c(t))/(1+ tg^2\alpha\, tg^2\varphi_c(t)), \tag{12}$$

where

$$\varphi_c(t)= tg^{-1}(\Gamma(E_1(t))/E_1(t)),$$

$$\omega_c(t)= d\varphi_c(t)/dt.$$

It is seen from (11) and (12) that instantaneous frequency and phase are not depend on energy parameters if $\alpha=0$. For this α $\varphi(t)=\beta$ and $\omega(t)=0$. If $E_1(t)$ and $E_2(t)$ are related by means of the Hilbert transform $\varphi(t)$ has uncertainty on β, and $\omega(t)$ is independent on β. Transform from (6) to (8) corresponds to the increase of β on $\pi/2$ angle, the change of α sign, and the Hilbert conjugation of temporal structure.

Setting E(t) in (6) as $E(\alpha,\beta)$ and $E^{\perp}(t)$ in (8) as $E^{\perp}(\alpha,\beta)$ we can realize the transform to the presentation of investigated signal on the basis of another polarization parameters (from old α_1,β_1 to new α_2,β_2):

$$E(\alpha_2,\beta_2)=\exp(i(\beta_2-\beta_1))(E(\alpha_1,\beta_1)\cos(\alpha_2-\alpha_1) + E^{\perp}(\alpha_1,\beta_1)\sin(\alpha_2-\alpha_1)), \tag{13a}$$

$$E^{\perp}(\alpha_2,\beta_2)=\exp(i(\beta_2-\beta_1))(E^{\perp}(\alpha_1,\beta_1)\cos(\alpha_2-\alpha_1) + E(\alpha_1,\beta_1)\sin(\alpha_2-\alpha_1)), \tag{13b}$$

It is easy convinced that (6) and (8) are also particular case of (13). The last relation can be rewritten in matrix form:

$$\begin{bmatrix} E(\alpha_2,\beta_2) \\ E^{\perp}(\alpha_2,\beta_2) \end{bmatrix} = e^{i(\beta_2-\beta_1)} \begin{bmatrix} \cos(\alpha_2-\alpha_1) & \sin(\alpha_2-\alpha_1) \\ -\sin(\alpha_2-\alpha_1) & \cos(\alpha_2-\alpha_1) \end{bmatrix} \begin{bmatrix} E(\alpha_1,\beta_1) \\ E^{\perp}(\alpha_1,\beta_1) \end{bmatrix} \tag{14}$$

or, briefly,

$$\mathbf{E}(\alpha_2,\beta_2)=Q(\alpha_1,\beta_1 ; \alpha_2,\beta_2)\, \mathbf{E}(\alpha_1,\beta_1), \tag{15}$$

where $Q(\alpha_1,\beta_1 ; \alpha_2,\beta_2)$ is unitary matrix. Scalar production of $\mathbf{E}^{(1)}(\alpha,\beta)$ and $\mathbf{E}^{(2)}(\alpha,\beta)$ vectors, corresponding to $E^{(1)}(t)$ and $E^{(2)}(t)$ signals , respectively, are written in following form:

$$(\mathbf{E}^{(1)}(\alpha_1,\beta_1,t),\mathbf{E}^{(1)}(\alpha_2,\beta_2,t))=\int_{-\infty}^{+\infty} \mathbf{E}^{(1)+}(\alpha_1,\beta_1,t)\cdot\mathbf{E}^{(2)}(\alpha_2,\beta_2,t)\,dt \tag{16}$$

where + is the Hermitian conjugation.

Relations (13) - (15) can be rewritten in affine (non-orthogonal) basis[1]. If in equations (6) and (8) temporal structures are real-valued functions, the orthogonality is presented when this pair has different temporal structures. They are real-valued functions for the ultra-wide-band signals (pulses). Thus polarization structure is independent from temporal structure in the sense of orthogonality if analyzed signals have components that matched on temporal structure.

This allows also to postulate next relations for signals unmatched on temporal structure:

$$E(t)=(G_1(t)\cos\alpha + i\,\Gamma(G_1(t))\sin\alpha - G_2(t)\sin\alpha + i\,\Gamma(G_2(t))\cos\alpha)e^{i\beta}, \tag{17}$$

i.e. we convert our relations to those with the two components having the identical polarization parameters and orthogonal on their polarization structure, but with different temporal structures $G_1(t)$ and $G_2(t)$. In order to equation (17) becomes identity it is necessary that $G_1(t)$ and $G_2(t)$ are associated with original $E_1(t)$ and $E_2(t)$ components of $E(t)$ by means of following relations:

$$G_1(t)=E_1(t)\cos\alpha\cos\beta - \Gamma(E_1(t))\sin\alpha\sin\beta + E_2(t)\cos\alpha\sin\beta - \Gamma(E_2(t))\sin\alpha\cos\beta, \tag{18a}$$

$$G_2(t)= -E_1(t)\sin\alpha\cos\beta + \Gamma(E_1(t))\cos\alpha\sin\beta - E_2(t)\sin\alpha\sin\beta - \Gamma(E_2(t))\cos\alpha\cos\beta, \tag{18b}$$

It is seen from (18a) and (18b) that if $E_1(t)$ and $E_2(t)$ are real-valued functions then $G_1(t)$ and $G_2(t)$ are real-valued function too. Relations (18a) and (18b) will be defined as the transformation of temporal structures of original components.

Relations (17), (18a) and (18b) can be written in matrix form:

$$\mathbf{E}(\alpha,\beta)=Q(\alpha,\beta,0,0)\,Q_t(\alpha,\beta,0,0)\,\mathbf{E}(0,0), \tag{19}$$

where

$$Q(\alpha,\beta;0,0)= e^{i\beta}\begin{bmatrix}\cos\alpha + i\sin\alpha\,\Gamma & 0\\ 0 & -\sin\alpha + i\cos\alpha\,\Gamma\end{bmatrix},$$

$$Q_t(\alpha,\beta;0,0)=\begin{bmatrix}\cos\alpha\cos\beta - \sin\alpha\sin\beta\cdot\Gamma & -\cos\alpha\sin\beta - \sin\alpha\cos\beta\cdot\Gamma\\ \sin\alpha\cos\beta + \sin\alpha\cos\beta\cdot\Gamma & -\sin\alpha\sin\beta + \cos\alpha\cos\beta\cdot\Gamma\end{bmatrix},$$

$$\mathbf{E}(0,0)=\left[E_1(t),\quad E_2(t)\right]^T,$$

$Q(\alpha,\beta;0,0)$ and $Q_t(\alpha,\beta;0,0)$ are unitary matrices.

Thus any pair of components of original signal admits the decomposition on two components which are orthogonal on polarization structure and have different temporal structures in general case. Figures 1 - 3 show circularly polarized signals with various temporal structures of components. At Fig.1 is presented convinient signal with sinusoidal temporal structure. Fig.2 presents signal with rectangular and sinusoidal temporal structures

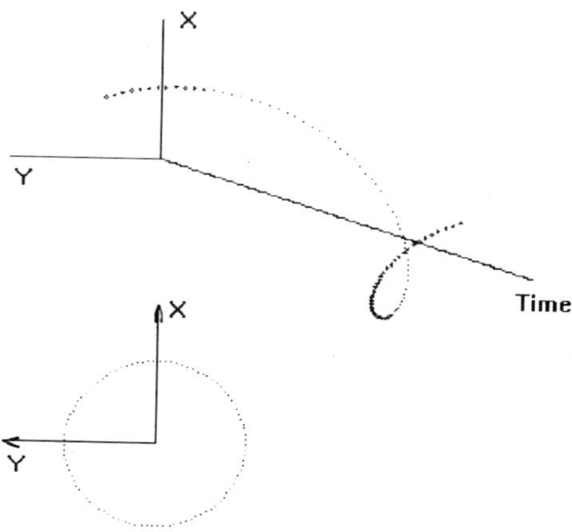

Figure 1. Circularly polarized signal with sinusoidal temporal structure.

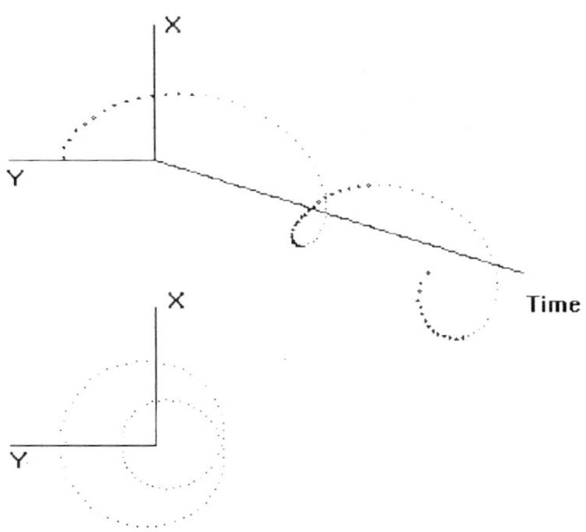

Figure 2. Circularly polarized signal with sinusoidal and triangular temporal structures of components.

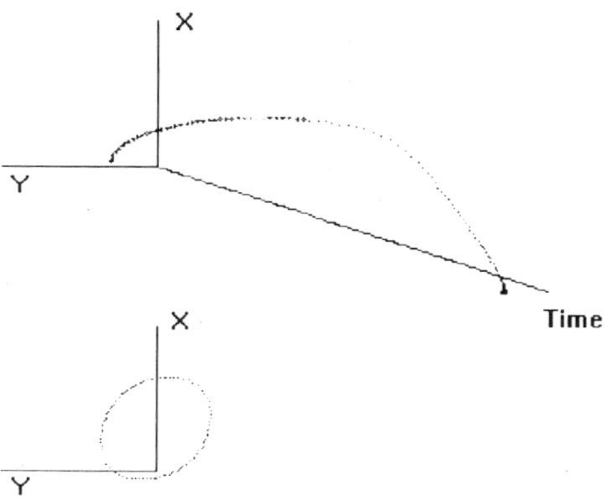

Figure 3. Circularly polarized signal with rectangular and triangular temporal structures of components.

of components. Fig.3 presents signal with triangular and rectangular temporal structures of components.

The procedure regularition of function is used for signals with discontinuities. This procedure defined as the convolution of the real-valued function f(x) with ω_ε (x) :

$$f_\varepsilon(x)=\int_{+\infty}^{-\infty} f(y)\omega_\varepsilon(x-y)dy,$$

where

$$\omega_\varepsilon = \begin{cases} C_\varepsilon \exp(\varepsilon^2/(\varepsilon^2-t^2), & |t|<\varepsilon \\ 0, & |t|\geq\varepsilon \end{cases}$$

and constant C_ε is chosen so, that:

$$\int_{+\infty}^{-\infty} \omega_\varepsilon(x)dx=1.$$

This presentation of ultra-wide-band signals permits also to describe transform of signals by various elements of radar channel. The process of radar scattering has important interest too.

Consider the case when radar target possesses following properties: linearity, translating invarianty (i.e. the delay of signal at same time causes the identical delay of response), reality (i.e. the real-valued action causes the real-valued response), original signal and response are integrable function. It is proved (in the theory of distribution [2]) that transfer characteristics matrix exist, which is connective matrix for incident and scattered waves:

$$\mathbf{E}^{sc}(t)=\int_{+\infty}^{-\infty} S(t-\tau)\mathbf{E}^{in}(\tau)d\tau. \tag{20}$$

Real-valued components of ultra-wide-band signals are provided by linear basis. For this reason we rewrite (20) as:

$$\mathbf{E}^{sc}(0,0;t)=\int_{+\infty}^{-\infty} S(0,0;t-\tau)\mathbf{E}^{in}(0,0;\tau)d\tau. \tag{21}$$

Elements of S(0,0;t) matrix represent the real-valued generalized function both regular and singular.

It is possible on above method to go from $\mathbf{E}^{sc}(0,0;t)$ and $\mathbf{E}^{in}(0,0;t)$ signals to their presentation $\mathbf{E}^{sc}(\alpha^{sc},\beta^{sc};t)$ and $\mathbf{E}^{in}(\alpha^{in},\beta^{in};t)$, i.e. we admit that incident and scattered waves have different polarization parameters. This allows to consider that transfer characteristics matrix is transformed on rule:

$$S(\alpha^{sc},\beta^{sc};\alpha^{in},\beta^{in};t) = Q(\alpha^{sc},\beta^{sc},0,0)\cdot Q_t(\alpha^{sc},\beta^{sc},0,0)\cdot S(0,0,t) \cdot Q_t^{-1}(\alpha^{in},\beta^{in},0,0)\cdot Q^{-1}(\alpha^{in},\beta^{in},0,0). \tag{22}$$

This matrix is in agreement with the equation (22) for the new representation:

$$\mathbf{E}^{sc}(\alpha^{sc},\beta^{sc};t)=\int_{+\infty}^{-\infty} S(\alpha^{sc},\beta^{sc};\alpha^{in},\beta^{in};t-\tau)\mathbf{E}^{in}(\alpha^{in},\beta^{in};\tau)d\tau. \tag{23}$$

where $\mathbf{E}^{sc}(\alpha^{sc},\beta^{sc};t)$ and $\mathbf{E}^{in}(\alpha^{in},\beta^{in};t)$ are transformed on (19).

Picking out pair in any set of components we can extend above relations to signals with arbitrary number of components. In particular, it is possible to describe antenna near-field having three components when ultra-wide-band signal components have different temporal dependencies (one from them is presented the derivation of antenna current, other reproduces current dynamic or integrates it).

REFERENCES

1. *Polarization of Signals in Large Complicate Transport Radio-electronic Systems,* A.I. Koslov, V.A. Sarytchev , ed., Transport Academy of Russia, St.-Petersburg (1994).
2. V.S. Vladimirov. *Distributions and Theirs Applications in Mathematics and Physics.* Nauka, Moscow (1984).

ANALYTIC METHODS FOR PULSED SIGNAL INTERACTION WITH LAYERED, LOSSY SOIL ENVIRONMENTS AND BURIED OBJECTS *

Leopold B. Felsen

Department of Aerospace and Mechanical Engineering and
Department of Electrical, Computer and Systems Engineering
Boston University
110 Cummington Street
Boston, MA 02215
Also: Polytechnic University (Emeritus), Brooklyn, NY 11201 USA

ABSTRACT

Inverse algorithms for processing data due to wave scattering from complex environments can be strengthened by incorporating relevant robust wave physics in the inversion scheme. Because of the complexity of the target-background environment for ground penetrating radar (GPR) applications, the GPR processing strategy generally relies heavily on numerical, model-based and statistical techniques. A model is proposed here which injects wave-based analytic techniques to reduce the size and(or) complexity of the overall problem. The model is structured around high resolution pulsed-beam propagators for transporting the incident signal to and from the target through lossy dispersive soil environments. Attention is given to those portions of an overall problem which are good candidates for wave-based analytic methods and those portions which are better served by other methods.

I. BACKGROUND PERSPECTIVES

The detection, classification and identification of hidden objects is a subject area which is steadily gaining importance for military as well as civilian applications. Here, "hidden" is taken as a generic designation that includes objects buried underground, embedded in concrete or other materials, hidden in foliage, etc., as well as objects submerged in water channels. Electromagnetic waves are used for interrogation, with emphasis on buried objects in the presence of background clutter. However, the proposed techniques could also be applied to acoustic coupling to buried objects, with the possibility of inducing elastic responses (e.g., resonances) in certain materials.

In the development and testing of inverse algorithms for processing data generated by

*Preliminary version presented at the AGARD Symposium on Remote Sensing: A Valuable Source of Information, held in Toulouse, France, April 22-25, 1996

scattering from a hidden target, it is becoming recognized that to improve accuracy, speed, and robustness, the underlying wave physics should be built into the inversion schemes; by thus enlarging the prior-knowledge base, the dimensionality of the subsequent processing is reduced. Parametric inversion of data implies extraction of characteristic parameters (footprints) which are discriminants for object classification and identification, and they are usually established by beginning with the forward problem for various model environments. *Wave-generated* footprints are best catalogued in the configuration (space-time) - spectrum (wavenumber-frequency) phase space because wave phenomena and their local features are most completely characterized by their configurational as well as spectral attributes; both are given equal weight in the phase space setting. While an awareness of the time-frequency phase space subdomain has begun to be felt within the electromagnetics and acoustics wave scattering communities because of the strong emphasis placed on it during the past decade by signal processors[1-10], the studies by Carin and Felsen[9,11,12] were among the first to apply phase space concepts also to the space-wavenumber domain. From these phase-space investigations, some important insights were gained concerning robustness in the presence of additive system noise and deterministic (target) -stochastic (clutter) interaction. Thus, before choosing a signal-processing strategy for a particular scattering scenario, it is advisable to perform a throrough parametrization of the underlying wave physics, thereby facilitating choices as to whether a particular parameter should be treated as deterministic or random.

The studies initiated by Carin and Felsen (see [9] and the references cited therein) have been continued, with Carin and his research group focusing primarily on the development of numerical modeling and model-based signal processing algorithms, and Felsen concentrating primarily on analytic techniques for classifying basic wave propagation and scattering issues. The latter are addressed here within the context of ground penetrating radar (GPR).

II. MODELING STRATEGY

A. Structure

To detect and classify buried targets by radar, the modeling strategy seeks to parametrize how various soil background environments modify the field incident on the target, and how these environments modify the field scattered from the target before it reaches the observer. If the effect of the background environment can be accounted for by a "good" propagation model, the target interrogation by the background-modified signal can proceed in the local environment near the target. The decomposition into the background domain and the target domain highlights the individual footprints of these distinct scattering phenomenologies, which must then be combined appropriately to parametrize the target-background interaction.

In essence, the propagation algorithm projects the signal from an initial surface S_I near the transmitter onto a surface S_T near the target (Fig. 1). The target domain encompasses the interior of S_T, and the scattering from the target in its local environment can be addressed analytically, when warranted, or purely numerically; in the latter case, the size of the numerical problem is much reduced over what it would have been if the entire problem, starting from S_I, were treated numerically. In this context, the analytic propagators (for which we choose beams in the frequency domain and pulsed beam wavepackets in the time domain, see Sec. IIE) could be regarded as long-range elements and might be embedded accordingly in numerical discretization codes.

More specifically, the modeling strategy advocated here is based on the following constructs:

1. Represent the incident field due to a given source "aperture" distribution as a superposition of beam basis fields.

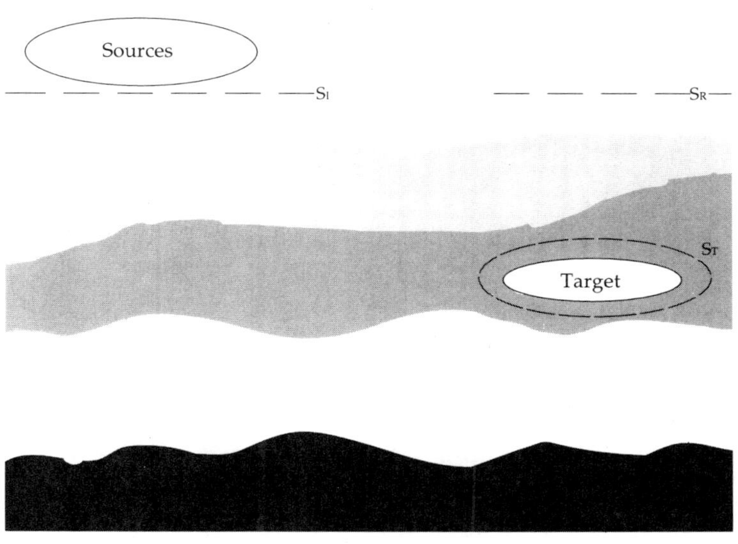

Figure 1. - Physical configuration of lossy, dispersive layers with a target. S_I: Initial field aperture; S_T : Target aperture; S_R : Received field aperture.

Rigorous algorithms to this effect are available.

2. Propagate these basis beams individually through the lossy dispersive soil environment into the vicinity of the target.

Establishing the rules of propagation by asymptotic and related techniques is partly known but requires new extensions (see Sec. IIIA).

Because the beams act like well-collimated local probes, they can be propagated similar to ray fields through heterogeneous media provided that the scales of variation in such media and on their boundaries are large compared to the beam width. However, unlike ray fields, beam fields do not fail near shadow boundaries, caustics, and other ray-optical transition regions.

3. Synthesize the resulting total field near the target by beam superposition.

This process projects the initial aperture distribution onto an "aperture" near the target, with full account taken of the distortion of the signal by the soil environment. The aperture distribution near the target furnishes the ambient incident field for target interrogation and classification; note that this aperture surface need not be planar but can be chosen to suit the target interrogation scheme.

4. Extract the target footprints.

The previous steps have accounted for the distortion of the original signal by the soil background, and therefore furnish the modified signal which is incident on the target from the equivalent aperture distribution near the target. The scattering from the target in its "local environment" represents one of the building blocks in the overall scheme. The local environment depends on the problem conditions. For example, it might incorporate the dominant interaction mechanisms of primary target-scattered fields with environmental features (like nearby interfaces) that induce secondary scattering. Note that for target classification and identification (ID), the classifier need not necessarily be the *shape* image of the target. *Target "footprints" in the phase space, such as resonances, are frequently more effective than information based on shape.* Because the problem scale for target ID has been much reduced by projecting from the initial data plane onto the aperture near the target, phase space modeling and footprinting of the target scattering process by direct numerical methods may be an attractive option, as noted earlier.

5. Construct the scattered fields at the receiver.
Here, the previous steps are followed in reverse order: the scattered field on the aperture near the target is decomposed into beams, the beams are backpropagated through the soil environment (the conventional designation of "backpropagation" is actually forward propagation in the reverse direction), and they are recombined at the receiver.

6. Calibration of the beam propagator algorithms.
The asymptotic beam propagator solutions need to be calibrated for accuracy. This is to be done by comparison with numerically generated reference data for selected model environments. We have previously pursued this strategy for various simple canonical scattering configurations ([9], etc.) which are cited in the list of references.

B. Implementation

As in all parametrization schemes, the implementation depends on the choice of the "parameters." This choice is generally problem dependent. Thus, in multilayered environments, wave objects that emphasize *individual* interface effects sample the layering via a succession of multiply reflected *progressing* waves or wavefronts, whereas wave objects that emphasize the *collective* effect of several interfaces sample the entire confining region via a series of *oscillatory* wavefields (modes or resonances). The *progressing* (ray or beam-type) wavefields furnish a *local* sensor of the layering whereas the *oscillatory* (mode or resonance) wavefields furnish a *global* sensor of the composite, which can, however, be synthesized formally by summing over all of the ray field multiples. The most versatile combination encompasses a self-consistent *local-global* (ray-mode or wavefront-resonance) *hybrid*[46-48]. For the pulsed signals which are of particular interest in our applications, the *progressing* wave formulation furnishes the natural basis, from which oscillatory events can by synthesized when appropriate.

C. Parametrizing complexity - the importance of scales

The GPR problem belongs to the category of *complex* problems. One of the characteristics of *complexity* is the presence of *many scales* that play a role in the parametrization. Relevant for the multilayer environment is the ratio of the scales that characterize the layers to the wavelength scales in the incident signal spectrum. With respect to resolution, large (d_n/l_{en}) favors local probing, while small (d_n/l_{en}) favors global probing; here, d_n is the nth layer thickness and l_{en} is the wavelength in the nth layer material. For the inverse problem, windowed transform processing of the scattered field data adds to the above the scale w representative of the window widths. When the target is included in the scenario, the scales of the target are added to those of the background environment and they influence the choice of the "most promising" target sensor. Finally, when scales in the environment or on the target are *irregular*, statistical parametrizations with *deterministic-stochastic* interactions, etc, play a role.

D. Learning the Rules

To address the issues raised in the preceding section it is prudent to escalate from "simplicities" to cumulative complexity. Judiciously chosen model problems provide the laboratory for quantification. These model problems should be analytically tractable and simulate conditions to be encountered in GPR applications. Details are given in Sec. III.

E. Beam parametrization of radiation from aperture distributions

Because the representation of radiation from frequency and time domain distributed aperture fields in terms of beams is a basic constituent in the problem strategy, we give here a brief review of the *continuous* beam basis representation; for the corresponding *discrete* representation on a Gabor lattice, see [13].

E.1 Formulation.

Beam propagation in a homogenous medium is addressed first; extensions for the case of a lossy dispersive, layered medium are discussed subsequently. In the frequency domain, suppressing a time dependence $\exp(-i\omega t)$, assume that field data $u_o(x_o)$ is specified along an aperture in the z=0 plane of a two-dimensional (x,z) coordinate space; x_o tags the x-coordinate in the aperture plane. The procedural steps enumerated next are straightforward (see, for example, [13]) and they lead to the final expression in (1) below. By applying the windowed Fourier transform along the x_o coordinate, one generates the phase space distribution $U_o(x_o',\xi_o')$, where ξ_o is the Fourier spectral wavenumber corresponding to x_o, and (x_o',ξ_o') represents the location of the window center in the (x_o,ξ_o) phase space. The phase space spectral distributions can be propagated away from the aperture plane to an observer plane $z > 0$ by applying the spectral propagator $\exp[i(k^2-\xi^2)^{1/2}z]$, where $k=\omega/c$ is the wavenumber in the medium and c is the wave propagation speed. The resulting phase space propagators excited by the transformed aperture field are beam-like fields $Y(x,z;x_o',\xi_o')$ which are localized around (x_o',ξ_o'). This leads to the following representation of the field u(x,z) and its spectrum $\hat{u}(\xi,z)\equiv\hat{u}(\xi_o,z)$[13],

$$u\,(x,z) = (\omega/2\pi)\, N_x^{-2} \int_{-\infty}^{\infty}\int_{-\infty}^{\infty} U_o(x_o',\xi_o') Y(x,z;x_o',\xi_o')\, dx_o' d\xi_o' \tag{1a}$$

$$\hat{u}\,(\xi_o,z) = (\omega/2\pi)\, N_x^{-2} \int_{-\infty}^{\infty}\int_{-\infty}^{\infty} U_o(x_o',\xi_o') \hat{Y}(\xi_o z;x_o',\xi_o')\, dx_o' d\xi_o' \tag{1b}$$

where N_x is a normalization constant, while the beam propagator $Y(x,z;x_o',\xi_o')$ and its spectrum $\hat{Y}(\xi_o,z;x_o',\xi_o')$ are given by

$$Y(x,z;x_o',\xi_o') = (\omega/2\pi) \int_{-\infty}^{\infty} d\xi_o \hat{Y}(\xi_o,z;x_o',\xi_o') \exp(i\omega\xi_o x) \tag{2a}$$

$$\hat{Y}(\xi_o,z;x_o',\xi_o')=\hat{w}(\xi_o-\xi_o')\exp[-i\omega[(\xi_o-\xi_o')x_o'-\zeta_o z]],\quad \zeta_o = [(1/c^2) - x_o^2]^{1/2} \tag{2b}$$

In (2b), for the windowed Fourier transform, $w(\xi_o) = (2\pi/\omega)^{1/2}\exp(-\omega\xi_o^2/2\alpha)$ is the spectrum of the Gaussian window $w(x_o) = \exp(-\omega\alpha x_o^2/2)$, with α defining the window width. The aperture domain relations are recovered from (1) and (2) by setting z=0. The beam axis, on which the beam propagator has its maximum, emerges from $x_o = x_o'$ in the aperture plane at an angle $\theta_o = \sin^{-1}(c\xi_o')$ relative to the z-axis. The representation in (1) describes the field as a continuous spectrum of shifted beams, whose origin and axis direction are determined by x_o' and ξ_o', respectively. The transform $U_o(x_o',\xi_o')$ in the z=0 plane specifies the excitation strength of these beams and provides the local matching of these beam fields to the aperture field u_o. The phase space window width determines the degree of confinement (collimation) of these excitation strengths to the vicinity of the axis of the beam propagators. By asymptotic (saddle point) evaluation of the spectral integral in (2a), with (2b), one obtains an explicit approximation Y_s for Y which behaves in the far zone of the (x,z) domain like a conventional Gaussian beam (Fig. 2). Saddle point asymptotics can also be applied to evaluation of the integrals that determine $U_o(x_o',\xi_o')$. With the integrand in (1a) thus approximated asymptotically, the field u(x,z) can be synthesized approximately as well. In asymptotic transition regions where saddle points are not isolated, uniform asymptotic techniques need to be utilized. Details pertaining to the above are documented in ref. [13]. This reference also contains the corresponding discretized representation based on the Gabor lattice in the phase space.

In the time domain, the asymptotic propagators in (2a) are pulsed beam wave packets which are locally matched to the space-time aperture distribution, and the rigorous transforms corresponding to the transform in (1) are the windowed forward and inverse radon transforms. The pulsed beams sample time resolved sections of the transverse beam profile in Fig. 2 at locations ct along the beam axis z_b, with spatial spread around z_b=ct determined by the pulse width. For details, see [14], which again contains also the Gabor representation.

Having thus parametrized the radiation from planar aperture distributions, the local matching of the asymptotic basis beams can be generalized to curved aperture distributions.

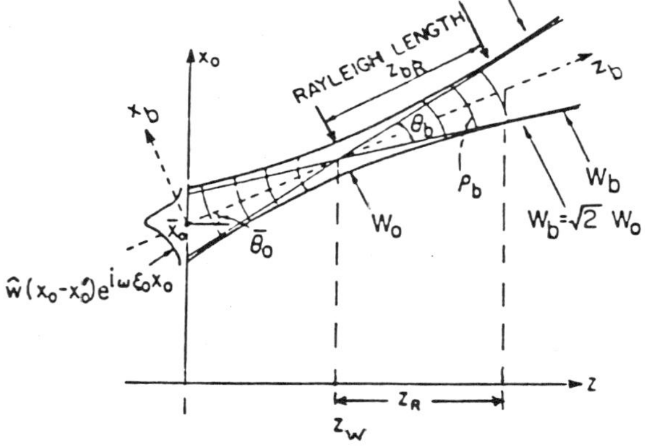

Figure 2 - Time harmonic beam field Y_s and beam parameters in the (x,z) configuration domain, phase matched to the windowed aperture profile in the $(x_o,0)$ plane. (x_b,z_b) are beam-based coordinates, z_w locates the beam waist of width W_o, ρ_b is the phase front radius of curvature, and $W_b=\rho_b\theta_b$ is the corresponding beam width. For pulsed beams, the wave packets are localized around the wavefronts which progress along the beam axis with speed c.

The above summary conveys that the formal frequency and time domain machinery for expanding aperture fields (either primary as produced by a source of radiation, or secondary as represented by scattering data from a target) in terms of a continuous or discrete distribution of beam-type basis fields is well established, as are the first-order asymptotic techniques to extract the wave physics from these formal representations. Thus, these algorithms can be used to calibrate the asymptotics against numerical reference data, and to assess the effects of window choice on the phase space resolution of the beam basis elements.

E.2. Applications and extensions

Having parametrized the launch conditions in free space, the basis beam propagators can be allowed to interact with various environments. In this endeavor, use can be made of the large reservoir of beam tracking algorithms that have been developed in various disciplines, especially in the regime based on paraxial, quasi-plane-wave, etc. approximations for the beam models. Of special interest is the rigorous complex source technique[13-34] whose power resides in the fact that it rigorously converts frequency and time domain Green's functions generated by elementary spatial and space-time sources, respectively, at *real* locations into steady state and pulsed beam excitations in the same environment when these source locations are made *complex*. The conversion applies to exact as well as asymptotic representations and has been utilized in a broad range of applications. Because it is rigorously based (i.e., the complex source beams are rigorous solutions of the wave equation), the method has made it possible to critically assess the effects of a *priori* assumptions in other models in the literature.

After individual phase space beams have been tracked through an environment, they need to be combined according to (1a) or its discretized Gabor version[13] for synthesis of the field u(x,z). Because these beams, which are generated by an extended aperture, arrive from different directions and have therefore sampled different portions of the environment, the asymptotically approximated individual beams may not be adequate and uniform methods may be needed. In this connection, the Gabor basis may offer advantages. In references [27] and [28], it has been demonstrated that narrow Gabor basis beams can explore complex noncanonical propagation environments[28] with remarkable accuracy.

III. PROVIDING THE TOOLS

With these observations as background, let us now consider application to the GPR problem. To provide tools for implementation, several "canonical" constituent problems require attention. Two basic problem categories are highlighted which are relevant not only for the GPR problem but also for other areas of application. Emphasis is on the time domain, especially for excitation by pulsed beams which model space-time resolved (i.e., wideband) wavepackets.

The first problem category is concerned with how lossy dispersive media, in bulk, affect the pulsed beam signal. The second category is concerned with how environments comprised of plane homogeneous layers affect the pulsed beam signal. Note again that beams in both the frequency and time domains a) are useful models individually for studying the wave phenomenology when spatially confined fields interact with environments, and b) can represent, by superposition, any prescribed excitation. For item b), the basis beam parameters can be chosen so as to yield well-collimated probes which can sample local inhomogeneities in an environment if the variations of these inhomogeneities are small over the beam width; with respect to localization, beams may be regarded as generalized rays (in fact, beams are complex rays[15-28]) which are not beset by the "catastrophes" that nonuniform ray theory encounters at caustics, shadow boundaries, etc. The outcomes from these investigations, i.e. exploration of the soil environment as such, are relevant for subsequent studies of buried targets in that environment.

A. Pulsed beam interaction with homogenous lossy dispersive materials

This is a basic electromagnetic problem that relates to materials of all types. A thorough study of transient fields in particular homogenous lossy dispersive media has been carried out by Oughston and collaborators and has been documented in a monograph[35]. This monograph presents theorems and rigorous alternative representations pertaining to general space-time dependent fields in such media (part I of [35]); the asymptotic reduction of the formal representation integrals via saddle point techniques has been carried out for the special case of pulsed plane wave excitation of a single resonance Lorentz medium (part II of [35]). The intricate asymptotics identifies and quantifies the relevant wave physics that gives rise to the distortions of the pulsed plane wave signal input, and furnishes valuable insight into the relation between the phenomenology and the model parameters as well as the input pulse shape for the plane wave case in that special medium. This still leaves the general *phenomenology* associated with a *pulsed beam* input (i.e., temporal and spatial localization) as a separate problem.

For the simplest nondispersive lossy model $\varepsilon=\varepsilon_r + i\sigma/\omega$, where ε_r and σ are positive real, the pulsed beam propagation problem has recently been addressed by Heyman, Tijhuis and Boersma[49], using the complex space-time source point approach. Their results furnish the exact and asymptotically approximated propagators in such a medium, and parametrize the effect of the loss factor σ on the pulse shape. For more general dispersive models, when these are required, it is unlikely that one can develop exact closed form expressions. One may now start with the pulsed beam formal spectral integrals which involve the transverse spatial wavenumber $\mathbf{k_t}$ and the temporal frequencies ω for lossy (i.e. complex) $\varepsilon(\omega)$. By performing sequential saddle point asymptotics, these integrals can be reduced to approximate forms that highlight the dependence on loss and beam parameters, thereby revealing the pulsed beam phenomenology for the well developed dispersive regime. For the precursors at short time scales, i.e., near the wavefronts, alternative asymptotics techniques are appropriate[36]. Such studies should suffice to extract generic phenomenology, which can be compared with the complex space-time ray asymptotics performed in [16], [50]. In the sequential reduction of the (k_t,ω) integrals, it is instructive to perform the ω-integration first. This represents the entire wave process in terms *of transient plane wave spectra* parametrized by the *spatial* wavenumbers, *directly* in the time domain. In this connection, the complex space-time source point technique[24,26,42], which deals directly with the time domain, is suggestive. After the problem in the unbounded homogeneous medium has been settled, adaptation to weakly inhomogeneous media can be handled by adiabatic asymptotics.

B. Synthesizing focusing beam and incident aperture profiles

Synthesis of input pulsed beam and aperture profiles for *focusing* the beam energy at a particular space-time locationin the lossy, dispersive environment is a problem of special importance. Recent studies have dealt with collimated ultra-wideband/short pulse aperture profiles in free space[37,43]. It is shown there how to synthesize the input profile by frequency-dependent beam apertures so as to render the Fresnel distance frequency insensitive. This yields better collimation than a constant-width aperture where that distance depends on frequency. Also relevant are previous asymptotic studies of plane wave pulse compression (i.e., space-time focusing) in lossless dispersive media[38].

The challenge is to extend these concepts to pulsed beam apertures in lossy dispersive media. It is suggestive to initially address the $(\varepsilon_r + i\sigma/\omega)$ model, for which the (pulsed beam parameter) - (medium - parameter) connection[49] takes on a simple form. For more general lossy medium models, the complex space-time ray asymptotics in [50] are appropriate, which reveal considerably more complex phenomenology, even for plane wave pulses. It remains to be seen what relevant approximate information can be extracted via pulsed beam asymptotics; due to the beam localization, the algorithms may actually simplify.

C. Pulsed beam interaction with layered lossy dispersive materials

In this model, lossy dispersive materials are arranged in homogenous parallel layers so as to simulate soil stratification. The objective is to explore how layering affects an incident pulsed beam signal; it is assumed that via the bulk studies in Sec. IIIA, the effects of loss and dispersion as such are understood. For normal incidence, the relevant mechanism is single and multiple reflection at the boundaries, with the multiply reflected field confined to the vicinity of the normal plane of incidence. This mechanism can be understood by first replacing the layers with a half space and establishing the reflection and transmission coefficients for a single interface; the multiple reflection mechanism is then built up sequentially. If the multiple reflections can no longer be resolved individually, collective summation into modal resonances in the plane-of-incidence cross section becomes an alternative descriptor. The extension from the bulk studies in Sec. IIIA to this normal incidence case is expected to be straightforward.

For oblique incidence of the pulsed beam signal, the phenomenology changes drastically. Multiple reflections between the layers now propagate laterally along the layers, thereby producing energy transport (guiding) in the lateral direction. Moreover, depending on the layer composition, a beam incident obliquely beyond the critical angle at an interface will produce a strong reflected and weak transmitted field which, under lossless conditions, would be evanescent (decay exponentially). The global (collective) effects of the layer-trapped multiples can by expressed in terms of *trapped* and *leaky modes*; the latter propagate along the layer direction and shed energy back into the exterior of the layered half space (the air region above ground can be regarded as an infinitely wide layer). The time *domain leaky modes* (which are sustained, damped oscillatory waveforms with time-dependent wavenumbers) are new wave objects which were first explored (to the best of our knowledge) in [39-41] (it should be recalled that extrapolation to the time domain of the frequency domain leaky mode phenomenology fails under short pulse conditions since the frequency dependent leakage angles then vary over a wide interval). In [39-41], the configuration consists of a *nondispersive, lossless* grounded dielectric slab excited by a pulsed horizontal dipole element inside the slab, with the fields observed outside the slab. The time domain leaky mode phenomenology, including the mechanisms of excitation and leakage, as derived by rigorously based asymptotics, is schematized in Fig. 3. The analysis revealed a curious anomalous leaky mode behavior when the problem conditions permit Brewster angle (perfect) transmission; at the instant of time when the leaky mode is matched to the Brewster angle, the radiated leaky mode field undergoes a highly singular transition which has not yet been traced out asymptotically but which can be avoided by employing a hybrid wavefront-resonance algorithm[41].

The results in [39-41] pertain only to the special problem conditions noted above. Therefore, much can be learned from this canonical configuration with respect to the GPR problem conditions. Further studies should include: A. *Lossless nondispersive case* - 1. replacing the pulsed dipole element by a pulsed beam; 2. allowing all combinations of inside-outside source and observer locations (although the radar is always in the exterior half space, energy can be coupled into a layer from the inside by virtual sources which model scattering from a strong soil inhomogeneity); 3. examining the effect of lateral waves excited by a wavefront incident at the critical angle (this regime was avoided in [39-41]; 4. creating a dielectric gap by having $\varepsilon_{int}>\varepsilon_{ext}$ (i.e., no trapped modes). B. *Lossy dispersive case* - introducing loss and dispersion into the items listed under A.

D. Including the target

The methodology for scattering from a target buried in the soil background has been detailed in Sec. II. The relevant problem sees the target in its *local* environment. The simplest local environment is a thick layer whose boundaries are so far removed from the target as to allow modeling in unbounded lossy space. The next escalations in "complexity" involve a target buried near the surface of a lossy half space, a target buried inside a layer with nearby

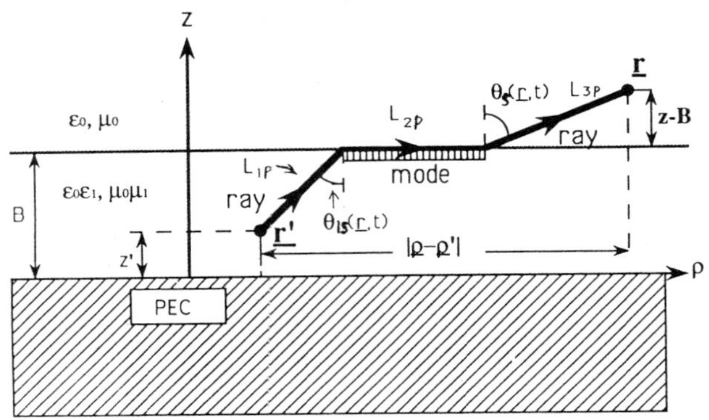

Figure 3. Physical configuration of horizontal electric current element located at **r**′ inside a grounded dielectric layer, with the observer **r** outside the layer. Also shown is the ray-mode interpretation of the space-time evolution of a leaky mode field excited by the pulsed dipole. Paths L_{1p} and L_{3p} describe trajectories of ray fields in the slab and exterior bulk media, respectively, whereas L_{2p} describes the trajectory of the pth leaky mode. The ray field wave vectors along the ray paths L_{1p} and L_{3p} are phase matched to the leaky mode wave vector along L_{2p}, with the space-time dependent leaky mode launch and detachment angles θ_{1s} and θ_s adjusted continuously so as to satisfy also the space-time dependent leaky mode dispersion relation.

boundaries, and so on. The matching from the background - corrected asymptotic incident field produced by the beam basis functions on the target domain boundary S_T to whatever basis is adopted for the "interior" problem can be handled to a lowest order of approximation by Kirchhoff-type assumptions wherein fields incident on one side of S_T furnish the initial conditions for outgoing fields on the other side, and vice-versa. It should be emphasized again that via the progressing pulsed - beam formalism adopted here, the asymptotic beam tracking can be performed, in principle, through any background (bulk material and/or interface) configuration provided that the previously stated "slow variability conditions" with respect to the beam widths are satisfied.

D.1 Weak-contrast targets — Born-type scattering. For weak nondispersive inhomogeneities in a homogeneous lossless nondispersive background recent studies have shown how frequency domain diffraction tomographic methods for object reconstruction can be generalized to interrogation by pulsed plane wave inputs[51], and thereby take advantage of the resolution capabilities of the short pulse signal. Going further, work now in progress shows that interrogation via *pulsed beams* optimizes the *local* reconstruction process[52, 53]. It is interesting to ask whether and how the procedure in [52, 53] can be applied to effect weakly-scattering-object reconstruction in a *lossy* environment.

IV. SUMMARY

There has been a growing awareness that recourse to phenomenology can play an important role when devising processing schemes for data from complex scattering environments. Without *some* phenomenological (i.e., wave physics) input into algorithms that rely entirely on non-wave-oriented statistical and other parametrizations, an opportunity may have been lost to reduce the complexity of the image processing by ignoring *a priori* information. If this premise is accepted for the overall GPR problem, the question is *where*, in the modeling chain, phenomenology makes sense, and what *measures* of that phenomenology (local or global, fine or coarse, etc.) should be utilized. The decision is evidently problem dependent, and is also influenced by the other parametrization schemes in the process.

The objective in this paper has been to parametrize the wave phenomenology for the GPR soil background environment by employing beam propagators as the basic synthesizing wave objects, separating the background domain from the target domain, and connecting the two domains via wave-based assumptions. Only "regular" background conditions have been considered, which do not require statistical measures. What portions of the overall problem can be phrased around this background and can then be perturbed by statistical measures depends on the overall problem conditions. The phenomenological hierarchy has been addressed here in detail, with due consideration of local vs. global characterizations, dispersive vs. nondispersive effects, etc., and the analytic framework for establishing the beam basis algorithms has been stated. Calibration of the analytic asymptotics against numerical reference solutions should be performed selectively. Extensions without calibration can then be carried out by local adaptation of the beam asymptotics to environments with smooth deviations from the test case. It should be noted that performing lossless-to-lossy dispersive generalization is usually nontrivial, and can give rise to substantial changes in phenomenology.

Clearly, implementation of the algorithm by beam tracking through the overall complex environment is computationally massive, and is *not* advocated here (unless it makes sense). Instead, by having listed the various anticipated phenomenologies and having stated how they can be analytically calibrated, an *intelligent* choice can be made as to "which aspect" and "what measure" of relevant wave physics should be embedded within the overall modeling strategy that includes non-wave-oriented (numerical and model-based) techniques. This decision requires the intimate coupling of the wave-oriented and non-wave-oriented portions. Only the former aspect has been discussed in this write-up, and the list of references cited here is limited accordingly. The latter aspect has been pursued in other investigations (not cited here), but a sample of those performed by L. Carin et al. has been documented in [54-56]. Recent examples for blending these alternative modeling strategies into an overall package with better potential than each alone may be found in [57-59].

V. ACKNOWLEDGEMENT

This work was sponsored in part by the US Air Force Office of Scientific Research.

VI. REFERENCES

1. Moghaddar and E.K. Walton, "Time-frequency distribution analysis of scattering from waveguide cavities," *IEEE Trans. Antennas Prop.*, vol. 41, pp. 677-679, May 1993.
2. L. Carin, L.B. Felsen, S.U. Pillai, D. Kralj, and W.C. Lee, "Dispersive modes in the time domain: analysis and time frequency representation," *IEEE Microwave and Guided Wave Letts.*, Jan. 1994.
3. D.R. Kralj, M. McClure, L. Carin, and L.B. Felsen, "Time domain wave-oriented data processing for scattering by nonuniform trunated gratings," J. *Optical Soc. America A*, vol. 11, pp. 2685-2694, Oct. 1994.
4. H. Kim and H. Ling, "Wavelet analysis of backscattering data from an open-ended waveguide cavity," *IEEE Microwave and Guided Wave Letts.*, vol. 2, pp. 140-142, April 1992.
5. H. Kim and H. Ling, "Wavelet analysis of radar echo from finite size targets," *IEEE Trans. Antennas Propagat.*, vol. 41, pp. 200-207, Feb. 1993
6. S.R. Cloude, P.D. Smith, A. Milne, D.M. Parkes, and K. Trafford, "Analysis of time do main ultrawideband radar signals," in Ultra-Wideband Short-Pulse Electromagnetics, Plenum: New York (H.L. Bertoni, L. Carin, and L.B. Felsen, Eds.), pp. 445-456, 1993.

7. G.C. Gaunaurd, H.C. Strifors, A. Abrahamsson, and B. Brusmark, "Scattering of short EM-pulses by simple and complex targets using impulse radar," in Ultra-Wideband Short-Pulse Electromagnetics, Plenum: New York (H.L. Bertoni, L. Carin, and L.B. Felsen, Eds.), pp. 437-444, 1993.
8. E.K. Walton and A. Moghaddar, "Time frequency-distribution analysis of frequency dis persive targets," in Ultra-Wideband Short-Pulse Electromagnetics, Plenum: New York (H.L. Bertoni, L. Carin, and L.B. Felsen, Eds.), pp. 423-436, 1993.
9. L. Carin and L.B. Felsen, "Wave-oriented data processing for frequency and time domain scattering by nonuniform truncated arrays," *IEEE Antennas and Propagat. Magazine*, vol. 36, pp. 29-43, 1994.
10. L. Cohen, "Time-frequency distributions - A review," *Proc. IEEE*, vol. 77, pp. 941-981, July 1989.
11. M. McClure, D.R. Kralj, T.-T. Hsu, L. Carin and L.B. Felsen, "Frequency domain scattering by nonuniform truncated arrays: wave-oriented data processing for inversion and imaging," *J. Optical Soc. America A*, vol. 11, pp. 2675-2684, Oct. 1994.
12. L.B. Felsen and L. Carin, "Wave-oriented processing of scattering data," *Elect. Letters*, vol. 29, pp. 1930-1932, Oct. 28, 1993.
13. B.Z. Steinberg, E. Heyman, and L.B. Felsen, "Phase space beam summation for time-harmonic radiation from large apertures," *J. Opt. Soc. Am. A*, vol. 8, pp. 41-59, 1991.
14. B.Z. Steinberg, E. Heyman, and L.B. Felsen, "Phase space beam summation for time-dependent radiation from large apertures: continuous parametrization," *J. Opt. Soc. Am. A*, vol. 8, pp. 943-958, 1991.
15. L.B. Felsen, "Complex-source-point solutions of the field equations and their to the propagation and scattering of Gaussian beams," Symposia Mathematica, Istituto Nazionale di Alta Matematica, vol. XVIII, Acad Press, London and New York, 40-56, 1976.
16. K.A. Connor and L.B. Felsen, "Gaussain pulses as complex-source-point solutions in dispersive media," *Proc. of the IEEE Issue on Rays and Beams*, Vol. 62, pp. 1614-1615, Nov. 1974.
17. S.Y. Shin and L.B. Felsen, "Gaussian beams in anisotropic media," *Applied Physics*, vol. 5, pp.239-250, 1974.
18. L.B. Felsen, "Rays, modes, and beams in optical fiber waveguides." *Optical and Quantum Electronics*, vol. 9, pp. 189-195, May 1977.
19. S.Y. Shin and L.B. Felsen, "Gaussian beam modes by multipoles with complex source points,," *IEEE Trans. Antennas Propagat.*, pp.189-195, May 1977.
20. G. Ghione, I. Montrosset and L.B. Felsen, "Complex ray analysis of radiation from large apertures with tapered illuminations," *IEEE Trans. on Antennas and Propagation*, vol. AP-32, pp.689-693, 1984.
21. L.B. Felsen, "Geometrical theory of diffraction, evanescent waves, complex rays and Gaussian beams," *Geophys. J. Roy.* Astron. Soc., vol. 79, pp.77-88 , 1984.
22. Y.Z. Ruan and L.B. Felsen, "Reflection and transmission of beams at a curved interface," *J. Opt. Soc. Am. A*, vol. 3, April 1986.
23. H. Ikuno and L.B. Felsen, "Complex rays in transient scattering from smooth targets with inflection points," *IEEE Trans. Antennas Propagat.*, vol. AP-36, pp.1272-1280, Sept. 1988.
24. E. Heyman and L.B. Felsen, "Propagating pulsed beam solutions by complex source parameter substitution," *IEEE Trans. Antennas Propagat.*, vol. AP-34, pp.1062-1065, 1986.
25. L.B. Felsen, "Systematic study of fields due to extended apertures by Gaussian beam discretization" *IEEE Trans. Antennas Propagat.*, vol. AP-37, pp.884-892, 1989.

26. E. Heyman and L.B. Felsen, "Complex source pulsed beam fields," *J. Opt. Soc. Am. A*, pp. 806-817, 1989.
27. J.Maciel and L.B. Felsen, "Gaussian beam analysis of propagation from an extended plane aperture distribution through dielectric layers: I- Plane layer,"*IEEE Trans. Antennas Propagat.*, vol. AP-38, pp.1607-1617, 1990.
28. J.Maciel and L.B. Felsen, "Gaussian beam analysis of propagation from an extended plane aperture distribution through dielectric layers: II- Circular cylindrical layer,"*IEEE Trans. Antennas Propagat.*, vol. 38, pp.1618-1624, 1990.
29. E. Heyman and L.B. Felsen, "Real and complex spectra- A generalization of WKJB seismograms," *Geophys. J. Roy. Astron. Soc.*, vol. 91, pp.1087-1126, 1987.
30. L.B. Felsen and S. Zeroug, "Beam parametrization of weak debonding in a layered aluminum plate," Review of Progress in Quantitative NDE, Plenum Press, pp.195-202, 1990.
31. L.B. Felsen and S. Zeroug, "Ultrasonic beam method for localized weak debonding in a layered plate," *J. Acoust. Soc. Am. A*, vol. 90, pp.1527-1538, 1991.
32. I.T. Lu, L.B. Felsen, J.M. Klosner, "Beams and modes for scattering from weak bonding flaws in a layered aluminum plate," *J. Acoust. Soc. Am.*., vol. 88, pp. 496-504, 1990.
33. I.T. Lu, L.B. Felsen and J.M. Klosner, "Beam-to-mode conversion in an aluminum plate for ultrasonic NDE applications," *ASME J. of Engineering Materials and Tech.*, vol. 112, pp.236-240, 1990.
34. I.T. Lu, L.B. Felsen, J.M. Klosner and C. Gabay, "Beams and modes for scattering from weak bonding flaws in a layered aluminum plate," , *J. Acoust. Soc. Am. A*, vol. 88, pp. 496-504, 1990.
35. K.E. Oughston and G.C. Sherman, "Electromagnetic Pulse Propagation in Causal Dielectrics," Springer-Verlag, New York, 1994.
36. L.B. Felsen, "Transients in dispersive media, Part I: Theory,"*IEEE Trans. Antennas Propagat.*, vol. AP-19, pp.424-432, 1971.
37. E. Heyman and T. Melamed, "Certain considerations in aperture synthesis of ultra-wideband/short-pulse radiation," *IEEE Trans. Antennas Propagat.*, vol. AP-42, pp.518-525, 1994
38. L.B. Felsen, "Asymptotic theory of pulse compression," *IEEE Trans. Antennas Propagat.*, vol. AP-19, pp.424-432, 1971.
39. L.B. Felsen and F. Niu, " Spectral analysis and synthesis options for short pulse radiation from a point dipole in a grounded dielectric layer," *IEEE Trans. Antennas Propagat.*, vol. AP-41, pp. 747-754, June 1993
40. F. Niu and L.B. Felsen, "Spectral analysis and synthesis options for short pulse radiation from a point dipole in a grounded dielectric layer," *IEEE Trans. Antennas Propagat.*, vol. AP-41, pp. 762-769, June 1993
41. F. Niu and L.B. Felsen, "Asymptotic analysis and numerical evaluation of short-pulse radiation from a point dipole in a grounded dielectric layer," *IEEE Trans. Antennas Propagat.*, vol. AP-41, pp.762-769, June 1993
42. E. Heyman, "Pulsed beam propagation in inhomogeneous medium,"*IEEE Trans. Antennas Propagat.*, vol. AP-42, pp.311-319, 1994.
43. A.M. Shaarawi, I.M. Besieris, R.W. Ziolkowski and S.N. Sedky, "Generation of approximate focus-wave-mode pulses from wideband dyanmic Gaussian apertures," J. Opt. Soc. Am. A, vol. 12, pp. 1954-1964, 1995.
44. E. Heyman, B.Z. Steinberg and L.B. Felsen, "Spectral analysis of focus wave modes," J. *Opt. Soc. Am. A*, vol. 4, pp. 2081-2091, 1987.
45. E. Heyman and L.B. Felsen, "Comments on 'Nondispersive waves: interpretation and causality'," *IEEE Trans. Antennas Propagat.*, vol. AP-42, pp.1668-1670, 1994.

46. L.B. Felsen, "Progressing and oscillatory waves for hybrid synthesis of source-excited propagation and diffraction," *IEEE Trans. Antennas and Propagat.*, AP-32, 775-796 , 1984.
47. I.T. Lu and L.B. Felsen, "Ray, mode and hybrid options for transient source-excited propagation in an elastic layer," *Geophys. J. Roy. Astron. Soc .* 86, 177-201 , 1986.
48. E. Heyman and L.B. Felsen, "Traveling wave and SEM representations for transient scattering by a circular cylinder," *J. Acoust. Soc. Am.* 79(2), 230-238 , 1986.
49. E. Heyman, A. Tijhuis and J. Boersma, "Spherical and collimated pulsed fields in conducting media," presented at the URSI Electromagnetic Theory Symposium, St. Petersburg, Russia, August 1995.
50. K.A. Connor and L.B. Felsen, "Complex space-time rays and their application to pulse propagation in lossy dispersive media," *Proceedings of the IEEE, Special Issue on Rays and Beams*, Vol. 62, No. 11, 1586-1598 , 1974.
51. T. Melamed, Y. Ehrlich and E. Heyman, "Short-pulse inversion of inhomogeneous media," submitted to *Inverse Problems*.
52. T. Melamed, E. Heyman and L.B. Felsen, "Local spectral analysis for short-pulse inversion of inhomogeneous media: Part I—spectral properties," *in preparation.*
53. T. Melamed, E. Heyman and L.B. Felsen, "Local spectral analysis for short-pulse inversion of inhomogeneous media: Part II—reconstruction," *in preparation.*
54. S. Vitebskij and L. Carin, "Moment-method modeling of short pulse scattering from , and the resonances of, a wire buried in a lossy, dispersive half space", *IEEE Trans. Antennas Propagat.*, vol. AP-43, Nov. 1995
55. S. Vitebskij, K. Sturgess and L. Carin, "Short pulse scattering from buried perfectly conducting bodies of revolution" *IEEE Trans. Antennas Propagat.*, to appear
56. S. Vitebskij and L. Carin, "Resonances of perfectly conducting wires and bodies of revolution buried in a lossy dispersive half space", submitted to *IEEE Trans. Antennas Propagat.*
57. L. Carin, L.B. Felsen and T.T. Hsu. "High frequency fields excited by truncated arrays of nonuniformly distributed filamentary scatterers on an infinite dielectric slab: Parameterizing (leaky-mode) - (Floquet-mode) interaction", *IEEE Trans. Antennas Propagat.* vol. 44, pp. 1-11, 1996
58. T.T. Hsu, M.R. McClure, L.B. Felsen, and L. Carin, "Wave-oriented processing of scattered field data from a plane-wave-excited finite array of filaments on an infinite dielectric slab", *IEEE Trans. Antennas Propagat.*, vol. 44, pp. 352-360, 1996
59. L. Carin, L.B. Felsen, D.R. Kralj, H.S. Oh, W.C. Lee, and S.U. Pillai, "Wave-oriented signal processing of dispersive time domain scattering data", submitted to *IEEE Trans. Antennas Propagat.*

SHORT-PULSE SCATTERING FROM AND THE RESONANCES OF BURIED AND SURFACE METAL MINES

Lawrence Carin and Stanislav Vitebskiy

Department of Electrical and Computer Engineering
Duke University
Durham, NC 27708-0291

INTRODUCTION

After several decades of research [1-19], buried and surface mine detection and identification remains a very difficult problem, with no single sensor providing sufficient performance (defined in terms of probability of detection and probability of false alarm) for reliable and repeatable system performance. Radar represents one of the oldest technologies for this problem, and there has been significant interest recently in ultra-wideband (UWB), short-pulse (SP) radar [20,21]. Such UWB-SP radar systems are designed to operate at a sufficiently low frequency to provide ground penetration, while retaining the wide bandwidth necessary for adequate down-range resolution. In addition to being used for detection, the late-time resonant signature scattered from buried mines has also been used for identification [1,2,5].

While UWB-SP radar systems have, under appropriate conditions, been found to be very effective tools for mine detection and identification [1,2,5], it is well known that under many other conditions such systems provide inadequate performance. For example, in high-clutter environments (e.g., surface roughness and rocks), for high-loss soils, and for low-RCS mines, radar systems often perform poorly. Thus, one must possess *a priori* insight into where UWB-SP radar systems are most likely to yield detection/identification benefits, thereby avoiding situations for which radar is likely to fail. Since system performance depends significantly on the highly variable electrical properties of soil, it is impossible to test system performance empirically for all possible conditions, and therefore accurate wave modeling is essential. Such theoretical studies, with comparison to experimental data, are the focus of this paper.

Although radar scattering from surface and buried mines represents a problem of long-standing interest, up until recently all studies of this problem were primarily experimental, due to problem complexity. However, with recent advances in computer speed and memory, it is now possible to develop accurate numerical models for UWB-SP scattering from and the resonances of buried and surface mines, which yield the predicted system performance alluded to above. A popular tool for such studies is the finite difference time domain (FDTD) algorithm [11], which allows scattering from rather general environments. However, the FDTD requires one to grid up the entire computational space, and therefore the available computer memory ultimately limits the complexity (size) of the problem to be studied. As an alternative, one can use the method of moments (MoM) [16-19], which only requires one to grid unknowns on the target surface (the surrounding media being characterized by an appropriate Green's function). The limitation of the MoM is that it can only be applied to environments for which the Green's function is known, such as lossy layered media. While this may be a severe limitation for some problems, for many others the surrounding environment is accurately so modeled. Moreover, the MoM, being based on an analytic evaluation of the Green's function, provides powerful insight that is not readily garnered from the FDTD, it being a purely numerical algorithm. In particular, poles in the MoM solution yield the complex resonant frequencies of buried targets. Furthermore, as will be discussed below, the MoM provides powerful insight into the polarimetric properties of scattering from realistic, three-dimensional mines, which can be exploited in polarimetric imaging and signal processing for mine detection/identification.

In the remainder of this paper, we provide a summary of our MoM algorithm, with appropriate references cited for additional details. It is explained that the target is modeled as a body of revolution (BOR), reducing algorithm complexity, with little loss of generality in the context of the mine problem (most mines can be modeled accurately as BORs). Additionally, important issues concerning the efficient evaluation of the half-space Green's function are addressed as well. Several sets of results are presented, with comparison to experimental data measured by the Army Research Laboratory (ARL). Finally, the results and impact of this research is summarized in the conclusions.

NUMERICAL ALGORITHM

Consider a nonmagnetic (μ_r=1) half-space composed of frequency-dependent permittivity $\varepsilon_0\varepsilon_r(\omega)$ and conductivity $\sigma(\omega)$; within this half-space is buried a perfectly conducting body of revolution (BOR), with its axis of rotation normal to the half-space interface (Fig. 1). A pulsed plane wave is assumed incident obliquely from the air region, and we are interested in calculating the scattered time-domain fields; the problem is analyzed in the frequency domain, with the time-domain fields calculated via a fast Fourier transform (FFT). For the short-pulse, wideband problem of interest here, we must develop as efficient a frequency-domain formulation as possible. As discussed below, this same algorithm can be modified for calculation of the complex target resonant frequencies.

To construct the Method of Moments (MoM) solution to this problem, we employ the mixed potential electric-field integral equation (EFIE) formulation of [15,16-19]. The boundary condition for vanishing tangential electric field on the surface of the perfect electric conductor (PEC) target yields the integral equation

$$\mathbf{n}\times[j\omega\int_S \bar{\mathbf{K}}_A(\mathbf{r}\,|\,\mathbf{r}')\bullet\mathbf{J}(\mathbf{r}')dS'+\nabla\int_S K_\phi(\mathbf{r}\,|\,\mathbf{r}')q(\mathbf{r}')dS']=\mathbf{n}\times\mathbf{E}^{inc}(\mathbf{r}),\quad \mathbf{r}\text{ on }S \tag{1}$$

which has the same form as integral equations used for scattering in free space; $\mathbf{J}(\mathbf{r}')$ and $q(\mathbf{r}')$ are the current and charge densities induced on the surface of the PEC target by the incident field $\mathbf{E}^{inc}(\mathbf{r})$, $\mathbf{r}$ and $\mathbf{r}'$ are the observation and source points, respectively, and $\mathbf{n}$ is a unit vector normal to the target surface S. The charge density $q(\mathbf{r}')$ in (1) is represented in terms of the electric current density $\mathbf{J}(\mathbf{r}')$ using the continuity relation. Among the different options for the Green's functions $\mathbf{K}_A$ and K_ϕ in (1) proposed by Michalski and Zheng [15], we have found "Formulation C" of their paper most preferable for scattering from a buried BOR. In this formulation the dyadic kernel $\mathbf{K}_A$ is written in the form

$$\bar{\mathbf{K}}_A(r|r') = (\mathbf{xx}+\mathbf{yy})K_A^{xx}+\mathbf{xz}K_A^{xz}+\mathbf{yz}K_A^{yz}+\mathbf{zx}K_A^{zx}+\mathbf{zy}K_A^{zy}+\mathbf{zz}K_A^{zz} \tag{2}$$

and reciprocity relations $K_A^{xz} = -K_A^{zx}$ and $K_A^{yz} = -K_A^{zy}$ apply (it is important to note that, for straight-wire targets and antennas [16], these reciprocity relations imply that the contributions from the cross terms of the dyadic cancel each other out in the computation of the MoM matrix components, and therefore one only requires components K_A^{xx} and K_A^{zz} for such problems – for the BOR problem, we have no such simplifications). Also, K_ϕ in (1) represents the scalar potential of the charge associated with a horizontal dipole (recall that the vertical and horizontal dipoles have different scalar potentials [15]). Explicit expressions for components of $\mathbf{K}_A$ and K_ϕ are given in [15] and therefore are not repeated here.

After appropriate manipulations, the half-space dyadic Green's function can be placed in a form appropriate for a BOR [17,18]. However, each component of the dyadic is represented in terms of slowly convergent Sommerfeld integrals, which, if not evaluated efficiently, severely limit the complexity and bandwidth of the problem to be considered. We have therefore applied the method of complex images [22,23], which performs an exponential fit to the spectral-domain Green's function, along a proper path in the complex wavenumber plane; the subsequent spectral integrals can then be integrated analytically using the Sommerfeld (Weyl) identity [22,23]. The numerical complexity is therefore transferred from a laborious and slowly convergent numerical evaluation of a oscillatory Sommerfeld integral, to the relatively efficient and simple task of providing an exponential fit to the spectral-domain half-space Green's function. We have used Prony's method to fit the parameters in the exponential model, although several other equally appropriate techniques are available [24-26].

The integral equation in (1) is solved numerically via the Method of Moments (MoM). In this well-known approach [16-19] the surface of the BOR is generated by rotating an arc (the "generating arc") about the axis of symmetry. The current on the BOR is represented by subsectional pulse basis functions, and the periodic azimuthal variation is represented by a Fourier series. The half-space Green's function components are also expanded in a Fourier series. As a result of the testing procedure, we obtain a matrix equation for each Fourier mode m

$$\begin{bmatrix} [Z_{tt}^m] & [Z_{t\phi}^m] \\ [Z_{\phi t}^m] & [Z_{\phi\phi}^m] \end{bmatrix} \begin{bmatrix} [I_t^m] \\ [I_\phi^m] \end{bmatrix} = \begin{bmatrix} [V_t^m] \\ [V_\phi^m] \end{bmatrix} \tag{3}$$

where the elements of the unknown current submatrices I are the basis-function coefficients, and the voltage submatrices V consist of the Fourier series coefficients of the tangential electric field on the BOR surface integrated with the testing functions. Expressions for V_t^m and V_ϕ^m (tangential and azimuthal components, respectively) are given in the literature [27] for the case of plane wave incidence in free space; for the problem considered here, these expressions must be modified to account for refraction at the interface.

The harmonics of the induced current are obtained by solving (3), which, with the half-

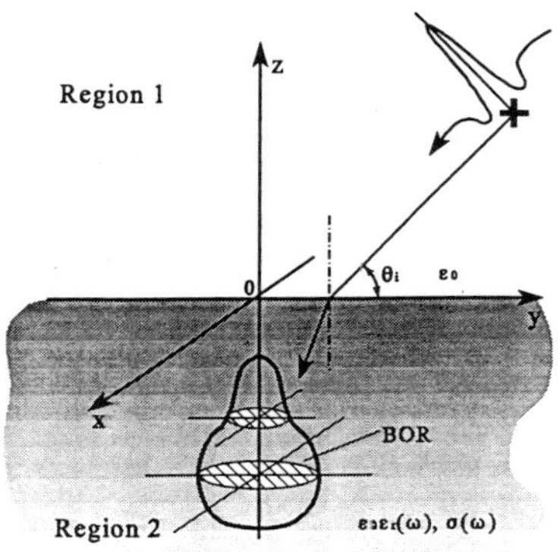

Figure 1. Schematic of short-pulse plane-wave scattering from a buried perfectly conducting body of revolution.

space Green's function, are then used to calculate the scattered field. The time-domain scattered fields are determined via FFT, taking into account the spectrum of the desired excitation pulse. The scattered fields vary strongly as a function of frequency while the components of the MoM matrix vary relatively slowly. Thus, to improve numerical efficiency, the MoM matrix is computed over a relatively coarse frequency grid, with the matrix components at intervening frequencies computed via linear interpolation [28].

As alluded to above, there are many analytical aspects to the MoM, particularly involving manipulation of the medium Green's function. This feature can lead to important phenomenological understanding. For example, if a target can be modeled as a BOR, as has been done for the mine problem considered here, it can be shown from (3) that there is no cross-polarized scattered field. In other words, if the incident field is vertically (V) polarized, so will be the scattered field, with the same property holding for horizontal (H) polarization. This is a very important property, in that many man-made targets (mines) can be accurately modeled as BORs, while most natural clutter (rocks) cannot. One can exploit this phenomenon, for example, in the context of polarimetric imaging of mine-field data. Mines should be visible for the VV and HH (co-polarized) images, and disappear (approximately) in the cross-polarized (HV or VH) images. Rocks, by contrast, should be visible in co- and cross-polarized imagery. While this important understanding is a direct consequence of the above analysis, its applications to measured polarimetric imagery is a subject for future research.

As discussed in the Introduction, there is also significant interest in resonance-based identification of buried targets (mines). However, it is well known that said resonances are often very low-Q, making them difficult to extract accurately from measured data. It is therefore of significant importance to perform calculations of such complex resonant frequencies, to quantify *a priori* which targets and under what conditions (e.g., soil properties) resonant-based identification represents a viable identification scheme. For this task, the MoM is *uniquely* qualified, since the FDTD cannot be applied to problems for which there is no source (incident field) – precisely the situation of interest when one considers the natural, source-free oscillatory modes characteristic of late-time resonances. For such problems the right side of (3) is set to zero (source-free) and non-trivial solutions for the (modal) currents are found at frequencies for which the determinant of the MoM impedance matrix vanishes. Although such matters are too complex

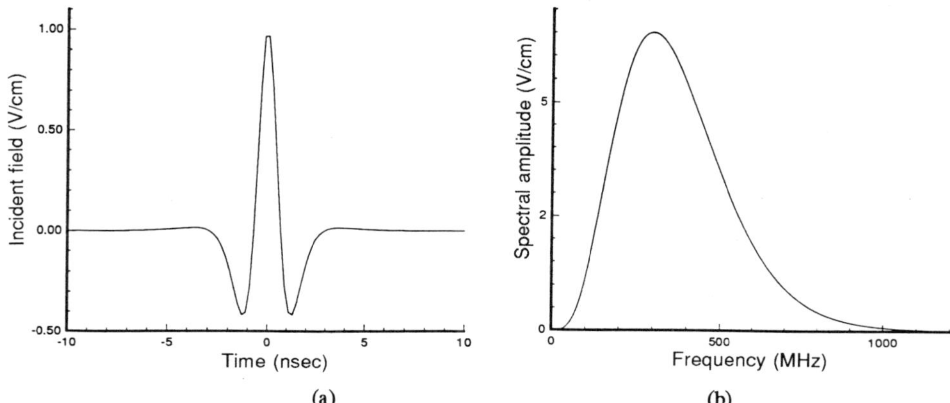

(a) (b)

Figure 2. Incident pulse and its spectrum used for all numerical examples. a) Incident pulse, b) Frequency spectrum

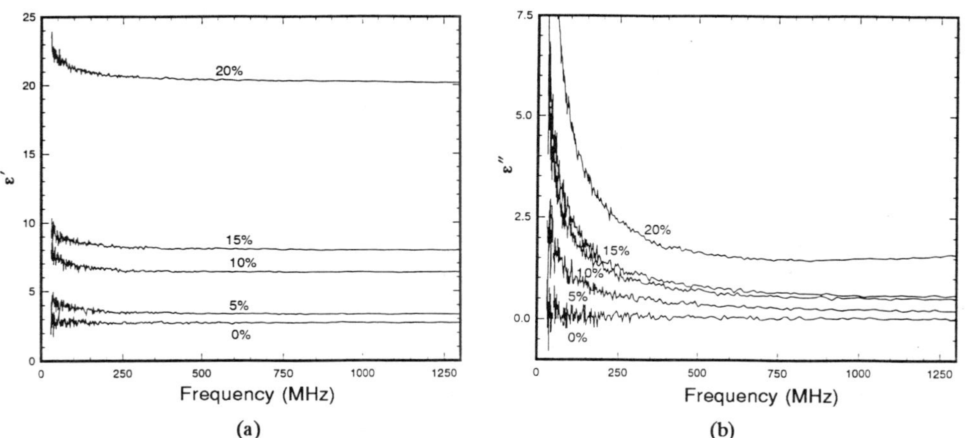

(a) (b)

Figure 3. Frequency-dependent complex permittivity $\epsilon_r = \epsilon_r' - j\epsilon_r''$ measured from a soil sample taken from Yuma, AZ. Results are plotted as a function of water content, by weight. a) ϵ_r', b) ϵ_r''.

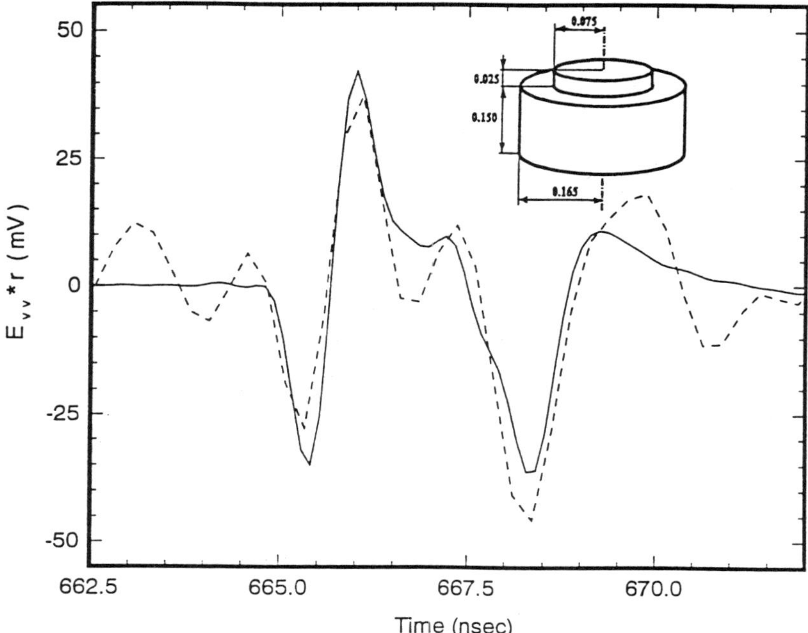

Figure 4. Comparison between theoretical (solid) and measured SAR (dashed) scattered fields from a surface anti-tank mine (shown inset, with units in meters).

to go into at length here, there are important considerations [18] which must be adhered to while computing the matrix parameters at *complex* frequencies (for the scattering problem discussed previously, all computations are at *real* frequencies).

EXAMPLE CALCULATIONS

Over the last few years we have published several papers on the aforementioned algorithm, for both UWB-SP scattering as well as resonant-frequency calculations [16-19]. Therefore, here we provide a summary of some the most important results. Considering first UWB-SP scattering, we consider pulsed plane-wave incidence, with the incident-pulse shape and spectrum shown in Fig. 2. This waveform was selected to be consistent with current UWB-SP technology, with the center frequency selected to provide significant ground penetration. The frequency-dependent (dispersive) soil parameters were measured with a network analyzer, and consisted of measuring the reflection coefficient from an open-circuit coaxial probe. Measurements were performed as a function of water content, by percentage weight, and are shown in Fig. 3 for the soil sample used in all subsequent scattering and resonance calculations, with results plotted in the form of complex relative permittivity $\epsilon_r=\epsilon_r'-j\epsilon_r''$.

The first example considered here is for scattering from the *surface* anti-tank mine (M20) schematized in Fig. 4. Results are presented for VV polarization, and a 20° incidence angle is considered (70° with respect to the air-ground normal); all calculations were performed using the 5% water data in Fig. 2. In Fig. 4 is shown a comparison between the computed (solid) and measured (dashed) data. The measured data represents the coherent superposition of several sensor positions (from a synthetic-aperture radar [29]). Therefore, one would anticipate that the

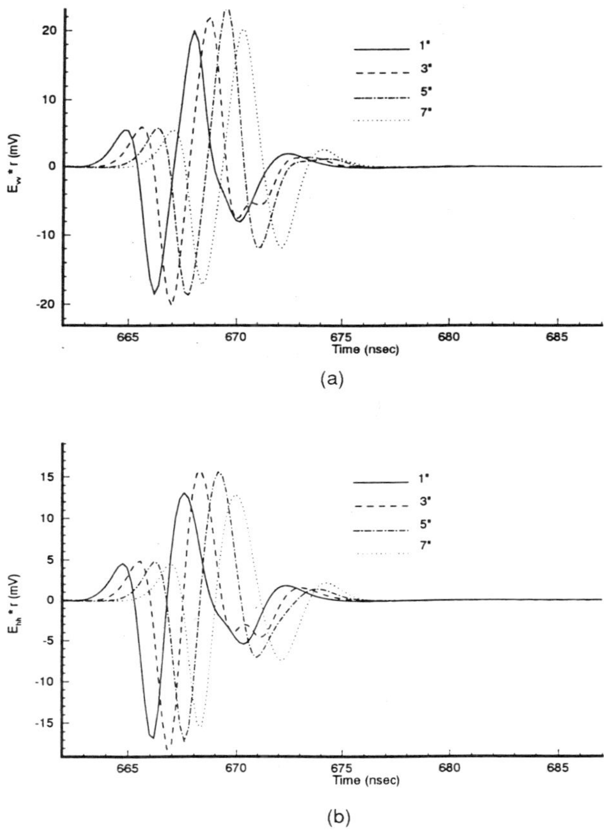

Figure 5. Normalized fields backscattered from a model anti-tank mine buried at depths of 2.54 cm, 7.62 cm, 12.7 cm, and 17.8 cm from the top of the target. a) VV, b) HH

imaging process will cause some blurring in the extracted signature. Further, in the theory we assume plane-wave incidence at a fixed angle and with a particular incident-pulse shape; in practice, the incident wave can only be approximated as planar, and the angle of incidence and incident-pulse shape can only be determined approximately. Nevertheless, we see in Fig. 4 that for VV polarization there is excellent agreement between theory and experiment. No effort was made to optimize the agreement between theory and experiment in Fig. 4; i.e., these results are typical of what we have found for several examples.

We next consider a model, disc-like, metal anti-tank mine with 38.1 cm diameter and 6.35 cm height. Results are shown in Fig. 5 for the VV and HH transient fields scattered from the anti-tank mine buried at depths of 2.54 cm, 7.62 cm, 12.7 cm, and 17.8 cm (1, 3, 5, and 7 inches, respectively) from the top of the target, with the excitation field incident at the Brewster angle (θ_B=28.25°). The most striking characteristic of these results is the similarity between the VV and HH fields (recall that there is no cross-polarized scattered field). Apparently the incident pulse does not have sufficient temporal (spatial) resolution to resolve features on the target. Interestingly, initially the peak scattered signal increases as the target depth increases, in contrast

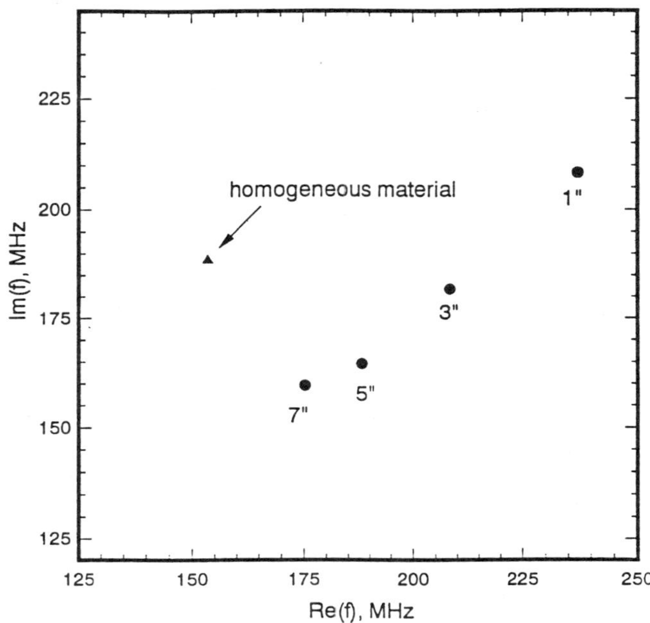

Figure 6. Complex resonant frequencies of the lowest-order resonant mode for the anti-tank mine considered in Fig. 5.

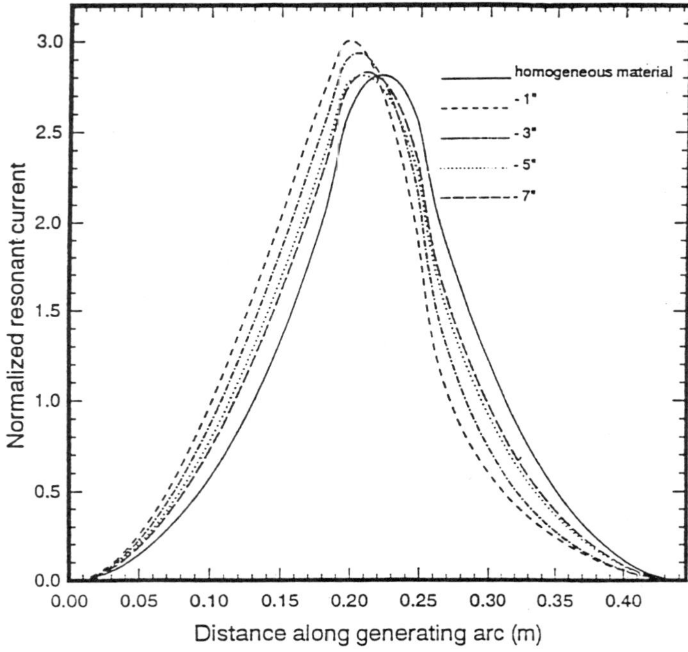

Figure 7. Normalized modal currents, as a function of target depth, corresponding to the resonant modes in Fig. 6. The currents are normalized such that they integrate to unity when integration is performed along the generating arc of the body of revolution.

with anticipation. We attribute this phenomenon to reverberations between the top of the target and the air-ground interface that constructively interfere with wavefronts scattered from the target. As expected, as the target depth is further increased, the peak scattered waveform starts to diminish (the reverberated waveform is attenuated and the temporal overlap of the primary and reverberated wavefronts diminishes).

Figure 5 does not show an obvious late-time resonant signature associated with the buried anti-tank mine. To explain this, we plot in Fig. 6 the complex resonant frequencies of the lowest-order SEM mode for the anti-tank mine considered in Fig. 5. For comparison, the resonant frequency of the lowest-order mode is also plotted for the same target situated in a homogeneous medium of the same electrical properties. Notice that the resonant frequency changes as a function of target depth. In Fig. 7 are plotted the normalized modal currents associated with the mode at each depth. There is a noticeable change in the modal current shape as the target depth is adjusted, with the currents concentrating under the target, nearer the high-dielectric soil, as the mine depth decreases. This latter phenomenon is consistent with the well-known concentration of fields in regions of relatively high dielectric constant.

The late-time resonant modes decay with time t as $\exp(-\omega_i t)$, for complex resonant frequency $\omega_r + j\omega_i$. After n periods the resonant fields decay by $\exp(-2\pi n\omega_i/\omega_r)$. Using the computed resonant frequencies from Fig. 6, after only one oscillation (n=1), the resonant signatures corresponding to depths of 2.54 cm, 7.62 cm, 12.7 cm, and 17.8 cm decay by a factor 0.0038, 0.0042, 0.0041, and 0.0026, respectively. These results explain the absence of a discernable resonant signature in Fig. 5, and indicate the extreme difficulty of resonance-based identification for buried targets of the type considered above.

CONCLUSIONS

A new Method of Moments (MoM) numerical algorithm has been utilized to examine the UWB-SP fields scattered from and the resonances of several targets buried in and placed atop soil. Results have been presented particularly for soil at Yuma, AZ, with account taken for dispersion and loss.

Results have only been presented for a small subset of targets and one soil type (albeit a relatively favorable, low-loss soil), but this study further substantiates the difficulty of radar-based detection and identification of buried and surface targets. For example, concerning the resonances of the buried anti-tank mine considered, the low-Q of such resonances, coupled with the soil loss, conspired to produce late-time resonant modes which decay extremely quickly with time, making virtually impossible resonance-based identification for such targets. However, resonant-based detection has proven useful for particular dielectric targets buried in frozen soil [5]. This dichotomy points to the need for modeling – which will yield *a priori* predictions of GPR performance – to assure that the radar is implemented under appropriate circumstances. It is highly unlikely that GPR will be an effective tool for *all* soil and target types, but when utilized properly, it can be an effective option.

Fortunately, the need for accurate modeling intersects with recent algorithmic developments, which now make possible the modeling of scattering from and the resonances of realistic three-dimensional targets buried in lossy, dispersive soils. In this paper, we have been concerned with UWB-SP radar, which involves incident signals with over 100% bandwidth. Up until very recently, it was virtually impossible to model the scattering of such waveforms from realistic buried targets, due to the complexity of computing the half-space Green's function (needed for the MoM). However, the recent development of the method of complex images, which efficiently computes the Sommerfeld integrals characteristic of the half-space dyadic Green's function, has been utilized here for several realistic and complicated targets of interest. Future studies will further utilize this algorithm to quantify anticipated radar performance as a function of soil and target type. Additionally, the predicted waveforms from our model can also be utilized in the development of matched filters for target detection.

The authors wish to thank Jeff Sichina, Karl Kappra, Marc Ressler and Francis Le of the Army Research Laboratory (Adelphi, MD) for providing the measured UWB-SP SAR data from Yuma Proving Grounds.

REFERENCES

1. L. Peters, Jr., and J. D. Young, "Applications of subsurface transient radars," in *Time-Domain Measurements in Electromagnetics*, E. K. Miller, Ed. New York: Van Nostrand Reinhold, 1986.
2. D. L. Moffatt and R. J. Puskar, "A subsurface electromagnetic pulse radar,' *Geophysics,* vol.41, pp. 506 - 518, June 1976.
3. M. E. Bechtel and A. V. Alogni, "Antennas and pulses for a vehicular-mounted mine detector," Calspan Corp., Buffalo, NY, Rep. no. MA-5366-E-1, 1974.
4. D.J. Daniels, D.J. Gunton, and H.F. Scott, "Introduction to subsurface radar," *Proc. IEE*, vol. 135, pt.F, no.4, pp. 278-320, Aug. 1988.
5. L. Peters, J.J. Daniels and J.D. Young, "Ground penetrating radar as a subsurface environmental sensing tool," *Proc. IEEE,* vol. 82, pp. 1802-1822, Dec. 1994.
6. R.W.P. King and C.W. Harrison, Jr., "The transmission of electromagnetic waves and pulses into the earth," *J. Appl. Phys.,* vol. 39, pp. 4444-4452, Aug. 1968.
7. J.A. Fuller and J.R. Wait, "Electromagnetic pulse transmission in homogeneous dispersive rock," *IEEE Trans. Antennas Propag.,* pp. 530-533, July 1972.
8. G.S. Smith and W.R. Scott, Jr., "A scale model for studying ground penetrating radars," *IEEE Trans. Geosci. Remote Sensing,* vol. 27, pp. 358-363, July 1989.
9. N. Osumi and K. Ueno, "Microwave holographic imaging of underground objects," *IEEE Trans. Antennas Propag.*, vol. AP - 33, pp. 152 - 159, Feb. 1985.
10. C. Liu and C. Shen, "Numerical simulation of subsurface radar for detecting buried pipes" *IEEE Trans. Geosci. Remote Sensing*, vol. GE - 29, pp. 795-798, Sept. 1991.
11. J.M. Bourgeois and G.S. Smith, "A fully three-dimensional simulation of a ground-penetrating radar: FDTD theory compared with experiment," *IEEE Trans. Geosci. Remote Sensing*, vol. GE - 34, pp. 36-44, Jan. 1996.
12. P.E. Wannamaker, G.W. Hohmann, and W.A. SanFilipo, "Electromagnetic modeing of three-dimensional bodies in layered earths using integral equations," *Geophys.,* vol. 49, pp. 60-74, Jan. 1984.
13. H.S. Chang and K.K. Mei, "Scattering of electromagnetic waves by buried and partly buried bodies of revolution," *IEEE Trans. Geosci. Remote Sensing*, vol. GE-23, pp. 596-592, 1985.
14. G. Kristensson and S. Strom, "Scattering from buried inhomogeneities - A general three dimensional formalism," in *Proc. 1977 URSI Symp. on Electromagnetic Wave Theory,* June 1977.
15. K.A. Michalski and D. Zheng, "Electromagnetic scattering and radiation by surfaces of arbitrary shape in layered media, Parts I and II," *IEEE Trans. Antennas and Propagat.,* vol. AP-38, pp. 335-352, March 1990.
16. S. Vitebskiy and L. Carin, "Moment-method modeling of short-pulse scattering from and the resonances of a wire buried inside a lossy, dispersive half space," *IEEE Trans. Antennas Propag.,* vol. AP-43, pp. 1303-1312, Nov. 1995.

17. S. Vitebskiy, K. Sturgess and L. Carin, "Short-pulse scattering from buried perfectly conducting bodies of revolution," *IEEE Trans. Antennas Propag.,*vol. AP-44, pp. 143-151, Feb. 1996.

18. S. Vitebskiy and L. Carin, "Resonances of perfectly conducting wires and bodies of revolution buried in a lossy, dispersive half space," *IEEE Trans. Antennas Propag.*, vol. AP-44, Dec. 1996.

19. S. Vitebskiy, L. Carin, M. Ressler and F. Le, "Ultra-wideband, short-pulse ground-penetrating radar: theory and measurement," to appear in *IEEE Trans. Geoscience and Remote Sensing*, 1997.

20. H.L. Bertoni, L. Carin, and L.B. Felsen (Eds.), Ultra-Wideband, Short-Pulse Electromagnetics, Plenum Co., NY, 1994.

21. L. Carin and L.B. Felsen (Eds.), Ultra-Wideband, Short-Pulse Electromagnetics II, Plenum Co., NY, 1995.

22. Y. L. Chow, J. J. Yang, D. G. Fang, and G. E. Howard, "A closed-form spatial Green's function for the thick microstrip substrate," *IEEE Trans. Microwave Theory Tech.,* vol. MTT-39, pp. 588 - 562, March 1991.

23. J. J. Yang, Y. L. Chow, D. G. Fang, "Discrete complex images of a three-dimentional dipole above and within a lossy ground," *IEE Proceedings-H,* vol. 138, no. 4, pp. 319 - 326, Aug. 1991.

24. M. Van Blaricum and R. Mittra, "A technique for extracting the poles and residues of a system directly from its transient response," *IEEE Trans. Antennas Propag.*, vol. 23, pp. 777-781, Nov. 1975.

25. Y. Hua and T.K. Sarkar, "Generalized pencil-of-function method for extracting poles of an EM system from its transient response," *IEEE Trans. Antennas Propag.*, vol. 37, pp. 229-234, Feb. 1989.

26. Y. Hua and T.K. Sarkar, "Matrix pencil method for estimating parameters of exponentially damped/undamped sinusoids in noise," *IEEE Trans. Acoustics, Speech, Sig. Proc.*, vol. 38, pp. 814-824, May 1990.

27. J.R. Mautz and R.F. Harrington, "Radiation and scattering from bodies of revolution," *Appl. Sci. Res.*, vol. 20, pp. 405-435, June 1969.

28. E.H. Newman, "Generation of wide-band data from the method of moments by interpolating the impedance matrix," *IEEE Trans. Antennas Propag.,* vol. AP-38, pp. 1820-1824, Dec. 1988.

29. M.A. Ressler and J.W. McCorkle, "Evolution of the Army Research Laboratory ultra-wideband test bed," in Ultra-Wideband Short-Pulse Electromagnetics 2, L. Carin and L.B. Felsen eds., Plenum Press, NY, pp. 109-123, 1995.

COMPARATIVE ANALYSIS OF UWB UNDERGROUND DATA COLLECTED USING STEP-FREQUENCY, SHORT PULSE AND NOISE WAVEFORMS

E. K. Walton and S. Gunawan

Ohio State University
ElectroScience Laboratory
1320 Kinnear Road
Columbus, Ohio 43212-1191

INTRODUCTION

One of the important questions in UWB underground radar systems is the choice of waveform. Clearly one must transmit and receive an electromagnetic signal with a large bandwidth in the frequency band below approximately 800 MHz to penetrate the ground, but the choice of the waveform is often determined by other considerations than underground propagation conditions such as cost and reduction of interference from (or to) other users of this crowded frequency band.

In this paper, we discuss three types of radar waveforms, the impulse, the noise and the step frequency. We show that all three types are linear transformations of each other and thus the raw processing results and signal to noise ratios are the same. We will discuss signal to noise after receiver detection as an important issue.

RADAR SYSTEMS

The three types of radar systems are shown in Figure 1. The impulse radar (a) is the simplest in concept, since it directly transmits a fast (IE: wide band) impulse and uses a voltage detector and high speed A/D to receive the signal. The noise radar (b) transmits a random noise waveform with a wide bandwidth and correlates a delayed version of the transmitted signal with the received signal. The correlation value as a function of the reference delay time can be shown to be the impulse response of the radar target (convolved with the radar antenna etc.). The step frequency radar transmits a carrier at a specific frequency, and measures the relative amplitude and phase of the received signal. The radar frequency is stepped through a set of frequencies over the band of interest. Impulse response profiles in this case are the result of Fourier transformation.

At the Ohio State University ElectroScience Laboratory, we have a long history of interest in ground penetration radars, and we have built UWB antennas and impulse

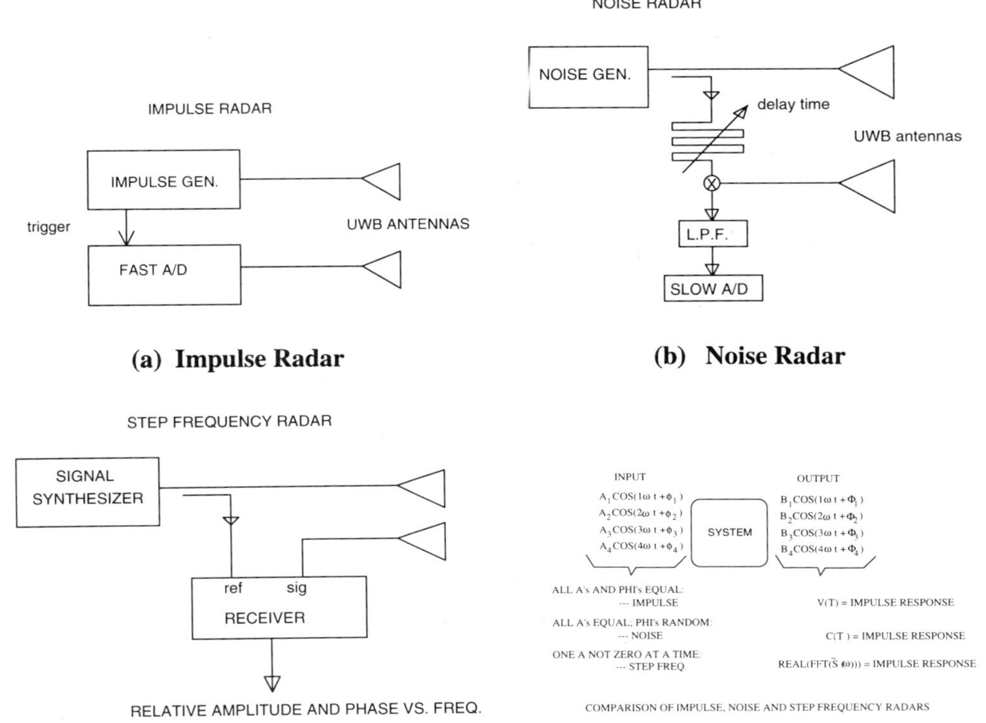

(a) Impulse Radar **(b) Noise Radar**

(c) Step Frequency Radar **(d) Generic Conceptual Diagram**

Figure 1. Basic block diagrams of the three types of radars along with a generic radar system.

radars. We also have network analyzer systems which we configured as radars for this test.

Finally, for this experiment, we built a noise radar operating in the 50 to 800 MHz frequency band for these comparisons. The noise radar uses a series of high gain amplifiers in the 50 to 1,000 MHz band and a mixer operating over the band from DC to approximately 750 MHz. The low pass filter is a two stage system, where the first stage is passive RF filtering and the second stage is a 1 Hz active low pass filter. The delay line is a pair of 10 step dual coaxial switches designed to switch in increasingly longer delays under computer control. The output of the low pass filter is A/D converted and stored in a computer file as correlation voltage as a function of delay line setting.

The performance of these three radar systems will be compared.

SPHERE EXPERIMENTS

The experimental systems were set up to penetrate a concrete building wall using a loaded crossed dipole antenna. All three radar systems used the same antenna in the same location, and the same set of targets and target positions. The geometry is shown in

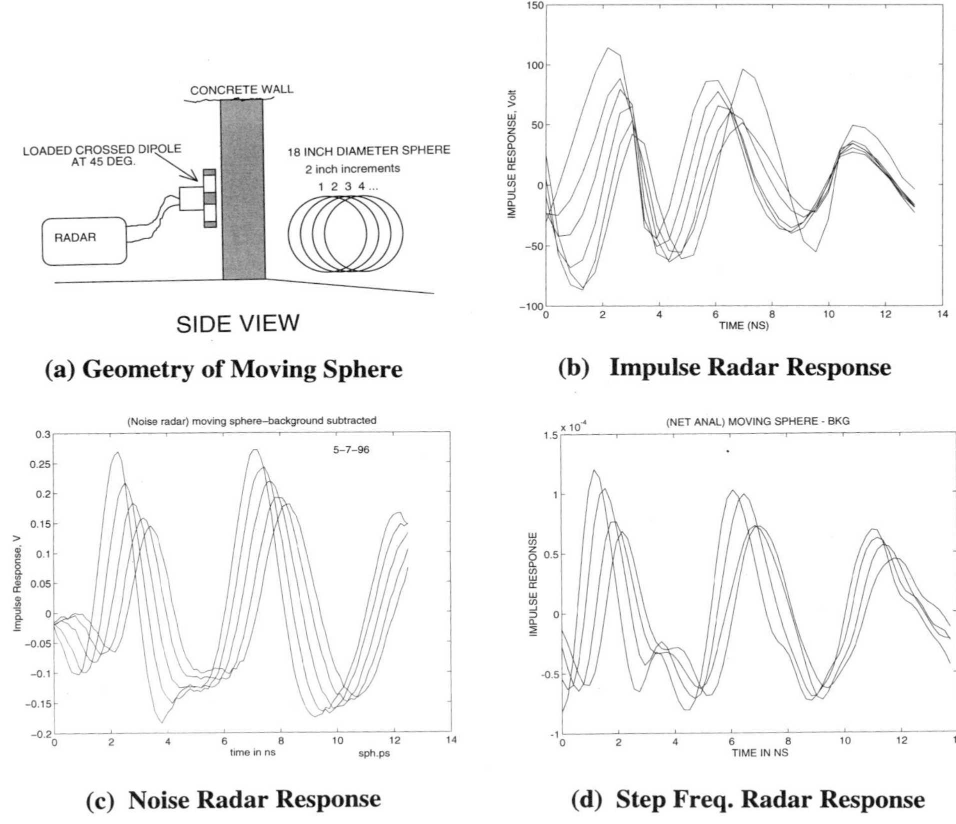

(a) Geometry of Moving Sphere **(b) Impulse Radar Response**

(c) Noise Radar Response **(d) Step Freq. Radar Response**

Figure 2. Moving sphere experiment.

Figure 2.a. Note that the sphere is offset from the centerline of the crossed dipole so that the polarization will not cancel. (The sphere has no cross-polarized RCS value.) The sphere is moved in 2 inch increments as the data are measured. In all cases, the background (no-sphere) data file is subtracted from the sphere data file so that the background scattering terms are suppressed.

Examples of the response waveforms are shown in Figure 2.b,c,and d. Figure 2.b. shows the curves measured using the impulse radar (after background subtraction). This is the directly digitized voltage response of the radiated impulse. Note that the waveforms become smaller and later in time as the sphere is moved away from the wall. The curves are the convolution of the sphere impulse response and the radar antenna and wall.

The response of the noise radar is shown in figure 2.c. Note that the curves are directly measured as the voltage output of the low pass filter as a function of delay time. Only background subtraction has been used. The resulting waveform has been shown to be the impulse response of the target convolved with the impulse response of the antenna and wall propagation values. The radar system response is ultra wide band and fairly flat, so the response waveforms as shown are approximately the impulse response of the sphere.

The impulse response of the sphere as measured using the network analyzer (NA) (a coherent step-frequency radar), is shown in figure 2.d. In this case, the NA was set up to transmit frequency steps from 0.1 MHz to 1,000 MHz. This frequency band was chosen to approximate the operational band of the noise radar. The amplitude and phase of the scattered signal (complex scattering values) were recorded at each frequency.

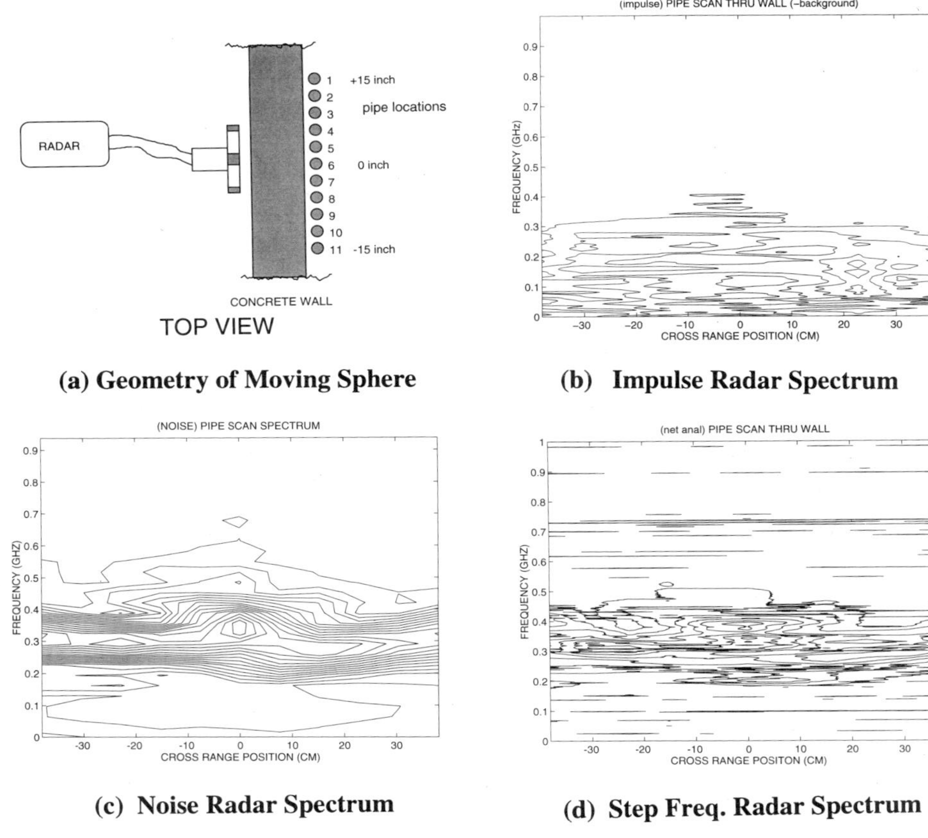

Figure 3. Frequency spectrum for moving pipe experiments.

Then, the inverse FFT was taken, and only the real parts retained and plotted here. Note that the result is the impulse response of the sphere (as convolved with the antenna and wall propagation values).

MOVING PIPE EXPERIMENTS

The moving pipe experiments were similar to the moving sphere experiment. In this case, a 12 inch long, ¾ inch diameter pipe was orientated vertically and moved horizontally along the opposite side of the wall from the small cross polarized dipole radar antenna. The geometry is shown in Figure 3.a.

The Frequency Spectra

At this point, we will show the frequency spectrum of the response waveforms as a function of cross-range position of the pipe for all three types of radars. The frequency response for the impulse radar system is shown in Figure 3.b. This was computed by taking the amplitude of the FFT of the of the impulse response data. Thus the plot is a

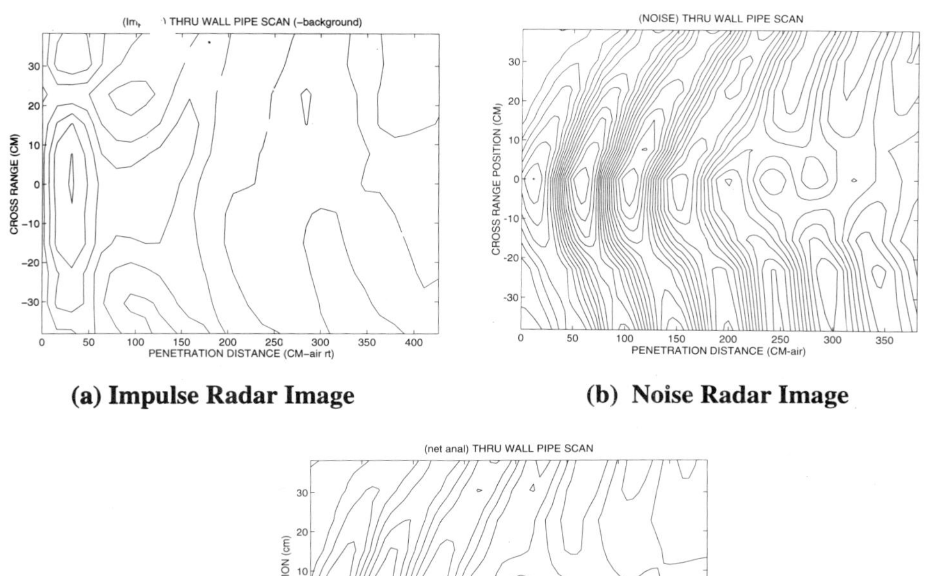

(a) Impulse Radar Image **(b) Noise Radar Image**

(c) Step Frequency Radar Image

Figure 4. Range vs. cross range images for moving pipe experiment.

contour plot showing amplitude as a function of cross range position and frequency. The same plot is shown in figure 3.c for the noise radar. In this case, the impulse response is simply proportional to the amplitude of the cross correlation (the voltage output of the low pass filter as a function of delay time). Thus the frequency spectrum is simply the amplitude of the FFT of this function. Once again, the plot is a contour plot giving amplitude as a function of cross range position and frequency.

The result for the Network analyzer is shown in figure 3.d. In this case, the spectral values are measured directly, and it is only necessary to retain the amplitude of the measured spectral values.

Note that the majority of the spectra from the pipe as measured by the NA and the noise radar is in the frequency band from 0.25 to 0.45 GHz. The NA curves look somewhat more "rough" than the noise radar curves because of the larger number of data points available over this frequency band for the NA. The impulse response spectral results (figure 3.b.) show a much lower response image because the output spectral response of the impulse radar used here is much lower than that of either the NA system or the noise radar system.

The Range/Cross Range Image

An image of the moving pipe in the cross-range versus down-range (impulse response) domain can now be created. In this case, the impulse response (directly measured or derived) is plotted as a function of pipe cross range and propagation delay time (here normalized to round trip propagation distance in cm). The first image is shown in figure 4.a. In this case, the impulse radar data is used, and directly plotted as a function of pipe cross range location and wall penetration distance (in cm, round trip). (Location (and time) zero is arbitrary.) Note the strong response near 40 cm at zero cross-range.

The image for the noise radar is shown in figure 4.b. In this case, the impulse response is directly proportional to the output voltage from the low pass filter as a function of delay line length. Note the increase in delay time as the pipe is moved off the center line (increasing distance).

The image for the NA is shown in figure 4.c. In this case, the impulse response is taken as the real part of the inverse FFT of the complex step frequency data. Note that since the bandwidth of the NA was set approximately to the same as that of the noise radar, the images for the NA data and the noise radar data were very similar.

CONCLUSIONS

We have demonstrated and compared the behavior of an impulse radar, a noise radar and a step frequency radar (NA) for the case of concrete wall penetration. We have seen the behavior of a sphere and a rod as a function of target movement.

It can be seen (exclusive of bandwidth effects) that all three radars are nearly identical in the quality of the presented data. Comparisons have been shown.

In the future, we expect to demonstrate RCS calibration and compensation. We also expect to have an impulse radar with a larger bandwidth so that the comparison is more direct. We also plan to study the effect of radar EMI on commercial systems (such as a TV set and an AM/FM radio.)

INDEX

Antennas, 43-204, 381-389, 405-421
  arrays, 15, 129-145
  biconical, 157
  circular helix, 190
  cross polarization, 200
  flared, 153
  impulse radiating (IRA), 43-63
  lens impulse radiating, 97-128
  photoconducting, 9
  reflector impulse radiating, 65-95, 424
  stripline, 197
  TEM horns, 22, 47, 97-128, 397-404, 407-408
Averaging (Crank-Nicholson), 269

Bayes hypothesis testing, 343
Bloch waves, 1-2
Blumlein, 31-38
Brewster angle, 108, 130, 386

Compact operator, 179
Constitutive parameters, 205-215, 231-246
CRB (Cramer Rao bounds), 351, 356
Creeping wave, 258

Deconvolution, 373-380
Dielectric
  loss, 413-415
  Lorentz model, 226, 254
  Debye model, 254
Diffraction, feed plate, 66
Dipole
  electric, 152
  combined, 97-105
Discretization, 266-267
Dispersion, 213
Doppler-shift, 351

Edge losses, 411-413
EFIE (electric-field integral equation), 263-270, 500
  conservation law, 265
Eigenvalue problem, 179
EM missile, 173
EM (beam) weapons, 279

Emission operator, 321
E-pulse, 327-334, 343

FCM (fuzzy C-means), 315
FDTD (finite-difference, time domain), 279, 331
Feyman diagram, 325
FIM (Fischer information matrix), 351
Fourier/Laplace transform (Also see time-frequency analysis.), 57, 69, 89-91, 140, 169, 219, 248, 321, 328, 462
  FFT (fast Fourier transform), 247, 374, 433, 500, 514
Fraunhofer zone, 283
Fresnel zone, 284

Gabor lattice, 489
Galerkin approximation, 263
Gaussian waveform, 202, 363
Glory wave, 257
GLRT (generalized likelihood ratio test), 345, 364
Green's function, 178, 501
Ground attenuation, 20, 22

Hilbert transform, 41, 477
Huygens
  principle, 200
  surface, 283

IRA (impulse radiating antenna) - See antennas.
ISAR (inverse synthetic aperture radar), 305

K-space entropy, 465

Laplace/Fourier transform (See Fourier/Laplace transform.)
Laser diode, 19
Lens, 43-56, 73-80, 97-128, 385-386
Lie group, 449
Lipschitz-Hankel integral, 249

Mellin transform, 352
Method of moments, 101, 185, 501
Mode converter, 108, 385, 391-396
Natural frequency (SEM pole), 327-342

Noise
Gaussian, 343, 372
signal processing in presence of, 343-372

Paraboloidal reflector, 43-46, 65, 73-74, 81, 89-95
Plasma, 247-254
Poisson random process, 362
Polarization, 207, 260
clutter matrix, 461
covariance matrix, 471
imaging, 447
Jones matrix, 448
Kennaugh matrix, 448, 470
Mueller matrix, 448, 470
radar, 461
Sinclair matrix, 448
Stokes vector, 469
Precursor (Brillouin, Sommerfeld), 227
Prepulse, 66
Principal value (Cauchy), 314
Propagation, 205-254
Pulser, 25-41, 381-389, 391

Radar, 21-23, 423-446, 511-516
polarimetric, 447-483
Radon transform, 353, 443
Rainbow, 257
RAM (radar absorbing material), 231
RCS (radar cross section), 295, 331
Reception operator, 321
Reciprocity, 122

Sampling detector, 431
SAR (synthetic aperture radar), 335-342
Scatterer (target), 255-304, 271-278
buried, 26, 485-516
dielectric, 255-262
Scattering (reflection) coefficient, 234, 243, 281
Schottky barrier effect, 10
Sea clutter, 435
SEM (singularity expansion method), 335, 343-344
Semiconductor switching, 1-24
Sensor, E-field, 407
SF_6, 19
Simulator (EMP), 129-130
Spectral filtering, 337
Spectral wave number, 489
Stationary phase, 356
Switches
gas, 381-383
photoconductive
GaAs, 17, 34, 383, 431
Si, 34

Target
identification, classification, discrimination, 26, 313-320, 327-350
imaging, 442
TE (wave or mode), 260
TEM (wave or mode)
spherical (conical transmission line), 65, 73, 81, 129, 397, 410
Time-frequency analysis, 3, 6, 305-326
wavelets, 321-326
window Fourier/Laplace transform, 489
TM (wave or mode), 226
Transmission coefficient, 79

Wavelet (spatial, physical), 332
Wigner distribution, 314